T0190238

Didier Astruc

Organometallic Chemistry and Catalysis

Grenoble Sciences

Grenoble Sciences pursues a triple aim:

- to publish works responding to a clearly defined project, with no curriculum or vogue constraints,
- to guarantee the selected titles' scientific and pedagogical qualities,
- to propose books at an affordable price to the widest scope of readers.

Each project is selected with the help of anonymous referees, followed by an average one-year interaction between the authors and a Readership Committee whose members' names figure in the front pages of the book. Grenoble Sciences then signs a co-publishing agreement with the most adequate publisher.

(Contact: Tél.: (33) (0)4 76 51 46 95 – Fax: (33) (0)4 76 51 45 79 –
E-mail: *Grenoble.Sciences@ujf-grenoble.fr*)

Scientific Director of Grenoble Sciences: Jean Bornarel, Professor at the Joseph Fourier University, France

Grenoble Sciences is supported by the French Ministry of Education and Research and the "Région Rhône-Alpes".

"Organometallic Chemistry and Catalysis" is an improved version of the French book published by Grenoble Sciences in partnership with EDP Sciences. The Reading Committee of the French version included the following members:

Yves Chauvin, Nobel prizewinner (2005), Honorary Scientific Director of the French Petrol Institute,

Robert Corriu, Member of the French Academy of Science and the French University Institute, professor at the University of Montpellier,

Jean-René Hamon, Senior Researcher at the CNRS, University of Rennes

Henri Kagan, Member of the French Academy of Science and the French University Institute, professor at he University of Paris-Sud, Orsay,

Paul Knochel, Professor at the Ludwig Maximilians Universität, München,

Bernard Meunier, Member of the French Academy of Science, Senior researcher at the CNRS in Toulouse,

Jean-François Normant, Member of the French Academy of Science, professor at the Pierre et Marie Curie University, Paris

Jean-Yves Saillard, professor at the University of Rennes

Jean-Pierre Sauvage, Member of the French Academy of Science, Senior researcher at the CNRS in Strasbourg,

Bernard Waegell, professor at the University of Aix-Marseille

Didier Astruc

Organometallic Chemistry and Catalysis

With 860 Figures and 27 Tables

Springer

Didier Astruc
Member of the Institut Universitaire de France
Institut des Sciences Moléculaires, Groupe Nanosciences Moléculaires et Catalyse
UMR CNRS N° 5255
Université Bordeaux I
351, Cours de la Libération
33405 Talence Cedex
France
email: d.astruc@ism.u-bordeaux1.fr

Originally published in French: Didier Astruc, Chimie Organométallique, EDP Sciences, Grenoble 2000, ISBN 978-2-86883-493-5

ISBN 978-3-662-50086-6 Springer Berlin Heidelberg New York
DOI 10.1007/978-3-540-46129-6

This work is subject to copyright. All rights are reserved, whether the whole or part of the material is concerned, specifically the rights of translation, reprinting, reuse of illustrations, recitation, broadcasting, reproduction on microfilm or in any other way, and storage in data banks. Duplication of this publication or parts thereof is permitted only under the provisions of the German Copyright Law of September 9, 1965, in its current version, and permission for use must always be obtained from Springer. Violations are liable for prosecution under the German Copyright Law.

Springer is a part of Springer Science+Business Media

springer.com

© Springer-Verlag Berlin Heidelberg 2007
Softcover re-print of the Hardcover 1st edition 2007

The use of general descriptive names, registered names, trademarks, etc. in this publication does not imply, even in the absence of a specific statement, that such names are exempt from the relevant protective laws and regulations and therefore free for general use.

Typesetting: Camera ready by Grenoble Sciences
Production: LE-TeX Jelonek, Schmidt & Vöckler GbR, Leipzig
Cover Design: Nicole Sauval, Grenoble, France
Front Cover Illustration: composed by Alice Giraud with extracts from
    E. Alonso, D. Astruc – Introduction of the Cluster Fragment $Ru_3(CO)_{11}$ at the periphery of Phosphine Dendrimers Catalyzed by the Electron-Reservoir Complex [$Fe^I Cp(C_6Me_6)$], J. Am. Chem. Soc. 2000, 122, 3222

Printed on acid-free paper   2/3180/YL   5 4 3 2 1 0

PREFACE

This text is a translation and thorough revision of the French book "Chimie Organométallique" that was published by EDP Sciences in 2000 and whose Spanish version, prepared by Carmen Claver and Beatriz Alonso, was published by Reverte in 2003. The catalysis part has been considerably extended, however. Not only has it been re-organized and complemented, but new chapters have been written on homogeneous catalytic hydrocarbon activation and on heterogeneous catalysis.

I would like to thank the colleagues who have corrected parts of the French Edition: Yves Chauvin (IFP and Villeurbanne) for the catalysis section, Jean-René Hamon (Rennes) who read the entire French version, Catherine Hemmert (LCC, Toulouse), who corrected the enzymatic catalysis chapter, Henri Kagan (Orsay) for general advice, Jean-Yves Saillard (Rennes) for the theoretical aspects and Bernard Waegel (Marseille) for remarks on the catalysis chapters. I am grateful to Alan H. Cowley (U.T. Austin), Andrew Wojcicki (Columbus), Jonathan Egbert and Travis J. Ebden (Seattle) for proofreading several chapters of the English text and to Cátia Ornelas (Talence) for proofreading the entire book.

Finally, it is a great pleasure to thank Jean Bornarel and Nicole Sauval for their editorial efforts as well as Sylvie Bordage (Grenoble Sciences) for her superb work in the book production and Theodor C.H. Cole (Springer) for excellent copy-editing.

Didier ASTRUC,

Talence, February 2007

CONTENTS

PART II
THE STOICHIOMETRIC REACTIONS OF TRANSITION-METAL COMPLEXES

PART III
THE MAIN FAMILIES OF ORGANOMETALLIC COMPLEXES

PART IV
CATALYSIS

PART V
APPLICATIONS IN ORGANIC SYNTHESIS

INTRODUCTION

1. ORGANIZATION OF THE TEXT

Organometallic chemistry is defined as the chemistry of metal-carbon bonds. More generally, it concerns the transformations of organic compounds using metals from the main groups, transition series, lanthanides and actinides.

It is necessary to understand, in first instance, the electronic structures of the complexes and the major bonding modes between metals and common ligands. This is the subject of the first part of the book. The two chapters of this first part are developed particularly from the pedagogical point of view in order to insure a good basic knowledge on the part of the student.

Once the electronic structure is understood, in the second part of the book we will examine reactivity of the complexes by scrutiny of selected stoichiometric reactions. The changes of the properties of a substrate by temporary complexation to a transition metal center, including stereochemical consequences, indeed provide many applications in organic chemistry. A second group of essential reactions is that involving the elementary steps of catalytic cycles.

In the third part of the book, we will survey the main families of organometallic complexes from both the structural and reactivity standpoints. The discussion of transition-metal complexes will be organized on the basis of ligand type, whereas that of other metals (alkali metals, rare earths, intermediate groups) will be examined according to the metal group.

The fourth part of the book is devoted to catalysis. The role of metals in their reactions with organic compounds can be stoichiometric or catalytic. The student will have carefully taken this important distinction into account. Chemists, especially those from industry, will seek transformation processes that involve metals in catalytic quantities, i.e. in small amounts (metal-to-substrate molar ratios much lower than one). These efforts are obviously driven by problems of cost, toxicity and sometimes corrosion. Catalytic processes mostly use transition metals, which makes this class of metals particularly important. Catalytic processes are numerous and very common in biology, industry and every-day operations in the laboratory. We will study the most important catalytic cycles with emphasis placed on homogeneous catalysis, because it is in this area that the mechanisms are mostly firmly established in this area. Emphasis is now placed not only on classic hydrogenation and carbonylation processes, but also on progress in the challenging catalytic activation of hydrocarbons that is the subject of a new chapter. Another new specific chapter is presented on heterogeneous catalysis, however, with a comparative view brought about by organometallic models.

The catalysis part of the book will also include a chapter on bio-organometallic chemistry, i.e. devoted to the main examples of enzymatic catalysis involving transition metal-carbon species.

The final chapter is devoted to the applications of metals in organic synthesis and will rely on the concepts developed in other parts of the book. We will summarize the various catalytic means of forming carbon-carbon bonds and in particular the use of palladium (Tsuji-Trost, Heck, Sonoganishara, Suzuki, Stille, Hiyama, Corriu-Kumada and animation reactions) and condensations of alkynes (Vollhardt's trimerization and Pauson-Khand reaction).

Exercises are presented at the end of each chapter along with a summary, and the answers to these exercises are gathered at the end of the book prior to the references.

2. FRONTIERS OF ORGANOMETALLIC CHEMISTRY

The definition of organometallic chemistry limiting this discipline to compounds with metal-carbon bonds is somewhat too strict. We prefer to extend the definition to include the transformation of organic compounds using metals. The compounds or intermediate species that are usually involved in these reactions contain one or several metal-carbon bonds, but the approach is not systematic. Some other species also play key roles in the reactions of organic compounds with metals such as those having metal-hydrogen (hydrides), metal-oxygen (oxo) and metal-nitrogen (imido) bonds. It is our objective to illustrate all the possibilities of organometallic chemistry in the broadest possible sense.

Organometallic chemistry is a part of coordination chemistry and inorganic chemistry. Although inorganic chemistry traditionnally involves metal-element bonds other than those to carbon, many families of compounds contain both metal-carbon and metal-element bonds. Moreover, intermediates in inorganic reactions may sometimes be organometallic species. The areas of application of organometallic and inorganic chemistry have been traditionally different, being focused on applications in the areas of organic synthesis and materials science, respectively. However, these demarcations are becoming fuzzier as these areas continue to develop in many directions. Thus, our presentation of ligands will be very broad, although the emphasis on carbon-based ligands will be preserved.

3. SITUATION OF THE BOOK WITH RESPECT TO TEACHING

The aim of the text is to serve as a support for the first elementary course on organometallic chemistry. The scope of the text goes far beyond, however, and should stimulate the students to think about new developments in addition to rising her or his basic culture in the area. Since the approach to the concepts is progressive, the book will also be useful for advanced courses, and selected references with their titles can serve as a starting point for a deeper knowledge of subject matters. This textbook has been mostly designed with the idea of serving as a pedagogical basis for students and teachers. However, up-to-date ideas are introduced in conjunction with the most recent advances in the field.

Comments, suggestions and constructive criticisms are welcome.

E-mail: *d.astruc@ism.u-bordeaux1.fr*

Internet: *http://u-bordeaux1.fr/lcoo/members/eastruc.htm*

4. *REFERENCE BOOKS AND OTHER SELECTED REFERENCES*

There are many organometallic chemistry books that are available, and students are strongly encouraged to become familiar with a number of them. Thus, important books in this and related areas are listed together with comments at the beginning of the reference section located at the end of this book. We have limited the reference section to a selection of approximately about 400 readily available key references.

HISTORY OF ORGANOMETALLIC CHEMISTRY

1760-1900: THE FIRST COMPLEXES

The boom in organometallic chemistry occurred essentially in the United States, England and Germany during the 3rd quarter of the XXth century. It is in a military pharmacy in Paris, however, that the discipline was born in 1760. Cadet, who was working on cobalt-containing inks, used arsenic-containing cobalt salts for their preparation. In this way, he discovered the fuming and ill-smelling Cadet's liquid which comprises a mixture of cacadoyl oxide and tetramethyldiarsine by carrying out the following reaction:

$$As_2O_3 + 4\,MeCO_2K \longrightarrow [(AsMe_2)_2O] + [Me_2As\text{-}AsMe_2] + ...$$

One of the key events of the XIXth century was the discovery of the first π complex, namely Zeise salt K[PtCl$_3$(η^2-C$_2$H$_4$)], in 1827, although the correct formula below was proposed much latter.

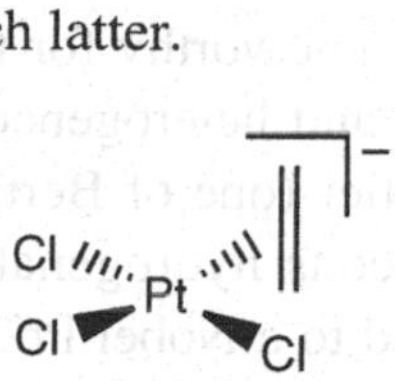

$$\left[\begin{array}{c} Cl\,\textit{//}_{\textit{//}} \quad {}^{\textit{\textbackslash}\textit{\textbackslash}}\!\!\|\| \\ Cl \blacktriangleright Pt \blacktriangleleft Cl \end{array} \right]^{-} \qquad K^+ \qquad \textbf{Zeise salt, 1827}$$

In the mid-1850s, Frankland synthesized several air-sensitive metal-alkyl complexes such as ZnEt$_2$ (1849), HgEt$_2$ (1852), SnEt$_4$ and BMe$_3$ (1860), the mercury and zinc complexes being immediately used to synthesize many other main-group organometallic complexes. For instance, Friedel and Crafts prepared several organochlorosilanes RnSiCl$_{4-n}$ from the reactions of SiCl$_4$ with ZnR$_2$ (1863). Shortly afterwards, Schützenberger, an Alsatian chemist, synthesized the first metal-carbonyl derivatives [Pt(CO)$_2$Cl$_2$] and [Pt(CO)Cl$_2$]$_2$ (1868-1870). Twenty years later, the first binary metal-carbonyl compounds appeared: [Ni(CO)$_4$] (Mond, 1890) and [Fe(CO)$_5$] (Mond and Berthelot, 1891). From 1893 onwards and over a period of twenty years, Werner developed the modern ideas of inorganic chemistry in his proposal according to which the Co^{3+} ion is surrounded by six ligands in an octahedral complex [CoL$_6$]$^{3+}$. These ideas were contrary to those proposed by the well-known chemists of that time such as Jorgensen, according to whom the ligands should be aligned in a chain with the metal at the terminus.

1900-1950: GRIGNARD, SABATIER, AND CATALYSIS IN GERMANY

It is at the turn of the XXth century that the major French contribution came into fruition with the work of Barbier and especially of his student Victor Grignard. Grignard's name is forever engraved in the history of chemistry for the reaction of magnesium with organic halides RX leading, by oxidative addition, to Grignard reagent RMgX that can alkylate carbonyl derivatives.

$$RX \xrightarrow[Et_2O]{} RMgX \xrightarrow{MgR'COR''} RR'R''COH$$

This discovery had an enormous impact, not only on organic chemistry, but also proved to be of considerable importance in transition-metal organometallic chemistry. For instance, in 1919, Hein synthesized what he believed to be a polyphenyl-chromium derivative $[(Cr(\sigma\text{-}Ph)_n]$:

$$CrCl_3 + PhMgBr \longrightarrow [Cr(\sigma\text{-}Ph)_n]^{0,+1} \qquad n = 2, 3 \text{ or } 4$$

However, some 36 years later, it was revealed that the compound posseses a π sandwich-type structure.

The first half of the XXth century was especially noteworthy for the emergence of catalysis. The distinction between homogeneous and heterogeneous catalysis was first delineated by the French chemist Paul Sabatier (one of Berthelot's students). Sabatier and Senderens' work on the heterogeneous hydrogenation of olefins to saturated hydrocarbons using nickel eventually led to a Nobel Prize shared in 1912 between Sabatier and Grignard. Many seminal discoveries, however, were made in Germany during this first half of the XXth century. In 1922, Fischer and Tropsch reported the heterogeneously catalyzed reaction between CO and H_2 (syngas) to form mixtures of linear alkanes and alkenes with few oxygenates as by-products (Fischer-Tropsch process, industrially developed in 1925). In 1938, Roelen discovered the catalysis by $[Co_2(CO)_8]$ of the hydroformylation of olefins by CO and H_2 (oxo process). From 1939 to the late 1940s, Reppe worked on the catalysis of transformation of alkynes (tetramerization to cyclo-octatetraene, 1948).

Curiously, there were only a few advances in the discovery of new complexes during this period. Besides iodotrimethylplatinum, $[PtMe_3I]$, synthesized by Pope at the beginning of the XXth century, the list includes $[Fe(\eta^4\text{-butadiene})(CO)_3]$ made by Reilhen in 1930 and Hieber's work on metal carbonyls from 1928 onwards (e.g. the synthesis of $[Fe(H)_2(CO)_4]$, 1931).

First σ-alkyl-metal complex	First diene-complex	First metal-hydride complex
Pope, 1909	Reilhen, 1930	Hieber, 1931

1950-1960: THE DISCOVERY OF FERROCENE AND THE BOOM OF ORGANOMETALLIC CHEMISTRY

It is in the 1950s that inorganic and organometallic chemistry boomed, especially in the United States. Henry Taube classified the inorganic complexes as inert or labile towards substitution reactions, depending on the nature of the ligand and oxidation state of the metal (1951). This work laid the foundation of molecular engineering that allowed him to carry out a crucial series of experiments: the catalysis of substitution reactions by electron or atom transfer, and the distinction between outer-sphere electron transfer and that proceeding by the inner sphere which is considerably faster (1953).

Structural bio-organometallic chemistry has its origin with the English chemist Dorothy Crowfoot-Hodgkin who established the X-ray crystal structure of vitamin B_{12} coenzyme between 1953 and 1961 using a primitive computer, resulting in her awarding the Nobel Prize in 1964. Consecutively, the legendary Harvard University chemist Robert B. Woodward achieved the total synthesis of this coenzyme in 70 steps with 99 co-workers from 1961 to 1972.

In England too, several chemists had noted, at the beginning of the 1950s, this very stable orange powder that formed when cyclopentadiene is passed through iron tubings. Some chemists had even filled pots with it in their laboratory. Keally, Pauson and Miller first published the compound in 1951, and Pauson reported the structure $[Fe(\sigma\text{-}C_5H_5)_2]$ in the journal *Nature*. In doing so, he made the same mistake as Hein, 32 years earlier. In the meantime, however, Sidgwick had published the 18-electron rule for transition-metal complexes in his book " *The Electronic Theory of Valency* " (Cornell University, Ithaca, 1927). Pauson's proposed σ structure for bis-cyclopentadienyl-iron only had 10 valence electrons and was soon challenged. At Harvard University, Willkinson and Woodward recognizing Sidgwick's rule, immediately understood that the Pauson formulation was incorrect. A few months later, they published the first sandwich structure, just before Fischer: bis-cyclopentadienyl-iron is ferrocene, an 18-electron complex, in which the two rings are perfectly parallel, π-bound to the iron atom and having aromatic properties (1952).

R.B. Woodward and Ferrocene

Robert Burns Woodward (1917-1979), who was a genious and had been a child prodigy, receiving his Ph.D. from MIT at age 20, has played a key role in the discovery of the sandwich structure of ferrocene. Ernst Otto Fischer (1918-) and Geoffrey Wilkinson (1921-1997) received the 1973 Nobel Prize for their pioneering work on the chemistry of the organometallic sandwich compounds. Although this Award is well recognized as being fully justified, the fact that Woodward was not included has been questionned. Woodward, a full Professor in the chemistry department at Harvard and Wilkinson, a first-year assistant Professor in the same department, joined their efforts in the beginning of 1952, after reading Peter Pauson's paper in *Nature* (published in England on December 15, 1951, and arrived in the U.S. about a month later) to carry out a series of experiments with Myron Rosemblum, a graduate student, and Mark Whittig, a post-doc, in order to show the sandwich structure. This, shortly thereafter led to the *JACS* communication of 1952. The proof for the sandwich structure was the dipole moment that was zero and the infrared spectrum showing only one type of C-H bond. It was Woodward, however, who recognized that ferrocene would behave like a typical aromatic compound, and Wilkinson later acknowledged that he had not considered this possibility. Later in 1952, another article from Woodward, Rosenblum and Whiting in *JACS* confirmed the aromatic properties of di-cyclopentadienyl-iron and proposed the name ferrocene.

On October 26, 1973, two days after the announcement of the Nobel Prize, Woodward mailed a letter from London to the Nobel Committee in which he stated: "*The notice in the Times (October 24, p. 5) of the award of this year's Nobel Prize in Chemistry leaves me no choice but to let you know, most respectfully, that you have – inadvertently, I am sure – committed a grave injustice... Indeed, when I, as a gesture to a friend and junior colleague interested in organometallic chemistry, invited Professor Wilkinson to join me and my colleagues in the simple experiments which verified my structural proposal, his reaction to my views was close to derision... But in the event, he had second thoughts about his initial scoffing view to my structural proposal and its consequences, and altogether we published the initial seminal communication that was written by me.*"According to Wilkinson, however, it was when he read the Pauson paper that he said to himself: "*Jesus-Christ it can't be that. It's a sandwich!*"

Indeed, there was room for another person in this Nobel Prize, but it is possible, as stated by Roald Hoffmann, that a strategic error of Woodward was not to expand his organometallic research to include other transition metals and instead concentrate on the aromatic chemistry of ferrocene. Actually, before Woodward and Wilkinson's work, William E. Doering at Columbia suggested in September 1951 the sandwich structure to Peter Pauson whose paper with the wrong structure had been sent to *Nature* on August 7, 1951. An article with the same wrong structure was submitted to the *Journal of the Chemical Society* by Miller, Tebboth and Tremaine from the British Oxygen Company on July 11, 1951 and published in England on March 24, 1952. Independently, Pfab and Fisher reported the first resolution of the X-ray crystal structure of ferrrocene in the beginning of 1952.

For a full account of the story, see: T. Zydowsky, *The Chemical Intelligencer*, Springer-Verlag, New York, 2000, p. 29. The author is grateful to Helmut Werner, a pioneer *inter alia* of π complexes including the first triple-sandwich compound, for bringing this article to his attention.

FeCp₂ : erroneous σ structure
Pauson, 1951

π-sandwich structure of ferrocene
Wilkinson and Fischer, 1952

Wilkinson did not make tenure in Harvard, and he moved back to London at Imperial College, fortunately for British inorganic chemistry. This structure awarded Wilkinson and Fischer the Nobel Prize in 1973. Cobaltocenium, isoelectronic with ferrocene, was also synthesized in 1952. Sandwich complexes with other metals shortly followed, opening the route to an organometallic chemistry of π complexes with poly-hapto ligands developed particularly in Oxford by Wilkinson's bright student, Malcolm L.H. Green. Some of these complexes such as [FeCp(CO)₂]₂ have a metal-metal bond and a non-rigid structure, carbonyls jumping rapidly from one metal to another: these are the fluxional complexes (Cotton and Wilkinson, 1955). Longuet-Higgins and Orgel predicted in 1956 that it would be possible to isolate 18-electron complexes of cyclobutadienyl, a highly instable anti-aromatic compound. Their prediction turned into reality only two years later with syntheses of such complexes using several transition metals.

Cotton and Wilkinson, 1955
Fluxional molecule

Hubel, 1958   Criegee, 1959
Stabilization of cyclobutadiene

Another very important discovery, in 1955, was that of olefin polymerization catalyzed by soluble titanium and aluminum compounds by Ziegler and Natta. These authors were awarded the Nobel Prize for chemistry in 1963. The stereospecific polymerization of propylene, which was also discovered by Ziegler and Natta, was attributed initially to the surface effect of the heterogeneous initiator, but this interpretation did not survive. Ziegler also discovered the hydroalumination of olefins, leading to tris-alkyl-aluminum, which is a very important industrial chemical. It was also during this period that Georg Wittig found the reaction that now bears his name (1953) and that Herbert C. Brown discovered the hydroboration of olefins (1956). They were jointly awarded the 1979 Nobel Prize. This decade ended with the report of the first oxidative addition reaction in transition-metal chemistry by Shaw and Chatt in 1959.

1961-1981: THE DISCOVERY OF MULTIPLE METAL-CARBON BONDS AND THE GOLDEN AGE OF CATALYSIS

Molecular compounds with metal-metal bonds had been known for a long time. Gold nanoparticles were known *inter alia* in Ancient Egypt for their decorative and therapeutic properties, and calomel was used by chemists in India in the XIIth century. The Hg-Hg bond in the mercurous ion was recognized at the beginning of the XXth century. It was in the 1960s, however, that a more complete development of metal clusters appeared, particularly for the metal-carbonyl clusters by Paolo Chini in Italy, Earl Muetterties and Lawrence Dahl in the United States and Lord Jack Lewis in England. At the same time, the syntheses, structures and properties of multiple metal-metal bonded compounds, including those with up to the quadruple bonds were studied, in particular by F. Albert Cotton.

The birth of an extraordinary family, the carboranes and metallocarboranes, whose father is Frederic Hawthorne, also occurred during this decade in the United States. Later, W.N. Liscomb obtained a Nobel Prize (1976) for the clarification of the structures of boranes.

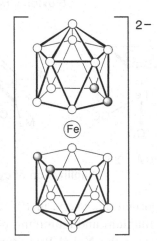

**Dicarbolyl iron(II)
analog of ferrocene**
M.F. Hawthorne, 1965

In 1962, Vaska published the famous 16-electron complex [Ir(CO)(Cl)(PPh₃)] that bears his name. It is well known for the reversible addition of dioxygen and for the series of other oxidative addition reactions that it undergoes with a large number of substrates, including H_2 at 25°C.

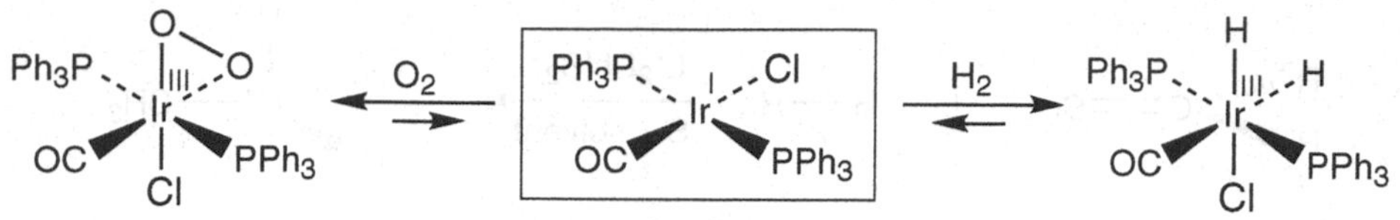

Vaska's complex and the oxidative addition of O_2 and H_2, 1965

In 1964, the first metal-carbene complex was published by E.O. Fischer and the catalyzed metathesis of olefins (below) was reported by Banks.

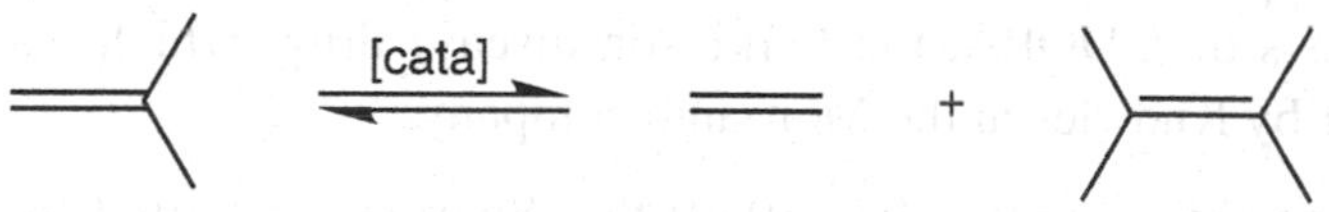

Although these two events were not connected at this time, several years later it turned out that they had a considerable mutual impact. Indeed, in 1971, Yves Chauvin, from the Institut Français du Pétrole near Paris, proposed that the mechanism of olefin metathesis proceeds by coordination of an olefin onto the metal center of a transition-metal alkylidene complex to form a metallacyclobutane that can decompose to a new metal-alkylidene and a new olefin. Subsequently, this mechanism was shown by Chauvin and others to be correct, and metal-alkylidene complexes, discovered three years later by Richard R. Schrock, now occupy a central place in organometallic chemistry and catalysis (*vide infra*).

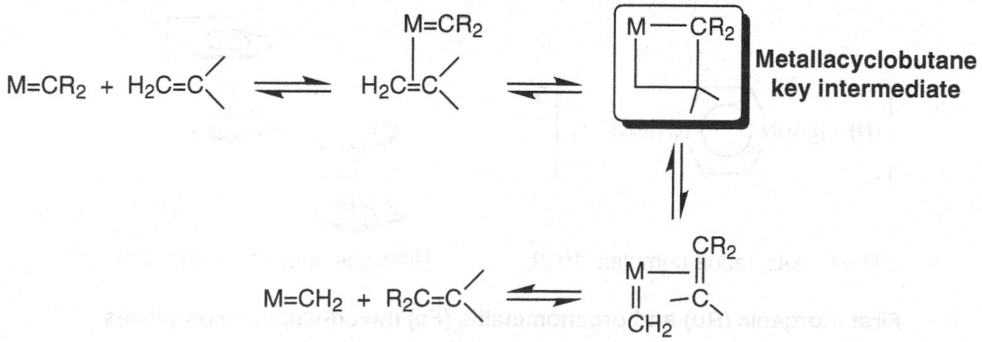

Mechanism proposed by Chauvin for olefin metathesis

In 1965, Wilkinson (together with his former Ph.D. student John Osborn), and Coffey independently discovered the first homogeneous olefin hydrogenation catalyst, Wilkinson's catalyst, [RhCl(PPh₃)₃]. In 1970, Henri Kagan reported in a patent the first efficient enantioselective asymmetric Rh[I] hydrogenation catalyst bearing his chelating chiral ligand DIOP of C_2-symmetry, and his work became widely known with his famous articles in *Chem. Commun.* and *J. Am. Chem. Soc.* in 1971 and 1972, respectively.

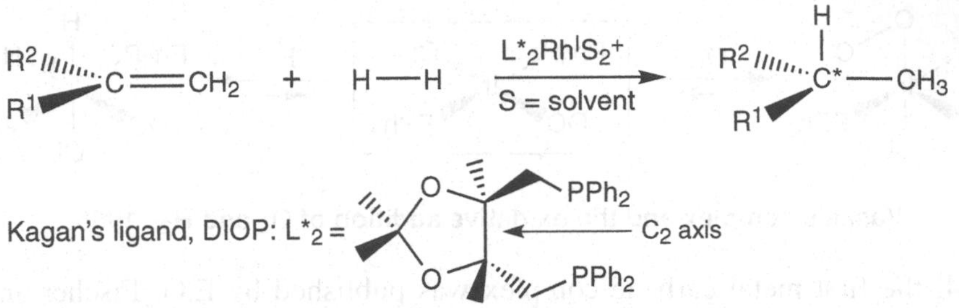

$$R^2\text{·····}C=CH_2 \quad + \quad H—H \quad \xrightarrow[\text{S = solvent}]{L^*_2Rh^IS_2^+} \quad R^2\text{·····}C^*—CH_3$$

Kagan's ligand, DIOP: L^*_2 =

PPh$_2$ ←————— C$_2$ axis

PPh$_2$

An important application of the Kagan ideas and type of chelate complexes permitted the synthesis of *L*-DOPA, the Parkinson disease drug, which was carried out five years later by Knowles at the Monsanto company.

It is also during this decade (1973-1980) that Sharpless reported the asymmetric epoxidation of allylic alcohols, a useful reaction in organic chemistry. The Nobel Prize in Chemistry was awarded to Knowles, Sharpless and Noyori in 2001 for asymmetric catalysis.

At the end of the 1960s, the concept of mixed valence appeared. H. Taube, on the experimental side, and N.S. Hush, with the theoretical aspects, made seminal decisive contributions. Taube characterized the first mixed-valence and average-valence complexes by varying the nature of the ligand L in the series $[(NH_3)_5RuLRu(NH_3)_5]^{5+}$. In organometallic chemistry, the first mixed-valence complex, biferrocenium monocation was published in 1970.

$$\left[(NH_3)_5RuN \bigcirc NRu(NH_3)_5 \right]^{5+}$$

Fe^{III}

Fe^{II}

The Creutz-Taube complex, 1969 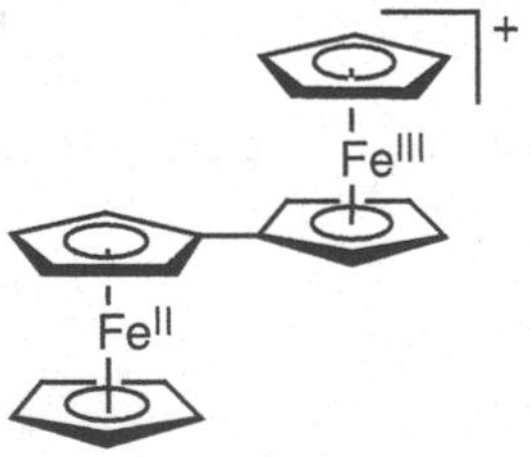   Biferrocenium, Cowan (1970)

First inorganic (Ru) and organometallic (Fe) mixed-valence complexes

A. Streitweiser and U. Mueller-Westerhoff reported uranocene, the first f-block sandwich complex, in 1968, and in 1970, Wilkinson found that metal-alkyl complexes that have less than 18 valence electrons are stable provided that they do not have β-hydrogen atoms. H. Werner and A. Salzer reported the first triple-decker sandwich complex in 1971.

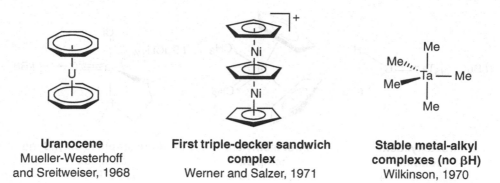

Uranocene
Mueller-Westerhoff
and Sreitweiser, 1968

**First triple-decker sandwich
complex**
Werner and Salzer, 1971

**Stable metal-alkyl
complexes (no βH)**
Wilkinson, 1970

During this same year, 1971, E.O. Fischer synthesized the first metal-carbyne complex and the Monsanto Company developed the rhodium-catalyzed carbonylation of methanol to acetic acid. In 1975, the DuPont de Nemours Company produced adiponitrile using the Ni^0-catalyzed hydrocyanation of butadiene.

Stabilized carbene complex
Fischer, 1964

Carbyne complex
Fischer, 1973

It is also in the early 1970s that Richard R. Schrock, a former student of John Osborn at Harvard, then at the DuPont de Nemours Company in Delaware, made the very important discoveries of the first stable, isolable methylene-, alkylidene- and nucleophilic alkylidyne complexes of transition metals. These classes of compounds later brought a new dimension to the field of the metathesis reactions. Thus, in 1974, Schrock reported the isolation and X-ray crystal structure of the neopentylidene complex $[Ta(CH_2CMe_3)_3(=CH\text{-}CMe_3)]$, the first transition metal carbene complex that is not stabilized by an heteroatom on the carbene. This complex also represented the discovery of the α-elimination reaction. In 1975, Schrock published the isolation and X-ray crystal structure of $[TaCp_2Me(=CH_2)]$, the first methylene complex and a transition-metal cousin of Wittig's reagent. Indeed, Schrock's "ylid" was shown to react by methylene transfer with aldehydes, ketones and esters. Other similar methylene transfer reagents followed, for instance, Tebbe's complex $[TiCp_2(\mu\text{-}Cl)(\mu\text{-}CH_2)AlMe_2]$ and later the species "$[TiCp_2(=CH_2)]$", which is now used and generated by heating $[TiCp_2Me_2]$. In 1976, Schrock reported the first synthesis of an alkylidyne complex $[Ta(CH_2CMe_3)(Me_2PCH_2CH_2PMe_2)(C\text{-}t\text{-}Bu)]$ in which the alkylidyne ligand possesses nucleophilic character.

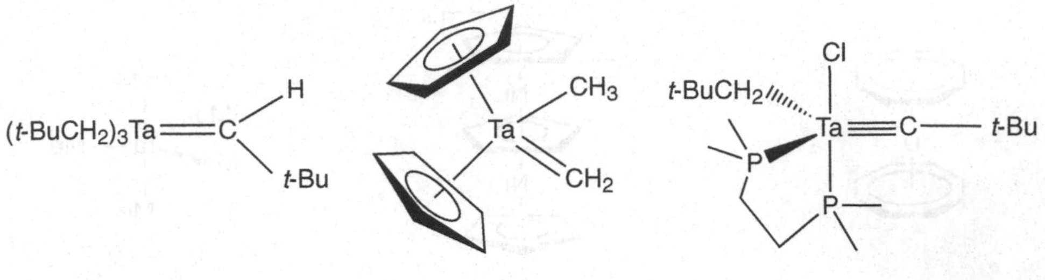

First alkylidene complex
Schrock, 1974

First methylene complex
Schrock, 1975

First nucleophilic alkylidyne complex
Schrock, 1976

Schrock also showed in 1980 that, importantly, some alkylidene complexes of tantalum and tungsten can metathesize olefins if they contain alkoxy ligands, and, in 1990, he reported a whole family of very active tungsten metathesis catalysts including chiral ones, which led him to disclose, in collaboration with Hoveyda, the first examples of enantioselective metathesis reactions.

The first application of Fischer-type metal-carbene complexes in organic synthesis also appeared in 1975 with the Dötz reaction:

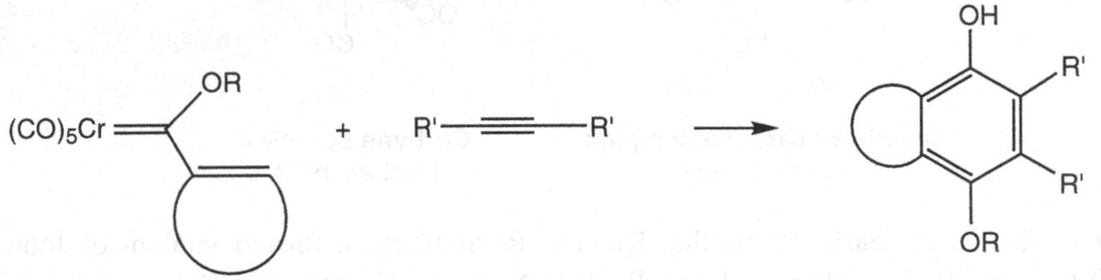

Other applications of the stoichiometric reactions of transition-metal complexes in organic synthesis, such as the Pauson-Khand reaction, were reported in the same year.

In 1977, H. Kagan published a practical synthesis of an excellent reducing agent, SmI_2, which found many applications in organic synthesis such as the pseudo-Barbier reaction, the Reformatsky reaction, the formation of pinacols, and the deoxygenation of epoxides and sulfoxides. Various reactions involving the formation of C-C bonds by treatment of SmI_2 with unsaturated substrates were subsequently developed by Kagan's group and many other research groups.

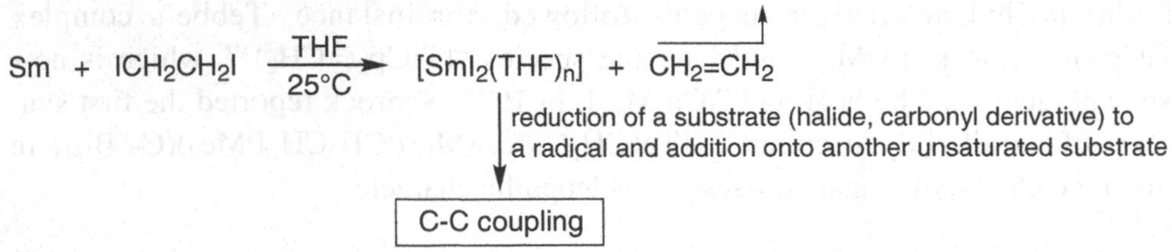

$$Sm + ICH_2CH_2I \xrightarrow[25°C]{THF} [SmI_2(THF)_n] + CH_2=CH_2$$

reduction of a substrate (halide, carbonyl derivative) to a radical and addition onto another unsaturated substrate

C-C coupling

In the 1970s, two important palladium-catalyzed reactions were disclosed by pioneers whose names are now associated with these reactions. Both reactions are widely used in organic chemistry.

Mizoroki-Heck reaction

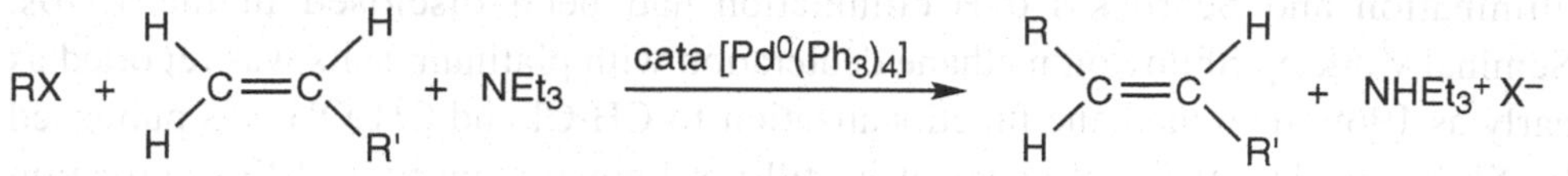

$$RX + \underset{H}{\overset{H}{}}C=C\underset{R'}{\overset{H}{}} + NEt_3 \xrightarrow{\text{cata } [Pd^0(Ph_3)_4]} \underset{H}{\overset{R}{}}C=C\underset{R'}{\overset{H}{}} + NHEt_3{}^+ X^-$$

R = allyl, vinyl, aryl, benzyl; X = halide, acetate, etc.

Tsuji-Trost reaction

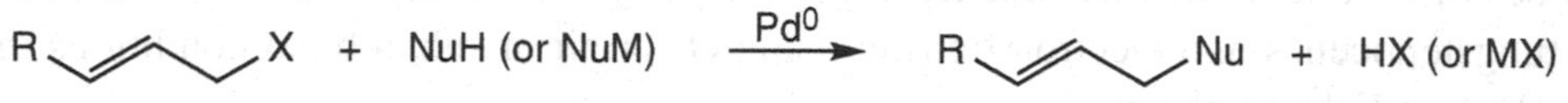

$$R\diagup\diagdown X + NuH \text{ (or NuM)} \xrightarrow{Pd^0} R\diagup\diagdown Nu + HX \text{ (or MX)}$$

X = OAc, OCO$_2$R, OH, OP(O)(OR)$_2$, Cl, NO$_2$, SO$_2$R, NR$_3{}^+$
NuH = ROH, RNH$_2$, etc. (or Nu$^-$ = OR$^-$, NHR$^-$, stabilized carbanion, etc.)

Among the first multiple C-C bond-forming reactions catalyzed by transition-metal complexes, mention should be made of Peter Vollhardt's very elegant [CoCp(CO)$_2$]-catalyzed synthesis of aromatics in the early 1970s. The later compounds are precursors to natural derivatives such as estrone (see Chap. 21) as well as non-natural derivatives (the 2+2+2 cycloaddition of TMSA below being sterically inhibited).

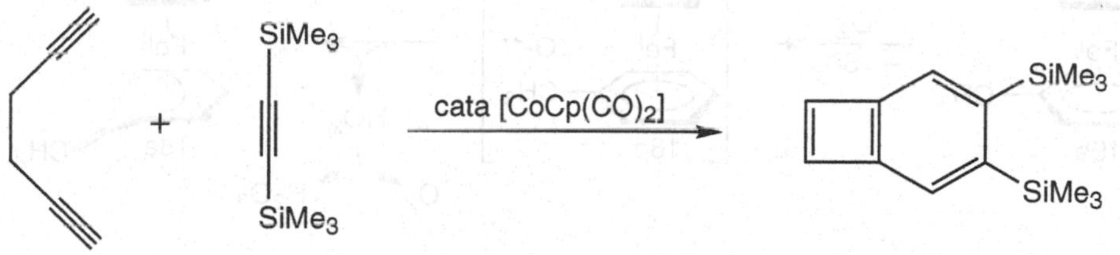

On the structural side, West published the first Si=Si double bond in 1981. The related triple bond needed to await A. Sekiguchi's disclosure in 2005.

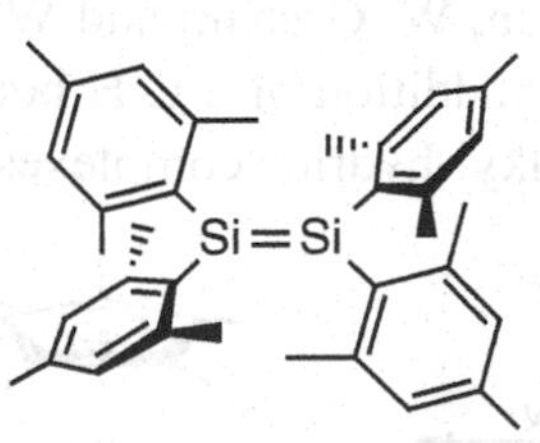

**Compound with a
silicon-silicon double bond**
West, 1981

In 1981, Roald Hoffmann and K. Fukui were awarded the Nobel Prize in Chemistry for the development of semi-empirical theoretical methods and their applications to organic, inorganic and organometallic chemistry. Hoffmann was also recognized for the development of the isolobal analogy, a very useful concept in molecular chemistry.

1970-1985: ACTIVATION OF C-H BONDS IN HYDRO-CARBONS, σ-BOND METATHESIS AND H₂ AS A LIGAND

Intramolecular examples of C-H activation of alkyl groups such as Wilkinson's β-H elimination and Schrock's α-H elimination had been disclosed in the 1970s. Seminal work by Shilov on methane deuteration with platinum salts was reported as early as 1969, then methane functionalization to CH_3Cl and CH_3OH was published by Shilov in the 1970s, although it is still not known even now if this activation proceeds by oxidative addition or σ-bond metathesis. In 1979, two research groups, those of Malcolm L.H. Green at Oxford and Didier Astruc at Rennes reported modes of C-H activation in toluene, respectively, by UV-induced oxidative addition in tungstenocenes and electron transfer using O_2 or air at subambient conditions in 19-electron Fe^I complexes.

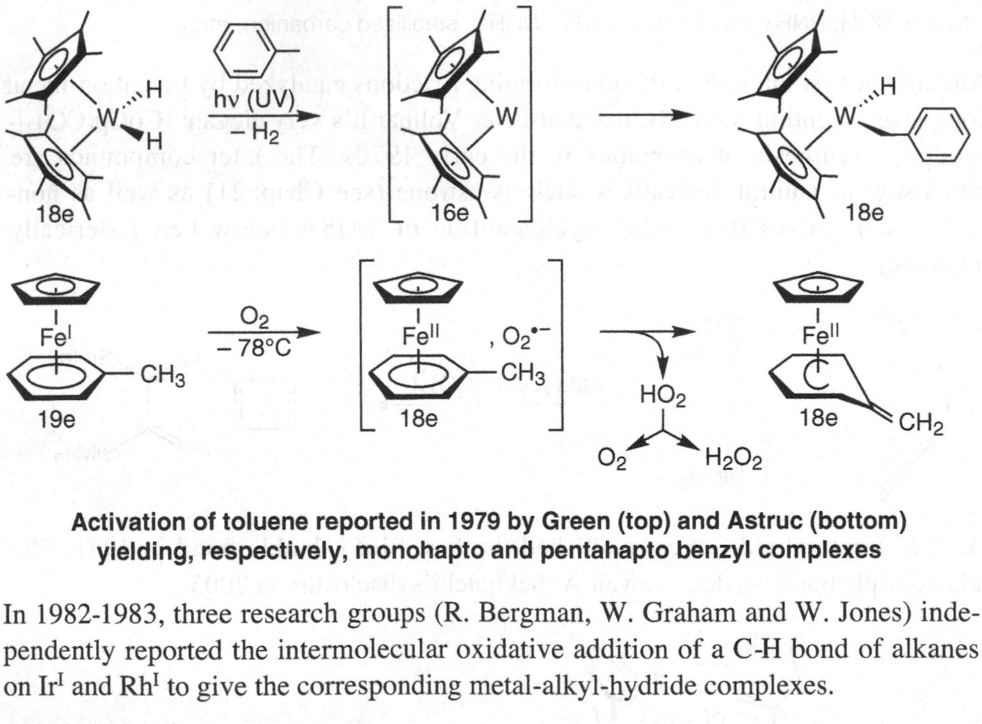

Activation of toluene reported in 1979 by Green (top) and Astruc (bottom) yielding, respectively, monohapto and pentahapto benzyl complexes

In 1982-1983, three research groups (R. Bergman, W. Graham and W. Jones) independently reported the intermolecular oxidative addition of a C-H bond of alkanes on Ir^I and Rh^I to give the corresponding metal-alkyl-hydride complexes.

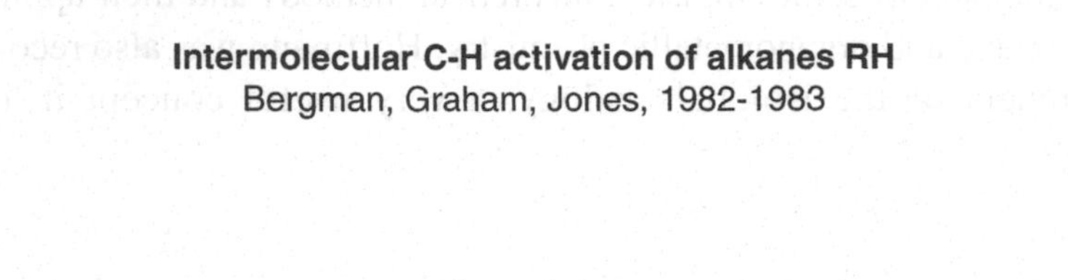

Intermolecular C-H activation of alkanes RH
Bergman, Graham, Jones, 1982-1983

In 1983, Maurice Brookhardt and Malcolm Green published their concept of the intramolecular agostic C-H bond in complexes that feature a Lewis-acidic transition-metal center, i.e. those that possess a vacant coordination site. During the same year, Patricia Watson at the Du Pont company and John E. Bercaw at Caltech independently reported the activation of methane by d^0 complexes of lanthanides and early transition metals (Lu and Sc respectively) according to the new σ-bond metathesis mechanism (an *intramolecular* version being Schrock's α-elimination, in other d^0 complexes, *vide supra,* which was developed by Tobbin Marks at Evanston with his seminal work in d^0 organoactinide chemistry also reported in 1982).

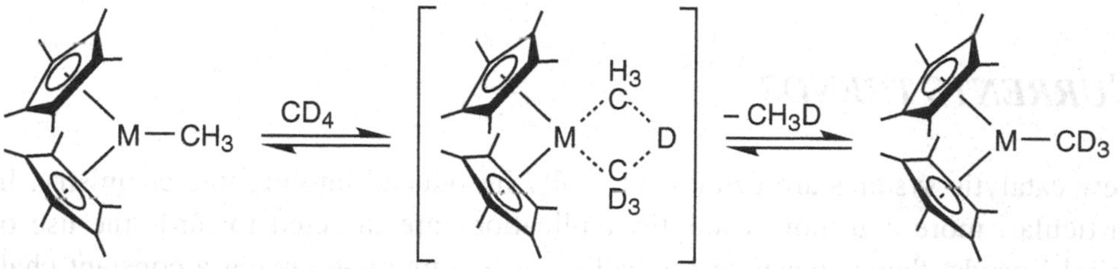

M = Lu (P. Watson) or Sc (J. Bercaw)

σ-bond metathesis reactions in CH₄ achieved in cyclohexane as a solvent at 70°C

In 1984, Gregory Kubas published another important discovery, namely that dihydrogen complexes are stabilized and isolable when the transition metal is sufficiently poor in terms of electronic density to avoid oxidative addition and the metal-dihydride formation.

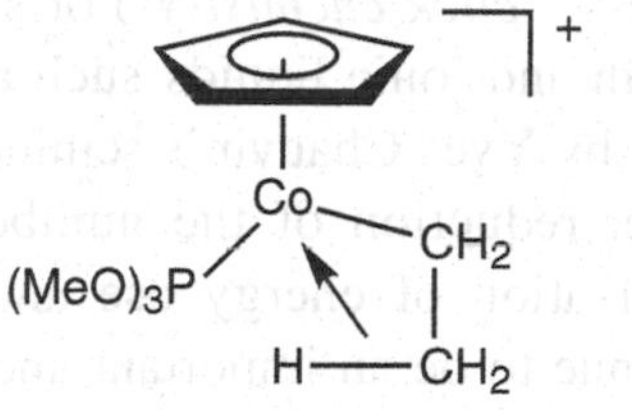

"Agostic"C-H-M bond
Brookhardt and Green, 1983

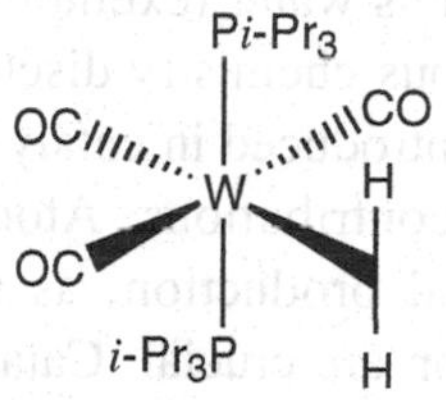

H₂ complex
Kubas, 1984

In 1985, Brintzinger published the first ansa-metallocene and demonstrated, along with Kaminsky, that propylene polymerization proceeds in a stereoregular fashion to yield isotactic polypropylene. This discovery disproved the view of Ziegler and Natta according to which polymer tacticity is induced by the structure of the surface of the heterogeneous initiator. Brintzinger and Kaminsky showed that polymer tacticity can be induced at a single reaction center. This event revolutionized the world of polymers and generated a sustained interest towards this research area.

Initiator complex for isotactic propylene polymerization - Brintzinger, 1985

CURRENTS TRENDS

New catalytic systems are being continually introduced into organic chemistry. In particular, more and more catalytic applications are directed towards the use of chiral ligands, the synthesis of optically active compounds being a constant challenge for the pharmaceutical industry. In the latter area, the discovery of the non-linear effect in asymmetric catalysis by Kagan represented a particularly important breakthrough.

It is becoming increasingly necessary to protect the environment by the development of new processes that are less polluting in terms of solvent emission and toxicity. The chemistry dedicated to these problems is called *"Green Chemistry"*. In this respect, the use of clays as solid supports or other environmentally friendly media such as water (exemplified by Sharpless' *"click chemistry"*) or supercritical CO_2, fluorous chemistry disclosed by Horvath and ionic liquids such as imidazolium salts introduced in catalysis in the 1980s by Yves Chauvin's seminal work are important contributions. Atom economy, i.e. reduction of the number of steps in industrial production, as well as minimization of energy use and chemical consumption are crucial. Catalysis will continue to be an important and integrated part of *"Green Chemistry"*. The recovery of catalysts and their re-use is essential for both economic and ecological reasons. The development of well-defined supported and metallodendritic catalysts represents an important step in this direction. High oxidation-state Lewis-acid catalysts have been discovered, and the concept of Lewis-acid catalyst (classic in organo-transition metal chemistry with 14- and 16-electron complexes) is increasingly used in organic synthesis, for instance by Kobayashi with activation of carbonyl oxygen atoms by rare earth complexes in aqueous solvents (water + THF or even water as the only solvent).

Fundamental research is more essential than ever in order to meet these challenges. A striking success story in organometallic chemistry was the discovery and use of metathesis catalysts that are compatible with organic functional groups. The recent development and use of the Schrock and Grubbs catalysts is particularly impressive.

Whereas a whole family of very active high oxidation-state molybdenum and tungsten catalysts were reported in 1990 by Schrock followed by chiral ones, the air-stable Grubbs ruthenium catalysts, which appeared later in the 1990s, have proved to be also particularly useful in organic chemistry.

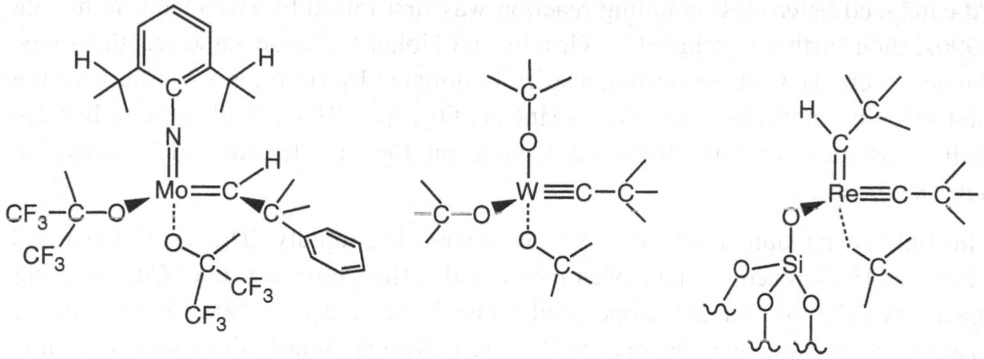

(1)   **(2)**   **(3)**

Schrock-type alkene- and alkyne metathesis catalysts
(1) Very reactive commercial prototype of the family of Schrock's metathesis catalysts, 1990 (can achieve RCM of tri- and tetrasubstituted olefins) – **(2)** Prototype of Schrock's alkyne metathesis catalysts, 1982 – **(3)** Basset's well-defined active catalyst for the metathesis of alkenes and alkynes, 2001.

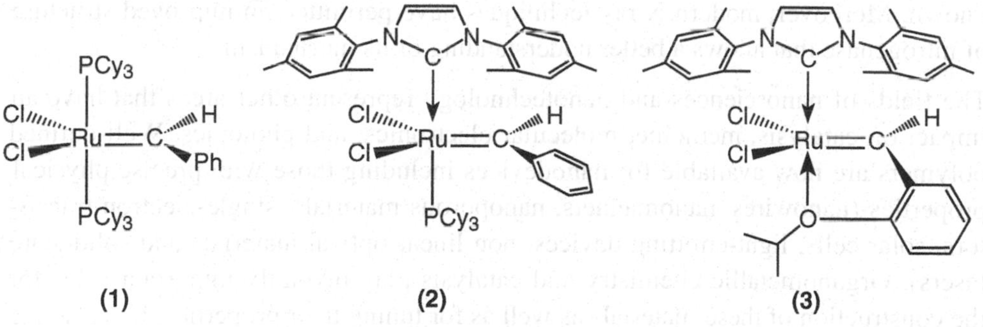

(1)   **(2)**   **(3)**

Air-stable Grubbs' type alkene metathesis catalysts
(1) Grubbs, 1st generation, 1995 - **(2)** Grubbs, Nolan, Herrmann, 2nd generation, 1999 - **(3)** Hoveyda, 2001. Hoveyda's first catalyst with PCy$_3$ instead of the diaminocarbene (1999).

The Schrock and Grubbs catalysts are complementary and their commercial availability continues to have a pronounced impact on synthetic organic chemistry and materials science. Yves Chauvin, Robert H. Grubbs and Richard R. Schrock have shared the 2005 Nobel Prize in chemistry.

Supported versions have also appeared, most notably Basset's well-defined silica-supported rhenium catalyst which is active for the metathesis of both alkenes and alkynes. Basset's concept of surface silica-supported catalysts was also successfully extended to alkane activation by σ C-H and C-C bond metathesis, and it is remarkable that a unified view is now proposed for both alkene and alkane metathesis mechanisms.

Improvements in ligand design for catalysis also appeared with Arduengo's stable *N*-heterocyclic carbenes (NHC) whose importance in olefin metathesis and Pd-catalyzed hetero C-C coupling reaction was first raised by Herrmann in the late 1990s, then further developed by Grubbs and Nolan for these same reactions (see the above chart). Carbene design, also early initiated by Herrman's synthesis of the first bridging methylene complex $\{[MnCp(CO)_2]_2(\mu_2\text{-}CH_2)\}$, is illustrated, besides NHC carbenes, by Guy Bertrand's work on the stabilization of *non-cyclic* heterocarbenes.

The field of palladium catalysis is also progressing rapidly. The Tsuji-Trost and Mizoroki-Heck reactions have been mentioned earlier, however other C-C coupling reactions (Miyaura-Suzuki, Sonogashira and Stille) are now extensively used in organic chemistry, and the search for non-polluting Sonogashira and Miyaura-Suzuki reactions is being actively pursued.

Another rapidly growing area is that of bioinorganic chemistry. In this field, modes of activation are found that sometimes resemble those known in organometallic chemistry. The case of vitamin B_{12} is no longer an isolated example. It is now known that enzymatic activation of H_2 resembles that established for inorganic transition-metal-dihydrogen complexes, the acidic properties of which are well known. Moreover, modern X-ray techniques have permitted an improved structure of nitrogenase that allows a better understanding of its mechanism.

The fields of nanosciences and nanotechnology represent other areas that have an impact on catalysis, medicine, molecular electronics, and photonics. Well-defined polymers are now available for nanodevices including those with precise physical properties (nanowires, nanomagnets, nanoporous materials, single-electron transistors, solar cells, light-emitting devices, non-linear optical materials and solid-state lasers). Organometallic chemistry and catalysis play pivotally important roles for the construction of these materials as well as for tuning their properties. It is also the best method of choice for the deposition of ultra-pure metals for micro-electronics. Nanoparticle catalysis is now growing very fast.

The famous inorganic anticancer drug *cis*-$PtCl_2(NH_3)_2$, named *cis*-platin and discovered by Rosenberg, has its organometallic equivalent, Cp_2TiCl_2, whose cancerostatic action was disclosed by H. Köpf and P. Köpf-Maier in 1979. This German group and many others have further developed this useful applied area of medicinal organometallic chemistry.

All these applications and emerging frontier disciplines have a profound impact on both fundamental and applied research. The molecular chemistry of metals is extraordinarily rich and varied, and one can be sure that it will always successfully contribute to the ever-increasing needs of modern society.

PART I

STRUCTURES OF THE TRANSITION-METAL COMPLEXES

Chapter 1
Monometallic transition-metal complexes

Chapter 2
Bimetallic transition-metal complexes and clusters

PART I

Chapter 1

MONOMETALLIC
TRANSITION-METAL COMPLEXES

The transition metals are, by definition, the elements with an incomplete d shell, and an empty last p shell (the valence one). These elements will need to complete more or less these subshells with electrons given or shared by the ligands in order to give rise to stable compounds. These electrons provided by the ligands allow the metal to reach more or less exactly the electronic structure of the rare gas following them on the same line of the periodic table. These notions will be refined later, but we will first examine the electron count given by the ligands to the transition metal.

We will use the convention that considers all ligands as neutral in transition-metal complexes. It is simple and close to reality for most transition metals. The ionic convention is less used, but appropriate for the scandium, lanthanide and actinide complexes. The characteristics of the complexes (see section 2) are the same whatever convention is used. In this chapter, we will examine the mononuclear complexes whereas the bi- and polymetallic complexes are described in Chap. 2.

1. THE LIGANDS

Given the convention according to which all the ligands are considered as neutral, there are two classes of ligands. First, there are ligands bringing one or several electron pairs to the metals; they are even ligands and are designated as L or L_n, n being the number of electron pairs given to the metal. Then, there are ligands bringing one electron or an odd number of electrons to the metal, i.e. the radical-type ligands; they are designated as X (one electron) or XL_n (odd number of electrons). The L or L_n ligands do not accept valence electrons from the metal to make the metal-ligand bond, because the bond involved is of the donor-acceptor type. On the other hand, the X or XL_n ligands require one valence electron from the metal to form the metal-ligand bond. Thus, the M-X bond resembles the covalent bond in organic chemistry as each partner brings one electron to form the bond electron pair. The triplet carbenes or alkylidenes (CR_2), oxo (oxene, O) and nitrido

(NR) are biradicals that form a double bond with the metal, and will therefore be considered as X_2 ligands, whereas the singlet carbenes give a pair of electrons to the metal, and they are thus considered as L ligands (see Chap. 9).

The number of electrons usually given to the metal by the most common ligands are the following:

1-electron, radical-type X ligands

▸ H, F, Cl, Br, I

▸ OH, OR, SR, NH_2, NR_2, PR_2, AsR_2

▸ CH_3, CR_3 (alkyl), Ph, Ar (aryl), $CH=CR_2$ (vinyl), $C≡CR$ (alkynyl), COR (acyl), SiR_3 (silyl), SnR_3, (stannyl)

▸ Metal-carbonyl radicals: $M(CO)_n$ (M being a metal with an odd number of valence electrons), $MCp(CO)_n$ (M being a metal with an even number of valence electrons)

▸ NO (bent bond with the metal) (see Chap. 7.5.4).

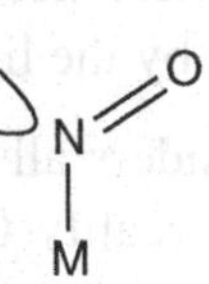

2-electron X_2 biradical-type ligands

▸ $=CH_2$, $=CR_2$ (carbenes or alkylidenes), $=C=CR_2$ (vinylidenes) (see Chap. 9.1)

▸ $=NR$ (amido), $=PR$ (phosphinidenes), $=O$ (oxo or oxene) (see Chap. 9.3)

▸ $-CH_2(CH_2)_nCH_2-$ (cycloalkyls), $-O-O-$ (peroxide), CO_2, CS_2

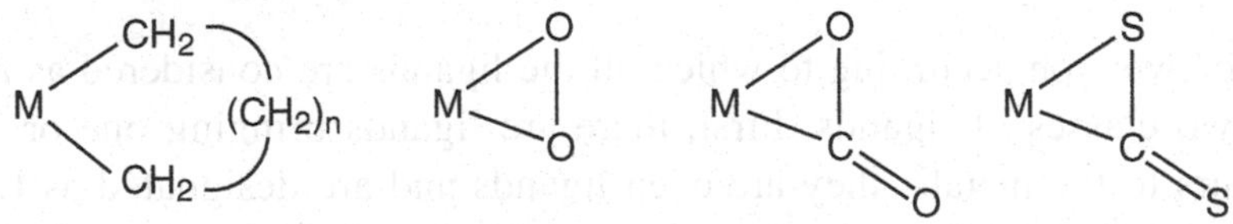

2-electron L ligands

They are:

▸ The donors of a non-bonding electron pair of an heteroatom

▸ H_2O, H_2S, ROH, RSH, R_2O, THF, R_2S

▸ NH_3, NR_3, PR_3 (phosphine), $P(OR)_3$ (phosphite), AsR_3 (arsine), SbR_3 (stilbine)

▸ N$_2$, O$_2$ (can also be L ligands *via* each of the two heteroatoms of the molecule), CO (carbonyl), CS (thiocarbonyl), (CH$_3$)$_2$CO, CH$_3$CN, RCN (nitrile), RNC (isonitrile), CH$_2$Cl$_2$ (non-bonding electron pair of a chlorine atom)

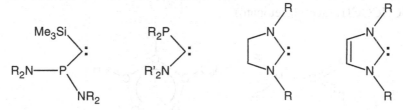

Triplet carbenes such as Bertrand's non-*N*-heterocyclic carbenes (non-NHCs) and Arduengo's *N*-heterocyclic carbenes (NHCs)

Bertrand's carbenes (non-NHCs)   **Arduengo's carbenes (NHCs)**

▸ The donors of a π electron pair

Alkenes, alkynes (can also be 4-electron L$_2$ ligands), double C=O bond or oxygen non-bonding electron pair of an aldehyde or a ketone (also C=S), O$_2$

X, Y = C, O, N, etc.

▸ The donors of a σ bond electron pair

H-H (dihydrogen ligand; see Chap. 8.6), H-SiR$_3$ (Si-H silane bond)

X = H, C, Si, etc.
Y = H

H-CR$_3$ (intramolecular: C-H bond belonging to another ligand of the same metal by another bond); the C-H-M bond is called "agostic" (see Chaps 6.3, 6.4 and 8.2)

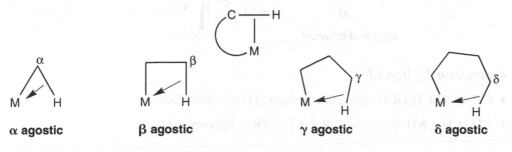

α agostic   β agostic   γ agostic   δ agostic

3-electron X₃ ligands (trivalent)

$\equiv$CR (carbynes or alkylidynes), $\equiv$N (nitrido), $\equiv$P (phosphido) (see Chap. 9.2)

3-electron radical LX ligands

▸ $CH_2{=}CH{-}CH_2$ (allyl), CHCHCH (cyclopropenyl) (see Chap. 9.4)

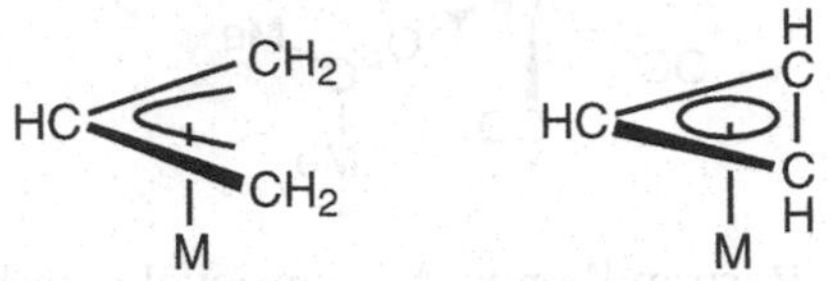

▸ CO_2R (carboxylato), O_2CNR_2 (carbamato), S_2CNR_2 (dithiocarbamato), $CH_3COCHCOCH_3$ (acetylacetonato)

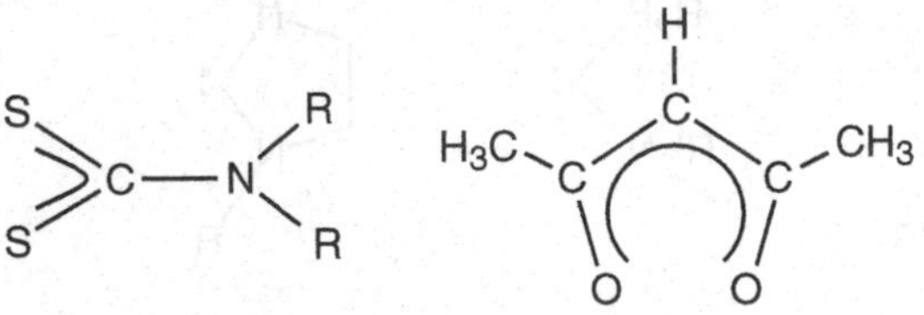

▸ NO (linear bond with the metal), can also be a bent 1-electron X ligand (see Chap. 7.5.4)

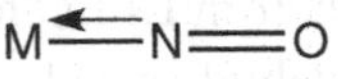

▸ The ligands halogeno, alkoxy (or aryloxy), alkylthio (or arylthio), amino, phosphido and NO that are all 1-electron ligands can also be 3-electron LX ligands when the metal is electron-deficient; i.e. has less than 18 valence electrons (*vide supra*)

▸ Analogously, the acyl and methoxy ligands are 3-electron ligands in electron-deficient early transition-metal complexes such as $[Ti(OCH_3)_4]$

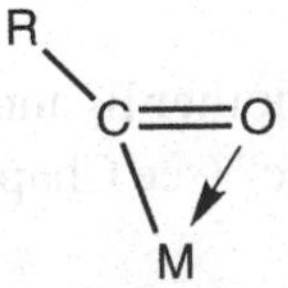

Acyl and thioacyl

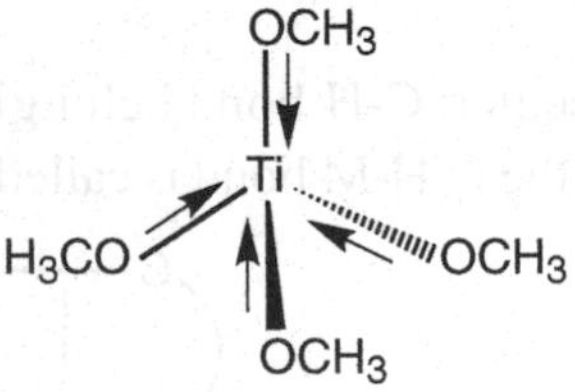

4-electron L₂ ligands

▸ $CH_3OCH_2CH_2OCH_3$ (dimethoxyethane, DME), disulfides

▸ $NH_2(CH_2)_nNH_2$ (diamines), $R_2P(CH_2)_nPR_2$ (diphosphines)

▶ $R_2As(CH_2)_nAsR_2$ (diarsines)

▶ $CH_2=CH–CH=CH_2$ (*cis*-1,3-butadiene), other dienes, $(CH)_4$ (cyclobutadiene) (see Chap. 10.2)

4-electron LX₂ ligands

▶ The oxo and imido ligands, that are X_2 ligands, can become LX_2 ligands when the metal is electron-deficient, i.e. has less than 18 valence electrons.

▶ The 2-electron imido ligand forms an angle with the metal, because the *p* orbital does not interact with the metal. On the other hand, in the 4-electron imido ligand, the *p* orbital of the nitrogen atom interacts with the vacant *d* metal orbital, and the M-N-R chain is linear, as for instance in $[Mo(NAr)_3]$.

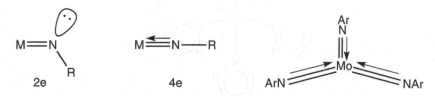

▶ Hawthorne has introduced the carborane ligands and prepared many metallo-carborane complexes, in particular the dianion *nido*-carborane $C_2B_9H_{11}^{2-}$ whose frontier orbitals are analogous to those of the cyclopentadienyl (see Wade's rules for the stereoelectronic structure of the carboranes, Chap. 2.2.4). In its neutral form, this ligand is counted as a 4-electron LX_2 ligand.

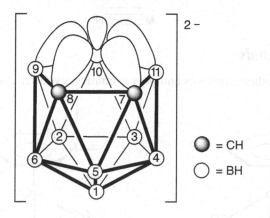

5-electron L₂X radical ligands (see Chap. 10.5)

▸ Dienyls

 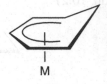

C₅H₅ (cyclopentadienyl, noted Cp)   **C₆H₇ (cyclohexadienyl)**

▸ Heteroatoms can be incorporated in rings, which gives rise to a large variety of ligands. Mathey's phospholyl C_4H_4P, comparable to the cyclopentadienyl, and Herberich's borabenzene, in which $5p$ electrons are delocalized in 6 contiguous orbitals are among the best known 5-electron L_2X ligands.

▸ $RB(C_3H_3N_2)_3$, tris-1- (pyrazolyl)borato, noted Tp, with R = H, alkyl, aryl.

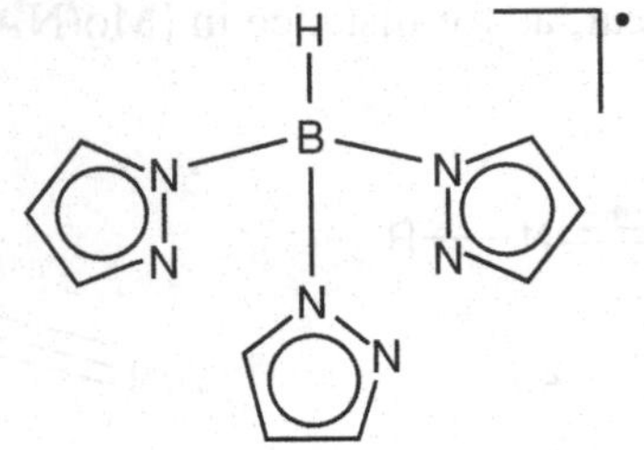

Tp

5-electron LX₃ ligands

The X_3 nitrido and phosphido ligands can aso become LX_3 if the metal is electron deficient

$$M\equiv\!\!\!\!-N$$

6-electron L₃ ligands

▸ C_6H_6 (benzene), other aromatics and polyaromatics, cyclic trienes, borazine

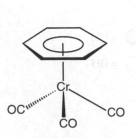 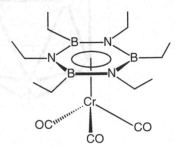

Benzene complex   **Borazine complex**

▶ C_5H_5N (pyridine), C_5H_5P (phosphinine), C_4H_4S (thiophene)– these three ligands are most often L ligands, however, *via* the lone pair of the heteroatom

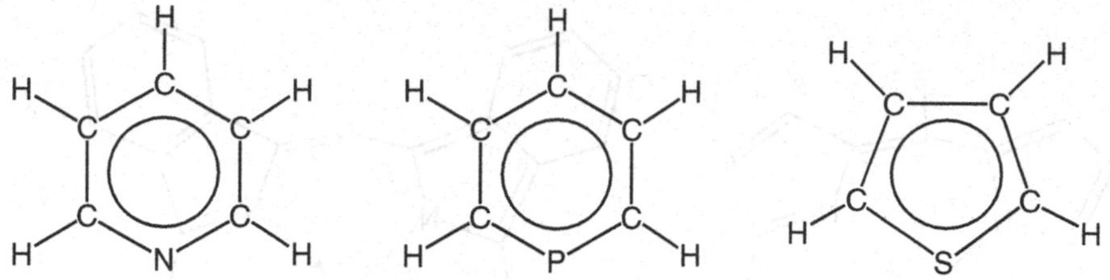

▶ Trisulfides, for instance $(CH_2CH_2S)_3$; triamines, for instance Wiegardt's triazacyclononane $(CH_2CH_2NR)_3$; triphosphines, for instance $CH(CH_2PPh_2)_3$

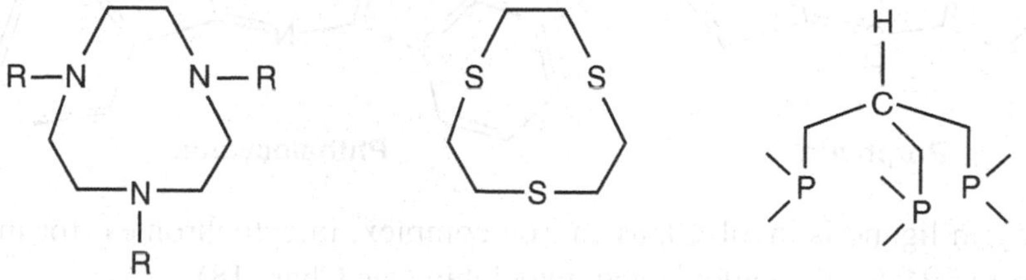

[9]ane   **[9]ane S$_3$**

▶ Tris(pyrazolyl)methanes and the little-developed tris(pyrazolyl)silanes are the neutral analogs of the widely known tris(pyrazolyl)borates

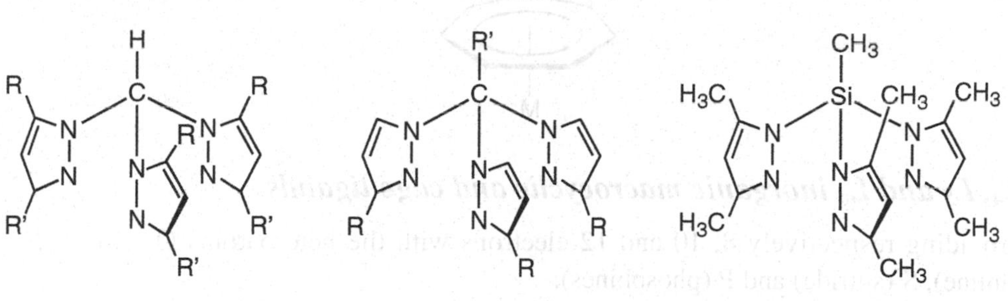

▶ Pincer *N*-heterocyclic carbenes recently introduced by Crabtree and Eisenstein

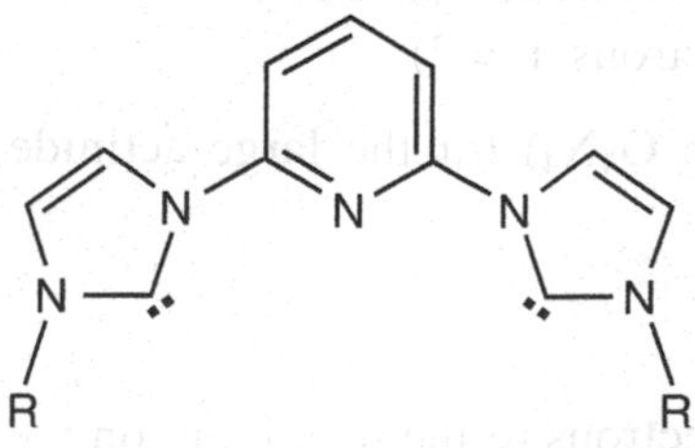

Various R, in particular R = *n*-Bu for solubility

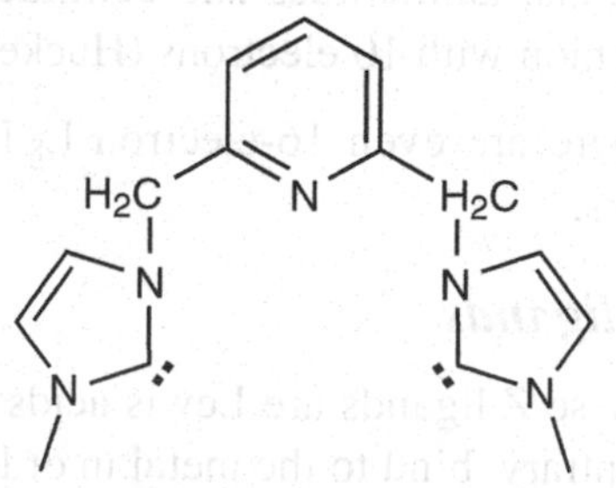

and other pyridine-linked pincer ligands (phosphines, amines, thiols)

6-electron L₂X₂ ligands (divalent)

N₄-ligands: porphyrins and phthalocyanins

Porphyrin   **Phthalocyanin**

The porphyrin ligand is involved, as an iron complex, in cytochromes (for instance in cytochrome P450), hemoglobin and myoglobin (see Chap. 18).

7-electron L₃X radical ligand

C_7H_7 (cycloheptatrienyl)

L₄, L₅ and L₆ inorganic macrocyclic and cage ligands

Providing respectively 8, 10 and 12 electrons with the heteroatoms O (ether), N (amine), S (sulfide) and P (phosphines).

Cyclooctatetraene C_8H_8 (COT) can be an 8-electron L_4 ligand. With the alkaline metals, lanthanides and actinides however, COT is found as the aromatic COT^{2-} dianion with 10 electrons (Hückel's rule: $4n + 2\pi$ electrons, n = 2).

There are even 16-electron L_8 ligands (for example O_4N_4) for the large actinide ions.

Z ligands

These Z ligands are Lewis acids that do not bring electrons to the metal, but on the contrary, bind to the metal in order to get their missing electron pair. Example: BH_3, $AlMe_3$.

In conclusion, the modes described here are the most frequent ones, but others can be found. The ligands can have variable coordination modes. If one hesitates, one can follow the 18-electron rule (vide infra). Such ambiguities exist for the ligands NO, polyenes and alkynes. NO is frequently found as either a bent 1-electron or a linear 3-electron ligand depending on the metal electron count. Alkynes often are 4-electron ligands but the 2-electron count is found when the metal has 18 valence electrons. Then, the second alkyne π orbital does not find a vacant metal orbital of adequate symmetry. Other coordination modes will be found in polynuclear compounds for the same ligands. Organometallic chemistry is very versatile and the coordination modes can be numerous for the same ligand.

Summary of ligands: most common modes

Ligands	Type	Number of electrons given to the metal
H, CH_3, Ph	X	1
C_2H_4, CO, PPh_3, O_2, N_2	L	2
π-allyl, π-enyl, dtc	LX	3
diolefin, diphosphine, diamine	L_2	4
Cp, dienyl	L_2X	5
porphyrinato	L_2X_2	6
arene, triolefin	L_3	6
cycloheptatriene	L_3X	7
cyclooctatetraene	L_4	8

Ambivalent ligands

Ligands		Type	Number of e	Condition
F, Cl, Br, I, OR, NR_2, PR_3, NO, allyl		X LX	1 3	according to metal need
carbenes	←CR_2 =CR_2	L X_2	2 2	electrophilic carbene nucleophilic carbene
oxo, imido, phosphinidene O, NR, PR		X_2 LX_2	2 4	according to metal need
carbyne	≡CH	X_3 LX	3 3	electrophilic carbyne nucleophilic carbyne
nitride	≡N	X_3 LX_3	3 5	according to metal need
alkyne	RC≡CR	L L_2	2 4	according to metal need

2. THE CHARACTERISTICS OF THE TRANSITION-METAL IN THE COMPLEXES

The four main characteristics of the transition-metal in a complex that allow to define this complex well are the Number of Valence Electrons (NVE), the Number of Non-Bonding Electrons (NNBE), the Oxidation State (OS) and the Coordination number (C).

2.1. THE NUMBER OF VALENCE ELECTRONS (NVE)

It is the total number of electrons in what is defined as the transition-metal valence shell, i.e. the group of ns, $(n-1)\,d$, np sublayers. It contains the electrons that are initially in the metal valence shell and the electrons brought by the ligands.

$$NEV = n_M + 2\,n_L + n_X - q$$

with: n_L = number of L ligands, n_X = number of X ligands, q = charge of the complex, n_M = number of electrons initially in the metal valence shell, i.e. 2 electrons from the ns sublayer and the electrons of the sublayer $(n-1)\,d$ which depends on the nature of the metal sublayer.

$n_M = 2 + y$ for a transition element having the electronic structure $ns^2\,(n-1)\,d^y\,np^0$.

y in d^y:	1	2	3	4	5	6	7	8	9
n_M:	3	4	5	6	7	8	9	10	11
	Sc	Ti	V	Cr	Mn	Fe	Co	Ni	Cu
	Y	Zr	Nb	Mo	Tc	Ru	Rh	Pd	Ag
	La	Hf	Ta	W	Re	Os	Ir	Pt	Au

Example: $[FeCp(CO)_2]^-$

Cp is an L_2X ligand and CO is an L ligand; the complex can be written in the form FeL_4X^- with $n_{Fe} = 8$.

$$NVE = n_{Fe} + 2\,n_L + n_X - q = 8 + 2 \times 4 + 1 - (-1) = 18$$

The NVE often is 18, which corresponds to the electronic structure $ns^2\,(n-1)\,d^{10}$ np^6 of the rare gas that follows the transition-metal on the same line of the periodic table. This "18-electron rule" is more a trend than a rule, and will be examined in detail later in this Chapter (section 4).

2.2. THE NUMBER OF NON-BONDING ELECTRONS (NNBE)

This is the number of electrons remaining in the valence electron shell after formation of the bonds with the ligands. Inorganic chemists often use the symbol d^n

for a metal atom or ion (or the complex itself) that has n non-bonding electrons on this metal in the complex (n = NENL).

$$NNBE = n_M - n_X - q = n_M - OS$$

For instance, $[FeCp(CO)_2]^-$, for which NNBE = $8 - 1 - (-) = 8$ is said to be a d^8 complex.

The value of NNBE is important in understanding the reactivity of a complex. For instance, the metals in complexes whose NNBE is 0 cannot be oxidized, and reactions on the metal in such complexes will be limited to those bringing electrons. The NNBE will also be useful to define the geometry of the complex, because the orbitals of the non-bonding electrons will have to be taken into account in the organization of space around the metal center.

Note - do not confuse d^y, given by the electronic structure $ns^2 (n - 1) d^y np^6$ of the atom and d^n indicating the NNBE (equal to n) of the metal in a complex. This double d nomenclature (atomic and molecular) is unfortunate. Thus, it is important to clearly distinguish these two meanings; n = y only if the oxidation state is 2. For instance, ferrocene, $[FeCp_2]$ is a d^6 complex, n = y = 2. On the other hand, $[Fe^0(CO)_5]$ is a d^8 complex (n = NNBE = 8) in which the iron atom has $6d$ electrons (y = 6).

2.3. THE OXIDATION STATE (OS)

The oxidation state (OS) is obvious when the complex contains only L or L_n ligands, because it is then equal to the charge q of the complex. For instance, in $[Fe(H_2O)_6]^{2+}$, it clearly appears, just by looking at the formula, that the OS is +2. Also, the OS is –2 in Collman's reagent $[Fe(CO)_4]^{2-}$. The OS is not so apparent, however, when the complex also contains X or X_n ligands, such as in $[FeCp(CO)_2]^-$, for example. It is then necessary to add the number n_X of X ligands to the charge q of the complex:

$$OS = n_X + q$$

Thus, for the above example of the monoanionic iron complex, OS = $1 + (-1) = 0$.

The transition metals all have numerous oxidation states, which accounts for the richness of their chemistry. For purely organometallic compounds, the oxidation states are low, zero or negative. On the other hand, purely inorganic complexes always have positive, even high oxidation states. For instance, in $[Fe(S_2CNMe_2)_3]^+$, each dithiocarbamato ligand is LX, and the complex is of the type $FeL_3X_3^+$; OS = $3 + 1 = +4$.

Let us recall that the OS is a formalism that we have decided to adopt for the whole area of inorganic and organometallic chemistry as a very practical form of language to communicate in the chemical community, and any controversy about oxidation states is meaningless. One can take the example of carbenes to illustrate this point.

Some carbenes (triplet ones) are clearly X_2 ligands whereas others (singlet ones) are rather L ligands, but in complexes, there is a whole area in between. The decision is clear-cut, but the reality is often not so obvious. CO is an L ligand, thus $[Ni(CO)_4]$ has OS = 0, but if it was considered an X_2 ligand, the oxidation state Ni would be 8, which would make an enormous difference. Is the metal so electron-rich that it deserves a 0 oxidation state or does the reality correspond to a somewhat larger OS? Another example is the family of the hydrides: ReH_9^{2-} is a Re^{VII} complex, but is the Re metal so electron-poor in this complex that it really deserves such a high oxidation state?

Note, however, that a strength of this formalism is that the OS cannot be higher than the number of the metal group. For instance, for Sc and the lanthanides, the OS cannot be larger than +3; for the Ti column, it cannot be larger than +4; for the V column, it cannot be larger than +5. For instance, this corroborates the fact that CO is conventionally an L ligand, not X_2: in $[Cr(CO)_6]$, Cr (OS = 0) could not have OS = +12, because the Cr atom can only provide 6 electrons.

2.4. THE COORDINATION NUMBER (C)

It is the number of occupied coordination sites on the transition-metal center. Thus, a L or X ligand occupies one site, a L_2 ligand (and most often a LX ligand) occupies two sites, and the L_3 or L_2X ligands occupies 3 sites. When the complex is of the form $ML_nX_p{}^q$, the coordination number is:

$$C = n_L + p_X$$

Example: $[FeCp(CO)_2]^-$

$[FeCp(CO)_2]^-$ is a FeL_4X^- complex: C = 4 + 1 = 5.

This simple formula does not work, however, when the complex contains one or several X_2 or X_3 ligands or a single-site LX ligand such as linear NO. Applying the formula gives a result in excess. In these cases, one must count the number of sites carefully.

Example: $[RuCp(CO)_2(=CH_2)]^+$

$[RuCp(CO)_2(=CH_2)]^+$ has a coordination number equal to 6 (3 for Cp, 2 for CO and 1 for the carbene).

Summary of the characteristics

$$NVE = n_M + 2\,n_L + n_X - Q$$

$$NNBE = n_M - n_X - q = n_M - DO$$

$$OS = n_X + q \text{ (mononuclear complexes)}$$

$$C = n_L + n_X \text{ (except if } X_2, X_3, \text{ single-site LX)}$$

Coordination number (C)

C			Example
2	linear	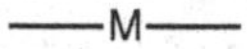 ——M——	$[NC–Ag–CN]^-$
3	trigonal	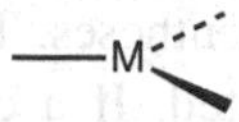	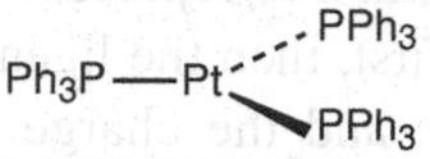 $Ph_3P—Pt$ with PPh_3 and PPh_3
	T shape	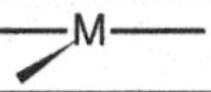 ——M——	$[Rh(PPh_3)_3]^+$
4	tetrahedron	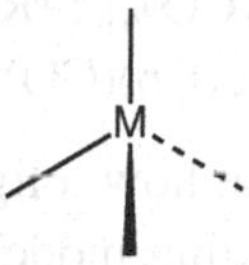	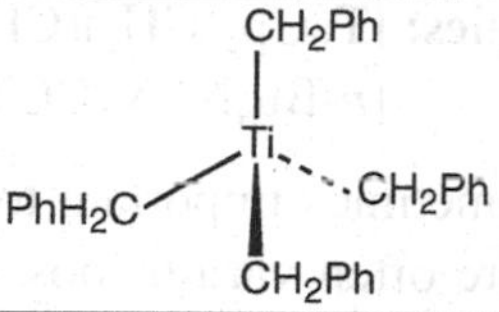 Ti with CH_2Ph, PhH_2C, CH_2Ph, CH_2Ph
	square planar	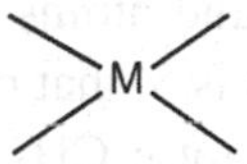	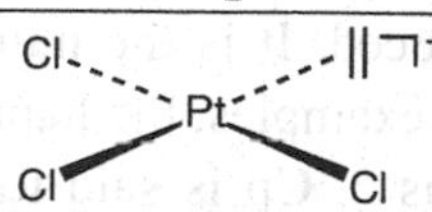 Cl, Cl, Pt, Cl
5	trigonal bipyramid	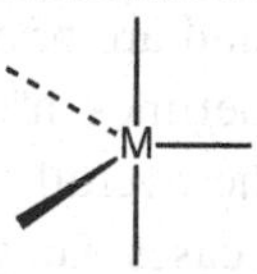	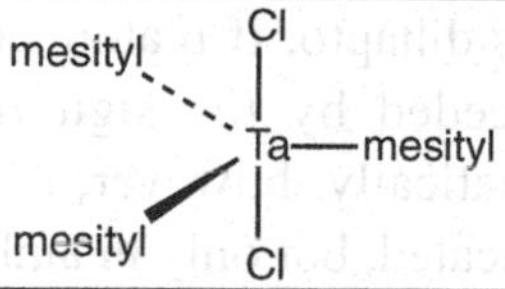 Ta with mesityl, mesityl, mesityl, Cl, Cl
	square-based pyramid	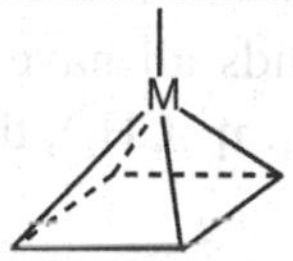	$[Co(CNPh)_5]^{2+}$
6	octahedron	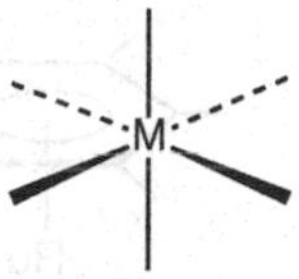	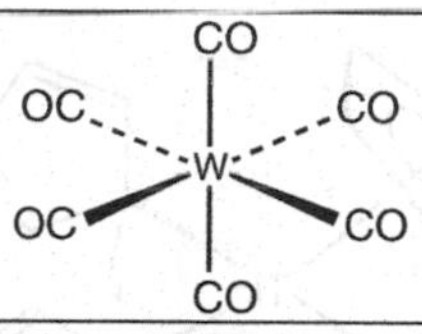 W with CO, OC, CO, OC, CO, CO
	pseudo-octahedral	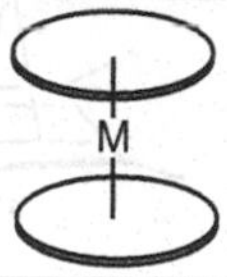 M	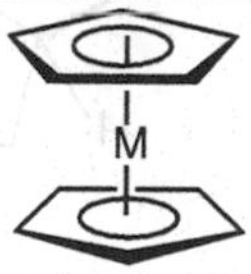 M
	antiprism	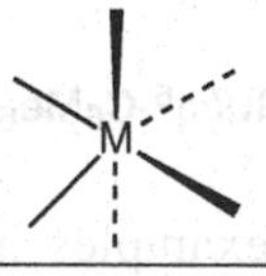	$[WMe_6]$
7	capped octahedron	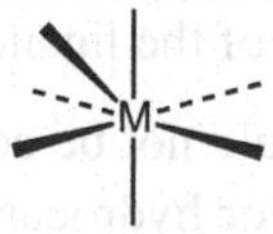	$[ReH(PR_3)_3(MeCN)_3]^+$
	pentagonal biprism	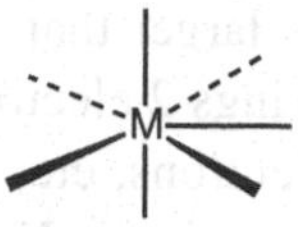	$[IrH_5(PPh_3)_2]$

3. HAPTICITY OF THE LIGANDS AND LINEAR WRITING OF THE COMPLEX FORMULAS

It is common to represent the complexes using linear writing. The metal should be written first, then the ligands in parentheses. Then, the complex is written between brackets, and the charge is indicated. If a counter-ion is written, however, the charges usually do not appear:

Examples: $[TaCp_2(CH_3)(CH_2)]$; $[V(CO)_6]^-$; $[RuCp(CO)_2(C_2H_4)]^+$;
$[n\text{-Bu}_4N][V(CO)_6]$; $K[Ru(Cp)(CO)(C_2H_4)]$

It is sometimes important to indicate how a ligand is bonded to the metal, because there are often various possible binding modes. This is why the ligand hapticity is introduced. It is the number of ligand atoms that are bonded to the metal. In the above examples, the hapticity of Cp is 5, that of CH_3, $=CH_2$ and CO is 1 and that of C_2H_4 is 2. Cp is said to be pentahapto; CH_3, $=CH_2$ and CO are monohapto, and C_2H_4 is dihapto. If n atoms of a ligand are bonded to the metal, the ligand formula is preceded by the sign η^n- (sometimes h^n- is also found). This is not done systematically, however, because the overall writing of the complex would be too complicated, but only in ambiguous cases, i.e. when the bonding mode is unusual.

For the above examples, the ligands all have their usual bonding modes in the complexes (η^5-Cp, η^1-CO, η^2-C_2H_4, η^1-CH_2), thus it is not necessary to emphasize it by the η^n- or h^n-writing.

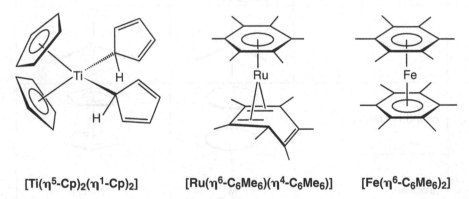

$[Ti(\eta^5\text{-Cp})_2(\eta^1\text{-Cp})_2]$ $[Ru(\eta^6\text{-C}_6\text{Me}_6)(\eta^4\text{-C}_6\text{Me}_6)]$ $[Fe(\eta^6\text{-C}_6\text{Me}_6)_2]$

Here (above) are some classic examples among many others for which it is indispensable to write the hapticity of the ligands.

Note 1 - The ligand hapticity should not be confused with the number of electrons given by the ligand to the metal. For hydrocarbons, these two numbers usually are identical, especially for hapticities larger than 1. The C_2H_4 ligand is bound to the metal by its 2 carbon atoms and brings 2 electrons, the Cp is bound to the metal by its 5 carbon atoms and brings 5 electrons, etc. The monohapto ligands can bring a variable number of electrons: 1 electron for X ligands such as CH_3, 2 electrons for

L ligands such as CO and for X_2 ligands such as the carbenes, 3 electrons for carbynes and linear NO, 4 electrons for the O ligand (LX_2) in d^0 complexes, and 5 electrons for the N ligand (LX_3) in d^0 complexes.

Note 2 - With the ionic convention that we are not using here for the transition-metal complexes, the X ligands are counted with a negative charge, i.e. X^-. Then they bring two electrons from X^- (just like L ligands) to the metal cation. The characteristics NVE, NNBE, OS and C remain the same whatever the convention. For instance, with the ionic convention in ferrocene, iron is considered as Fe^{2+} and has thus 6 valence electrons. Each Cp^- ligand brings 6 electrons. NVE = 6 + (2 × 6) = 18; NBBE = 6; OS = 2; C = 6 . Our convention is justified with transition-metal complexes, because they are essentially covalent (for instance, ferrocene is pentane soluble and the metal does not bear a charge), but this is no longer the case for the Sc, Ln and Ac complexes (see Chap. 12.4) that are essentially ionic and for which the ionic convention is preferred. For instance, in uranocene, $[U(COT)_2]$, the dianionic COT^{2-} ligands are genuine 10-electron ligands for the ion U^{4+} (see Chap. 12.4).

4. THE "18-ELECTRON RULE":
TENDENCIES AND EXCEPTIONS

The NVE is often equal to 18 for transition-metal organometallics and for many inorganic complexes. This "18-electron rule" should be better viewed as a strong tendency than a rule, but it is followed by a majority of complexes (despite many exceptions, *vide infra*). The 18-electron electronic structure often brings a good stability for complexes. For instance, this is the case for the metal carbonyl complexes, possibly the largest family. Transition-metal sandwich complexes are more stable in the 18-electron count than in others. The 18-electron electronic structure is also found most of the time in complexes containing a mixture of carbonyls, hydrocarbons, carbenes, hydrides, etc.[1.1-1.3]

The 18 electrons correspond to the filling of the 9 molecular orbitals (one pair for each of them) coming from the 9 atomic orbitals of the transition-metal {5(n – 1) *d* orbitals, 1n *s* orbital and 3n *p* orbitals}. Some of these 9 molecular orbitals are bonding, and some others are non-bonding or antibonding. The interaction of atomic orbitals with the ligand orbitals of the same symmetry usually gives rise to occupied molecular orbitals and to unoccupied antibonding orbitals.[1.4]

The NVE is sometimes not 18, however, for various reasons:

▸ The more at the right of the periodic table the metal is located, the more *d* electrons it has, and it will be easier to complete its coordination sphere with 18 valence electrons. On the other hand, the early transition metals may have an average or weak tendency to fulfill the 18-electron shell. It can even happen that,

	Mo	= 6
MoL_6	6 L	= 12
		18
	Cr	= 6
CrL_6	6 L	= 12
		18
	Re	= 7
ReL_5X	5 L	= 10
	X	= 1
		18
	Fe	= 8
FeL_5	5 L	= 10
		18
	Ta	= 5
TaL_4X_5	4 L	= 8
	5 X	= 5
		18
	Co	= 9
$CoL_4X_2^+$	4 L	= 8
	2 X	= 2
	– (+ 1)	= – 1
		18
	Fe	= 8
FeX_6^{4-}	6 X	= 6
	– (– 4)	= + 4
		18

for steric reasons, the valence shell be far away from 18 electrons. The NEV can come down to 8, for instance in $[Ti(CH_2Ph)_4]$ or 10 in $[TaMe_5]$, 12 in $[Cr(CH_2SiMe_3)_6]$. In all these complexes, however, the metal tries by all means to accept more electrons by binding in ways that are not included in the formal electron count (agostic interactions, distal π interactions with aromatic rings).

▶ "Noble" transition metals (2nd and 3rd periods of transition metals: Ru and Os, Rh and Ir, Pd and Pt) whose complexes are excellent catalysts, often have a square-planar structure and NVE = 16. This is due to the fact that the p_z orbital cannot be occupied in the square-planar geometry, because it has a high energy. Thus, the complexes are stable with the 16-electron count.

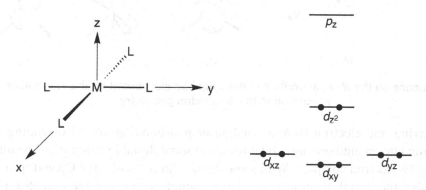

Their remarkable catalytic properties may well be attributed to chemical reactivity directed towards occupancy of this higher energy level.

▶ For the latest transition metals, i.e. those of the Cu column, the tendency is to have commonly 14 valence electrons in organometallic complexes of MLX structure (M = Cu, Ag, Au, L = phosphine, X = halogeno for example).

Whereas for the 16-electron square-planar complexes, a p orbital is too high in energy, these linear 14-electron compounds, cannot easily fill the p_x and p_y orbitals (z being the symmetry axis of the molecules) for geometrical reasons. Indeed, the more one goes to the right of the periodic table, the higher is the gap between the energy level of the p orbitals and the energy level of the next orbitals when the ligand field increases (which is the case for the carbon ligands). The high-energy level p_x and p_y orbitals often remain vacant for this reason.

▶ On the contrary, inorganic ligands that have a weak ligand field lead to form dicationic octahedral complexes ML_6^{2+} (M = Cu and Zn) having NVE = 21 and 22 respectively. The reason is as follows:

The $5d$ metal orbitals are split, under the influence of the ligand field, into $3t_{2g}$ bonding orbitals that are degenerate (i.e. of the same energy) and 2 degenerate antibonding e_g^* orbitals. In 18-electron complexes such as $[Fe(H_2O)_6]^{2+}$ or $[Co(NH_3)_6]^{3+}$, only the 3 bonding t_{2g} orbitals are occupied. Since the inorganic ligand field is weak, however, the 2 antibonding e_g^* orbitals are not of very high energy and can also be occupied. In this way, it is possible to accommodate up to 4 more electrons, in addition to 18.

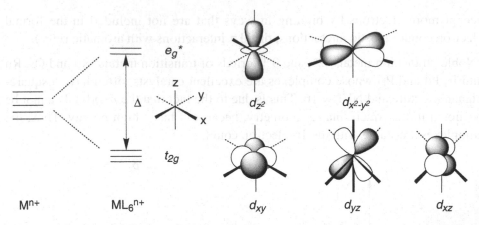

Influence on the *d* metal orbitals of the 6 ligands distributed on the x, y, z axes according to the octahedral geometry

▸ Removing one electron from a bonding or non-bonding orbital or adding one electron into an antibonding orbital results in some destabilization in the resulting 17- or 19-electron complex. With some ligands (in particular the Cp and η^6-arene ligands), the destabilization is not large enough to prevent the complex from existing, and it can sometimes be isolated. It is thus possible to find odd NVE numbers. Such complexes have at least one unpaired electron and are paramagnetic. It is also possible to have two or several unpaired electrons on the same metal when the ligand field is not too strong, i.e. for inorganic complexes, but also sometimes for purely organometallic complexes:

Examples:

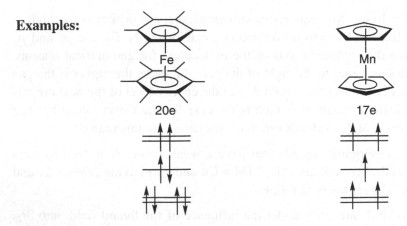

Several experimental techniques provide access to the magnetic susceptibility of a complex that is related to the number n of unpaired electrons according to the formula:

$$\mu_{theoretical} = \sqrt{n(n + 2)}\ \mu_B \qquad \mu_B \text{ is the Bohr magneton.}$$

For example, for n = 1, $\mu_{theoretical}$ = 1.73 μ_B ; for n = 2, $\mu_{theoretical}$ = 2.83 μ_B ; etc.

In conclusion, departures from the 18-electron rule are much more frequent than the exceptions to the octet rule (NVE = 8) in organic and main-group inorganic chemistry.

5. BONDING BETWEEN THE TRANSITION METAL AND THE π-ACCEPTOR (CO, C$_2$H$_4$) AND π-DONOR LIGANDS (HALOGENO, ALKOXY, AMINO)

CO and C$_2$H$_4$ are two among the most important ligands in organometallic chemistry: they are representative of the type of bond encountered. They are L ligands, and as such, they coaxially give a pair of electrons to the metal: this is the σ bond of a p orbital of CO or of a π orbital of C$_2$H$_4$ towards a vacant d metal orbital. It is this bond that is formally taken into account in the characteristics of the complexes. There is another bond that is not considered in the electron count, but which is also very important: it is the lateral backbonding of π type from a filled d metal orbital to a vacant antibonding orbital of CO or C$_2$H$_4$. This π bond is in the opposite direction to the σ bond, as indicated by its name. It partly compensates the σ bond and allows the metal to give back part of its excess of electron density and therefore to exist in a low or even negative oxidation state.

It is this backbonding that makes the big difference between inorganic and organometallic chemistry. For inorganic complexes, the backbonding cannot exist, because of the too high energy level of the antibonding N and O orbitals (the P and S atoms are weak π acceptors). The fact that there is no backbonding with O and N ligands explains, for instance, why complexes such as M(H$_2$O)$_6$ or M(NH$_3$)$_6$ (M = Cr, Mo or W) do not exist even with NVE = 18. These complexes would have OS = 0, and the metal would be too electron-rich, since backbonding would be absent. On the other hand, the only solution to compensate the enormous electronic gain provided by the inorganic ligands is that the metal be oxidized at least to the oxidation state +2. The complexes [Fe(H$_2$O)$_6$]$^{2+}$ and [Fe(NH$_3$)$_6$]$^{2+}$ are stable whereas the 18-electron complex [Fe(CO)$_6$]$^{2+}$ is extremely fragile. Indeed, the metal is very electron-poor in the latter complex because of the loss of electron density through oxidation to Fe^{II}, and therefore the π backbonding to all the CO ligands decreases.

The Dewar-Chatt-Duncanson model that takes into account this bonding mode, including the π backbonding with the CO and C$_2$H$_4$ ligands, is represented below for classic metal-carbonyl and metal-ethylene bonds. Of course, the other unsaturated hydrocarbons bind the transition-metals according to the same π backbonding model.

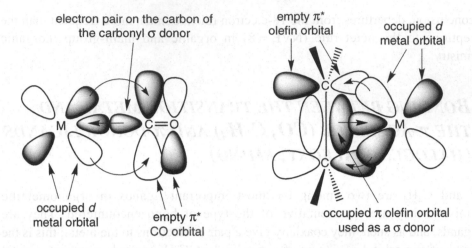

The consequences of this backbonding are experimentally observable. X-ray crystal structures show that the C-O and C-C bonds of CO and C_2H_4 respectively are longer when these molecules are bonded to a transition metal than when they are free, which means a decrease of the bond order upon coordination. This bond lengthening is due to the population of the antibonding ligand orbital in the complex. Also, the infrared frequency of the carbonyl is lower for the CO ligand in the metal-carbonyl complex than for free CO, as a consequence of the decrease of the C-O bond order upon coordination.

If the π backbonding becomes even more important, the lengthening of the C-C bond will be such that this bond will reach the length of a single bond. In this extreme case, it is appropriate to change its representation. The system should then no longer be viewed as a metal-olefin complex, but as a metallacyclopropane. This reasoning can be extended to metal-alkyne complexes and to the bond between the carbonyl of ketones and very oxophilic early transition-metals such as Sc and Ti.

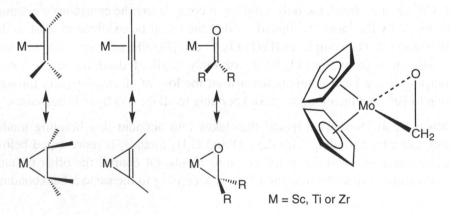

It is important to emphasize that the π backbonding is of course only possible if the transition metal has available d electrons in the complex, i.e. if the NNBE is not zero. In d^0 complexes, this π backbonding cannot take place, which considerably destabilizes metal-ligand bonds that can then only be very weak and, in fact, occur very rarely. This is the case, for instance, for Sc, lanthanide and actinide complexes that are most often in the oxidation state +3.

The opposite effect is also well known. Some ligands provide π electron density from the p electron pair of an heteroatom as ligand: these ligands are π donors. It is necessary to have a geometrical overlap between the p ligand orbital and the empty d metal orbital involved in the π bonding. Therefore, the available p ligand orbital must be perpendicular to that engaged as a σ donor with the metal. This interaction proceeds well with halogeno, alkoxy, amino, oxo and amido ligands and d^0 metals such as Sc^{III}, Ti^{IV} or Ta^V. One may notice that the early transition-metal chemistry contains a plethora of such bonds in complexes that would have a NVE much lower than 18 valence electrons if this π bonding was not taken into account.

Examples: $[ScCp^*{}_2Cl]$, $[TiCp_2Cl_2]$, $[Ta(mesityl)Cl_4]$, $[Ta(OAr)_5]$

Ar = 2,5-diisopropylphenyloxy.

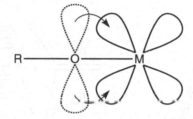

On the other hand, if the metal has non-bonding d electrons, their orbitals and the p orbital of the ligand exert mutual repulsion against each other, which leads to some destabilization characterized by an increase of the energy level of these orbitals. This does not prevent the compounds to exist with halides, alkoxy or amido ligands in late transition-metal complexes, but interesting effects are sometimes encountered. For instance, the complex $[RuCp^*(PMe_3)(Cl)]$ is stabilized, due to the π donor chloride that gives a non-bonding p electron pair into the vacant d orbital. Thus, the compound reaches NVE = 18 by counting the chloro as a 3-electron LX ligand:

6. *MOLECULAR ORBITAL DIAGRAMS*

Molecular orbital calculations allow to derive a molecular orbital diagram resulting from the interaction of metal orbitals with the ligand orbitals of corresponding symmetry.[1,4] The molecular orbital diagram of the complex contains bonding and antibonding molecular orbitals resulting from these interactions and non-bonding metal orbitals that do not interact with ligand orbitals. Let us take, for instance, the simplest case of an octahedral ML_6 complex. Six metal orbitals, i.e. the s, p and two d orbitals interact with the six orbitals of the six ligands in the six directions of the octahedron, whereas the three other metal d orbital remain non-bonding. If L is a π acceptor such as CO in a metal hexacarbonyl complex $[M(CO)_6]$, M = Cr, Mo or W, the three non-bonding metal d orbital are those involved in the π backbonding with the CO antibonding π^* orbitals. As a result, these three d orbitals will then become bonding. The energy difference between the two groups of d orbitals, called Δ or 10 dQ, will consequently be increased (diagram on the right below).

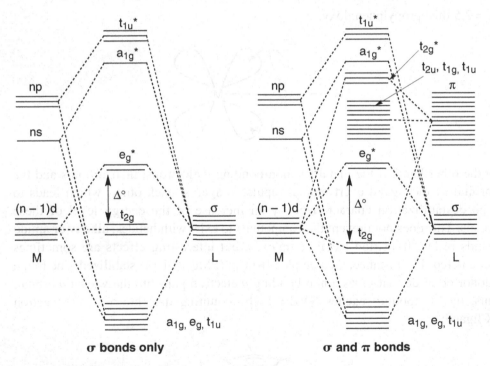

Molecular orbital diagram for a octahedral ML6 complex (simplified)

SUMMARY OF CHAPTER 1
THE MONOMETALLIC TRANSITION-METAL COMPLEXES

1 - The ligands (considered as neutral by convention)

Even ligands L_n bringing one or several electron pairs to the metal (e.g., ethylene).

Odd ligands (radicals) L_nX bringing an odd number of electrons to the metal (e.g., H).

2 - The formal characteristics of the transition-metal in the complexes

Number of valence electrons $NVE = n_M + 2 n_L + n_X - q$.

Number of non-bonding valence electrons $NNBE = n_M - n_X - q = n_M - DO$.

Oxidation state $OS = n_X + q$ (mononuclear complexes).

Coordination number $C = n_L + n_X$ (except if there are X_2, X_3 or single-site LX ligand in the complex).

3 - Hapticity

Number n of ligand atoms bonded to the metal in the complex (notation: η^n-ligand).

Example: the cyclopentadienyl ligand Cp usually is pentahapto (in this case, the haptivity is most of the time not indicated), but it can also be monohapto, and, rarely, even trihapto.

Linear notation of the complexes: $[NiCp_2]$ for nickelocene, but in the following ambiguous case: $[Fe(\eta^5-C_5H_5)(CO)_2(\eta^1-C_5H_5)]$.

One Cp is pentahapto, and the other is monohapto, so that $NVE = 18$.

4 - 18-electron rule: tendencies and exceptions

In *most* complexes, the number of valence electrons, NVE, is 18, but there are many other possibilities.

Examples: $[Ti(CH_2Ph)_4]$ (NVE = 8), $[VCp_2]$ (NEV = 15), $[RhCl(PPh_3)_3]$ (NVE = 16), $[Fe(C_6Me_6)_2]$ (NVE = 20), $[AgCl(PPh_3)]$ (NEV = 14).

5 - π Acceptor (CO, ethylene) and π donor (halogeno, alkoxy, amido) ligands

π Acceptor: backbonding from the filled d metal orbital to antibonding ligand orbital.

π Donor: π bond from the filled p orbital of the ligand to the empty metal d orbital.

6 - Molecular orbitals

Molecular orbital calculations give a detailed description of the ligand-metal orbital interactions and resulting energy diagram of the complex.

EXERCISES

1.1. Draw developed (three-dimensional) formula of the following complexes, give their $ML_nX_p^q$ form, and determine their numbers of valence electrons (NVE), numbers of non-bonding electrons (NNBE), oxidation states (OS) and their coordination numbers (C):

$[FeCp_2]^+$, $[RhCl(PPh_3)_3]$, $[Ta(CH_2CMe_3)_3(CHCMe)_3]$, $[ScCp^*_2(CH_3)]$, $[HfCp_2Cl_2]$, $[W(H)(CO)_5]^-$, $[Os(NH_3)_5(\eta^2-C_6H_6)]^{2+}$, $[Ir(CO)(PPh_3)_2(Cl)]$, $[ReCp(O)_3]$, $[PtCl_4]^{2-}$, $[PtCl_3(C_2H_4)]^-$, $[CoCp_2]$, $[Fe(\eta^6-C_6Me_6)_2]$, $[AuCl(PPh_3)]$, $[Fe(\eta^4-C_8H_8)(CO)_3]$, $[Ru(NH_3)_5(\eta^1-C_5H_5N)]^{2+}$,

$[(Re(CO)_4(\eta^2\text{-phen})]^+$ (phen = phenanthroline $C_{12}H_8N_2$),
$[(FeCp^*(CO)(PPh_3)(CH_2)]^+$, $[Ru(bpy)_3]^{2+}$ (bpy = η^2-2,2'-bipyridine $C_{10}H_8N_2$).

1.2. The complex $[FeCp^*(\eta^2\text{-dtc})_2]$ (dtc = dithiocarbamate S_2CNMe_2) exists in equilibrium between two forms: $[FeCp^*(\eta^2\text{-dtc})_2]$ and $[FeCp^*(\eta^2\text{-dtc})(\eta^1\text{-dtc})]$. Draw this equilibrium in writing the developed complex formulas, and calculate the NVE, NNBE, OS and C for each of them. Explain why these two forms are in equilibrium.

1.3. Draw the developed structure of the complex $[Ir(\eta^2\text{-O}_2)(CO)(PPh_3)_2]$ and show the various possibilities of coordination of the O_2 ligand and their consequences on NVE, NNBE, OS and C. Knowing that this is an Ir^{III} complex, give the correct coordination mode of O_2, the type of this ligand and the O-O bond order.

1.4. Are the following complexes stable? Justify your answer.
$[CoCp_2]^+$, $[CoCp_2]^-$, $[V(CO)_7]^+$, $[Cr(H_2O)_6]$, $[Ni(H_2O)_6]^{2+}$, $[ReH_9]^{2-}$, $[Zr(CH_2Ph)_4]$, $[Zr(CH_2Ph)_4(CO)_2]$, $[Cr(CO)_5]^{2-}$, $[Cu(Cl)(PPh_3)]$, $[Pt(PPh_3)_3]$, $[Ir(CO)_2(PPh_3)(Cl)]$, $[ScCp^*_2(CH_3)]$, $[NbCp_2(CH_3)_3]$, $[FeCp^*(dtc)_2]^+$.

1.5. Knowing that the diphenylacetylene complex $[W(CO)(PhC_2Ph)_3]$ is quite stable, what is the number of electrons provided by each alkyne ligand to the metal?

1.6. What is the stability order of the following metallocenes: $[TiCp_2]$, $[VCp_2]$, $[CrCp_2]$, $[MnCp_2]$, $[FeCp_2]$, $[CoCp_2]$, $[NiCp_2]$?

1.7. Represent schematically the splitting of the d orbitals under the influence of a pseudo-octahedral ligand field in manganocene. How are the valence electrons distributed in the orbitals? Why? Is the complex high spin or low spin? Calculate the theoretical value of its magnetic susceptibility. With $Cp^* = \eta^5\text{-C}_5Me_5$ instead of Cp, is the ligand field weaker or stronger? What are you expecting for the complex $[MnCp^*_2]$ in terms of the number of unpaired electrons?

1.8. Given the series of complexes $[MnCp(CO)_2L]$, with L = CO, CS, THF, PPh_3, C_2H_4, arrange these complexes in order of decreasing electron density on Mn. How could you experimentally check this? What is the most easily oxidizable complex?

1.9. What are the ligand hapticities in the complex $[Cr(CpCH_2CH_2Cp)(CO)_2]$? Draw the three-dimentional formula.

Chapter 2

BIMETALLIC TRANSITION-METAL COMPLEXES AND CLUSTERS

In organometallic and inorganic chemistry, there is a large variety of bi-[2.1] and polymetallic [2.2-2.5] structures, with or without metal-metal bonds, with or without bridging ligands and with a number of metal atoms that can vary between two and several hundred. The modern tools and techniques such as X-ray crystallography and electronic microscopy have recently allowed a remarkably accurate analysis of the molecular and intermolecular arrangements. On the other hand, the imagination of chemists and the need to search new materials for catalysis and nanotechnology have led to an explosion of research on molecular, supramolecular and macromolecular arrangements involving metal complexes. Here, we will give an introduction to this chemistry, especially concerning the structures of complexes. Their properties and reactivities should lead to applications in the future. The behavior of clusters resembles more and more that of metallic surfaces as the number of metal atoms increases. We are then in the world of nanoparticles. A comparison of the reactivity of mono- and bimetallic complexes, clusters and nanoparticles with that of metal surfaces is of great interest to improve our understanding and design of heterogeneous catalysts.

We will define the rules that allow us to understand and construct bimetallic complexes, then we will provide the most salient features of metallic clusters including electron counting (limits of the 18-electron rule and Wade's rules) and the isolobal analogy.

The 18-electron rule and electron count using the NVE, NBBE and C already described for the monometallic complexes are also applicable to bimetallic complexes and clusters containing up to 4 or 5 metal atoms. From 6-metal clusters on, these rules hardly apply and often fail, and Wade's rules, that must be used, will be considered in detail. Concerning the oxidation number, we prefer to skip its use for metal-metal bonded bi- and polymetallic complexes, because it would often be confusing.

When it is necessary to emphasize, in its linear formula, the fact that a ligand bridges or caps n metal centers, one introduces the symbol μ_n, immediately before the symbol η.

Examples:

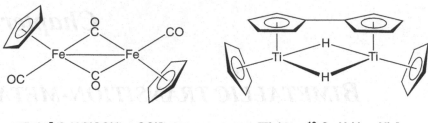

$$[Fe(\eta^5\text{-}C_5H_5)(CO)(\mu_2\text{-}CO)]_2 \qquad [Ti_2(\mu_2,\eta^{10}\text{-}C_{10}H_8)(\mu_2\text{-}H)_2]$$

1. THE METAL-METAL BOND IN BIMETALLIC COMPLEXES

1.1. GENERAL DESCRIPTION OF THE METAL-METAL BONDS

The metal-metal bond most often concerns two identical metals and even two iden-tical inorganic or organometallic metal-ligand fragments. This is the "general" case that we will describe here. Each of the two metals has d_{z^2}, d_{xy}, d_{xz}, d_{yz} and $d_{x^2-y^2}$ orbitals involved in the interaction with its neighbor. A coaxial (σ) interaction and eventually two lateral (π and δ) interactions represented below will result, depen-ding on the number of valence electrons available in these orbitals. The d_{z^2} orbitals of each metal coaxially overlap to make the σ bond. The d_{xz} orbitals of each metal can laterally overlap to give rise to a π bond. The d_{yz} orbitals do similarly. Finally, the d_{xy} orbitals can overlap according to another lateral mode to form the δ bond, and it is the same for the $d_{x^2-y^2}$ orbitals of each metal (see below) .

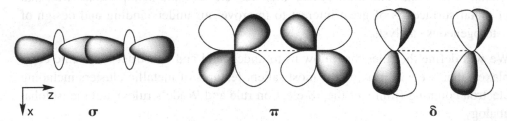

This latter type of overlap corresponding to the δ bond is much weaker than the other ones, and the energy of the molecular orbital involved is thus higher than those of the σ and π molecular orbitals. This means that, although this δ molecular orbital is of a bonding type, it is the last populated one in the molecular orbital diagram. Note that there are two δ molecular orbitals, but the second one, $\delta_{x^2-y^2}$, has a too high energy to be populated, because of its strong sensitivity to the ligand field.

The order of the metal-metal bond can now be deduced from this molecular orbital diagram, and it is equal to $1/2(n_1 - n_2)$ where:
n_1 = number of electrons in the bonding orbital (σ, π, δ);
n_2 = number of electrons in the antibonding orbitals (π^*).

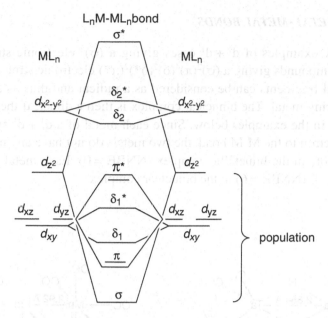

General molecular orbital diagram for the metal-metal bond in bimetallic complexes
(with halogeno ligands, the σ orbital is of higher energy than the π orbital)

The sequence of filling of the orbital levels shown below defines in this way the order of the metal-metal bond, as in organic and molecular inorganic chemistry.

n	Electronic structure	Bond type			Bond order
		σ	π	δ	
2	$(\sigma)^2$	1	0	0	1
4	$(\sigma)^2(\pi)^2$	1	1	0	2
6	$(\sigma)^2(\pi)^4$	1	2	0	3
8	$(\sigma)^2(\pi)^4(\delta)^2$	1	2	1	4
10	$(\sigma)^2(\pi)^4(\delta)^2(\delta^*)^2$	1	2	0	3
12	$(\sigma)^2(\pi)^4\,(\delta)^2(\delta^*)^2(\pi^*)^2$	1	1	0	2
14	$(\sigma)^2(\pi)^4(\delta)^2(\delta^*)^2(\pi^*)^4$	1	0	0	1

Thus, it is possible to have a single metal-metal bond with metals each bringing 1 or 7 valence electrons, a double bond with metals each bringing 2 or 6 electrons, a triple bond with metals bringing 3 or 5 electrons, and a quadruple bond with metals each bringing 4 electrons. The distance between the two metals can vary between 2 Å and 3.5 Å. For a given system, the metal-metal bond is shorter as the bond order is higher. However, the nature of the ligands (in particular bridging or non-bridging) and of the metal also has a strong influence on the intermetallic distance, and a comparison between different systems is therefore not appropriate. Finally, the δ bond of quadruple bonds is so weak that it does not imply a significant shortening of the metal-metal distance within a given series.

1.2. SINGLE METAL-METAL BONDS

There are a few examples of $d^1 + d^1$ cases giving a $(\sigma)^2$ electronic structure and many $d^7 + d^7$ compounds giving a $(\sigma)^2(\pi)^4(\delta)^2(\delta^*)^2(\pi^*)$ electronic structure.[2,4] Each of the two metal fragments can be considered as a radical and thus as an X ligand for the neighboring metal. The binuclear complex is then a dimer, if the two halves are the same as in the examples below. Since each metal of a $d^1 + d^1$ system gives its available electron to the M-M bond, the two metals do not have any non-bonding electrons available in the bimetallic complex (NNBE = 0). Each metal of a $d^7 + d^7$ system becomes d^6 (NNBE = 6) in the binuclear complex.

Examples:

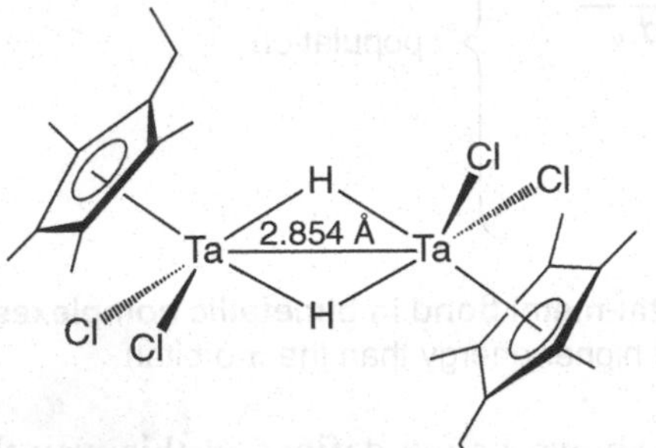

Before the Ta–Ta bond formation: $d^1 + d^1$ Before the Mn–Mn bond formation: $d^7 + d^7$
In the Ta–Ta complex: d^0 In the Mn–Mn complex: d^6

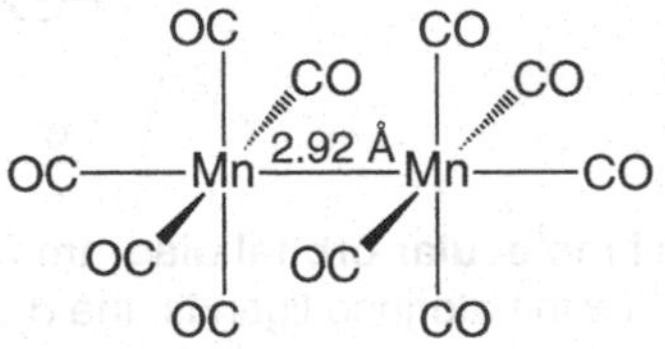

The complexes such as the above d^6 dimanganese complex follow the 18-electron rule, and each of the two d^7 manganese radicals taken separately before the Mn-Mn bond formation has 17 valence electrons. On the other hand, each of the d^0 Ta atoms has NVE = 16 in the above complex, and each d^1 Ta fragment has 15 valence electrons before the Ta-Ta bond formation.

The strength of the M-M single bonds is lower than that of the C-C bonds and depends on the steric bulk around the metal center. For instance, in the $[(CO)_5Mn-Mn(CO)_5]$ dimer for which the carbonyl ligands are all terminal, it is only 28 ± 4 kcal·mol^{-1} (117 ± 17 kJ·mol^{-1}), and the Mn-Mn distance (2.93 Å) is very high because of the filling of the antibonding levels and the absence of bridging carbonyl ligand added to the octahedral environment. These data are those of a weak bond. In the single metal-metal bonded dimer $[Cp(CO)_3Cr-Cr(CO)Cp]$, it is even weaker: only 17 kcal·mol^{-1} (71 kJ·mol^{-1}) because of the additional steric bulk of the Cp ligand. Under these conditions, there is an equilibrium between the monometallic d^7 radical and the d^6 dimer which favors the monomer as the temperature increases. Here is an analogy with organic chemistry: hexaphenylethane is also in equilibrium with Gomberg's bulky radical Ph$_3$C·. The reactivity of these organometallic dimers will often be that of the d^7 radical.

$$[Cp(CO)_3Cr–Cr(CO)_3Cp] \rightleftharpoons 2\ Cr(CO)_3Cp\cdot$$

$$Ph_6C_2 \rightleftharpoons 2\ Ph_3C\cdot$$

In order to strengthen the metal-metal bond, it is necessary:
▸ to go down in the periodic table,
▸ to have a multiple metal-metal bond,
▸ to introduce bridging ligands.

1.3. DOUBLE METAL-METAL BONDS

There are many compounds belonging to the families $d^2 + d^2$ leading to the $(\sigma)^2(\pi)^2$ electronic structure as well as $d^6 + d^6$ leading to the $(\sigma)^2(\pi)^4(\delta)^2(\delta^*)^2(\pi^*)^2$ electronic structure.

Examples:[2.4]

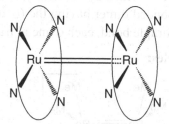

Before the W=W bond formation: $d^2 + d^2$
In the W=W complex: d^0

Before the Fe=Fe bond formation: $d^6 + d^6$
In the Fe=F complex: d^4

The NNBE is zero in the dimer resulting from the W=W bond formation from two d^2 tungsten fragments, since these fragments have engaged their two non-bonding electrons in the W=W bond. The two d^6 iron fragments give a Fe=Fe bond in which the iron atoms are now d^4. The complexes (porphyrin)M=M(porphyrin) (M = Ru or Os; parallel porphyrinic frames) discovered by Collman also belong to this latter family.

Before the Ru=Ru bond formation: $d^6 + d^6$
In the Ru=Ru complex: d^4

1.4. TRIPLE METAL-METAL BONDS

They are provided by the families $d^3 + d^3$ leading to a $(\sigma)^2(\pi)^4$ electronic structure and $d^5 + d^5$ leading to the $(\sigma)^2(\pi)^4(\delta)^2(\delta^*)^2$ electronic structure.

Examples:

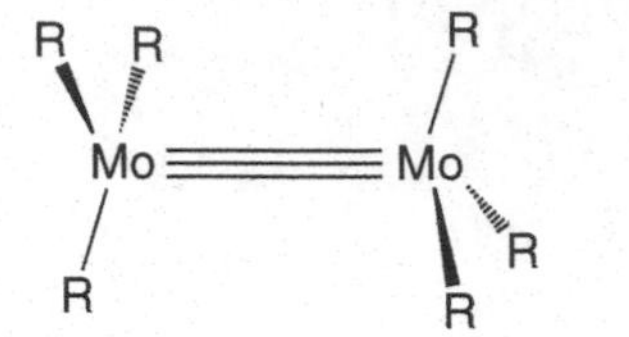

R = CH$_2$SiMe$_3$: d$_{Mo-Mo}$ = 2.17Å
R = NMe$_2$: d$_{Mo-Mo}$ = 2.21Å

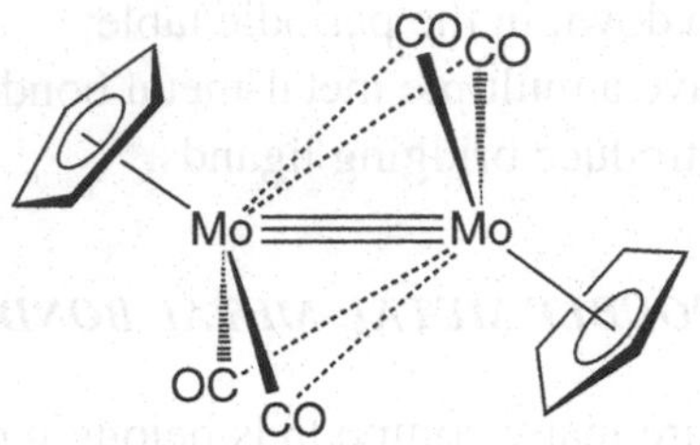

Semi-bridging carbonyls

Before the metal-metal bond formation: d^3 + d^3 Before the metal-metal bond formation: d^5 + d^5
In the Mo$_2$ complex: d^0 In the Mo$_2$ complex: d^2

After the formation of the triple bond, the two d^3 metals of the [MoR$_3$] fragments become d^0 in the triply metal-metal bonded dimer. On the other hand, the d^5 fragments [MoCp(CO)$_2$] give rise to a triply bonded Mo dimer in which the Mo atoms are d^2.

It is easy to determine the bond order of metal-carbonyl dimers, because they follow the 18-electron rule. This remains true if the dimer contains both Cp and carbonyl ligands. For instance, the dimer [Cp(CO)$_2$Mo-Mo(CO)$_2$Cp] can be written under the form [Mo(CO)$_2$Cp(X)$_y$] if y is the Mo-Mo bond order (each Mo being a X$_y$ ligand for the Mo neighbor), i.e., MoL$_4$X$_{y+1}$.

NEV = 18 = n$_{Mo}$ + 2 n$_L$ + n$_X$ + q = 6 + (2 × 4) + y + 1 = 15 + y; thus, y = 3.

The molybdenum-molybdenum triple bond in this complex is 2.2 Å long, whereas the length of a single Mo-Mo bond is 2.78 Å in the dimer [Cp(CO)$_3$Mo-Mo(CO)$_3$Cp] which is its precursor. The shortening is indeed very large.

1.5. QUADRUPLE METAL-METAL BONDS

They are provided by the family of d^4 metal fragments and lead to quadruply metal-metal bonded dimer having the (σ)^2(π)^4(δ)^2 electronic structure. After formation of the quadruple bond, each d^4 metal fragment becomes d^0 in the dimer.

Examples:

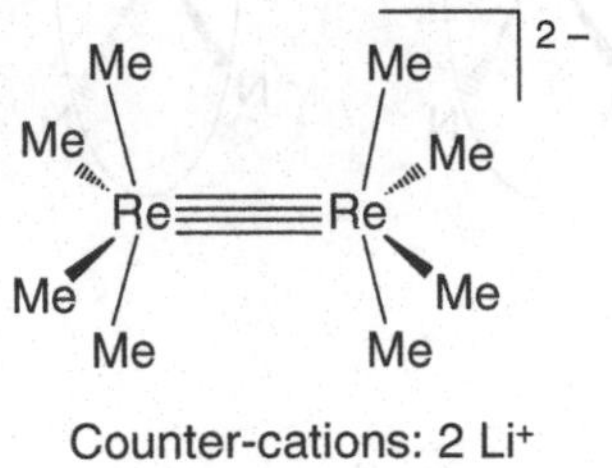

Counter-cations: 2 Li$^+$ 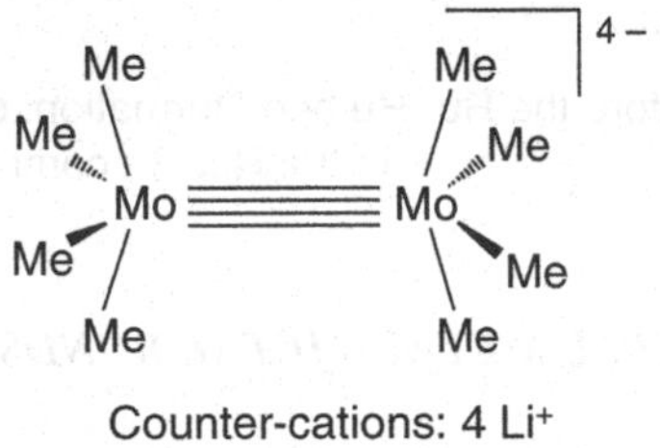   Counter-cations: 4 Li$^+$

Before the quadruple metal-metal bond formation: d^4 + d^4
In the dinuclear complex: d^0

In the dimer $[Re_2Cl_8]^{2-}$, the NVE is 16 if all the Cl are counted as X ligands. Each rhenium fragment must be counted as monoanionic, and can be described in the form ReX_{4+y}^-.

$$NEV = n_{Re} + 2\,n_L + n_X + q = 7 + 0 + 4 + y - (-1) = 12 + y = 16; \text{ thus, } y = 4.$$

The Re-Re distance (2.24 Å) is lower than the Re-Re distance in the metal itself (2.71 Å), but the strength of the quadruple rhenium-rhenium bond is only 85 ± 5 kcal·mol^{-1} (355 ± 20 kJ·mol^{-1}) including 5 kcal·mol^{-1} (21 kJ·mol^{-1}) for the fourth bond (δ bond), which is as weak as a hydrogen bond.

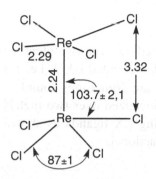

Structure of the ion $[Re_2Cl_8]^{2-}$

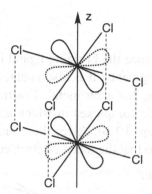

Scheme showing the formation of the δ bond in $[Re_2Cl_8]^{2-}$ by overlap of the d_{xy} metal orbitals of the two metals

Let us recall that the chloro and alkoxy ligands are π donors that can be counted as LX ligands if the metal has vacant d orbitals of suitable symmetry. For instance, in $[(CH_3O)_3MoMo(OCH_3)_3]$, one would have, for each Mo, NEV = 12 if each methoxy is counted only as an X ligand, the molybdenum-molybdenum bond being a triple bond. If each methoxy ligand is counted as LX, one finds more logically NVE = 18 for each Mo.

The work by F.A. Cotton and his group has provided a broad range of data in this area and a remarkable description of multiple bonds including the quadruple bond.[2.1]

1.6. THE BRIDGING LIGANDS

A large number of the ligands that are mentioned as terminal ligands X, L, X_2, LX, L_2, L_2X, etc. of mononuclear complexes in Chap. 1 can also serve as bridging ligands in metal-metal bonded bimetallic complexes. The electron count must be different, however, as follows.

▸ **The X ligands become LX** - The X ligands of monometallic complexes become LX for the group of the two metals in the bimetallic complexes. The X ligand brings the two metals close to each other, and the two metals almost always form a metal-metal bond. It is necessary to distinguish among the X ligands that have a non-bonding pair of electrons and that are already LX ligands in electron-deficient monometallic complexes (halogeno, hydroxy, alkoxy, thiolato, amido, phosphido, nitrosyl), those that have a pair of π electrons available on a nearby atom (vinyl, allyl, alkynyl) and those that do not have a pair of non-bonding π electrons (H, alkyl).

▸ **The X ligands having a pair of non-bonding electrons** use this pair to become formally L ligands for the second metal while remaining X ligand for the first metal (let us recall that, in monometallic complexes, these X ligands become LX with electron-deficient metals).

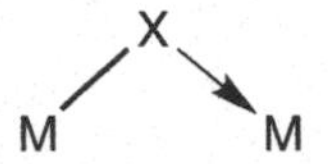

In fact, since the bridging ligand is symmetrically located between two identical metal fragments, it is also counted symmetrically, and is counted LX/2, i.e. 1.5 electrons, for each metal. Fractional electron counts are sometimes found in bimetallic complexes in which a single electron is delocalized over two metals (see Chap. 3.1.6). Also, these are usually, two bridging LX ligands as shown below, so that the metal electron counts are no longer fractional.

Example:

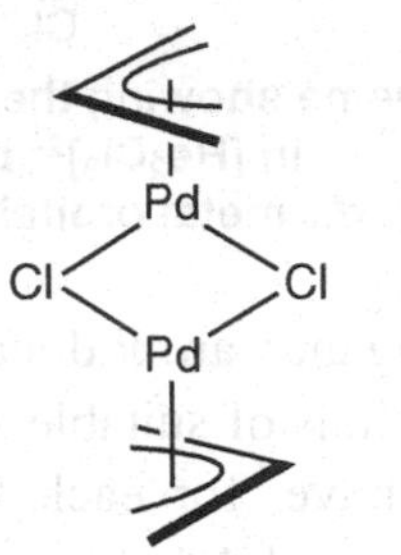

Pd(II)

In the above complex, each Pd fragment is in the form PdL_2X_2, $NVE = 10 + 4 + 2 = 16$.

Another means for two metal centers to avoid a fractional electron count is to have an odd number of charge for the bimetallic complex in which the two metals are bridged by a single LX bridging ligand.

Example:

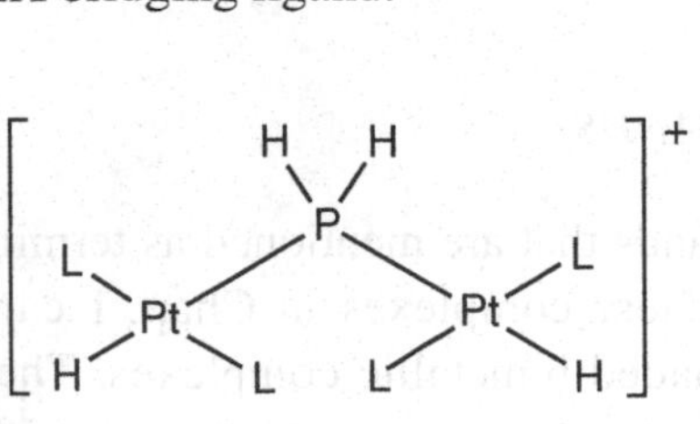

Each Pt fragment is written as $PtL_{2.5}X_{1.5}^{0.5+}$ with $NVE = 10 + 5 + 1.5 - 0.5 = 16$.

▸ **The X ligands having a multiple bond or a heteroatom in close proximity** can bridge the two metals in an unsymmetrical way with an X ligand for one metal and an L ligand for the other metal.

Example:

$$
\begin{array}{c}
R \\
C = O \\
(CO)_3Re \longleftarrow W(CO)_3\{P(OMe)_3\} \\
Ph_2P \qquad PPh_2 \\
C \\
H_2
\end{array}
$$

▸ **Two-electron three-center electron-deficient bonds -** The X ligands of mono-metallic complexes that do not have a non-bonding electron pair such as the hydrides and the alkyls can play the role of LX ligands bridging two metals, because the M-X σ bond electrons can serve as a ligand for the second metal nearby. The bridging ligand is symmetrical, so that the ligand is counted as LX altogether, i.e. LX/2 (1.5 electron) for each metal. For instance, protonation of singly metal-metal bonded bimetallic anions gives a bridging hydride in which the three atoms M-H-M are bound in a triangle together with only a pair of electrons (which originally served to bind the two metals). One should be very careful, because the writing in publications may be misleading due to the difficulty of choosing a formalism representing the two-electron three-center bond. When one compares the metal-metal distances before and after protonation (below), one may be convinced of a reduction of the metal-metal bond order upon protonation.

Example:

$$
\underset{2.97\ \text{Å}}{\left[(CO)_5Cr \longrightarrow Cr(CO)_5 \right]^{2-}} \xrightarrow{\ H^+\ } \underset{3.41\ \text{Å}}{\left[(CO)_5Cr \cdots\cdots Cr(CO)_5 \atop H \right]^{-}}
$$

The above complex containing the bridging hydride follows the 18-electron rule without counting the metal-metal bond if the hydride is counted as LX/2 for each metal. Each Cr atom is noted $CrL_{5.5}X_{0.5}^{0.5-}$. The above dashed representation is often used, but since it most often means 1/2 σ bond (i.e. 1 electron), it seems that there are 3 electrons altogether, which is in excess by 1 electron. The representations below are also encountered, but the formulas IV represent the only correct formalism.

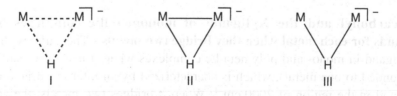

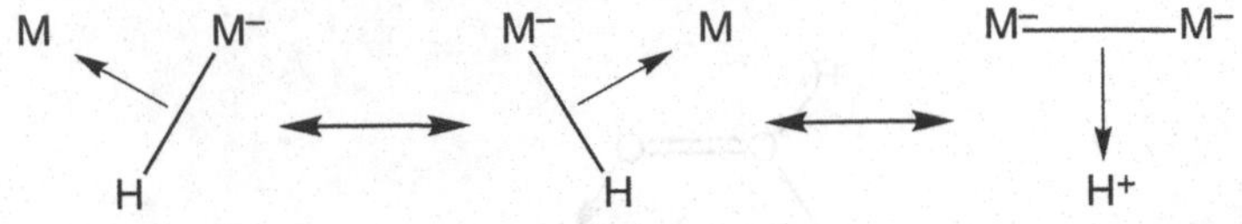

The two-electron, three-center bond is also found in hydrides and alkyls of Li, B, Al, Be, lanthanides and actinides. For instance, in diborane B_2H_6, the structure must allow each boron atom to satisfy the octet rule by using a neighboring B-H σ bond as a ligand.

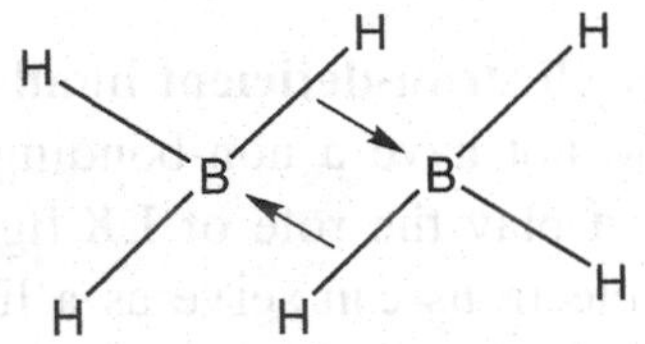

In section 2.4, we will examine, in more details, the Wade's rules that rationalize the electron count in polyboranes, carboranes and metallocarboranes.

The alkyl groups are used, but more rarely, as LX bridging ligands in the same way as described above for the bridging hydrides. One has then a two-electron three-center M-C-M system as below:

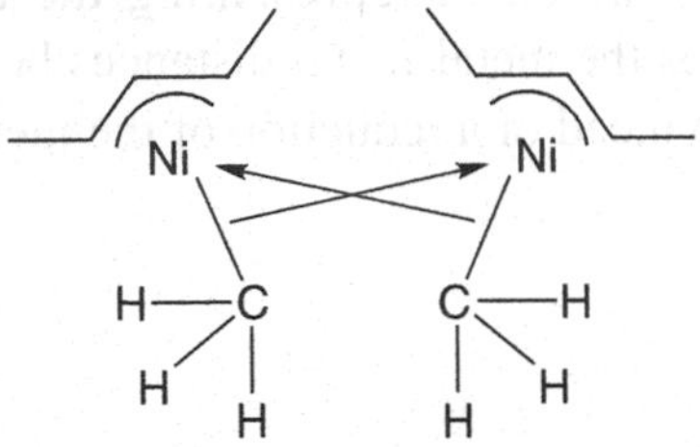

There are also situations in which the C-H bond of a methyl, alkyl or benzyl ligand X serves as an L ligand for the second metal, creating a dissymmetrical bridge as described above for the vinyl or alkynyl. These three-center M-C-H bonds are *agostic* bonds:

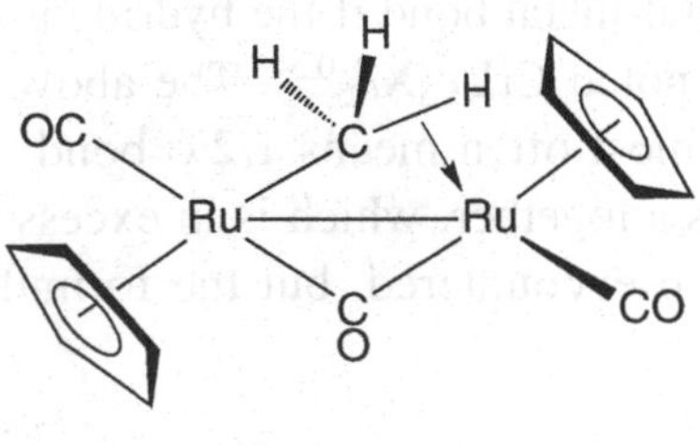

▸ **The carbonyl and the X_2 ligands of monometallic complexes become X ligands for each metal when they bridge two metals** - The carbonyl ligand is an L ligand in mono- and polymetallic complexes when it is a "terminal" ligand (i.e. bonded to one metal), which is characterized by an intense infrared absorption band in the region of 2000 cm^{-1}. When it bridges two metals, the carbonyl becomes a X_2 ligand, i.e. an X ligand for each of the two bridged metal centers.

Its structure is that of a dimetallaketone, and its strong infrared absorption is now found around 1650 cm^{-1} as for organic ketones. The CS, CSe and CTe ligands behave analogously. Other bridging X$_2$ ligands that are X ligands for each of the bridged metals are those that are already known ligands in monometallic complexes such as carbenes or alkylidenes, vinylidenes, nitrido, phosphido, oxo and sulfido.

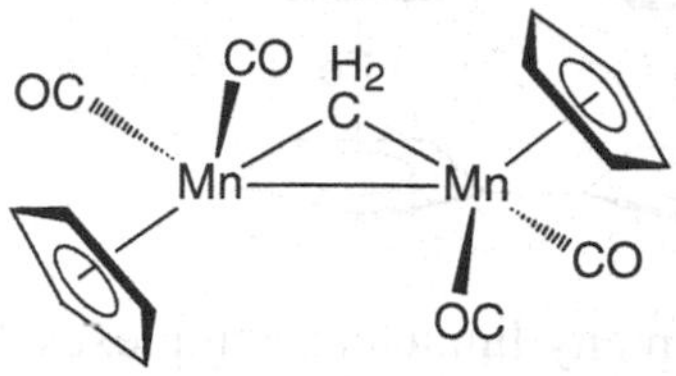

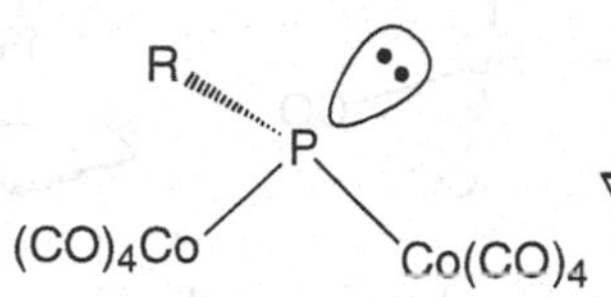

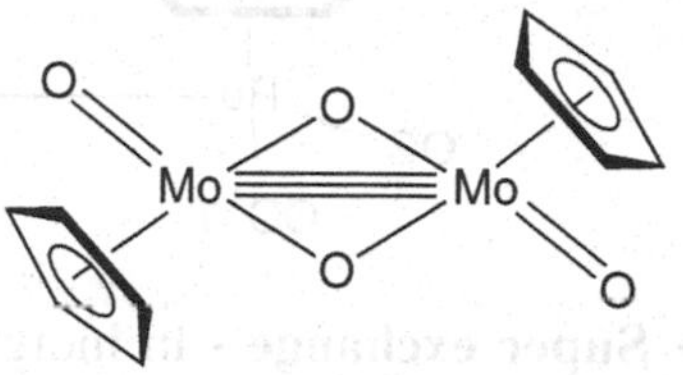

▸ **The LX and L$_2$X ligands of mononuclear complexes can bridge two metals:** then, they share the given electrons between the two metals (respectively LX/2, i.e. 1.5 electrons and LX$_2$/2, i.e. 2.5 electrons for each metal).

This is the case for the allyl ligand, for chelating inorganic LX ligands and even occasionally for the cyclopentadienyl (see the example below).

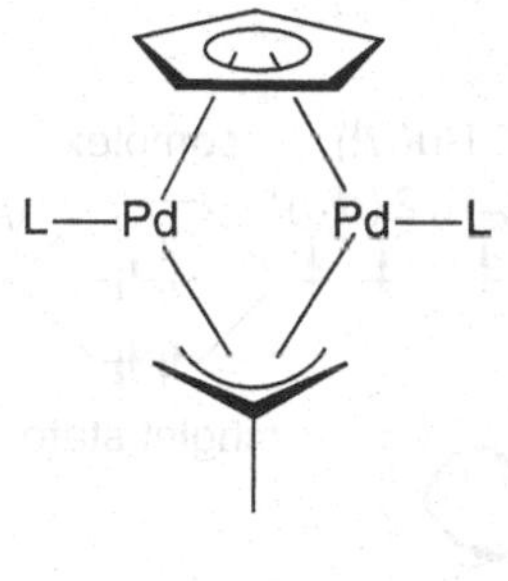

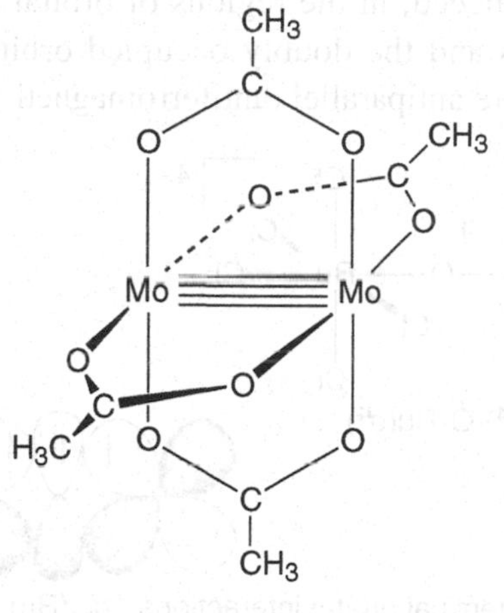

In the above complexes, the metals are noted: PdL$_{2.5}$X, NVE = 10 + 5 + 1 = 16, and MoL$_2$X$_6$, NVE = 6 + 4 + 6 = 16.

▸ **The L$_2$ ligands of the monometallic complexes can bridge two metals**, the L$_2$ being distributed as one L for each metal:

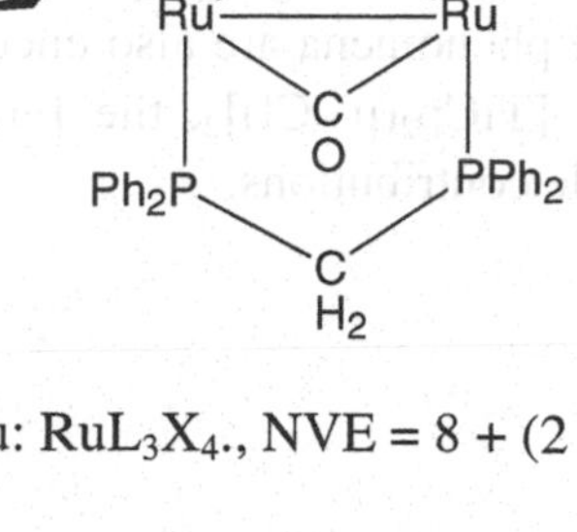

Electron count for each Ru: RuL$_3$X$_4$., NVE = 8 + (2 × 3) + 4 = 18.

▸ **The benzene and other aromatic ligands, the cyclooctatetraene (L$_4$), the fulvalenediyl (L$_4$X$_2$, see below)** can bridge two metals with or without a metal-metal bond.

For each metal, half of the available ligand electrons are counted.

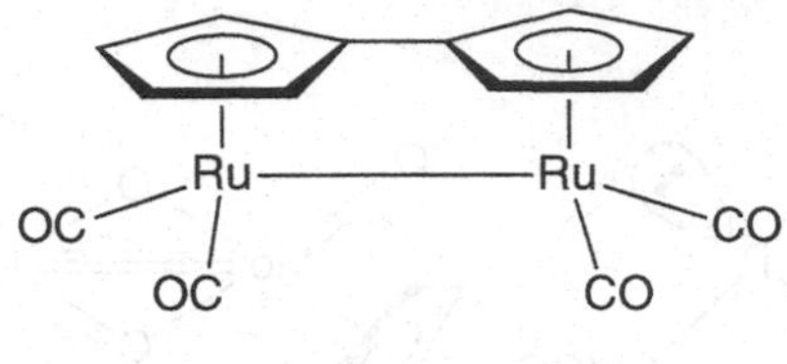

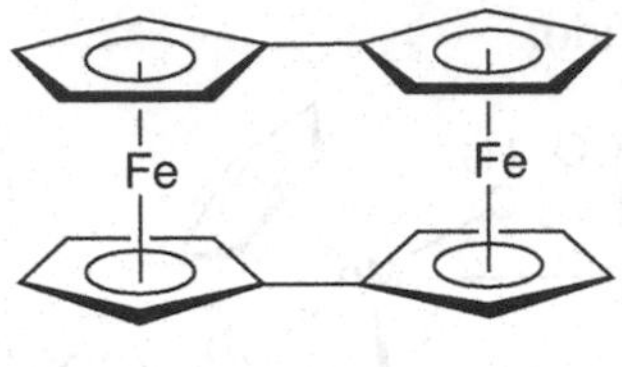

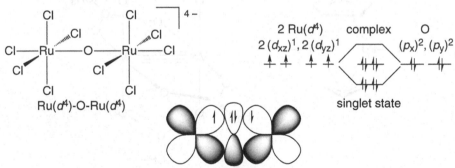

▸ **Super-exchange** - In inorganic chemistry, there are many binuclear complexes that are bridged by a ligand such as O, O$_2$, N$_2$, CN, pyrazine, 4,4'-bipyridine, etc. without an M-M bond. Sometimes, electron pairing occurs through a ligand such as O without a direct chemical bond between the two metals: this phenomenon is called super-exchange. For instance, in [Cl$_5$Ru(μ_2-O)RuCl$_5$]$^{4-}$ (figures below), each metal taken separately would have an electronic configuration $(d_{xy})^2 (d_{xz})^1 (d_{yz})^1$, i.e. with two unpaired electrons. The dinuclear complex is diamagnetic, however. Indeed, in the regions of orbital overlap between the d orbitals of the two metals and the doubly occupied orbitals of the O ligand, the spins of the d orbitals are antiparallel (antiferromagnetic coupling, scheme below):

Identical orbital interactions : d_{xz}(Ru)-p_x(O)-d_{xz}(Ru) in the xz plane
and d_{yz}(Ru)-p_x(O)-d_{yz}(Ru) in the yz plane

Antiferromagnetic coupling

This is also true when the interacting d orbitals are coplanar (d_{xz} for each of the two Ru and d_{yz} for each of the two Ru). On the other hand, if the interacting d orbitals are orthogonal (d_{xz} for one of the two metals and d_{yz} for the other one), the spins of these d orbitals near the bridging ligand are parallel (ferromagnetic coupling, triplet state). These phenomena are also encountered in organometallic complexes. For instance, in [TiCp$_2$(μ_2-Cl)]$_2$, the ferro- and antiferromagnetic contributions have comparable contributions.

TYPES OF MOLECULAR MAGNETISM
(see ref. 11.4)

A molecule or ion is:

▸ **diamagnetic** when all the electrons are coupled. Ex.: CH_4: ⥮

▸ **paramagnetic** when at least one electron is alone. Ex.:

   $[FeCp_2]^+$: ↑ (spin = 1/2); $[Fe(\eta^6\text{-}C_6Me_6)_2]$ and O_2 in the ground state: ↑ ↑ (spin = 1)

▸ **ferromagnetic** when two electronically coupled centers have one or several electrons with parallel spins: this phenomenon occurs when the orbitals of the electrons of the two centers are orthogonal (Kahn's concept [11.4]);

▸ **ferrimagnetic** when the coupled centers have opposed spins with different spin numbers, so that the resulting spin is not zero;

▸ **antiferromagnetic** when two centers have the same number of spins, but of opposed signs (antiparallel); when this coupling becomes very strong, the two centers form a chemical bond, and the compound is then diamagnetic. As the intensity of the coupling depends on temperature, the diamagnetic-antiferromagnetic thermal transition are common.

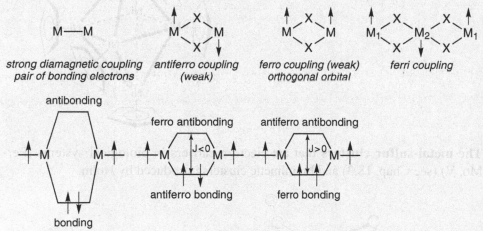

M——M	M⟨X⟩M	M⟨X⟩M	M_1⟨X⟩M_2⟨X⟩M_1
strong diamagnetic coupling pair of bonding electrons	*antiferro coupling (weak)*	*ferro coupling (weak) orthogonal orbital*	*ferri coupling*

Experimentally, the magnetism is recorded by the variation of the magnetic susceptibility χ as a function of the temperature [T in Kelvin, or usually $\chi T = f(T)$].

In order to have a magnetic material, it is necessary to induce the alignment of the spins in a whole three-dimensional network (first along a chain, for instance using ferrimagnetism, then using interchain interactions). Otherwise, the molecular spins take different orientations in the solid network, and the overall component is zero.

A complex can be low spin, if its ligand field is strong or high spin, if this field is weak. In intermediate cases such as that of manganocene, $[MnCp_2]$, (see Chap. 1.4), a **spin transition** (spin "cross-over") can be observed (thermal, photochemical, etc.). When a lattice-induced hysteresis phenomenon is observed, these transitions can lead to promising applications in the field of signals and molecules with memories.

2. CLUSTERS

2.1. DIFFERENT TYPES OF CLUSTERS

The main types [2.2-2.5] are:

▸ **The metal carbonyl clusters** - They have been characterized by X-ray diffraction with nuclearities up to the range 30 to 40, by the pioneering work of Chini.[2.2] They are neutral or, mono-, di-, or trianionic, and these anions can be protonated to give respectively mono-, di- or trihydrides. They have been much studied by organometallic chemists who have used them as models of the Fischer-Tropsch reaction and replaced some carbonyls by unsaturated hydrocarbon ligands. Wade has compared them to metalloboranes, which allowed him to propose a new, correct electron count. Metallocarboranes, derived from the latter, were introduced by Hawthorne.

▸ **The metal cyclopentadienyl clusters**
(with or without carbonyls)

Example (see also section 2.2):

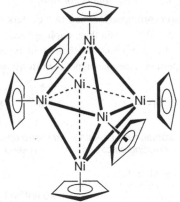

▸ **The metal-sulfur clusters** that are electron carriers in biological systems (Fe, Mo, V) (see Chap. 18.3) and biomimetic clusters introduced by Holm.

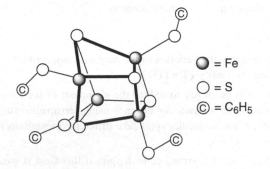

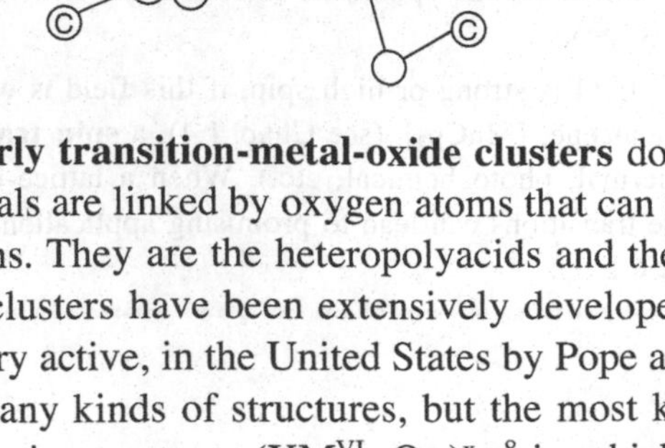

$\bigcirc$ = Fe
$\bigcirc$ = S
$\copyright$ = C_6H_5

$$[Fe_4S_4(SC_6H_5)_4]^{2-}$$

▸ **The early transition-metal-oxide clusters** do not have metal-metal bonds, but the metals are linked by oxygen atoms that can also occupy non-bridging terminal positions. They are the heteropolyacids and their salts, the heteropolymetallates. These clusters have been extensively developed in France by Souchay's school now very active, in the United States by Pope and in Germany by Müller.[2.6] They have many kinds of structures, but the most known ones in oxidation catalysis are Keggin structures $(XM^{VI}_{12}O_{40})^{x-8}$ in which the central atom $X = Si^{IV}$, Ge^{IV},

P^V, As^V, etc. is located at the center of a XO_4 tetrahedron surrounded by 12 octahedrons MO_6 (M = Mo or W) having common vertices and summits. The very robust structure includes 12 M = O terminal ligands in addition to the terminal oxo ligands.

▸ **The metal-sulfide clusters -** Some, such as $[PbMo_6S_8]$ (Chevrel phases, introduced by Sergent and Chevrel) are superconductors.[2.7]

▸ **The octahedral early transition-metal-halide clusters** $[Mo_6Cl_8]^{4+}$ and $[M_6X_{12}]^{n+}$ (M = Nb, Ta, Re; X = F, Cl, I; n = 2, 3, 4)[2.8] contain both strongly anchored "inner" halide ligands that bridge metals and substitution-labile "outer" chloride ligands that are terminal in the six axial/equatorial positions and give rise to a rich hexasubstitution chemistry.

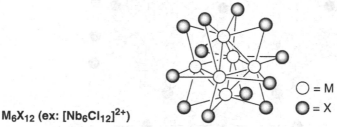

M_6X_{12} (ex: $[Nb_6Cl_{12}]^{2+}$)

○ = M
● = X

It is also possible to design mixed inorganic-organometallic systems by appropriate decoration of such clusters, as in the case of $Mo_6Br_{14}^{4-}$ below. The 6 apical Br ligands can be substituted by triflates, then by functional pyridines to synthesize dicationic cluster-cored ferrocenyl dendrimers:

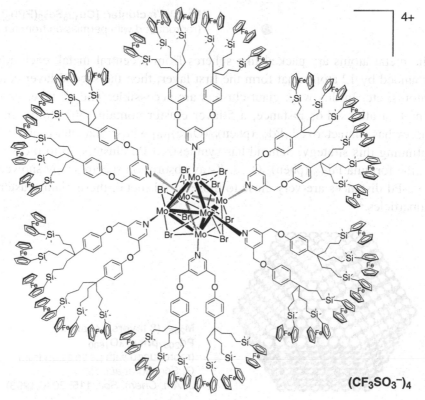

$(CF_3SO_3^-)_4$

▶ **High-nuclearity clusters and nanoparticles -** X-ray diffraction has allowed to characterize clusters up to 30 to 40 metal atoms, for instance $[Pt_{38}(CO)_{44}H_2]^{2-}$ and $[Au_{29}(PPh_3)_{14}Cl_6)Cl_2]$. Larger clusters such as $[Au_{55}(PPh_3)_{12}Cl_6]$ and $[Cu_{146}Se_{73}(PPh_3)_{30}]$ are more difficult to characterize by X-rays, but they are still precisely defined. Electronic microscopy [2.5] now is the standard technique to study these very large clusters.

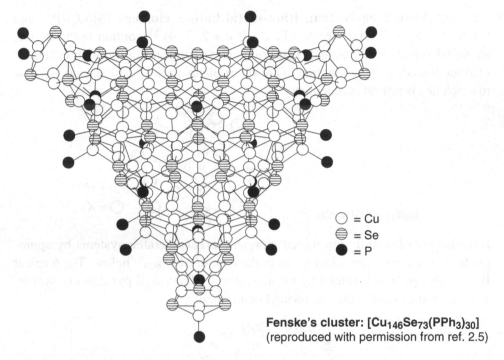

○ = Cu
⊜ = Se
● = P

Fenske's cluster: [Cu$_{146}$Se$_{73}$(PPh$_3$)$_{30}$]
(reproduced with permission from ref. 2.5)

If the metal atoms are packed like spheres from a central metal, each atom is surrounded by 12 atoms that form the first layer; then the second layer contains 42 atoms, etc. In this way, giant clusters are accessible. The nth layer contains $(10\,n^2 + 2)$ atoms. For instance, a 5-layer cluster contains 561 atoms. Indeed, Moiseev has characterized $[Pd_{561}(phen)_{60}(OAc)_{180}]$ whose diameter is about 25 Å. Continuing this strategy, Schmid has synthesized Pd clusters up to the 8th layer with the formula $Pd_{2057}(phen)_{84}O_{1600}$. X-ray absorption spectroscopy showed that the Pd-Pd distances are very close to those of Pd metal: these giant clusters are nanoparticles.[2.5]

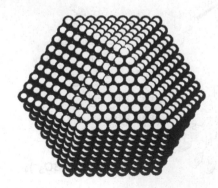

M$_{2057}$ (8 layers); example:
Pd$_{2057}$phen$_{84}$O$_{1600}$
(reproduced with permission from
G. Schmid *et al.*,
J. Am. Chem. Soc. **115**, 2046, 1993)

Colloids or nanoparticles have been known since antiquity (gold ones). They can be stabilized in water, for instance by reduction of an aqueous solution of a metallic salt in the presence of polyvinyl alcohol (PVA), but do not have a very precise nuclearity unlike the above giant clusters. Their size ranges from 2 to 100 nm. Nanoparticles play an important role in catalysis because their surface is very active. They can be stabilized by long-chain thiolate ligands, and then become like molecular compounds. The frontier between large clusters and nanoparticles is fuzzy, possibly only related to the accuracy of the nuclearity and polydispersity of the samples. These objects belonging to nanochemistry will go a long way, partly because they link the molecular and solid-state worlds. Indeed, the collective oscillation of the conducting 6s electrons of gold nanoparticle cores larger than 3 nm, a quantum size effect, gives rise to the surface plasmon band absorption around 520 nm (deep red color) that leads to multiple applications, *inter alia*, in optoelectronics and biological sensing. For instance, the superb stained-glass windows of French Gothic cathedrals owe their red colors to the plasmon band of inserted gold nanoparticles.

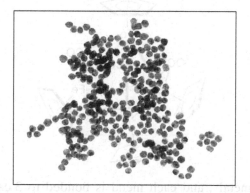

**Transmission electron micrograph (TEM) of a monodisperse
14-nm hydrosol prepared by reduction of HAuCl₄ by citrate,
recorded by Dr D. Weitz, Exxon Research Corporation**
(reproduced with permission from ref. 2.5)

2.2. LIGANDS BRIDGING 3 OR 4 METALS

▸ **X₃ ligands capping three metals -** The trivalent ligands such as nitrido N, phosphido P, carbynes or alkylidynes CR can cap three metals as an X ligand for each of the three metals. The alkylidyne tricobalt nonacarbonyl complexes are the most classical. The NO ligand that can be a LX ligand in monometallic complexes can become such a X₃ ligand when it caps three metals in clusters.

The 15-electron organometallic fragments having at least three non-bonding electrons such as $M(CO)_3$ (M = Co, Rh, Ir) or $M(CO)_4$ (M = Mn, Tc, Re) can also be X₃ ligands in capping three 17-electron metal fragments. They complete their NVE to 18 in this way. A similar reasoning can be applied to the capping chloro ligand which can be a L₂X ligand when its caps 3 metals in clusters.[2.3, 2.4]

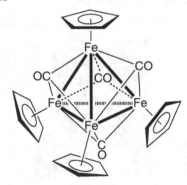

> ► **X_4 ligands capping four metals** - The carbide ligand C can cap four metals and so become X_4. Other examples are organometallic 14-electron fagments such as $M(CO)_3$, (M = Fe, Ru, Os) or $M(CO)_2$, (M = Cr, Mo, W).[2.3, 2,4]

> ► **Carbonyl and phosphinidene ligands capping 3 or 4 metals** - A CO ligand contributes $2/n$ electrons to the NVE of a metal atom, n being the number of metal atoms in the cluster that are capped by the carbonyl.

Example: $[Fe_4Cp_4(CO)_4]$

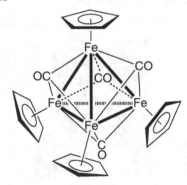

Each CO caps 3 metals, and each metal is bonded to 3 carbonyls that bring: $3 \times 2/3 = 2e$, X_2. Each Fe is also bonded to three neighboring Fe atoms that count as 3 X ligands and bring 3e. Each Fe may be written as: FeL_2X_6, i.e. NVE = 18.

Note - Many clusters exist in several oxidation states because of the degeneracy of their HOMOs and LUMOs (several orbitals with the same energy due to symmetry) and the steric protection by the ligands of the metal-based redox centers. In particular, this is the case for the metal-cyclopentadienyl clusters. For instance, the above cluster has been isolated in the forms $[FeCp(CO)]_4^n$ with n = –1, 0, +1 and +2. A consequence, in cyclic voltammetry, is the cascade of several chemically reversible waves (no decomposition of the oxidized and reduced species: the anodic and cathodic intensities are the same) that are also electrochemically reversible (fast electron transfer, which means that there is no significant structural change in the course of electron transfer; the potentials of the anodic and cathodic waves differ by $58/n$ mV at 20°C and do not change when the sweep rate varies).

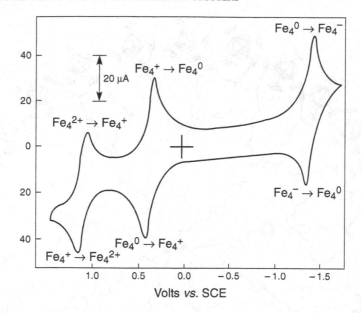

Cyclovoltammogram of [FeCp(CO)]$_4$ in acetonitrile
Electrode: Pt, supporting electrolyte: [n-Bu$_4$N][PF$_6$]; potentials measured using the
saturated calomel electrode (SCE) as a reference (reproduced with permission from
J.A. Ferguson and T.J. Meyer, *J. Am. Chem. Soc.* **94**, 3409, 1972)

The robustness of both the neutral and cationic Fe$_4$ clusters led to the use of the
chemically and electrochemically reversible cluster redox couple [FeCp(CO)$_4$]$^{+/0}$
(see above) to sense electrochemically anions such as adenosine triphosphate
ATP^{2-}, a DNA fragment, in metallodendrimers whose tethers are terminated by
this cluster. A single cyclic-voltammetry wave is observed for this redox system
in the dendrimers because the cluster units are far enough from one another to
render to extremely weak and unobservable electrostatic factors. The advantages
of dendrimers are that they better recognize ATP^{2-} due to dendritic effects and
they easily and strongly adsorb on electrodes due to their large size to make
modified electrodes functioning as re-usable ATP sensors (the principle of ATP
recognition is based on the cathodic shift of the wave due to the [FeCp(CO)$_4$]$^{+/0}$
redox system observed by cyclic voltammetry), see the dendrimer-cluster p. 66.

2.3. THE LOCALIZED ELECTRON COUNT IN CLUSTERS AND ITS LIMITS

▸ **Number of valence electrons (NVE)** - For small clusters, the 18-electron rule
remains valid. It is possible to count electrons as we have already done it above
for [FeCp(CO)]$_4$. The most common CO ligand in clusters is a L ligand when it
is terminal, X ligand for each metal when it bridges 2 metals, 2X/3 when it caps
3 metals and X/2 when it caps 4 metals. Each metal is an X ligand for the
neighboring metal. For instance, in [Ru$_3$(CO)$_{12}$], all the carbonyls are terminal,
and each Ru is represented as RuL$_4$X$_2$ since it bears 4 CO's and 2 Ru.

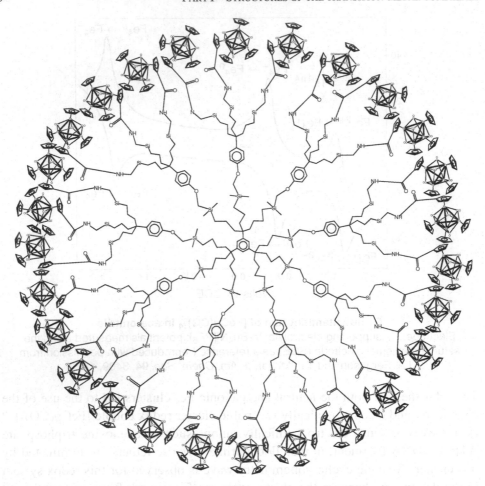

On the other hand, in [$Fe_3(CO)_{12}$], one of the Fe's bears 4 terminal CO's whereas the two other Fe's bear only 3 CO's. The first Fe is represented as FeL_4X_2 and the two other Fe's are represented as FeL_3X_4. The 3 Fe's follow the 18-electron rule.

▸ **Number of cluster electrons (NCE)** - The number of cluster electrons (NCE) is also defined for each cluster. It is the total number obtained by adding the valence electrons of all the metals of the cluster to the number of electrons brought by the ligands. The electrons in the metal-metal bonds are not added.

$$NEC = x_{nT} + 2\,n_L + n_X + q$$

Example: $M_3(CO)_{12}$ (M = Fe or Ru), NEC = $(3 \times 8) + (12 \times 2) = 48$.

The NEC does not detail the bonding mode of the carbonyl ligands, i.e. if they are terminal, bridging or capping. Since the 18-electron rule is fulfilled, the NEC is equal to x times 18 if x is the number of metals minus the 2 y electrons of the y metal-metal bonds:

$$NEC = 18 \, x - 2 \, y.$$

In the above example, $NEC = (18 \times 3) - (2 \times 3) = 48$.

This electron count works for small clusters, but generally breaks down for hexanuclear clusters because of electronic delocalization. For example, in $[Os_6(CO)_{18}]$:

$$NCE = (6 \times 8) + (18 \times 2) = 84.$$

This cluster has a monocapped trigonal bipyramidal structure in which the 18-electron rule is obeyed for only two of the six Os atoms. The situation is worse for the localized electron count and for the 18-electron rule that do not work at all in the very stable octahedral 86-electron dianion (see below). On the other hand, Wade's theory that uses the polyhedral skeleton electron pairs very well rationalizes the electron counts in large organometallic clusters.[2.9]

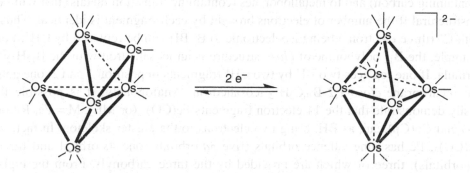

Note - In the above cathodic reduction, the electron transfer is slow because of the deep structural change. The cyclovoltammogram reflects this situation by the fact that the peak potentials E_{pc} and E_{pa} do not remain constant when the scan rate is varied (this potential peak shift is 30 mV when the scan rate is multiplied by 10), and the $E_{pc} - E_{pa}$ value is much larger than $58 \, mV \cdot n^{-1}$ (n = number of exchanged electrons, here: 2) obtained at 20°C with redox systems for which the electron transfer is fast on the electrochemical time scale. The variation of $E_p - E_{pa}$ as a function of the scan rate gives access to the heterogeneous electron transfer rate constant between the cathode and the cluster in solution (see for instance chap. 2 of ref. 3.3, pp 104 and 174).

2.4. WADE'S RULE: POLYBORANES, CARBORANES, METALLOCARBORANES AND TRANSITION-METAL CLUSTERS

Wade has taken into account the fact that transition metals are electron-deficient in average-size clusters in order to rationalize their geometries and electron counts. He proposed that the metals must share their available electrons to form the cluster skeleton. Wade also noticed that the situation is analogous for polyborane clusters, a very rich and varied family of general formula $B_nH_n^{x-}$, the starting point of this analysis.[2.9] In these compounds, each boron atom, that has 3 valence electrons, introduces one electron in the bond with H, which leaves 2 available electrons with 3 orbitals to contribute to the formation of the cluster.

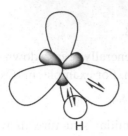

B(3e)-H(1e)
BH fragment:
2 electrons and 3 orbitals are available.

A borane $B_nH_n^{2-}$ has n + 1 electron pairs to form the cluster. For example, one has, for n = 6, an octahedron $B_6H_6^{2-}$ with one BH at each vertex and 14 electrons in the cluster skeleton. Such a structure is called "*closo*" due to the fact that all the vertices of the polyhedron are occupied. From the boranes, one can jump to carboranes (containing carbon) and to metalloboranes (containing transition metals) that will be isostructural if the number of electrons brought by each fragment is identical. Thus, with C^+ (three electrons) being isoelectronic to B, BH can be replaced by CH^+. For example, the B_{12} polyborane of *closo* structure is an icosahedron with the $B_{12}H_{12}^{2-}$ formula. If one replaces two BH by two CH^+ fragments in position 1 and 2, one gets the *o*-carborane *closo*-1,2-$B_{10}C_2H_{12}$ icosahedron. Analogously, it is possible to easily demonstrate that the 14-electron fragments $Fe(CO)_3$ (or ML_3, M = Fe, Ru or Os) and Co(Cp) can, as BH, bring two electrons to the cluster skeleton. In fact, in $Fe(CO)_3$, Fe has nine valence orbitals (five $3d$ orbitals, one $4s$ orbital and three $4p$ orbitals), three of which are provided by the three carbonyls. From the eight metal valence electrons, the filled orbitals (six electrons) are involved in the π-backbonding (which is equivalent, in some way, to consider that CO is a X_2 ligand and to say that Fe only has two non-bonding electrons in $Fe(CO)_3$. Three orbitals and two electrons remain available in the cluster, exactly as B in BH. The 14-electron fragment CoCp, isoelectronic with $Fe(CO)_3$, behaves in the same way. The fragments BH, CH^+, $M(CO)_3$ (M = Fe, Ru, Os) and M'Cp (M' = Co, Rh, Ir) are called *isolobal* (see section 2.3).

For a fragment $ML_nX_p^q$, the number T of electrons given to the cluster is:

T = n_M + 2n_L + n_X – 12, n_M being the number of valence electrons of the metal.

The total number S of electrons brought to the cluster by the different fragments is:

S = Σ T + Σ l – z, l being the number of electrons brought by the ligands that are not counted in T, for instance eventually the bridging ligands (H: one electron, CO: two electrons, Cl: three electrons, etc.), z being the algebraic charge.

The number n_s of vertices of the polyhedron is given by:

$$n_s = S/2 - 1.$$

The 6 BH of the $B_6H_6^{2-}$ octahedron can be replaced by 6 $Os(CO)_3$ fragments. The octahedron $[Os_6(CO)_{18}]^{2-}$ is then exactly obtained, whereas its structure could not be explained by the localized NCE model. The 6 $Os(CO)_6$ fragments contribute for a total of 12 electrons to the cluster, which makes a total S of 14 electrons resulting in the double negative charge. The number n of vertices of the polyhedron is:

$$n = S/2 - 1 = 6.$$

Another isoelectronic example is $[Co_6(CO)_{14}]^{4-}$ which can also be written as $[\{Co(CO)_2\}_6(CO)_2]^{4-}$. For the fragment $Co(CO)_2$: $T = 9 + 4 - 12 = 1$ and altogether: $S = (6 \times 1) + (2 \times 2) + 4 = 14$, i.e. $n = 14/2 - 1 = 6$.

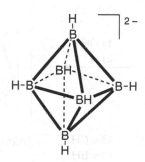

We have seen that the general formula of close-shell boranes, i.e. whose polyhedral vertices are all occupied, (called *"closo"*), is $B_nH_n^{2-}$ ($n = 6 - 12$). If a BH^{2+} fragment is removed from the borane, no electron is removed from the cluster, and thus the cluster remains unchanged, except for one less vertex. The general formula of the boranes whose structure is derived from the *"closo"* boranes, but with one less summit, is $B_nH_n^{4-}$. This family is called *"nido"*. For instance, if a BH vertex is removed from the octahedron $B_6H_{12}^{2-}$, the tetragonal pyramid $B_5H_5^{4-}$ is obtained.

If a second BH^{2+} fragment is removed, a new family called *"arachno"* is obtained, and its structure is derived from the preceding one by elimination of two vertices. Its general formula is $B_nH_n^{6-}$. In general, the number n of vertices of a borane of formula $B_xH_y^{z-}$ is:

$$n = 1/2 \, (x + y + z) - 1 \quad \text{for a } closo \text{ cluster,}$$
$$n = 1/2 \, (x + y + z) - 2 \quad \text{for a } nido \text{ cluster,}$$
$$n = 1/2 \, (x + y + z) - 3 \quad \text{for an } arachno \text{ cluster.}$$

It is also possible, in the same way, to find carboranes that are isostructural with these two families by carrying out the same exercise of replacement of BH by CH^+. For instance, if a BH^{2+} in position 1,2 of the *o*-carborane *closo*-1,2-$C_2B_{10}H_{12}$ icosahedron (*vide supra*), is removed, the *nido*-7,8-$C_2B_9H_{11}^{2-}$ is obtained. It is possible to synthesize this *nido* dianionic carborane, analogous to the cyclopentadienyl anion with six π electrons (see Chap. 1), from the *closo* icosahedron. Both are excellent ligands. In its neutral form, it should be counted as a LX_2 ligand.

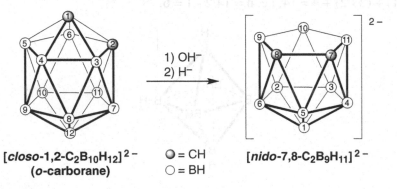

$[$*closo*-1,2-$C_2B_{10}H_{12}]^{2-}$ ◐ = CH $[$*nido*-7,8-$C_2B_9H_{11}]^{2-}$
(*o*-carborane)   ○ = BH

Hawthorne has synthesized many exceptionally stable sandwich complexes containing carborane ligands.[2.10] For example, the dianionic Fe^{II} complex below is the carborane equivalent of ferrocene.

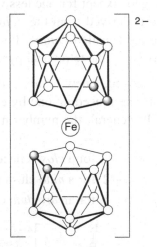

In the same way, a BH fragment can be replaced by a 14-electron metallic fragment such as FeL_3. For instance, the cluster $[Fe_5(CO)_{15}C]$ is well known. It can be written $[(Fe(CO)_3)_5C]$ in which the C ligand (carbido, X_4 ligand) is inserted between the 4 Fe of the tetragonal bipyramid basis:

$$T = 8 + (3 \times 2) - 12 = 2 \text{ and } S = (5 \times 2) + 4 = 14 \text{ with } n_s = 14/2 - 1 = 6$$

i.e. an octahedron with one less summit, since there are only 5 Fe (the 4 carbon valence electrons must be counted for their contribution to S).

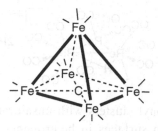

In conclusion, the bimetallic complexes and the clusters display fascinating structures that have been beautifully rationalized by comparison with the borane families, creating yet another area of chemistry.

2.5. THE REACTIVITY OF CLUSTERS

A considerable amount of work has been carried out on the reactivity of clusters, especially organometallic ones, that can be complex, versatile and very rich.[2.11] Let us only examine a few main trends. *Structural changes* in the cluster geometry can occur simply upon adding electrons, as already shown in section 2.3 with $Os_6(CO)_{18}$. In other cases such as $[FeCp(CO)]_4$ (see section 2.2) or giant metal clusters (nanoparticles), various redox changes can occur *without structural change*. Electron-transfer-chain catalysis, detailed in Chap. 5.3.1, is a very practical, useful and general method to cleanly exchange a carbonyl ligand by another less electron-withdrawing ligand such as a P donor. One needs a catalytic amount of an initiator whose oxidation potential is more negative than the reduction potential of the cluster, for instance the "electron-reservoir" complex $[Fe^ICp(\eta^6\text{-}C_6Me_6)]$: [2.12]

Another method to carry out carbonyl substitution in metal-carbonyl clusters is to use the nucleophile Me_3NO that is well known to provoke the liberation of CO from transition-metal complexes, and this method has been used to substitute CO by MeCN in the analogous cluster $[Os_3(CO)_{12}]$ giving $[Os_3(CO)_{11}(MeCN)]$.[2.13]

Protonation of clusters is another simple reaction that yields metal hydride clusters in which the hydride ligands are bridging two metals. For instance, the binary anionic cluster isoelectronic to $[Os_3(CO)_{12}]$ is $[Re_3(CO)_{12}]^{3-}$. This trianion can undergo three successive protonation reactions ultimately yielding a neutral triply bridging trihydride cluster.[2.14]

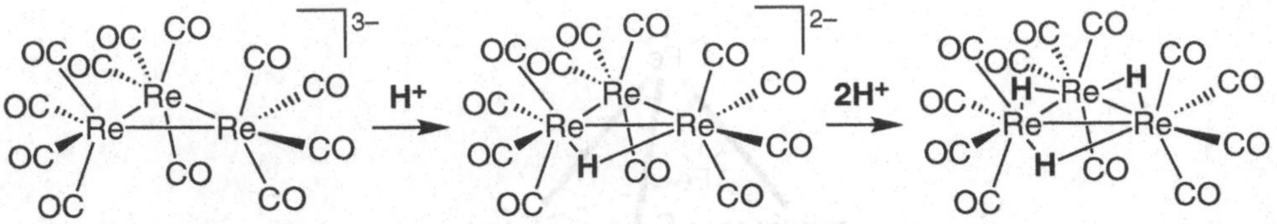

The reactivity of metal-carbonyl clusters with unsaturated hydrocarbons gives an idea of what may happen on surfaces in heterogeneous catalysis, although the conditions and pathways are different, because surfaces of heterogeneous catalysts are not saturated with ligands whereas clusters are (see Chap. 20). Many bond activations have been shown (CO, PPh$_3$, alkenes, alkynes, etc.), a simple one being the reaction of ethylene leading to double C-H activation. Such bond activations are common and rather facile because of the proximity of the metals in the clusters: [2.15]

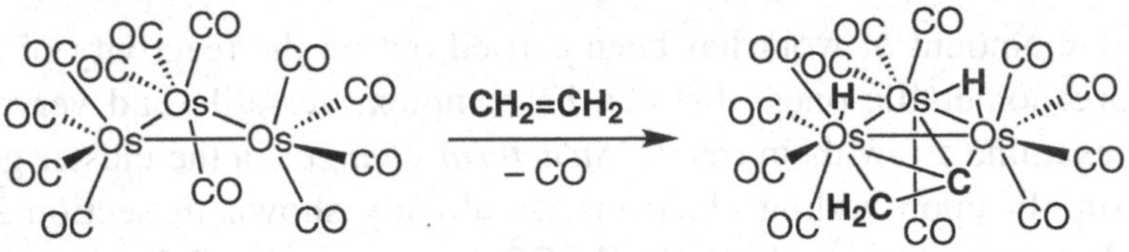

Related clusters have been used in catalysis because of this enhanced potential for bond activation; for instance [Ru$_3$(CO)$_{11}$]$^-$ is active in hydroformylation. Their role is unclear, however, because it is not known whether the catalytic activity is due to the cluster itself or to an easily generated monometallic fragment.[2.16]

Octahedral inorganic metal-halide clusters can also undergo single or hexa ligand substitution of terminal halide ligands, which allows to introduce alkoxy, aryloxy and pyridine ligand *via* the intermediate mono or hexa-triflate complexes. This leads to use them as cores and to decorate then with molecular motifs presenting sensor or catalytic functions.[2.17] Likewise, giant clusters, i.e. nanoparticles which are stabilized for instance by thiolate ligands can undergo ligand substitution to a certain extent or direct synthesis (both paths shown below), which allows introducing interesting properties at the periphery, even organometallic clusters that have been appropriately functionalized: [2.18]

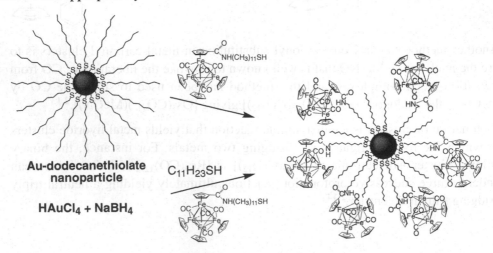

Au-dodecanethiolate nanoparticle

C$_{11}$H$_{23}$SH

HAuCl$_4$ + NaBH$_4$

3. THE ISOLOBAL ANALOGY

In comparing fragments such as $CH_3^{\bullet}$ and $Mn(CO)_5^{\bullet}$, chemists have noticed their similarity for a long time. These two fragments each have a single electron in a σ orbital pointing in the direction opposite to that of the substituents. They both rapidly dimerize and have the same radical-type chemistry.

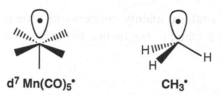

d^7 Mn(CO)$_5^{\bullet}$ CH$_3^{\bullet}$

Hoffmann has called these two fragments *isolobal* (etymologically: *with the same lobe*) and spread the definition. This powerful concept, well enlightened by Hoffmann, has allowed to bring closer together different types of molecular structures of both organic and inorganic chemistries.

Two fragments are isolobal if they have the same number of frontier orbitals, the same total number of electrons in these orbitals and if the symmetry, energy and shape properties of these orbitals are close (look alike). The convention indicating the isolobality of two fragments is the following:

$$Mn(CO)_5^{\bullet} \quad \longleftrightarrow \quad CH_3^{\bullet}$$

▸ If a fragment is isolobal with one of two isolobal fragments, it is also isolobal with the other one. For example $Cr(CO)_5^-$ and $Fe(CO)_5^+$ are isoelectronic with $Mn(CO)_5^{\bullet}$; they are thus isolobal to $CH_3^{\bullet}$ and, of course, isolobal with each other.

▸ Since the main quantum number has only little influence on the shape of the frontier orbital, $Tc(CO)_5^{\bullet}$ and $Re(CO)_5^{\bullet}$ also are isolobal with $CH_3^{\bullet}$.

▸ Cp^- being equivalent to 3 CO, for example $FeCp(CO)_2^{\bullet}$ is isolobal with $Fe(CO)_5^+$, and thus also with $Mn(CO)_5$.

▸ Any 17-electrons ML_5 fragment such as $Mn(PR_3)_5$ is considered to be isolobal with $Mn(CO)_5^{\bullet}$.

$$CH_3^{\bullet} \quad \longleftrightarrow \quad MnL_5^{\bullet} \quad \longleftrightarrow \quad Cr(CO)_5^- \quad \longleftrightarrow \quad FeCp(CO)_2^{\bullet}$$

The divalent fragments $Fe(CO)_4$ and CH_2 (carbene) are isolobal:

d^8 Fe(CO)$_4$ CH$_2$

The CH_3^+, CH^-, SiR_2 and SnR_2, fragments are thus also isolobal with $Fe(CO)_4$, as well as, for example, $Ru(CO)_4$, $Co(CO)_4^+$ and $Mn(CO)_4^-$.

Finally, the trivalent fragments $Co(CO)_3$, CH (carbyne), P and As are isolobal.

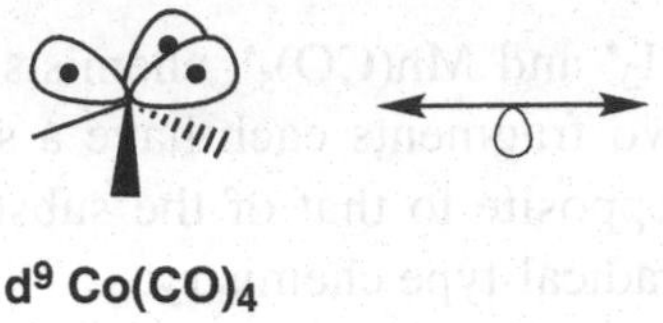

$$d^9 \; Co(CO)_4 \qquad\qquad CH$$

The interest in isolobal analogy mainly concerns its structural implications. The term isolobal is not only applied to fragments, but it is spread to compounds made of isolobal fragments.

Examples:

▸ Ethane (two $CH_3^{\bullet}$ fragments) is isolobal with $[Mn_2(CO)_{10}]$ (two $Mn(CO)_5^{\bullet}$ fragments isolobal with $CH_3^{\bullet}$):

$$H_3C\text{-}CH_3 \qquad\longleftrightarrow\qquad (CO)_5Mn\text{-}Mn(CO)_5$$

▸ Cyclopropane (three CH_2 fragments) is isolobal with $[Ru_3(CO)_{12}]$ (three $Ru(CO)_4$ fragments isolobal to CH_2) and with Herrmann's bridging carbene $[\{RhCp(CO)\}_2(\mu_2\text{-}CH_2]$ (the two $RhCp(CO)_2$ fragments are isolobal with the third one, CH_2).

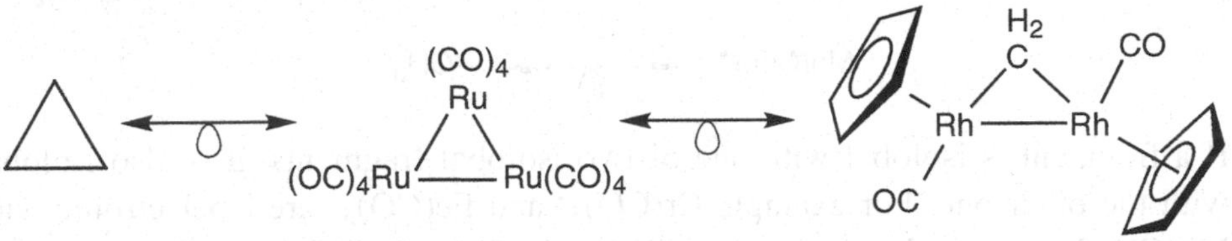

▸ Tetrahedrane (four CR fragments, for example R = CH_3) is isolobal with its inorganic analogs P_4, As_4 and $[Ir_4(CO)_{12}]$ (four $Ir(CO)_4$ fragments isolobal with CH, P and As). It is also isolobal with organometallic analogs having the hypothetical structure $[Co_4(CO)_{12}]$ in which one or two $Co(CO)_3$ fragments are replaced by isolobal carbynes fragments P or As. These isolable compounds are well known and very stable:

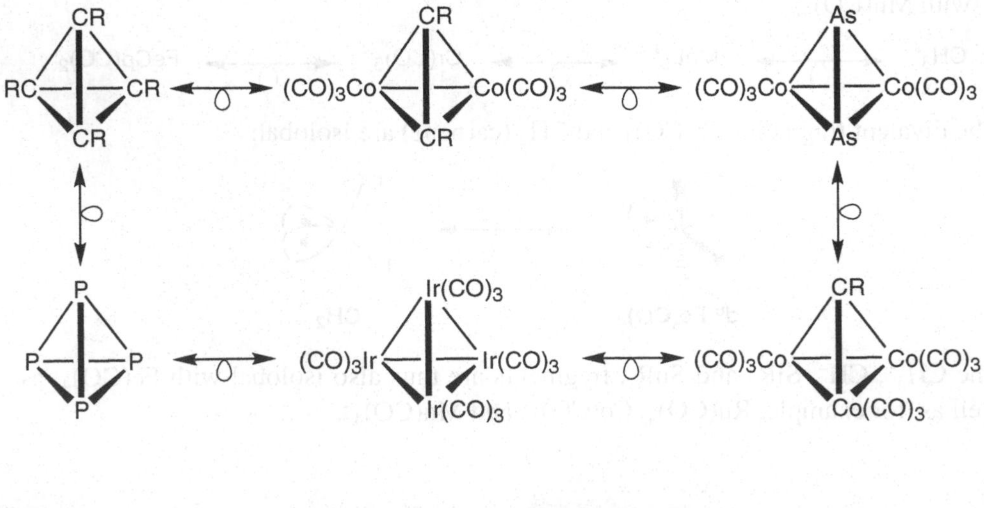

Thus, isolobal analogy is a Lego game between organic, inorganic and organometallic chemistries that allows to rationalize existing structures and to predict other ones. Some of the predicted isolobal structures are stable, others are not.

For example, ethylene $CH_2=CH_2$ is isolobal with the instable structures $(CO)_4Fe=CH_2$, $(CO)_4Fe=Fe(CO)_4$ and $(CO)_4Mn=PH_2$. On the other hand, both of the following isolobal structures of spirocyclopentane type are known.

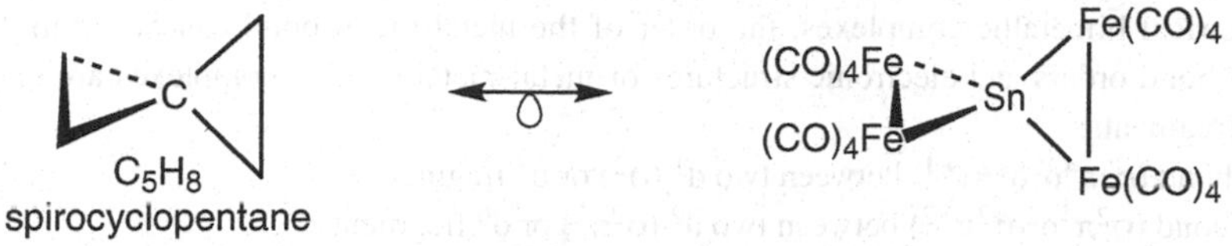

In section 2, we have examined the analogy between the boranes, carboranes and metallocarboranes according to Wade's rules. The BH and CH^+ fragments are isolobal. These two fragments are also isolobal with ML_3 fragments (M = Fe, Ru, Os) or M'Cp (M' = Co, Rh, Ir). Wade had thus already used the isolobal analogy for a good purpose before the latter bore its name.

SUMMARY OF CHAPTER 2
BIMETALLIC TRANSITION-METAL COMPLEXES AND CLUSTERS

1 - Bimetallic complexes

In symmetrical bimetallic complexes, the order of the metal-metal bond can be up to 4. The following bond orders and electronic structures of metal-metal bonded complexes are obtained from the fragments:

▸ simple bond ($\sigma^2\pi^4\delta^2\delta*^2\pi*^4$) between two d^1 (σ^2) or d^7 fragments;
▸ double bond ($\sigma^2\pi^4\delta^2\delta*^2\pi*^2$) between two d^2 ($\sigma^2\pi^2$) or d^6 fragments;
▸ triple bond ($\sigma^2\pi^4\delta^2\delta*^2$) between two d^3 ($\sigma^2\pi^4$) or d^5 fragments;
▸ quadruple bond ($\sigma^2\pi^4\delta^2$) between two d^4 ($\sigma^2\pi^4$) or d^5 fragments.

In bridged bimetallic complexes, the bridging ligand can be:

▸ H, alkyl, aryl (deficient 2-electron, 3-center bond);
▸ a LX ligand (halogeno, alkoxy, amino); also L_2 (diphos), L_2X (Cp), L_4X_2 (fulvalene), etc.;
▸ ligand providing an electronic and magnetic super-exchange between the two metals.

2 - Clusters

Various types: metal-carbonyls, metal-Cp, biomimetic M-S clusters (analogs to biological electron-transfer mediators), polyoxometallates, oxides, sulfides, halides, nanoparticles.

Electron counts in clusters:

Localized electron count up to about 5 metals

Number of cluster electrons NEC = $x\, n_M + 2\, n_L + n_X + q$, x being the number of metals in the cluster. If the 18-e rule is fulfilled: NEC = $18\, x - 2\, y$, y being the number of metal-metal bonds.

Wade's rule for boranes, carboranes, metallacarboranes and transition-metal clusters

Each BH, CH^+, $Fe(CO)_3$, $Os(CO)_3$ or CoCp brings 2 electrons and 3 orbitals, thus a borane $B_6H_6^{2-}$, with the analog $[Os_6(CO)_{18}]^{2-}$, has $2\,n + 2$, i.e. 14 e^-. A $ML_nX_p^q$ fragment brings $T = n + 2\,n_L + n_X + q$ electrons to the cluster to which it binds. For the cluster, the total e^- number is $S = \Sigma T + \Sigma l - z$ (Σl being the number of e^- brought by the other ligands).

The number of vertices of the polyhedron is $n_s = S/2 - 1$, for example $14/2 - 1 = 6$ in the above cases. If one, resp. 2 BH (or $Fe(CO)_3$, etc.) vertices are removed from a borane, carborane or metalloacarborane, one jumps from a *closo* structure to a *nido*, resp. *arachno* structure. For example, the *nido*-7,8-$C_2B_9H_{11}^{2-}$ carborane is obtained to removing a BH vertex from a *closo*-1,2-$C_2B_{10}H_{12}$; it is analogous to the cyclopentadienyl ligand and forms with Fe^{II} a dianionic sandwich, analogous to ferrocene.

3 - The isolobal analogy

The fragments having the same number of frontier orbitals, the same total number of electrons in these frontier orbitals and frontier orbitals of analogous shapes and symmetries (but not necessarily identical) are isolobal. Examples:

CH_3 ⟷ $Mn(CO)_5$ CH_2 ⟷ $Fe(CO)_4$ CH ⟷ $Fe(CO)_3$

The term isolobal also applies to compounds made of isolobal fragments:

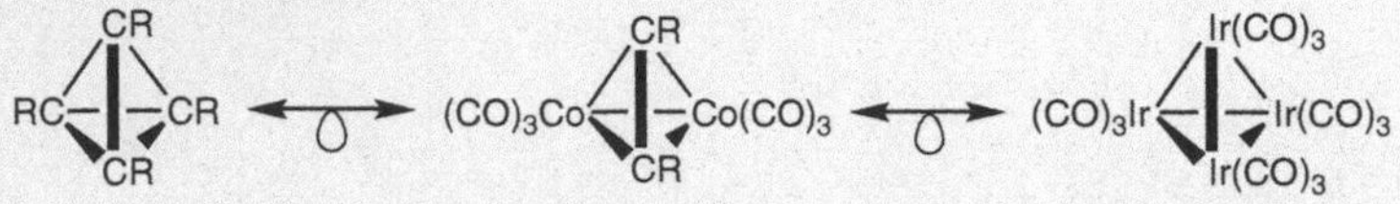

EXERCISES

2.1. Draw structures of the complexes $[FeCp(CO)(\mu\text{-}CO)]_2$, $[CrCp(CO)_3]_2$, $[W(CO)_4(\mu_2\text{-}SCH_2Ph)]_2$ and $[\{Rh(CO)(PPh_3)\}_2(\eta^5,\eta^5,\mu_2\text{-}fulvalenyl)]$, and determine the NVE, NNBE and C of the metals in these complexes.

2.2. Determine the order of the metal-metal bonds in the following binuclear complexes $[(RhCp)(\mu_2\text{-}CO)]_2$, $[CrCp^*(CO)_2]_2$, $[(FeCp^*)_2(\mu_2\text{-}CO)_3]$, in such a way that the metals have NVE = 18.

2.3. Determine the number of electrons (NCE) of the clusters $[Re_4H_4(CO)_{12}]$, $[Fe_6\{C(CO)_{16}\}]^{2-}$ and $[NiCp]_6$ by adding the metal valence electrons and the electrons provided by the ligands on one hand and by the localized electron count on the other hand. Do these clusters follow the localized electron count?

2.4. Find boranes analogous to the three clusters in question **2.3**. In other words, do these clusters follow Wade's rule?

2.5. What are the numbers of Pd atoms in the nanoparticles containing respectively 6 and 7 Pd atom layers around the central metal, the number of neighboring atoms around each Pd atom being 12?

2.6. a. Calculate the Mo oxidation state in the octahedral cluster $Mo_6Cl_{14}^{2-}$ $(Cs^+)_2$.
   b. Deduce the number of core electrons on the cluster using the ionic convention. Thus, consider that the Mo-Cl bonds are ionic to carry out this calculation (which is far from reality since the 8 inner Cl ligands capping Mo_3 triangles cannot undergo substitution reactions).
   c. Calculate the total number of electrons of the cluster (NEC).

PART II

THE STOICHIOMETRIC REACTIONS
OF TRANSITION-METAL COMPLEXES

Chapter 3
Redox reactions, oxidative addition and σ-bond metathesis

Chapter 4
Reactions of nucleophiles and electrophiles with complexes

Chapter 5
Ligand substitution reactions

Chapter 6
Insertion and extrusion reactions

Part II

The Stoichiometric Reactions of Transition-Metal Complexes

Chapter 3

REDOX REACTIONS, OXIDATIVE ADDITION AND σ-BOND METATHESIS

Transition-metal complexes have an extraordinary reactivity due to their ability to change oxidation states, often having up to eight oxidation states. For instance, for iron, oxidation states are known from –II in Collman's reagent, $[Fe(CO)_4]^{2-}$, up to +V in some complexes of the type $[Fe(porphyrin)(X)(=X'_2)]$ (X = halogeno, $X'_2 = O$, NR, CR_2). There are different types of redox reactions of transition-metal complexes. We will distinguish, as Henry Taube already did in the early 1950s,[3.1] the most simple electron-transfer reactions, that proceed by outer sphere, from atom transfer reactions that proceed by inner sphere. These reactions are essential, not only from the fundamental point of view, but also for their applications in various fields: molecular materials, organic synthesis, catalysis and understanding of biological processes. Organometallic chemists are familiar with another type of redox reactions: oxidative addition and its reverse, reductive elimination, both reactions playing a central role in catalysis. Oxidative addition, however, cannot occur in d^0 complexes that do not have available electrons in their valence shell. Activation of metal-alkyl bonds in such complexes must occur by σ-bond metathesis. This reaction is seemingly not of redox type, but it is equivalent to an oxidative addition followed by a reductive elimination. Thus, it will be discussed in this chapter in order to compare these two processes.

1. OUTER-SPHERE ELECTRON TRANSFER

1.1. DEFINITION AND EXAMPLES

An electron-transfer reaction occurs by outer sphere between two molecules or ions if these two entities have adequate redox potentials and do not form intermediate bonds in the course of this reaction, i.e. if the electron is transferred from distance through the solvent cage. This distance can vary between a few Angstroms to a few tens of Angstroms depending on the steric bulk and the nature of the reaction medium.

The following example led Miller [3.2] to find one of the very first molecular magnets (i.e. in which all the spins are aligned from one molecule to the next):

$$[Fe^{II}(C_5Me_5)_2] + TCNE \longrightarrow [Fe^{III}(C_5Me_5)_2]^{\bullet+}, TCNE^{\bullet-}$$

The neutral 19-electron complexes of the $[Fe^I Cp(arene)]$ family are electron-reservoirs that are able to reduce many substrates by outer sphere: [3.3]

$$n\ [Fe^I Cp(C_6 Me_6)] + C_{60} \longrightarrow n\ [Fe^{II} Cp(C_6 Me_6)]^+, C_{60}^{\ n-} \quad (n = 1 - 3)$$

1.2. THERMODYNAMICS: WELLER'S EQUATION [3.3]

The thermodynamics of an outer-sphere electron-transfer reaction is represented by the Weller equation that provides the free enthalpy $\Delta G°$ as a function of the standard redox potentials of the donor and the acceptor and an electrostatic factor containing the sum d of the radiuses of the donor and acceptor, their charges Z_D and Z_A, the dielectric constant ε of the medium and the ionic strength factor f (often approximated to one).

$$\Delta G° \ (kcal \cdot mol^{-1}) = 23.06\ [(E°_D - E°_A) + (Z_A - Z_D - 1)\ e^2\ f/\varepsilon\ d]$$

$$= 23.06\ [E°_D - E°_A] + 331.2\ [(Z_A - Z_D - 1)(f/\varepsilon\ d)]$$

If the acceptor bears a +1 charge, and if the donor is neutral, which is very often the case (as in Miller's example above) the electrostatic factor is nil $(Z_A - Z_D - 1 = 0)$, and Weller's equation is then reduced to the simple form:

$$\Delta G° \ (kcal \cdot mol^{-1}) \approx 23\ (E°_{D/D+} - E°_{A/A-})$$

In the other cases, it is essential to examine if the electrostatic factor is important, for instance in weakly polar solvents that have small dielectric constants (ε low) and when the reacting species are small (d low).

1.3. KINETICS: THE MARCUS' EQUATION [3.3,3.4]

The kinetics of an outer-sphere electron transfer reaction:

$$red_1 + ox_2 \xrightleftharpoons[]{K_{12}} ox_1 + red_2$$

is expressed by the (simplified) Marcus' equation connecting the activation energy $\Delta G^{\ddagger}$ to the free enthalpy $\Delta G°$ (above) of this reaction:

$$\Delta G^{\ddagger} = \lambda/4\ (1 + \Delta G°/\lambda)^2$$

λ is a positive constant; it is the reorganization energy $\Delta G^{\ddagger}$ of the system when $\Delta G°$ is nil, i.e. for the self-exchange (system 1 = system 2); for example:

$$[FeCp^*_2]^+ + [FeCp^*_2] \xrightleftharpoons[]{K_{11}} [FeCp^*_2] + [FeCp^*_2]^+ \quad (K_{11} = 1)$$

The Marcus equation expresses, *inter alia*, the fact that, if the thermodynamics is moderately favorable $(0 < \lambda < \Delta G°)$, the electron-transfer reaction is fast ($\Delta G^{\ddagger}$ weak). If the thermodynamics is unfavorable ($\Delta G° > 0$), the reaction is very slow, or sometimes even so slow that it is not observable, except if the redox reaction is displaced towards the product for any reason, for instance when there is an irreversible follow-up reaction.

The rate constant k of the electron-transfer reaction is connected to $\Delta G^{\ddagger}$ by Eyring's equation:

$$k = K Z e^{-\Delta G^{\ddagger}/RT}$$

K is the adiabaticity $(0 < K < 1)$, i.e. the orbital interaction between the donor and the acceptor that determines the probability of the electron transfer; Z is a constant representing the collision frequency of the molecules $(Z = 6 \times 10^{11}$ at 25°C).

A practical means, using another form of Marcus equation, that allows to estimate the order of magnitude of the rate constant k_{12} of a new reaction, consists in using the self-exchange rate constants k_{11} and k_{22} of the donor and of the acceptor, and the equilibrium constant K_{12} of the redox reaction:

$$k_{12} \sim (k_{11} \, k_{12} \, K_{12})^{1/2}$$

1.4. ELECTRON TRANSFER FROM A PHOTO-EXCITED STATE

The photo-excited states of some complexes have a sufficiently long life time to give rise to electron-transfer reactions. The latter usually are very exergonic, thus fast (which is indispensable to contribute avoiding the back electron transfer, also very fast). This exergonic character is provided by the fact that the photo-excited state is both a much better oxidant and a much better reductant than the ground state. This increase of redox power is approximatively equal to the photo-excitation energy. The classic example is that of $[Ru^{II}(bipy)_3]^{2+}$ (bipy = 2,2'-bipyridine) for which the photo-excited state has a life time of 0.6 μs. Its photo-excitation energy in the visible region is 2.12 eV, the oxidation potential $Ru^{II/III}$ is 1.02 V vs. SCE, and the reduction potential $E°$ $(Ru^{II/I})$ is –1.59 V vs. SCE in the ground state. For the photo-excited state, these potentials become respectively:

$$1.02 - 2.12 = -1.10 \text{ V} \quad \text{and} \quad -1.59 + 2.12 = +0.53 \text{ V } vs. \text{ SCE.}$$

For instance, cobaltocenium $[Co^{II}Cp_2]^+$ whose reduction potential $E°$ $(Co^{II/III})$ is –0.89 V vs. SCE can be reduced exergonically by the photo-excited state of $[Ru^{II}(bipy)_3]^{2+}$:

$$[Ru(bipy)_3]^{2+*} + [Co^{III}Cp_2]^+ \longrightarrow [Ru(bipy)_3]^{3+} + [Co^{II}Cp_2]$$

$\Delta G° = -1.10 + 0.89 = -0.21 \text{ V} = 0.21 \times 26 = -5.46 \text{ kcal} \cdot \text{mol}^{-1} = -22.32 \text{ kJ} \cdot \text{mol}^{-1}$

1.5. ELECTRON TRANSFER INDUCED BY IRRADIATION
OF A CHARGE-TRANSFER COMPLEX

If the electron transfer between two species is thermodynamically slightly unfavorable, but if these species have geometries that facilitate their interaction by orbital overlap (in particular with the planar geometry), these two species can assemble in the solid state to form a "charge-transfer complex" with a slight electronic transfer (corresponding to a small fraction of electron, for instance of the order of 0.1 e⁻) from the donor to the acceptor. This "charge-transfer complex", even though it is weakly bound, gives rise to an absorption band in the visible region (it is colored). Its irradiation at the wavelength corresponding to the absorption band leads to a photo-induced electron transfer between the donor and the acceptor parts of this "charge-transfer complex". In other words, the energy missing in the ground state for electron transfer is provided by the photo-excitation at this wavelength.

Example: [3.5]

$$[CoCp_2] + [Co(CO)_4] \xrightarrow{\Delta} [Co^+Cp_2, Co^-(CO)_4] \xrightarrow{\Delta}\!\!\!\!\times [CoCp_2]^\bullet + [Co(CO)_4]^\bullet$$

$$\downarrow hv$$

$$[CoCp_2]^\bullet + [Co(CO)_4]^\bullet \xrightarrow{PR_3} \text{reaction}$$

The back electron transfer in the ground state subsequent to irradiation is very fast, because it is exergonic, and it does occur in the absence of a follow-up reaction. It can only be inhibited by a fast follow-up reaction with which it competes. In the above example, the use of a phosphine provokes a carbonyl substitution by the phosphine on the Co^{-I} center that is irreversible and faster than the back electron transfer (see Chap. 4).

1.6. INTRAMOLECULAR ELECTRON TRANSFER
IN MIXED-VALENCE COMPLEXES

An important potential application of electron transfer is the intramolecular electron transfer that could lead chemists to fabricate molecular wires. This rich field has been pioneered by Taube and Creuz who have discovered, in 1969, the first mixed-valence complex, the so-called Creuz-Taube ion, a diruthenium complex in which the Ru ions have oxidation states between 2 and 3 and are bridged by a pyrazine ligand:

$$\left[(NH_3)_5RuN\bigcirc NRu(NH_3)_5 \right]^{5+}$$

By using a whole range of bridging ligands, Taube and his coworkers could show that, depending on the ligands, some complexes of this family are average-valence complexes with a singly-occupied orbital delocalized over the two metals (above case) whereas other complexes are genuine mixed-valence complexes (see below) in which the electron is transferred between the two metals at a rate that is measurable using spectroscopic techniques:[3.6]

$$\left[(NH_3)_5Ru^{III} N\bigcirc\!\!-\!\!\bigcirc NRu^{II}(NH_3)_5 \right]^{5+} \rightleftharpoons \left[(NH_3)_5Ru^{II} N\bigcirc\!\!-\!\!\bigcirc NRu^{III}(NH_3)_5 \right]^{5+}$$

Beyond the infrared frequency (of the order 10^{13} per second), it is considered that a faster electron transfer cannot be taken into account, because this is the domain of delocalized orbitals.

Robin and Day's classification of mixed-valence complexes

The frequency of infrared spectroscopy marks the frontier between average-valence complexes (class 3 below) and mixed valence complexes (class 2 below), and this technique is easily accessible. The unsymmetrical complexes belong to class 1 (below).

Class 1 - the two redox centers have a different environment and the valence is localized whatever the time scale;

Class 2 - the two redox centers have identical environments, but the valence is localized at the infrared time scale;

Class 3 - the valence is delocalized on the infrared time scale (10^{-13} s).

For example, with the biferrocenium cation below, both infrared bands of the Fe^{II} and Fe^{III} complexes are observed: it belongs to the class 2. On the other hand, with two arene ligands (below, right), the monocation shows infrared absorptions intermediate between those of Fe^{II} and Fe^{I} complexes:[3.7] it belongs to the class 3.

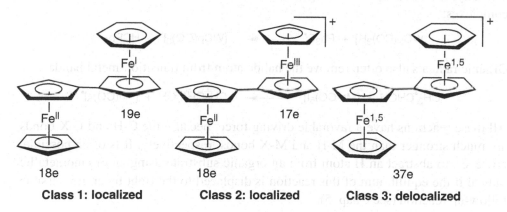

| **Class 1: localized** | **Class 2: localized** | **Class 3: delocalized** |

2. INNER-SPHERE ELECTRON- AND ATOM TRANSFER

This mechanism allows a much faster electron-transfer mechanism than by the outer-sphere mechanism, because it involves an intermediate in which electron transfer is considerably facilitated by a chemical bond between the donor and the acceptor. The first example designed by Henry Taube in 1953 is also the pioneering example of molecular engineering in chemistry. The Co^{III} oxidant has a potentially bridging Cl ligand and the Cr^{II} reductant has six labile H_2O ligands. The H_2O substitution on the substitution-labile Cr^{II} center by the Cl ligand of Co^{III} that bridges both metals very much fastens the reaction: the same reaction with a sixth NH_3, a non-bridging ligand, instead of Cl, leads to an outer-sphere electron-transfer reaction that is a billion times slower!

$$[Co^{II}(H_2O)_6]^{2+} + 5\ NH_4^+$$

start here $\uparrow + 5\ H_3O^+$

$[Co^{III}(NH_3)_5Cl)]^{2+} + [Cr^{II}(H_2O)_6]^{2+} \longrightarrow [Co^{II}(NH_3)_5(H_2O)]^{2+} + [Cr^{III}(H_2O)_5Cl)]^{2+} + 5\ NH_4^+$

$\downarrow - H_2O$ $\uparrow + H_2O$

$[(NH_3)_5Co^{III}\text{-}Cl{\rightarrow}Cr^{II}(H_2O)_5]^{4+} \longrightarrow [(NH_3)_5Co^{II}{\leftarrow}Cl\text{-}Cr^{III}(H_2O)_5]^{4+}$

$k = 6 \times 10^5\ M^{-1} \cdot s^{-1}\ (25°C)$

Note that the transfer of the Cl atom from Co^{III} to Cr^{II} in the intermediate corresponds to an electron transfer in the opposite direction. Overall, the addition of an atom to a metal corresponds to an increase by one unit of its oxidation state, since this atom is one more X ligand for the metal. The Cl atom is transformed from an X to a L ligand for Co in the intermediate complex, and it is the reverse for chromium.

$$[(NH_3)_5Co^{III}\text{–}Cl{\rightarrow}Cr^{II}(H_2O)_5]^{4+} \rightleftharpoons [(NH_3)_5Co^{II}{\leftarrow}Cl\text{–}Cr^{III}(H_2O)_5]^{4+}$$

Atom transfer reactions are very common, in particular *inter alia* with the H atom. The trityl radical is the usual reagent that removes the H atom from metal-hydride complexes:

$$[WCp(CO)_3H] + Ph_3C^\bullet \longrightarrow [WCp(CO)_3]^\bullet + HCPh_3$$

Organic radicals also often remove the halide atom from transition-metal halides.

$$CH_3{}^\bullet CHOH + [Mn(CO)_5I] \longrightarrow CH_3CH(I)OH + [Mn(CO)_5]^\bullet$$

All these reactions have a favorable driving force, because the C-H and C-X bonds are much stronger than the M-H and M-X bonds, respectively. It is often possible, however, to abstract an H atom from an organic substrate using an organometallic radical if the equilibrium of this reaction is displaced to the right by an irreversible follow-up reaction (see Chap. 5).

It is thus useful, in this context, to indicate the strengths of a few M-H bonds in the following table:

Dissociation energy of the element-H bond (E-H) for various element hydrides

E-H bond	Dissociation energy of E-H in $kcal \cdot mol^{-1}$ ($kJ \cdot mol^{-1}$)
$PhCH_2$–H	85 (355)
Et_3Si–H	90 (376)
Bu_3Sn–H	74 (309)
$(CO)_5Mn$–H	60 (251)
$(CO)_4Co$–H	58 (242)
$Cp(CO)_3Cr$–H	61 (255)
$Cp(CO)_3Mo$–H	66 (276)

The photo-excited states of some inorganic complexes are able to abstract an H atom from alkanes. Finally, high-oxidation-state late transition-metal oxo complexes such as the active $Fe^V=O$ or $Fe^{IV}–O^{\bullet}$ species of cytochrome P450 methane mono-oxygenase enzyme, are also able to abstract an H atom from alkanes, which then leads to their hydroxylation (see Chap. 18). It is also possible, in some cases, to remove an H atom from H_2 (see Chap. 15):

$$2\,[(TMP)Rh^{II}CO] + H_2 \longrightarrow 2\,[(TMP)Rh^{III}C(O)H]$$

The consequences of an electron transfer or atom transfer generating a 17-electron species from a less reactive 18-electron complex will be examined in the following paragraphs and chapters.

3. OXIDATIVE ADDITION, REDUCTIVE ELIMINATION AND σ-BOND METATHESIS

3.1. OXIDATIVE ADDITION:
DEFINITION AND CONDITIONS OF APPLICATION

The oxidative addition of a substrate A-B by a complex results in the cleavage of the A-B bond to form two new bonds M-A and M-B, A and B becoming X ligands for the metal. In other terms, the metal inserts between A and B. The opposite reaction, reductive elimination, consists in producing an A-B molecule from the two X ligands A and B in the complex M(A)(B):

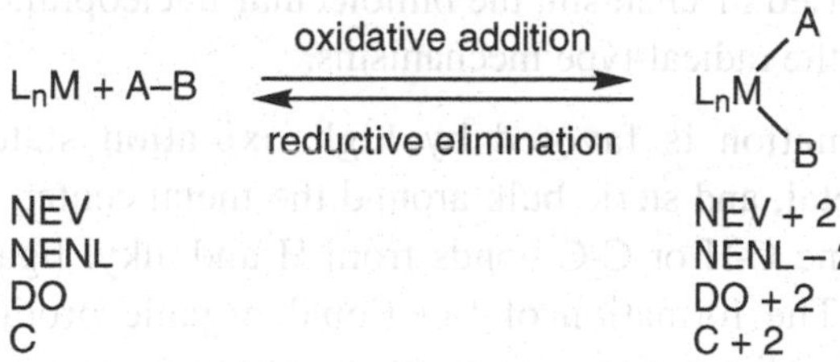

Example of oxidative addition:

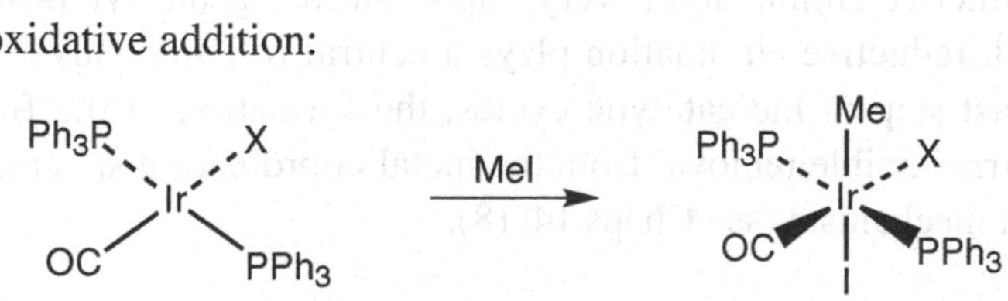

Example of reductive elimination:

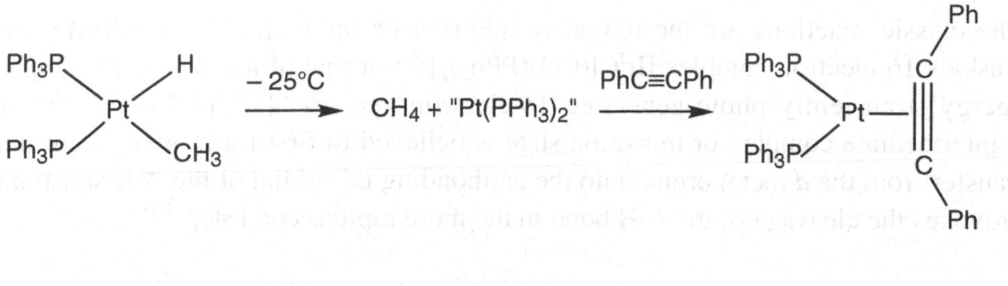

The variations of the characteristics of the complex that are noted above bring constraints for the reaction. The complex undergoing the oxidative addition must have:

▶ at most 16 valence electrons (the resulting complex must not bear more than 18 valence electrons on the metal); see the exception of the ionic mechanism;

▶ at least two non-bonding electrons, because they will be required in order to form the two bonds with A and B;

▶ a low oxidation state (since this oxidation state must increase by two units in the reaction);

▶ a coordination number (C) at most equal to 4 (the square-planar complexes are ideal, because they form octahedral complexes upon oxidative addition); see also the exception of the ionic mechanism.

The oxidative addition reaction has been known for a long time with main-group elements:

$$Mg + RX \longrightarrow R–Mg–X$$

$$PCl_3 + Cl–Cl \longrightarrow PCl_5$$

It is now extremely frequent in transition-metal chemistry in which it plays a key role in catalytic processes. Usually, it involves organic halides and substrates of the RH type (R = H, CN, $SiMe_3$), these categories of substrates being frequently involved in catalysis. Other substrates whose activation would be very valuable such as alkanes sometimes undergo stoichiometric oxidative addition, most often with late, third-row transition metals. The mechanism varies and depends on the polarity of the A-B bond and the nature of the complex. We will distinguish the classic 3-center concerted mechanism, the bimolecular nucleophilic substitution, the ionic mechanism and the radical-type mechanisms.

The reductive elimination is favored by high oxidation states, oxidation or protonation of the metal, and steric bulk around the metal center. It often involves the formation of alkane C-H or C-C bonds from H and alkyl ligands or two alkyl ligand respectively. The formation of functional organic products from ligand fragments by reductive elimination is very important in organic synthesis (see Parts 4 and 5). Indeed, reductive elimination plays a central role in catalysis, because it allows, as the last step of the catalytic cycles, the formation of the final reaction product and its irreversible removal from the metal coordination sphere, most often by the concerted mechanism (see Chaps 14-18).

3.2. THE CONCERTED THREE-CENTER MECHANISM

The classic reactions are the oxidative addition of *inter alia* H_2 or $HSiMe_3$ on Vaska's 16-electron complex $[Ir^ICl(CO)(PPh_3)_2]$ [3.8] or that of alkanes on the (high-energy) transiently photo-generated 16-electron species "$[Ir^ICp^*(PMe_3)]$".[3.9] A σ intermediate complex or transition state is believed to first form, then electronic transfer from the *d* metal orbital into the antibonding σ* orbital of the A-B substrate provokes the cleavage of the A-B bond in the more rapid second step.[3.10]

$$L_nM + A\text{–}B \longrightarrow \left[L_nM\overset{A}{\underset{B}{-}} \right] \longrightarrow L_nM\overset{A}{\underset{B}{\diagdown}}$$

16e, M(0)   18e, M(0)   18e, M(II)

The kinetics is second order ($v = k\,[ML_n][A\text{-}B]$) and the activation entropy is negative ($\Delta S^{\ddagger}$ of the order of –20 eu), which corresponds to an ordered transition state such as that indicated above. When the substrate is H_2, electron-withdrawing ligands (or a positive charge) can remove enough electron density from the metal to let the reaction stop at the stage of the σ H_2 complex. On the other hand, such a complex is not isolable in the neutral form when the ligands are electron-releasing such as with trialkylphosphines, because the metal center is then too electron rich. The σ complexes [M(RH)] analogous to the σ [M(H$_2$)] complexes are not known as stable (because of the combined weakness of the bond and bulk of the R group), although thermally sensitive methane complexes have been spectroscopically characterized around –90°C. When the C-H bond is part of a ligand, however, the agostic σ M-H-C bonds are stable and have often been characterized by X-ray diffraction. In these agostic C-H-M bonds, H is closer to the metal than C, again for steric reasons.[3.10] It is believed that the transition state of alkane oxidative addition involve a σ complex in which the C-H bond is bonded to the metal (see below), but with a M-C distance somewhat larger than the C-H one.[3.10]

$$M\overset{H}{\underset{H}{-\!\mid}} \qquad\qquad M\overset{H}{\underset{CH_3}{-\!\mid}} \qquad\qquad M\overset{H}{\underset{SiMe_3}{-\!\mid}}$$

Vaska's complex gives many oxidative additions [3.8], according to this mechanism. On the other hand, it does not give oxidative addition with alkanes, because the modest difference of energy between the stable starting complex and the final complexes (in particular the Ir-C + Ir-H bond dissociation energies) would not compensate the large C-H bond dissociation energy of alkanes.

The oxidative C-H oxidative addition of alkanes is indeed more difficult to achieve than the above reactions, because of the very high dissociation energy of the C-H bonds in alkanes (see table, section 2). For instance, the oxidative addition of methane on Vaska's complex would be endergonic by 8 kcal·mol^{-1}, which explains that it has not been observed. In order to carry out this oxidative addition of alkanes, it is necessary to start from a high-energy 16-electron species. In 1982, the three research groups of Bergman, Graham and Jones have independently discovered this reaction. The photochemical irradiation of the 18-electron dihydride complexes [M^{I}(η^5-C$_5$R$_5$)(H)$_2$L$_2$], M = Ir or Rh, L = CO or PMe$_3$, in the presence of an alkane, gives H$_2$ and the alkyl-hydride complexes [M^{III}(η^5-C$_5$R$_5$)L(H)(R)]. It was shown that the 16-electron species [M^{I}(η^5-C$_5$R$_5$)L$_2$] is generated after photo-induced reductive elimination of H$_2$. This intermediate species is extremely reactive and provokes the alkane oxidative addition:

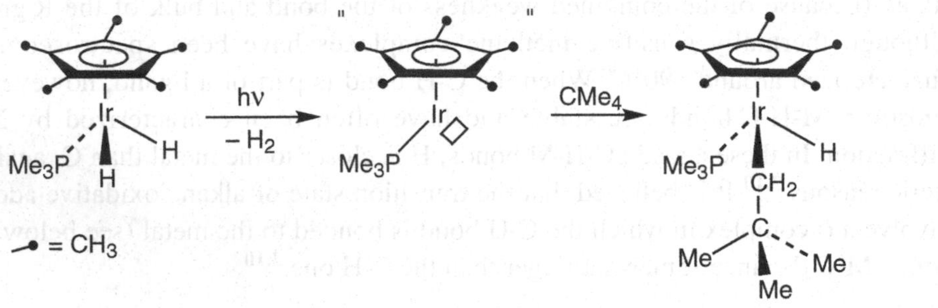

In the above scheme, the experiment carried out in cyclohexane *d*-12 led to [Cp*IrPMe$_3$(D)(Cy-*d*-11)], proving that the photolysis did not involve a 17e radical species [Cp*IrPMe$_3$(H)] that would further react with cyclohexane *d*-12. The related reaction with *p*-xylene yielded a mixture of Ir^{III}-aryl and Ir^{III}-benzyl complexes with the kinetic product ratio of 3.7:1 showing that the aromatic C-H bond reacted faster than the benzylic C-H bond. This was also against a radical mechanism that would have yielded the reverse selectivity.

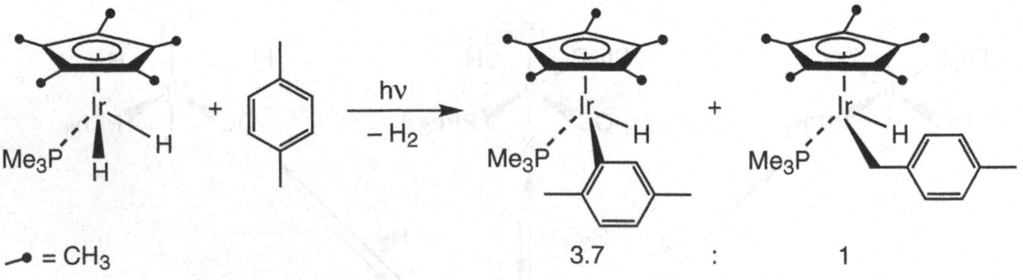

In the same vein, with propane, the primary C-H bond reacted twice as fast as the secondary despite its high bond dissociation energy, meaning antiradical reactivity. Whereas *radical H-abstraction* occurs from the *weakest C-H bond of hydrocarbons* (selectivity: tertiary C-H > secondary C-H > primary C-H), this new activation mode proceeding with the *opposite selectivity* using C-H oxidative addition by organometallics would allow functionalization of terminal carbons yielding highly valuable linear functional products (see Chap. 18).[3.9a]

Although intermolecular oxidative addition of alkanes is difficult, intramolecular oxidative addition has been known for a much longer time, and is indeed favored by the entropic assistance.

Example:

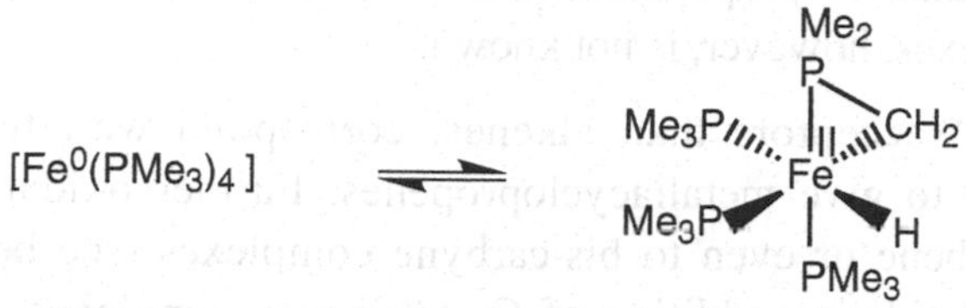

$$[Fe^0(PMe_3)_4] \rightleftharpoons$$

Finally, it is possible to achieve *oxidative addition of substrates that have a double bond A=B, in particular alkenes and O_2, or a triple bond in alkynes.*

The ability of double bonds to behave as L ligands to form π complexes is well known. These ligands are subjected to backbonding that is not conventionally taken into account in the formal representation or in the electron count. When this backbonding becomes too important, however, it must be considered that the ligand has undergone oxidative addition, which is much more pronounced with O_2 than with olefins. The side-on dioxygen complexes are indeed often formulated as peroxide complexes. Double oxidative addition of O_2 leads to high-oxidation-state metal-dioxo complexes (see below).

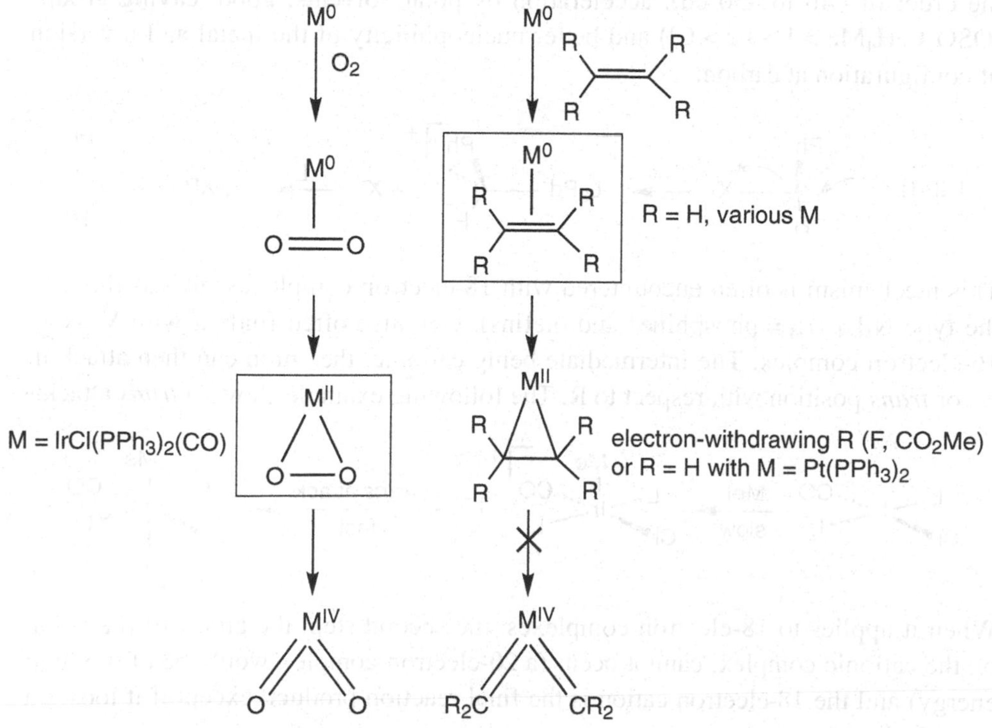

Even for ethylene, the bonding mode can sometimes be considered as resulting from oxidative oxidation when it is bonded to very electron-rich metal centers such as Ni^0, Pd^0 and Pt^0 bearing phosphine ligands. This tendency is of course stronger when the olefin is electron poor, because all the ligand orbitals then have lower energies, and the $\pi*$ orbitals are more easily accessible by the d metal orbitals. For instance, tetrafluoroethylene and fullerene (C_{60} et C_{70}) π metal complexes must be considered as metallacyclopropanes. Double oxidative addition of olefins yielding bis-carbene complexes, however, is not known.

Alkynes, better $\pi*$ acceptors than alkenes, correspond well to the model of oxidative addition to give metallacyclopropenes. Further oxidative addition of alkynes to bis-carbene or even to bis-carbyne complexes (see below) would be analogous to the oxidative addition of O_2 to di-oxo complexes. Such oxidative addition is not yet known, however.

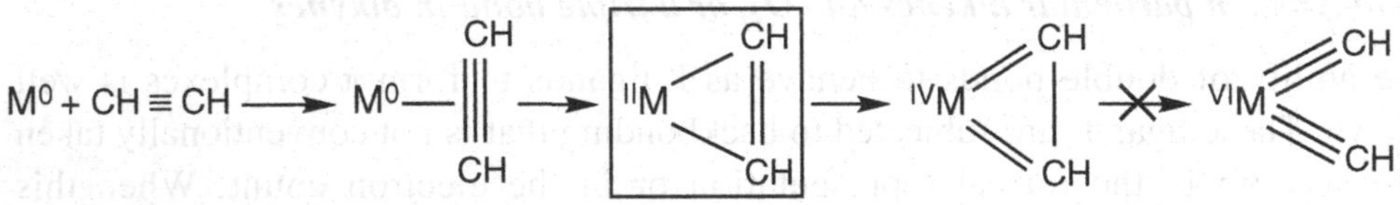

3.3. THE SN₂ MECHANISM

It applies to strongly polar RX substrates and presents all the characteristics of the organic SN₂ mechanism: second-order kinetics, negative activation entropy ($\Delta S^{\ddagger}$ of the order of –40 to –50 eu), acceleration by polar solvents, good leaving groups ($OSO_2C_6H_4Me > I > Br > Cl$) and better nucleophilicity of the metal and inversion of configuration at carbon:

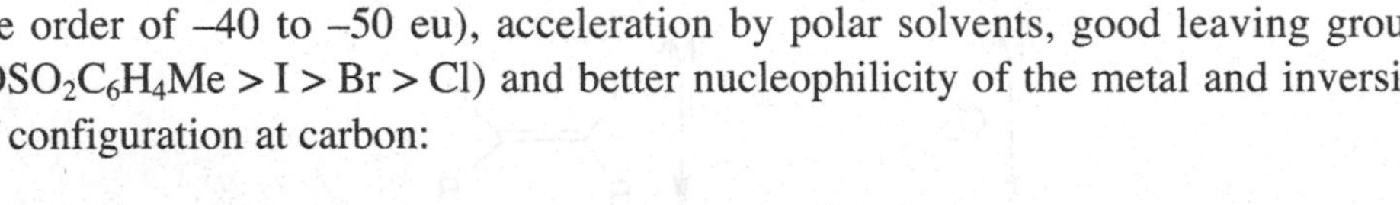

This mechanism is often encountered with 18-electron complexes such as those of the type NiL_4 (L_4 = phosphines and olefins). One also often finds it with Vaska's 16-electron complex. The intermediate being cationic, the anion can then attack in *cis* or *trans* position with respect to R. The following example shows a *trans* attack:

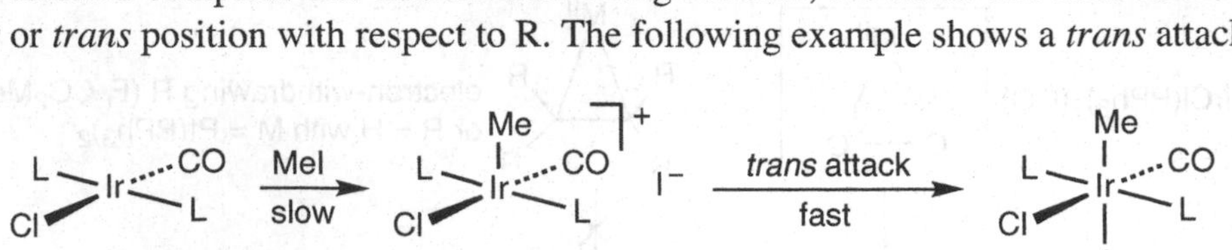

When it applies to 18-electron complexes, the second step, the attack of the anion on the cationic complex, cannot occur (a 20-electron complex would be of too high energy) and the 18-electron cation is the final reaction product, except if it looses a labile L ligand.

$$[Ir^I Cp(CO)(L)] + MeI \longrightarrow [IrCp(CO)(L)(Me)]^+ \ I^-$$

A particular case of SN_2 mechanism that is remarkable is that of the oxidative addition of RX substrates on two non-bonded metals in binuclear cyclic Au^I complexes (note the unusual reversibility of both steps):

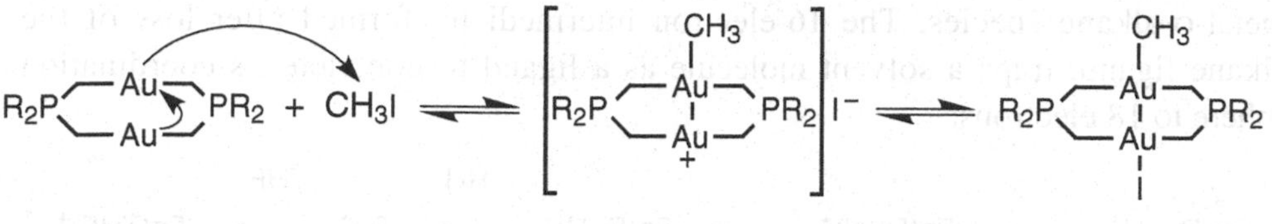

Many other cases of oxidative addition are known with metal-metal-bonded binuclear complexes, such as for instance the synthesis of methylene-bridged binuclear complexes:

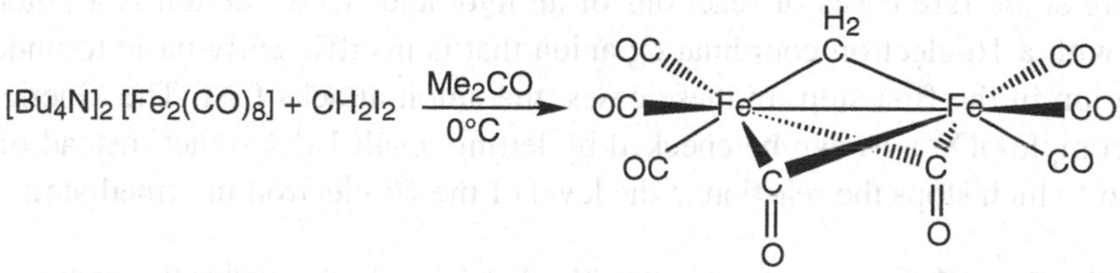

3.4. THE IONIC MECHANISM

The above mechanism can be applied to the case of the attack of 18-electron complexes by totally dissociated electrophiles in a polar solvent. This is particularly the case of hydracids. Thus, the simple protonation of an 18-electron complex to give a hydride is considered as an oxidative addition:

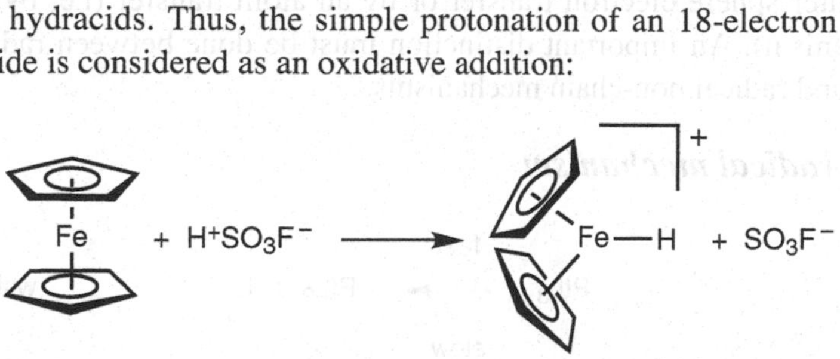

If, instead, a 16-electron complex is used or if the complex looses an L ligand in the course of the protonation, and if the anion is a potential ligand, it will coordinate in the second step to the 16-electron cation, which brings us back to the previous case:

$$[PtL_4] + H^+Cl^- \xrightarrow{\ -L\ } [Pt(H)L_3] + Cl^- \longrightarrow [Pt(H)(Cl)L_3] \quad \text{with } L = PPh_3$$

In all these cases, the second-order kinetics is defined with respect to the slow protonation step:

$$v = k \, (\text{complex})(H^+).$$

The protonation of a 16-electron complex by an acid having a "non-coordinating" anion such as $B(C_6F_5)_4^-$ is of considerable interest in catalysis, because it very much enhances the Lewis-acid properties of the metal (see Part IV).

The protonation of a metal-alkyl complex usually gives a very instable metal-alkyl-hydride immediately leading to the reductive elimination of the alkane *via* the metal-σ alkane species. The 16-electron intermediate, formed after loss of the alkane ligand, traps a solvent molecule as a ligand to complete its coordination sphere to 18 electrons:

$$
\begin{array}{ccccccccc}
 & & & & & -\,RH & & THF & \\
Fp–R + H^+ & \longrightarrow & Fp(R)(H)^+ & \longrightarrow & Fp(R–H)^+ & \longrightarrow & Fp^+ & \longrightarrow & Fp(THF)^+ \\
18e,\ Fe^{II} & & 18e,\ Fe^{IV} & & 18e,\ Fe^{II} & & 16e,\ Fe^{II} & & 18e,\ Fe^{II}
\end{array}
$$

with Fp = [CpFe(CO)$_2$]

There are some rare cases of reactions of an hydracid, whose anion is a potential ligand, with a 16-electron coordinating anion that is insufficiently basic to undergo protonation in the first step. In these cases, the anion attacks first. The kinetics is: v = k (complex)(X$^-$), as can be checked by letting a salt Li$^+$X$^-$ react instead of the hydracid (which stops the reaction at the level of the 16-electron intermediate):

$$[Ir^I(cod)L_2]^+ + H^+Cl^- \longrightarrow [Ir^I(Cl)(cod)(L_2)] + H^+ \longrightarrow [Ir^{III}(H)(Cl)(cod)(L_2)]^+$$

3.5. RADICAL MECHANISMS

There are several variations.[3.11] These mechanisms are found with the halides RX, start by an outer-sphere electron transfer or by an atom transfer (i.e. by an inner-sphere mechanism). An important distinction must be done between radical chain mechanisms and radical non-chain mechanisms.

Non-chain-radical mechanism

$$
\begin{array}{ccc}
 & fast & \\
PtL_3 & \longrightarrow & PtL_2 + L \qquad\qquad with\ L = PPh_3 \\
 & slow & \\
PtL_2 + RX & \longrightarrow & Pt(X)L_2^{\bullet} + R^{\bullet} \\
 & fast & \\
Pt(X)L_2^{\bullet} + R^{\bullet} & \longrightarrow & Pt(X)(R)L_2
\end{array}
$$

The reaction is favored by the basicity of the metal, the transferability of X that goes along its polarizability (RI > RBr > RCl) and the stability of the radical favoring its formation (tertiary R > secondary R > primary R).

In all the preceding examples, the two fragments R and X add to the same metal. This is no longer the case, however, if one starts from a 17-electron complex:

$$[Co(CN)_5]^{3-} + RX \longrightarrow [Co(X)(CN)_5]^{3-} + R^{\bullet}$$

$$R^{\bullet} + [Co(CN)_5]^{3-} \longrightarrow [Co(R)(CN)_5]^{3-}$$

Chain-radical mechanism

It has been discovered by Osborn [3.11] with an electron-rich Ir^I analog of Vaska's complex (L = PMe$_3$, see below) and ethyl- or benzyl bromide:

initiation: $\qquad\qquad\quad$ $Q^{\bullet} + RX \longrightarrow QX + R^{\bullet}$

propagation: $\quad$ $R^{\bullet} + [Ir^I(Cl)(CO)L_2] \longrightarrow [Ir^{II}(R)(Cl)(CO)L_2]$

$\qquad\qquad$ $[Ir^{II}(R)(Cl)(CO)L_2] + RX \longrightarrow [Ir^{III}(R)(X)(Cl)(CO)L_2] + R^{\bullet}$

termination: $\qquad\qquad\quad$ $2 R^{\bullet} \longrightarrow R_2$

In the course of the reaction, the α carbon atom of the radical looses the stereochemistry that it could have had. The chain reaction can be blocked by a radical inhibitor such as 2,6-di-*t*-butylphenol, because the latter gives a relatively stable ArO$^{\bullet}$ radical.

$$ArOH + R^{\bullet} \longrightarrow ArO^{\bullet} + RH$$

Another strong indication that contributes to show the intermediacy of R$^{\bullet}$ radicals consists in testing the reaction with 1-hexenyl bromide as a RX substrate. It is indeed known that the 1-hexenyl radical very rapidly rearranges to give the cyclopentylmethyl radical:

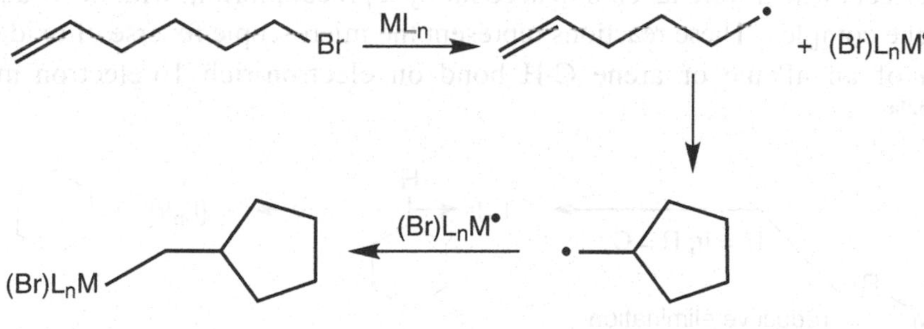

We will come back to the chain reaction in Chap. 5, but let us already note that there are various ways to initiate such reactions by generating radicals that enter the chain: redox reagents, electrodes, light, etc.

3.6. REDUCTIVE ELIMINATION

The concerted three-center mechanism is the most common. It is exactly symmetrical to that already described for oxidative addition, according to the microreversibility principle (see section 3.1). It could indeed be shown that the reaction is intramolecular. For instance, in the following example, no "cross-product" was obtained:

$$\left\{ \begin{array}{c} [WCp_2(H)(CD_3)] \\[2mm] [WCp_2(D)(CH_3)] \end{array} \right\} \xrightarrow{\Delta \text{ or } h\nu} CH_3D + CD_3H + [WCp_2] \ .$$

The two A and B ligands must be located in *cis* position in order to undergo the reductive elimination. The following example shows that prior photoisomerization to the *cis* derivative is necessary before reductive elimination takes place:

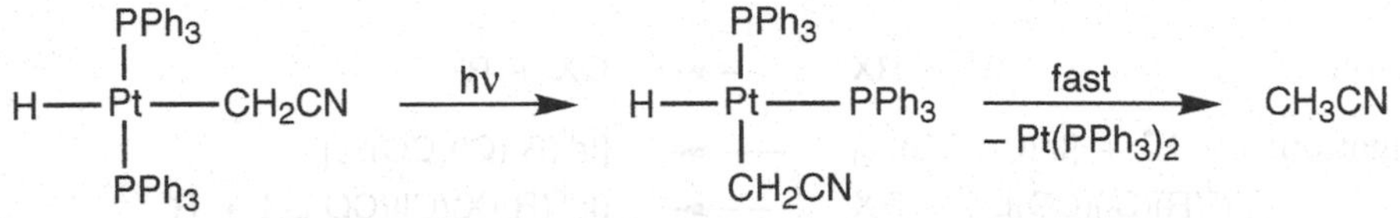

Reductive elimination is favored by high oxidation states of the metal that can eventually be reached by oxidation. For instance, Kochi has shown that the $[Fe^{II}(bipy)_2Et_2]$ complex is stable, and that its 2-electron oxidation provoked the selective, irreversible formation of butane resulting from reductive elimination of the two ethyl groups from the a Fe^{IV} intermediate.[3.5]

$$[Fe^{II}(bipy)_2Et_2] - 2\,e \longrightarrow [(Fe^{IV}(bipy)_2Et_2]^{2+} \longrightarrow n\text{-}C_4H_{10} + [(Fe^{II}(bipy)_2]^{2+}$$

Another means to decrease the electron density on the metal center often is the prior decoordination of a donor ligand.

The reductive elimination from metal-alkyl-hydride complexes $L_nMR(H)$ (M = Ir, R = Cy or M = Rh, R = Ph) giving the alkane or arene occurs with inverse kinetic isotope effects which were taken into account by a preequilibrium with the σ-alkane or η^2-arene complex. These reactions represent the microscopic reverse of oxidative addition of an alkane or arene C-H bond on electron-rich 16-electron metal centers:[3.9a]

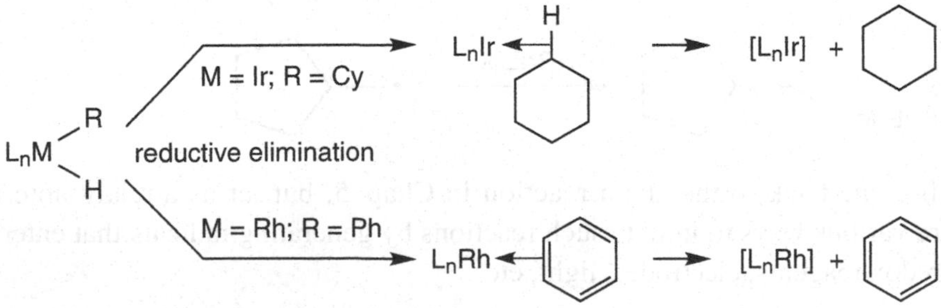

Reductive elimination reactions involving the intermediacy of a σ-alkane or η^2-arene complex

Reductive elimination can also occur on two metal centers:

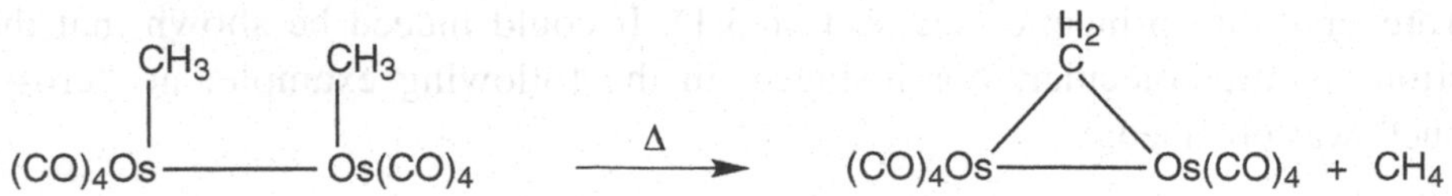

Radical mechanisms are also found in reductive elimination. An important example is the radical-chain mechanism of the dehydrogenation of cobalt carbonyl hydride:

$$2 \, [HCo(CO)_4] \longrightarrow [Co_2(CO)_8] + H_2$$

The number of carbonyl atoms on cobalt being uncertain, the CO ligands are not represented in the following mechanistic scheme of the reductive elimination:

initiation: $CoCo \rightleftharpoons 2 \, Co^{\bullet}$

propagation: $Co^{\bullet} + HCo \longrightarrow HCoCo^{\bullet}$

$$HCoCo^{\bullet} + HCo \longrightarrow HCoCoH + Co^{\bullet} \xrightarrow{-H_2} CoCo \rightleftharpoons 2 \, Co^{\bullet}$$

The reductive elimination will most often be found in Part IV as the last step (irreversible) of catalytic mechanism.

3.7. σ-BOND METATHESIS

3.7.1. Hydrogenolysis: H-H activation

It is possible to let H_2 or H^+ react with a d^0 metal-alkyl complex in a reaction that seems to be an oxidative addition of H_2 followed by a reductive elimination of alkane.

$$M–R + H_2 \longrightarrow M–H + R–H \qquad (\text{example: } M = ZrCp_2Cl)$$
$$M–R + H^+ \longrightarrow M^+ + R–H$$

It is known, however, that oxidative addition is impossible with d^0 metal complexes, because they do not have the required non-bonding electrons to form the two bonds with the new ligands resulting from this reaction. As shown in the scheme below (left), an intermediate H_2 complex is first formed, then a square, 4-center transition state forms that can then follow two topologically symmetrical paths: come back to the starting σ H_2 complex or proceed forward to the transient σ alkane complex that finally releases the alkane.[3.13] This mechanism follows σ-bond metathesis.

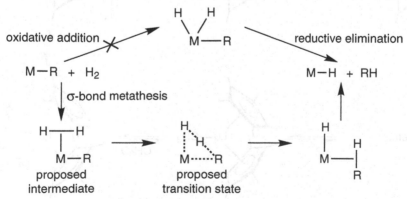

Hydrogenolysis by σ-bond metathesis plays an important role in the decoordination of polymers from metals (see Chap. 6).

3.7.2. Protonation and deprotonation

In the case of protonation, the σ alkane complex is directly formed by protonation of the M-R bond, before liberation of the alkane.

$$M-R + H^+ \longrightarrow M^+\!\!-\!\!\overset{R}{\underset{H}{|}} \longrightarrow M^+ + RH$$

When the complex has at least two non-bonding electrons, this hydrogenolysis process of the metal-alkyl bond to give the metal-hydride usually occurs by oxidative addition followed by reductive elimination. The protonation of the d^6 complex [FeCp(CO)$_2$R], for instance, follows this latter mechanism that we have examined in section 3.4. The reverse σ-bond metathesis of that noted in the above equation, i.e. formation of an alkane complex followed by its deprotonation to give a metal-alkyl species, is a mechanistic possibility that has been envisaged to explain the C-H activation of methane by platinum chloride (*vide infra*).

3.7.3. Methane C-H activation by [MCp*$_2$(CH$_3$)] (M = Lu or Sc)

σ-bond metathesis also applies to the reaction of d^0 complexes with methane, which leads to its activation. Patricia Watson at Du Pont and John Bercaw at Caltech have indeed shown, using deuterium labeling, that [MCp*$_2$CH$_3$] (M = Lu or Sc) can activate CH$_4$ (or CD$_4$) in this way by exchange of methyl groups in the symmetrical square transition state below (this σ-bond metathesis is degenerate, i.e. only isotope exchange occurs without other chemical transformation). Only methane can be activated, not longer alkanes, so that these reactions can be carried out in cyclohexane at 70°C. [3.13]

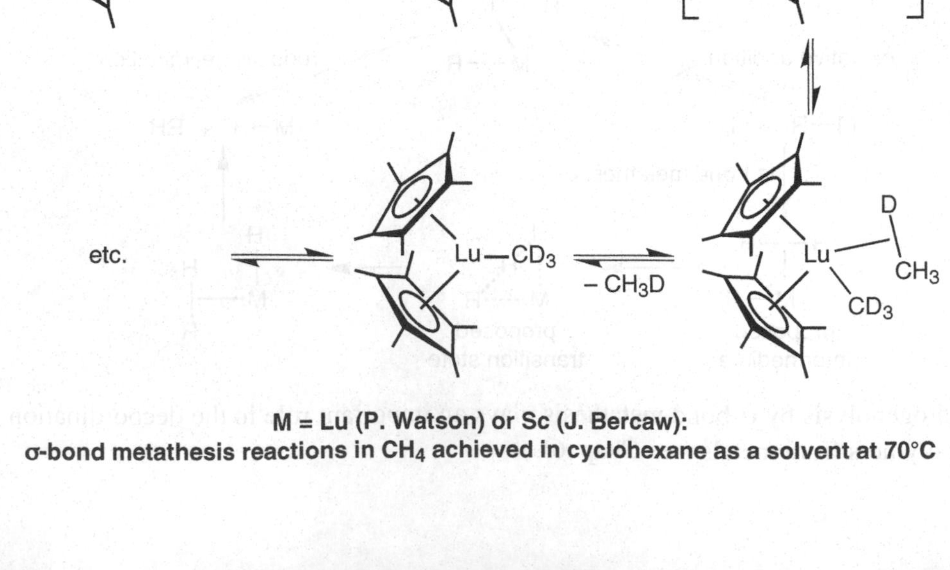

M = Lu (P. Watson) or Sc (J. Bercaw):
σ-bond metathesis reactions in CH$_4$ achieved in cyclohexane as a solvent at 70°C

3.7.4. Conclusion on σ-bond metathesis: α-, β- and γ-elimination

C-H bond activation of a ligand in which a C atom is located in α, β or γ position *vs.* the metal involves σ-bond metathesis in many cases, in particular for α and γ-elimination. These aspects are dealt with in Chaps 3 (stoichiometric reactions leading to metal-alkylidene complexes: section 5), 15 (Ziegler-Natta polymerization: section 1) and 20 (silica-supported alkane metathesis catalysts: section 5).

The relationship between agostic α, β or γ C-H-M bonds (see Chap. 1.1) and α, β or γ-elimination is crucial, because the coordination of an agostic α, β or γ C-H bond to a metal center as a 2-electron L ligand is the first step of α, β or γ C-H activation (by oxidative addition or σ-bond metathesis).[3.15]

Both oxidative addition and σ-bond metathesis are core subjects of organo-transition metal chemistry and catalysis. They also provide the unique regioselective intermolecular alkane activation modes at the terminal (less substituted) C atom contrary to radical-type and cationic C-H activation (see Chap. 19).

3.8. σ C-H ACTIVATION OF ALKANES BY Pt[II] AND Ir[III]: OXIDATIVE ADDITION OR σ-BOND METATHESIS?

In the early 1970s, Shilov showed that a mixture of $PtCl_2$ and H_2PtCl_6 activates and functionalizes methane and other alkanes (see Chap. 17 for the catalytic aspects including the role of H_2PtCl_6).[3.14a] This activation mode starts by the formation of an acidic σ alkane complex. The latter can be directly deprotonated to give a Pt-methyl complex or eventually (but not necessarily) proceed to oxidative addition, as proposed by Shilov, to $Cl_2Pt^{IV}(R)(H)$ before deprotonation. RD is formed in the presence of D^+Cl^-.[3.14a] With various alkanes, the factors influencing the rate of the reaction ($CH_4 < C_2H_6 < C_3H_8$) are the steric bulk and the strength of the activated C-H bond, the latter being larger in methane than in the methyl groups of the other alkanes. The mechanism has been recently intensively studied and largely debated.[3.14b]

Karen Goldberg's group has isolated stable Pt[IV] alkyl-hydride complexes by reaction of alkanes with Pt[II] complexes bearing nitrogen ligands, demonstrating oxidative addition of alkanes to Pt[II].[3.14c]

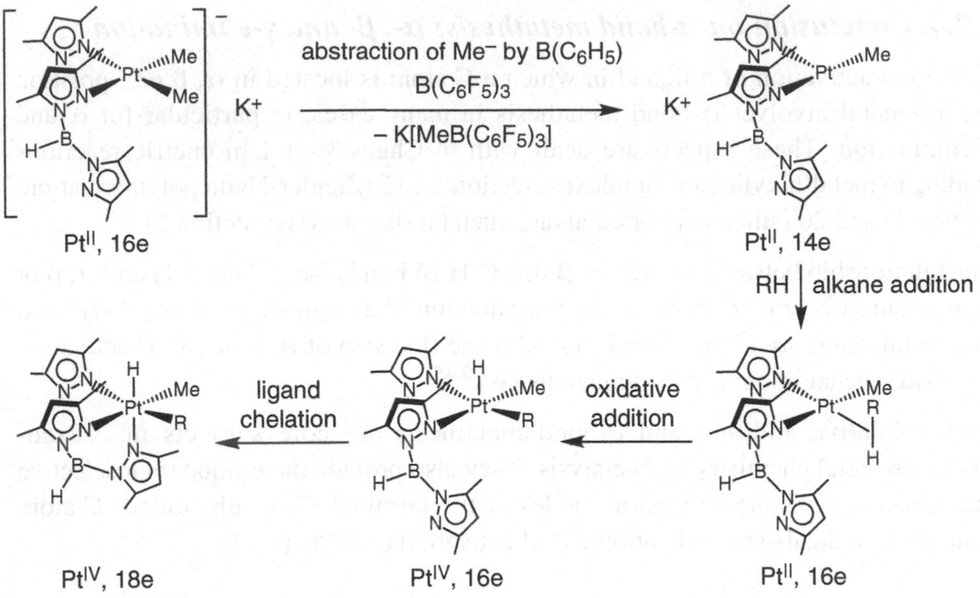

TpMe2 = hydridotris (3,5-dimethylhydridotris(3,5-dimethyl)pyrazolyl) borate
(see Chap. 1, section 1)

The microscopic reverse reaction, reductive elimination from [Tp^{Me2}Pt^{IV}Me$_2$(H)], was observed upon heating this compound in benzene at 110°C to form [Tp^{Me2}Pt^{IV}Me(Ph)(H)]. Reductive elimination of a second equivalent of methane then leads to C-H oxidative addition of a second equivalent of benzene to yield [Tp^{Me2}Pt^{IV}(Ph)$_2$(H)]. Thermolysis of [Tp^{Me2}Pt^{IV}Me$_2$(D)] at 60°C leads to scrambling into the methyl ligands, without loss of methane, in accord with a σ-CH$_4$ complex intermediate before reductive elimination of methane.[3.14d]

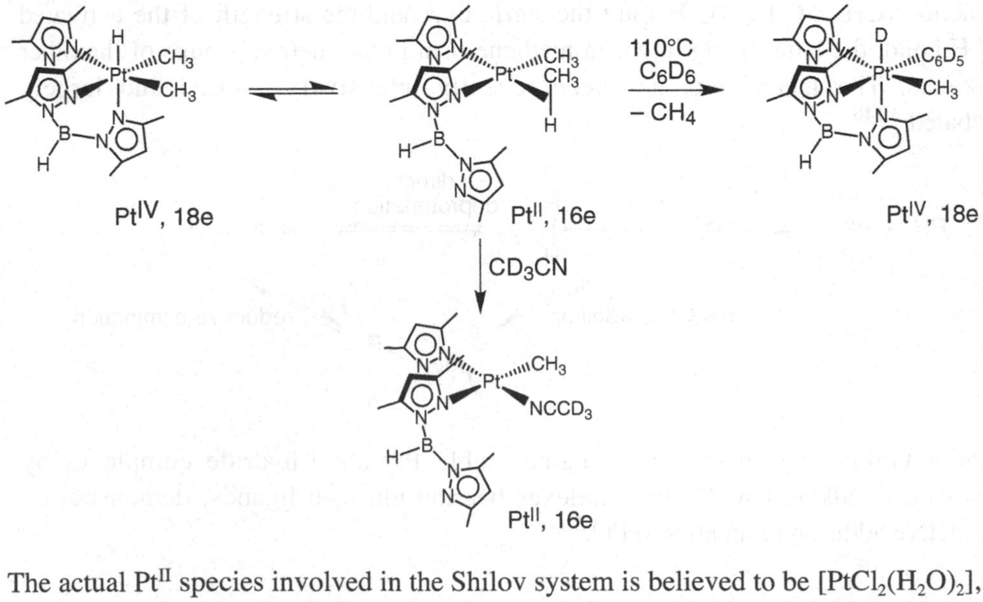

The actual Pt^{II} species involved in the Shilov system is believed to be [PtCl$_2$(H$_2$O)$_2$], however, and calculational studies on the reaction of methane with this species indicated that the barrier to both mechanistic pathways were very similar:[3.14d]

Change from H_2O ligand to nitrogen donors alters the electronic density on Pt and might be responsible for a switch (if any) from σ-bond metathesis to an oxidative addition mechanism in Pt^{II} activation of alkanes.

The same dichotomy of mechanism exists in the alkane activation by Bergman's 16e cationic methyl-Ir^{III} complexes. One might think that, as Pt^{IV}, Ir^V is a very high oxidation state that is eventually difficult to reach, in which case σ-bond metathesis would be an alternative pathway avoiding the high oxidation state. These methyl-Ir^{III} complexes are generated by reversible ionization of polar neutral methyl-Ir^{III}-triflate complexes or by facile dissociation of the weakly coordinated dichloromethane ligand in 18e cationic methyl-Ir^{III}-CH_2Cl_2 complexes (if the counter-anion is a "non-coordinating" perfluorinated tetraarylborate $B(Ar_f)_4^-$). They react with alkanes to form an intermediate methyl-Ir^{III}-alkane cation that yields methane and the Ir^{III}-alkyl cation. The arguments for an oxidative addition mechanism rather than σ-bond metathesis are:

▶ the isolation of a cyclometallated Ir^V complex in the case of oxidative addition of a Si-H bond to Ir^{III} (see equation below),

▶ C-H reductive elimination from this cyclometallated Ir^V complex upon reaction with MeCN, and

▶ DFT calculation carried out on the oxidative addition of methane to $[IrCp^*(PMe_3)(Me)]^+$.

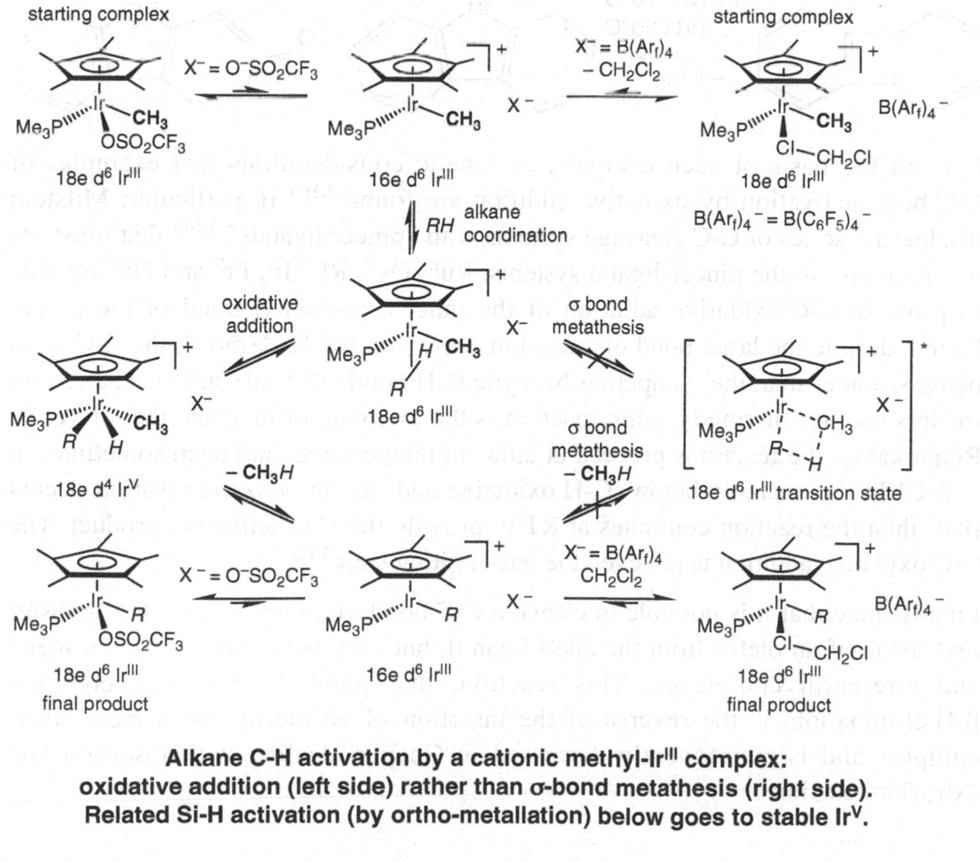

Alkane C-H activation by a cationic methyl-Ir^{III} complex: oxidative addition (left side) rather than σ-bond metathesis (right side). Related Si-H activation (by ortho-metallation) below goes to stable Ir^V.

18e d^6 Ir^{III} 18e d^4 Ir^V

3.9. OXIDATIVE ADDITION OF C-C BONDS

The oxidative coupling of a C-C bond to give a bis-alkyl metal complex is much less common than C-H oxidative addition. For instance, no oxidative coupling of a C-C bond of linear alkane is known, because C-H activation in alkanes is preferred for steric and energetic reasons. It is common, however, for cyclopropane yielding metallacyclobutane, because, understandably, strain energy of cyclopropane is released upon C-C activation. Jones showed that various electron-rich noble-metal complexes such as Ni^0 or Pd^0 complexes insert into the strained 4-membered ring of biphenylene leading to non-strained five-membered Ni^{II} or Pd^{II} metallacycles upon C-C oxidative addition, an energetically-favored reaction in these particular cases: [3.14e]

It is on the basis of such energetic or kinetic considerations that examples of C-C bond activation by oxidative addition are found.[3.16a] In particular, Milstein disclosed a series of C-C cleavage systems with "pincer ligands" [3.16b] that illustrate this problem. In the pincer-ligand systems, Ru^{II}, Os^{II}, Rh^I, Ir^I, Pt^{II} and Ni^{II} are able to provoke C-C oxidative addition of the inner aryl-methyl bond of the pincer ligand, despite the large bond dissociation energy of 102 kcal·mol^{-1}, the C-C bond being stronger than the competing benzylic C-H bond (88 kcal·mol^{-1}). The reason for this counter-thermodynamic reaction is the evolving of methane that drives it. Remarkably, the reactions proceed at ambient temperature, and even sometimes at −78°C! In the example below, C-H oxidative addition in benzylic position occurs first, then the reaction continues at RT to provide the C-C activation product. The C-C oxidative addition is not even the rate-limiting step.[3.16c]

Finally, note that it is possible to cleave a C-C bond of a ligand by β-alkyl transfer (extrusion of an olefin from the alkyl ligand), but only with early transition-metal and rare-earth complexes. This reaction, that parallels the more common β-H elimination, is the reverse of the insertion of an olefin into a metal-alkyl complex and is consequently discussed in Chap. 6 dealing with insertion and extrusion reactions.

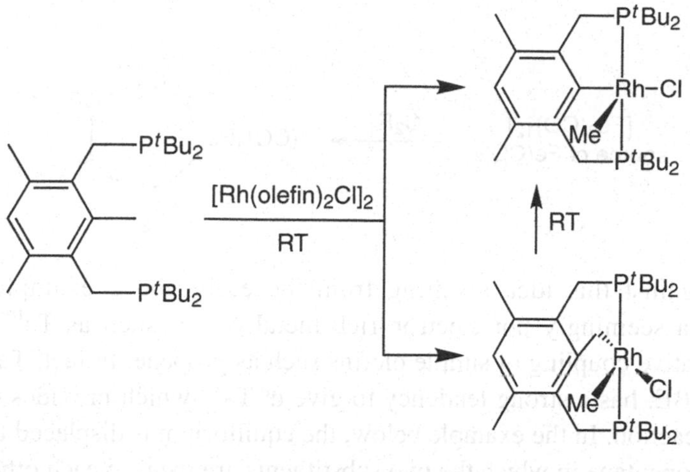

3.10. OXIDATIVE COUPLING AND REDUCTIVE DECOUPLING

Oxidative coupling consists in coupling two ligands by formation of a covalent bond between them. This reaction goes along with the increase of two units of the metal oxidation state (which justifies its name) and the decrease of its NVE and NNBE by two units. The coordinance remains unchanged. The opposite reaction, reductive decoupling, is rarely encountered.

The classic case of oxidative coupling is that of two alkene ligands giving a metallacyclopentane: [3.14]

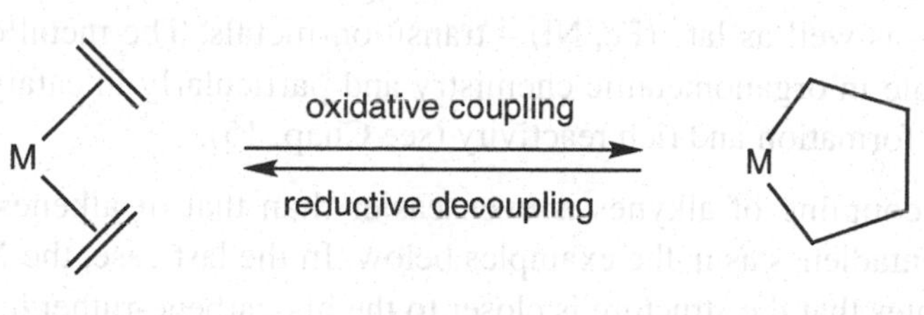

It is also possible to show that oxidative coupling is a form of oxidative addition. Indeed, if one writes the oxidative addition of each olefin to a metallacyclopropane followed by a single reductive elimination of two alkyl groups from the two metallacyclopropanes, the same result is obtained as show below:

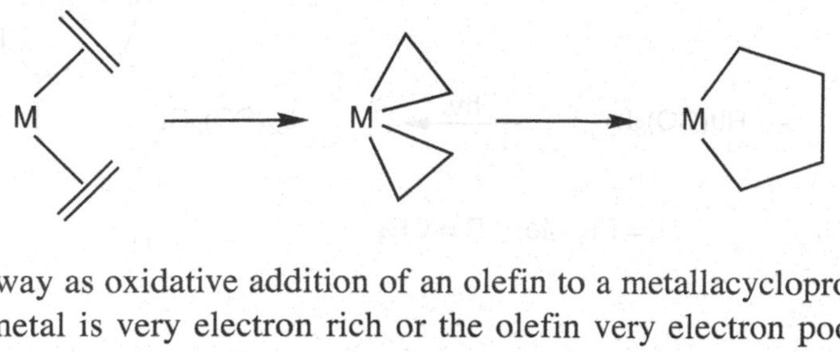

In the same way as oxidative addition of an olefin to a metallacyclopropane occurs only if the metal is very electron rich or the olefin very electron poor, oxidative coupling of two alkenes usually requires the same conditions.

Example:

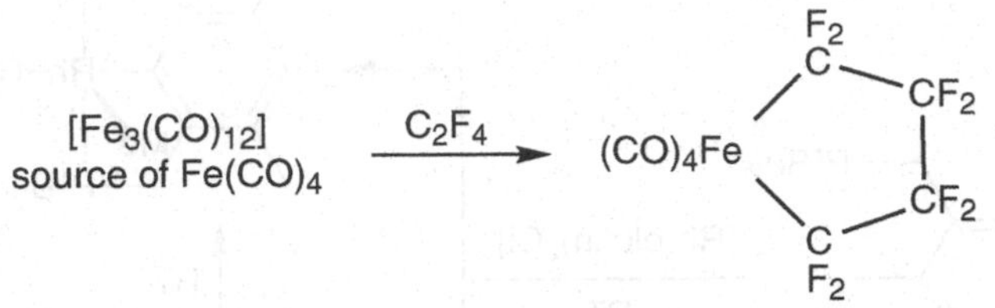

$$[Fe_3(CO)_{12}] \xrightarrow{\ C_2F_4\ } (CO)_4Fe$$
source of $Fe(CO)_4$

Somewhat against this idea resulting from the early above example, Schrock showed that a seemingly not electron-rich metal center such as Ta^{III} is able to undergo oxidative coupling of simple olefins such as propene. In fact, Ta^{III}, that has only two NNBE, has a strong tendency to give d^0 Ta^V, which provides the driving force of the reaction. In the example below, the equilibrium is displaced towards the d^0 tantalacyclopentane in which the two substituents are *trans* to each other.[3.17]

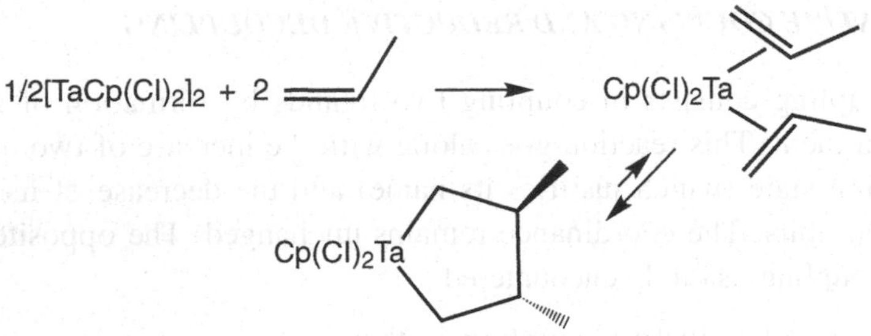

$$1/2[TaCp(Cl)_2]_2 + 2 \quad \longrightarrow \quad Cp(Cl)_2Ta$$

$$Cp(Cl)_2Ta$$

There are other examples of oxidative coupling of metal-olefin complexes with early (Ti, Ta) – as well as late (Fe, Ni) – transition-metals. The metallocycles play an important role in organometallic chemistry and particularly in catalysis because of their ease of formation and rich reactivity (see Chap. 15).

The oxidative coupling of alkynes is even easier than that of alkenes; it leads to metallacyclopentadienes as in the examples below. In the last case, the X-ray crystal structure indicates that the structure is closer to the bis-carbene-ruthenium form than to the ruthenacyclopentadiene mesomer form, and Nishihara showed that this chemistry can be extended to the synthesis of conjugated polymeric ruthenacycles starting from 1,3-butadiyne.[3.18]

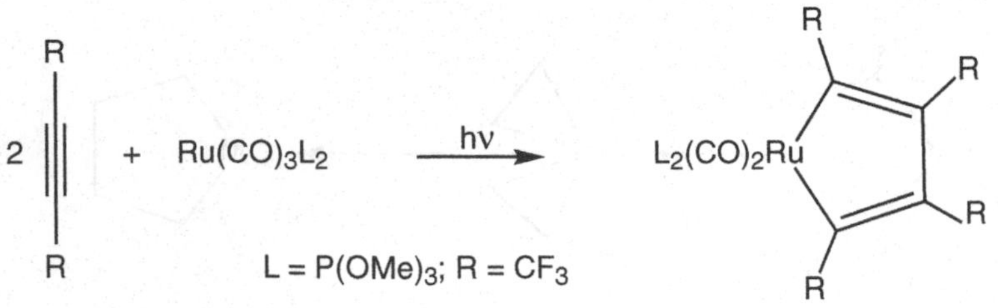

$$2 \quad + \quad Ru(CO)_3L_2 \xrightarrow{\ h\nu\ } L_2(CO)_2Ru$$

$$L = P(OMe)_3;\ R = CF_3$$

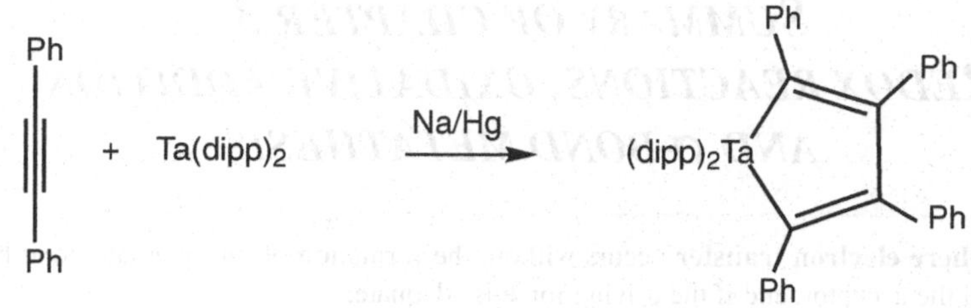

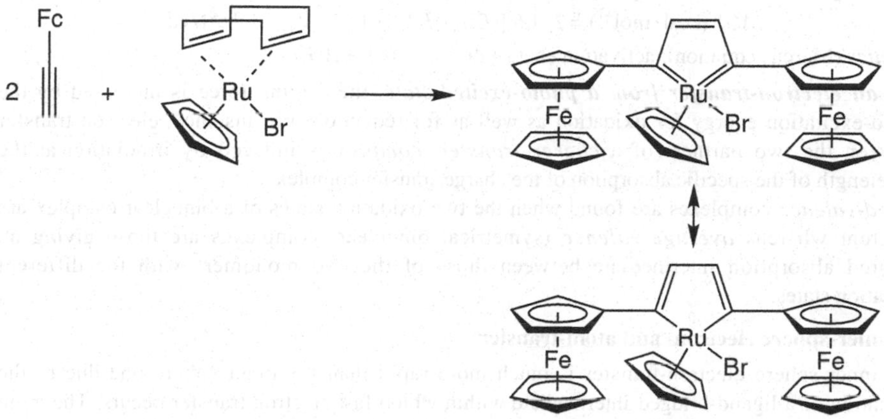

The coupling of two identical ligands located in *cis* positions in a metal complex is also known for other ligands such as CO, alkylidenes and isonitriles (coupling to diaminoalkynes).[3.19] Note that the coupling of two alkylidenes can give the metallacyclopropane or even the olefin, each step going along with a decrease of two units of the metal oxidation state; a reductive coupling is now involved. The opposite reaction is the oxidative addition of an olefin that has already been examined in section 3.2.

SUMMARY OF CHAPTER 3
REDOX REACTIONS, OXIDATIVE ADDITION
AND σ-BOND METATHESIS

1 - Outer-sphere electron transfer occurs without the formation of intermediate bond between the donor and the acceptor, and if the driving force is adequate.

Thermodynamics (Weller's equation):

$$\Delta G° \ (kcal \cdot mol^{-1}) = 23.06 \ [(E°_D - E°_A) + (Z_A - Z_D - 1)] \ e^2 f / \varepsilon d$$

Kinetics (Marcus equation): activation energy $\Delta G^{\ddagger} = \lambda/4 \ (1 + \Delta G°/\lambda)^2$

For an electron-transfer from a photo-excited state, the driving force is increased by the photo-excitation energy for oxidation as well as for reduction. For instance, electron transfer between the two partners of a *charge-transfer complex* is induced by irradiation at the wavelength of the specific absorption of the charge-transfer complex.

Mixed-valence complexes are found when the two oxidation states of a binuclear complex are different whereas *average-valence* (symetrical binuclear) complexes are those giving an infrared absorption intermediate between those of the two monomers with the different oxidation states.

2 - Inner-sphere electron- and atom-transfer

The inner-sphere electron-transfer is much more rapid than the outer-sphere one due to the formation of a ligand-bridged intermediate within which fast electron transfer occurs. The atom transfer $A^{\bullet} + BH \rightarrow AH + B^{\bullet}$ depends on the relative energies of the A-H and B-H bonds and eventually of a follow-up reaction.

3 - Oxidative addition, reductive elimination and σ-bond metathesis

Three-center mechanism *via* a transient σ complex:

There are also SN_2, ionic and radical (non-chain and chain) mechanisms.

Reductive elimination proceeds according to the mechanism reverse of the oxidative addition mechanism (microreversibility principle) from $M^{II}(A)(B)$ to give M^0 and A-B.

Oxidative coupling - Example: $M^n(\eta^2\text{-}C_2H_4)_2$ gives a M^{n+2} metallacyclopentane and, more commonly, addition of 2 equiv. alkyne to M^n yields M^{n+2} metallacyclopentadiene complexes or the bis-carbene mesomers.

σ-bond metathesis is equivalent to an oxidative addition followed by a reductive elimination and occurs in d^0 complexes (oxidative addition being impossible). Ex.: reactions of H_2 or H^+ with d^0 complexes (hydrogenolysis or protonation of d^0 metal-alkyl bonds and alkane activation).

C-C oxidative addition is unknown with linear alkanes, but can be designed in specific systems in which it is driven by the thermodynamics (cyclopropane releasing ring strain by giving metallacyclobutane) or by displacing the reaction (for instance if methane is produced).

EXERCISES

3.1. Is it possible to oxidize decamethylferrocenium ($E° = +1.5$ V vs. ECS; radius ~ 3 Å) using the ion $Co(III)W_{12}O_{40}^{5-}$ ($E° = +1.0$ V vs. ECS, radius ~ 3 Å):
a. in acetonitrile ($\varepsilon = 35$)?
b. in acetic acid ($\varepsilon = 6.2$)?

3.2. Can $[Ru(II)bpy_3]^{2+}$ oxidize ferrocene ($E° = +0.4$ V vs. ECS) in acetonitrile:
a. in the absence of light?
b. in the presence of sunlight? See the numerical values in section 3.1.4.

3.3. What are the reaction products of the protonation of $[MCp(PMe_3)_2(H)]$, M = Fe vs. Os in THF and the difference of reactivity between the two metals?

3.4. What is the reaction product of $[LuCp*_2CD_3]$ with CH_4?

Chapter 4

REACTIONS OF NUCLEOPHILES AND ELECTROPHILES WITH COMPLEXES

These reactions concern 18-electron complexes, the electronic configuration that brings the necessary robustness to late organotransition metal complexes. Such compounds usually are air stable, thermally robust, and they are very useful in the perspective of applications in organic synthesis. The nucleophilic and electrophilic reactions also involve the most simple reagents: the proton and the hydride. When the nucleophile is bulky, it becomes a base, and when the electrophile becomes bulky, it becomes an abstractor of hydride. Thus, the abstractions of H^+ and H^-, common in organometallic chemistry, allow simple interconversions between electrophilic and nucleophilic complexes. This practice is common with the goal of functionalizing a ligand depending on the nature, nucleophilic or electrophilic, of the available reagent. For applications in organic synthesis, see Chap. 19.

1. NUCLEOPHILIC REACTIONS

1.1. NUCLEOPHILIC ADDITION

The nucleophilic additions have a driving force that is all the larger as the organo-metallic fragment on which the attacked ligand is coordinated is more electron-withdrawing. This electron-withdrawing property of the metal fragment can be due to a positive charge or to electron-withdrawing ligands, typically carbonyls. The factors that determine the nucleophilic attack on the coordinated ligand are:
▸ the orbital control,
▸ the charge control,
▸ the steric control.

When the metal fragment is positively charged, the charge control often (but not always) plays the predominant role. In the absence of positive charge or when this charge is attenuated by only hydrocarbon ligands that are donors, the orbital control predominates. The steric control always plays an important role, however. It induces the stereochemistry of the addition, because nucleophilic attacks on ligands occur in *exo*, i.e. on the other side of the metal with respect to the ligand. Steric effects sometimes also induce the regioselectivity of the reactions.

Hydrocarbon ligands from η^2 to η^7

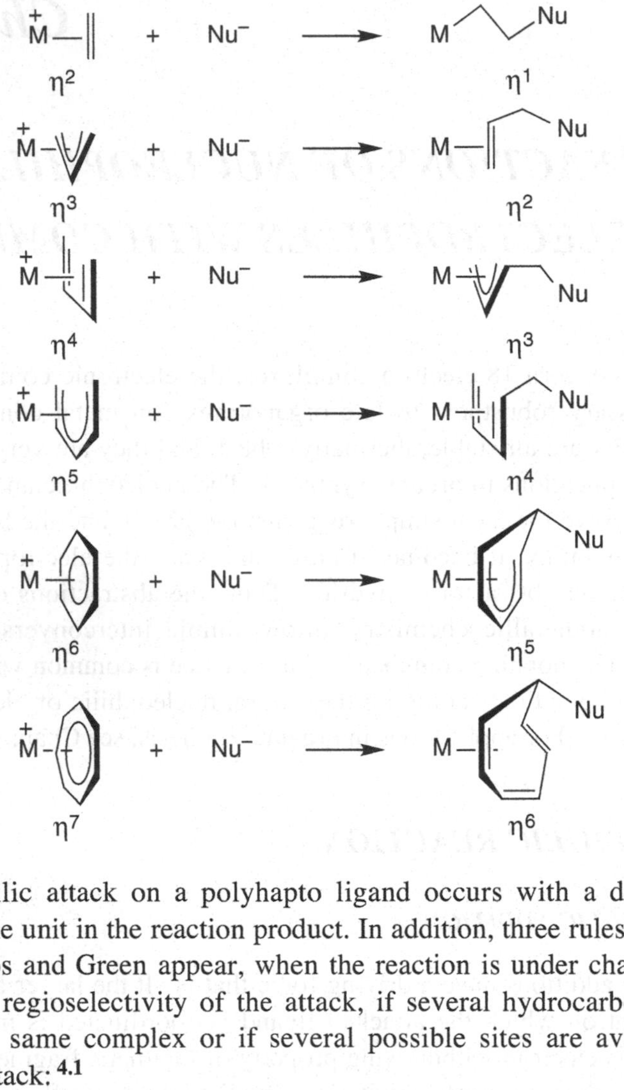

The nucleophilic attack on a polyhapto ligand occurs with a decrease of the hapticity by one unit in the reaction product. In addition, three rules established by Davies, Mingos and Green appear, when the reaction is under charge control, to determine the regioselectivity of the attack, if several hydrocarbon ligands are present in the same complex or if several possible sites are available for the nucleophilic attack: [4.1]

1. Even ligands are attacked preferably to odd ligands.
2. Open ligands are attacked preferably to cyclic ligands.
3. The attack preferably occurs on the terminal carbon of the ligand in the polyenes and open polyenyls.

For instance, in the cationic molybdenum complex below containing the following ligands: benzene (even, closed), butadiene (even, open) and allyl (odd, open), the nucleophilic attack will occur in *exo* on the terminal butadiene carbon. The first rule indicates that an even ligand (benzene or butadiene) should be preferably attacked; the second rule says that the open butadiene ligand is attacked preferably to the closed benzene one, and the third rule states that a terminal butadiene carbon is preferred:

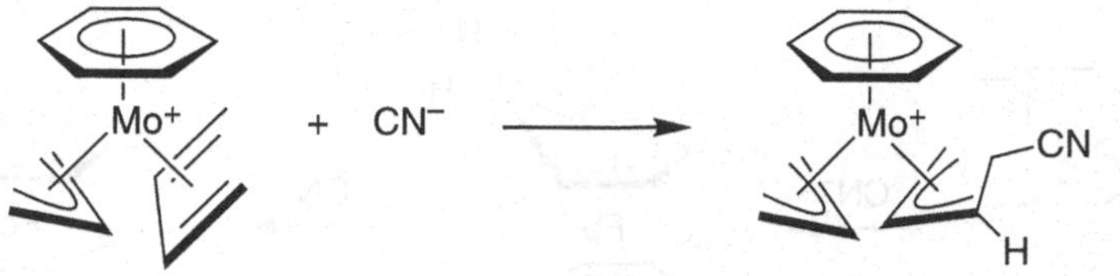

In the following rhodium complex, the first rule does not apply in the absence of even ligand, the second rule gives preference to the cyclohexadienyl ligand, and the third rule to one of its terminal carbon atoms:

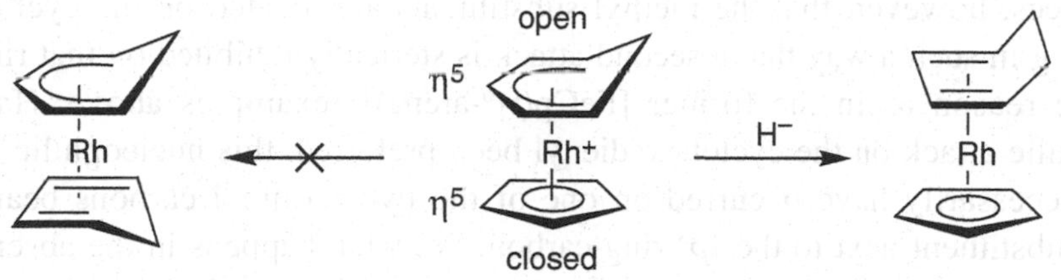

These rules are meant to apply in the order: rule 1 before rule 2 before rule 3, to all cationic complexes, and they work most of the time, but not always.[4.2]

In the mixed cation $[FeCp(\eta^6\text{-}C_6H_6)]^+$, rule 1 indicates that nucleophilic attack must occur on the benzene ligand, and it always does so:

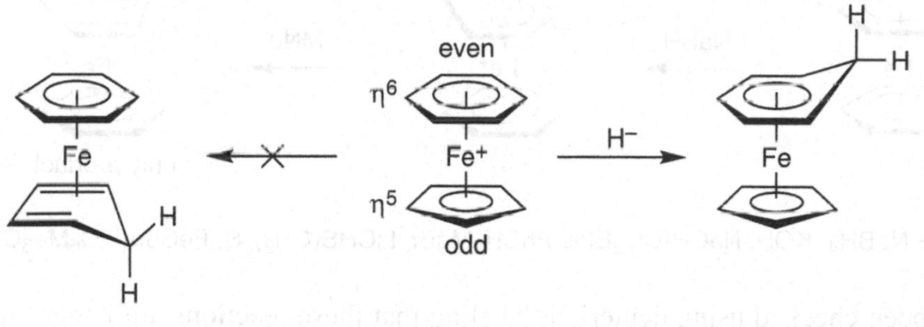

In the analogous hexamethylbenzene complex, the steric control dominates, and the preferred attack occurs on the Cp ligand, although rule 1 gives priority to the arene ligand:

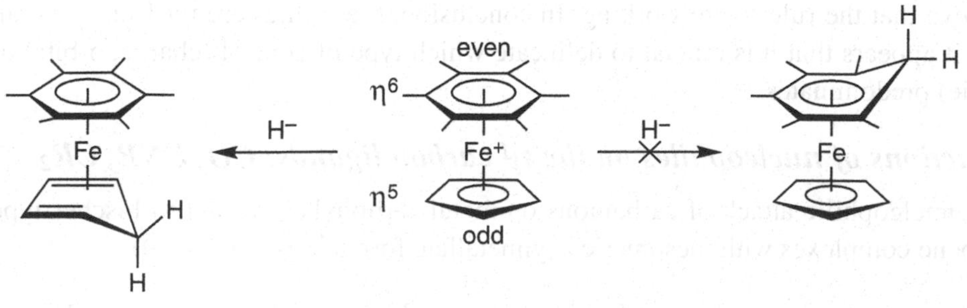

Let us now examine the nucleophilic attacks on the cations $[Fe(arene)_2]^{2+}$. The first attack gives a monocationic cyclohexadienyl-benzene complex. The second nucleophilic attack can then occur either on the arene or the cyclohexadienyl ligand. When arene = mesitylene, the rules seemingly work very well since the even mesitylene ligand is attacked rather than the odd cyclohexadienyl (rule 1).

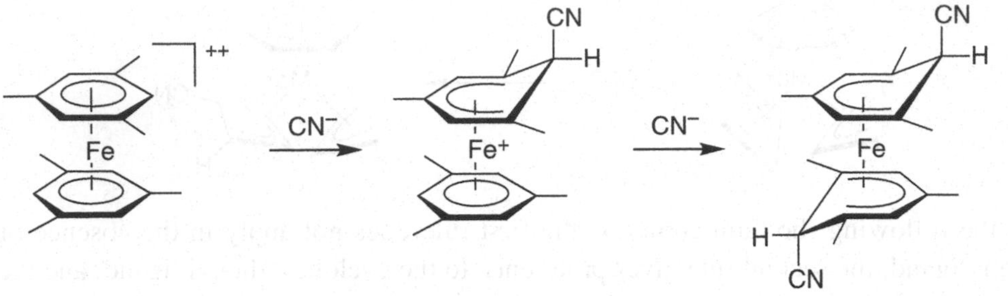

One notices, however, that the methyl substituents are located on the cyclohexadienyl ring in such a way that a second attack is sterically inhibited on that ring for the same reason as in the former $[FeCp(\eta^6\text{-arene})]^+$ examples above. Had the nucleophilic attack on the cyclohexadienyl been preferred, this nucleophilic attack should necessarily have occurred on one of the two terminal carbons bearing a methyl substituent next to the sp^3 ring carbon. So, what happens in the absence of the methyl substituents on the parent complex $[Fe(\eta^5\text{-}C_6H_7)(\eta^6\text{-}C_6H_6)]^+$? Indeed, the reactions of this complex systematically lead, with various nucleophiles, to the attack of the cyclohexadienyl which is contrary to the rule.[4.2]

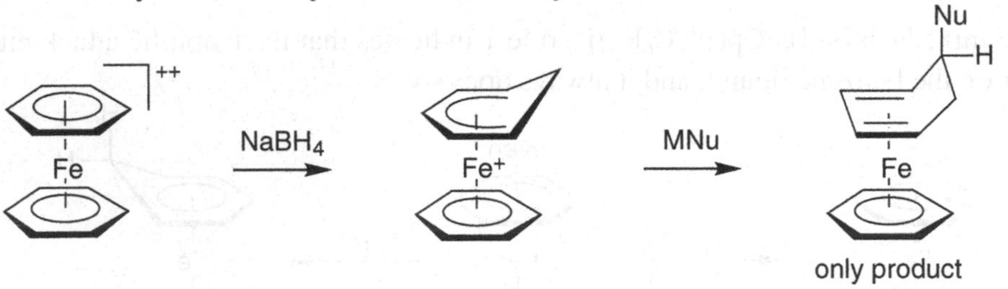

$MNu = NaBH_4, KOH, NaCH(CO_2Et)_2, PhCH_2MgBr, LiCHS(CH_2)_3S, FeCp(\eta^5\text{-}C_6Me_5CH_2)$

It has been checked using deuterium labeling that these reactions are under kinetic control, and do not result from a rearrangement. On the other hand, this latter reaction is well explained by the orbital control. This also shows that, in the case of the mesitylene ligand above, it was the predominance of steric control that let believe that the rules were working. In conclusion, these rules are far from general, and it appears that it is crucial to delineate which type of control (charge, orbital or steric) predominates.

Reactions of nucleophiles on the η^1 carbon ligands: CO, CNR, CR_2

The nucleophilic attack of carbanions on metal carbonyls have led to Fischer-type carbene complexes with mesomeric acylmetallate forms [4.3] (see Chap. 9).

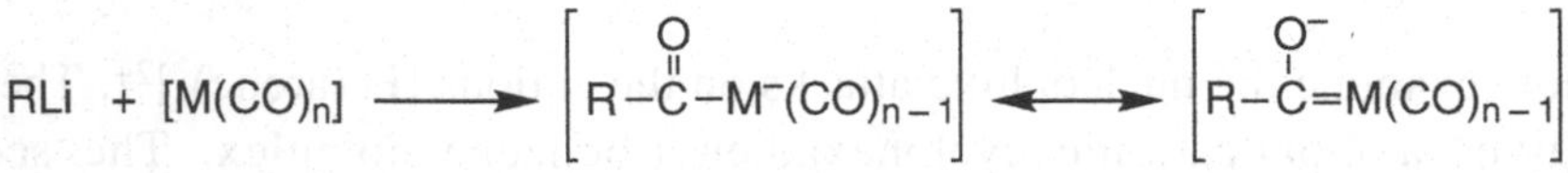

The Fischer-type carbene complexes themselves can undergo nucleophilic attacks: [4.4a]

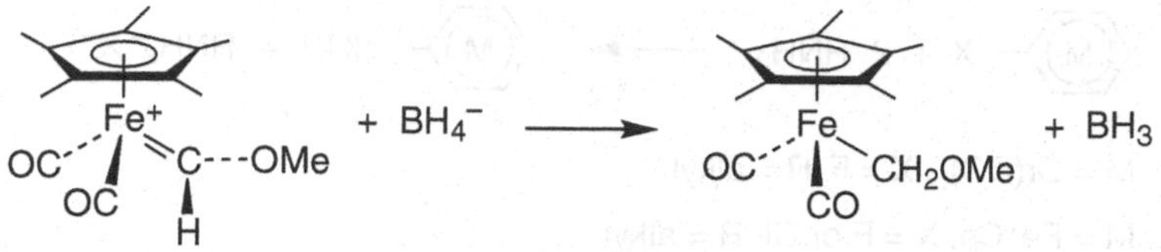

Thus, the CO ligand can be reduced to methyl; this is a kind of monometallic model for the heterogeneous Fischer-Tropsch reaction, i.e. reduction of CO by H_2 to alkanes using iron metal as catalyst (see Chap. 20.3).

$$CO + H_2 \xrightarrow[\text{Fe}]{\text{heter. cat.}} CH_3(CH_2)_nH + CH_3(CH_2)_nOH + H_2O$$

The ligand CNR, isoelectronic to CO, is also very easily attacked by nucleophiles, even neutral ones; for instance, the attack by aniline leads to a carbenic complex: [4.4b]

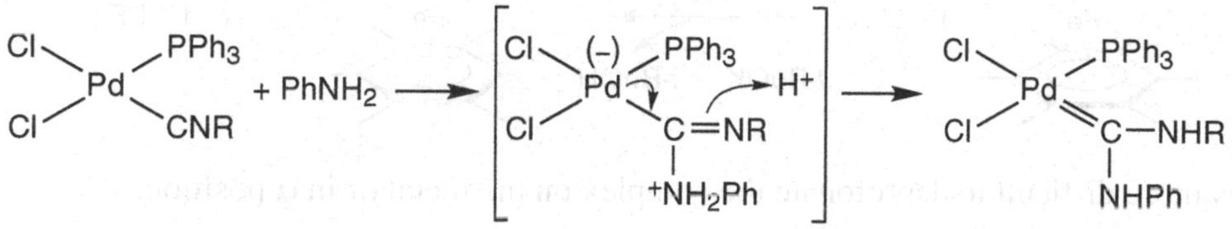

Nucleophilic substitutions

The addition of a nucleophile on an aromatic coordinated to a 12-electron organo-metallic fragment such as $Cr(CO)_3$, $FeCp^+$, $Mn(CO)_3^+$ yields a cyclohexadienyl complex. [4.5] The hydrogen atom located in *endo* position being not a good leaving group (as a hydride), such a complex is stable. In the case of a halogenoarene, however, it is the *ipso* carbon bearing the halogen that is attacked by oxygen, nitrogen and sulfur nucleophiles. The *endo* halogen is then a good leaving group as a halide. The aromatic halide substitutions, very difficult in organic chemistry, can occur under very mild conditions by temporary complexation of the halogenated aromatic by a 12-electron organometallic fragment. [4.6-4.10] These nucleophilic substitutions are all the easier as the organometallic fragment is more electron-withdrawing. Thus, the following activation order is observed:

$$Cr(CO)_3 < CpFe^+ < Mn(CO)_3^+.$$

For instance, a primary amine cannot react with $[Cr(\eta^6\text{-}C_6H_5Cl)(CO)_3]$, and it is necessary, in this case, to use the amide anion NHR^-. On the other hand, the reaction occurs with the amine itself when the aromatic is coordinated to the $CpFe^+$ fragment, except in the case of the less reactive aromatic amines such as aniline.

With the latter substrates, the nucleophilic substitution reactions can be carried out using the strongly activating fragment $Mn(CO)_3^+$.

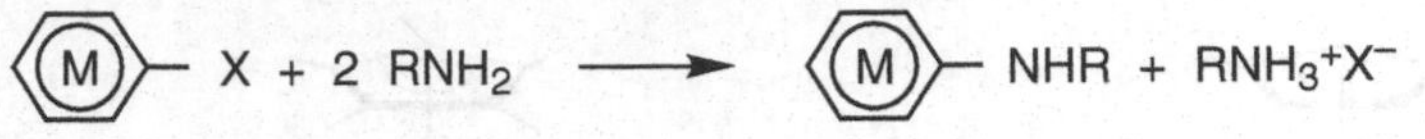

$M = Cr(CO)_3$, $X = F$, $R = $ alkyl

$M = Fe^+Cp$, $X = F$ or Cl, $R = $ alkyl

$M = [Mn(CO)_3]^+$, $X = F$ or Cl, $R = $ alkyl or aryl

1.3. DEPROTONATIONS

The cationic organometallic complexes containing a CH_3, NH_2 or OH in β position with respect to the metal can be easily deprotonated:

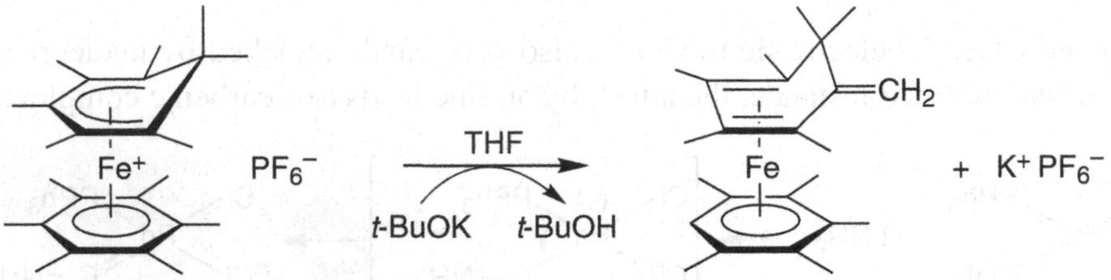

It is more difficult to deprotonate the complex on the metal or in α position:

$$ReH_7L_2 + BuLi \longrightarrow LiReH_6L_2 + BuH$$

$$Cr(C_6H_6)(CO)_3 + BuLi \longrightarrow Cr(C_6H_5Li)(CO) + BuH$$

$$[TaCp_2(CH_3)_2]^+ BF_4^- + Me_3P=CH_2 \longrightarrow TaCp_2(CH_3)(=CH_2) + Me_4P^+ BF_4^-$$

1.4. SIDE ELECTRON-TRANSFER REACTIONS

The orbital overlap between the two reactants has so far been considered as being adequate for all the nucleophilic reactions. For instance, if the charge difference between them is too large, the nucleophilic reactions cannot occur, and electron transfer from the nucleophile to the electrophile dominates:

$$Nu^- + El^+ \longrightarrow Nu^\bullet + El^\bullet$$

Such electron transfer is usually minimized by carrying out reactions at low temperature, electron-transfer reactions being favored thermally.

2. REACTIONS OF ELECTROPHILES

2.1. ELECTROPHILIC ADDITIONS

Electrophiles can add:

▸ on an insaturation conjugated with the ligand (in β position with respect to the metal):

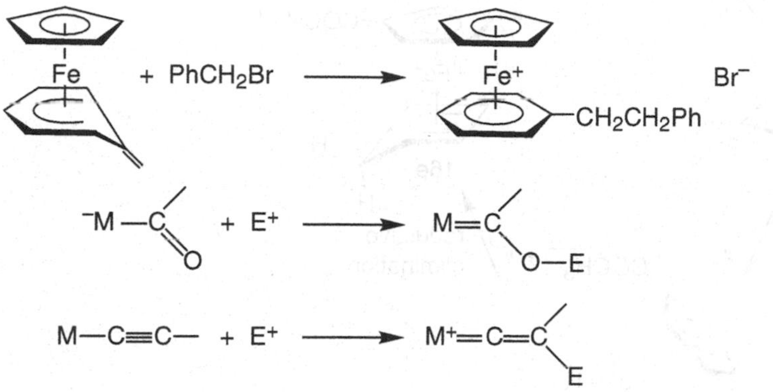

▸ on γ position with respect to the metal:

▸ on the metal (which corresponds to an oxidative addition; see Chap. 3.3):

$$[Fe(CO)_4]^{2-} (Na^+)_2 + CH_3I \longrightarrow [CH_3Fe(CO)_4]^- Na^+ + Na^+I^-$$

2.2. ELECTROPHILIC SUBSTITUTION

Electrophilic substitution can occur on a hydrocarbon ligand of neutral 18-electron organometallic complexes, the most well-known examples being ferrocene derivatives. These reactions can either involve direct ligand attack with *exo* stereochemistry or "ricochet" on the metal before ligand attack with *endo* stereochemistry.[4.11] Ferrocene itself is very reactive towards electrophiles due to its electron richness. For instance, it is acetylated 3×10^6 times faster than benzene. The mechanism involves direct *exo* attack of a Cp ring with "hard" electrophiles such as the acylium cation CH_3CO^+ or attack of iron forming an iron (IV) intermediate followed by *endo* transfer to a Cp ring with "soft" electrophiles such as Hg^{II}.[4.12] The mechanism and regioselectivity of ferrocene acylation reactions are represented below:

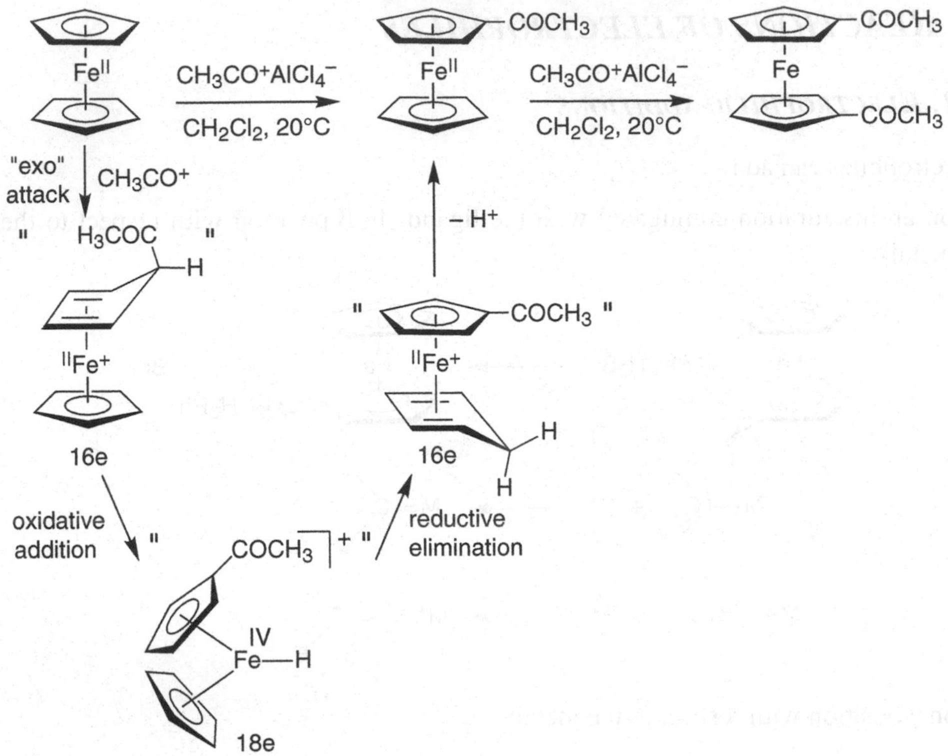

In the presence of an excess of acylium reagent, the acylated ring is deactivated by the electron-withdrawing acyl group. The electronic effect is partly transmitted through the iron atom, but the other ring is less deactivated than the acylated one and can also be selectively acylated. This second electrophilic attack is thus slower than the first one, which allows to selectively obtain the monoacylated derivative if the acylium reagent $CH_3CO^+ AlCl_4^-$ is added dropwise to the solution of ferrocene. Even in the presence of excess reagent, the reaction stops after the second electrophilic substitution. The 1,1'-diacylferrocene can also be selectively obtained by dropwise addition of the ferrocene solution to a solution of the acylium reagent in slight excess with respect to the 2-1 stoichiometry. Such a good selectivity cannot be obtained in the electrophilic reactions of alkyl halides because the electron-releasing alkyl substituents enhance the ferrocene reactivity towards further reactions. The result is a useless complex mixture of polyalkylated ferrocene derivatives.

Various other electrophilic substitutions are known in the ferrocene series, but there is an important limitation concerning the use of electrophiles that are good oxidants, because ferrocene is easily oxidized to ferrocenium cation. For instance, with sulfuric acid, only oxidation is obtained:

$$2 [FeCp_2] + H_2SO_4 \longrightarrow 2 [FeCp_2]^+, SO_4^= + H_2 \nearrow$$

2.3. ABSTRACTION REACTIONS

Hydride abstraction

The trityl cation, in the form of the salt $Ph_3C^+BF_4^-$, is the classic reagent to abstract, for instance, a hydride in β position with respect to the metal:

$$M-CH_2CH_2R + Ph_3C^+ BF_4^- \longrightarrow [M(\eta^2-CH_2=CHR)]^+ BF_4^- + Ph_3CH$$

Example: $M = FeCp(CO)_2$

This abstraction can sometimes occur in α leading to a metal-alkylidene complex:[4.12]

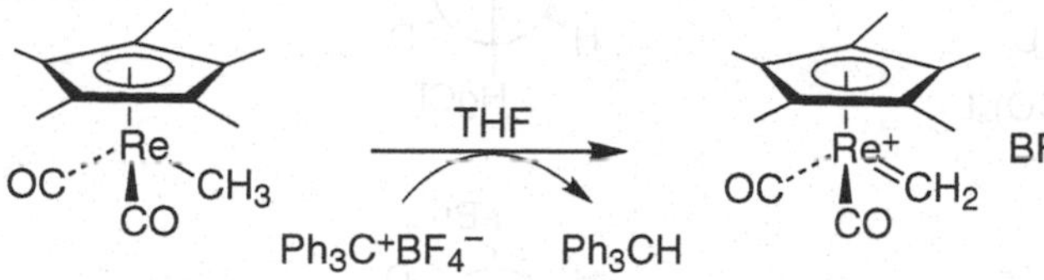

Abstraction of an inorganic anion

Similarly, it is possible to abstract other inorganic anions such as F^- or OH^- using Ph_3C^+:

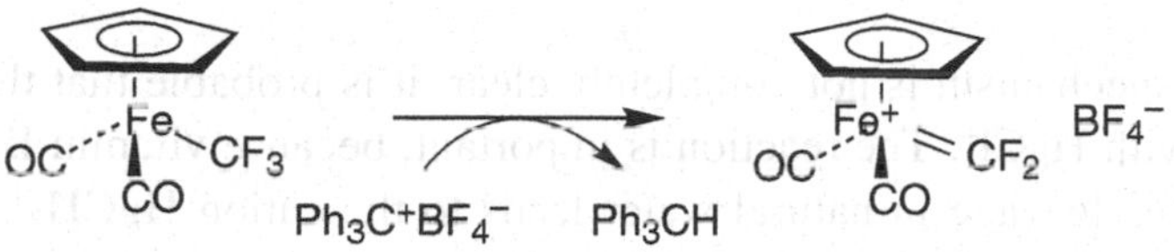

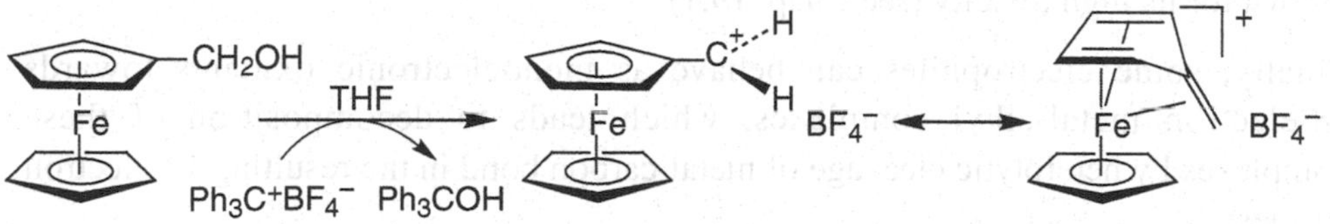

The compound obtained by hydroxide abstraction from hydroxymethylferrocene can be viewed as an 18-electron hexahapto fulvene complex rather than as a genuine carbocation. The bending angle of the *exo*-cyclic double bond towards iron is 21°, which clearly shows that this double bond is coordinated to iron.

Abstraction of an alkyl group

The abstraction of a carbanion CH_3^- by Ph_3C^+ from a neutral Ta-CH$_3$ complex provided a cationic precursor (below) of Schrock's tantalum-methylene complex:

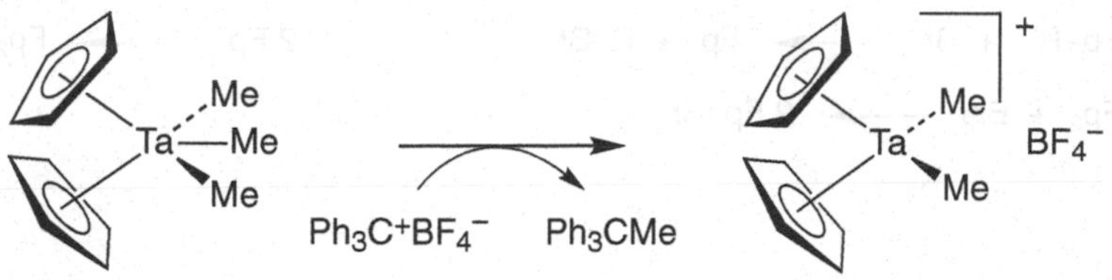

The reaction of $HgCl^+$ with metal-alkyl complexes led to the cleavage of the metal-carbon bond. This electrophile can attack the carbon atom or the metal atom depending on the electron richness of the metal. If the metal is electron poor (case of Mn below), the attack occurs on the carbon atom, which leads to inversion of configuration at carbon. On the other hand, if the metal is sufficiently basic (case of Fe below), the attack occurs at the metal or on the metal-carbon bond, which leads to retention of configuration at carbon.

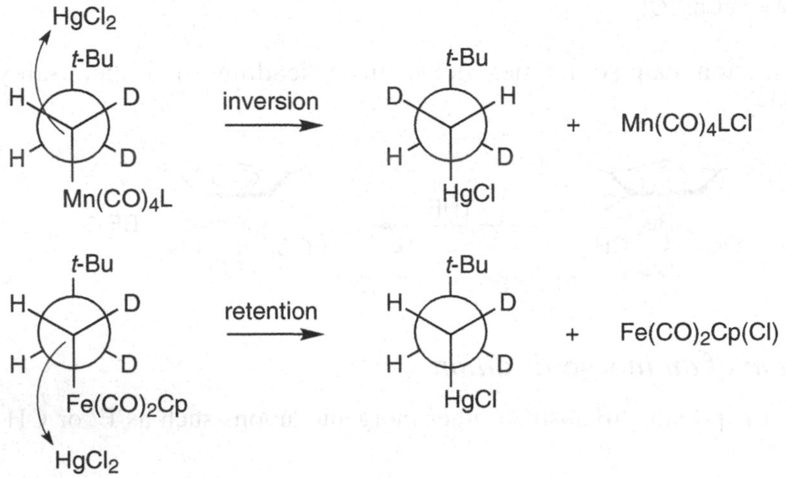

Although the mechanism is not completely clear, it is probable that the complexes react in fact with $HgCl^+$. The reaction is important, because vitamin B_{12} has such a Co-CH$_3$ whose cleavage in natural water leads to the cation $HgCH_3^+$ that is well-known for its high toxicity (see Chap. 19.1).

Finally, some electrophiles can behave as monoelectronic oxidants towards 18-electron metal-alkyl complexes, which leads to decomposition of these complexes by heterolytic cleavage of metal-carbon bond in the resulting 17-electron species:

$$M\text{–}R - e^- \longrightarrow M\text{–}R^{\cdot+} \longrightarrow \text{``}M^+\text{''} + R^\cdot \quad \text{or} \quad M^\cdot + \text{``}R^+\text{''}$$
$$\quad\text{18e} \qquad\qquad\quad \text{17e} \qquad\qquad \text{16e} \qquad\qquad\quad \text{17e}$$

Example:

$$Fp\text{–}R + Br_2 \longrightarrow Fp\text{–}Br + R\text{–}Br \qquad \text{avec } Fp = [FeCp(CO)_2]; R = CH_3$$

Mechanism:

$$Fp\text{–}R + Br_2 \longrightarrow Fp\text{–}R^{\cdot+} + Br^- + Br^\cdot \qquad\qquad 2\,Br^\cdot \longrightarrow Br_2$$

$$Fp\text{–}R^{\cdot+} + Br^- \longrightarrow Fp^\cdot + R\text{–}Br \qquad\qquad\qquad 2\,Fp^\cdot \longrightarrow Fp_2$$

$$Fp_2 + Br_2 \longrightarrow 2\,Fp\text{–}Br$$

SUMMARY OF CHAPTER 4
REACTIONS OF NUCLEOPHILES AND ELECTROPHILES

1 - Nucleophilic attacks

Additions on unsaturated hydrocarbon ligands

▸ with decrease by one unit of the hapticity of the ligand (example: η^3-allyl → η^2-olefin)

▸ *endo* stereochemistry of the attack with respect to the metal

▸ if the reaction is under charge control: the Mingos-Davies-Green rules apply with the following priority order: – even ligand before odd ligand,

– open ligand before closed cyclic ligand,

– terminal carbon of open ligand before inner carbon.

These rules do not apply if orbital control conflicts with charge control (then orbital control has priority; steric control can also change the priorities).

Reduction of activated η^1 ligands. Example of CO (model of heterogeneous Fischer-Tropsch process).

Aromatic nucleophilic substitution of halide

$M(\eta^6\text{-ArX}) + Nu^-$ or $NuH → M(\eta^6\text{-ArNu}) + X^-$; activating power of the 12 e^- fragments:

$$Mn(CO)_3^+ > CpFe^+ > Cr(CO)_3$$

Deprotonation of hydride MH, of methyl ligand (α) to $=CH_2$, or in β position (exocyclic to π hydrocarbon ligand).

2 - Electrophilic reactions

Addition: in β position (with increase of the hapticity by one unit), in γ, or on the metal.

Substitution: ferrocene chemistry: $FcH + E^+ → FcE + H^+$. Hard E attacks the ligand (*exo*), soft E attacks the metal first, then jumps to *endo* ligand position. Mono- and 1,1'-disubstitution are selective with acyl.

Abstraction: using Ph_3C^+, of H^- ($-CH_3 → =CH_2$ or alkyl → alkene), F^- ($-CF_3 → =CF_2$), OH^- (in β position).

In all these reactions of nucleophiles and electrophiles with metal complexes, electron transfer from a reducing nucleophile or to an oxidizing electrophile competes and sometimes inhibits the reaction.

EXERCISES

4.1. Forecast the two sites of H^- attack (by $NaBH_4$) on $[CoCp(C_6H_6)]^{2+}$ supposing that the reactions are under charge control.

4.2. What is the product of the reaction of $Ph_3C^+BF_4^-$ with the alcohol $[MoCp(dppe)(CO)(CHOHCH_3)]$?

4.3. Is the electrophilic reaction of $EtCl + AlCl_3$ with ferrocene a good synthetic method to prepare ethylferrocene ?

4.4. What is the product resulting from the deprotonation of $[FeCp(\eta^6\text{-}C_6Me_6)]^+$ and that resulting from the reaction of this deprotonated complex with $SiMe_3Cl$?

Chapter 5

LIGAND SUBSTITUTION REACTIONS

1. INTRODUCTION

The ligand substitution reactions play a fundamental role in organometallic chemistry, because they are systematically involved in the syntheses of complexes and in catalysis. As in organic chemistry, they can occur according to a "pairwise process" or a single-electron-transfer mechanism involving radicals (paramagnetic species). For each of these two categories, ligand substitution can occur associatively or dissociatively, which corresponds respectively to the SN_2 or SN_1 mechanisms in organic chemistry. The analogy between these two fields continues with the distinction, for single-electron-transfer mechanisms, to proceed according to a chain- or non-chain mechanism. What distinguishes organometallic from organic substitution reactions is the weakness of metal-ligand bonds compared to the strength of most organic bonds. For instance, in 18-electron metal-carbonyl complexes, despite their reputation of stability, the strength of metal-carbonyl bonds is only from $25 \, kcal \cdot mol^{-1}$ for $[Ni(CO)_4]$ to $46 \, kcal \cdot mol^{-1}$ for $[W(CO)_6]$. The metal-carbonyl bond is in the average of metal-ligand bonds strength whose order is the following: $C_5Me_5 > Cp > C_6Me_6 > C_6H_6 > CO \sim PMe_3 \sim C_2H_4 > PPh_3 > Py > CH_3CN > THF \sim CH_3COCH_3$.

The metal-cyclopentadienyl bond strength, for instance, reaches $118 \, kcal \cdot mol^{-1}$ in cobaltocenium, a very robust 18-electron complex.

In addition, the metal-ligand bonds tend to become stronger as one goes down in a column of the periodic table. Finally, the strength of a metal-metal bond is of the same order of magnitude as that of a metal-carbonyl bond.

We will successively examine the two pairwise mechanisms – dissociative and associative – then those involving organometallic radicals.

2. "PAIRWISE" MECHANISMS

2.1. DISSOCIATIVE MECHANISM

The dissociative mechanism is that usually encountered for the ligand-substitution reaction of 18-electron complexes.[5.1,5.2]

A typical case is that of the reactions of metal-carbonyl complexes with phosphines studied *inter alia* by Basolo.

Slow step:

$$ML_n \xrightarrow{\ \ k\ \ } ML_{n-1} + L$$

Fast step:

$$ML_{n-1} + L' \longrightarrow ML_{n-1}L'$$

The kinetics is of the type: rate = k $[ML_n]$, as for organic SN_1 reactions. The activation energy is close to that of a M-L bond (*vide supra*) and the enthalpy variation is positive ($\Delta S^\ddagger$ of the order of 10 to 15 eu), because the transition state is less ordered than the initial one.

The stereochemistry of the reaction, observable when the starting complex has a metal-centered chirality, can vary. Brunner has shown that, if the recombination step is faster than the rearrangement of the 16-electron transition state or intermediate, the reaction occurs with retention of configuration at the metal center. In the opposite case, racemization is observed with the following mechanism (the rate being inversely proportional to [PPh3], the dissociation of this ligand must be involved):

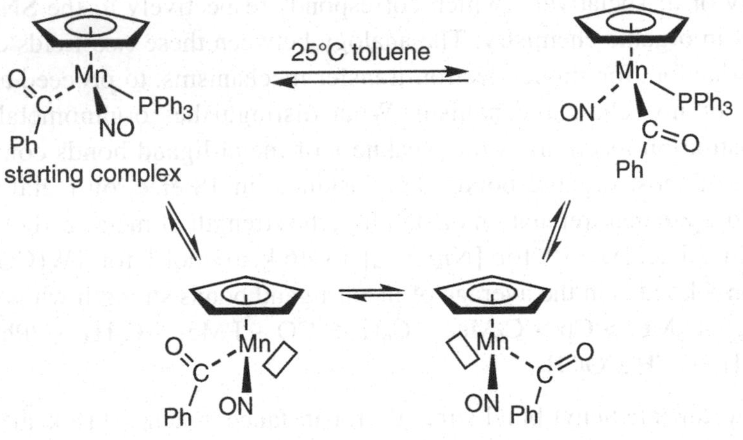

The dissociative mechanism applies to the L ligands, but not the X, LX or L_2X ligands. The exchange of odd ligands only occurs if a Lewis acid such as $AlCl_3$ or Ag^+ able to abstract X^-, LX^- or L_2X^- is involved or if the reagent carries a reactive anion (see section 3). The exchange of L ligands can also be facilitated by the presence of reagents such as Me_3NO in the case where L = CO, but also using various other means such as photochemistry, ultrasounds, microwaves (in the context of the dissociative mechanism). Electrochemistry and other redox processes involve electron-transfer mechanisms (*vide infra*). A practical technique that takes advantage of the dissociative mechanism consists in preparing a complex containing one or several labile ligands (see the bond strengths in section 1) such as NH_3 or N_2 or solvent molecules such as THF, CH_3CN or CH_3COCH_3. These ligands are easily substituted by other ligands according to the dissociative mechanism, because the energies of their ligand-metal bonds are weak.

2.2. ASSOCIATIVE MECHANISM

This is most of the time encountered in ligand-substitution reactions of 16-electron complexes,[5.3] the classic cases being those involving the square-planar d^8 Pd^{II}, Pt^{II} and Rh^I complexes. The kinetics (rate = k $[ML_n]$ $[L']$) and mechanism are analogous to those of organic SN_2 reactions.

Slow step:
$$ML_n + L' \xrightarrow{k} L'ML_n$$

Fast step:
$$L'ML_n \longrightarrow L'ML_{n-1} + L$$

For example, in square-planar complexes, the pentacoordinate intermediate has a trigonal bipyramid geometry. The substitution depends on the *"trans effect"* discovered by Chernaev in 1920. This *"trans effect"* of a ligand is defined by its ability to provoke the dissociation of the ligand located in *trans* position. Ligands having a strong *"trans effect"* are those having a strong σ bond with the metal such as H^- or CH_3^- and those having a strong π bond such as CO and C_2H_4. The consequence of this strong bond is the weakening of the bond between the metal and the *trans* ligand. This phenomenon is called the *"trans influence"*, of thermodynamic nature, and must not be confused with the *"trans effect"* that is kinetic. The *"trans effect"* is the consequence of the *"trans influence"*. The most spectacular application of this *"trans effect"* is the synthesis of *"cis* platinum", a well-known antitumor agent, whereas the *trans* isomer has no action. This synthesis uses the fact that Cl has a much stronger *"trans effect"* than NH_3.

We have seen that 18-electron complexes undergo ligand substitution by the dissociative mechanism, whereas 16-electron complexes do so by the associative mechanism. This distinction results from the fact that, if 18-electron complexes would undergo ligand substitution following an associative mechanism, the 20-electron transition state involved would be of much too high energy. There are cases, however, for which the 20-electron transition state can be avoided. The 18-electron complexes having a flexible ligand, i.e. a ligand that can behave either as an X or as an LX ligand, can undergo ligand substitution following an associative mechanism that involves an 18-electron transition state. Suffices therefore that the ligand be of LX type in the starting complex and X type in the transition state. This can be the case for the ligands Cl, OR, NR_2, PR_2, NO, allyl, carboxylato, etc. The same possibility exists for some polyhapto- or chelating ligands such as indenyl whose transition from η^5 to η^3 is easier than for Cp, because the benzo group of the indenyl ligand recovers the aromatic stabilization when the ligand becomes η^3. Even the Cp can occasionally give rise to this phenomenon of hapticity reduction from η^5 to η^3 in the transition state.

η^5-indenyl   proposed η^3-indenyl

Similarly:

$$[Mn(CO)_4(NO\ linear)],\ (18e) + L \longrightarrow [Mn(CO)_4(NO\ bent)],\ (16e)$$

$$[Mn(CO)_4(NO\ bent)],\ (16e) + L \longrightarrow [Mn(CO)_3(NO\ linear)L],\ (18e)$$

Finally, it is important to note that the associative and dissociative mechanisms are two extreme schemes, and there is a continuum of possibilities between these two defined mechanistic schemes.

2.3. PHOTOCHEMICAL LIGAND SUBSTITUTION

This technique is much used synthetically,[5.4,5.5] in particular for metal-carbonyl-[5.4] and metal-arene complexes:

$$[Fe(CO)_5] + 2\ PPh_3 \xrightarrow{h\nu_{(uv)}} [Fe(CO)_3(PPh_3)_2] + 2\ CO$$

$$[FeCp(C_6H_6)]^+ + 3\ P(OMe)_3 \xrightarrow{h\nu_{(vis)}} [FeCp\{(P(OMe)_3\}_3]^+ + C_6H_6$$

By irradiation of metal carbonyls in THF as solvent, it is possible to obtain THF complexes that are very useful for further facile thermal (20°C) THF exchange reactions with a large variety of ligands:

$$[W(CO)_6] \xrightarrow{h\nu} [W(CO)_5(THF)] + CO$$

$$[W(CO)_5(THF)] + L \xrightarrow{THF} [W(CO)_5(L)] + THF$$

Photo-excited states have lifetimes of the order of 10^{-10} to 10^{-6} seconds, and it is essential that these excited states be reactive before returning to the ground states. The irradiation of $[W(CO)_6]$ transfers a d electron into a d_σ orbital that is antibonding with respect to the M-CO bond. Under these conditions, the cleavage of the M-CO bond is faster than the return to the ground state. From $[W(CO)_5L]$, it is also possible to achieve a photochemical substitution reaction of CO or L. The degeneracy of the two orbital levels d_σ (or e_g, antibonding) and d_π (or t_{2g}, bonding) of the octahedral field in $[W(CO)_6]$ is lifted. The L ligand in axial position (along the z axis) usually having a weaker ligand field than CO, the d_{z^2} orbital is of lower energy than $d_{x^2-y^2}$. Thus, irradiation at $\nu_1 = 400$ nm in pyridine leading to the antibonding σ^*_z level labilizes the L ligand whereas irradiation at $\nu_2 = 250$ nm in pyridine leading to σ^*_{xy} labilizes a CO ligand.

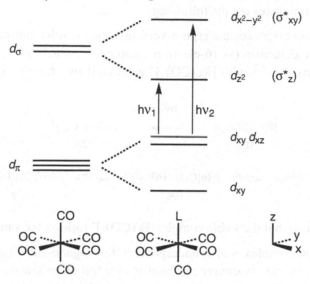

Although the photochemical CO or arene substitution by a phosphine is the most common among the light-induced organometallic reactions, other ones are known such as the photo-cleavage of M-H bonds of dihydrides (see Chap. 3.3.2) and metal-metal bonds (following section 2.4).

2.4. 17- AND 19-ELECTRON COMPLEXES

Some of these organometallic radicals are sufficiently stable to be isolable;[5,6] in particular, many stable 17-electron complexes are known. As organic radicals, organometallic radicals are indeed stabilized by bulk around the metal center that inhibits radical reactions. When they are not sterically stabilized, their main property, beside the redox ones, is the very fast intra- or intermolecular interconversion between the 17-electron and 19-electron forms. The combination of a 17-electron complex with a 2-electron L ligand to give a 19-electron species is schematized below:

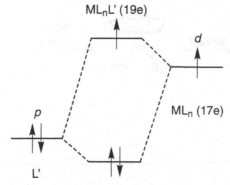

Metal-centered 19e complex formed by reaction of a 2e L ligand with a 17e species

The dissociation of an L ligand from a 19-electron species is provoked by the localization of the 19th electron in an orbital that is antibonding with respect to the M-L bond. The consequences of the very rapid interconversion between 17-electron and 19-electron complexes are the following:

▸ The 17-electron complexes undergo a very fast ligand substitution according to an associative mechanism (as 16-electron complexes), which has been shown for the first time by Poë [5.7] with $[Re(CO)_5]^•$ generated by photolysis of the dimer $[Re_2(CO)_{10}]$.

$$[Re_2(CO)_{10}] \xrightarrow{\ h\nu\ } 2\,[Re(CO)_5]^•$$
$$18e \qquad\qquad\qquad 17e$$

$$[Re(CO)_5]^• + PPh_3 \rightleftharpoons [Re(CO)_5(PPh_3)]^{•*} \rightleftharpoons [Re(CO)_4(PPh_3)]^• + CO$$
$$17e \qquad\qquad\qquad\quad 19e \qquad\qquad\qquad\qquad 17e$$

Basolo has shown that the stable complex $[V(CO)_6]^•$ follows the same process.

▸ The 19-electron complexes also undergo very fast ligand-substitution reactions according to the same associative mechanism (second-order kinetic low), because they are in very fast pre-equilibrium (eventually intramolecular) with the 17-electron complexes. In the intramolecular case, the 17- and 19-electron forms can even sometimes be mesomer forms.

In the example below, the 17-electron species $[FeCp(PPh_3)_2]$ formed in the course of the substitution reaction undergoes both reversible L ligand substitution and irreversible H-atom abstraction by the solvent to yield the final reaction product $[FeCp(PPh_3)_2H]$.

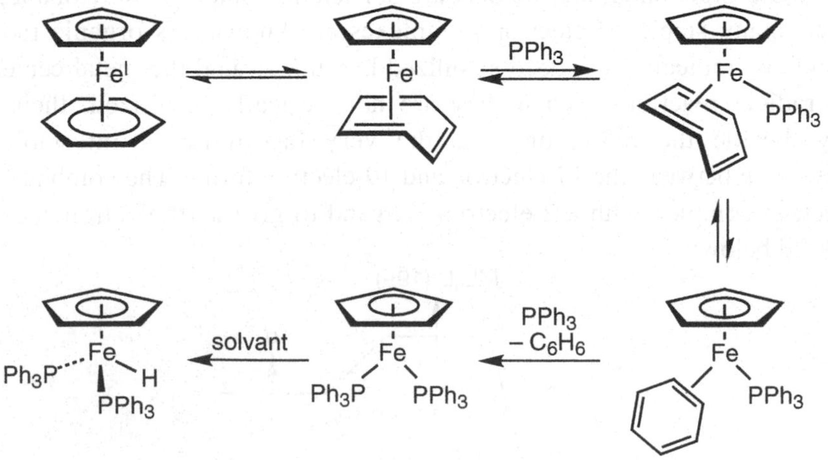

3. ELECTRON-TRANSFER-CHAIN
AND ATOM-TRANSFER-CHAIN MECHANISMS

Henry Taube [5.8] showed in 1954 that it is possible to initiate a ligand-substitution reaction by using a catalytic quantity of a redox reagent, a radical or a physical means (such as light). The radicals, generated using one of these initiation modes, react according to a chain mechanism to yield the product resulting from ligand substitution. Whatever the mode of initiation, we will distinguish two types of chain mechanisms, both disclosed by Taube, electron-transfer-chain and atom-transfer-chain.[5.9] These mechanisms both usually involve 17-electron and 19-electron intermediates in the chain (in some of Taube's examples, 15-electron Cr species are also involved). The variations of the modes of initiation may lead to either mechanism. We will envisage, as the initiation mode, electron transfer for the electron-transfer-chain mechanism and atom transfer for the atom-transfer-chain mechanism. These cases are indeed the most frequent ones, but various initiation types can be used for either mechanism.

3.1. ELECTRON-TRANSFER-CHAIN (ETC) MECHANISM

Let us take the typical case of a robust 18-electron complex and an L ligand that would react together at a negligible rate. The technique consists in initiating the ligand substitution by introducing into the reaction medium a catalytic quantity of an oxidant or reductant by electrochemistry or using a redox reagent behaving as a reservoir of electrons or electron holes. This initiation leads to 17-electron and 19-electron species that exchange ligands at rates 10^6 to 10^9 times faster than that of the starting 18-electron complex.[5.9-5.11] If A is the starting complex and B is the reaction product, C and D the incoming and leaving ligands respectively, the chain mechanism can be written in a linear or cyclic way as follows. The representation of the schemes are rather similar whatever the initiation type, oxidation or reduction.

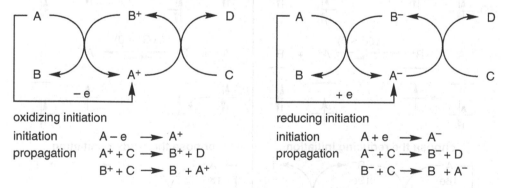

oxidizing initiation

initiation	$A - e$	$\longrightarrow A^+$
propagation	$A^+ + C$	$\longrightarrow B^+ + D$
	$B^+ + C$	$\longrightarrow B + A^+$

reducing initiation

initiation	$A + e$	$\longrightarrow A^-$
propagation	$A^- + C$	$\longrightarrow B^- + D$
	$B^- + C$	$\longrightarrow B + A^-$

The price to pay for the considerable kinetic gain is the high risk of side reactions of the very reactive radical intermediates in the chain. Not only is it necessary that the free enthalpy of the overall propagation cycle be negative, but also that each of the two steps, the ligand substitution and the cross redox steps also have a negative free enthalpy. It is difficult to evaluate the thermodynamic data of the ligand substitution between transient paramagnetic intermediates, but it is easy to organize the cross redox step in such a way that it be thermodynamically favorable. Suffices therefore to compare the redox potentials of A and B, or even only to compare the electron donicities of the incoming and leaving ligand. The only difference between the complexes A and B being the nature of the exchanged ligand(s), their molecular diagrams including their HOMOs and LUMOs are closely related, and in particular the relative energy orders of the HOMOs and LUMOs of A and B are the same. The consequence is that there is only one correct mode of initiation of the ETC reaction if a favorable thermodynamics is expected for the cross redox step: reductive initiation if B is more electron rich than A (i.e. C is a better donor than D), and oxidizing initiation in the opposite case. According to the simplified Marcus' equation (see Chap. 3.1.3), it is necessary to have a (moderately) favorable driving force to have a fast cross redox step. The mode of initiation, oxidation or reduction, should thus be chosen according to this simple criterion. This molecular engineering works well in about 80% of the cases with high turnover numbers for the redox catalyst (or coulombic efficiency if the catalyst is an electrode: number of reactions induced by ± one electron). This molecular engineering neglects the chemical step that most often is more or less isoergonic for a classic ligand substitution reaction. {There are cases that are not envisaged here, however, for which the chemical step is energetically demanding and has to be taken into account (chelation, other organometallic ETC catalyzed reactions)}.

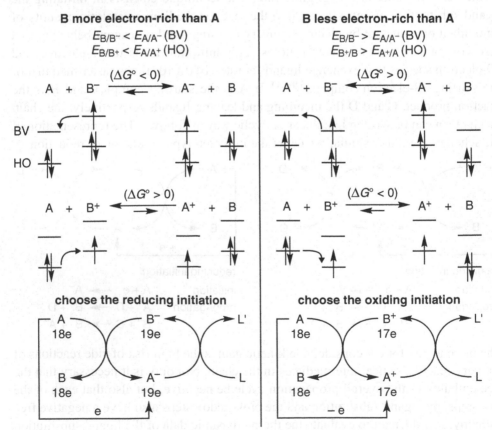

In the above scheme, a ligand L is exchanged by L' in A to give B. The relative donicities of A and B, that determine the type of initiation, directly depend on the π-acceptor properties of L and L': the following order of the relative donicities is opposite to that of the π-acceptor properties:

CO < unsaturated hydrocarbons < $P(OR)_3$ < PR_3 < ligands O, N, S (amines, ethers...)

After the choice (oxidation or reduction) of the initiation type, it remains to determine the strength of the redox reagent, i.e. its standard redox potential. This choice must be carried out referring to the adequate redox potential of A. The electron-transfer reaction between the initiator and A must be exergonic or isoergonic; even a very slightly endergonic electron transfer can be tolerated. In the latter case, the unfavorable equilibrium is displaced to the right by the chain reactions, but if the initiation step is too endergonic, initiation cannot occur. The standard redox potentials of the main classic redox reagents are as follows:

Oxidation:	**oxidant**	$E°$ *vs* **ECS**	$E°$ *vs* **[FeCp₂]**	**solvent**
	$[FeCp*(C_6Me_6)^{2+}]\,[SbCl_6^-]_2$	1.4 V	1.03 V	CH_2Cl_2
	$[NO^+][PF_6^-]$	1.37 V	1.00 V	CH_2Cl_2
	$[Ag^+][BF_4^-]$	1,05 V	0.65 V	CH_2Cl_2
	$[FeCp_2^+][PF_6^-]$	0.4 V	0.00 V	CH_2Cl_2
	I_2	0.4 V	–0.14 V	MeCN
	O_2	–0.82 V		MeCN

Reduction:	**reductant**	$E°$ *vs* **ECS**	$E°$ *vs* **[FeCp₂]**	**solvent**
	Na	–2.72 V	–3.04 V	THF
	$[Na^+(naphtalene)^{•-}]$	–1.98 V	–2.30 V	THF
	$[Na^+(Ph_2CO)^{•-}]$	–1.86 V	–2.30 V	THF
	$[Fe^ICp*(C_6Me_6)]$	–1.75 V	–2.24 V	THF
	$[Fe^ICp(C_6Me_6)]$	–1.54 V	–2.01 V	THF
	$[CoCp_2]$	–0.86 V	–1.33 V	THF

Kochi[5.11] has shown that, in the 18-electron complex $[MnCp(CO)_2(MeCN)]$, the substitution of MeCN by other less donating ligands (i.e. better π acceptors) such as PPh_3 is initiated by anodic oxidation or by chemical oxidants such as ferrocenium cation. The process occurs with coulombic efficiencies that reach 10^3.

$$[MnCp(CO)_2(MeCN)] + PPh_3 \xrightarrow{\text{cat. ox.}} [MnCp(CO)_2(PPh_3)] + MeCN$$

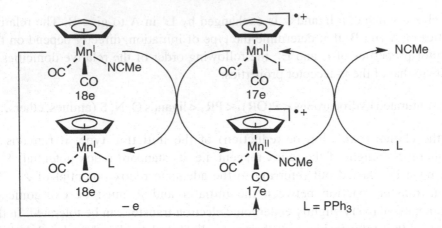

In the complexes [FeCp(η^6-arene)]$^+$ (arene = benzene, toluene, mesitylene, etc.), arene exchange is carried out by three ligands PR$_3$ (R = Me, OMe) that are not as good π acceptors as the arene. This exchange thus requires reducing initiation, for instance using an electron-reservoir complex [Fe^{I}Cp(η^6-arene)]$^{\bullet}$ whose self-electron transfer is of course isoergonic. The coulombic efficiency is also larger than 100 (see also Chap. 11): [5.12-5.14]

$$[FeCp(\eta^6\text{-}C_6H_6)]^+ + 3\ P(OMe)_3 \xrightarrow{\text{cat. red.}} [FeCp(CO)_2\{P(OMe)_3\}]^+ + C_6H_6$$

In this case, the initiator can be the species involved in the chain reaction, because it is sufficiently stable. Many of these 19-electron complexes are moderately stable in THF and can be indeed prepared, which is not usual for 19-electron complexes especially with such negative redox potentials (see the above table and Chap. 11).

The exchange of one or several CO by phosphanes in metal-carbonyl clusters also proceeds by reducing initiation using either [Na$^+$(Ph$_2$CO)$^{\bullet -}$] or, more conveniently, the prototypal electron-reservoir complex [Fe^{I}Cp(η^6-C$_6$Me$_6$)]$^{\bullet}$:

$$M_n(CO)_m + PR_3 \xrightarrow{\text{cat. red.}} M_n(CO)_{m-1}(PR_3) + CO$$

This technique is so efficient, in particular for the substitution of the first CO, that it has been used to cleanly carry out in a few minutes at room temperature the substitution of a CO in $[Ru_3(CO)_{12}]$ by a phosphine on each branch of dendrimers containing 32 or 64 phosphine termini. Thus, dendrimers containing 32 or 64 $[Ru_3(CO)_{11}(phosphine)]$ clusters at the dendritic periphery could be easily prepared:

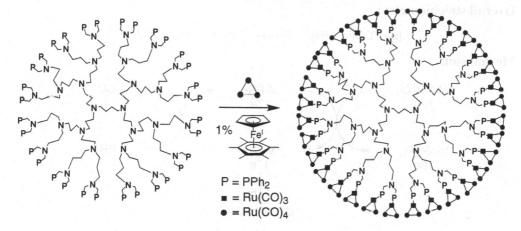

P = PPh$_2$
■ = Ru(CO)$_3$
● = Ru(CO)$_4$

Besides ligand substitution reactions, various other organometallic reactions have been carried out according to the ETC mechanism (insertion, carbene transfer, isomerization, etc.) [5.15,5.16]. Moreover, it is possible to couple ETC catalysis with organometallic catalysis in order to largely improve the selectivity and efficiency of catalytic reactions such as the tungsten-catalyzed polymerization of terminal alkynes (see Chap. 14.4).[5.17]

3.2. ATOM-TRANSFER-CHAIN MECHANISM

This mechanism resembles the preceding one, but an atom, instead of an electron, is exchanged in the cross redox propagation. Taube's pioneering example [5.8,5.9] concerns the incorporation of radioactive Cl^- into $[Pt^{IV}Cl_6]^{2-}$, the initiation being most often carried out by irradiation (various means have been used):

Overall stoichiometry:

$$[Pt^{IV}Cl_6]^{2-} + 6\,{}^*Cl^- \xrightarrow{h\nu} [Pt^{IV\,*}Cl_6]^{2-} + 6\,Cl^-$$

Mechanism:

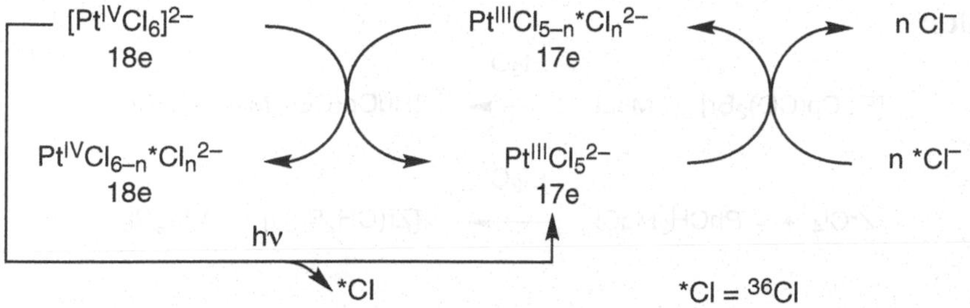

$$[Pt^{IV}Cl_6]^{2-} \quad 18e$$
$$Pt^{III}Cl_{5-n}{}^*Cl_n{}^{2-} \quad 17e$$
$$n\,Cl^-$$

$$Pt^{IV}Cl_{6-n}{}^*Cl_n{}^{2-} \quad 18e$$
$$Pt^{III}Cl_5{}^{2-} \quad 17e$$
$$n\,{}^*Cl^-$$

$$h\nu$$

$${}^*Cl$$

$${}^*Cl = {}^{36}Cl$$

Brown has shown that metal-carbonyl-hydride complexes such as $[Re(H)(CO)_5]$ can undergo CO substitution by a phosphine (PBu_3) according to the H-atom-transfer-chain mechanism.[5.18] A classic type of initiation to introduce the radical species into the chain is to photolyze the metal-carbonyl dimer, which generates the reactive 17-electron metal-carbonyl monomer:

Overall stoichiometry:

$$[H–MCO] + PR_3 \longrightarrow [H–M–PR_3] + CO$$

Mechanism:

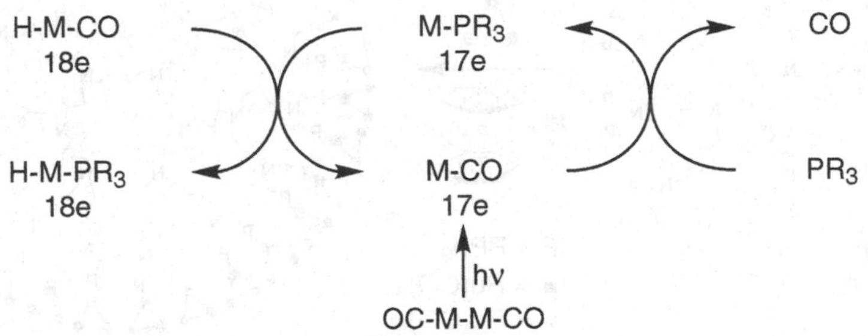

Here again, it is possible to forecast if the redox propagation step is thermodynamically favorable by comparing the relative strengths of the M-H bonds in the starting and final compounds. The M-H bond is indeed strengthened after CO substitution by a phosphine (because electron density is increased), which justifies the exergonicity of the cross redox step and the fact that this overall reaction works well.

Finally, these atom-transfer-chain reactions that are common in substitution reactions are, as ETC reactions, also known for other organometallic reactions.[5.9,5.19]

4. SUBSTITUTION OF XL$_n$ LIGANDS (n = 0 - 2)

The ligand substitution reactions that bring to the metal an odd number of electrons, i.e. X (1 electron), LX (3 electrons) and L_2X (5 electrons), are also very useful for the synthesis of organometallic complexes. The main ligands concerned are those found in starting materials, i.e. halogeno (Cl and Br), cyano (CN), acetylacetonato (acac) and cyclopentadienyl (Cp). The incoming ligands can be alkyls, aryls, enyls or polyenyls or various inorganic ligands. The reagents that carry these potential ligands are mostly alkali, magnesium, zinc and aluminum reagents:

Examples:

$$[RuCp(CO)_2Br] + MeLi \xrightarrow{Et_2O} [RuCp(CO)_2Me] + LiBr$$

$$ZrCl_4 + 4\ PhCH_2MgCl \xrightarrow{Et_2O} [Zr(CH_2Ph)_4] + 4\ MgCl_2$$

$$TaCl_5 + ZnMe_2 \xrightarrow{\text{toluene}} [TaCl_3Me_2] + ZnCl_2$$

$$K_3[Cr(CN)_6] + 6\,NaC{\equiv}CH \xrightarrow[-40°C]{NH_3} \underset{\text{explosive}}{K_3[Cr(C{\equiv}CH)_6]} + 6\,NaCN$$

$$[Fe(\eta^5\text{-}C_5Me_5)(acac)] + LiCp \xrightarrow{\text{THF}} [Fe(\eta^5\text{-}C_5Me_5)(\eta^5\text{-}Cp)] + Li(acac)$$

When the metal of the reagent is in a high oxidation state and easily reducible, for instance with the halogeno ligands, the lithium and Grignard reagents (very polar) tend to reduce the metal instead of alkylating it. This problem is, for instance, crucial in the organometallic chemistry of early transition metals. It is then necessary to use very mild (slightly polar) reagents such as the dialkylzinc complexes. For instance, with MCl_5 (M = Ta or Nb), $ZnMe_2$ allows to carry out up to the disubstitution, but not further. For the third and further chloride substitution by alkyls up to the pentamethylation, the more polar Grignard reagent CH_3MgI is adequate.

SUMMARY OF CHAPTER 5
LIGAND SUBSTITUTION REACTIONS

1 - Pairwise mechanisms

▸ dissociative, SN_1 type (18e complexes) or associative, SN_2 type (16e complexes);

▸ photochemical CO or arene substitution in 18e complexes;

▸ very fast ligand L substitution in 17e and 19e complexes due to rapid interconversion of these species in the absence of bulk;

2 - Chain mechanisms: $ML + L' \rightarrow ML' + L$ by electron-transfer-chain (ETC) or atom-transfer-chain (ATC)

ETC catalysis is very efficient: it is used in substitutionally inert 18e complexes, the kinetic gain being enormous (10^6 to 10^{10}) at the 17e - 19e level.

An exergonic cross redox propagation step should be planned. Therefore, if L' is a better donor than L: initiate by reduction; if L' is a less good donor than L: initiate by oxidation.

ATC catalysis (H, Cl, etc. tranferred in the propagation step) is initiated by a radical and requires that the M-H or M-Cl bond in the final product be stronger than in the starting complex.

3 - Substitution of XL_n ligands (n = 0 - 2)

Examples: Cl^-, Br^-, CN^-, $acac^-$, etc. by alkyl, allyl or Cp, etc. proceed thermally.

EXERCISES

5.1. The 19-electron complex $[Fe^ICp(\eta^6\text{-toluene})]$ undergoes uncatalyzed arene substitution by $P(Me)_3$ ligands. What is the final reaction product? The kinetic studies of these reactions show that they proceed according to the associative mechanism. Why does the mechanism imply a pre-dissociation to the 17-electron species $[Fe^ICp(\eta^4\text{-arene})]$? (J. Ruiz *et al.*, *J. Am. Chem. Soc.*, **1990**, *112*, 5471).

5.2. How would you best proceed to selectively replace a CO ligand by a phosphine ligand in the cluster $[CH_3Co_3(CO)_9]$? [5.9]

5.3. At room temperature, how would you synthesize $[MnCp(CO)_2\{P(OMe)_3\}]$ from $[MnCp(CO)_2(NCMe)]$? [5.10]

Chapter 6

INSERTION AND EXTRUSION REACTIONS

1. INTRODUCTION

The insertion of an unsaturated molecule (potentially L ligand) into an M-R bond (R = X ligand: H, alkyl, hydroxy, alcoxy, amino, etc.), is very common:

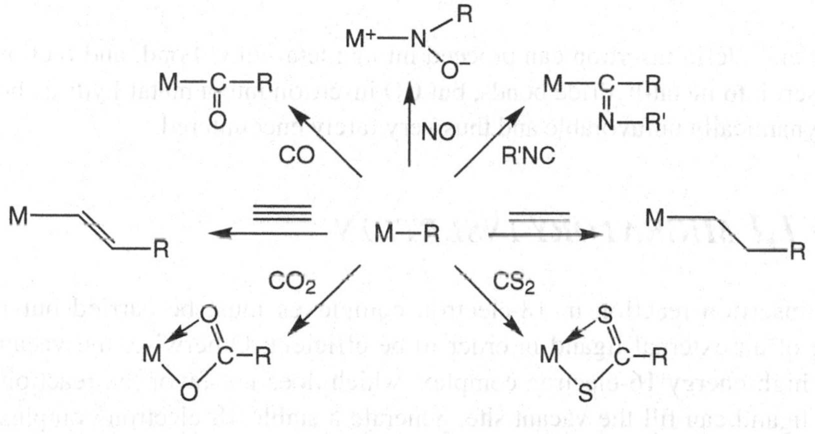

General scheme summarizing the insertion reactions (R = H or alkyl)
Pre-coordinated ligands that are instable in the free state such as $=CH_2$ or CS can also be inserted.

The insertion reaction leads to a new species M-L-X by formation of a covalent bond between L and X. The new ligand X' is a 1-electron radical ligand, as X. The insertion goes along with a decrease of two units of the NVE and one unit of the coordination number (liberation of one coordination site). The NNBE and the OS remain unchanged. The reverse reaction is the extrusion or, with metal-alkyl complexes, the β-elimination.

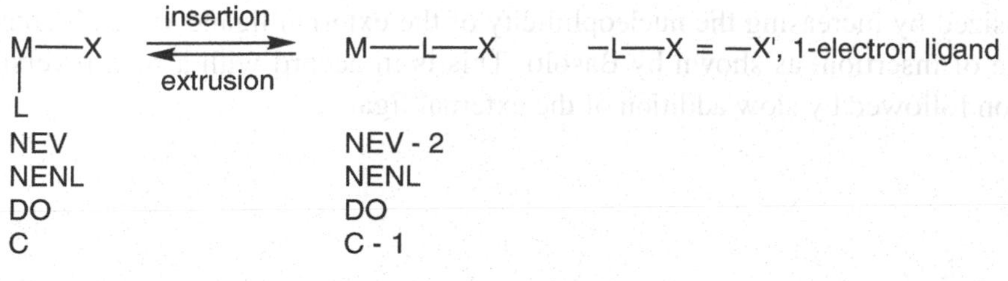

Insertion is a very important reaction in synthesis and catalysis concerning especially two ligands: CO and ethylene (or olefins).

There is a fundamental difference between the topologies of CO and olefin insertions. Carbon monoxide, a η^1 ligand, leads to a 1,1 insertion, i.e. the metal and the X fragment are both bonded to the first atom of the CO ligand, the carbon atom.

Insertion of ethylene, a η^2 ligand is, on the other hand, a 1,2 insertion, i.e. the metal is bonded to the first ethylene carbon in the insertion product, whereas the X fragment is bonded to the second ethylene carbon atom.

Both CO and olefin insertion can proceed into a metal-alkyl bond, and olefins also easily insert into metal-hydride bonds, but CO insertion into a metal-hydride bond is thermodynamically unfavorable and thus very rarely encountered.

2. CO 1,1 MIGRATORY INSERTION

The CO insertion reaction in 18-electron complexes must be carried out in the presence of an external ligand in order to be efficient. Otherwise, the vacant site leaves a high-energy 16-electron complex, which does not favor the reaction. The external ligand can fill the vacant site, generate a stable 18-electron complex, and therefore displace the overall reaction towards the insertion product.

For example, Calderazzo [6.1] has shown that, if ^{13}CO is introduced into the reaction medium, it is found as a ligand in the insertion product, not in the acylated ligand. This was observed by infrared spectroscopy using the shift of the carbonyl band with the isotopic labeling and the difference between the CO ligand absorption around 2000 cm^{-1} and metallaketone absorption around 1650 cm^{-1} (see Chap. 7.3):

$$[Mn(CO)_5CH_3] + {}^{13}CO \longrightarrow [Mn(CO)_4({}^{13}CO)(COCH_3)]$$

The importance of the external ligand on the insertion reaction can also be emphasized by increasing the nucleophilicity of the external ligand which increases the rate of insertion, as shown by Basolo. This is in accord with a fast, reversible insertion followed by slow addition of the external ligand.

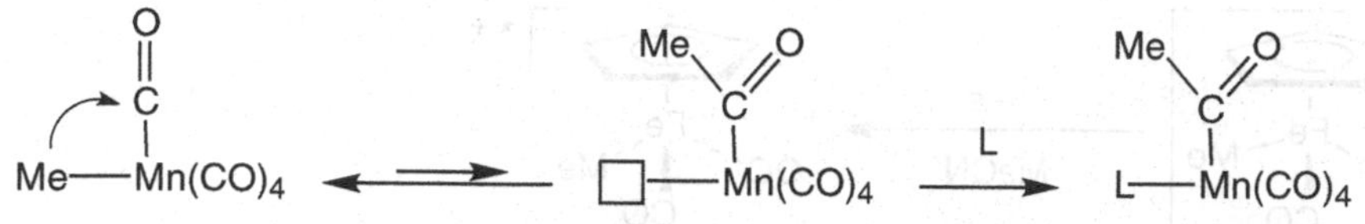

The above CO insertion into the metal-alkyl bond was shown by isotopic labeling to be in fact a migration of the *cis* alkyl ligand onto the CO ligand, which provided the name of *migratory insertion* for this reaction.

Since the insertion is all the faster as the external ligand is more nucleophilic, it is also considerably accelerated by factors that increase the electrophilicity of the metal center. Two types of reagents have been shown to be efficient in favoring the insertion: acids and oxidants. Protons or Lewis acids can bind to the acyl oxygen atom, withdrawing electron density from the metal as shown by writing the mesomeric carbenic form: [6.2]

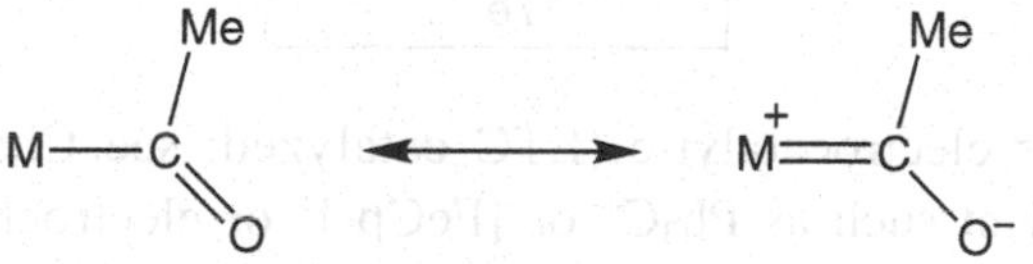

For instance, insertion of CO in the M-H, M-COR and M-CF$_3$ bonds is usually thermodynamically unfavorable (on the other hand, extrusion of CO from M-CHO, M-COCOR and M-COCF$_3$ bonds is favorable). Oxophilic metal complexes, however, play this role of Lewis acids. The additional resulting driving force can then make the CO insertion possible into a thorium-hydride bond [6.3] (see Chap. 12), whereas this was impossible with transition-metal-hydrides:

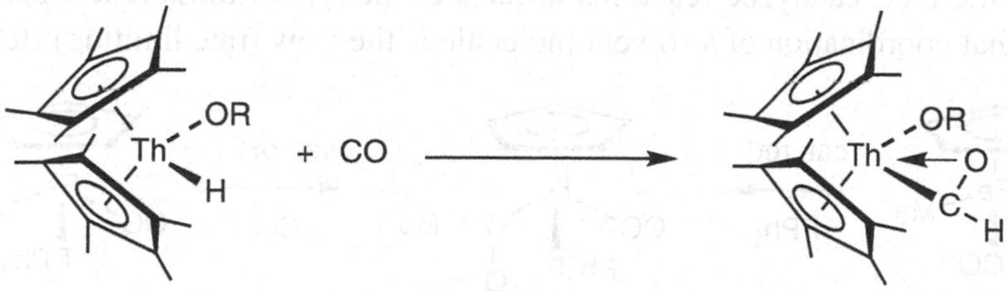

Likewise, monoelectronic oxidation of a neutral 18-electron metal complex to a 17-electron cationic complex before insertion also considerably facilitates this reaction. Whereas the 18-electron complex [FeCp(CO)$_2$Me] only very slowly undergoes CO insertion into the Fe-CH$_3$ bond at ambient temperature, the 17-electron cation rapidly gives the acetylated complex in MeCN at $-78°C$, the solvent playing the role of the nucleophile that coordinates to the iron center.[6.4]

The two alternative mechanistic possibilities shown below are:
▸ coordination of MeCN to form a 19-electron species before insertion, and
▸ direct insertion in the 17-electron species to give a 15-electron intermediate that then traps MeCN:

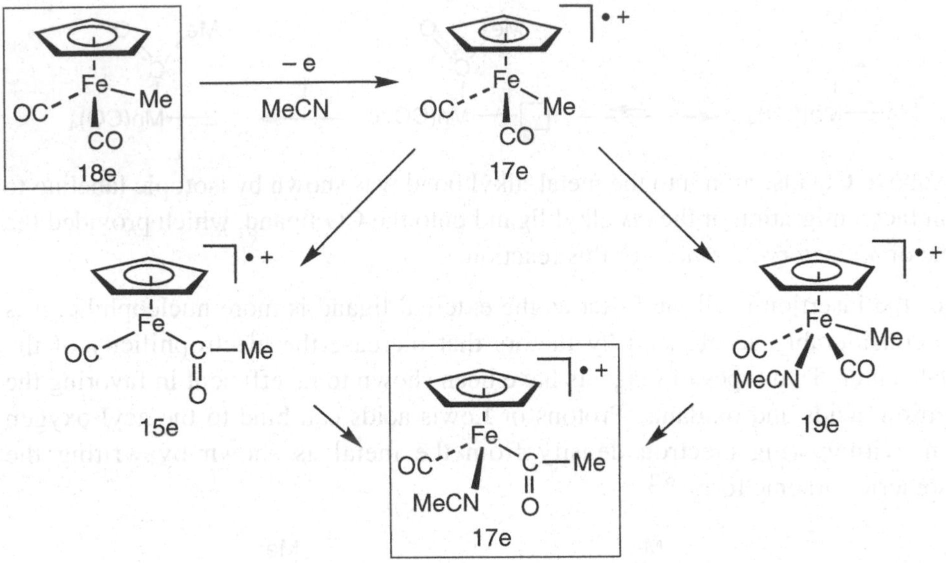

This reaction can be electrocatalytic (ETC catalyzed; see Chap. 5.3.1) using a monoelectronic oxidant such as Ph_3C^+ or $[FeCp_2]^+$ or electrochemically with an anode. In these ETC catalyzed insertion reactions represented below, the final product is more electron-rich or more electron-poor than the starting complex depending on whether the external ligand is PPh_3 or CO, respectively. Thus, reductive initiation is necessary in the first case, whereas it is oxidizing initiation that is necessary in the second one (see Chap. 5.3.1). The turnover is higher in the oxidizing initiation with CO as the external ligand than in the other case, because there are less reactive radical intermediates in the chain. The nucleophilicity of the solvent of the ETC catalyzed reactions influences the rate of these reactions, which indicates that coordination of a solvent molecule is the slow (rate limiting) step.

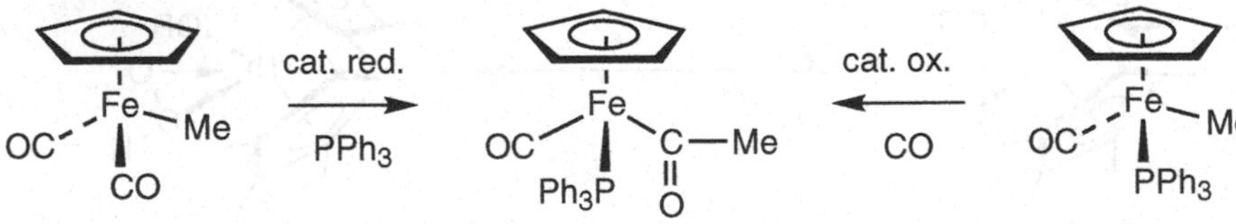

Kochi [6.5] has shown that redox reagents or an electrode can considerably accelerate extrusion reactions that are thermodynamically favorable but sometimes very slow such as that of CO from metal-formyl complexes. These redox reagents introduced in catalytic amounts efficiently initiate ETC- or ATC-catalysis variations of the extrusion mechanism:

$$\text{a -} \qquad [(CO)_5ReRe(CO)_4(CHO)]^- \quad \underset{}{\overset{[e^-]}{\rightleftharpoons}} \quad [(CO)_5ReRe(CO)_4H]^- + CO$$
$$\text{Rdt: 100\%}$$
$$E_{1/2} = -2.1 \text{ V/SCE (irrev.)} \qquad\qquad E_{1/2} = -2.6 \text{ V/SCE (irrev.)}$$

b -

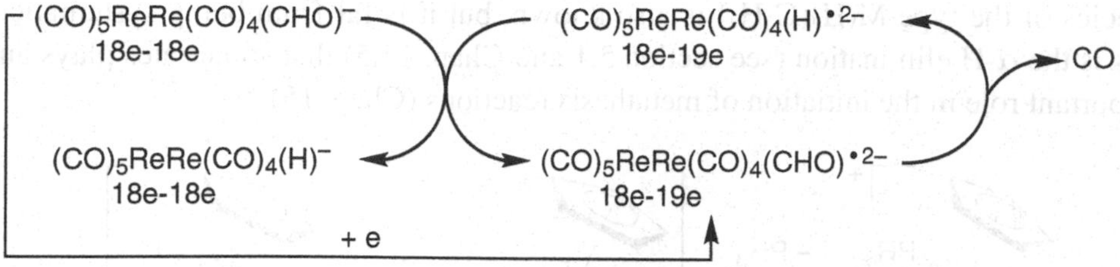

ETC catalysis mechanism initiated by a cathode

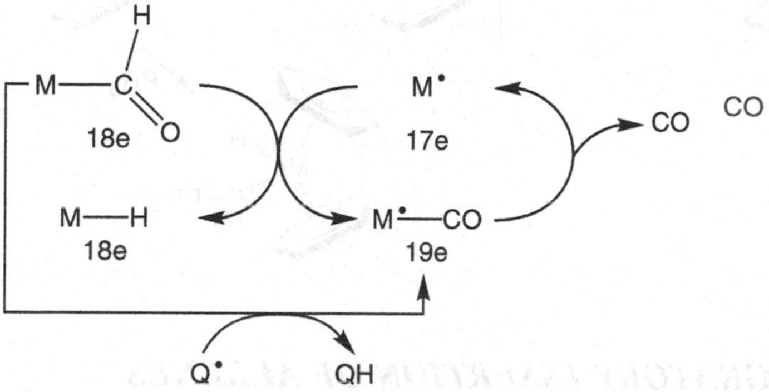

$M^{\bullet} = [(CO)_5ReRe(CO)_4]^{\bullet-}$, $[Cr(CO)_5]^{\bullet-}$, $[W(CO)_5]^{\bullet-}$, $[Mn(CO)_3(PPh_3)_2]^{\bullet}$, $[Re(CO)_4Br]^{\bullet-}$, $[(CH_3)CORe(CO)_4]^{\bullet-}$, $[Cp^*Ru(CO)L]^{\bullet}$ (L = CO or PMe$_2$Ph).

ATC-catalysis mechanism initiated by AIBN

3. METHYLENE INSERTION AND EXTRUSION

Since carbene ligands $=CH_2$ and $=CR_2$ have electronic structures related to that of CO, it is expected that methylene and carbene insertions are possible. There are only very few studies, however, because carbene ligands are more fragile than CO.

Example:

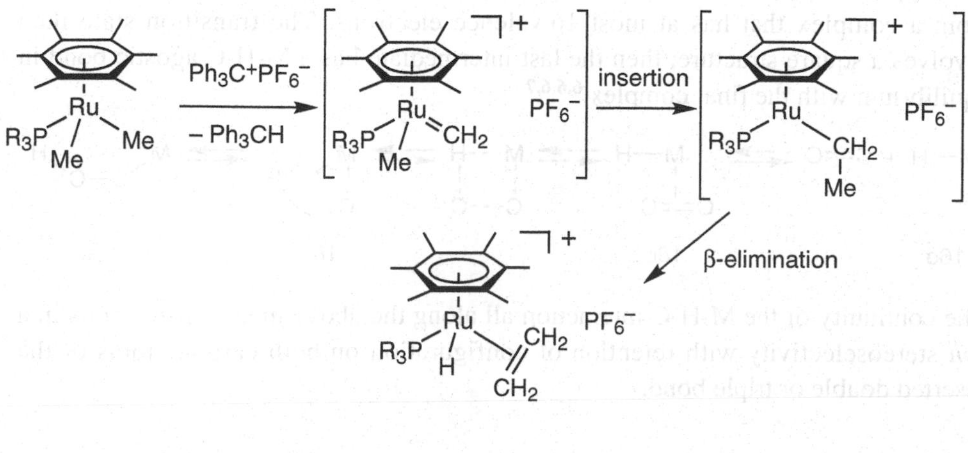

The reverse reaction, extrusion of a methylene from a metal-alkyl complex to give a species of the type $M(H)(C_2H_4)$ is also known, but it is far from being systematic: this is the α-H elimination (see section 5.1 and Chap. 18.5) that sometimes plays an important role in the initiation of metathesis reactions (Chap. 15).

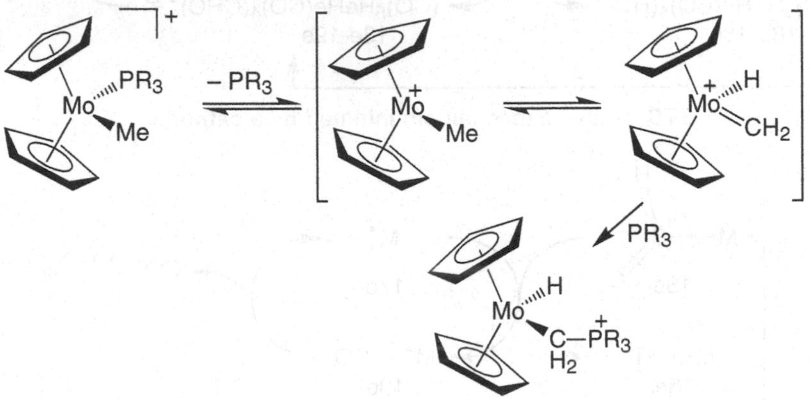

4. 1,2 MIGRATORY INSERTION OF ALKENES AND ALKYNES INTO M-H BONDS

Contrary to CO, alkenes very easily insert into a metal-hydride bond to give a metal-alkyl complex, and the reaction is reversible. Indeed, depending on the conditions, the metal-alkyl complex can give rise to the extrusion reaction, also called β-elimination, yielding the metal-alkene-hydride complex. This β-elimination reaction is the mechanism of decomposition of metal-alkyl complexes that have at least one β hydrogen and at most 16 valence electrons on the metal.

$$M\text{–}H + C\text{=}C \rightleftharpoons M\text{–}C\text{–}C\text{–}H$$

The insertion reaction, as its reverse, involves a metal-olefin-hydride species, and thus requires, in one direction or the other, a vacant site. It is thus necessary to start from a complex that has at most 16 valence electrons. The transition state then involves a square structure, then the last intermediate has a M-H-C agostic bond in equilibrium with the final complex.[6.6,6.7]

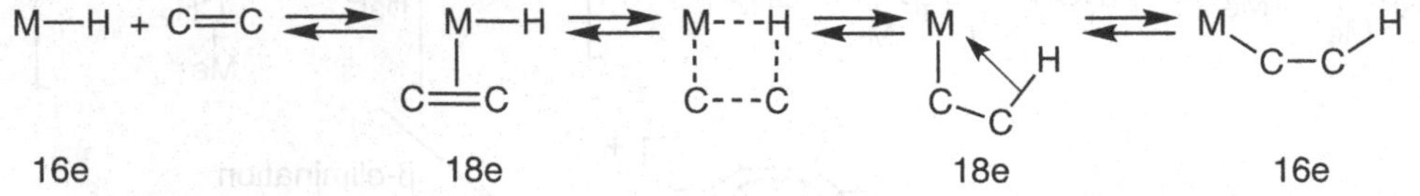

The continuity of the M-H-C interaction all along the above mechanism results in a *syn* stereoselectivity with retention of configuration on both carbon atoms of the inserted double or triple bond.

Example:

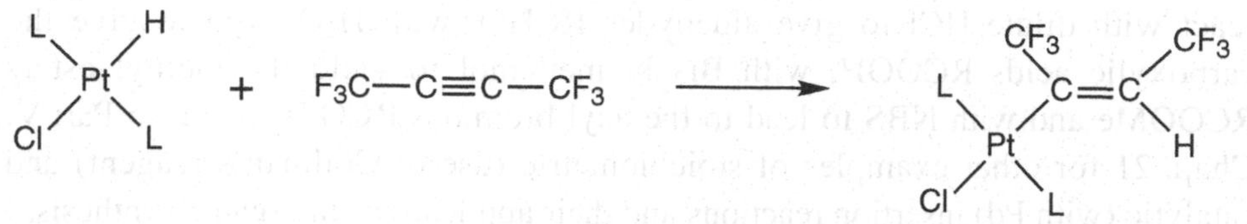

The insertion-β-elimination equilibrium is displaced in the direction of the insertion for alkenes and alkynes bearing electron-withdrawing substituents. Thus, in the examples above and below, the metal-fluoroalkyl complexes are stable, even with 16 valence electrons and one β hydrogen atom.

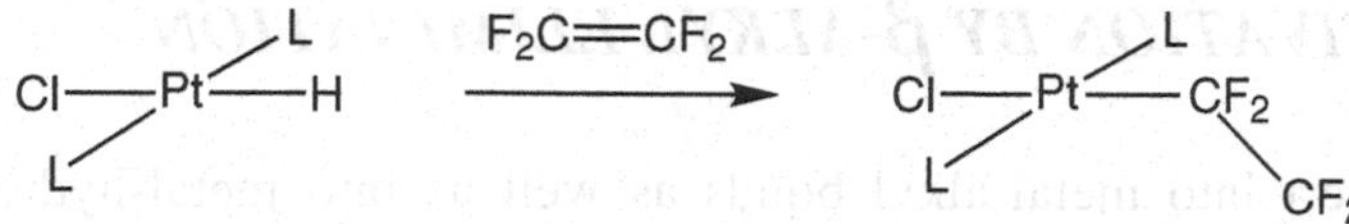

The presence of electron-withdrawing substituents is not always an indispensable condition to make the insertion quantitative. Schwarz [6.8] has indeed established an important application of alkene insertion in the Zr-H bond of the so-called Schwarz reagent, [ZrCp$_2$(H)(Cl)].

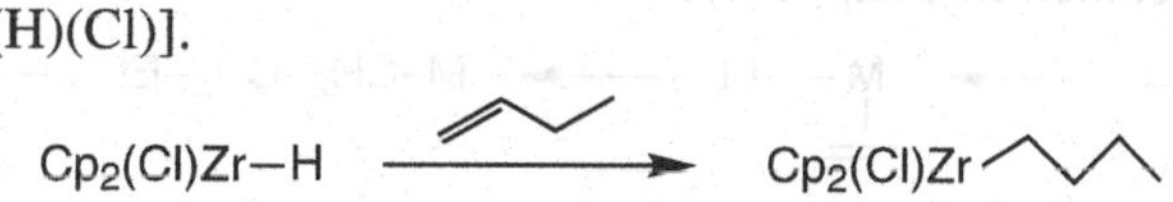

$$Cp_2(Cl)Zr-H \longrightarrow Cp_2(Cl)Zr \diagdown\diagup\diagdown$$

The regiospecificity of the above reaction is antiMarkownikov, which yields a primary alkyl complex. In addition, if the Schwarz reagent reacts with 2-butene, the insertion product is an instable secondary alkyl complex, this kinetic instability being presumably due to the steric bulk. This latter complex rearranges to give a primary alkyl complex by β-H elimination followed by regiospecific insertion of the intermediate 1-butene.

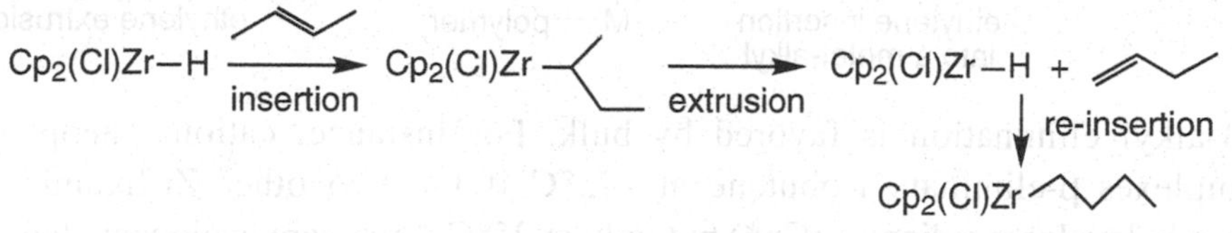

▸ The Zr-alkyl complexes [ZrCp$_2$(R)(Cl)] obtained by olefin insertion react with various electrophiles with retention of configuration at the carbon atom in the course of the decomplexation, which leads to the functionalization of the olefins; for example, the reaction with *N*-bromosuccinimide (NBS) or I$_2$ gives the halide RX:

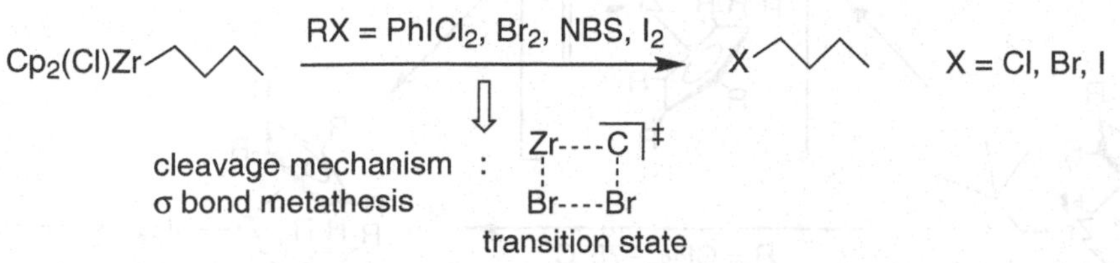

$$Cp_2(Cl)Zr \diagdown\diagup\diagdown \xrightarrow{RX = PhICl_2, Br_2, NBS, I_2} X \diagdown\diagup\diagdown\diagup \qquad X = Cl, Br, I$$

cleavage mechanism :
σ bond metathesis

$$\left[\begin{array}{c} Zr \text{----} \overline{C} \\ \vdots \qquad \vdots \\ Br \text{----} Br \end{array} \right]^{\ddagger}$$

transition state

▸ The oxidation of [ZrCp$_2$(R)(Cl)] using H$_2$O$_2$/H$_2$O or *t*-BuOOH gives the alcohol ROH. These complexes [ZrCp$_2$(R)(Cl)] also very easily undergo CO insertion into the Zr-alkyl bond to give Zr-acyl complexes [ZrCp$_2$(COR)(Cl)]. The latter react with dilute HCl to give aldehydes RCHO, with H$_2$O$_2$/H$_2$O to give the carboxylic acids RCOOH, with Br$_2$ in methanol to yield the methyl esters RCOOMe and with NBS to lead to the acyl bromides RCOBr. See also Part V, Chap. 21 for other examples of stoichiometric (use of Collman's reagent) and catalytic (with Pd) insertion reactions and their applications in organic synthesis.

5. INSERTION OF ALKENES INTO METAL-ALKYL BONDS AND THE REVERSE REACTION: C-C ACTIVATION BY β-ALKYL ELIMINATION

Olefins can insert into metal-alkyl bonds as well as into metal-hydride bonds provided that a vacant coordination site is available on the metal center. This reaction can thus be repeated even when the alkyl ligand lengthens, which makes the proposed mechanism a Ziegler-Natta olefin polymerization. This type of polymerization is detailed in Chap. 15.1:

$$M\text{—}Et + \; = \; \longrightarrow \; M\text{—}Et \; \longrightarrow \; M\text{–}CH_2\text{–}CH_2\text{–}Et \; \longrightarrow \; etc.$$

Although the most common reverse reaction, β-elimination (extrusion of an olefin) is β-H elimination forming a terminal olefin, β-alkyl elimination is sometimes observed with d^0 metal-alkyls of the early transition metals and the rare-earth metals. This reaction is the basis for depolymerization of the polymers.

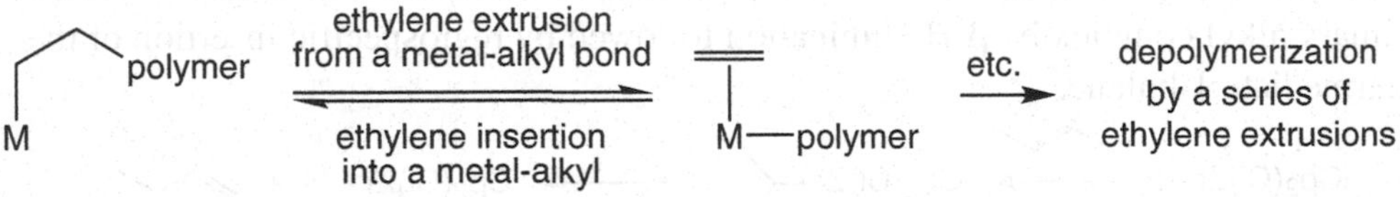

The β-alkyl elimination is favored by bulk. For instance, cationic neopentyl Zr complexes β-eliminate isobutene at –75°C if the two other Zr ligands are pentamethylcyclopentadienyls (Cp*) but only at 25°C if they are cyclopentadienyls, [ZrCp$_2$(CH$_2$CMe$_3$)]$^+$ being stable in solution at 0°C:

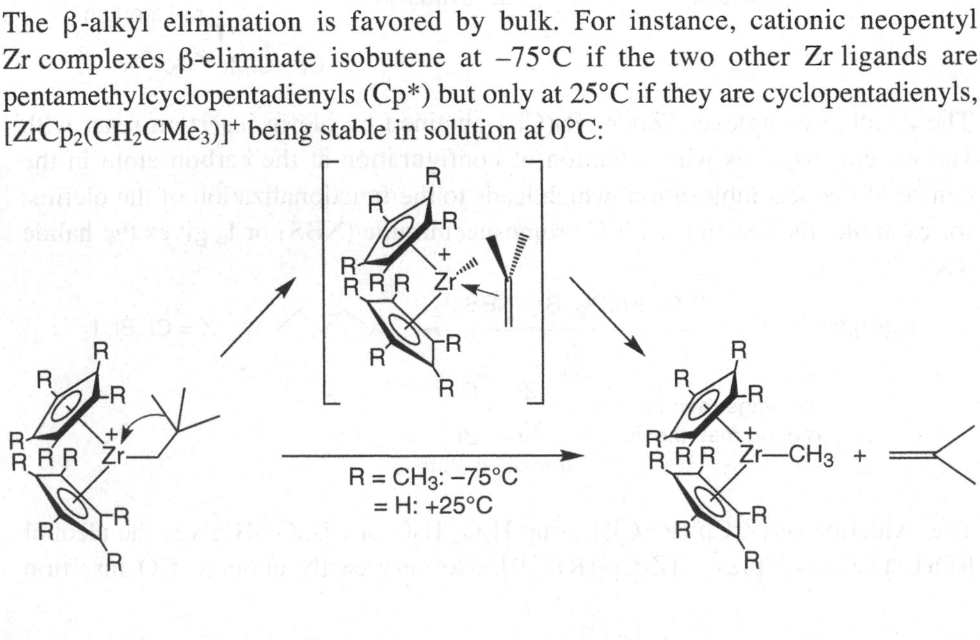

R = CH$_3$: –75°C
= H: +25°C

In such Zr complexes, β-H and β-methyl elimination reactions are in competition when both H and methyl are present on the β carbon atom, and it has been observed that increased bulk favors the β-methyl elimination. In $[ZrCp*_2(CH_2CHMeR)]^+$, β-methyl elimination is 10 times faster than β-H elimination, and it is even 50 times faster in the analogous Hf complex. On the other hand, in the less bulky complex $[ZrCp_2(CH_2CHMeR)]^+$, β-H elimination is 100 times faster than β-methyl elimination. The reason for this trend is that the steric interactions between bulky ligands and the two substituents of the β carbon atom destabilize the transition state of the β-H elimination to a larger extent than that of the β-methyl elimination.[6.15b]

6. α- AND γ-ELIMINATION

In section 4, we have seen that the β-H elimination is the main general decomposition pathway of transition-metal-alkyl complexes that have a β hydrogen and at most 16 valence electrons. Consequently, 18-electron metal-alkyl complexes and other metal-alkyl complexes without β hydrogen are relatively stable except if an L ligand is removed by heating or photolysis, in which case decomposition by β-elimination is again possible. There are other decomposition pathways, however, for complexes lacking β hydrogens: β-alkyl elimination mentioned above for early transition-metal and rare-earth complexes, but also α- and γ-elimination, concerning complexes that have hydrogens in these positions.

6.1. α-ELIMINATION

The neopentyl ligand is one of these ligands lacking β hydrogens. During an attempt to synthesize $[Ta(CH_2CMe_3)_5]$, Schrock[6.16] obtained the first alkylidene complex by α-elimination due to the steric bulk around the metal. The tantalum atom being in a d^0 electronic configuration, this α-elimination is well explained by σ-bond metathesis that intervenes either after the first substitution of Cl by CH_2CMe_3, or after the second one.

Overall:

$$[Ta(CH_2CMe_3)_3Cl_2] + 2\ LiCH_2CMe_3 \longrightarrow [Ta(CH_2CMe_3)_3(=CHCMe_3)] + CMe_4 + 2\ LiCl$$

Mechanism:

The steric bulk necessary around the Ta atom to induce α-elimination can also be provoked by addition of a phosphine:

$$[TaCp(CH_2Ph)_2Cl_2] + 2\ PMe_3 \longrightarrow [TaCp(=CHPh)Cl_2(PMe_3)_2] + PhCH_3$$

Many other examples have been disclosed by Schrock's group in this family of complexes, showing that α-elimination is rather systematic when bulk conditions are met, and has become a major component of the chemistry of early transition-metals.[6.16] It is in particular a remarkable route to early transition-metal alkylidene complexes.

There are also cases for which α-elimination is possible starting from a metal-methyl complex.[6.16,6.17] If the metal has at least two d electrons, the path from a ligand-methyl complex to a methylene-hydride complex involves α-C-H agostic interaction with the metal followed by C-H oxidative addition. Thus, overall, it corresponds to methylene extrusion as mentioned in section 3.[6.17] Such a case is also known from Bercaw's studies of d^0 [Cp*$_2$Ta(=CH$_2$)(H)] in equilibrium with d^2 [Cp*$_2$TaCH$_3$] that can be trapped by CO: [6.17b]

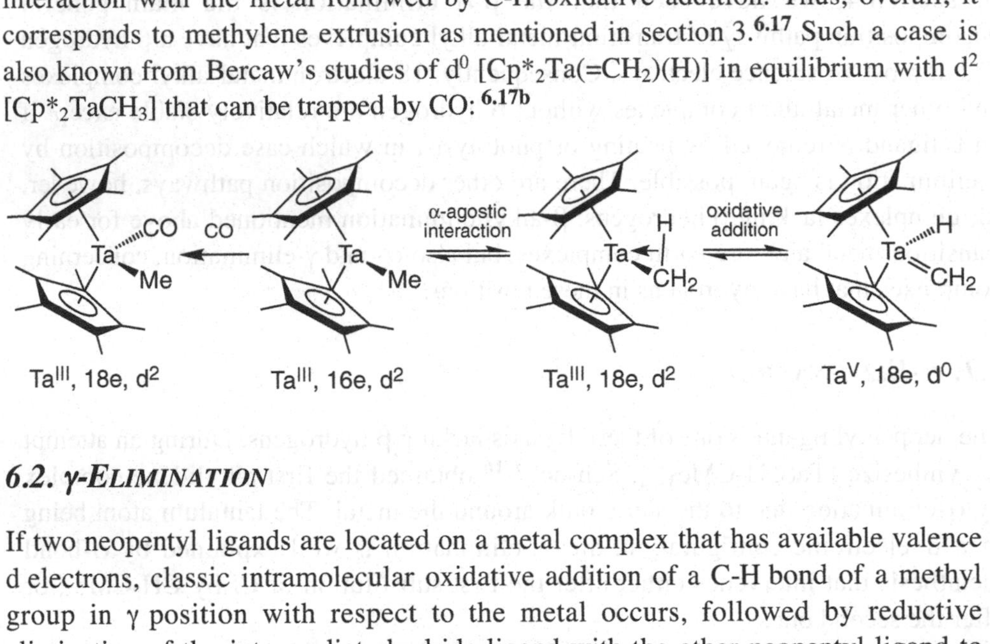

Ta^{III}, 18e, d^2 Ta^{III}, 16e, d^2 Ta^{III}, 18e, d^2 Ta^V, 18e, d^0

6.2. γ-ELIMINATION

If two neopentyl ligands are located on a metal complex that has available valence d electrons, classic intramolecular oxidative addition of a C-H bond of a methyl group in γ position with respect to the metal occurs, followed by reductive elimination of the intermediate hydride ligand with the other neopentyl ligand to form neopentane: [6.18]

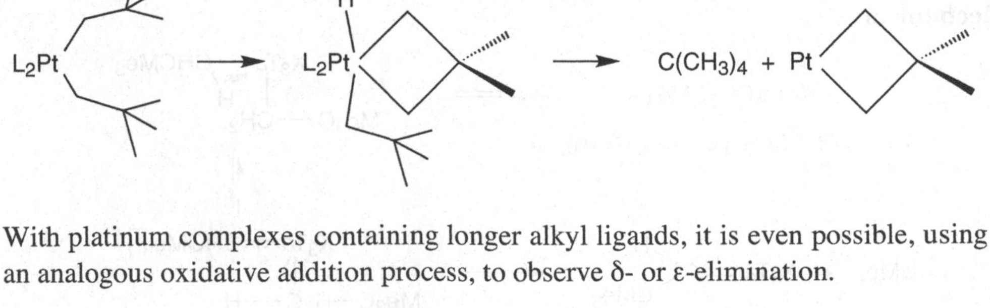

With platinum complexes containing longer alkyl ligands, it is even possible, using an analogous oxidative addition process, to observe δ- or ε-elimination.

Addition of a phosphine to a tantalum-methyl-mesityl complex (below) also leads to γ-elimination rather than α-elimination, which yields methane and a dimethyl-benzylidene complex.[6.19] The σ-bond-metathesis mechanism can again take this γ-elimination into account:

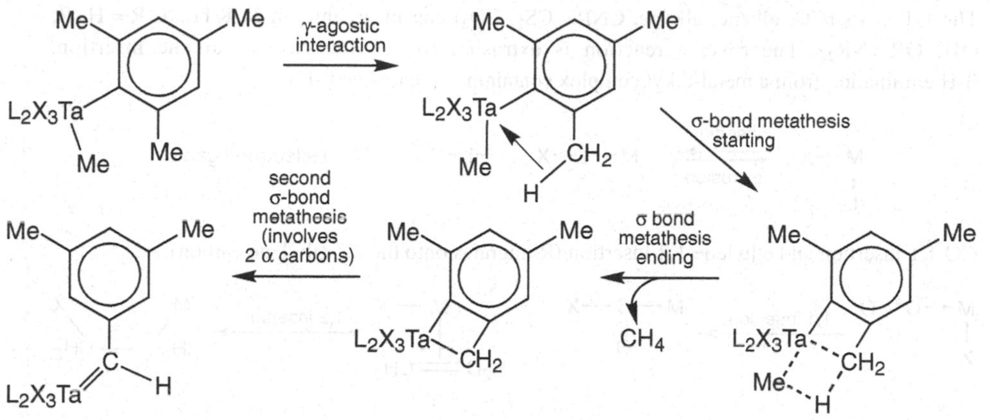

With the analogous benzyl-mesityl complex, α-elimination from benzyl (rather than γ-elimination from mesityl) is observed to give a benzylidene complex and mesitylene.[6.19]

$$[Ta(mesityl)(CH_2Ph)Cl_3] + 2\ PMe_3 \longrightarrow [Ta(=CHPh)(PMe_3)_2Cl_3] + mesitylene$$

The seminal work by Tobbin Marks group at Northwestern on rare-earth alkyl complexes also has examples of γ-elimination in d^0 complexes that parallel Schrock's reactions with d^0 Ta and Nb neopentyl complexes. A neopentyl ligand removes an H atom in γ position of another neopentyl ligand to yield a dimethylmetallacyclobutane. Again the reaction proceeds by σ-bond metathesis.[6.20]

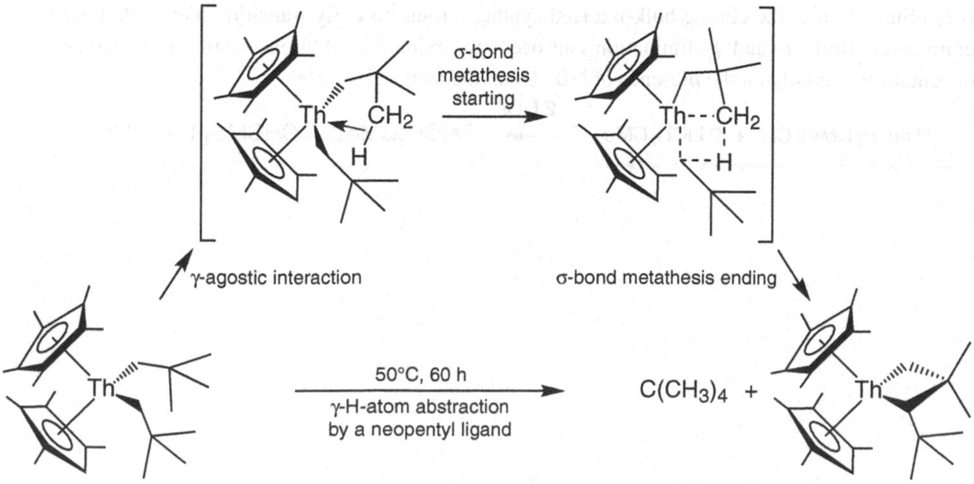

SUMMARY OF CHAPTER 6
INSERTION AND EXTRUSION REACTIONS

1 - Insertion and extrusion

The L ligands (CO, alkene, alkyne, CNR', CS_2, CO_2) can insert into an M-R bond (R = H, R, OH, OR, NR_2). The reverse reaction is extrusion (or, in the case of alkene insertion, β-H elimination from a metal-alkyl complex containing at least one β-H).

$$\underset{L}{\overset{}{M}}-X \underset{\text{extrusion}}{\overset{\text{insertion}}{\rightleftarrows}} \quad M-L-X \quad -L-X = -X',\ 1\text{-electron ligand}$$

CO 1,1-insertion and ethylene 1,2-insertion (R migrates onto the 2nd ethylene carbon):

$$\underset{X}{\overset{}{M}}-C{=}O \xrightarrow{\text{1,1 insertion}} \underset{O}{\overset{}{M}-C-X} \qquad \underset{H_2C{=}CH_2}{\overset{}{M}-X} \xrightarrow{\text{1,2 insertion}} \underset{H_2C-CH_2}{\overset{M\qquad X}{}}$$

CO insertion into an M-alkyl bond is a migratory insertion liberating the site of the alkyl group. Such an insertion is much accelerated by stoichiometric or catalytic (ETC) single-electron oxidation.

CO insertion is possible in an M-alkyl bond, but energetically unfavorable in M-H (except with rare earths).

The reverse, CO extrusion is favorable: M-CHO ⟶ (CO)M-H.

Alkene insertion can occur either in an M-H bond or in an M-alkyl bond (see also Chap. 14).

2 - Decomposition of metal-alkyl complexes

The extrusion of an alkene from a metal-alkyl (β-H elimination) explains the instability of metal-alkyl complexes having both NEV < 18 and at least one β-H atom. β methyl abstraction is also known, but only with early transition-metals and rare-earth complexes.

In the absence of β-H atom, other decomposition routes are viable: α- and γ-elimination. The α-H elimination is the classic bulk-induced synthetic route to early-transition-metal-alkylidene complexes. Both α- and γ-elimination can occur by oxidative addition or σ-bond metathesis depending essentially on the presence of NBVEs on the metal. Example:

$$[Ta(CH_2CMe)_3Cl_2] + 2\ LiCH_2CMe_3 \xrightarrow{-\ 2\ LiCl} [Ta(CH_2CMe_3)_3(=CHCMe_3)] + CMe_4$$

EXERCISES

6.1. a. What is the product resulting from the reaction of $[FeCp(CO)_2CH_3]$ with $[FcPPh_2]$ (Fc = ferrocenyl)?

b. How is it possible to accelerate this (otherwise slow) reaction?

6.2. What is the reaction product between Schwarz reagent $[ZrCp_2(Cl)(H)]$ and cyclopentadiene?

6.3. What is the major mode of decomposition, if any, of the transition-metal complexes $[Ta(\eta^1\text{-aryl})_nCl_{5-n}]$ with the following aryl ligands:

a. phenyl;

b. mesityl;

c. pentafluorophenyl?

6.4. Why is it possible to synthesize the thermally stable complexes $[TaMe_5]$ and $[Ta(CH_2Ph)_5]$, but not $[Ta(CH_2CMe_3)_5]$?

6.5. Which gas or gases can be characterized upon thermal decomposition of $[Ta(CH_3)_5]$? Propose mechanisms.

PART III

THE MAIN FAMILIES OF ORGANOMETALLIC COMPLEXES

Chapter 7
Metal carbonyls and complexes of other monohapto L ligands

Chapter 8
Metal-alkyl and -hydride complexes
and other complexes of monohapto X ligands

Chapter 9
Metal-carbene and -carbyne complexes
and multiple bonds with transition metals

Chapter 10
π Complexes of mono- and polyenes and enyls

Chapter 11
Metallocenes and sandwich complexes

Chapter 12
Ionic and polar metal-carbon bonds:
alkali and rare earth complexes

Chapter 13
Covalent chemistry of the organoelements
of frontier (11, 12) and main (13-16) groups

Chapter 7

METAL CARBONYLS AND COMPLEXES
OF OTHER MONOHAPTO *L* LIGANDS

1. INTRODUCTION

The family of metal carbonyls is one of the very most important ones in inorganic and organometallic chemistry. This is due to the interest in their structures, applications in organic and organometallic synthesis and essential role in catalysis. The remarkable ability of CO to give π backbonding is responsible for the stability of metal carbonyls with oxidation states that are nil (neutral metal carbonyls) or negative (anions). A large variety of mixed complexes containing carbonyls and other ligands is also available, the best known being those that also contain phosphines, cyclopentadienyls, arenes or other unsaturated hydrocarbon ligands. Because the carbonyl ligands are strongly bonded to transition metals, the variations of the frequency (around 2000 cm^{-1} for terminal CO) of the intense infrared carbonyl absorption allow to measure the electronic effects of the other ligands.[7.1]

The low-nuclearity binary metal carbonyls always have 18 valence electrons on the metal except the 17-electron radical [V(CO)$_6$] whose structure is stabilized by the octahedral geometry that largely contributes to prevent radical-type reactions. The 18-electron complex [Mn(CO)$_6$]$^+$ is stabilized for the same structural reason, metal carbonyl cations being otherwise instable.

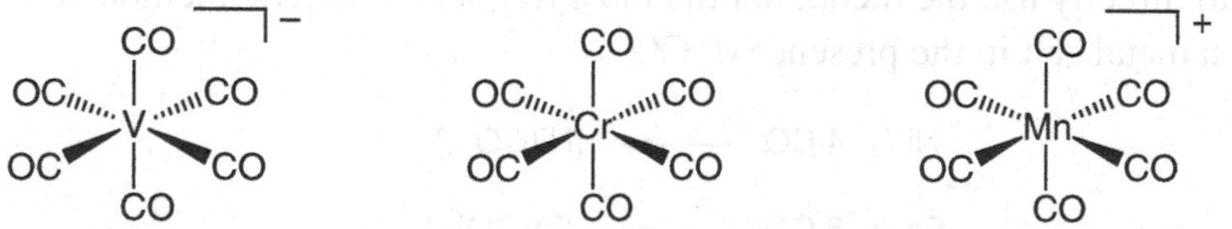

Three stable isoelectronic mononuclear octahedral metal carbonyls
The favorable octahedral geometry also stabilizes the neutral 17e radical [V(CO)$_6$]$^{•}$ and [Mn(CO)$_6$]$^+$, a rare cationic metal carbonyl.

The high-nuclearity metal-carbonyl clusters [M$_6$(CO)$_{16}$] and beyond often escape the 18-electron rule to obey Wade's rules (see Chap. 2). Thus, mononuclear neutral metal carbonyls are known for the metals that have an even number of d electrons, whereas neutral dimers with a single metal-metal bond are known for metals that have an odd number of d electrons. Metal-carbonyl clusters are essentially known in the columns of iron, cobalt and nickel (groups 8-10). They are all the more stable as one goes down in the columns, because the strength of the metal-metal bond increases in this direction. The most current metal carbonyls are compiled in the table below.

Reduction of all the neutral dinuclear metal carbonyls yields monoanionic mononuclear metal carbonyls such as $[Mn(CO)_5)]^-$ that are very stable, whereas reduction of mononuclear neutral metal carbonyls gives stable dianions such as Collman's reagent $[Fe(CO)_4]^{2-}$. Some trianions such as $[Nb(CO)_5]^{3-}$ are known to be thermally stable and extremely air-sensitive. Ellis has synthesized a series of multi-anionic metal carbonyls under extreme conditions.[7.2] Anions of metal-carbonyl clusters are also known.

Stable binary metal carbonyls

$V(CO)_6$*	$Cr(CO)_6$*	$Mn_2(CO)_{10}$*	$Fe(CO)_5$*	$Co_2(CO)_8$*	$Ni(CO)_4$*
			$Fe_2(CO)_9$*	$Co_4(CO)_{12}$*	
			$Fe_3(CO)_{12}$*	$Co_6(CO)_{16}$	
	$Mo(CO)_6$*	$Tc_2(CO)_{10}$	$Ru(CO)_5$		
			$Ru_3(CO)_{12}$*	$Rh_4(CO)_{12}$*	
	$W(CO)_6$*	$Re_2(CO)_{110}$*	$Os(CO)_5$		
			$Os_3(CO)_{12}$*	$Ir_4(CO)_{12}$*	
				$Ir_6(CO)_{16}$	

* commercially available

2. SYNTHESES

The synthesis of metal carbonyl complexes proceeds industrially. It is sometimes possible to directly use the metal, but the most frequent synthetic method consists in reducing a metal salt in the presence of CO:

$$Ni + 4\,CO \longrightarrow [Ni(CO)_4]$$

$$Fe + 5\,CO \longrightarrow [Fe(CO)_5]$$

$$CrCl_3 + Al + 6\,CO \longrightarrow [Cr(CO)_6] + 1/2\,Al_2Cl_6$$

$$WCl_6 + 2\,Et_3Al + 6\,CO \longrightarrow [W(CO)_6] + 2\,Al_2Cl_6 + 3\,C_4H_{10}$$

$$6\,Mn(OAc)_2 + 4\,Et_3Al + 30\,CO \longrightarrow 3\,[Mn_2(CO)_{10}] + 4\,Al(OAc)_3 + 6\,C_4H_{10}$$

$$Re_2O_7 + 17\,CO \longrightarrow [Re_2(CO)_{10}] + 7\,CO_2$$

$$2\,CoCO_3 + 2\,H_2 + 8\,CO \longrightarrow [Co_2(CO)_8] + 2\,CO_2 + 2\,H_2O$$

In the laboratory, some syntheses classically employ photochemistry or redox reagents. The use of alkali metals leads to metal carbonyl anions starting from neutral metal carbonyls.

$$2\ [Fe(CO)_5] \xrightarrow{CH_3COOH,\ h\nu} [Fe_2(CO)_9] + CO$$

$$[Fe(CO)_5] + 2\ OH^- \longrightarrow [HFe(CO)_4]^- + HCO_3^-$$

$$3\ [HFe(CO)_4]^- + 3\ MnO_2 \longrightarrow [Fe_3(CO)_{12}] + 3\ OH^- + 3\ MnO$$

$$[Fe(CO)_5] + 2\ Na/Hg \longrightarrow [Fe(CO)_4]^{2-}(Na^+)_2 + CO + 2\ Hg$$

3. BINDING MODES

3.1. CLASSIC MODE: TERMINAL CO

In the first chapter, it has been noted that CO is a σ donor (two-electron L ligand) and a π acceptor. The corresponding orbital interactions are the following:

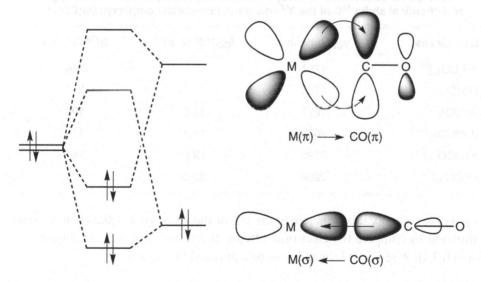

The two mesomer forms below (a neutral- and a ionic one) can be used to describe the formalism of the M-C-O chain:

$$M \overset{\pi}{\underset{\sigma}{\rightleftharpoons}} C{=}\overset{\cdot\cdot}{\underset{}{O}} \longleftrightarrow |M \overset{-}{\underset{\sigma}{\longleftarrow}} \overset{-}{C}{\equiv}O|^+$$

The σ donation from the carbonyl to the metal weakens the electron density at the carbon atom, and the backbonding enriches both the carbon and the oxygen atoms of the carbonyl. The overall result is that the carbon atom is more electron poor upon coordination, and the oxygen atom is more electron rich. The ligand is thus polarized upon coordination, and undergoes nucleophilic attacks on the positively polarized carbon atom. The more important the π backbonding from the metal, the more populated is the π^* antibonding carbonyl orbital, the more the C-O bond is weakened and lengthened, the more the metal-carbon bond is strengthened and shortened. It is possible to correlate the bond length determined by X-ray diffraction. For instance, for $[Mo(CO)_3(NR_3)_3]$, the length of a single Mo-C bond would be 233 pm, and the experimental length, 193 pm, indicates a strong double bond character, i.e. important π backbonding.

The most practical method for these correlations is infrared spectroscopy, however, because the weakening of the C-O bond, provoked by the π backbonding, results in the lowering of the absorption frequency. The frequency domain is between 1820 and 2150 cm^{-1} for a terminal carbonyl. The correlation between the vibration frequency and the bond order is possible, because the elongation frequency of the C-O bond is, in first approximation, independent from the other vibrations of the molecule. Let us examine, for instance, the decrease of the π backbonding (characterized by the C-O force constant) when the charge on the metal center decreases upon going to the right of the periodic table for a series of octahedral metal complexes of the 5d transition metals:

Absorption frequencies ν_{CO} (asymmetric elongation), force constant f_{CO} and chemical shifts ^{13}C of the 3rd-row transition-metal complexes $[M(CO)_6]^x$

Complexes	$\nu_{CO}(f_{1u})$/cm^{-1}	f_{CO}/10^2 N·m^{-1}	$\delta(^{13}C)$, ppm
$[Hf(CO)_6]^{2-}$	1757		244
$[Ta(CO)_6]^-$	1850		211
$W(CO)_6$	1977	17.0	192
$[Re(CO)_6]^+$	2085	18.1	171
$[Os(CO)_6]^{2+}$	2190	19.8	147
$[Ir(CO)_6]^{3+}$	2254	20.8	121

It is possible to apply this principle to the use of the carbonyl frequency as a means of allowing to compare the electronic effects of another ligand L, for instance in $[Mo(CO)_3L_3]$. A series of L ligands can be compared in this way:

Infrared asymmetric elongation frequencies (wavenumbers) of the carbonyl absorption in Mo(CO)$_3$L$_3$ complexes

Complex	ν_{CO} (cm^{-1})
$[Mo(PF_3)_3(CO)_3]$	2055, 2090
$[Mo(PCl_3)_3(CO)_3]$	1991, 2040
$[Mo\{P(OMe)_3\}_3(CO)_3]$	1888, 1977
$[Mo(PPh_3)_3(CO)_3]$	1835, 1934
$[Mo(NCMe)_3(CO)_3]$	1758, 1898
$[Mo(Py)_3(CO)_3]$	1746, 1888

PF$_3$ is the best π acceptor. Pyridine and acetonitrile are not π acceptors, but only σ donors. The order of donicities (opposite to that of π acceptors) is the following:

$$PF_3 < PCl_3 < P(OMe)_3 < PPh_3 < CH_3CN < Py.$$

In a complex containing carbonyl ligands, the number of absorption bands corresponding to the asymmetric elongation of the carbonyl depends on the local symmetry around the metal center. This symmetry can even be directly deduced by counting the number of bands in the carbonyl infrared region.

Number and modes of the active infrared bands (ν_{CO}) in connection with the local symmetry in the complexes [M(CO)$_n$] and [M(CO)$_n$L$_m$]

complex	number and modes of IR active bands	symmetry	complex	number and modes of IR active bands	symmetry
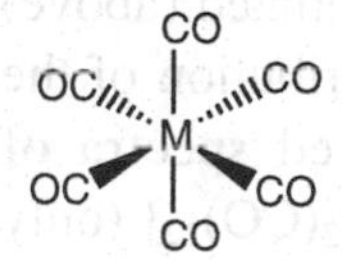 M(CO)$_6$ octahedron	1 T$_{1u}$	O$_h$	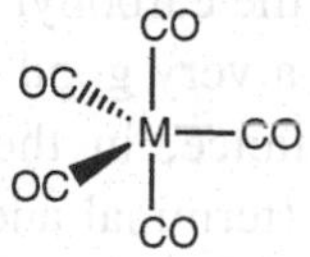 M(CO)$_5$	2 A"$_2$ + E'	D$_{3h}$
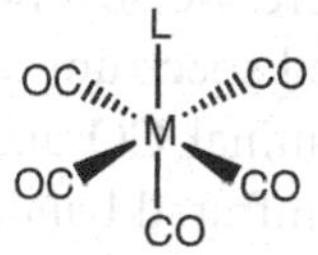 M(CO)$_4$L	3 2A$_1$ + E	C$_{4v}$	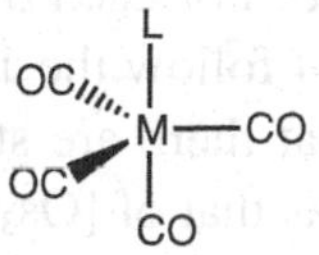 M(CO)$_4$L	3 2A$_1$ + E	C$_{3v}$
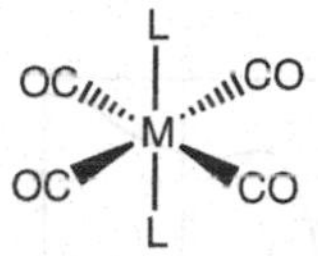 M(CO)$_4$L$_2$ (trans)	1 E$_u$	D$_{4h}$	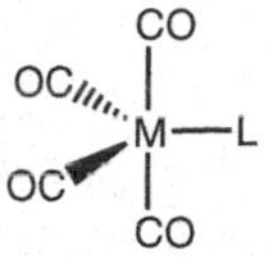 M(CO)$_3$L$_2$	4 2A$_1$ + B$_1$ + B$_2$	C$_{2v}$
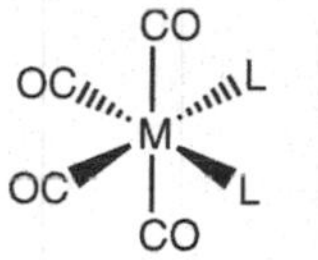 M(CO)$_3$L$_2$	4 2A$_1$ + B$_1$ + B$_2$	C$_{2v}$	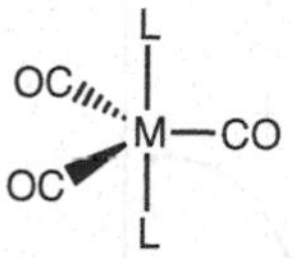 M(CO)$_2$L$_3$	1 E'	D$_{3h}$
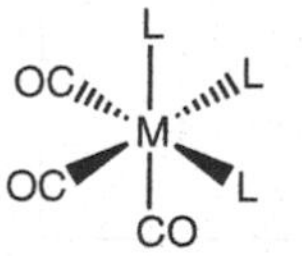 fac-L$_3$M(CO)$_3$	2 A$_1$ + E	C$_{3v}$	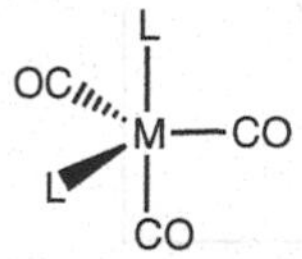 M(CO)$_2$L$_3$	3 2A' + A"	C$_s$
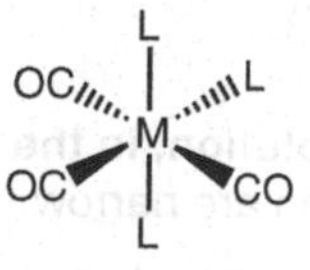 mer-L$_3$M(CO)$_3$	3 2A$_1$ + B$_2$	C$_{2v}$	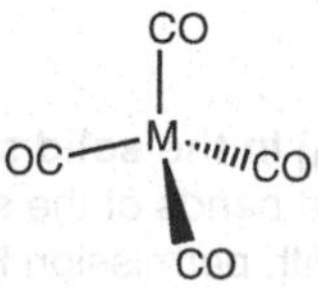 M(CO)$_4$ tetrahedron	1 T$_2$	T$_d$
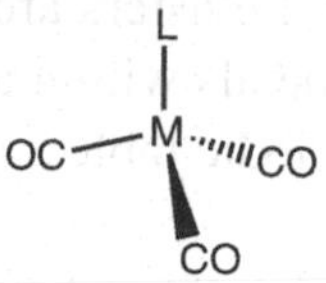 M(CO)$_3$	1 T$_{1u}$	O$_h$	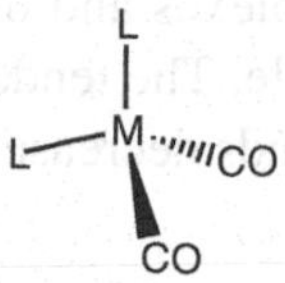 M(CO)$_2$L$_2$	2 A$_1$ + B$_1$	C$_{2v}$

3.2. OTHER MODES: BRIDGING AND CAPPING CO

The main binding modes of CO are represented below together with their characteristic infrared carbonyl absorption frequencies. They include, in addition to those of the terminal mode (section 3.1), the frequencies for CO bridging two or three metals:

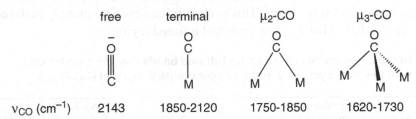

free	terminal	μ_2-CO	μ_3-CO
ν_{CO} (cm^{-1}) 2143	1850-2120	1750-1850	1620-1730

The vibration frequencies of the carbonyl ligands are so well differentiated (above) that infrared spectroscopy is a very good method to diagnose the attribution of the binding CO mode. For instance, in the figure below, the infrared spectra of $[Fe_2(CO)_9]$ in the solid state (terminal and bridging CO) and of $[Os_3(CO)_{12}]$ (only bridging CO) are represented. For each of these two compounds, there are several possibilities of structures that follow the 18-electron rule. The infrared spectrum of $[Fe_2(CO)_9]$ clearly shows that there are strong bands in both the terminal CO and bridging CO regions, whereas that of $[Os_3(CO)_{12}]$ only shows strong infrared bands in the terminal CO region.

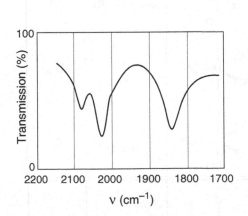

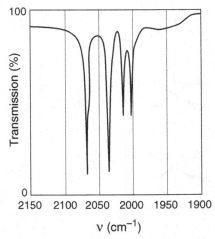

Infrared spectra of $[Fe_2(CO)_9]$ in the solid state and of $[Os_3(CO)_{12}]$ in solution, in the carbonyl region. Note that the bands of the spectrum taken with the solution are narrow. (Reproduced with permission from reference 10.19, p. 651)

The structures of these complexes and of several other metal-carbonyl clusters are gathered in the following table. The tendency of carbonyls to bridge metals within a family of analogous compounds decreases as one goes down in the periodic table.

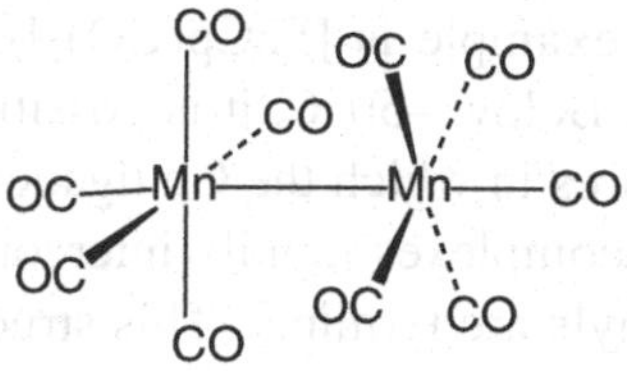

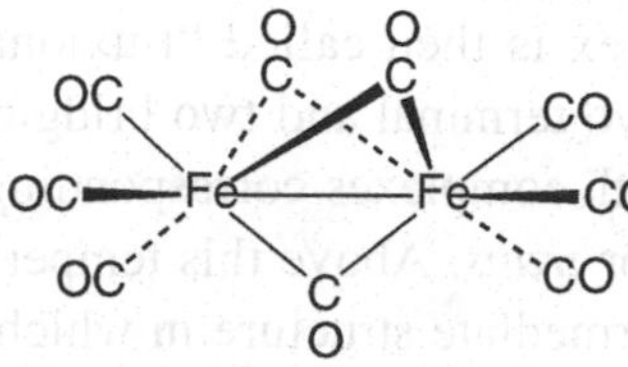

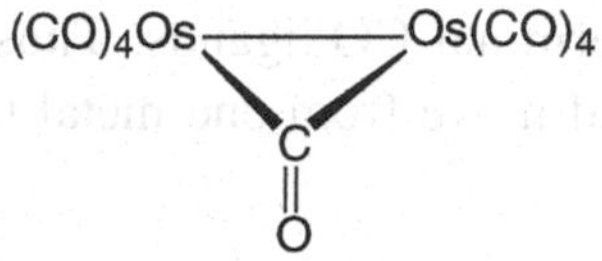

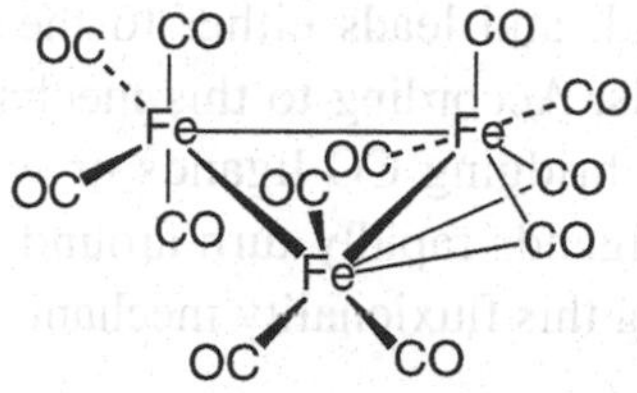

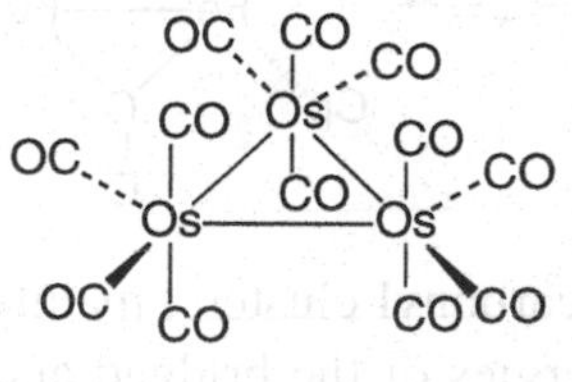

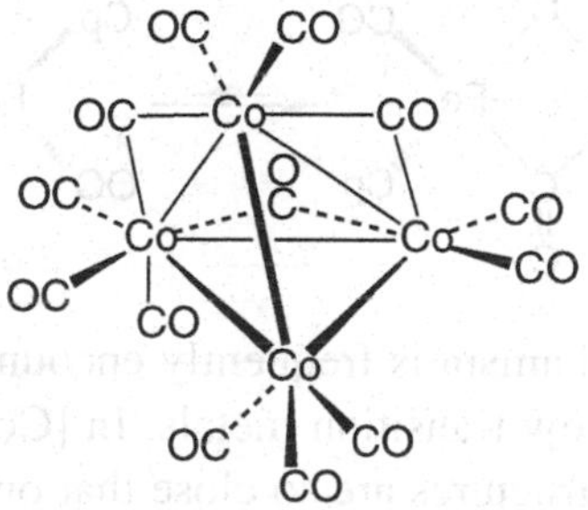

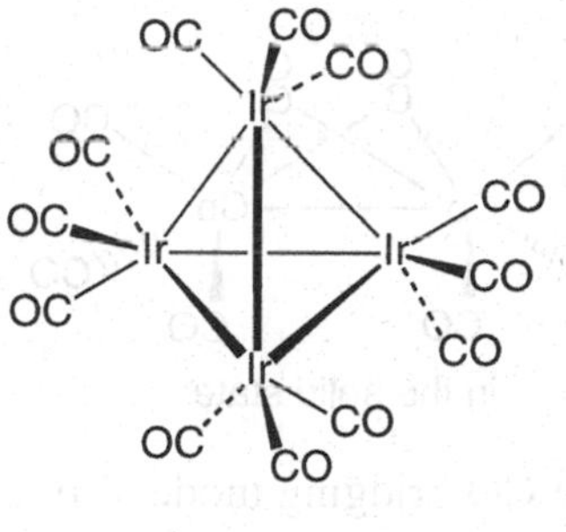

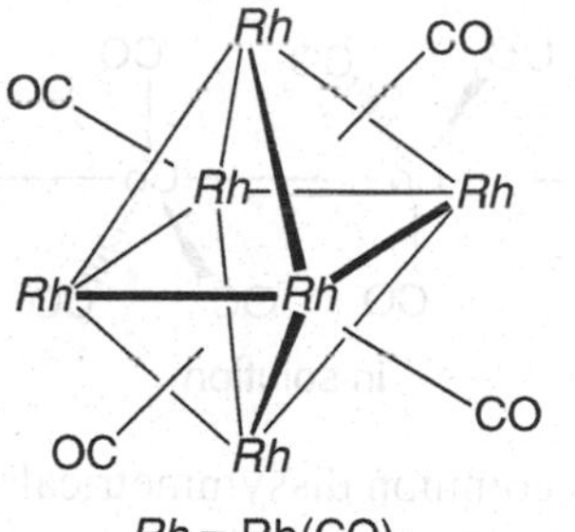

$(CO)_4Os——Os(CO)_4$

$Rh = Rh(CO)_2$

$Rh_6(CO)_{16} = Rh_6(CO)_{12}(\mu_3\text{-}CO)_4$

The CO ligand can bridge two metals in several ways. The most known one is the symmetrical mode in which CO brings one electron to each of the two metals. The CO ligand can easily move from one metal to the next *via* this bridging mode whose energy is close to that of the terminal CO:

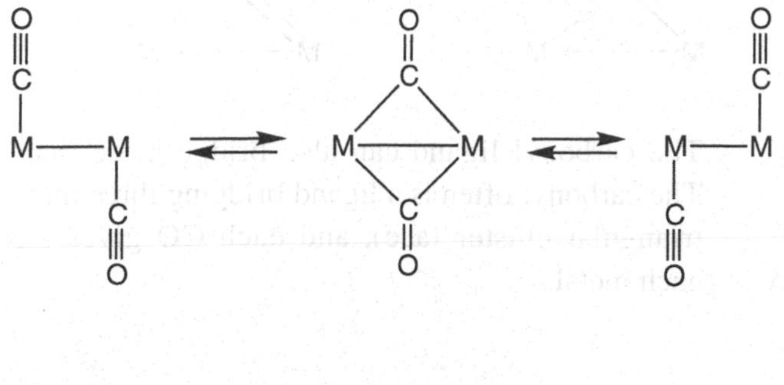

The complex is then called "fluxional". A classic example is $[FeCp(CO)_2]_2$ that contains two terminal and two bridging CO ligands. Below –50°C, it is possible to observe both complexes corresponding to the structures in which the Cp ligands are either *cis* or *trans*. Above this temperature, the two complexes rapidly interconvert *via* an intermediate structure in which all the carbonyls are terminal. This structure is detectable and represents 1% of the complex. The rotation is free around the Fe-Fe bond, and leads either to the *cis* or *trans* Cp complex upon bridging two CO ligands. According to this mechanism, the "closing" can occur either with the originally bridging CO ligands or with originally terminal CO ligands. Thus, the four CO ligands rapidly turn around the complex and move from one metal to the other using this fluxionality mechanism.

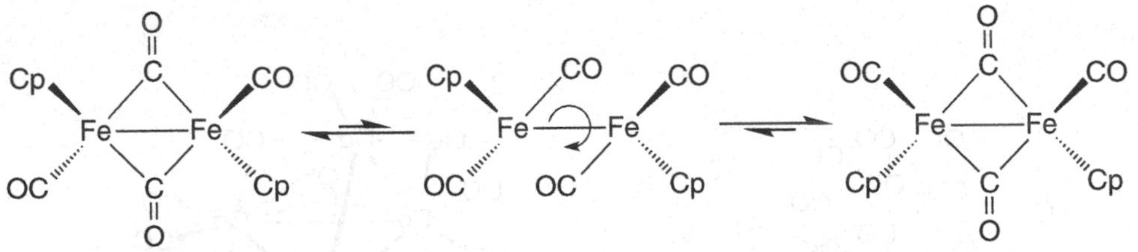

This mechanism is frequently encountered in metal-carbonyl clusters, mainly with the first-row transition metals. In $[Co_2CO)_8]$, the energies of the bridged and non-bridged structures are so close that one is found in the solid state and the other one in solution:

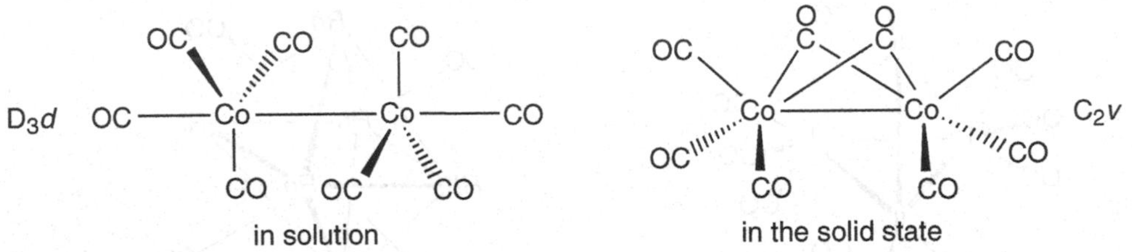

in solution   in the solid state

Three less common dissymmetrical variations of the CO bridging mode also can be considered. The carbonyl is an L ligand for one of the two metals, and the C-M bond, the C-O bond or the oxygen atom is an L ligand for the other metal, as for bridging alkyls. In the first case, it is a 2-electron, 3-center bond, whereas in the two latter cases, one has a genuine 4-electron L_2 donor. The oxygen atom of the CO is indeed a good ligand for classic Lewis acids such as $AlMe_3$.

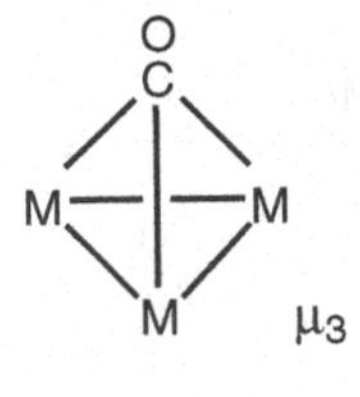

The carbonyl ligand can also bridge more than two metals. The carbonyl often is a ligand bridging three metals (capping a triangular cluster face), and each CO gives 2/3 electron to each metal.

Some exceptional binding modes can also be found. For instance, in the cluster $[NbCp_3(CO)_7]$, one of the carbonyls is an L ligand capping a triangle of three metals, but it is also an L ligand giving each of the two oxygen lone pairs to two of the three metals. Overall, this carbonyl is a 6-electron donor for the three metals:

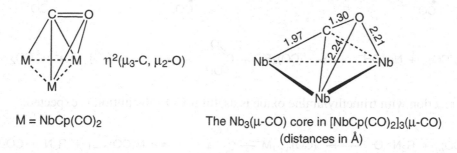

M = NbCp(CO)₂

The Nb₃(μ-CO) core in $[NbCp(CO)_2]_3(\mu\text{-}CO)$
(distances in Å)

4. REACTIONS

4.1. LIGAND SUBSTITUTION REACTIONS (see Chap. 5)

These reactions are useful in organometallic synthesis, the metal carbonyls being commercial and inexpensive:[7.2-7.5]

$$[Fe(CO)_5] + 2\ L\ (or\ L_2) \xrightarrow{\ C_6H_6\ (reflux)\ } [Fe(CO)_3L_2] + 2\ CO$$

L = phosphane, L_2 = diene, etc.   *trans* if L = PR_3

$$[Cr(CO)_6] + arene \xrightarrow{\ n\text{-}Bu_2O\ } [Cr(arene)(CO)_3] + 3\ CO$$

$$[FeCp^*(CO)_3]^+ + CH_3CN\ (solvent) \xrightarrow{\ h\nu_{vis}\ } [FeCp^*(CH_3CN)_3]^+ + 3\ CO$$

The substitution of several CO ligands often leads to the coupling of two incoming ligands or to various rearrangements:

$$Fe(CO)_5 + 2\ H\text{–}C\equiv C\text{–}H \longrightarrow \quad + CO$$

4.2. NUCLEOPHILIC ATTACKS

These lead to Fischer-type metal-carbene complexes that are mesomers of formyl, acyl and metallocarboxylic acid complexes (the latter decompose to hydrides). See also Chap. 5.

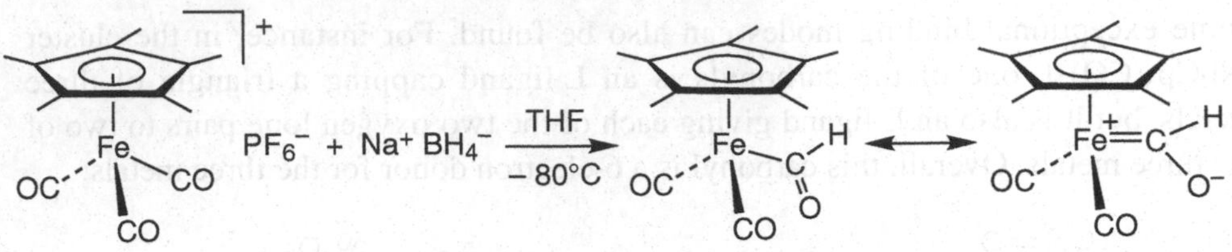

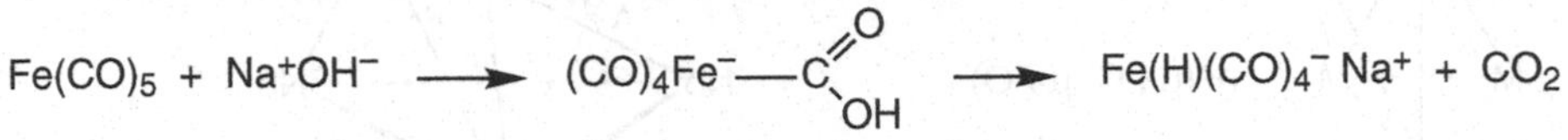

$$Fe(CO)_5 + Na^+OH^- \longrightarrow (CO)_4Fe^- -C\overset{O}{\underset{OH}{}} \longrightarrow Fe(H)(CO)_4^- \, Na^+ + CO_2$$

The reaction with trimethylamine oxide is useful if CO substitution is expected:

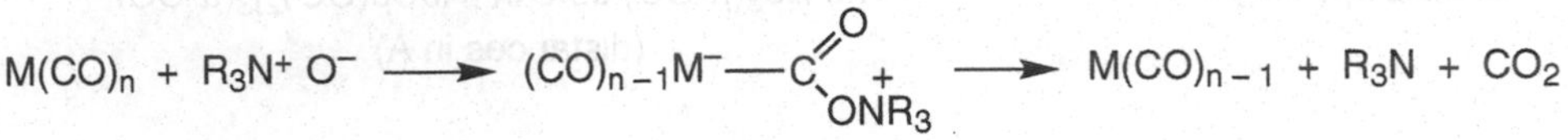

$$M(CO)_n + R_3N^+O^- \longrightarrow (CO)_{n-1}M^- -C\overset{O}{\underset{ONR_3}{}}^+ \longrightarrow M(CO)_{n-1} + R_3N + CO_2$$

4.3. *ELECTROPHILIC ADDITION ON THE OXYGEN ATOM*

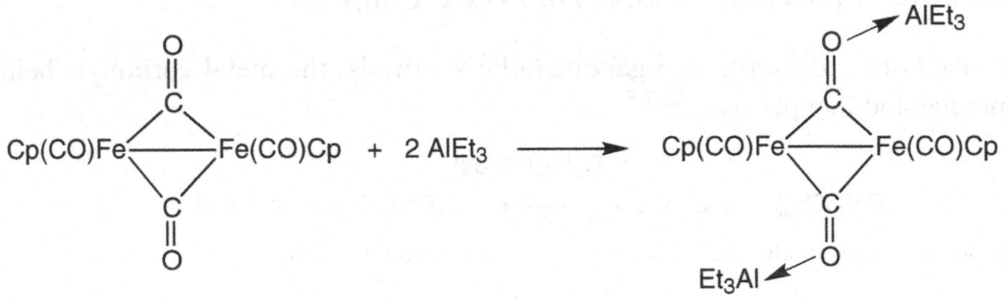

4.4. *INSERTION (see Chap. 6)*

$$[FeCp(CO)_2CH_3] + PPh_3 \longrightarrow [FeCp(CO)(PPh_3)COCH_3]$$

4.5. *OXIDATION*

$$[FeCp^*(CO)_2]_2 + Br_2 \xrightarrow{\text{CH}_2\text{Cl}_2,\ 20°C} 2\,[FeCp^*(CO)_2Br]$$

$$[Re^ICp^*(CO)_3] + H_2O_2\,(30\%) \longrightarrow [Re^{VII}Cp^*(=O)_3] + 3\,CO_2$$

4.6. *REDUCTION*

$$[VCp(CO)_4] + Na/Hg \longrightarrow [VCp(CO)_3]^- \, Na^+ + Hg$$

4.7. *DISMUTATION*

$$[L_nM-ML_n] \xrightarrow{PR_3} [M(PR_3)_3]^+ \, [L_nM]^-$$

The mechanism follows a chain process[7.6] (ETC catalysis, see Chap. 5).

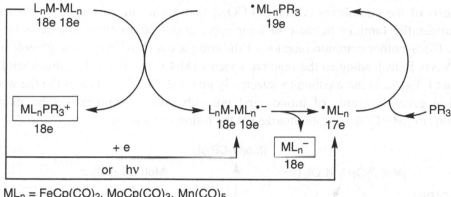

ML_n = $FeCp(CO)_2$, $MoCp(CO)_3$, $Mn(CO)_5$

4.8. EXAMPLES OF IRON CARBONYL AND MOLYDENUM CARBONYL

The chemistry of $[Fe(CO)_5]$, one of the best known and representative of metal carbonyls, is summarized in the following scheme. Contrary to metal carbonyls that react with arenes to give metal-arene-tricarbonyl complexes, metal-pentacarbonyls do not react with simple aromatics. This is due to robustness of the $M(CO)_3$ fragment, M = Fe, Ru or Os that do not give up a CO ligand, except in some intramolecular processes such as the mild thermal decomposition of $[Fe(\eta^4-C_5H_6)(CO)_3]$ to H_2, CO and $[FcCp(CO)_2]_2$.

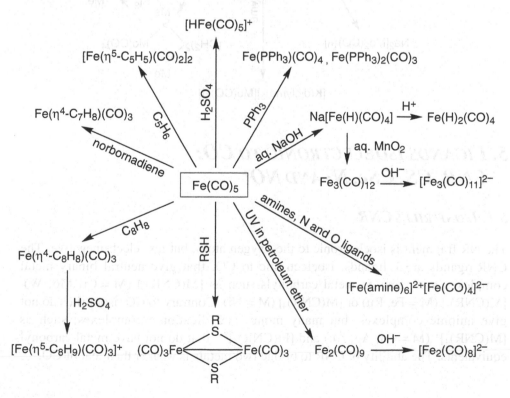

All of the three 18-electron octahedral metal carbonyls [M(CO)$_6$], M = Cr, Mo and W are very much used. Hexacarbonyl chromium is the starting material for the synthesis of the complexes [Cr(arene)(CO)$_3$] that form one of the most studied organometallic families because of their applications in aromatic synthesis (see Chap. 10). Another common reaction of the complexes [M(CO)$_6$] is the photolysis by UV-vis light leading to the reactive species [M(CO)$_5$]. The latter immediately traps an L ligand in the medium to selectively give [M(CO)$_5$L], whereas the thermal reaction gives mixtures of mono- and polysubstituted complexes. The main reactions of [Mo(CO)$_6$] are summarized in the following scheme:

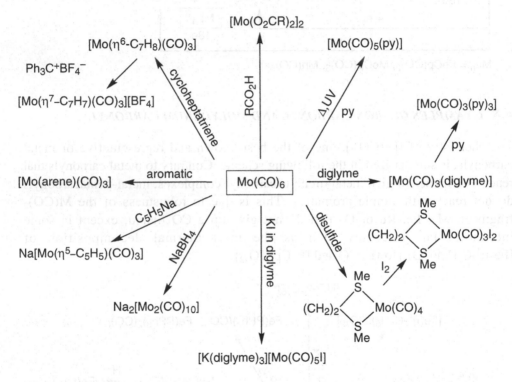

5. LIGANDS ISOELECTRONIC TO CO: CNR, CS, CSe, N$_2$ AND NO$^+$

5.1. ISONITRILES CNR

The NR fragment is isoelectronic to the oxygen atom, but less electronegative. The CNR ligands are L ligands, isoelectronic to CO, that give neutral binary metal complexes analogous to metal carbonyls such as [M(CNR)$_6$] (M = Cr, Mo, W), [M(CNR)$_5$] (M = Fe, Ru) or [M(CNR)$_4$] (M = Ni). Contrary to CO, isonitriles do not give anionic complexes, but many mono- and dicationic complexes, such as [M(CNR)$_4$]$^+$ (M = Cu, Ag, Au) and [Pt(CNR)$_4$]$^{2+}$, that do not have metal carbonyl equivalents. The abillity of CNR to bridge two metals is weaker than that of CO.

The lone pair on carbon is antibonding for the C-N bond. Upon coordination, it is given as a σ bond to the metal. The antibonding character is then suppressed, which leads to the strengthening of the C-N bond (in CO, this lone pair is not antibonding for the C=O bond). The π backbonding exists as in metal carbonyls and reduces the strength of the C-N bond. These two effects, σ and π, are opposite to each other.

The C-N bond is either stronger than in non-coordinated CNR, or weaker, depending on the effect that dominates, whereas the C-O bond is always weaker in metal-carbonyl complexes than in free CO. The infrared absorption frequency is thus either lower than that of free isonitrile (2136 cm^{-1} for p-$CH_3C_6H_4NC$), for instance 2065 cm^{-1} with [$Ni(CNR)_4$] or higher as in [$Ag(CNR)_4$]$^+$ (2186 cm^{-1}).

The M-C-N-R chain is usually linear, which shows the delocalization of the lone pair of the nitrogen atom and the essential weight of the zwitterionic mesomer form. When the π backbonding is strong with a basic metal, however, it is possible to find a bent isonitrile ligand, i.e. in which the nitrogen lone pair is not delocalized, as for instance in [$NbCl(CO)(CNR)(dmpe)_2$].[7.7]

$$\overset{\ominus}{M}-C\equiv\overset{\oplus}{N}-R \qquad\qquad M=C=N\diagdown_R$$

a   b

The isonitrile metal complexes can be synthesized by reactions of isonitriles with metal carbonyls (substitution reactions) or, more rarely, by alkylation of metal-cyanide complexes.

Thus, the isonitrile ligand easily undergoes insertion reactions, even into metal-hydride bonds. The nucleophilic additions onto the ligand carbon atom are also facile. For instance, amines readily add to yield diaminocarbene complexes:

$$[Pd(CNMe)_4]^{2+} \quad\xrightarrow[\text{2) HBF}_4]{\text{1) H}_2\text{NNH}_2}\quad$$

5.2. THIO- AND SELENOCARBONYL CS AND CSe

The complexes containing these ligands are rare and usually do not contain more than one or two of these ligands CS and CSe. The only known binary complex is [$Ni(CS)_4$]. The σ donor as well as the π acceptor properties increase in the following order:

$$CO < CS < CSe.$$

CS and CSe being not stable except at very low temperature in matrixes, the complexes are most often synthesized from CS_2 or CSe_2. The S or Se atom is abstracted from a CS_2 or CSe_2 intermediate using PPh_3:

$$[MCp(CO)_2] + CS_2 + PPh_3 \longrightarrow [MCp(CO)_2(CS)] + SPPh_3$$

M = Mn, Re

Another route consists in letting react a mono- or binuclear carbonylmetallate dianion with thiophosgene:

$$[Fe(CO)_4]^{2-}(Na^+)_2 + Cl_2CS \longrightarrow [Fe(CO)_4CS] + 2\,Na^+Cl^-$$

The CS is an excellent bridging ligand. In binuclear complexes containing both CO or NO and one or two CS ligands, the latter preferentially occupies the bridging positions:

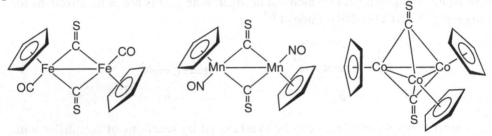

The infrared absorption frequencies are respectively 1273 cm^{-1} for free CS, between 1160 and 1410 cm^{-1} for terminal CS, between 1100 and 1160 cm^{-1} for μ_2-CS and between 1040 and 1080 cm^{-1} for μ_3-CS.[7.8,7.9]

5.3. DINITROGEN N_2

The dinitrogen molecule is inert because of the very strong bond between the two atoms. Only strongly reducing metals such as Li and Mg were known to reduce N_2, giving nitrides:

$$6\,Li + N_2 \longrightarrow 2\,Li_3N$$

Since 1965, many N_2 complexes with transition metals have been synthesized,[7.10-7.12] and the chemists have attempted, with modest success, to mimic nitrogenase enzyme reducing N_2 to NH_3 under ambient conditions.[7.12]

N_2, formally isoelectronic to CO, is an L ligand, but the energy of the lone pair engaged towards the metal in the σ bond is lower than that of carbon in CO complexes, because nitrogen is more electronegative than carbon. The result is that this σ bond is much less efficient than in CO complexes (N_2 is a poor σ donor). The π^* orbital that accepts the π backbonding from the d metal orbital is of lower energy than that of CO, which is favorable for a good π backbonding. This orbital is equally distributed on the two nitrogen atoms, however, and therefore a less good overlap with the d metal orbital is offered than the π^* orbital in CO that is located on carbon. The π backbonding being dominant, only the very basic metals can give N_2 complexes. The monometallic M-N≡N and the bimetallic complexes M-N≡N-M are known:

$$[MoCl_3(THF)_3] \xrightarrow[\text{dppe}]{\text{Mg, THF, } N_2} [Mo(N_2)(dppe)_2]$$

$$[Fe(H_2)(H)_2(PPh_2Et)_3] \xrightarrow{N_2} [Fe(N_2)(H)_2(PPh_2Et)_3]$$

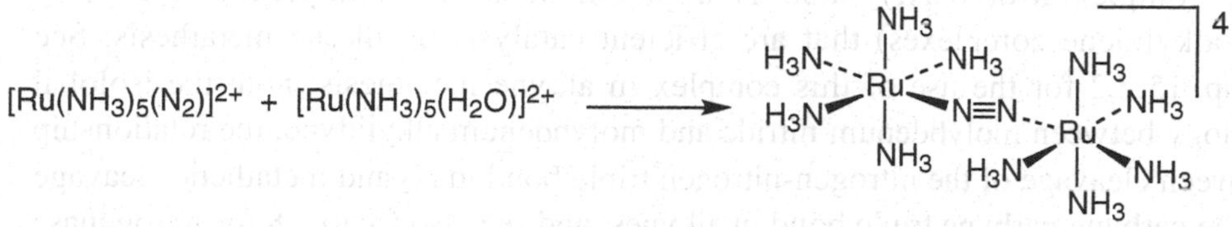

$$[Ru(NH_3)_5(N_2)]^{2+} + [Ru(NH_3)_5(H_2O)]^{2+} \longrightarrow$$

When the π backbonding becomes very (too) important, it is necessary to change the formalism and to adopt the writing M-N=N-M, as in the following complex:

When the π backbonding becomes very (too) important, it is necessary to change the formalism and to adopt the writing M-N=N-M, as in the following complex:

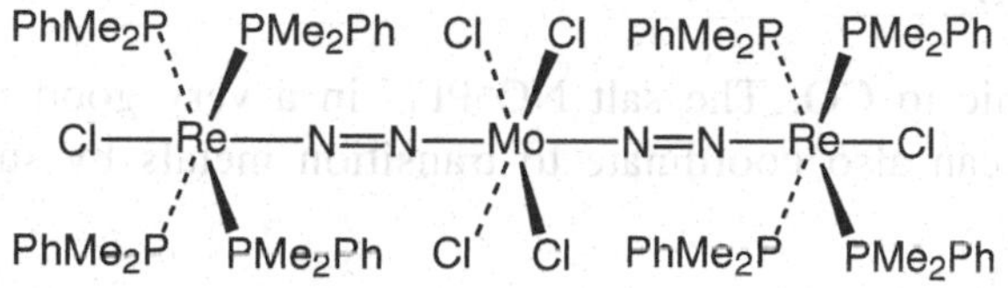

The complexation of N_2 polarizes the molecule:

$$M-N≡N \longleftrightarrow M=N^+=N^- \longleftrightarrow M^+-N=N^-$$

The consequences of this polarization are the following:

▸ the presence of a strong infrared band between 1920 and 2150 cm^{-1} due to the elongation of N-N, whereas the free molecule does not absorb in the infrared region because of its symmetry, but only in Raman. In the M-N=N-M complexes, this frequency is lower (for instance 1680 cm^{-1} in the above trimetallic complex).

▸ the reactivity of the terminal nitrogen atom towards electrophiles, in particular the proton. Only the very basic metals give rise to this reaction, because they sufficiently polarize the dinitrogen ligand. Thus, this protonation is only possible in a minority of complexes:

$$[W(N_2)_2(dpe)_2] + 2\,HCl \longrightarrow [WCl_2(=N-NH_2)(dpe)_2] + N_2$$

There are rare examples of dinitrogen complexes that can be protonated to NH_3. In these complexes, the metal is in a low oxidation state that can be raised to a high oxidation state.[7.12b]

$$[W(N_2)_2(PMe_2Ph)_4] \xrightarrow{H_2SO_4,\ MeOH} W^{VI} + 3\,NH_3 + N_2$$

Cummins complex $Mo(NRAr)_3$ is an interesting one, because it allows both the formation of a dinitrogen complex and the cleavage of the very strong nitrogen-nitrogen bond in N_2 ($942\ kJ \cdot mol^{-1}$) at room temperature to form two nitrogen-molybdenum triple bonds ($680\ kJ \cdot mol^{-1}$ each):[7.12c]

$$N_2 + Mo(NRAr)_3 \longrightarrow (ArN)_3Mo—N\equiv N—Mo(NRAr)_3$$

red-orange   purple

$$\xrightarrow{RT} \quad 2\ N\equiv Mo(NRAr)_3$$

golden

The complex $Mo(NRAr)_3$ also is a precursor of Mo complexes (including Mo-alkylidene complexes) that are efficient catalysts of alkyne metathesis. See Chap. 15.2.2 for the use of this complex in alkyne metathesis; note the isolobal analogy between molybdenum nitride and molybdenum alkylidyne, the relationship between cleavage of the nitrogen-nitrogen triple bond in N_2 and metathetic cleavage of the carbyne-carbyne triple bond in alkynes, and see also Chap. 18 for nitrogenase and the proposed mechanism for nitrogen fixation.

5.4. NITROSYL NO

NO^+ is isoelectronic to CO. The salt $NO^+PF_6^-$ in a very good monoelectronic oxidant, but NO^+ can also coordinate to transition metals by substitution of a carbonyl ligand:

$$[FeCp_2] + NO^+PF_6^- \xrightarrow{CH_2Cl_2} [FeCp_2]^+ PF_6^- + NO$$

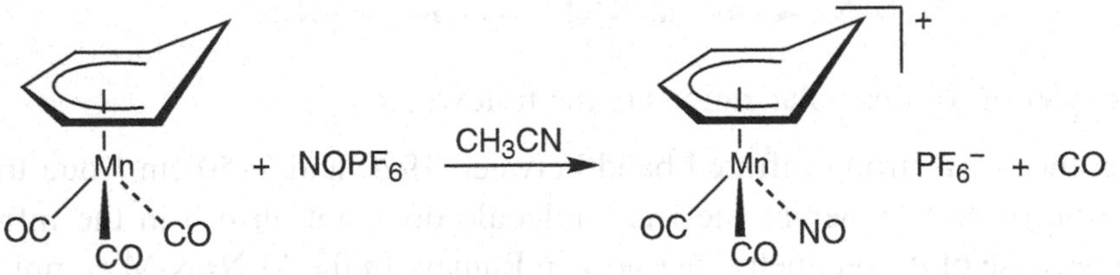

In the latter case, the starting complex and the reaction product are isoelectronic (although the formal oxidation states are different). The interest in the above reaction is that the cyclohexadienyl ligand can undergo a nucleophilic addition in the cationic reaction product to give an *exo*-substituted cyclohexadiene complex, whereas the neutral starting compound does not react, because it is not enough electrophilic (see Chap. 4).

In the above NO complex, NO is a 3-electron LX ligand in which the Mn-N-O chain is linear. This is the same for many other complexes such as $[Cr(NO)_4]$, $[Fe(CO)_2(NO)_2]$ and $[Co(CO)(NO)]$. The NO ligand is also a 1-electron X ligand when the 18-electron rule dictates it. Then, the chain is bent with an angle between 120 and 140° in order to allow the nitrogen lone pair out of the metal coordination sphere. The 18-electron rule usually imposes the direction of the equilibrium:

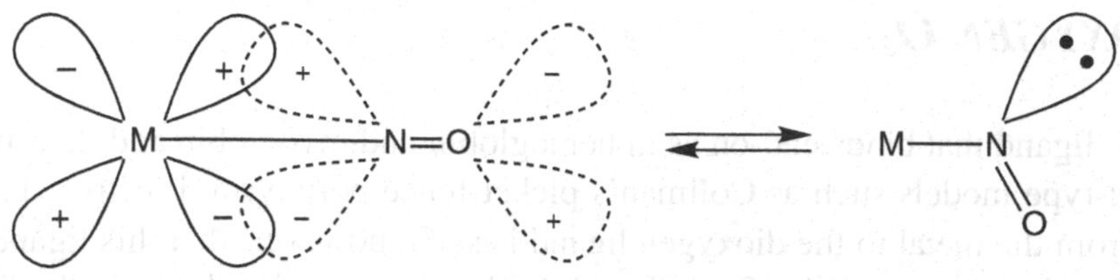

Between 16 and 18 valence electrons, one cannot decide for sure, however, because some 16-electron complexes are known with a bent NO ligand.

Example:

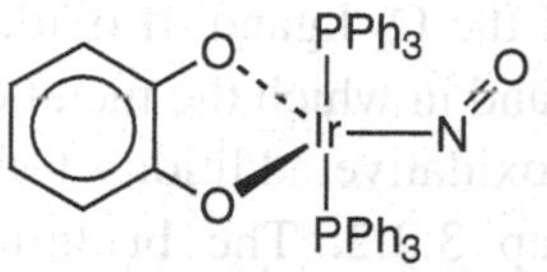

The distinction between the two coordination modes cannot be done using infrared spectroscopy, although bent NO has a lower infrared frequency than linear NO, because the regions of absorption of these two ligand modes overlap. The distinction can be made using ^{15}N NMR (deshielding from 350 to 700 ppm for linear NO) or by X-ray diffraction.

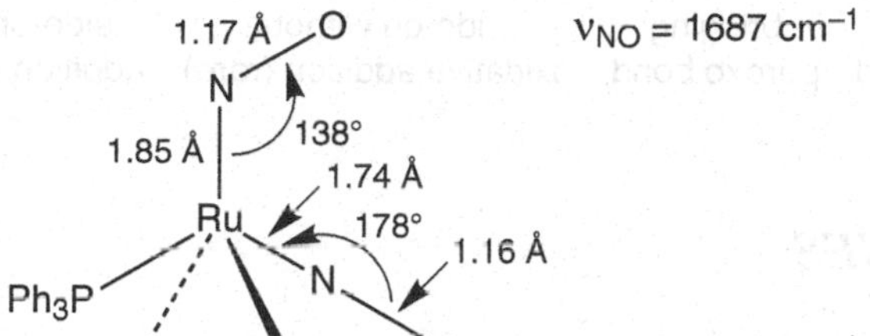

$\nu_{NO} = 1687\ cm^{-1}$

$\nu_{NO} = 1845\ cm^{-1}$

NO can also be a 3-electron bridging LX ligand as in $[CrCp(NO)(\mu_2\text{-}NO)]_2$. The bridging NO ligand ($\nu_{NO} = 1518\ cm^{-1}$) is easily distinguished from terminal NO ($\nu_{NO} = 1677\ cm^{-1}$) by its very low infrared frequency.[7.13]

As CO, coordinated NO can easily undergo nucleophilic attacks in some cationic complexes, whereas alkylation of anionic NO complexes can be followed by insertion into the metal-alkyl bond induced by an external potential ligand:

$$[CoCp(NO)]^- \xrightarrow{\ RI\ } [CoCp(R)(NO)] \xrightarrow{\ PPh_3\ } [CoCp(NOR)(PPh_3)]$$

The importance of NO gas as a neurotransmitter in mammalian brain has been revealed in 1990, which could greatly increase the interest in NO coordination chemistry. Also, the complex $Na_2[Fe(CN)_5(NO)]$ has been used for a long time to reduce the pressure during surgical operations.

6. DIOXYGEN O₂

O_2 is an L ligand that binds end-on as in hemoglobin and myoglobin and their bulky porphyrin-type models such as Collman's picket-fence porphyrin. Electron density transfer from the metal to the dioxygen ligand is such, however, that this ligand can be considered as a superoxide ($O_2^{\bullet -}$) ligand. In the absence of bulk above the $Fe-O_2$ site, this species irreversibly binds to Fe-porphyrin to give a μ_2-peroxo dimer intermediate [porphyrin-Fe^{III}-O-O-Fe^{III}-porphyrin] that decomposes to [porphyrin-Fe^{III}-O-Fe^{III}-porphyrin]. Side-on O_2 complexes are also known ($M-\eta^2-O_2$), most often with oxidative addition of the O_2 ligand. If oxidative addition occurs, the O_2 ligand becomes a peroxo X_2 ligand in which the metal oxidation state has increased by two units. The more facile oxidative addition of side-on O_2 than ethylene has already been stressed in Chap. 3.3.2. The bridging modes $M-\eta^1, \mu^2-O_2$ and $M-\eta^2, \mu^2-O_2$ are also known.

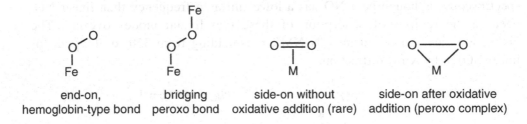

| end-on, hemoglobin-type bond | bridging peroxo bond | side-on without oxidative addition (rare) | side-on after oxidative addition (peroxo complex) |

7. PHOSPHANES

The phosphanes PR_3 are divided into phosphines (R = alkyl or aryl) and phosphites (R = OR'). The phosphines are among the most important ancillary ligands in inorganic and organometallic chemistry,[7.14] in particular in catalysis. The most frequent and less expensive ones are PPh_3 and dppe. Indeed, it is possible to control the stereoelectronic properties of the phosphines by choosing the electronic and bulk properties of the R groups. Chiral phosphines play a central role in asymmetric catalysis, phosphorus being a stereochemically stable atom, contrary to nitrogen in amines. The phosphines are good σ donors, *via* the lone pair of the phosphorus atom and mediocre π acceptors at least for the most common phosphines (i.e. trialkyl and triaryl phosphines). It is the σ^*(P-R) orbital that is the π acceptor. With electron-withdrawing substituents on phosphorus, the energy of the σ^*(P-R) orbital is lower than in trialkyl- and triarylphosphines. The consequence is that phosphines PX_3 (R = X: halogen), and to a lesser extent the phosphites, are excellent π acceptors. For example, PF_3 is as good π acceptor as CO, and binary 18-electron complexes are known such as [$M(PF_3)_4$], M = Ni, Pd or Pt. The π acceptor properties follow the order:

$$PMe_3 = P(NR_2)_3 < PAr_3 < P(OMe)_3 < P(OR)_3 < PCl_3 < PF_3 = CO.$$

We already know that the infrared frequencies of the CO absorption are excellent criteria of comparison (see section 2). Another criterion is provided by the pK_a values of the conjugated acids HPR_3^+ in the following table established by Tolman.[7.7] They confirm that the trialkylphosphines are better donors than triarylphosphines, and that a tertiary phosphine ligand is a better donor than the corresponding secondary (and *a fortiori* primary) phosphine.

These pK_a are 2.5 units lower than those of the corresponding ammoniums, i.e. the amines are better donors than phosphines (because of the absence of π backbonding in amine complexes).

pK_a values of the conjugated acids HPR_3^+ of the phosphines PR_3

Phosphine	pK_a	Phosphine	pK_a
PCy_3	9.70	PMe_2Ph	6.50
PMe_3	9.65	PCy_2H	4.55
PEt_3	8.69	PBu_2H	4.51
PPr_3	8.64	PPh_3	2.73
PBu_3	8.43	$P(CH_2CH_2CH)_3$	1.37
$P(i\text{-}Bu)_3$	7.97	PPh_2H	0.03
$P(CH_2CH_2Ph)_3$	6.60	$PBuH_2$	−0.03

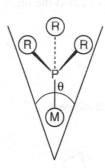

The other important property of phosphines is their steric bulk defined by Tolman[7.7] as the cone angle starting from the metal to include all the space occupied by the substituents.

Phosphane cone angles

Phosphane ligand	Cone angle θ	Phosphane ligand	Cone angle θ
PH_3	87	PH_2Ph	101
$P(OCH_2)_3CR$	101	PF_3	104
$P(OMe)_3$	107	PMe_3	118
PMe_2Ph	122	$PPh_2PCH_2CH_2PPh_2$	123
PEt_3	132	PPh_3	145
$PPh_2(t\text{-}Bu)$	157	PCy_3	170
$PPh(t\text{-}Bu)_2$	170	$P(t\text{-}Bu)_3$	182
$P(C_6F_5)_3$	184	$P(mesityl)_3$	212

The ability of phosphanes to dissociate from a metal center is directly related to their cone angle:

$$PMe_3 < PMe_2Ph < PMePh_2 < PEt_3 < PPh_3 < P(i\text{-}Pr)_3 < P(Cy)_3.$$

This modulation is of key importance in catalysis in order to control the liberation of a phosphine from a metal center and create a vacant site for the activation of substrates.

^{31}P NMR (spin: 1/2; isotopic abundance: 100%) is a very useful tool allowing to study complexes of phosphorus ligands although the relationship between the chemical shift and the electronic density or the steric bulk is not obvious for phosphanes.

Phospha-alkenes $RP=CR'_2$ can give σ complexes (P ligands) and π complexes (the ligand is the double bond). Interconversion is sometimes observed between these two types of complexes as in $[Pt(PPh_3P)_2(ArP=CPh_2)]$ due to the low energy difference between the σ and π phosphorus orbitals. π complexes of diphosphenes $RP=PR$ are also known in which the P atoms are electrophilic.

Phosphorus-containing heterocycles are a special class of phosphorus L ligands that has been developed by Mathey, in particular phosphirene, 1*H*-phosphole, 2*H*-phosphole, phosphinine and phosphanorbornadiene. The corresponding chelating ligands BIPNOR and 2,2'-biphosphinine are known as well and have proven to be useful in catalysis as also the organometallic phosphorus ligands (phosphametallocenes).

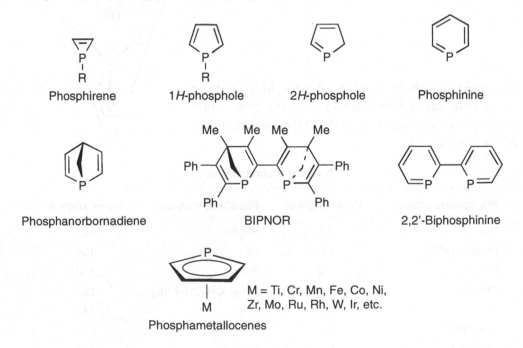

Phosphirene   1*H*-phosphole   2*H*-phosphole   Phosphinine

Phosphanorbornadiene   BIPNOR   2,2'-Biphosphinine

M = Ti, Cr, Mn, Fe, Co, Ni, Zr, Mo, Ru, Rh, W, Ir, etc.

Phosphametallocenes

8. WEAK L LIGANDS AND THEIR ROLE TO GENERATE ORGANOMETALLIC LEWIS ACIDS

When a coordination site is vacant, one can say that the metal has Lewis-acid properties (a term borrowed from organic chemistry) essential for polymerization and more generally in catalysis. This property will be more or less pronounced depending on the energy of the d metal orbital. If the metal has a strong tendency to complete its coordination sphere even with weak ligands, the coordination can occur with very weak ligands such as N_2, a solvent molecule through the lone pair of an heteroatom (ether, nitrile, chlorinated solvents) or even so-called "non-coordinating" anions (BF_4^-, PF_6^-, etc.). This situation is very much looked for, precisely because of the resulting catalytic properties mentioned above. Thus, chemists have proposed anions that are as little coordinating as possible (tetra-perfluoroarylborates such as $B(C_6F_5)_4^-$, Michl'permethylated carborane anion $[closo\text{-}CB_{11}Me_{12}]^-$) in order to enhance and protect the Lewis-acid properties of the metal center in these catalysts. The protonation of an 18- or 16-electron metal-alkyl complex and the hydride abstraction using reagents containing one of these special anions are the classic ways to generate active cationic organometallic Lewis-acid catalysts.

Finally let us recall that the agostic C-H-M bond and the ligands H_2 and $H\text{-}SiR_3$ also are weak ligands through their σ bond.

Note - In inorganic chemistry, the most common monodentate L ligands are water, amines, ethers, thiols and sulfides (Chap. 1.1). They are all good donors, because they are not π acceptors. Thus, they are weakly bound to metals, and, as such, also frequently used in organometallic complexes for facile substitution chemistry (see Chap. 5) and catalysis (see Parts IV and V).

SUMMARY OF CHAPTER 7
METAL CARBONYLS AND COMPLEXES
OF OTHER MONOHAPTO L LIGANDS

1 - The metal carbonyls are one of the essential inorganic families. They are known for all the transition metals as stable neutral or anionic mono- and polynuclear complexes (with NVE = 18 for the low-nuclearity complexes except [V(CO)$_6$]; Wade rule applies for high nuclearities; see Chap. 2).

2 - Synthesis: from the metals (or salt + reductant) and CO under pressure

3 - Bonding mode

Terminal CO: C lone pair given to the metal (σ bond) and π backbonding.

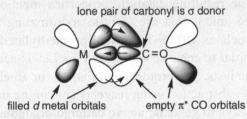

lone pair of carbonyl is σ donor

filled *d* metal orbitals   empty π* CO orbitals

The π backbonding is characterized by a weakening of the C=O bond that leads to a lowering of the infrared carbonyl absorption (1820-2150 cm^{-1}) compared to free CO (2143 cm^{-1}) and to low-field chemical shifts in ^{13}C NMR (around 200 ppm for neutral metal carbonyls).

CO bridging two or several (n) metals in bi- and polymetallic complexes (noted μ_2-CO or μ_n-CO respectively). μ_2-CO: 1750-1850 cm^{-1} ; μ_3-CO: 1620-1730 cm^{-1}.
The bridging CO mode can take various forms, and CO can migrate from one metal to the next (fluxionality).

4 - Reactions: ligand substitution, nucleophilic or electrophilic addition, insertion, oxidation or reduction, dismutation. Examples of the most common metal carbonyls: Fe(CO)$_5$ and Mo(CO)$_6$.

5 - Other L ligands: CNR, CS, CSe, N$_2$, O$_2$, PR$_3$ and P(OR)$_3$

Isonitriles CNR: the neutral and cationic binary complexes M(CNR)$_n$ are known, but not the anions.

Thiocarbonyls CS and *selenocarbonyls CSe* are instable, but stabilized by complexation. In bridged bimetallic complexes containing both CO and CS, CS is the bridging ligand.

Dinitrogen N$_2$ forms many complexes, but M-N$_2$ bonds are labile. Presence of a strong infrared band between 1920 and 2150 cm^{-1} (M-N$_2$) and 1680 cm^{-1} (M-N=N-M). A few metallic fragments are basic enough to sufficiently polarize N$_2$ for protonation and reduction to NH$_3$.

Nitrogen monoxide NO is a 3e LX ligand (M-N=O, linear) or a 1e LX ligand (M-N=O, bent), most often depending on the need of the metal (18e rule).

5 - Dioxygen O$_2$: reversibly binds hemoglobin, myoglobin and their models (bent Fe-O-O); dihapto side-on (M-η^2-O$_2$) peroxide complexes, resulting from oxidative addition of O$_2$, are also known. Superoxo- and peroxo ligands bridging two metals (μ^2) can be η^1 or η^2.

Phosphines PR$_3$ and phosphites P(OR)$_3$, L ligands, play a key role in catalysis, their steric (cone angle) and electronic (*pKa*) properties being adjustable by variations of R.

Other O, S, and N-based inorganic L ligands are only σ donors, not π acceptors, and are thus weak ligands overall. They can be easily substituted.

EXERCISES

7.1. The complex $[(FeCp(CO)_2]_2$, also called Fp_2, shows infrared carbonyl absorptions bands at 1760 and 2000 cm^{-1}; what can be deduced for its structure?

7.2. Fp_2 is in equilibrium in solution with an isomeric structure of close, although slightly higher energy? What is this structure?

7.3. How would you call structures such as Fp_2 for which the equilibrium of question 2 is easily achieved? What is the consequence for the carbonyl ligands?

7.4. The photolysis of Fp_2 leads to the loss of one carbonyl ligand, and the new complex obtained shows only one carbonyl infrared band at 1760 cm^{-1}. What is the structure of this new complex?

7.5. Heating Fp_2 to 180°C leads to a tetranuclear complex that has lost 2 CO per Fp_2 complex and shows a strong band at 1620 cm^{-1} in the infrared spectrum. What is its structure, the number of valence electrons of each of the four Fe atoms and the number of electrons of the cluster (NCE)?

METAL-ALKYL AND -HYDRIDE COMPLEXES AND OTHER COMPLEXES OF MONOHAPTO X LIGANDS

1. DIFFERENT TYPES OF sp^3 METAL-CARBON BONDS

The bond between tetrahedral carbon and transition metals mainly exists in the form of terminal metal-alkyl complexes such as $[W(CH_3)_6]$.[8.1] Metal-alkyl complexes are now numerous. It has been understood that they are stable if the β-climination path is blocked either because the complexes have 18 valence electrons,[8.2] or have less valence electrons, but no β hydrogen.[8.1] Complexes of other tetrahedral carbon are also known in which two, three or four metal substituents are metal fragments. They are respectively bridging alkylidene, alkylidyne and carbide complexes, below:

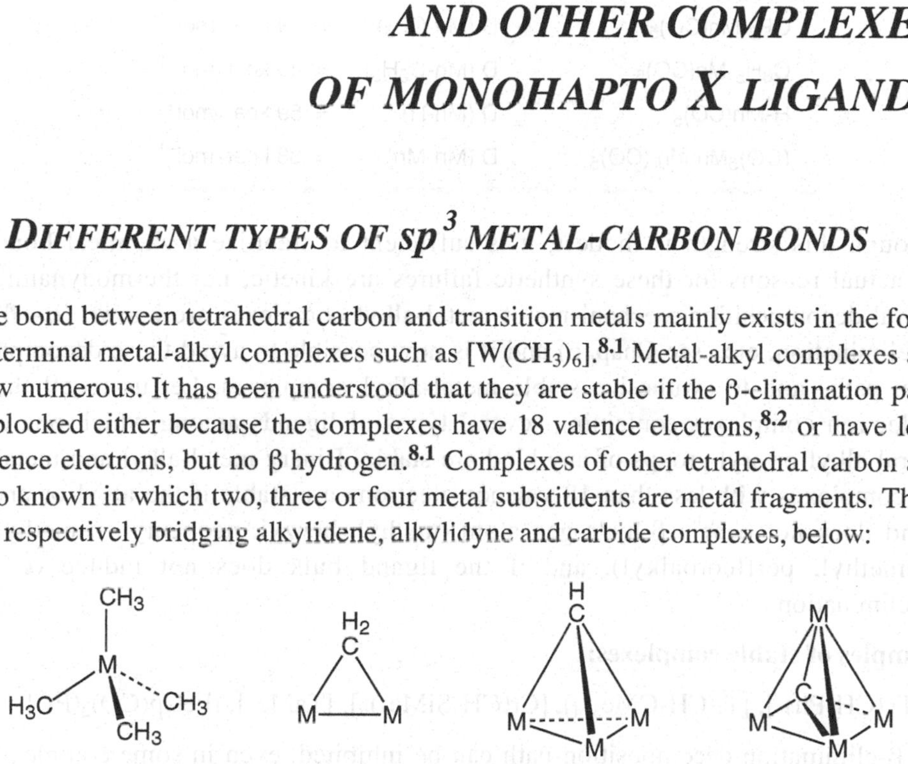

Finally, an alkyl group can also occasionally bridge two or three metals in various modes. Metal-alkyl complexes are examined in this chapter, terminal and bridging alkylidene and alkylidyne complexes being studied in the following chapter.

2. STABILITY OF METAL-ALKYL COMPLEXES

The interest of chemists for transition-metal-alkyl complexes in the 1950s [8.2a] was partly due to the fact that attempts to synthesize them had failed during the first half of the XXth century. These failures had been attributed for a long time to the supposed intrinsic weakness of the metal-carbon σ bond. It is now known that the strengths of these bonds are in the range between 30 and 65 kcal·mol^{-1},[8.3] an increase of strength being observed upon going down in the periodic table as well as with perfluorinated ligands such as C_6F_5.[8.4]

Dissociation energy D of various M-C and some other M-elements bonds

PhCH$_2$-Mn(CO)$_5$	D (Mn-CH$_2$Ph)	= 29 kcal·mol^{-1}
Ti(CH$_2$Ph)$_4$	D (Ti-C)	= 62 kcal·mol^{-1}
Zr(CH$_2$Ph)$_4$	D (Zr-C)	= 74 kcal·mol^{-1}
CH$_3$-Mn(CO)$_5$	D (Mn-CH$_3$)	= 36 kcal·mol^{-11}
CH$_3$-Re(CO)$_5$	D (Re-CH$_3$)	= 52 kcal·mol^{-1}
CF$_3$-Mn(CO)$_5$	D (Mn-CF$_3$)	= 49 kcal·mol^{-1}
C$_6$H$_5$-Mn(CO)$_5$	D (Mn-C$_6$H$_5$)	= 49 kcal·mol^{-1}
H-Mn(CO)$_5$	D (Mn-H)	= 59 kcal·mol^{-1}
(CO)$_5$Mn-Mn (CO)$_5$	D (Mn-Mn)	= 38 kcal·mol^{-1}

Although this strength is modest, it is sufficient to stabilize a chemical bond. The actual reasons for these synthetic failures are kinetic, not thermodynamic. Several decomposition mechanisms of metal-alkyl complexes are known (α-, β-, γ-H elimination, etc., see Chap. 6) and it is now possible to avoid these decomposition paths and to synthesize stable metal-alkyl complexes. Let us recall that 18-electron complexes containing several types of ligands among which one or several alkyl or aryl groups of any kind are stable. Binary metal-alkyl and metal-aryl complexes with less than 18 valence electrons are stable if the alkyl or aryl ligand does not bear a β hydrogen atom (methyl, benzyl, neopentyl, trimethyl-silylmethyl, perfluoroalkyl), and if the ligand bulk does not induce α or γ-H elimination.

Examples of stable complexes:

[Ta(CH$_2$Ph)$_5$], [Ti(CH$_2$CMe$_3$)$_4$], [Cr(CH$_2$SiMe$_3$)$_6$], [TaMe$_5$], [FeCp(CO)$_2$(Et)].

The β-elimination decomposition path can be inhibited, even in some complexes that have both less than 18 valence electrons and β hydrogen atoms, when the steric constraints inhibit the approach of the β H such that it be coplanar with the M-C-C plane: this is the case for the following complexes that are stable:

Cr(1-adamantyl)$_4$ Nb(6-norbornyl)$_4$ PtL$_2$CHCH$_2$CH$_2$CH$_2$

PtH(C≡H)L$_2$ Cr(CMe$_3$)$_4$

We also know from Chap. 3 that reductive elimination in 18-electron metal-alkyl-hydride complexes yielding an alkane and lower oxidation-state complexes is favorable for the first line of the transition metals if this lower oxidation state is accessible.

There are still some factors that stabilize metal-alkyl complexes:

The agostic C-H-M bond

Discovered by M. Brookhardt and M.L.H. Green,[8.5] the agostic bond formally consists in using the electron pair of a C-H bond of one of the pre-coordinated ligands as an additional L ligand if a d metal orbital is empty, i.e. if the metal has otherwise less than 18 valence electrons. The entropic factor allowing to form stable agostic complexes is essential here, because otherwise C-H bonds of alkanes are too bulky to form stable alkane complexes whereas H_2 complexes are numerous. This difficulty is due to the unfavorable steric bulk around the tetrahedral carbon and the high energy of the σ^* orbital of the C-H bond (see Chap. 8.6). Agostic bonds are structurally characterized by X-ray diffraction, the β carbon atom being located closer to the metal than expected if the agostic bond was absent. The β hydrogen interaction with the metal, often difficult to see among the X-ray data, is sometimes characterized by neutron diffraction of a deuterium-labelled complex.

Example of agostic bond:

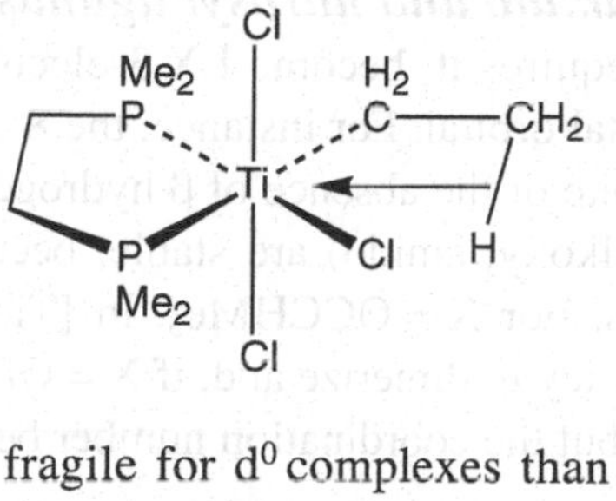

$$
\begin{array}{c}
Me_2 \quad\quad \overset{\displaystyle Cl}{\underset{\displaystyle|}{}} \quad H_2 \\
P\cdots \quad\quad C\!-\!CH_2 \\
\quad\quad\;\; Ti \longleftarrow \\
P \quad\quad Cl \quad H \\
Me_2 \quad\quad \underset{\displaystyle Cl}{\overset{\displaystyle|}{}}
\end{array}
$$

The agostic bond is more fragile for d^0 complexes than for d^n complexes ($n \neq 0$), because the former do not have available d electrons for π backbonding into the π^* orbital of the C-H bond. The agostic bond is also found in binuclear complexes in which an alkyl group such as CH_3 dissymmetrically bridges two metals. The methyl is an X ligand for the first metal and gives, as an L ligand, the doublet of one of its C-H bonds to the second metal:

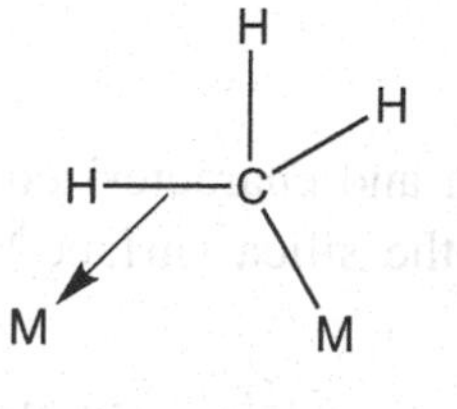

$$
\begin{array}{c}
H \\
| \\
H\!-\!C\!-\!\overset{\displaystyle H}{} \\
\swarrow \quad \searrow \\
M \quad\quad M
\end{array}
$$

The coordination of an instable metal-alkyl complex such as $[TiMe_4]$ by L, or better, L_2 ligands (bipy, dppe), allows to reach a NVE closer to 18 and a coordination number 5 or 6, which stabilizes the complex.

Hexacoordination is a very favorable factor for the stability of a complex, which compensates other eventually unfavorable factors. Thus $[WMe_6]$ (trigonal prismatic) is stable, despite NVE = 12 only (compare to the situation of octahedral $[V(CO)_6]$, stable in spite of NVE = 17).

The benzyl ligands, in the complex [Ti(CH$_2$Ph)$_4$], form angles Ti-C-C from 84 to 86° only instead of 109° awaited for a *sp^3* carbon. This shows that titanium interacts with the phenyl to formally accept a doublet of π electrons in the closest proximity from the phenyl sextet of π electrons. This allows to increase the NVE that, with only 4 X ligands, would otherwise only be 8. Each benzyl ligand acts independently in the same way.

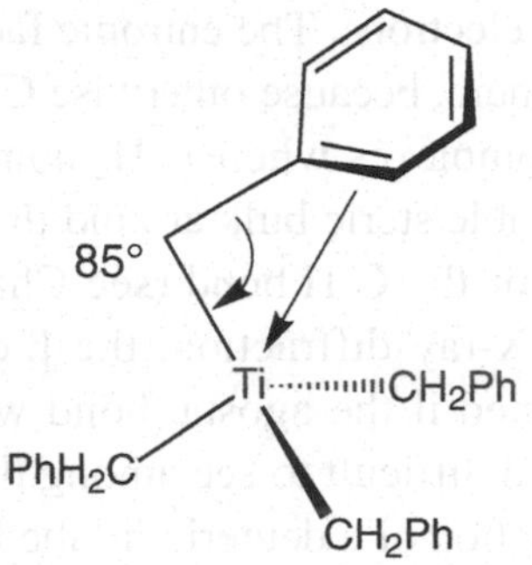

The halogeno, alkoxy, amido and nitrosyl ligands are, as alkyls, X ligands, but they can, if the NVE requires it, become LX 3-electron ligands by giving the *p* lone pair to a vacant *d* metal orbital. For instance, the 8-electron complex [TiMe$_4$] decomposes at – 40 °C in spite of the absence of β hydrogen whereas the complexes [TiMeX$_3$] (X = halogeno, alkoxy, amido) are stable, because they have, with this count, 14 valence electrons. For X = OCCHMe$_3$, in [TiMeX$_3$], the complex is a monomer with a weak tendency to dimerize and, if X = OEt, the complex exists as a dimer with also NVE = 14, but the coordination number becomes 5:

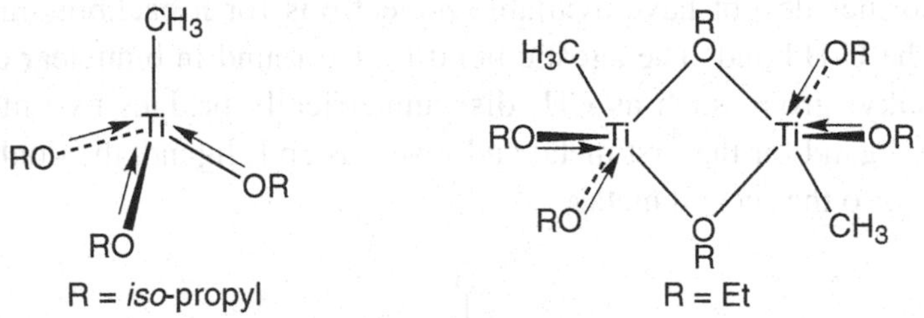

Likewise, Basset has discovered and characterized *inter alia* hydride complexes of the type [Zr(H)(OSiO-)$_3$] at the silica surface [8.7] for which the NVE can be considered to be 14.

There are also many stable binary complexes with these ligands, their number being sometimes restricted.[8.8] For instance Zr(O*t*-Bu)$_4$ and Mo(NMe$_2$)$_4$ are respectively 16-electron and 18-electron complexes (and not 8- and 10-electron complexes, respectively!), which is better in accord with their stability. With this LX mode, the chains M-N-R and M-N-O are linear in monometallic complexes and the bond angle M-O-R tends to open up to 180°. The alkoxy ligands,[8.8] as the alkyls, undergo β-elimination under the same conditions as the latter (β-H atom and less than 18 valence electrons). Thus, the transition metal is a good oxidant of primary alcohols to aldehydes and of secondary alcohols to ketones, especially in the presence of a base, and the alcohols are good reductants of transition-metal complexes to transition-metal hydrides:

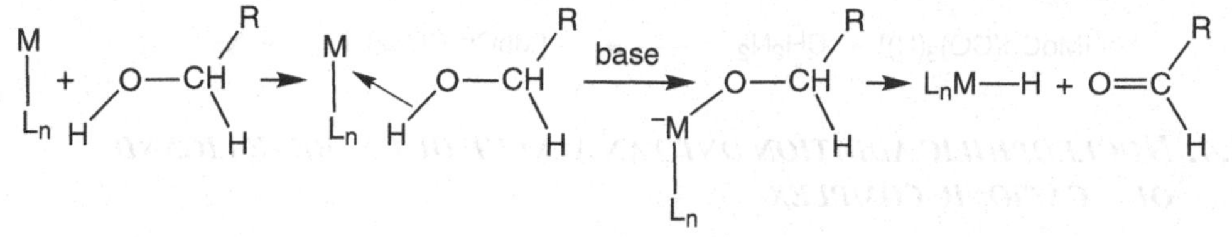

3. SYNTHESIS OF METAL-ALKYL COMPLEXES

3.1. METAL-HALIDE + MAIN-GROUP ORGANOMETALLIC (TRANSMETALLATION)

A Grignard reagent or an organometallic alkali is generally used if there is no risk of electron transfer leading to the unproductive reduction of the transition metal. For instance, with Nb^V and Ta^V, it is indispensable to use a very soft alkylating reagent (zinc, cadmium or tin reagent) in order to avoid the reduction of the metal penta-halide. After two or three alkylation reactions, the risk of side reduction does not exist any longer, and it is then necessary, on the contrary, to use a powerful alkylating agent in order to carry out additional alkylation reactions.[8.1,8.7,8.8]

$$[TiCl_4] + 4\ PhCH_2MgCl \longrightarrow [Ti(CH_2Ph)_4] + 4\ MgCl_2$$

$$TaCl_5 + [Zn(Mes)_2] \longrightarrow [Ta(Mes)_2Cl_3] + ZnCl_2$$

$$[Ta(Mes)_2Cl_3] + Li(Mes) \longrightarrow [Ta(Mes)_3Cl_2] + LiCl$$

3.2. METALLATE OR 16-ELECTRON COMPLEX + ALKYL HALIDE OR ALKANE (OXIDATIVE ADDITION)

$$[FeCp^*(CO)_2]_2 \xrightarrow{Na/Hg} 2\ [FeCp^*(CO)_2]^-\ Na^+ \xrightarrow{CH_3OCH_2Cl} 2\ [Fe(Cp^*CO)_2(CH_2OCH_3)]$$

$$[FeCp(CO)_2]^-\ Na^+ + C_6F_6 \longrightarrow [FeCp(CO)_2(C_6F_5)] + Na^+F^-$$

The above reaction is a nucleophilic aromatic substitution facilitated by the electron-withdrawing effect of the five fluorine atoms on the phenyl ring.

$$[Ir^IBrL_2(CO)] + CH_3Cl \xrightarrow{trans\ addition} [Ir^{III}(Br)L_2(CO)(CH_3)(Cl)]$$

Oxidative addition of alkanes with reactive late third-row transition-metal complexes gives stable alkyl-hydride complexes:

$$[IrCp^*(H)_2(PMe_3)] \xrightarrow{CMe_4,\ h\nu} [IrCp^*(H)(CH_2CMe_3)(PMe_3)] + H_2$$

3.3. OLEFIN AND CARBENE INSERTION INTO METAL HYDRIDES

$$[FeCp(CO)_2(H)\,] + CH_2=CH-CH=CH_2 \longrightarrow [FeCp(CO)_2(CH_2-CH=CH-CH_3)]$$

$$[MoCp(CO)_3(H)] + CH_2N_2 \longrightarrow [MoCp(CO)_3(CH_3)] + N_2$$

3.4. NUCLEOPHILIC ADDITION ONTO AN ALKENE OR CARBENE LIGAND OF A CATIONIC COMPLEX

$$[FeCp(CO)_2(CH_2=CMe_2)]^+ + NaBH_4 \longrightarrow [FeCp(CO)_2(t\text{-Bu})]$$

$$[ReCp(NO)(PPh_3)(=CH_2)]^+ + EtLi \longrightarrow [ReCp(NO)(PPh_3)(CH_2CH_2CH_3)]$$

4. PROPERTIES OF METAL-ALKYL COMPLEXES

The binary complexes $[MR_n]$ have oxidation states reaching the NNBE, i.e. high. The metal-alkyl complexes are easily characterized in ^{13}C NMR by the ^{13}C chemical shift of the carbon bonded to the metal at higher field than for the corresponding alkane. A coupling between this α carbon or the H atoms to which it is bonded and the metal, and also eventually with the ^{31}P nucleus (100%) of phosphorus ligands, is observed with some metals that have a spin (^{103}Rh, 100%; ^{183}W, 14%; ^{187}Os, 1.6%; ^{195}Pt, 34%; ^{199}Hg, 17%).

We have seen that the main reason for the instability of metal-alkyl complexes is elimination of an α-, β- or γ-H atom. The two other properties of d^n metal-alkyl complexes ($n \neq 0$) are:

▸ **The insertion of CO, NO, SO₂, olefins and alkynes** in metal-alkyl bonds (see Chap. 6). Dioxygen can also insert:

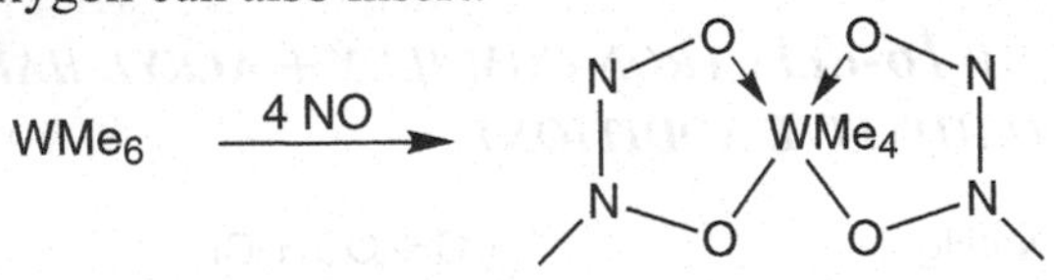

$$[WMe_6] \xrightarrow{O_2} [W(O)Me_6]$$

▸ **The reaction with electrophiles,** often is an oxidative addition followed by a reductive elimination with the alkyl ligand (impossible with d^0 complexes). With the proton, the alkane is formed and, with functional electrophiles, the alkyl ligand can be functionalized in this way (see Chap. 3).

$$[Fe(CO)_4(CH_3)]^- \, Na^+ + PhCOCl \longrightarrow \text{"}Fe(CO)_4(CH_3)(COPh)\text{"} + Na^+, Cl^-$$

$$\longrightarrow PhCOCH_3 + \text{"}Fe(CO)_4\text{"}$$

$$FeCp(CO)_2(C_2H_5) + H^+ \, BF_4^- \xrightarrow{Et_2O} \text{"}[FeCp(CO)_2(C_2H_5)(H)][BF_4]\text{"}$$

$$\longrightarrow [FeCp(CO)_2(Et_2O)][BF_4] + C_2H_6$$

$$[FeCp(CO)_2(C_2H_5)] + Br_2 \longrightarrow [Fe(Cp(CO)_2Br] + C_2H_5Br$$

The main property of metal-alkyl complexes is the metathesis of σ bonds encountered in the hydrogenolysis, protonation and reaction with methane (see Chap. 3).

5. METALLOCYCLES

Metallacycloalkanes are a particular class of metal-alkyls, but they can be synthesized in the same way as the ordinary alkyl complexes by using classic methods:

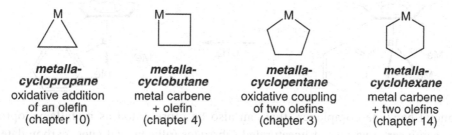

Other specific synthetic routes have been described in the chapter mentioned below the following formula:

metalla-cyclopropane	metalla-cyclobutane	metalla-cyclopentane	metalla-cyclohexane
oxidative addition of an olefin (chapter 10)	metal carbene + olefin (chapter 4)	oxidative coupling of two olefins (chapter 3)	metal carbene + two olefins (chapter 14)

▸ The metallocycles have specific properties in addition to those indicated above for the metal-alkyl complexes:
   a. the reaction giving rise to their formation is reversible, and it is possible to get back the starting products;
   b. their two major modes of evolution other than (**a**) are (see Chaps 9.1 and 15):

▸ Reductive elimination to give the cycloalkane (for instance, metallacyclopentane gives cyclobutane, etc.)

▸ β-elimination followed by reductive elimination to give the terminal olefin (for instance, metallacyclopentane gives 1-butene).

Unsaturated metallocycles also exist. They are formed according to the same mode as those described above by replacing the alkenes by the alkynes (see Chap. 15): metallacyclopropenes, metallacyclobutenes and metallacyclobutadienes (intermediates in alkyne metathesis and alkyne polymerization), metallacyclopentadienes (intermediates of alkyne trimerization, see Chap. 21) and metallabenzenes (see Chap. 15) are obtained in this way. Finally, heterometallocycles are formed analogously by reactions of metal-oxo and metal-imido complexes that are isoelectronic to metal-carbene complexes. Their role is crucial as intermediates in metathesis (Chap. 15) and oxidation catalysis (Chap. 17).

6. *METAL-ARYL, -VINYL, -ALKYNYL, -SILYL, AND -STANNYL COMPLEXES*

6.1. ARYL

The metal-aryl bond is somewhat more robust than the metal-alkyl bond, and binary metal-aryl complexes are more stable than the corresponding metal-alkyl complexes. For instance, $TiPh_4$ decomposes (by reductive elimination) only above 0°C, whereas $TiMe_4$ decomposes (giving CH_4 by α-elimination) at -40°C. These comparisons are thus based on kinetic factors.[8.1]

$$[TiPh_4] \xrightarrow[-C_6H_4]{} Ph_3Ti{-}H \longrightarrow 1/2\ Ph{-}Ph + \text{"}TiPh_2\text{"} \longrightarrow \text{etc.}$$

The phenyl complexes often decompose as metal-alkyl complexes according to the β-H-elimination mechanism. The benzyne intermediate can sometimes be stabilized as shown by Schrock with a tantalum complex:[8.5]

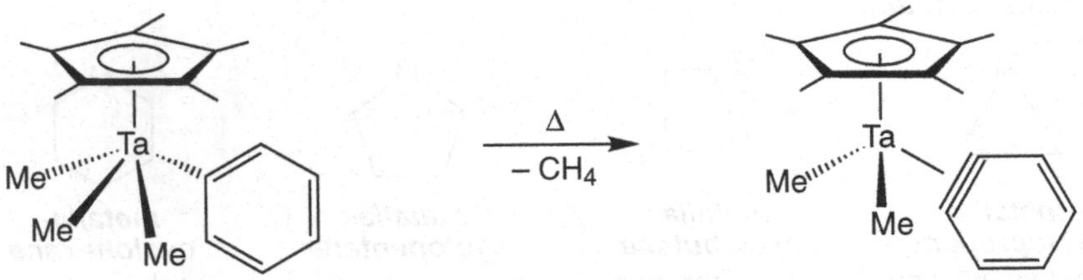

The metal-benzyne complexes, that can also be represented as metallacyclopropenes, can insert a variety of unsaturated substrates (alkenes, alkynes, carbonylated derivatives, nitriles, etc.), if their NVE is lower or equal to 16, to give 5-membered metallocycle precursors of organic heterocycles:

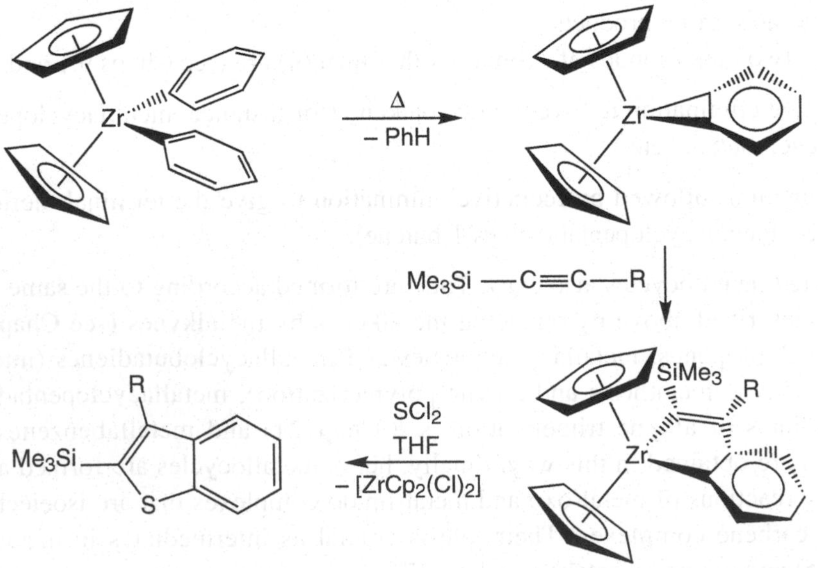

Although 18-electron complexes containing one or several phenyl ligands are stable, binary metal-aryl complexes are only stable if the β-H-elimination reaction is blocked by replacing both *ortho* H atoms of the phenyl ligands by substituents. This condition is perfectly achieved in the mesityl ligand whose 10-electron niobium and tantalum complexes $[M(\eta^1\text{-mesityl})_nCl_{5-n}]$, M = Nb or Ta, n = 1, 2, 3 are stable:[8,9]

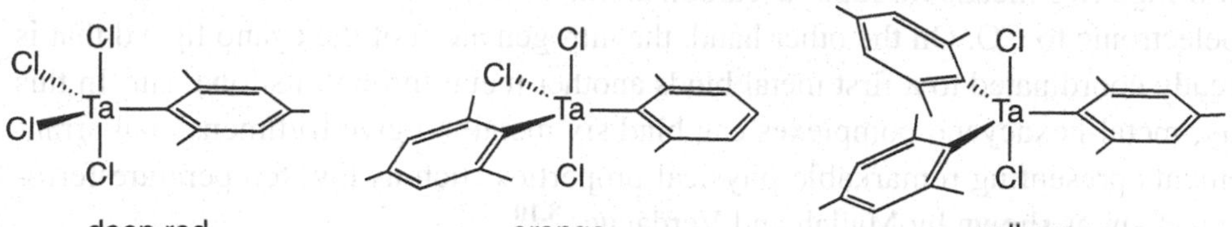

deep red,
spontaneous
ignition in air

orange,
smoothly hydrolyzed
in air in one minute

yellow,
air stable

6.2. VINYL

The vinyl complexes are accessible by transmetallation, oxidative addition of a vinyl halide, addition of an acid on a neutral alkyne complex, insertion of an alkyne in a metal hydride, reduction of a vinylidene complex or nucleophilic attack of a cationic alkyne complex:

$MCl_n + n\ Li(RC=CR'_2) \longrightarrow M(RC=CR'_2)_n$ ex: M = Zr, R = Ph, R' = Me, n = 4

$L_nM + RCH=CHBr \longrightarrow L_nM(Br)(CH=CHR)$ ex: ML_n = Pt(PPh$_3$)$_3$, R = Ph

$L_nM(RC\equiv CR) + HX \longrightarrow L_nM(X)(CR=CHR)$ ex: ML_n = Pt(PPh$_3$)$_2$, R = Me, X = Cl

$L_nM–H + HC\equiv CR \longrightarrow L_nM–CH=CHR$ ex: ML_n = ZrCp$_2$(Cl), R = *n*-Bu

$[L_nM=C=CR_2]^+ + H^- \longrightarrow L_nM–CH=CR_2$ ex: ML_n = FeCp(dppe), R = Me

$L_nM(RC\equiv CR)^+ + M'R' \longrightarrow L_nM–CR=C(R)(R')$ ex: ML_n = FeCp(CO)(PPh$_3$), R = Me, M'R' = Ph$_2$Cu(CN)Li$_2$, R' = Ph

The vinyl ligand can also coordinate to a second metal through its double bond:

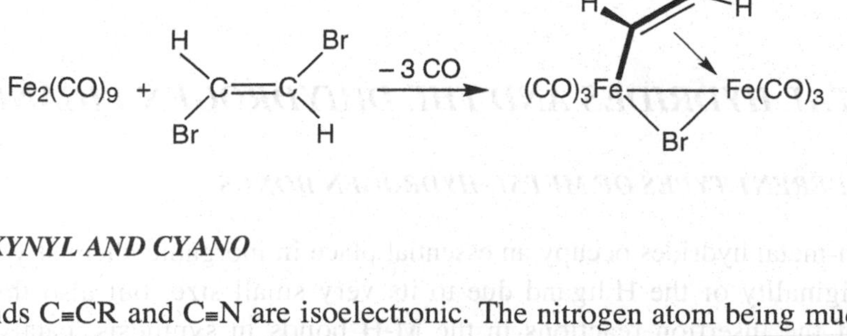

6.3. ALKYNYL AND CYANO

The ligands C≡CR and C≡N are isoelectronic. The nitrogen atom being much more electronegative than carbon, CN^- is much less basic than $C\equiv CR^-$. Thus, CN stabilizes the polyanionic complexes such as the metallocyanides $[Fe(CN)_6]^{4-}$ and

$[Mo(CN_8)]^{3-}$, whereas the alkynyl complexes usually stabilize more or less negatively charged complexes. Another consequence is that the metallic acetylides are easily hydrolyzed, contrary to the cyano complexes. It is also possible to isolate acids suchs as $[H_4Fe(CN)_6]$, the protons being not bonded to the metal, but to the nitrogen atoms through hydrogen bonds M-CN–H–NC-M. The cyano ligand does not bridge two metals through its carbon atom, as CO does so well, although CN^- is isoelectronic to CO. On the other hand, the nitrogen atom of the cyano ligand that is already coordinated to a first metal binds another metal through its lone pair. In this way, metal-hexacyano complexes can bind six metals to give tridimensional arrangements presenting remarkable physical properties such as low-temperature ferromagnetism as shown by Mallah and Verdaguer.[8.10]

Whereas the complexes MCN (M = Cu, Ag, Au) form infinite linear polymer chains, their homologs MC≡CR (R = H or alkyl) give chains in which the π bond is coordinated as an L ligand to a metal acetylide (these structures are broken by addition of a stronger L ligand such as PPh_3 to give a monomer such as $[Au(-C≡CR)(PPh_3)]$. The binary complexes $[M(C≡CR)_n]$ are usually explosive, even those mentioned above that are resistant to hydrolysis. Polyacetylide complexes (mesomers of polyallenylidenes), models of conducting wires have been pioneered by the groups of Lapinte and Gladysz: $[L_nM-(C≡C-)_nML_n]$.

Examples: L_nM = FeCp*(dppe) [8.11] or ReCp*(CO)(NO) [8.12].

6.4. SILYL AND STANNYL

There are many metal-silyl complexes because of the robustness of the metal-silyl bond.[8.13] Several reasons explain this stability:
▸ The metal-silicon bond is much longer than the metal-carbon bond.
▸ The β-elimination does not occur because of the instability of the Si=C bond.
▸ The π backbonding of the vacant d orbitals of silicon strengthens the bond.

The complexes with a $M-SiR_3$ bond are known for R = Me, but also with R = alkyl, aryl, hydroxy, chloro and fluoro.

The complexes with a $M-SnR_3$ bond are known for R = alkyl or chloro. With several $SnCl_3$ ligands, they are active in catalysis, because of their labilization due to the strong *trans* influence, which leads to the liberation of their coordination sites.

7. METAL HYDRIDES AND THE DIHYDROGEN LIGAND

7.1. DIFFERENT TYPES OF METAL-HYDROGEN BONDS

Transition-metal hydrides occupy an essential place in inorganic chemistry because of the originality of the H ligand due to its very small size, but also the applications of the insertion reactions in the M-H bonds in synthesis, catalysis and polymers science.[8.14] The H ligand is most of the time terminal, but many complexes are known in which the H ligand bridges two or three metals or even more. The H atoms are sometimes interstitial in inorganic clusters such as $[Rh_{13}(H)_3(CO)_{24}]^{2-}$.[8.15] It is possible to count electrons brought to the metal or to two metals. In bimetallic complexes, the bridging hydride is an X (1-electron) ligand for a metal, and this M-H bond is a 2-electron L ligand for the second metal (Chap. 2). Beyond two metals, the count can be carried out only for the whole cluster (NCE, Chap. 2).

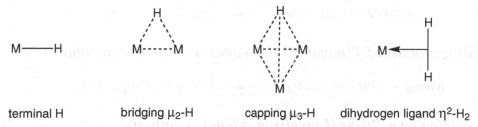

| terminal H | bridging μ_2-H | capping μ_3-H | dihydrogen ligand η^2-H_2 |

There are many metal-carbonyl-hydride complexes with one or several hydride ligands, due *inter alia* to their easy formation by protonation of metal-carbonyl anions. According to the same principle, various other families of hydrides can be synthesized. The only stable binary complex, $[ReH_9]^{2-}$, is well known for its exceptionally high valence (C = 9), due to the very small size of the hydrogen atom. There also exist a whole category of high-coordinance, high-oxidation-state poly-hydride complexes. Since Kubas' discovery of the H_2 ligand in 1984,[8.16] the structures of many of these complexes have been re-examined and most often re-assigned with one or several dihydrogen ligands.

7.2. SYNTHESIS OF THE HYDRIDES

Many routes are known:

Protonation

$$[CoCp(PMe_3)_2] + NH_4^+PF_6^- \xrightarrow{Et_2O} [CoCp(H)(PMe_3)_2]^+ PF_6^- + NH_3$$

$$[FeCp_2] + H^+SO_3F^- \xrightarrow{Et_2O} [FeCp_2(H)]^+ SO_3F^-$$

$$[Fe(CO)_4]^{2-} [Na^+]_2 + CH_3CO_2H \xrightarrow{H_2O} [Fe(CO)_4(H)]^- Na^+ + CH_3CO_2^- Na^+$$

Hydriding

$$[FeCp^*(CO)_2Br] + Na^+ BH_4^- \xrightarrow{THF} [FeCp^*(CO)_2(H)] + Na^+ Br^- + BH_3$$

$$[ZrCp_2Cl_2] \xrightarrow{Et_2O} [ZrCp_2(H)Cl] + Na^+ Cl^-$$
$$+ Na^+ [AlH_2(OCH_2CH_2OMe)_2]^- \qquad + [AlH(OCH_2CH_2OMe)_2]$$

H atom transfer from the medium

$$[Fe^ICp(arene)] + 2\, PR_3 \xrightarrow{\text{THF or toluene}} [Fe^{II}Cp(PR_3)_2(H)] + arene$$

R = Ph, Me

Oxidative addition of H_2 on 16-electron complexes

$$[IrCl(CO)(PPh_3)_2] + H_2 \longrightarrow [IrCl(CO)(PPh_3)_2(H)_2]$$

Hydrogenolysis of d^0 metal-alkyl complexes by σ-bond metathesis

$$[WMe_6] + 3\, PMe_2Ph + 6\, H_2 \longrightarrow [WH_6(PMe_2Ph)_3] + 6\, CH_4$$

Hydriding by n-Bu_3SnH involving σ-bond metathesis

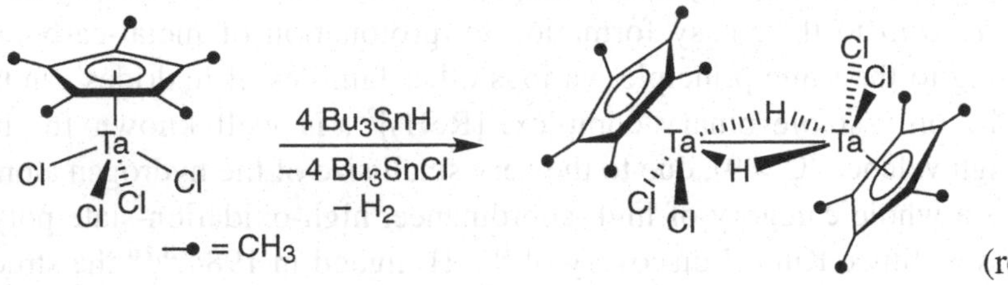

$$\xrightarrow[\substack{-\,4\,Bu_3SnCl \\ -\,H_2}]{4\,Bu_3SnH}$$

—● = CH_3

(ref. **8.17a**)

Decomposition of a ligand

$$Fe(CO)_5 + OH^- \longrightarrow Fe(CO)_4(COOH)^- \longrightarrow [(Fe(CO)_4(H)]^- + CO_2$$

Deprotonation of H_2 complexes

$$[FeCp(dppe)(H_2)]^+ PF_6^- + NEt_3 \longrightarrow [FeCp(dppe)H] + [NHEt_3]^+ PF_6^-$$

7.3. PROPERTIES OF THE HYDRIDES

The hydride ligands are easily detected by ^1H NMR, because they are usually found between 0 and –20 ppm, this region being otherwise empty (only hydrides of d^0 and d^{10} transition-metal hydride complexes are found at lower field). The hydride can be coupled with the metal when the latter has a spin 1/2 and with the phosphorus atom of phosphanes coordinated to the same metal in *cis* ($J = 15$ to 30 Hz) or *trans* ($J = 90$ to 150 Hz) position. The infrared stretching M-H band near 1500 to 2200 cm^{-1} is not intense enough to be reliable. The hydrogen atoms are also detected with difficulty by X-ray because of their very small size. Neutron diffraction is a much better technique, although it is a heavier one and requires relatively large crystals (at least one mm^3).

We have already seen the insertion of olefins into the M-H bond leading to M-alkyl complexes in Chap. 6, and repeated insertions further leading to olefin polymerization (this reaction is important and will be developed in Chap. 15)

The hydride complexes can be either good hydride donors (hydridic property) or acids (in which case they are still called hydrides, but do not deserve this name). This hydridic vs. acidic property depends on the charge, nature of the ligands and location of the metal in the periodic table. $[VH(CO)_6]$ and $[CoH(CO)_4]$ are strong acids. The acidity decreases as one goes down in the periodic table, or if one or several carbonyls are replaced by more donor ligands. There is also, of course, a *continuum* of intermediate situations where some hydrides, such as for instance $[WCp(CO)_3(H)]$, are able to behave as H^+ or H^- donors depending on the substrate.

pK_a values of transition-metal hydride complexes in various solvents

Hydride	H_2O	MeOH	CH_3CN
$[HV(CO)_6]$	strong acid	–	–
$[HV(CO)_5PPh_3]$	6.8	–	–
$[HW(CO)_3Cp]$	–	8.00	16.1
$[HW(CO)_2(PMe_3)Cp]$	–	–	26.6
$[HMn(CO)_5]$	7.1	–	15.1
$[HRe(CO)_5]$	–	–	21
$[H_2Fe(CO)_4]$ (pK_1)	4.00	–	11.4
$[H_2Ru(CO)_4]$	–	–	18.7
$[H_2Os(CO)_4]$	–	15.2	20.8
$[HFe(CO)_2Cp]$	–	–	19.4
$[HFe(CO)_2(\eta^5\text{-}C_5Me_5)]$	–	–	26.3
$[HRu(CO)_2Cp]$	–	–	20.2
$[HCo(CO)_4]$	strong acid	strong acid	8.4
$[HCo(CO)_3P(OPh)_3]$	4.95	–	11.4
$[HCo(CO)_3PPh_3]$	6.96	–	15.4
$[H_4Ru_4(CO)_{12}]$	–	11.7	–
$[H_2Ru_4(CO)_{13}]$	–	11.1	–

The charge plays a key role for the acidity and hydride-transfer property of metal-hydride complexes. Tilset has shown that the monoelectronic oxidation of a metal hydride lowers its pK_a by 20.6 units for a variety of structures!

The H^- donor property of transition-metal hydrides can be paralleled with that of n-Bu_3SnH which is very much used in organic chemistry in spite of the toxicity of molecular tin compounds:

$$[FeCp^*(CO)_2H] + PhCH_2Cl \longrightarrow [FeCp^*(CO)_2Cl] + PhCH_3$$

$$[SnBu_3(H)] + PhCH_2Cl \longrightarrow [SnBu_3Cl] + PhCH_3$$

The hydrides can even be titrated by CCl_4 using this property:

$$[M–H] + CCl_4 \longrightarrow [M–Cl] + CHCl_3$$

The parallel between main-group hydrides and transition-metal hydrides also allows to consider the H^- transfer mechanism that can be a direct H^- transfer or an electron transfer followed by an H atom transfer. For instance, it could be shown by isolating the electron-transfer intermediate that the reduction by several hydrides of the complexes $[FeCp(arene)]^+$ proceeds by electron transfer followed by H atom transfer (see Chap. 11).

Likewise, concerning the reactions of the type $[MH] + RX \longrightarrow [MX] + RH$, it is often difficult to distinguish between the mechanisms: **(a)** hydride transfer, **(b)** electron transfer followed by H atom transfer (non chain) and **(c)** atom-transfer chain (ATC, see Chap. 4).

a.　　　$[M–H] + R–X \longrightarrow [M]^+ + H–R + X^- \longrightarrow [M–X] + R–H$

b.　　　$[M–H] + R–X \longrightarrow [M]^+ + (H^\bullet, R^\bullet) + X^- \longrightarrow [M–X] + R–H$

Example:

$$[ArCH_2Mn(CO)_4PAr_3] + [HMn(CO)_5] \longrightarrow ArCH_3 + [Mn_2(CO)_9PAr_3]$$

c. When the radicals are stoichiometrically generated, for instance spontaneously (reaction **(b)** above), they can undergo the following propagation cycle:

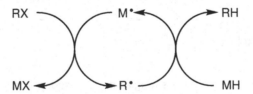

Example:

$$[Mn(CO)_4(PPh_3)H] + CF_3I \longrightarrow [Mn(CO)_4(PPh_3)I] + CF_3H$$

In the same way, hydrides can undergo the substitution of CO by a phosphine much more rapidly than the other complexes $[M(CO)_nR]$ ($R \neq H$) because of this ATC mechanism detailed in Chap. 4.

In the mechanism mentioned above, the metal hydride transfers an H atom to an organic radical. This reaction is useful, even in stoichiometric (non-chain) reactions, because it allows to easily generate an organometallic or inorganic radical. It is thermodynamically favorable, because the C-H bond is stronger than the M-H bond (see the table of E-H bond energies, Chap. 5).

$$[M–H] + Ph_3C^\bullet \longrightarrow [M]^\bullet + Ph_3CH$$

It is as difficult to distinguish between direct H⁻ transfer and electron transfer followed by H atom transfer (H⁻ = e⁻ + H•) as it is to distinguish between direct H atom transfer and electron transfer followed by proton transfer (H• = e⁻ + H⁺). It is possible to favor the electron-transfer reaction by using an anode or a powerful monoelectronic oxidant that is not an H atom acceptor. We have noted above that the acidity of a metal hydride is enormously increased by monoelectronic oxidation. Indeed, the monoelectronic oxidation of a neutral 18-electron metal-hydride complex to a 17-electron radical cation M-H•⁺ is systematically followed by the protonation of the neutral hydride by the very acidic cationic 17-electron metal hydride to give the 18-electron dihydride or dihydrogen complex (see section 7.4) and the deprotonated 17-electron neutral radical. The latter traps a solvent molecule as an L ligand, and the transient 19-electron species formed is oxidized to the 18-electron cation:[8.17b]

$$[MH] - e^- \longrightarrow [MH]^{•+} \xrightarrow{[MH]} M^• + [MH_2]^+$$
$$\quad 18e \qquad\qquad 17e \qquad\qquad 17e \quad 18e$$

$$[M]^• + S \longrightarrow [MS]^• \xrightarrow{-e^-} [MS]^+$$
$$\quad 17e \qquad\qquad 19e \qquad\qquad 18e$$

Overall:

$$2\,[MH] - 2e^- + S \longrightarrow [MH_2]^+ + [MS]^+$$

Example:

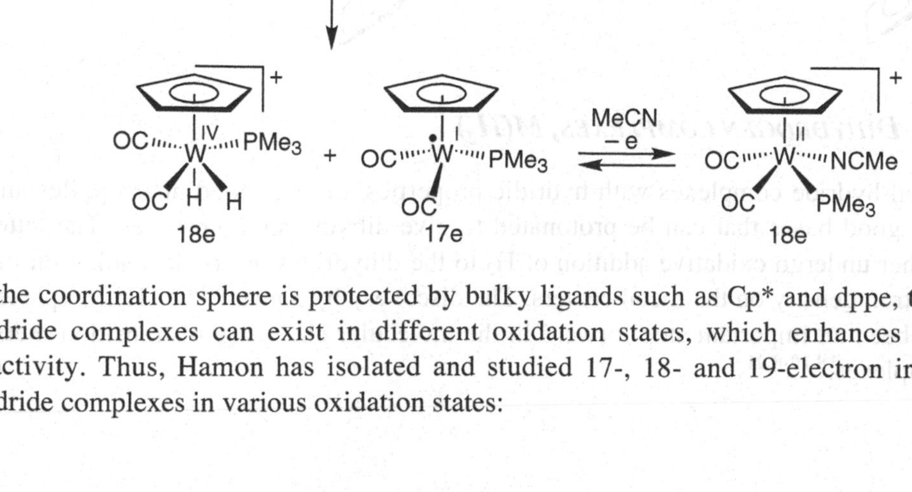

If the coordination sphere is protected by bulky ligands such as Cp* and dppe, the hydride complexes can exist in different oxidation states, which enhances its reactivity. Thus, Hamon has isolated and studied 17-, 18- and 19-electron iron hydride complexes in various oxidation states:

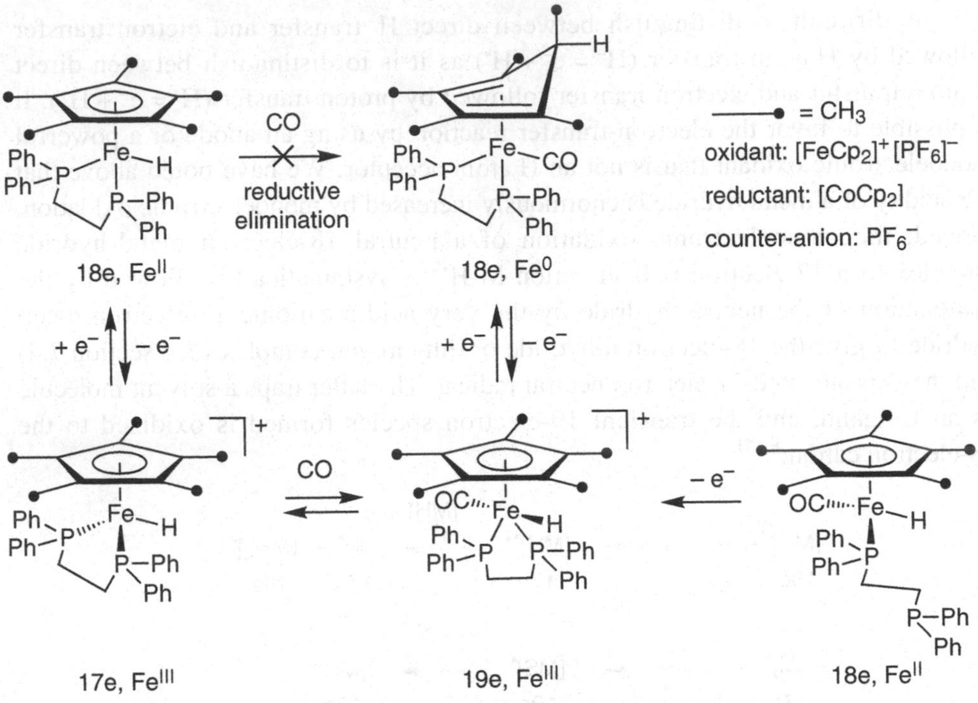

**Square-redox scheme for the CO-induced reductive elimination
of hydride from an iron-hydride complex [8.22]**

Finally, a last important property of the M-H bond is its role as a ligand to give bimetallic hydrides, given its ability to bridge metals (see Chap. 2). Therefore, a 16-electron species must be generated for instance by removing a CO ligand in the second metal complex:

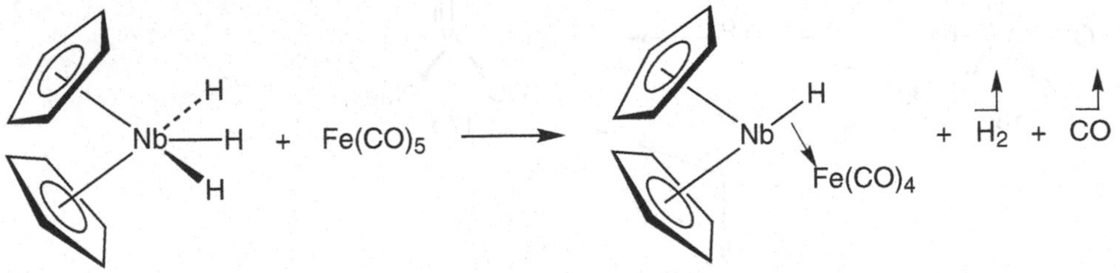

7.4. DIHYDROGEN COMPLEXES, M(H₂)

Metal-hydride complexes with hydridic properties are thus good nucleophiles and also good bases that can be protonated to give dihydrogen complexes. The latter further undergo oxidative addition of H_2 to the dihydrides or not depending on the electron density on the metal centers. The discovery by Kubas [8.16] of H_2 complexes has been an important step forward in the molecular chemistry of transition-metal complexes.[8.19-8.21]

$$P i\text{-}Pr_3$$

First H$_2$ complex,
characterized as such by Kubas in 1984

The complexes of σ bonds have indeed original properties and present a great interest from a theoretical point of view as intermediates of (a) oxidative addition and (b) heterolytic H$_2$ cleavage.

a.   M + H—H   ⟶   M←| ⟶   M⟨$\begin{array}{c}H\\H\end{array}$

b.   M—H + H$^+$ ⇌ M$^+$←|

Formally, the H-H σ bond is given to the metal as a 2-electron L ligand in the NVE count. The d_{z^2} metal orbital is in good position to accept this bond. There is also π backbonding, as for olefins, from the filled d metal orbital to the antibonding ligand orbital. Since the ligand orbital is a σ* and not a π* orbital, it is of much higher energy, and consequently, the π backbonding is weaker than for olefins. It is significant, however, and contributes to the bond stabilization. When it becomes too important, this π backbonding provokes the dissociation of the H$_2$ bond to give the dihydride complex, i.e. oxidative addition has occurred. The amount of π backbonding can be evaluated by the H-H distance in the H$_2$ complexes, because the electron density in the σ* antibonding orbital weakens the H-H bond, thus lengthens it. This distance is usually in the range 0.82 Å to 1 Å, whereas it is 0.74 Å in free H$_2$.

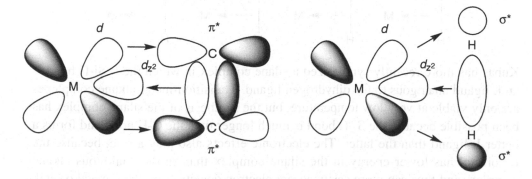

Comparison between the metal-olefin bond (left) and metal-H$_2$ bond (right)

For instance, in Kubas' first complex above, the H-H distance, determined by neutron scattering, is 0.84 Å. A band is observed in the infrared spectrum at 2695 cm^{-1}, attributed to the elongation of the H-H bond. This band usually is observed in the infrared spectrum between 2300 and 2900 cm^{-1} for the

H_2 complexes, but not always. In ^1H NMR, the H_2 ligand resonates between 0 and −10 ppm (often broad), which does not allow the distinction with the dihydride complexes. In order to characterize an H_2 complex (besides neutron diffraction), it is necessary to synthesize the H-D complex which presents, in NMR, a coupling constant $^1J_{H-D}$ between 20 and 34 Hz (for example 33.5 Hz for Kubas' complex above). Indeed, the metal-hydride-deuteride complex only shows a coupling constant $^2J_{H-D}$ of about 1 Hz. Many H_2 complexes are now known with the majority if not all transition metals. They are among the weakest ligands, and can be easily displaced by other ligands, even weak ones such as N_2. In polyhydrides, some couples of H atoms are in fact H_2 ligands, often in fluxional exchange with the other hydride ligands, which reduces the H-D coupling. The H_2 complex is sometimes in equilibrium with the dihydride, which is the case for Kubas' complex. In other cases, the H_2 complex generated at low temperature by protonation of a hydride such as [FeCp*dppe(H)], thermally yields the dihydride, as shown by Hamon and Lapinte: [8.22]

$$[Fe\text{–}H] + H^+ \xrightarrow{-80°C} [Fe(H\text{–}H)]^+ \xrightarrow{20°C} [Fe(H)_2]^+ \quad Fe = Fe\ Cp^*(dppe)$$

The protonation of a polyhydride can have a dramatic effect on the rearrangement of hydride ligands to dihydrogen ligands, because of the considerable lowering of the electron density on the metal due to the additional cationic charge.

Example:

$$[Ir^V H_5 (PCy_3)_2] + H^+ \longrightarrow [Ir^{III}(H_2)_2(H)_2(PCy_3)_2]^+$$

In conclusion, let us stress the analogy between the different σ bonds as ligands for transition metals. Besides the dihydrogen ligand, we already know the agostic C-H-M bond and the M-H-M bond.

Kubas has more recently synthesized a silane complex in which the σ Si-H bond is an L ligand analogous to the dihydrogen ligand. Transition-metal-alkane complexes are only viable at very low temperature, but the synthesis of the silane complex has been possible because the Si-H bond is much longer than the C-H bond and forms a better L ligand than the latter. The electronic effects also play a role, because the σ* orbital has lower energy in the silane complex than in the analogous alkane complex, and thus can more easily accept electron density from the d metal orbital. Finally, it is possible that analogous bonds can form between transition metals and other σ X-H bonds, with X = B, P, S, Sn, etc. For example, such B-H-M bonds have been characterized crystallographically in polypyrazolylborate complexes.

8. *ALKOXY AND AMIDO COMPLEXES*

Early transition metal, lanthanide and actinide alkoxy and amido complexes are common, and they often are stable because of the interaction between the filled *p* orbital of the O or N atom of the ligand and an empty *d* metal orbital. The alkoxy and aryloxy ligands play a crucial role in the catalytic properties of group 5-7 metal-alkylidene and metal-alkylidyne complexes for the metathesis of simple, double and triple bonds.[15.2] On the other hand, the behavior of late transition-metal alkoxy and amido complexes is less known. Many of them are stable, however, in spite of the possible repulsion between the filled *d* orbital and the *p* orbital of the heteroatom. The metal-heteroatom bonds are *robust*, and the main characteristic of these is that they are *strongly polar* and possess a significant ionic character. They exhibit *nucleophilic reactivity* and sometimes form strong bonds to proton donors (they even deprotonate relatively weak acids).[8.23,8.24]

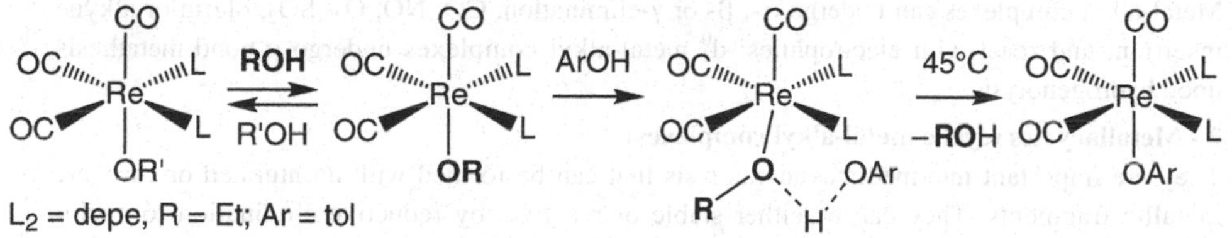

L_2 = depe, R = Et; Ar = tol

Even the parent hydroxy and amido complexes are stable; they are available by classic routes.[8.24]

SUMMARY OF CHAPTER 8
METAL-ALKYL AND -HYDRIDE COMPLEXES
AND OTHER COMPLEXES OF MONOHAPTO X LIGANDS

1 - Metal-alkyl complexes

They are stable if they do not have β hydrogen or if NVE = 18. Binary metal-alkyl complexes MR_n are thus known for R = Me, CH_2Ph, mesityl, CH_2CMe_3, CH_2SiMe_3. They are stable if the coordination sphere is not too crowded, otherwise α-elimination proceeds to give Schrock-type alkylidene complexes.

They are synthesized as follows: $MX_n + n\ M'R \longrightarrow MR_n + n\ M'X$ (M' = alkali metal or MgX, etc.). Metal-alkyl complexes with one alkyl ligand are synthesized as above, from L_nMX, from $M^- + RX$, by oxidative addition or olefin insertion into a M-H bond or by nucleophilic addition onto an olefin or carbene ligand.

Metal-alkyl complexes are characterized in ^{13}C NMR by upfield signals compared to alkanes.

Metal-alkyl complexes can undergo α-, β- or γ-elimination, CO, NO, O_2, SO_2, olefin or alkyne insertion, and react with electrophiles; d^0 metal-alkyl complexes undergo σ bond metathesis upon hydrogenolysis.

2 - Metallacycles (cyclic metal-alkyl complexes)

They are important intermediates in catalysis that can be formed with unsaturated or carbenic metallic fragments. They can be either stable or reactive, by reductive elimination to cyclo-alkanes or by β-elimination followed by reductive elimination to terminal olefins.

3 - Other organometallic complexes with σ M-C bonds

• metal-aryl complexes decompose by β-elimination to metal-benzynes except for C_6F_5 or mesityl in which there is no β hydrogen;

• metal-vinyl complexes are known with a large variety of syntheses and reactions.

• metal-alkynyls produce models of organic conductors: $M(CC)_nM$

• metal-cyanides $M(CN)_6^{4-}$ give polymetallic magnetic complexes $M(CNM')_6{}^x$.

4 - M-Si or M-Sn analogs: stable metal silyl and stannyl $M-ER_3$ (E = Si, Sn; R = Me, Cl).

5 - Metal hydrides (M-H bonds) and dihydrogen complexes $M(H_2)$

Various types:

terminal H   bridging μ_2-H   bridging μ_3-H   dihydrogen complex η^2-H_2

Metal hydrides, important in catalysis, have various pK_as and can be synthesized in many ways. They are characterized by ^1H-NMR shifts between 0 and –20 ppm, which distinguishes them from other complexes except for H_2 complexes $M(H_2)$. In the latter, the H-H distance is between 0.84 Å and 1 Å (compare: free H_2: 0.74 Å).

The complexes $M(H_2)$ are met when the metal center is too electron-poor to undergo oxidative addition of H_2 to dihydride. The complexes $M(H)_2$ or $M(HD)$ and $M(H_2)$ are distinguished in NMR by the coupling between H and D: $^1J_{HD}$ = 20 to 34 Hz in M(HD), 1 Hz in M(H)(D). The $M(H)_2$ or $M(HD)$ and $M(H_2)$ are sometimes in equilibrium:

EXERCISES

8.1. Determine the NVEs of the metals in the following complexes and conclude on their thermal stability:

$[Zr(CH_2CH_2Ph)_4]$; $[Re(CH_2CH_3)_5]$; $[MoCp(CO)_3(CH_2CH_3)]$; $[TiCp_2(CH_2CH_3)_2]$; $[Cr(CH_2Ph)_6]$; $[Nb(CH_3)_5]$; $[TaPh_5]$; $[MoCp_2(CH_2CH_2CH_3)_2]$.

8.2. Write the homolytic decomposition of $[CH_3–M(CO)_5]$, M = Mn or Re. Which complex, the Mn or Re one, is more stable towards this decomposition mode? Why?

8.3. Determine and explain the order of stability of the complexes $[TiR_4]$ with R = CH_3, Ph, CH_2Ph, OCH_3 or OPh.

8.4. Is β-elimination faster with an alkyl ligand such as CH_2CH_3 or with an alkoxy ligand such as OCH_3 (or with the substituents of the exercise 8.3 above)? For instance, compare the relative stabilities of $[TaCp*(C_2H_5)_4]$ and $[TaCp*(OMe)_4]$.

8.5. In the series of complexes $[LnM(H_2)]$ and $[LnM(H)_2]$, with M = Cr, Mo and W, the H_2 and dihydride complexes have close energies, and are thus in equilibrium. Indicate in which direction this equilibrium shifts when one goes down in the periodic table, and explain.

Chapter 9

METAL-CARBENE AND -CARBYNE COMPLEXES AND MULTIPLE BONDS WITH TRANSITION METALS

The complexes containing a double carbon-element bond, i.e. metal-carbenes M=C [9.1], metal-oxo M=O [9.17] and metal-imido M=NR [9.17,9.18] are key species that are important from a theoretical point of view, in biochemistry, organic synthesis and catalysis of metathesis, dimerization, polymerization and oxidation of olefins (see Part IV). Analogously, the complexes with a triple metal-carbon bond M≡C, i.e. metal-carbynes, catalyze the metathesis and the polymerization of alkynes and are potentially, as metal-nitrides M≡N, the source of a very rich chemistry, although still less developed than that of metallocarbenes.

1. METAL-CARBENE COMPLEXES

1.1. STRUCTURE

The M=C bond can be polarized in one way or the other given the large flexibility of electronic properties of transition-metal fragments. [9.1-9.4]

$$M^+ - C^- \quad \longleftrightarrow \quad M = C \quad \longleftrightarrow \quad M^- - C^+$$

One can distinguish the cationic or neutral complexes of electrophilic carbene of Pettit type, [9.1-9.3] that tend to behave somewhat as carbocations coordinated to transition metals, and the neutral metal complexes of nucleophilic carbene complexes of Schrock [9.2] type that tend to react as ylids.

It has been suggested that complexes of electrophilic carbenes be considered as the combination of a metal fragment with a singlet carbene, an L ligand, whereas the complexes of nucleophilic carbenes are the combination of a metal fragment with a triplet carbene, an X_2 ligand (scheme below). [9.6] Free methylene is a triplet in the ground state and has a low-energy singlet excited state. Carbene-bearing electron-withdrawing substituents such as Fischer's and Arduengo's carbenes are singlets. However, Pettit's iron-methylene species has a strongly electrophilic methylene, which is explained by the positive charge and electron-withdrawing carbonyl ligands.

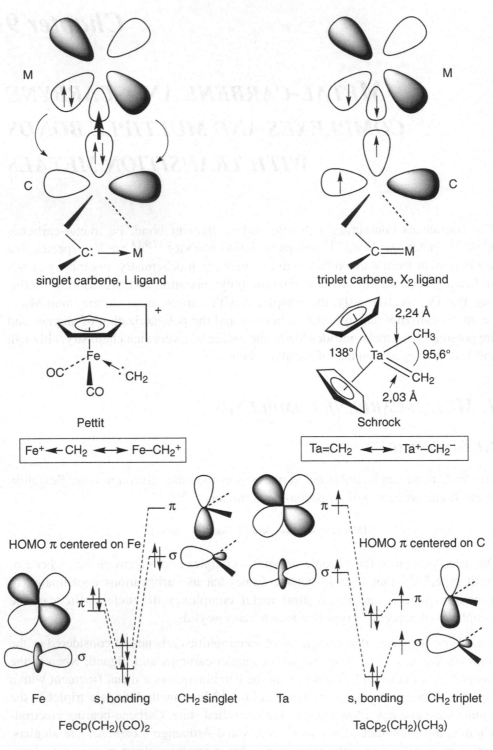

Two methylene complexes of opposite polarities

Pettit's methylene complex could be characterized only by its reactivity or spectroscopically upon attenuating the positive charge by introducing phosphines (instead of the two CO's) or C_5Me_5 (instead of Cp). These ligands strengthen the $Fe^+=CH_2$ bond, which is characterized by a lower value of the rotation barrier around this bond (for instance, $\Delta G^{\ddagger} = 10.6\ kcal \cdot mol^{-1} = 44.3\ kJ \cdot mol^{-1}$ for $[Fe(C_5Me_5)(CO)(PPh_3)(=CH_2)]^+$, which remains weak).[9.4]

It is Schrock's methylene complex that was isolated and characterized first, this complex being more stable partly because of a weaker polarity, which allowed the determination of its X-ray crystal structure that confirmed the $Ta=CH_2$ double bond. This double bond is also apparent in ^1H NMR, and the two methylene protons remain non-equivalent up to at least 100°C, the decomposition temperature ($\Delta G^{\ddagger} > 22\ kcal \cdot mol^{-1}$ or $92\ kJ \cdot mol^{-1}$). When the methylene bears one or two alkyl substituents, the carbene complexes are **alkylidene complexes**.[9.2] The electron-releasing property of these substituents strengthens the triplet structure of the carbene and the nucleophilicity of the metal-alkylidene complex, although it does not turn a Pettit-type carbene complex nucleophilic.

The family of complexes of nucleophilic carbenes or alkylidenes is essentially composed with d^0 complexes. For instance, the complexes $[M(CH_2t\text{-}Bu)_3(=CHt\text{-}Bu)]$, M = Nb or Ta formally only have 10 valence electrons. The X-ray diffraction structures of Schrock's alkylidene complexes that have less than 18 valence electrons show that the M-C-H angle is weak and that the C-C-H angle is abnormally high, i.e. there is an agostic M=C-H bond that increases the NVE by two units:

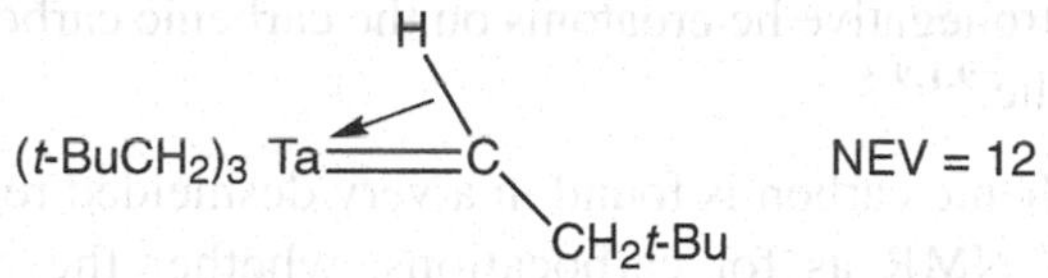

$$(t\text{-BuCH}_2)_3\ Ta\equiv\!\!\!=\!\!\!=\!C\underset{CH_2t\text{-Bu}}{\overset{H}{\diagup}} \qquad NEV = 12$$

There are two other important categories of metal-carbene complexes:

The metal-carbene complexes stabilized by the presence of an heteroatom, most often oxygen (but sometimes also N or S) or even two heteroatoms.[9.1, 9.3]

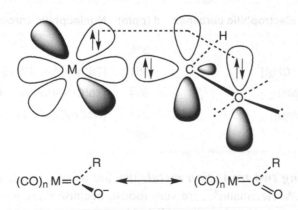

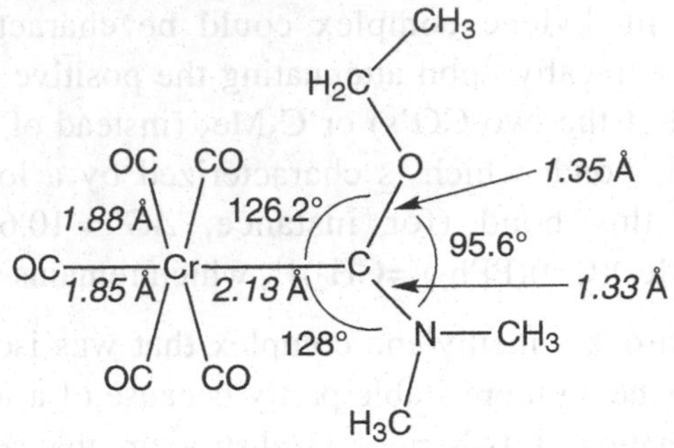

The structure shows the complex with bond lengths and angles: OC–Cr 1.88 Å, OC 1.85 Å, Cr–C 2.13 Å, angle 126.2°, angle 128°, angle 95.6°, C–O 1.35 Å, C–N 1.33 Å, with N–CH₃ and additional methyl groups.

[Cr(CO)₅{C(OEt)NMe₂}]

The M-C-heteroatom conjugation is characterized by:

▸ the shorter C-heteroatom bond than a single bond, thus double-bond character;

▸ the metal-carbon bond that is also approximately half-way between a single bond M-C and a double bond such as M=C=O (therefore the writing M←CO and M←CR(OR') is preferred). The carbene ligands are better donors and/or less good acceptors than CO, which is reflected in infrared spectroscopy by a lowering of about 150 cm⁻¹ of the CO absorption wavenumbers when a carbonyl is substituted by a carbene ligand of this type (for instance 1953 cm⁻¹ for [Cr(CO)₅{=C(OMe)Me}] and 2108 cm⁻¹ for [Cr(CO)₆)]).

These complexes, synthesized by E.O. Fischer in 1964, form a category called Fischer carbenes. They were the first successful approach to the synthesis and stabilization of metal-carbene complexes. They gave a rich chemistry (see Part V) because of their stability and ease of access. The presence of carbonyl ligands on the metal and of electronegative heteroatoms on the carbenic carbon makes this type of carbene electrophilic.[9.1,9.3]

The signal of the carbenic carbon is found in a very deshielded region between 200 and 400 ppm in ^{13}C NMR as for carbocations, whether the carbene ligand is electrophilic or nucleophilic. The electrophilic carbenes are the most deshielded ones, however:

Chemical shifts (δ_{ppm}) of the carbenic carbons in ^{13}C NMR

Carbocations and electrophilic carbenes	d (ppm)	Nucleophilic carbenes	d (ppm)
Me₃C⁺	336		
[FeCp*(CO)(PPh₃)(=CH₂)]⁺	351	[TaCp₂(Me)(=CH₂)]	224
[Cr(CO)₅{=C(Ph)(OMe)}]	354	[Ta(CH₂t-Bu)₃(=CHt-Bu)]	250
[Cr(CO)₅{C(Me)(NMe₂)}]	271	[NbCp₂Cl(=CHt-Bu)]	299
[Cr(CO)₆]	226		

Carbenes bridging two transition metals that are usually bonded to each other, synthesized by W.A. Herrmann[9.7], are very robust, because there is no double bond M=C. They can be formulated as dimetallacyclopropane and have been considered as molecular models of the interaction of carbenes with metal surfaces. For example, in the heterogeneous reactions such as the Fischer-Tropsch reduction of CO by H_2, bridging carbenes between two metal centers have been proposed to form as intermediates at the surface of the catalyst (see Chap. 20).

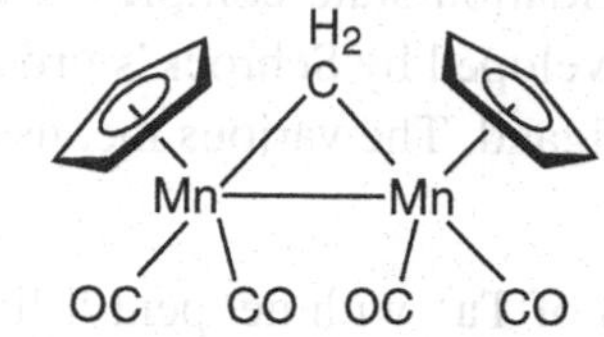

There are also complexes in which the carbene bridges a transition metal and a main group element such as aluminum. For instance, Tebbe's complex [9.8a] is well known, because it is especially reactive, being in equilibrium with the terminal methylene complex:

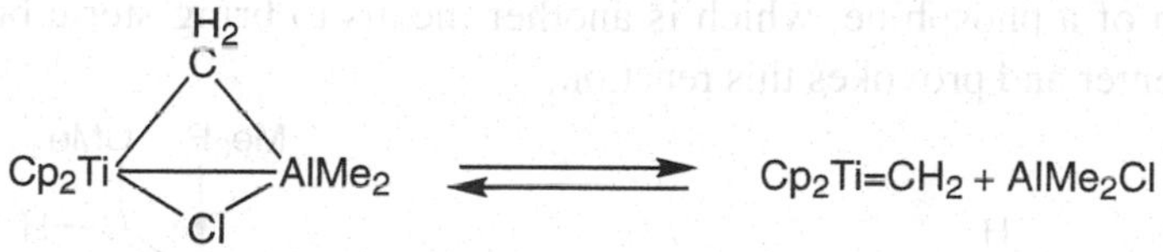

Another common class of metal-carbene complexes is that of the vinylidene complexes (whose structure compares to that of allenes),[9.8a-c] isomers of metal-alkyne complexes.[9.8c] A well-known example is the vinylidene complex $[Ru(PPh_3)_2Cl_2(=C=CHPh)]$, the first unimolecular ruthenium catalyst of olefin metathesis discovered by Grubbs in 1992.[9.8d] This class is extended to the allenylidenes and cumulenylidenes.[9.8a,b,e,f]

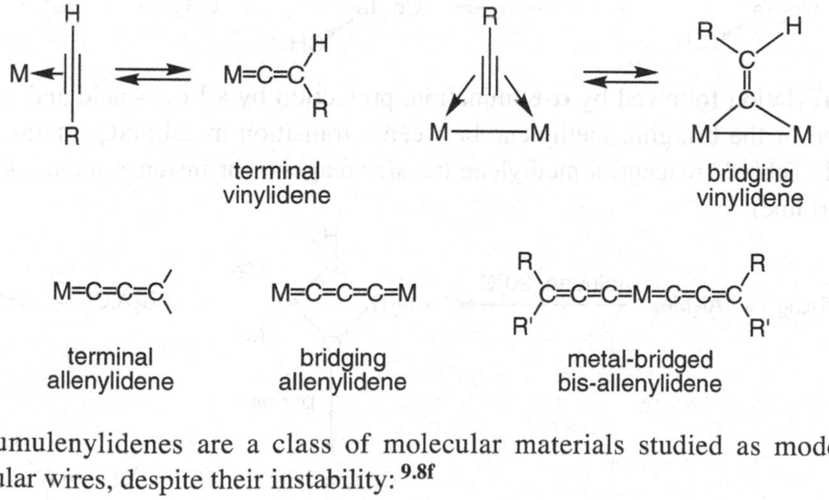

terminal
vinylidene

bridging
vinylidene

$M=C=C=C$

terminal
allenylidene

$M=C=C=C=M$

bridging
allenylidene

metal-bridged
bis-allenylidene

The cumulenylidenes are a class of molecular materials studied as models of molecular wires, despite their instability: [9.8f]

1.2. SYNTHESIS

Complexes of nucleophilic carbenes

The synthesis of these high oxidation-state complexes of nucleophilic alkylidene ligands has been extensively developed by Schrock's group and usually involves the α-H elimination from an alkyl ligand. The various means to achieve this elimination are:[9.2]

▸ the tentative penta-alkylation of Ta^V with neopentyl ligands. This reaction led to the discovery of the α-elimination reaction (see Chap. 6.5).

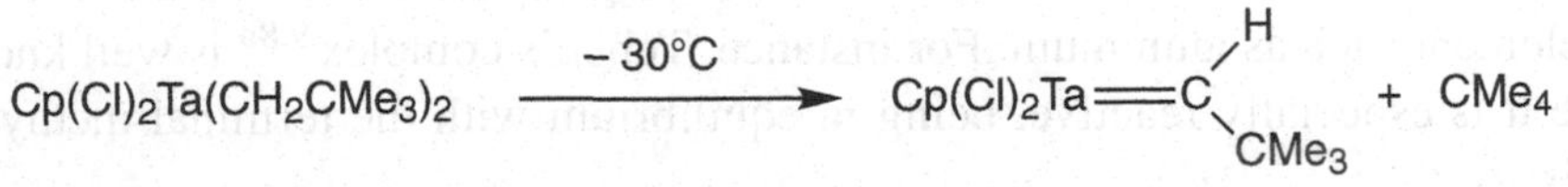

$$Cp(Cl)_2Ta(CH_2CMe_3)_2 \xrightarrow{-30°C} Cp(Cl)_2Ta{=}C{\overset{H}{\underset{CMe_3}{}}} + CMe_4$$

▸ the addition of a phosphine, which is another means to bring steric bulk around the metal center and provokes this reaction,

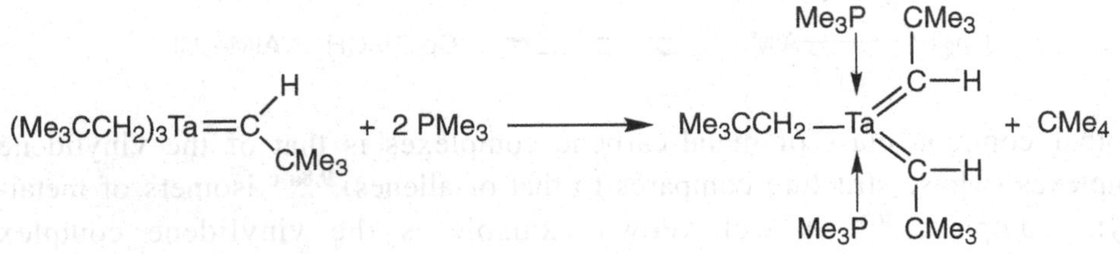

$$(Me_3CCH_2)_3Ta{=}C{\overset{H}{\underset{CMe_3}{}}} + 2\ PMe_3 \longrightarrow Me_3CCH_2{-}Ta{\begin{array}{c} Me_3P\ \ CMe_3 \\ {=}C{-}H \\ {=}C{-}H \\ Me_3P\ \ CMe_3 \end{array}} + CMe_4$$

▸ the deprotonation of a cationic methyl complex,

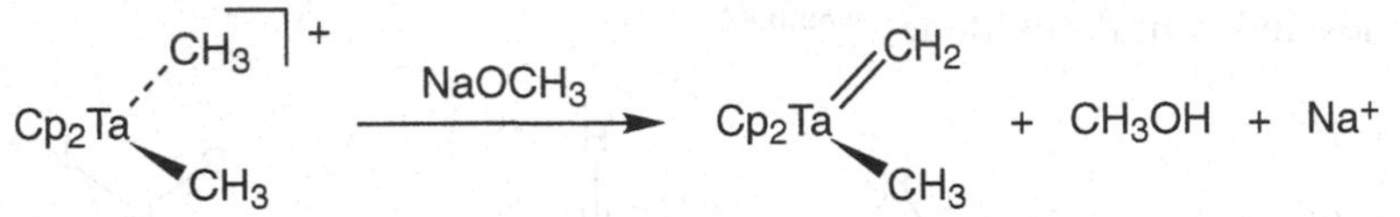

$$\left[Cp_2Ta{\overset{CH_3}{\underset{CH_3}{}}}\right]^+ \xrightarrow{NaOCH_3} Cp_2Ta{\overset{CH_2}{\underset{CH_3}{}}} + CH_3OH + Na^+$$

▸ the alkylation followed by α-elimination, protection by a Lewis acid and deprotection of the bridging methylene between a transition metal and a main-group metal (Tebbe's reagent, a methylene transfer reagent, for instance upon addition of pyridine).

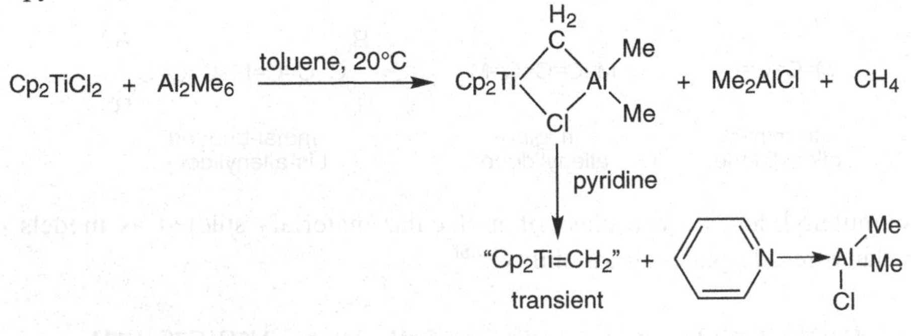

$$Cp_2TiCl_2 + Al_2Me_6 \xrightarrow{toluene,\ 20°C} Cp_2Ti{\begin{array}{c} H_2 \\ C \\ \diagdown\ \diagup \\ Cl \end{array}}Al{\overset{Me}{\underset{Me}{}}} + Me_2AlCl + CH_4$$

$$\downarrow \text{pyridine}$$

"$Cp_2Ti{=}CH_2$" + [pyridine]${-}Al{\overset{Me}{\underset{Cl}{\overset{|}{-}Me}}}$
transient

▸ the dehydrohalogenation of a bulky complex,

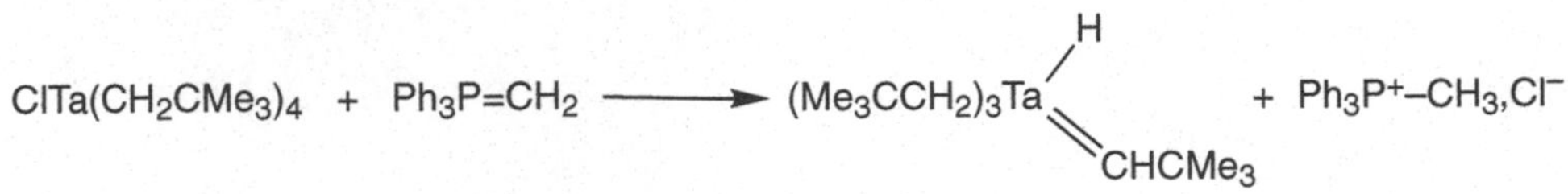

$$ClTa(CH_2CMe_3)_4 + Ph_3P{=}CH_2 \longrightarrow (Me_3CCH_2)_3Ta{\overset{H}{\underset{CHCMe_3}{}}} + Ph_3P^+{-}CH_3,Cl^-$$

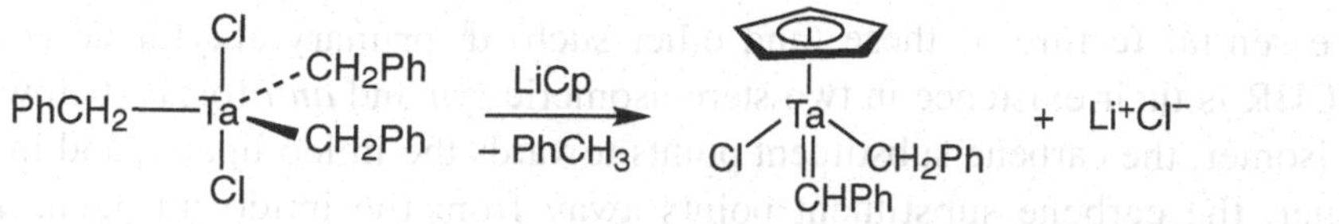
PhCH2—Ta(Cl)2(CH2Ph)2
LiCp
– PhCH3
+ Li+Cl–

▸ the transmetallation of an alkylidene ligand,

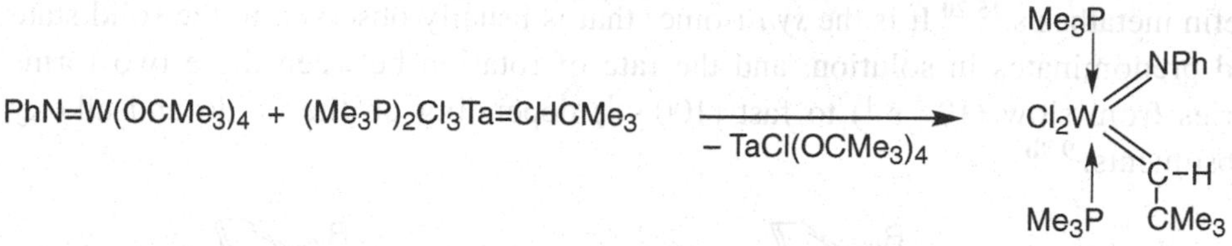
PhN=W(OCMe3)4 + (Me3P)2Cl3Ta=CHCMe3
– TaCl(OCMe3)4
Me3P
NPh
Cl2W
C-H
Me3P
CMe3

Several routes are known to Schrock's Mo- and W-alkylidene complexes that catalyze the metathesis of alkenes, including those starting from metal-carbyne complexes (see section 3.3). A practical one was discovered starting from Na_2MoO_4 and produced a large family of catalysts including enantiopure ones: [9.9]

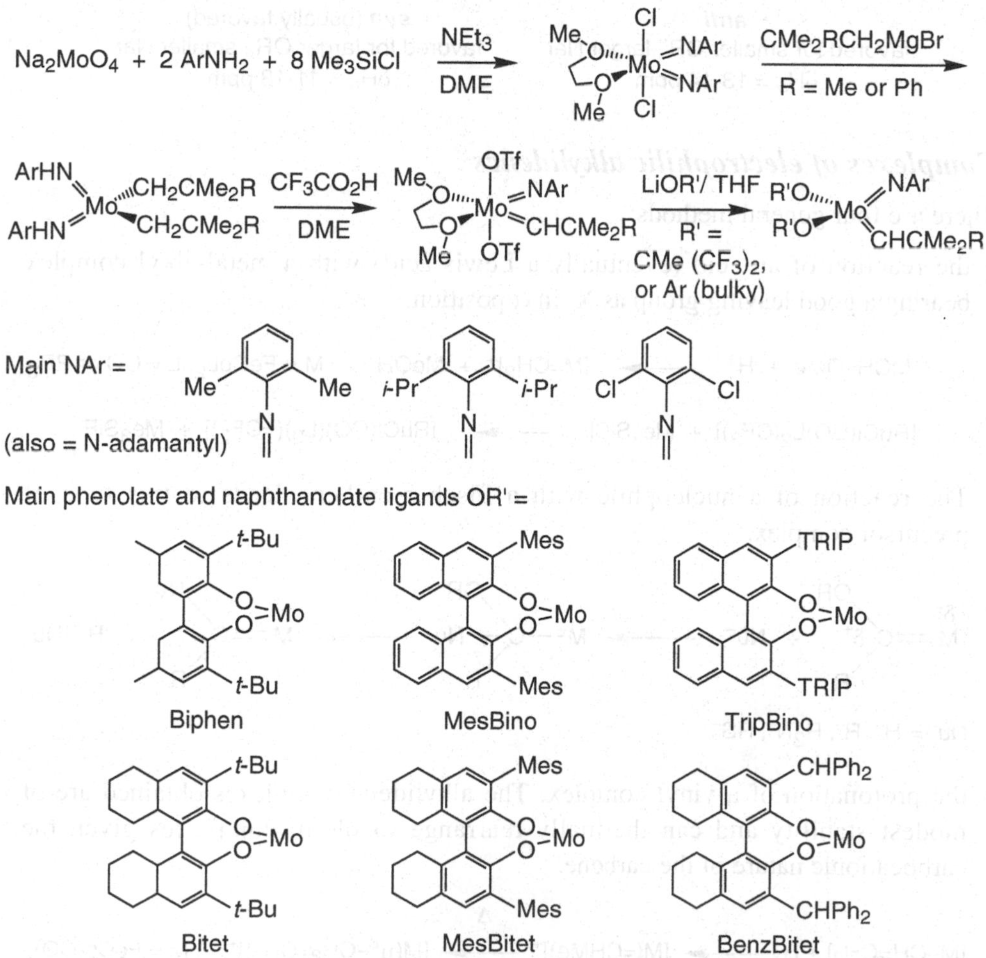
Na2MoO4 + 2 ArNH2 + 8 Me3SiCl
NEt3
DME
CMe2RCH2MgBr
R = Me or Ph
CF3CO2H
DME
LiOR'/ THF
R' =
CMe (CF3)2,
or Ar (bulky)
Main NAr =
(also = N-adamantyl)
Main phenolate and naphthanolate ligands OR' =
Biphen
MesBino
TripBino
Bitet
MesBitet
BenzBitet

Synthesis of the main family of Schrock's molybdenum-imido-alkylidene complexes that catalyze olefin metathesis

The routinely used imido and diolate ligands are indicated below the scheme.

An essential feature of these (and other such) d^0 primary alkylidene complexes M=CHR is their existence in two stereoisomeric *syn* and *anti* forms (below). In the *syn* isomer, the carbene substituent points towards the imido ligand, and in the *anti* isomer, the carbene substituent points away from the imido nitrogen. The two isomers are in an equilibrium that brings insight into the intimate mechanism of olefin metathesis.[15.20] It is the *syn* isomer that is usually observed in the solid state and predominates in solution, and the rate of rotation between these two forms varies from slow (10^{-5} s^{-1}) to fast (100 s^{-1}) depending on the imido and alkoxy substituents: [9.9b]

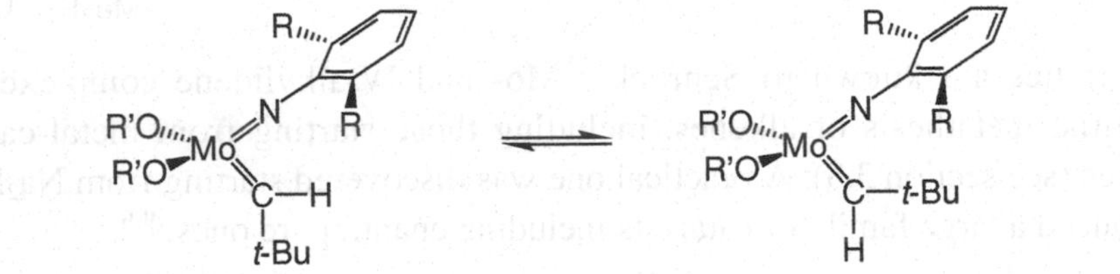

anti	**syn** (usually favored)
favored for smaller OR', larger Nar	favored for larger OR', smaller Nar
$\delta H_\alpha = 13\text{-}14$ ppm	$\delta H_\alpha = 11\text{-}13$ ppm

Complexes of electrophilic alkylidenes

There are four general methods:

▸ the reaction of an acid (eventually a Lewis acid) with a metal-alkyl complex bearing a good leaving group as X$^-$ in α position.

$$MCH_2OMe + H^+ \longrightarrow [M=CH_2]^+ + MeOH \qquad M = FeCpL_2, L = CO \text{ or } PR_3$$

$$[RuCl(CO)L_2(CF_3)] + Me_3SiCl \longrightarrow [RuCl(CO)(L_2)(=CF_2)] + Me_3SiF$$

The reaction of a nucleophile with a Fischer carbene leads to this type of precursor complex.

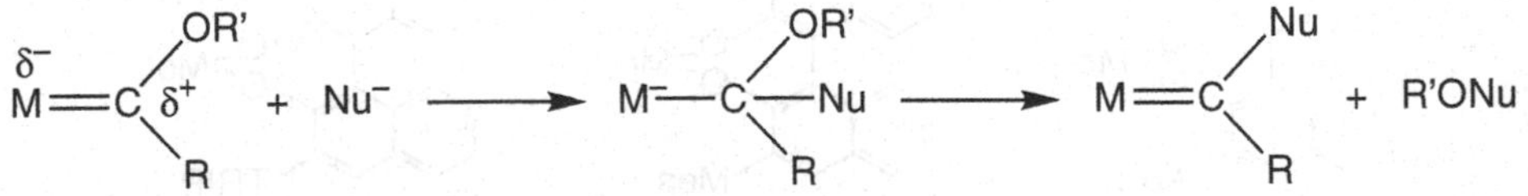

$Nu^- = H^-, R^-, R_2N^-, RS^-$

▸ the protonation of a vinyl complex. The alkylidene complexes obtained are of modest stability and can thermally rearrange to olefin complexes given the carbocationic nature of the carbene.

$$[M–CH=CH_2] + H^+ \xrightarrow{\quad} [M(=CHMe)]^+ \xrightarrow{\;\Delta\;} [M(\eta^2–CH_2=CH_2)]^+ \qquad M = FeCp(CO)_2$$

▸ the reaction of a diazoalkane or another carbene source on a complex bearing a potential site under the form of a weakly coordinated ligand. Roper's osmium-methylene complex is at the frontier between electrophilic and nucleophilic (see the reactions).

$$[Os(PPh_3)_3(Cl)(NO)] \xrightarrow[- PPh_3, - N_2]{CH_2N_2} [Os(PPh_3)_2(Cl)(NO)(CH_2)]$$

It is this method that also allowed Herrmann to synthesize the first bridging methylene complex.

$$[CpMn(CO)_2(THF)] \xrightarrow[- THF, - N_2]{CH_2N_2} Cp(CO)_2Mn \overset{CH_2}{\triangle} Mn(CO)_2Cp$$

▸▸ In the same line, Mansuy showed that CCl_4 is a source of dichlorocarbene ligand, which also shows how CCl_4 could be metabolized by cytochrome P450 and revealed for the first time the biochemical dimension of metallo-carbenes:[9.10]

▸▸ In 1995, Grubbs reported the synthesis of the family of olefin metathesis catalysts $[RuCl_2(=CHR')(PR_3)]$ with a variety of R and R' groups. The key synthetic method was the alkylidene transfer from various diazoalkanes to $[RuCl_2(PPh_3)_3]$ at –78°C. Subsequent phosphine exchange of PPh_3 by PCy_3 could be carried out either at room temperature from isolated $[RuCl_2(PPh_3)_2(=CHPh)]$ (that is relatively instable in solution) or by adding PCy_3 in situ at –50°C after generation of the Ru-benzylidene complex. This yielded Grubb's metathesis catalyst $[RuCl_2(PCy_3)_2(=CHPh)]$ of the first generation whose X-ray crystal structure revealed a distorted square pyramidal coordination.[15.11a] Herrmann, who pioneered the use of Arduengo's NHCs in catalysis, then synthesized another olefin metathesis catalyst, $[RuCl_2(NHC)_2(=CHPh)]$, resulting from the exchange of both phosphines with the bis-cyclohexyl NHC, from $[RuCl_2(PCy_3)_2(=CHPh)]$. Subsequently, in 1999, substitution of only one PCy_3 by Arduengo's stable bis-mesityl NHC was reported by the Grubbs, Nolan, Herrman and Fürstner groups yielding the Grubbs olefin metathesis catalyst of second generation. Variation of the bis-mesityl NHC structure (saturated vs. unsaturated bis-mesityl NHC, achiral or chiral) produced a family of very efficient and stable metathesis catalysts bearing both a PCy_3 and a bis-mesityl NHC.[15.3,15.11b] This family of Ru-benzylidene complexes has later been extended with many variations, in particular by the introduction of chelating benzylidene ligands (see Chap. 15).

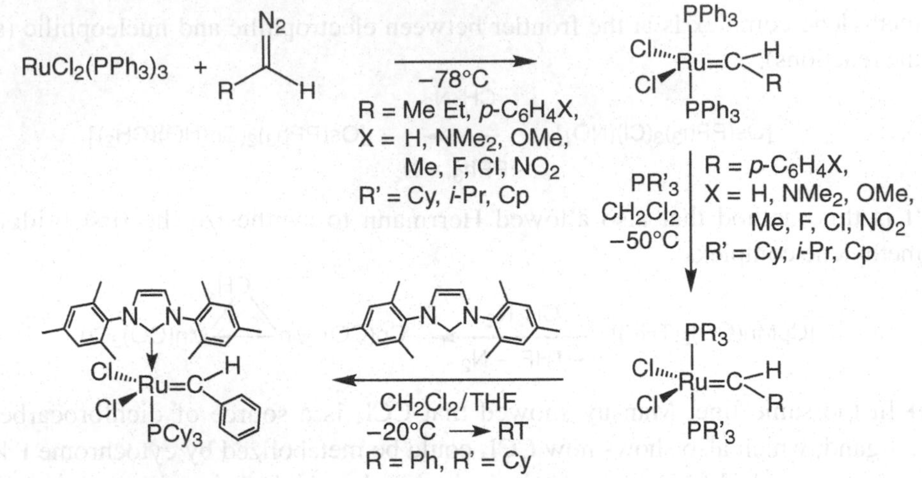

NHC generated from NHC, HCl and *t*-BuOK in THF at −20°C,
added to a toluene solution of Ru = C

▸ The cross-metathesis reaction (see Chap. 15 for mechanistic understanding) of the Ru-benzylidene complexes also yields other alkylidene complexes quantitatively in the presence of a tenfold excess terminal olefin. The methylene complex [RuCl$_2$(PCy$_3$)$_2$(=CH$_2$)] could also be made (only) in this way in the presence of 100 psi ethylene at 50°C in CD$_2$Cl$_2$ and isolated as a red-purple, air-stable solid that decomposes in CH$_2$Cl$_2$ or C$_6$H$_6$ solution.[15.11a]

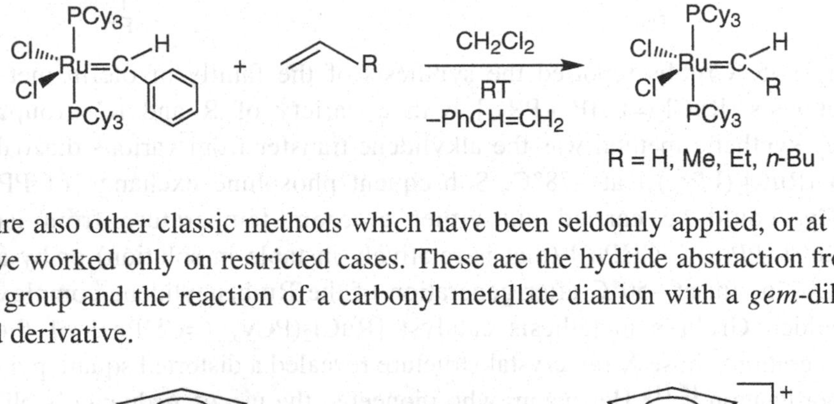

R = H, Me, Et, *n*-Bu

There are also other classic methods which have been seldomly applied, or at least that have worked only on restricted cases. These are the hydride abstraction from a methyl group and the reaction of a carbonyl metallate dianion with a *gem*-dihalogenated derivative.

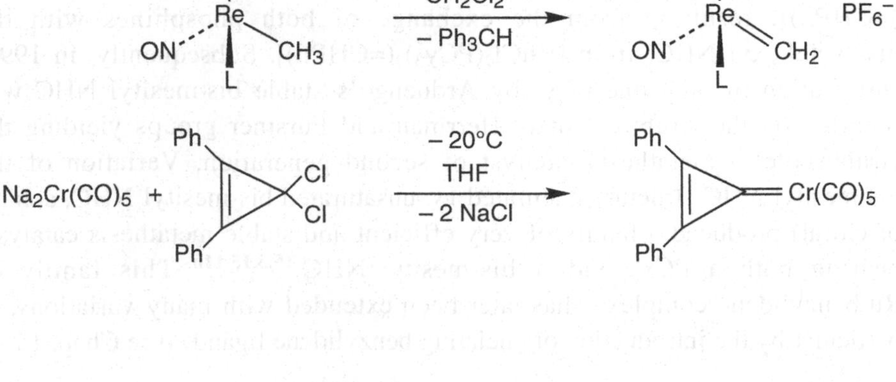

Electrophilic carbenes stabilized by heteroatom(s)

In 1964, Fischer showed that heteroatom-stabilized transition-metal carbenes are accessible by the reaction of a metal carbonyl with a powerful anionic nucleophile (typically a carbanion or an amide). The anion obtained can also be formulated as an acylmetallate. It is stabilized by the reaction of an electrophile:

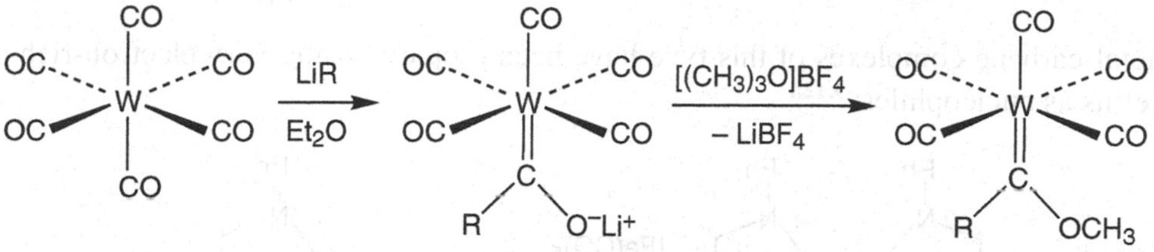

The anionic metal-carbene intermediate having an acylmetallate mesomeric form, the follow-up alkylation can occur either on the oxygen atom as above leading to the Fischer carbene or on the metal to give a metal-acyl-alkyl intermediate. The latter undergoes reductive elimination to give the dissymmetrical ketone as with Collman's reagent (see Chap. 21). The rationalization of this competition between O- and M-alkylation is provided by Pearson's theory of hard and soft acids and bases (HSAB theory) well known in organic chemistry (C- *vs.* O-alkylation of enolates). Thus, O-alkylation (hard) is favored by the use of HMPA as the solvent, NMe_4^+ as counter-cation M^+ instead of Li^+, a large alkylating agent and a slightly reactive leaving group (toluene sulfonate rather than a soft anion such as iodide). These problems have been enlightened by Semmelhack with the use of $[Fe(CO)_5]$:[9.11]

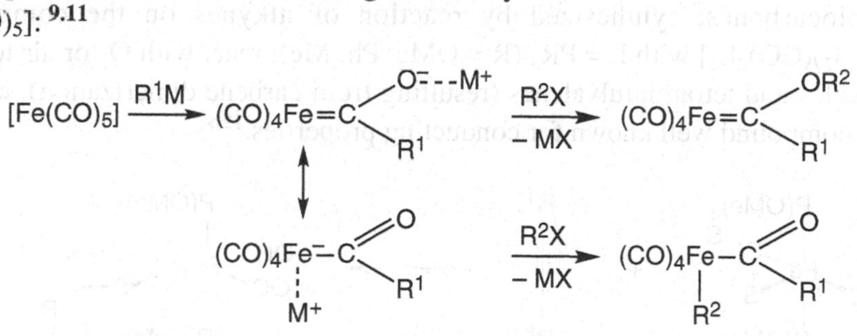

The neutral complexes of acyl, produced by acylation of a monoanionic metallate, can react with powerful electrophiles to give hydroxycarbene complexes:

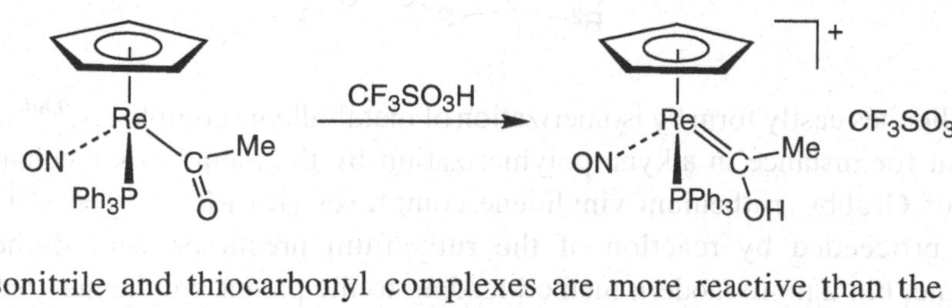

The isonitrile and thiocarbonyl complexes are more reactive than the carbonyl ligand towards nucleophiles and can even react with neutral nucleophiles such as amines and alcohols:

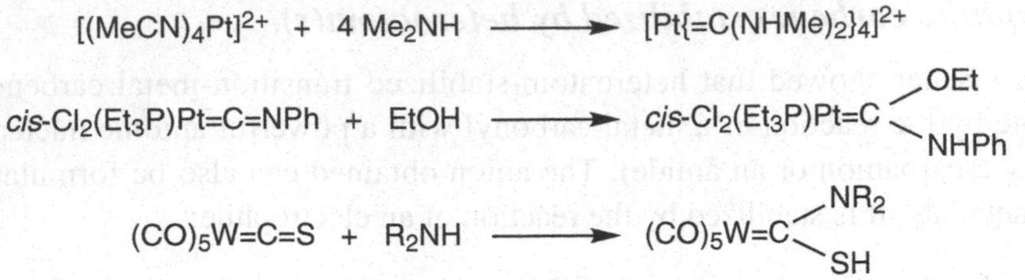

$$[(MeCN)_4Pt]^{2+} + 4\ Me_2NH \longrightarrow [Pt\{=C(NHMe)_2\}_4]^{2+}$$

$$cis\text{-}Cl_2(Et_3P)Pt=C=NPh + EtOH \longrightarrow cis\text{-}Cl_2(Et_3P)Pt=C \overset{OEt}{\underset{NHPh}{\big\langle}}$$

$$(CO)_5W=C=S + R_2NH \longrightarrow (CO)_5W=C \overset{NR_2}{\underset{SH}{\big\langle}}$$

Metal-carbene complexes of this type have been prepared using very electron-rich olefins as nucleophiles: [9.12a]

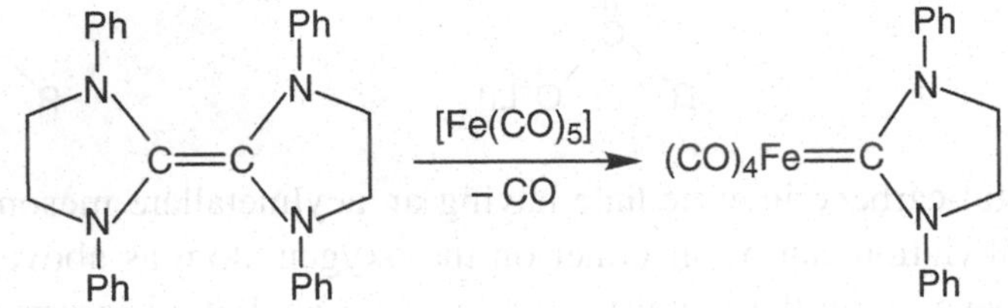

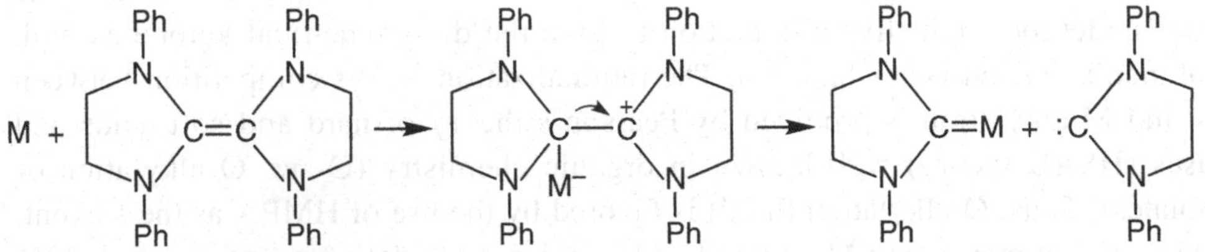

M = Fe(CO)$_4$ or other 16-electron fragment

Iron dithiocarbenes, synthesized by reaction of alkynes on the complexes [Fe^0(η^2-CS$_2$)(CO)$_2$L$_2$] with L = PR$_3$ (R = OMe, Ph, Me), react with O$_2$ or air to give metallocycles and tetrathiafulvalenes (resulting from carbene dimerization), a class of planar compound well known for conducting properties.[12b]

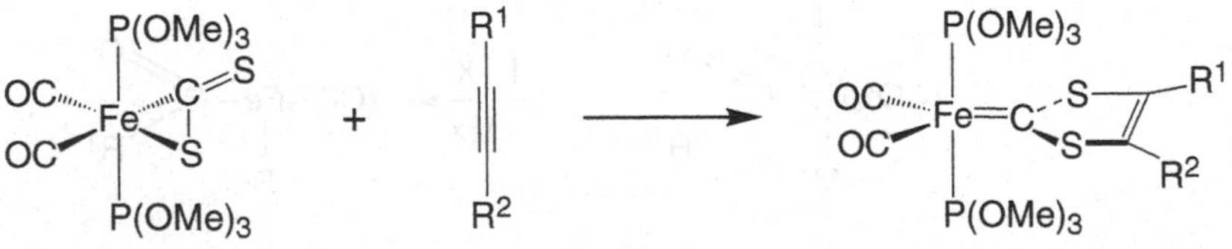

R^1, R^2: H, CO$_2$Et, p-ClC$_6$H$_4$, CHO, CO$_2$Me, p-O$_2$NC$_6$H$_4$, COCH$_3$, C$_6$H$_5$, CHO

tetrathiafulvalenes:

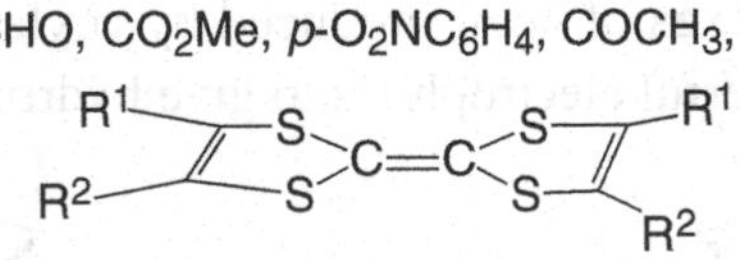

Metal vinylidenes easily form by isomerization of metal-alkyne complexes,[9.8d] which is important for instance in alkyne polymerization by the metathesis mechanism. Synthesis of Grubbs' ruthenium vinylidene complexes [Ru(PR$_3$)$_2$Cl$_2$(=C=CHPh)] originally proceeded by reaction of the ruthenium precursor with diphenyl-cyclopropene, but a more modern method leading to the parent vinylidene complex involves benzylidene exchange for the vinylidene in the metathesis with 1,3-butadiene: [9.8f]

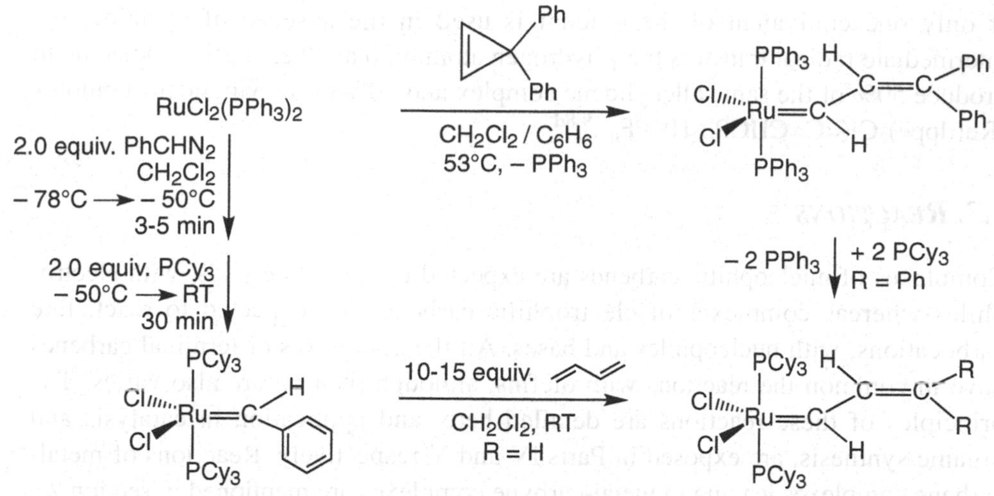

The cationic complexes of vinylidene [9.8a-c] can be obtained by protonation of neutral acetylide complexes. Indeed, the mesomer form of these acetylides is the zwitterionic vinylidene. These cationic vinylidene complexes react with ethanol to give ethoxycarbenes. For this reaction, it is possible to start directly from neutral acetylide complexes that react with ethanol in acidic medium:

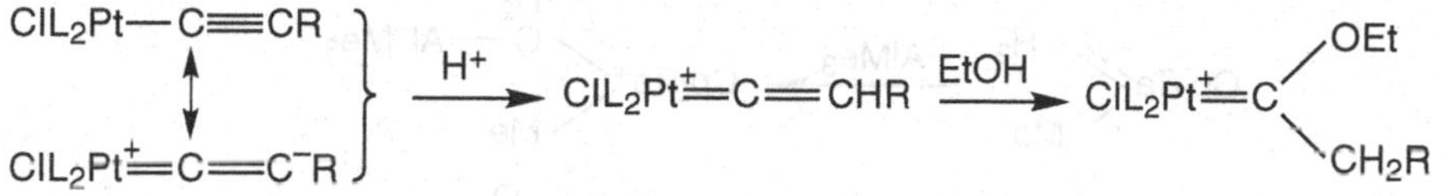

The allenylidene complexes are readily synthesized since the discovery by Selegue of a straightforward general method proceeding by activation of propargylic alcohols.[9.8g]

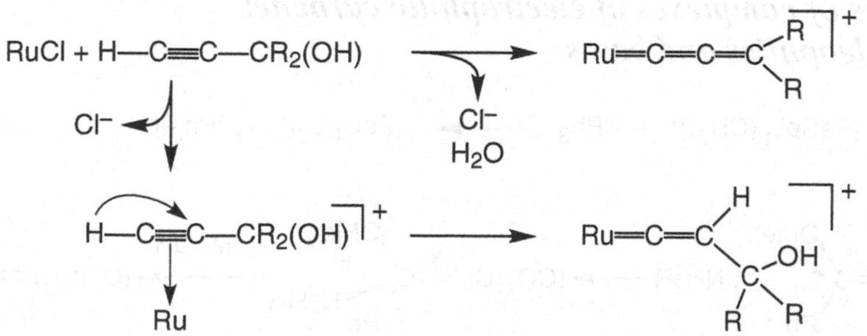

Another more recent synthesis is by oxidation using two equivalents of ferrocenium salt in the presence of pyridine of alkynyl complexes having a γ hydrogen atom: [9.8d]

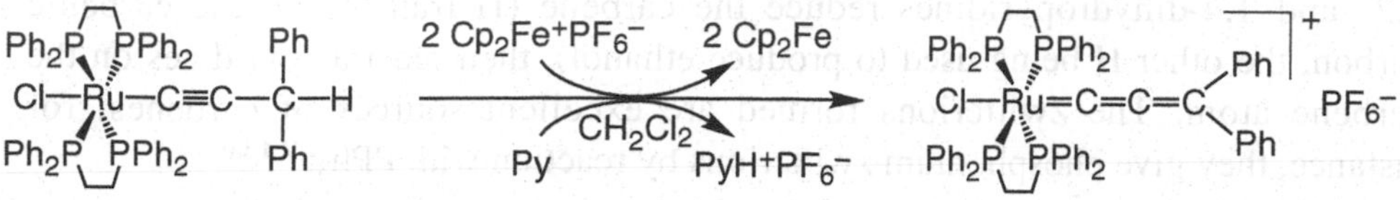

If only one equivalent of the oxidant is used in the absence of pyridine, the intermediate radical transfers the γ hydrogen atom onto another starting molecule to produce 50% of the same allenylidene complex and 50% of the vinylidene complex $[Ru(dppe)_2Cl(=C=CHCPh_2H]^+PF_6^-$.[9.8d]

1.3. REACTIONS

Complexes of nucleophilic carbenes are expected to react, like ylids, with electrophiles whereas complexes of electrophilic carbenes are expected to react, like carbocations, with nucleophiles and bases. All the complexes of terminal carbenes have in common the reactions with olefins, although their nature also varies. The principles of these reactions are detailed here, and application in catalysis and organic synthesis, are exposed in Parts IV and V respectively. Reactions of metal-carbene complexes leading to metal-carbyne complexes are mentioned in section 2.

Reactions of complexes of nucleophilic carbenes with electrophiles and acids

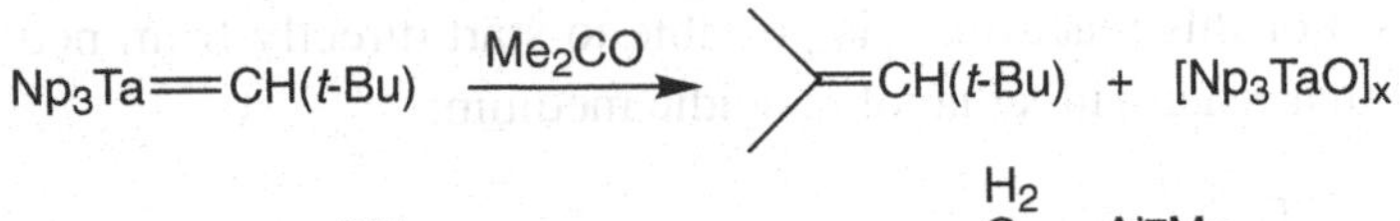

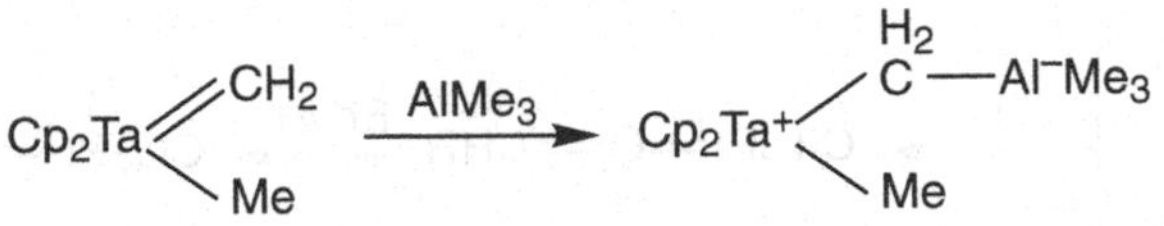

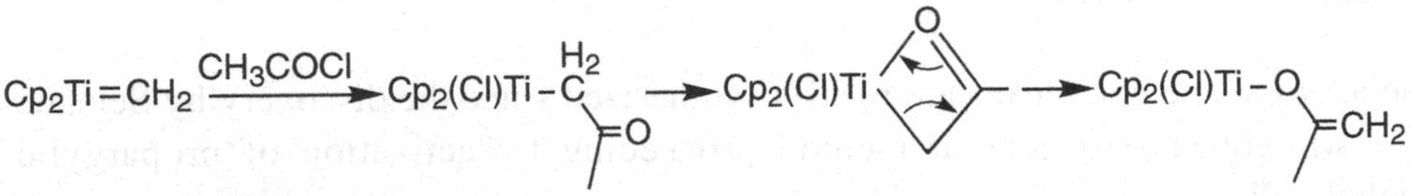

Reactions of complexes of electrophilic carbenes with nucleophiles and bases

$$[FeCpL_2(CH_2)]^+ + PPh_3 \longrightarrow [FeCpL_2(CH_2PPh_3)]^+ \qquad L = CO \text{ or } PPh_3$$

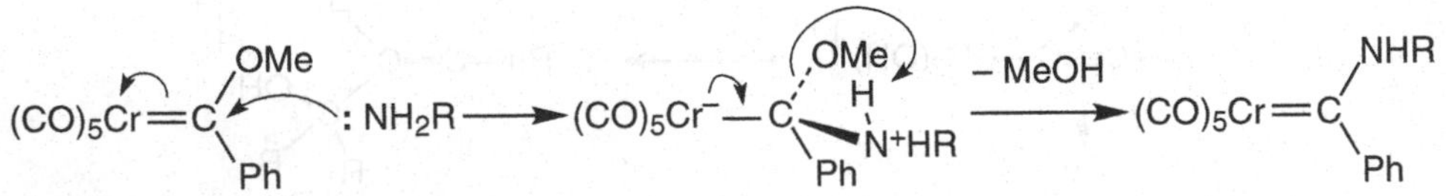

This reaction is analogous to that allowing to synthesize a non-stabilized carbene complex by letting react a Fischer carbene with a hydride (see section 1.2).

1,2- and 1,4-dihydropyridines reduce the carbene (H transfer on the carbenic carbon, the other H being used to produce ethanol), then react as pyridines on the carbene atom. The zwitterions formed are excellent sources of carbenes; for instance, they give phosphonium zwitterions by reaction with PPh_3:[9.13a]

R = (CH$_2$)$_n$R', n = 2-8, R' = Ph, OPh, CH=CH$_2$

n-butyllithium deprotonates Fischer carbenes, and the anion formed reacts with benzaldehyde to yield a complex of carbene that is conjugated with a double bond.

Interconversion between metal-alkyl, -carbene and -carbyne complexes; α-elimination and its reverse, insertion of carbenes into a M-X bond

In the preparative section 3.2 devoted to metal-carbene complexes, it is shown how the α-elimination reaction from high oxidation state early-transition-metal-alkyl complexes is one of the general methods of synthesis of Schrock's Ta and Nb alkylidene complexes. The other direction, formation of an alkylidene from an alkylidyne complex, can also be a valuable route to metal alkylidenes. For instance, Schrock's arylamino-tungsten-carbynes can be isomerized to imido-tungsten-carbene by using a catalytic amount of NEt$_3$ as a base. These compounds are precursors of olefin metathesis catalysts by substitution of the two Cl ligands by bulky alkoxides (dimethoxyethane then decoordinates for steric reasons), and this route was extended to Mo complexes: [9.13c]

Caulton and Eisenstein have examined the interconversion of metal-alkyl, metal-alkylidene and metal-alkylidyne species using combined experimental and DFT theoretical approaches. This useful strategy has yielded examples of such inter-conversions. For instance, in the first reaction below, Me$_3$SiCF$_3$ is a source of nucleophilic CF$_3$ ligand. It is proposed that the transient Ru-CF$_3$ bond does form and further rearranges by α-F elimination producing the difluorocarbene-metal-fluoride product. The latter reacts with CO which provokes the reverse reaction, insertion of the carbene into the Ru-F bond leading to the Ru-CF$_3$ product. [9.13d]

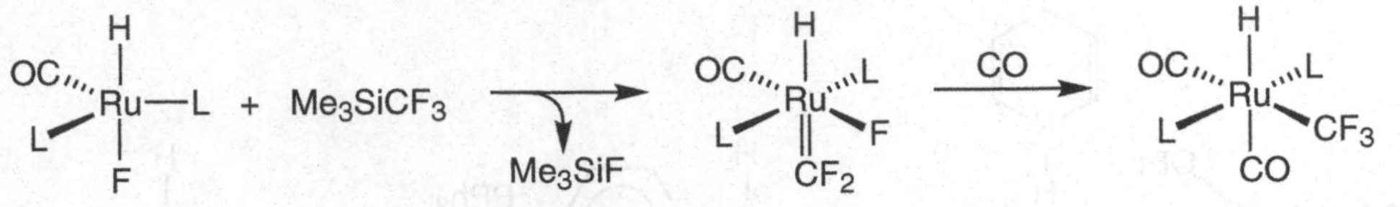

In another example, DFT calculations determined whether a hydrido-metal-carbyne was thermodynamically preferred to the inserted metal-carbene product. The answer to this kind of question was shown to very much depend on the electronic effects of the metal and ligands. Designed, successful experiments in both directions are the following: [9.13d]

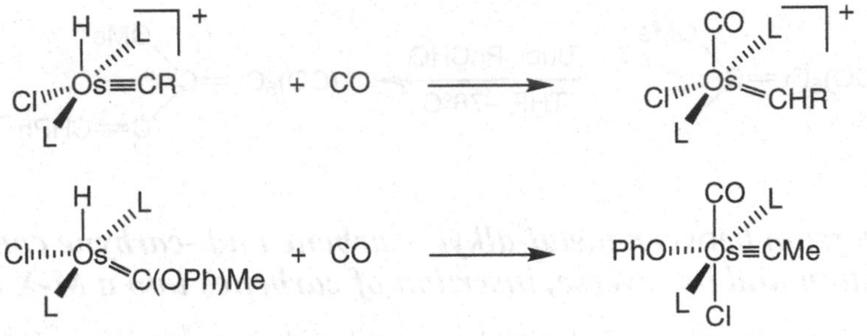

Reactions of carbene complexes with olefins and various modes of evolution of metallocycles

The reactions of metal-carbene complexes with olefins are important in catalysis and organic synthesis. The first reaction disclosed by Pettit in 1966 involved transient $[FeCp(CO)_2(=CH_2)]^+$ that reacted with styrene to give phenylcyclopropane resulting from methylene transfer from the metallacarbocation to the olefin, a reaction that was shown later to be general with iron-methylene complexes of this series.[9.5, 9.6]

However, the reactions with less polar metal-carbene complexes most often lead to coordination of the olefin if the metal has a free coordination site, then formation of a metallacyclobutane that can undergo various types of reactions:

▸ The metallacyclobutane formation being reversible, it is possible to retrieve the starting materials.

▸ The metallacyclobutane has a symmetric topology. Thus it can, analogously to the above process, shift in the perpendicular direction to give the other olefin and the other metal-carbene: this is the Chauvin mechanism of the olefin metathesis reaction (see Chap. 15.2).

▸ Reductive elimination from the metallacyclobutane can give the cyclopropane and liberate the metal that can in turn enter catalytic cycles (see Chap. 15.2).

▸ The β-elimination to a metal-allyl-hydride species that further reductively eliminates to give the olefin and the free metal center that can also be available for new catalytic processes.

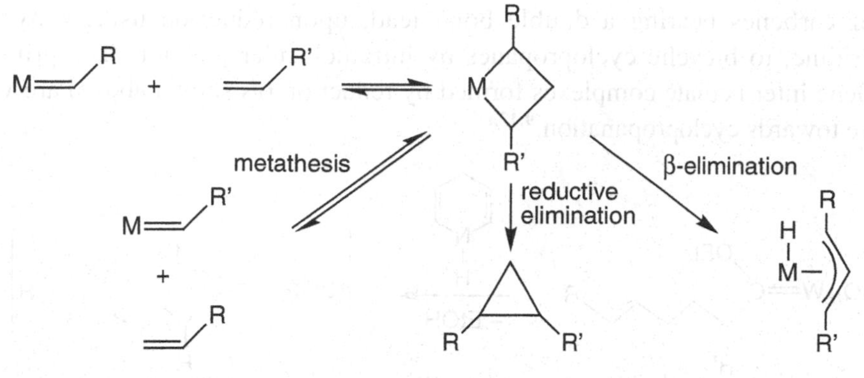

Formation and various modes of evolution of metallacyclobutanes

The mechanism of formation of the metallacyclobutane from the metal-carbene complex and the olefin is different depending on whether the metal complex has 18 valence electrons:

▸ Only if the metal is not saturated can the olefin coordinate before leading to the metallacyclobutane: [9.2]

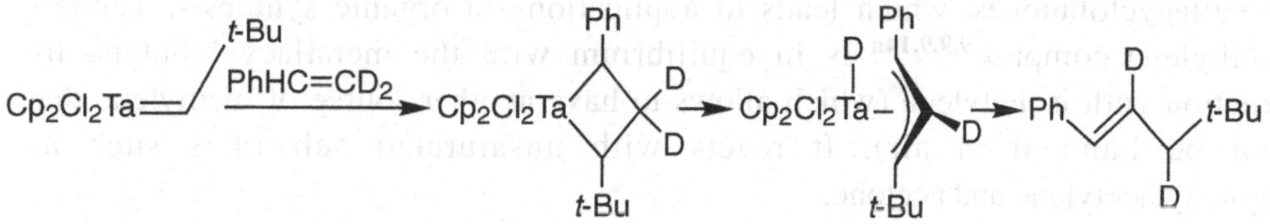

▸ If the metal already has 18 valence electrons, which is the case for all the Fischer carbenes, the electrophilic carbene carbon is attacked by the olefin to develop a zwitterionic intermediate before ring closure.[9.1,9.3] From the 18-electron metalla-cyclobutane, β-elimination, that requires a free coordination site, cannot occur. Among the above reactions, only the metathesis and reductive elimination of the metallacyclobutane to cyclopropane can be observed, as in the two following examples:

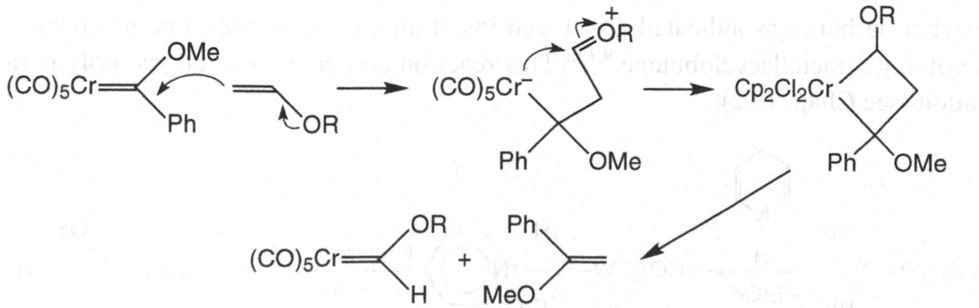

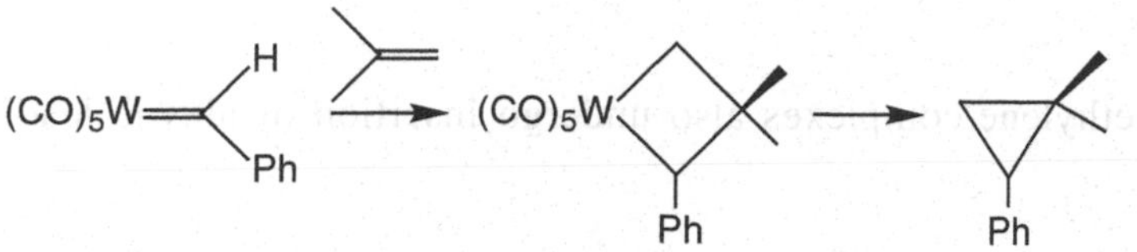

Fischer carbenes bearing a double bond lead, upon reduction using dihydro-1,4-pyridine, to bicyclic cyclopropanes by intramolecular reaction. The primary alkylidene intermediate complexes formed by reduction (as shown above) are very reactive towards cyclopropanation.[9.13a]

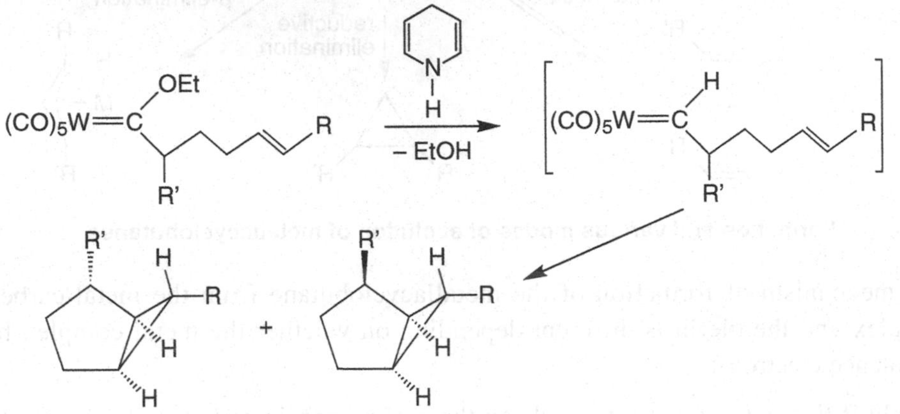

In the same way, the reactions of metal carbene complexes with alkynes give metallacyclobutenes, which leads to applications in organic synthesis. Tebbe's methylene complex[9.9,9.14a] is in equilibrium with the metallacyclobutane by reaction with isobutylene (which allows to have another source of methylene that can be handled in air). It reacts with unsaturated substrates such as diphenylacetylene and acetone.

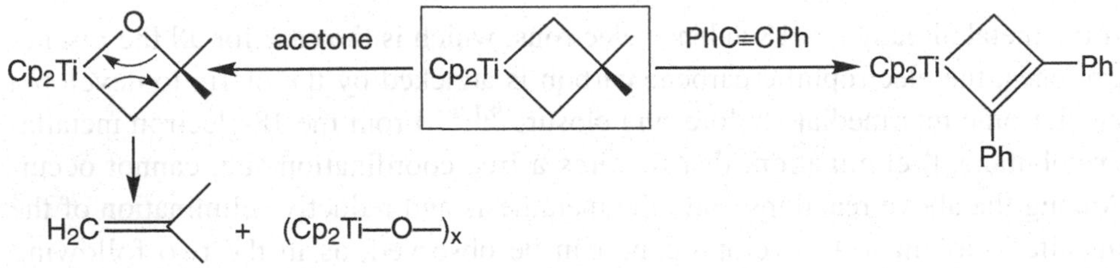

Likewise, zwitterionic complexes obtained by reaction of 1,4-dihydropyridine on Fischer carbenes as indicated above can insert an alkyne by the same mechanism involving a metallacyclobutene.[9.13a] This reaction compares with alkyne polymerization (see Chap. 15.2).

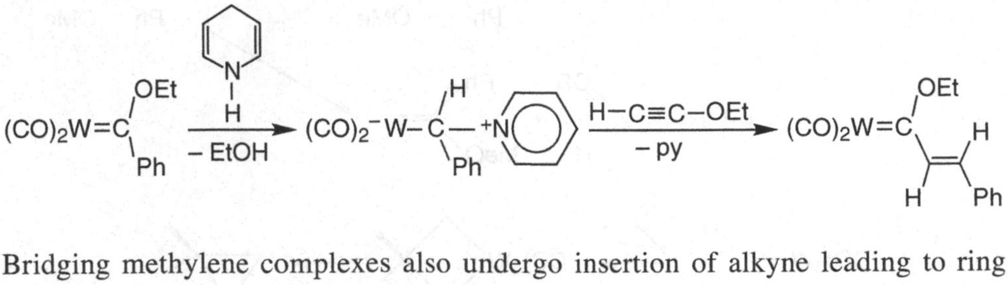

Bridging methylene complexes also undergo insertion of alkyne leading to ring extension:

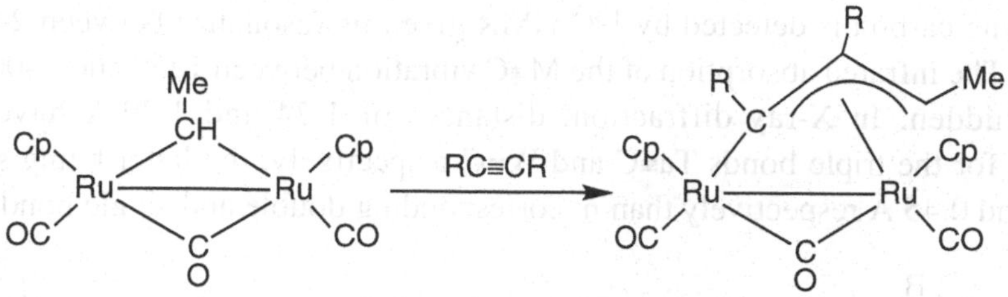

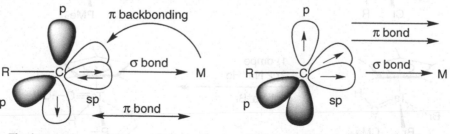

Other formations of metallocycles are possible with various unsaturated substrates. With Roper's osmium methylene complex,[9.4] the carbenic carbon reacts with a nucleophilic substrate (CO) as well as with an electrophilic one (SO_2):

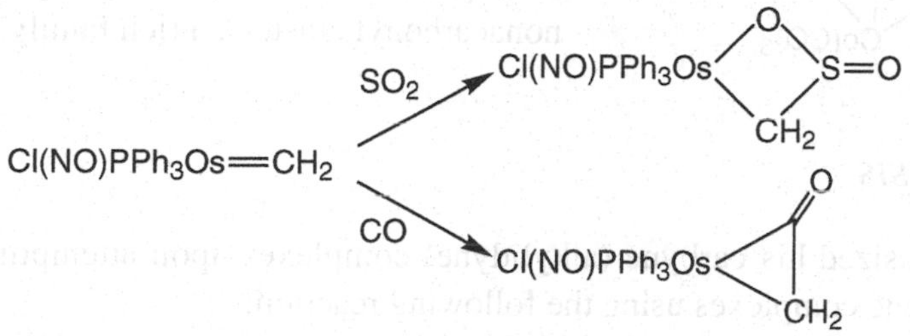

Finally, the redox reactions of metal-carbene complexes lead to the transfer of H atom, alkyl group or carbene ligand.[19.14b]

2. *METAL-CARBYNE COMPLEXES*

To Fischer carbenes correspond Fischer carbynes that are 18-electron complexes of electrophilic carbyne, and to Schrock carbenes correspond Schrock carbynes that are d^0 high-oxidation-state early transition-metal complexes of nucleophilic carbyne, often with a less-than-18-electron count. This latter category is important in alkyne metathesis, and all carbyne complexes show a rich chemistry.

2.1. STRUCTURE

The carbyne ligand CR can be considered either as an LX ligand, or as an X_3 ligand. In Fischer carbynes, the carbyne ligand is counted LX whereas in Schrock's carbynes, the carbyne ligand should be viewed as X_3. This distinction follows the same logic as for metal-carbene complexes, i.e. is derived from the electronic structure of the free carbyne:

Fischer carbyne (RC is a LX ligand) **Schrock carbyne (RC is a X_3 ligand)**

The carbyne carbon is detected by ^{13}C NMR given its resonance between 200 and 350 ppm. The infrared absorption of the M≡C vibration between 1250 and 1400 cm^{-1} is often hidden. In X-ray diffraction, distances of 1.74 and 1.79 Å have been measured for the triple bonds Ta≡C and W≡C respectively, the latter being shorter by 0.15 and 0.45 Å respectively than in corresponding double and single bonds.

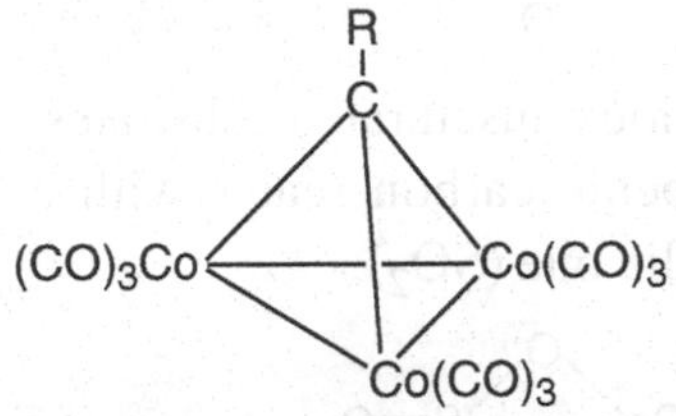

The carbyne ligands can also cap three metals as part of a metal cluster (see Chap. 2). It is then much less reactive. The best known example is Seyferth's alkylidyne tricobalt nonacarbonyl clusters, a rich family.

2.2. SYNTHESIS

Fischer synthesized his carbyne (alkylidyne) complexes upon attempting to make halogenocarbene complexes using the following reaction:

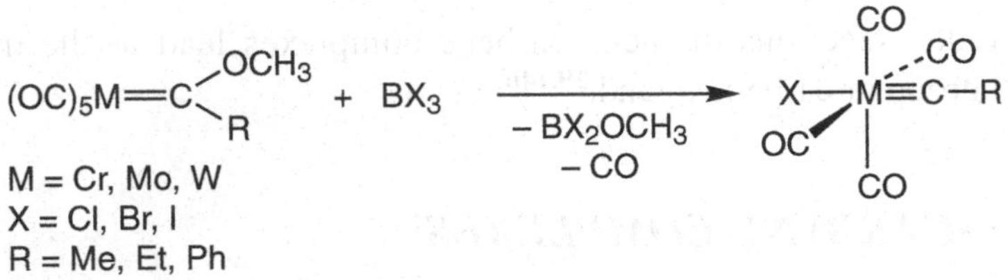

$M = Cr, Mo, W$
$X = Cl, Br, I$
$R = Me, Et, Ph$

Schrock alkylidyne complexes [9.16] are obtained by deshydrohalogenation of primary alkylidene complexes $L_n(X)Ta=C(H)R$ upon reaction with a base or reducing agent. In both cases, the reaction is carried out in the presence of phosphine in order to fill the sites that are left vacant after removing the halide ligand. When the tantalum complex has two halogeno ligands, intermediate d^2 tantalum (III) is formed and undergoes oxidative addition of the C-H bond that was agostic in the d^0 starting complex. This is another mechanism for the α-elimination. Finally, it is possible to obtain W^{VI} metal-carbyne complexes by hexaalkylation of WCl_6 along with double α-elimination or oxidative decarbonylation of a Fischer carbyne.

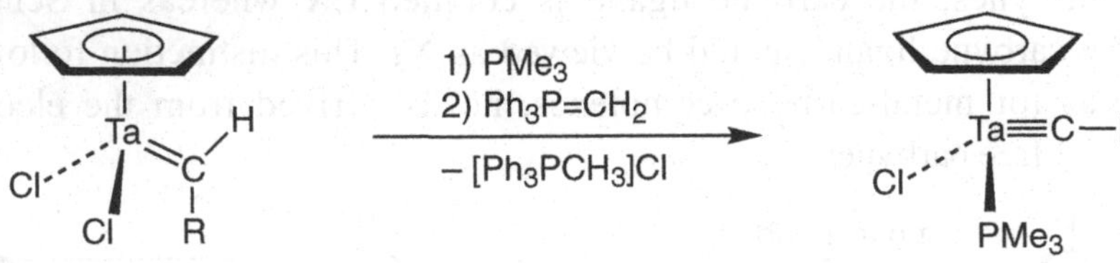

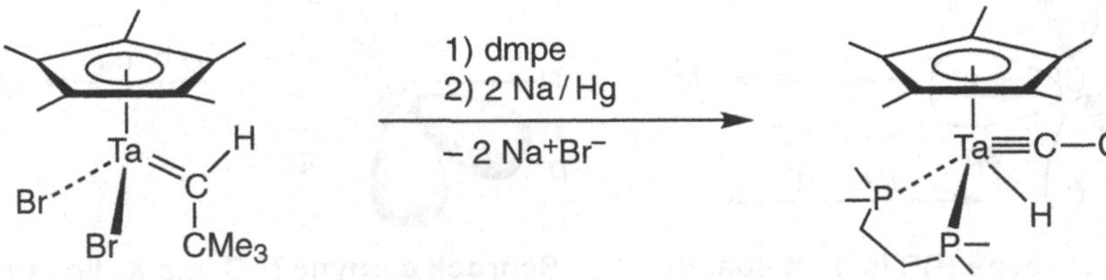

$$[(t\text{-}BuO)_3W \equiv W(Ot\text{-}Bu)_3] + t\text{-}BuC \equiv Ct\text{-}Bu \longrightarrow 2\,[(t\text{-}BuO)_3W \equiv Ct\text{-}Bu]$$

$$WCl_6 + 6\,LiCH_2SiMe_3 \longrightarrow [(Me_3SiCH_2)_3W \equiv CSiMe_3] + 6\,LiCl + 2\,Me_4Si$$

$$[Br(CO)_4W \equiv CMe] \xrightarrow{\ Br_2,\ DME\ } [Br_3(DME)W \equiv CMe] + 4\,CO$$

2.3. REACTIONS

The three types of reactions of metal-carbyne complexes are metathesis whose catalysis aspects are developed in Chap. 15.2, the reactions of the carbynic carbon with electrophiles and nucleophiles depending on the polarity, and the heterocycle formations with the following substrates:

$$Cl_3(Et_3PO)W \equiv C\text{--}CMe_3 + PhC \equiv CPh \xrightarrow[\ 1\ h\ ]{\text{toluene}} \left[\begin{array}{c} Cl_3(Et_3PO)W = C\text{--}CH_3Me_3 \\ | \qquad\qquad | \\ PhC = CPh \end{array} \right]$$

$$\downarrow$$

$$Cl_3(Et_3PO)W \equiv CPh + PhC \equiv C\text{--}CMe_3$$

$$L_n(CO)_mM \equiv CR + R'Li \longrightarrow [L_n(CO)_mM = CRR']^- Li^+$$

$$(t\text{-}BuO)_3W \equiv Ct\text{-}Bu + 2\,HCl \longrightarrow (t\text{-}BuO)_2Cl_2W = CHt\text{-}Bu + t\text{-}BuOH$$

$$\begin{array}{c} L \\ OC \cdots | \\ \quad\;\; Os \equiv C\text{--}Ar \\ Cl \;\; | \\ L \end{array} \xrightarrow{\ SO_2\ } \begin{array}{c} \qquad\; Ar \\ L \;\;\; | \\ OC \cdots | \;\; C \\ \quad\;\; Os \;\;\;\; S=O \\ Cl \;\; | \;\;\; O \\ L \end{array}$$

3. MULTIPLE BOND BETWEEN AN HETEROATOM AND A TRANSITION METAL: OXO, IMIDO, PHOSPHINIDENE AND NITRIDO LIGANDS

The complexes with a double bond M=O and M=NR, i.e. metal-oxo and metal-imido are isoelectronic to metal-carbene complexes and show, as the latter, a rich chemistry, all the more so as the binary complexes $M(=O)_n$ and $M(=NR)_n$ are well known.[9.17,9.18] The phosphinidenes M=PR and the nitrido complexes M≡N, isoelectronic to metal carbynes, are also known. All these ligands can bridge two or three metals, in which case there is no multiple bond.

3.1 EXISTENCE AND STRUCTURE

An essential difference between multiple metal-oxygen and metal-nitrogen bonds and multiple metal-carbon bonds is the larger electronegativity of oxygen and nitrogen that causes a systematic polarity of the bonds with a metal such as $M^{\delta+}=O^{\delta-}$ (see Chap. 16.5.1) or $M^{\delta+}=N^{\delta-}$:

$$M=O \longleftrightarrow M^+{-}O^- \qquad M=NR \longleftrightarrow M^+{-}NR^-$$

We have seen that metal-carbene complexes polarized in this way, $M^{\delta+}=C^{\delta-}$, are Schrock carbenes. Thus, the reactivity of the M=O and M=N complexes resembles that of early transition-metal complexes of nucleophilic alkylidene complexes, which has consequences in organic chemistry and catalysis (see Part IV). It is indeed possible to transfer the O and NR ligands in the same way as alkylidene ligands of early transition-metal complexes were transferred.

A consequence of the strong polarization of the M=O bond is the lack of such a double bond in early transition metal (groups 3 to 5), lanthanide and actinide complexes because of systematic M-O-M bridging. In transition-metal complexes, the oxo ligand is very basic and can be protonated in hydoxy or alkylated in alkoxy ligand:

$$M{-}O{-}R \xleftarrow{\text{RCl}} M^+{-}O^- \xrightarrow{\text{HCl}} M^+{-}O{-}H$$

On the other hand, when the number of d electrons becomes too large, i.e. on the right of the periodic table, there is, in the metal-oxo complex, an important repulsion between these filled d non-bonding orbitals and the non-bonding orbitals of the heteroatom. The consequence is that M=O bonds are instable at the right of the iron column. Between these two sides of the periodic table, the most stable metal-oxo (M=O) complexes are found along a diagonal from V to Os, the most common ones being the Mo family. The M=O complexes are most often d^0, i.e. do not have non-bonding electrons that would have to face the non-bonding oxygen p lone pairs. Some d^2 complexes are known, however, and even some d^4 ones, but not in octahedral geometry because the latter would maximize this destabilization. For instance, $[Re(=O)X(RC{\equiv}CR)_2]$ is tetrahedral.

The same observation can be made for the imido complexes M=NR, an extremely rich family. Indeed, nitrogen being not as electronegative as oxygen, the above constraints are less severe. Thus the M=NR family spreads from the titanium column to the iron one and even to iridium. Here again, there is a predilection for d^{0-2} complexes, but it is possible to raise the NNBE until d^6, for instance in $Cp^*Ir \leqq NAr$, with a linear geometry, below.

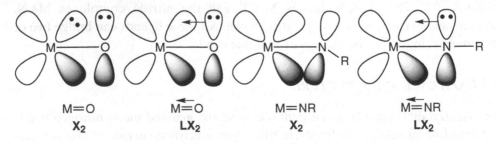

The second important feature in the structure of M=O and M=NR complexes is the duality of behavior concerning the number of electrons given to the metal. Indeed, a lone pair of the heteroatom can be given to the metal, if the latter is electron deficient after forming the double bond M=O or M=NR. In this case, O and NR are LX_2 ligands (usually noted $M\overset{\text{s}}{=}O$ or $M\equiv O$). On the other hand, if the metal already has NVE = 18, it does not accept this lone pair, and O and =NR are only X_2 ligands. This situation usually corresponds for the imido ligand to a bent M=N-R geometry allowing to orientate the nitrogen lone pair away from the coordination sphere of the metal. The classic example is that of the Mo complex (on the side) that contains both a NPh ligand of LX_2 type and another NPh ligand of X_2 type in order to allow molybdenum to follow the 18-electron rule.

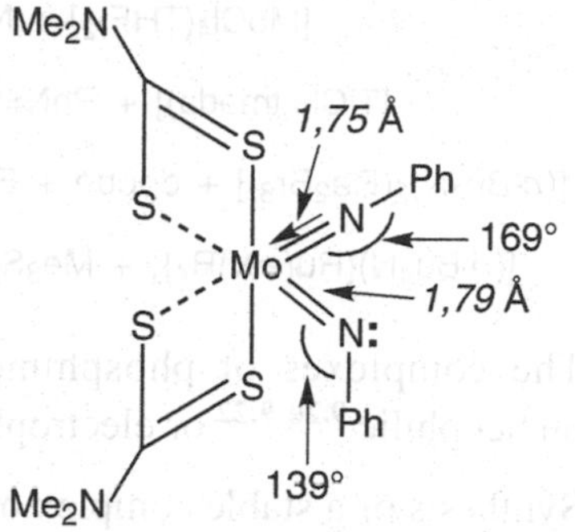

The same considerations govern the coordination chemistry of the nitrido N ligand that can be a 3-electron X_3 ligand or a 5-electron LX_3 ligand depending on the need of the metal.

3.2. SYNTHESES

The metal-oxo complexes are often obtained by reaction of air or water with an oxophilic metal precursor located along the V-Os diagonal. There are other more sophisticated methods as, for instance, the remarkable oxidative addition of a carbonyl from W^{II} to W^{VI}.

$$ReMe_6 \xrightarrow{\ O_2\ } ReOMe_4$$

$$Cp^*Re(CO)_3 \xrightarrow{\ O_2,\ h\nu\ or\ H_2O_2\ } Cp^*ReO_3 + 3\,CO$$

$$Np_3W\equiv Ct\text{-}Bu + H_2O \longrightarrow Np_2W(\equiv O)(\equiv CHt\text{-}Bu) + NpH$$

$$Np = Me_3CCH_2$$

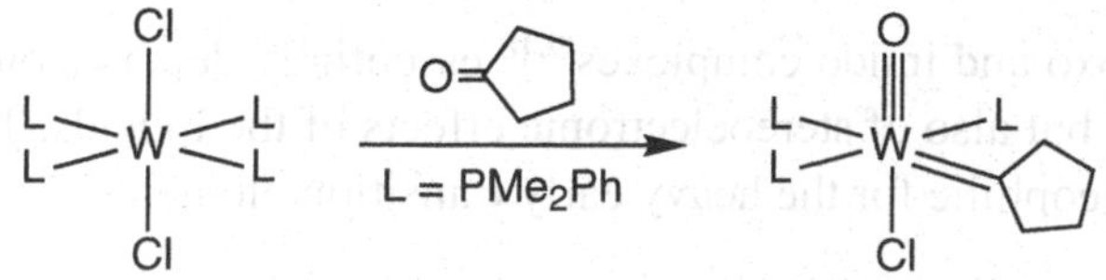

It is possible to synthesize imido complexes by cleavage of a N-H or N-Si bond of an amine, upon complexation with an oxide or a metal chloride, by metathesis, by reaction with a nitrile, by electrophilic attack of a nitride or by other reactions of various nitrogen-containing precursors:

$$[OsO_4] + H_2N\text{-}t\text{-}Bu \longrightarrow [Os(=N\text{-}t\text{-}Bu)O_3] + H_2O$$

$$[NbCpCl_4] + (MeSi_3)_2NMe \longrightarrow [NbCp(=NMe)Cl_2]$$

$$[(n\text{-}Bu_4N)_2(Mo_6O_{19})] + Ph_3P=Ntol \longrightarrow [(n\text{-}Bu_4N)_2\{Mo_6O_{18}(=Ntol)\}] + OPPh_3$$

$$[NbCp(=NMe)Cl_2] + [TaCp^*(=NAr)Cl_2] \rightleftharpoons [NbCp(=NAr)Cl_2] + [TaCp^*(=NMe)Cl_2]$$

$$[MoCl_4(THF)_2] + N_3tol \longrightarrow [Mo(=Ntol)Cl_4(THF)]$$

$$[TiCl_2(tmeda)] + PhN=NPh \longrightarrow [Ti(=NPh)Cl_2(tmeda)]$$

$$[(n\text{-}Bu_4N)_2(Re_2Br_8)] + dppbe + EtCN \longrightarrow [Re(=N\text{-}n\text{-}Pr)Br_3(dppbe)]$$

$$[(n\text{-}Bu_4N)\{Ru(\equiv N)R_4\}] + Me_3SiOTf \longrightarrow [Ru(=NSiMe_3)R_4] \quad \text{with } R = CH_2SiMe_3$$

The complexes of phosphinidene [9.19-9.22] can, as the carbene complexes, be nucleophilic [9.20-9.22] or electrophilic,[9.19] terminal [9.20] or bridging.[9.21,9.22]

Synthesis of a stable complex of nucleophilic phosphinidene:

$$[MCp_2(H)Li] + ArPCl_2 \longrightarrow [MCp_2(=PAr)]$$
with M = Mo, W and Ar = $1,3,5\text{-}t\text{-}Bu_3C_6H_2$

Marinetti et Mathey [9.19] have shown that the complexes of electrophilic phosphinidene, although they are not isolable, have a rich chemistry resembling that of carbenes:

The nitrido ligand can be obtained for instance by oxidation of coordinated NH_3:

$$[Os^{III}(NH_3)_6]^{3+} \xrightarrow{\;Ce^{4+}\;} [Os^{VI}(\equiv N)(NH_3)_5]^{3+}$$

3.3. REACTIONS

The reactivity of oxo and imido complexes [9.18] essentially depends on the metal and its oxidation state, but also of stereoelectronic effects of the ligands. The heteroatom ligand is very nucleophilic for the heavy early transition elements:

$$[TaCp^*(=N\text{-}t\text{-}Bu)H_2(PMe_2Ph)] + 4 MeOH \longrightarrow [TaCp^*(OMe)_4] + t\text{-}BuNH_2 + 2 H_2 + PMe_2Ph$$

An impressive consequence of the pronounced polarity of the Zr=NR bond is the activation of methane discovered with the complexes [Zr(alkyl)(NHR)$_3$]. The latter loses the alkane by α-elimination upon heating, producing the Zr=NR(NHR)$_2$

species that is very reactive because of the polarity of the Zr=N bond and the electronic deficiency of Zr.

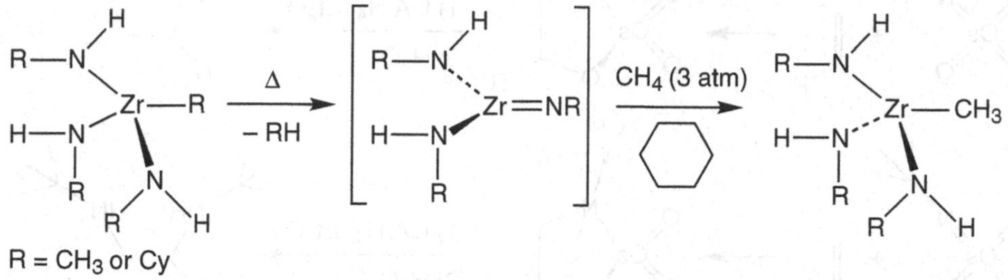

R = CH₃ or Cy

The allylic alcohols react with bis-imido Mo and W complexes leading to allylated amines after hydrolysis. This reaction is a model for the C-N bond formation step in the aminolysis of propene. However, it is not known whether the allyl group is directly captured by the imido ligand or by the oxo ligand followed by O → N migration according to an electrocyclic concerted mechanism. Both hypotheses are represented below:

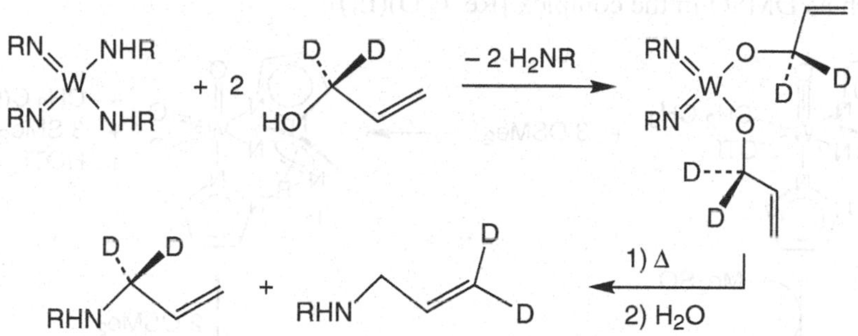

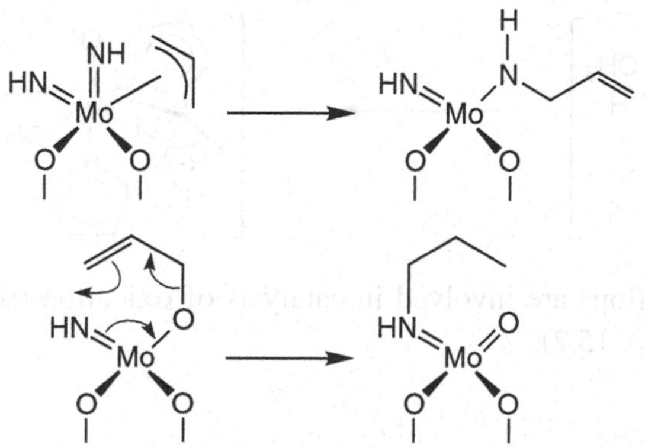

The charge of the heteroatom that is doubly bonded to the metal decreases by going towards the top right of the periodic table. With the complexes [OsO₄] and [Os(=NR)O₃], the O and NR ligands react with the double bond of non-activated alkenes to undergo dihydroxylation, ammoxidation and diamination of the olefins.

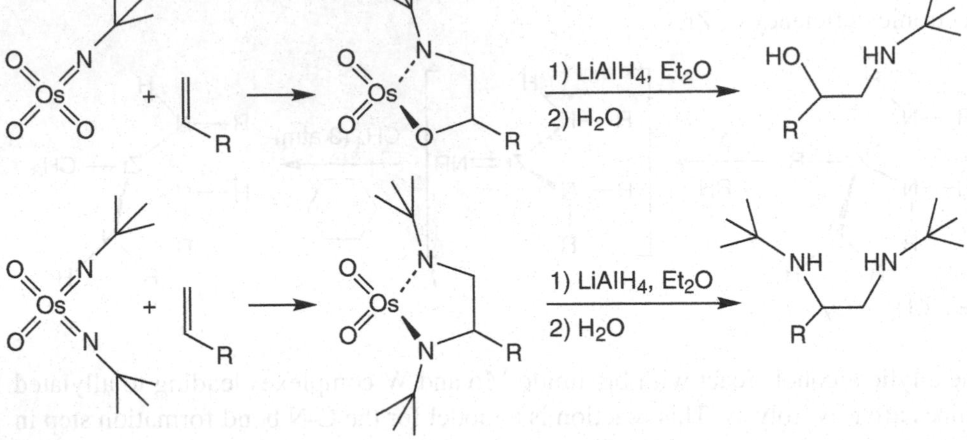

These reactions have been made catalytic by Sharpless and their mechanism is controversial (see Chap. 16.5).

It is also possible to oxidize an ethyl ligand to acetaldehyde using an oxygen atom donor such as DMSO in the complex $[Re^V(=O)(Et)]$:

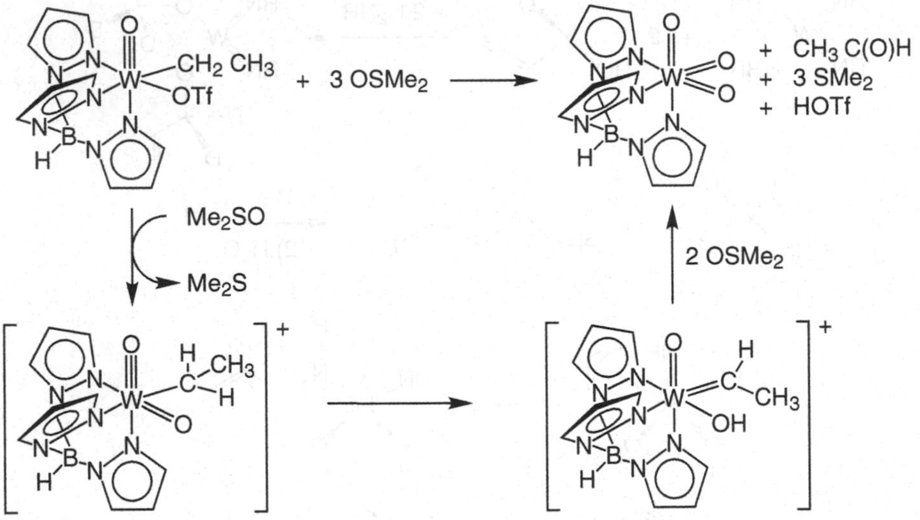

Some of these reactions are involved in catalysis of oxidation (see Chap. 16) and metathesis (see Chap. 15.2).

SUMMARY OF CHAPTER 9
METAL-CARBENE AND CARBYNE COMPLEXES AND OTHER MULTIPLE BONDS WITH TRANSITION METALS

1 - Metal-carbene and carbyne complexes occupy a central place in synthesis and catalysis. Metal-carbene or alkylidene complexes can have an electrophilic singlet carbene (Pettit type) or a nucleophilic triplet carbene (Schrock type, often d^0).

Structure:

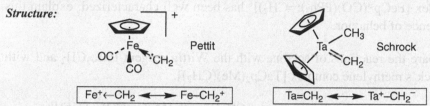

Fisher-type metal-carbene complexes are stabilized by one or two heteroatoms:

The carbene ligand can also bridge two metal.

Syntheses:

Schrock-type carbene complexes: α-elimination reaction (penta-neopentylation of $TaCl_5$, addition of a phosphine, deprotonation of a cationic methyl complex, deprotection of a methylene, dehydrohalogenation) and, more rarely, transmetallation of a carbene.

Pettit-type electrophilic alkylidene complexes: reaction of an acid with a methoxy ligand, protonation of a vinyl complex, reaction of a diazo or another source of carbene on a complex that has a vacant coordination site or a labile ligand.

Fischer-type metal-carbene complexes: reaction of a carbanion on a metal carbonyl; the anions obtained can be protonated or alkylated on the oxygen atom.

Vinylidene complexes $M=CH_2=CHR$: protonation of metal-alkynyl complexes.

Reactivity

• with olefins, yielding metallocycles that can react in various ways: metathesis, reductive elimination (giving a cyclopropane), and β-H elimination followed by reductive elimination (giving an olefin).

• with electrophilic or nucleophilic reagents, depending on their polarity. Example: Schrock-type metal-alkylidenes react like ylids (methylenation of ketones by $Cp_2Ti=CH_2$).

2 - Metal-carbyne complexes $M\equiv CR$ are less known than metal carbenes. The carbyne can also be nucleophilic (Schrock type) or electrophilic (Fischer type). Fischer-type metal-carbynes are obtained by reaction of BF_3 on a neutral Fischer-type metal-carbene complex, whereas Schrock carbynes are often obtained by deshydrohalogenation of a Schrock-type metal-carbene complex. They catalyze alkyne metathesis and, in particular, give heterocycles with unsaturated substrates.

3 - The other complexes containing a multiple bond between the transition metal and an element are the metal-oxo **M=O**, metal-imido **M=NR**, metal-phosphinidene **M=PR** and metal-nitride **M≡N** complexes, all of them being very polar, and the heteroatom ligand is nucleophilic. It is even possible to activate CH_4 with $[ZrR(CH_3)(=NHR)_3]$, R = methyl or cyclohexyl.

EXERCISES

9.1. Which of the two carbene complexes $[TaCp_2(Me)(=CH_2)]$ and $[FeCp^*(PPh_3)(CO)(CH_2)]^+$ reacts with PMe_3? With $AlMe_3$? Write and explain the reactions involved.

9.2. The methylene complex $[FeCp(CO)_2(=CH_2)]^+$ is not isolable whereas the complex $[FeCp^*(CO)(PPh_3)(=CH_2)]^+$ has been well characterized: explain this difference of behavior.

9.3. Compare the reactions of acetone with the Wittig reagent PMe_3CH_2 and with Schrock's methylene complex $[TaCp_2(Me)(CH_2)]$.

9.4. What is the product of the reaction of CH_2N_2 with $[MnCp(CO)_2(THF)]$?

9.5. Compare and explain the relative stabilities of the binary oxo complexes CrO_4^{2-}, MnO_4^- and FeO_4.

9.6. Is it possible to form a metal-carbene or a metal-carbyne complex from an alkene or an alkyne?

9.7. Draw a catalytic cycle of hydroamination of diphenylacetylene by an arylamine $ArNH_2$ (Ar = 2,6-$Me_2C_6H_3$) catalyzed by $Cp_2Zr(NHAr)_2$ and involving imido and azametallacyclobutane intermediates.

Chapter 10

π COMPLEXES
OF MONO- AND POLYENES AND ENYLS

The complexation of unsaturated hydrocarbons by transition metals is a powerful activation method that plays a fundamental role in their stoichiometric and catalytic transformation. The bonds formed are governed by π backbonding. Olefins, dienes and arenes are usually 2-electron L, 4-electron L_2, and 6-electron L_3 ligands respectively, and alkynes are 2-electron L or 4-electron L_2 depending on the needs of the metal. The odd-electron ligands are the 3-electron LX allyl radical and 5-electron L_2X dienyl radicals. Cyclopentadienyl and arene sandwich complexes will be examined in Chap. 11.

1. METAL-OLEFIN COMPLEXES

1.1. STRUCTURE

The π bond from the olefin to the metal is the σ bond (L ligand), whereas the filled d metal orbitals give back an electron density from the side into the antibonding π* ligand orbitals, which makes the π backbonding (see Chap. 2). [10.1, 10.2]

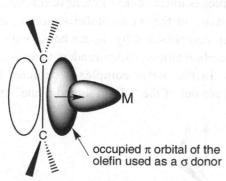

σ bond from the π olefin doublet to the metal in ITS d_z2 orbital

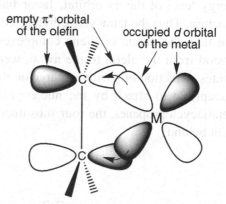

Lateral retrodonation from the filled d metal orbital into the antibonding π* olefin orbital

This bond requires non-bonding d electrons in the metal valence shell. Thus, d^0 complexes such as those of Sc^{III}, Ti^{IV} or V^V cannot give stable olefin complexes. In the classic metal-olefin complexes, the backbonding is not taken into account in the formal electron count (NVE and NNBE), which is justified in the majority of the complexes. With electron-rich late transition metals in low oxidation states (Ni^0, Pd^0 and Pt^0 with phosphine ligands) and/or electron-poor olefins, the backbonding is so important that it is formalized as an oxidative addition of the olefin, i.e. giving a metallacyclopropane. Between these two conventions, the actual situation often is intermediate:

$$M^0 \leftarrow \| \quad\longleftrightarrow\quad \overset{\text{II}}{M}\triangleleft$$

Therefore, X-ray crystallography that allows to accurately determine the exact C-C distance is useful in order to give a measure of the π backbonding. For instance, in Zeise salt $K\{PtCl_3(\eta^2\text{-}C_2H_4)\}.H_2O$, it is 1.37 Å, hardly larger than in free C_2H_4 (1.33 Å). On the other hand, in $[Pt(PPh_3)_2(\eta^2\text{-}C_2H_4)]$, it is 1.43 Å, which indicates a structure close to the platinacyclopropane. The lowering of the C=C bond order can also be monitored by the lowering of its absorption frequency in infrared spectroscopy. For instance, free C_2H_4 absorption is located at 1623 cm^{-1}, whereas in Zeise salt the absorption band of coordinated C_2H_4 is located at 1516 cm^{-1}. The π backbonding partly hinders the rotation about the axis of the metal-olefin bond, and the measure of this rotation barrier is another means to estimate the degree of π backbonding. The range of rotation barriers is rather broad, from 10 to 25 kcal·mol^{-1} (48 to 105 kJ·mol^{-1}), which shows again the flexibility of the π backbonding. The orientation of the olefin in the metal coordination sphere is represented below in the most stable conformation:

Note that the electron-withdrawing substituents on the olefin, by lowering the energy level of the $\pi*$ orbital, favor backbonding and the metallacyclopropanic structure. Thus the tetracyanoethylene complexes much more often have this structure than the parent ethylene complexes. Also, in the metal-olefin structure, the σ bond from the olefin to the metal, weakly compensated by the $\pi*$ backbonding, creates a fraction of positive charge on the carbon atoms, which renders them more susceptible to attack by the nucleophiles. In the olefin complexes depicted as metallacyclopropanes, the four substituents are out of the "olefin ligand plane" and point behind:

This situation fits to electron-poor double bonds of fullerenes such as C_{60}, for instance, given its truncated icosahedral structure (soccer ball). Thus C_{60} and C_{70} form 16-electron metallacyclopropanic-type complexes with the "electron-rich" 14-electron metal fragments $M(PEt_3)_2$ (M = Ni, Pd, Pt): [10.3-10.5a]

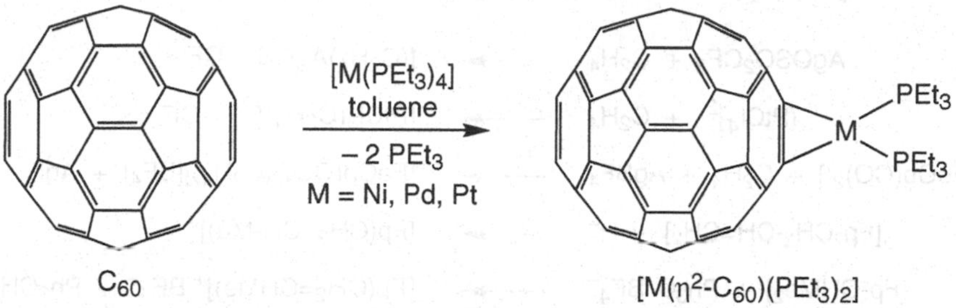

$$C_{60} \xrightarrow[\substack{- 2\,PEt_3 \\ M = Ni,\,Pd,\,Pt}]{\substack{[M(PEt_3)_4] \\ toluene}} [M(\eta^2\text{-}C_{60})(PEt_3)_2]$$

Likewise, fullerenes C_{60} and C_{70} form metallacyclopropanic complexes with various 16-electron fragments that tend to form 18-electron complexes. For example, C_{60} binds three $[Cp^*Ru(MeCN)_2]^+$ fragments, despite the cationic nature of this group, by reaction of a 10-fold excess of the 18-electron precursor $[Cp^*Ru(MeCN)_3]^+$ $[O_3SCF_3]^-$.

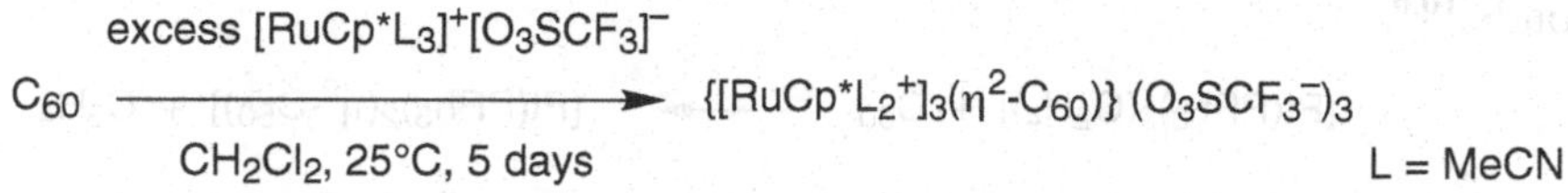

$$C_{60} \xrightarrow[\substack{CH_2Cl_2,\,25°C,\,5\ days}]{\substack{excess\ [RuCp^*L_3]^+[O_3SCF_3]^-}} \{[RuCp^*L_2^+]_3(\eta^2\text{-}C_{60})\}\,(O_3SCF_3^-)_3 \qquad L = MeCN$$

This precursor is well known to very easily give, with aromatics, complexes of the type $[RuCp^*(\eta^6\text{-arene})]^+$, even when the aromatic bears a double bond in a substituent. Thus, this reaction is a good example of the inability of C_{60} to coordinate to the metal in a hexahapto fashion, and on the contrary of its strong tendency to give dihapto complexes.[10.3]

Nakamura has complexed a cyclopentadienyl ring of C_{60} by CpFe, which yields ferrocene-fullerenes. For instance, the parent compound (R = H) can be obtained in 25% yield upon heating C_{60} with $[FeCp(CO)_2]_2$ as the CpFe source. This ferrocene-fullerene can be further alkylated, and a ruthenocene analog is also known with R = Me.[10.5b]

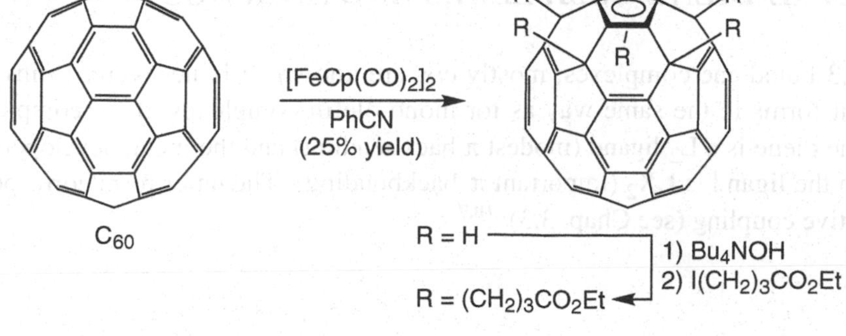

$$C_{60} \xrightarrow[\substack{PhCN \\ (25\%\ yield)}]{[FeCp(CO)_2]_2}$$

R = H $\xrightarrow[\substack{2)\ I(CH_2)_3CO_2Et}]{\substack{1)\ Bu_4NOH}}$ R = $(CH_2)_3CO_2Et$

1.2. SYNTHESIS

The olefin complexes are usually obtained by addition or substitution, eventually using a substrate able to abstract X^- such as Ag^+ or a reductant. It is also possible to protonate an allyl complex or to abstract H^- from an alkyl complex:

$$AgOSO_2CF_3 + C_2H_4 \longrightarrow [(C_2H_4)AgOSO_2CF_3]$$

$$[PtCl_4]^{2-} + C_2H_4 \longrightarrow [PtCl_3(C_2H_4)]^- + Cl^-$$

$$[FeCp(CO)_2I] + C_2H_4 + AgBF_4 \longrightarrow [FeCp(CO)_2(C_2H_4)][BF_4] + AgI$$

$$[Fp-CH_2-CH=CH_2] + H^+ \longrightarrow [Fp(CH_2=CH-Me)]^+$$

$$[Fp-CHMe_2] + Ph_3C^+BF_4^- \longrightarrow [Fp(CH_2=CHMe)]^+ BF_4^- + Ph_3CH$$

1.3. REACTIONS

The main reactions of olefin complexes are the facile exchange of the olefin ligand with another ligand, the insertion of the olefin ligand into an M-H or M-alkyl bond (see Chap. 5), the addition of nucleophiles [10.5] and the activation of allylic C-H bonds.[10.6]

$$[Pt(PPh_3)_2(C_2H_4)] + C_{60} \longrightarrow [Pt(PPh_3)_2(\eta^2-C_{60})] + C_2H_4$$

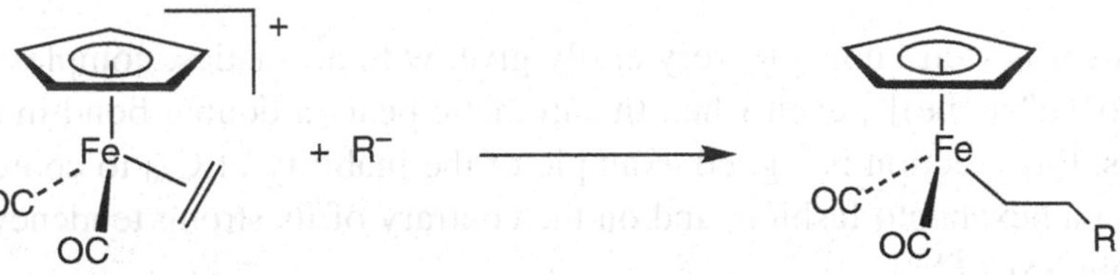

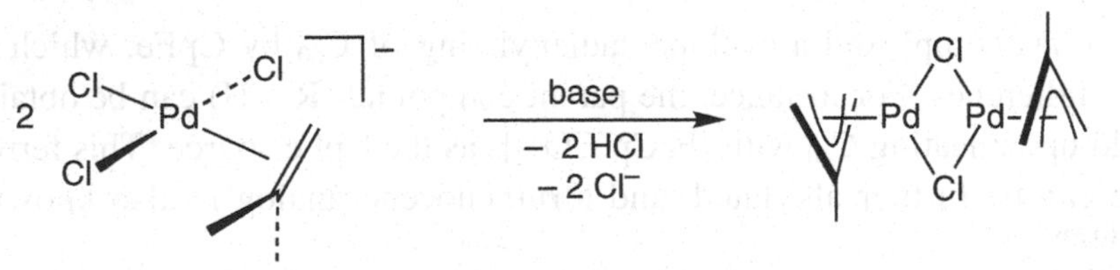

2. METAL-DIENE COMPLEXES AND ANALOGS

Many 1,3-butadiene complexes, mostly *cis*, are known. It is necessary to involve two limit forms in the same way as for mono-olefin complexes, the π complex in which the diene is a L_2 ligand (modest π backbonding) and the metallacyclopentene in which the ligand is LX_2 (important π backbonding). The latter form corresponds to oxidative coupling (see Chap. 3.3): [10.7]

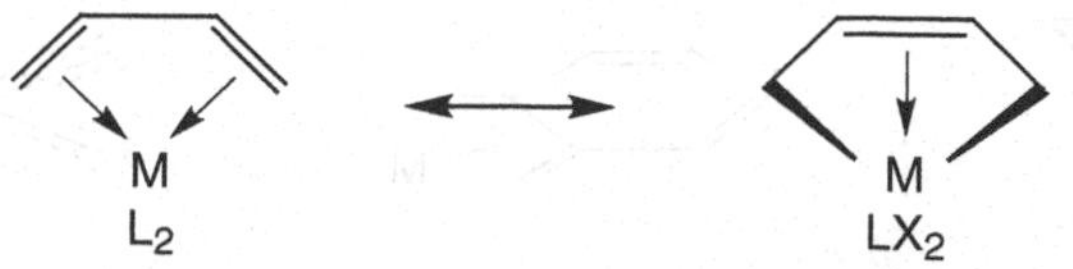

The syntheses are related to the ones used for the mono-ene complexes. The metal-vapor technique of synthesis involving very low pressure of metals has been used to make binary complexes such as molybdenum (0) tris-1,3-butadiene.

The diene ligand can be protonated or acylated. Subsequently, a free coordination site is liberated and can then be filled by a ligand present in the reaction medium or by the oxygen atom of the acyl group. In the absence of such a ligand, this site is occupied by a C-H bond of the dienyl ligand, which provides an *agostic* bond:

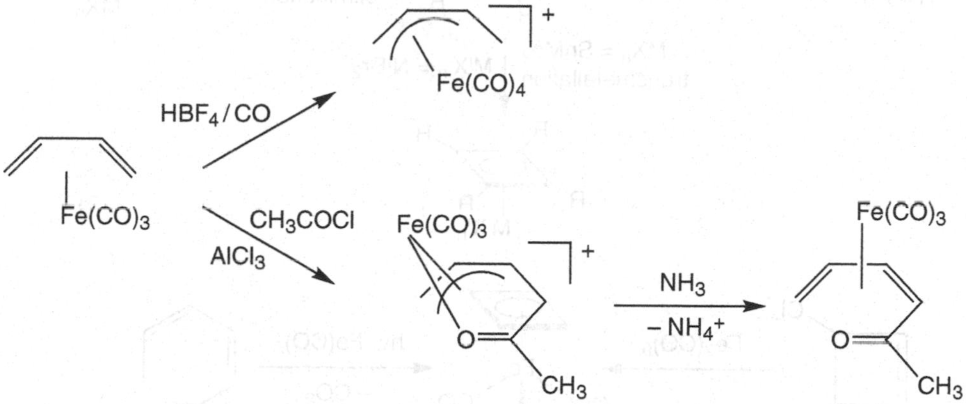

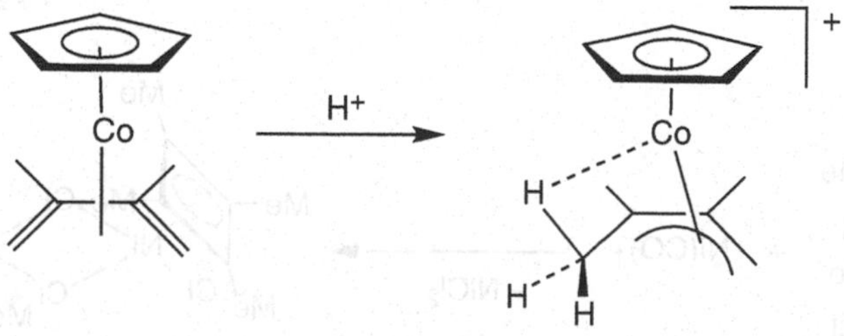

The complex $[Fe(CO)_3(\eta^4\text{-}1,3\text{-cyclohexadiene})]$, useful in organic synthesis, is obtained by complexation-isomerization of the 1,4-isomer by reaction with $[Fe(CO)_5]$. There are also other robust 18-electron complexes of unsaturated hydrocarbons as L_2 ligands that are related to dienes but are instable in the free state: cyclobutadiene, orthoxylylene (orthoquinodimethane) and trimethylenemethane (below).

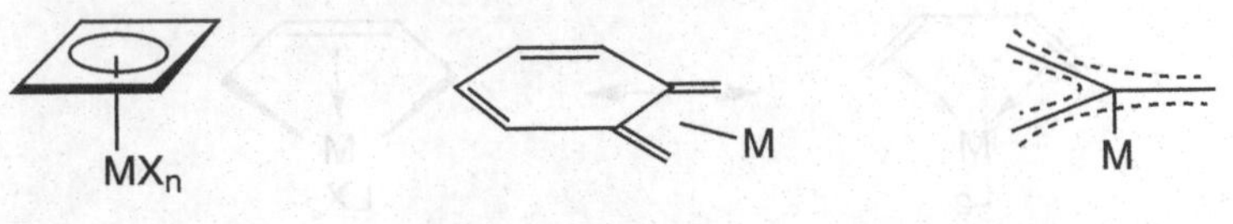

M-η⁴-cyclobutadiene   **M-η⁴-orthoxylylene**   **M-η⁴-trimethylenemethane**

The cyclobutadiene complexes are accessible from dichlorocyclobutene, α-pyrone or alkynes. Disubstituted alkynes often react with transition metals to give tetrasubstituted metallacyclopentadienes that usually yield tetrasubstituted π-cyclobutadiene complexes of the type shown above by reductive elimination. Main-group metallacyclopentadienes (Al, Sn) can transmetallate the ligand onto transition metals that also rearrange to form π-cyclobutadiene complexes. Cyclobutadiene-iron-tricarbonyl undergoes electrophilic substitutions such as acylation of the ring, and carbocations are stabilized on the ring substituent in exocyclic α position (see Chap. 11).

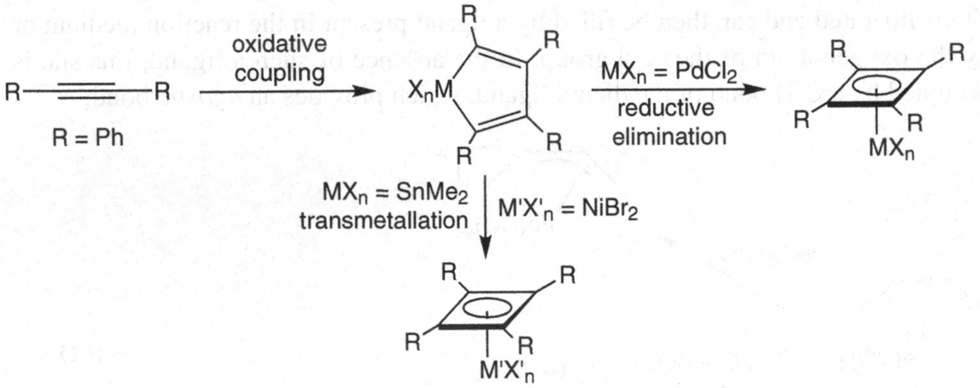

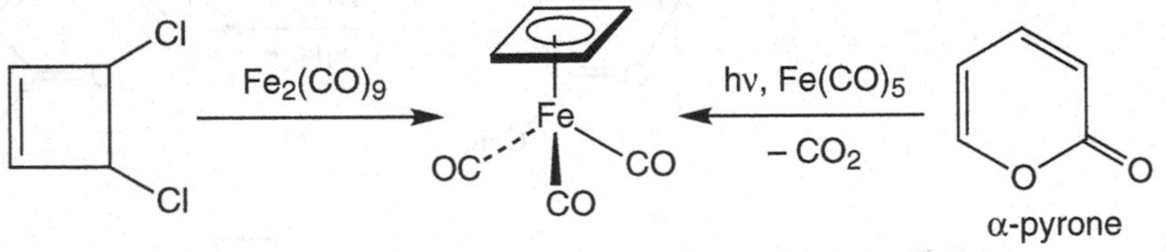

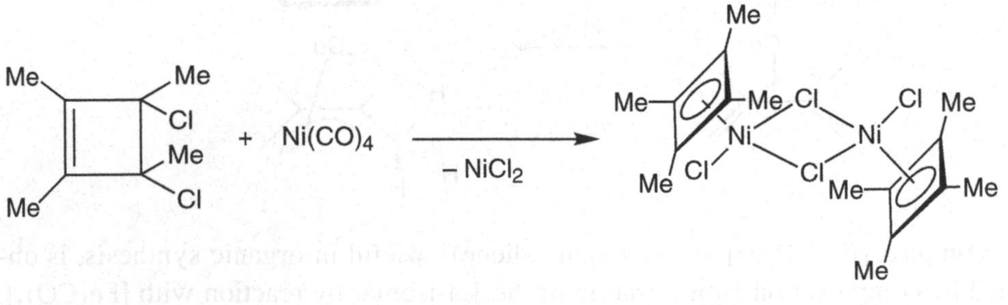

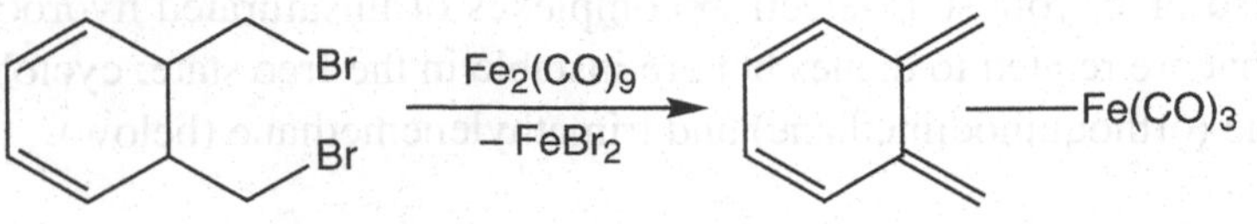

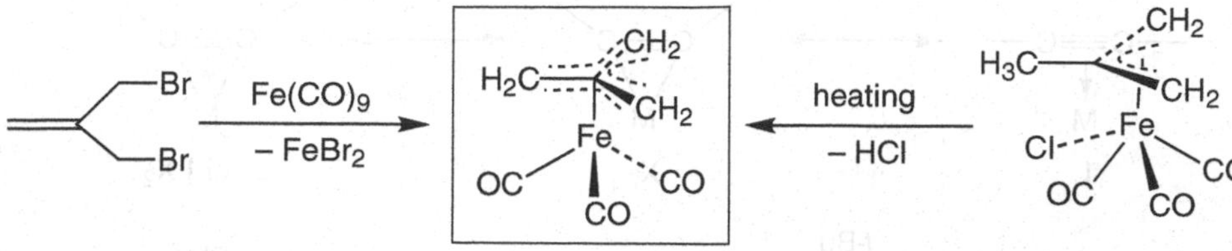

For orthoxylylene, the intracyclic coordination is also known. In the following synthesis, the 20-electron sandwich complex $[Fe(C_6Me_6)]_2$ transfers an electron to O_2 to form the intermediate superoxide radical anion $O_2^{\bullet-}$ which deprotonates the 19-electron cation. This electron transfer/proton transfer sequence is repeated a second time to give the final 18-electron orthoquinodimethane product directly obtained by contact of the starting 20-electron Fe complex with O_2: [10.8]

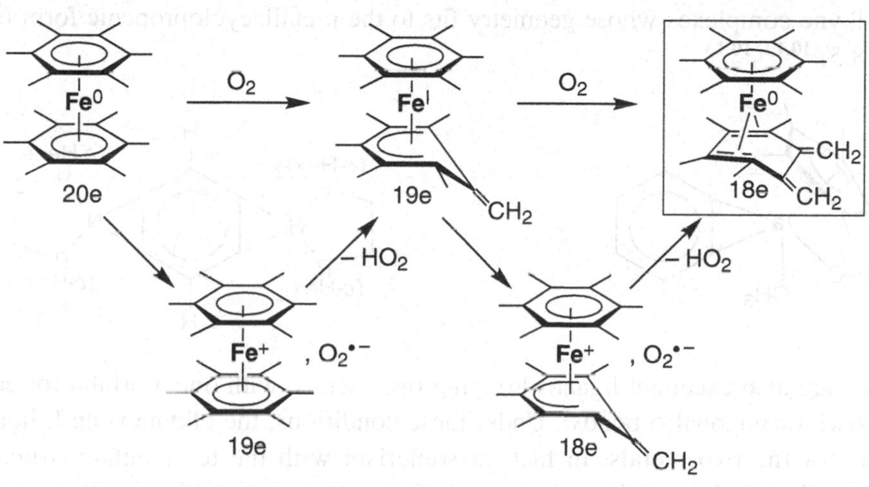

3. METAL-ALKYNE COMPLEXES

The organometallic chemistry of alkynes is very rich, including oxidative addition and coupling, insertion and metathesis and leads to a variety of applications in organic synthesis such as functionalization, C-H and C-C bond cleavage and formation, dimerization, oligomerization and polymerization of alkynes (see Parts IV and V)

Alkynes being much easier to reduce than alkenes (much lower energy of the π* LUMO), the π backbonding of metal d orbitals is much more marked than for alkenes. The adequate formalism for the metal-alkyne bond is thus most often that of a metallacyclopropene. Alkynes usually are 2-electron L ligands. In addition, when the metal is still electron deficient, there is another π (lateral) 2-electron bond towards a vacant d metal orbital from the π orbital orthogonal to that donating the first π doublet forming the σ bond.[10.9-10.11] For example, in the complex $[W(CO)(PhC≡CPh)_3]$, two alkyne ligands are 4-electron donors, whereas the third one is a 2-electron ligand, which brings the NVE to 18.

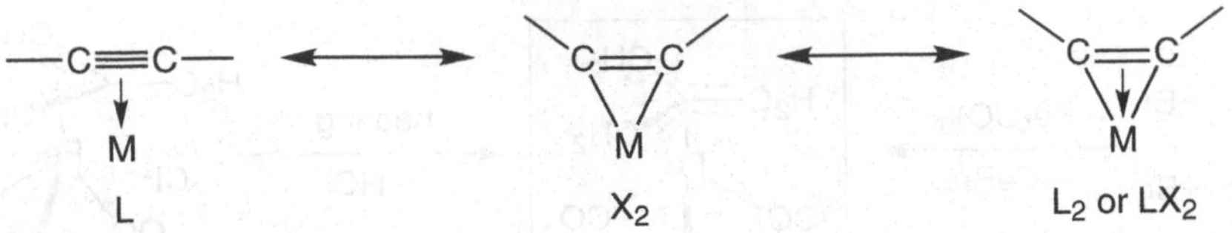

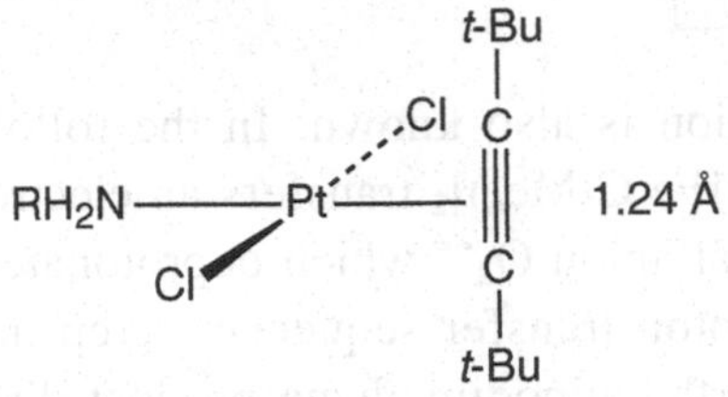

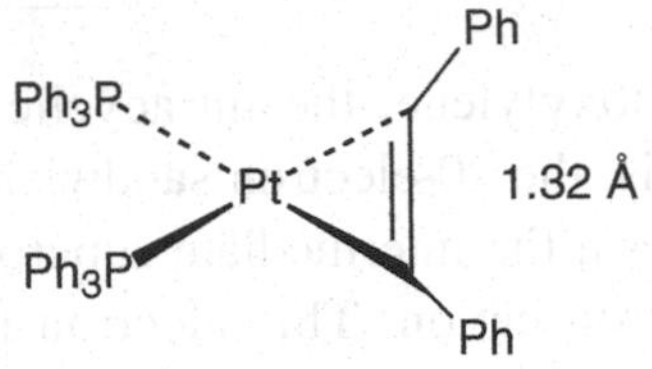

The complexes of benzyne (an instable species in the free state) are remarkable cycloalkyne complexes whose geometry fits to the metallacyclopropenic form (see Chap. 8.5).[10.12, 10.13]

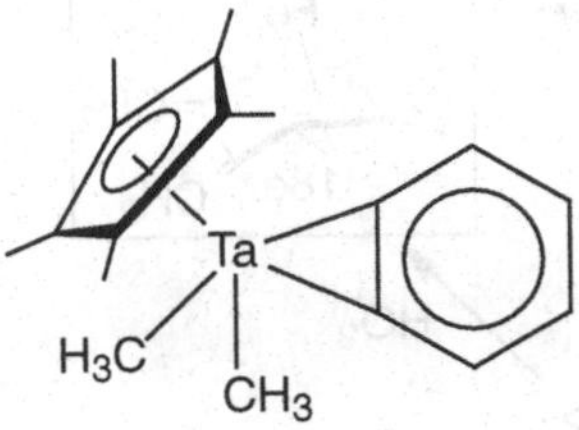

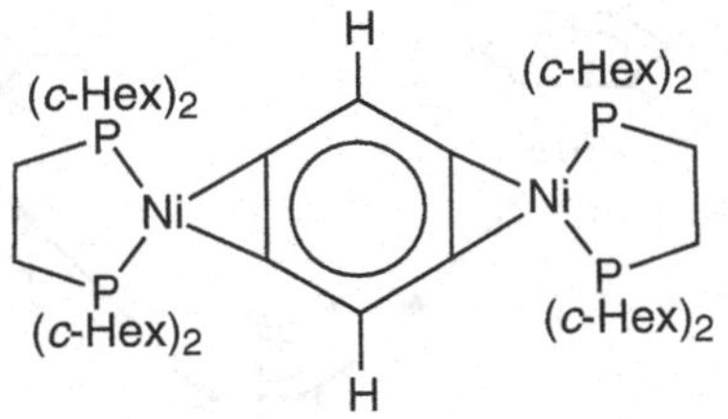

Alkynes are also excellent ligands bridging two metals, with one π orbital for each metal (two orthogonal σ bonds). Under these conditions, the alkyne is an L ligand for each of the two metals. In fact, mesomerism with the tetradentate structure corresponding to the metallacyclopropanic form of monometallic complexes must be taken into account as well (Ni_2C_2 tetrahedron, bottom right formula below).

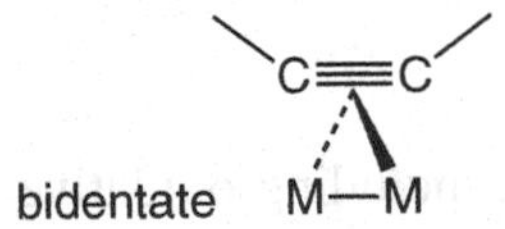

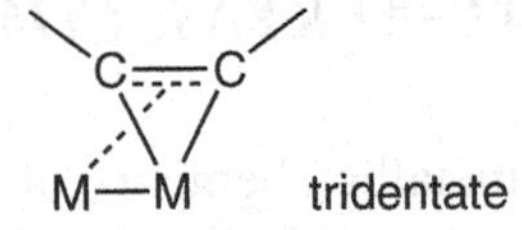

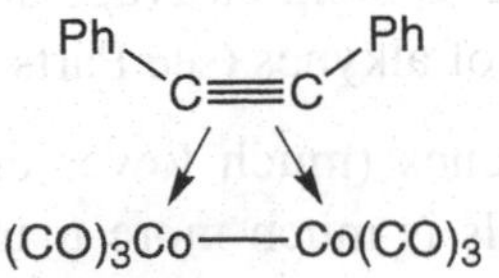

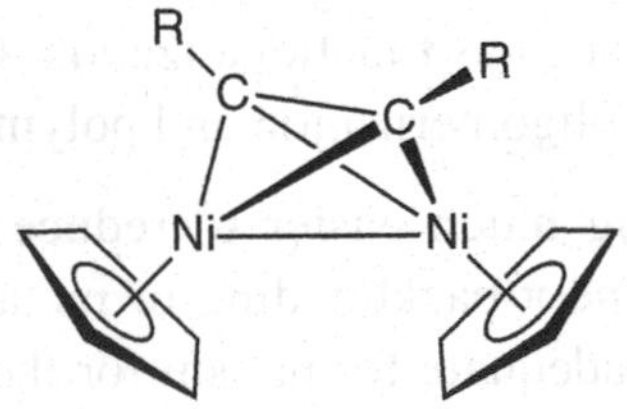

The complexation of alkynes by the fragment $Co_2(CO)_6$, achieved by reaction with $[Co_2(CO)_8]$, is often used to protect the alkynes, a process that is very useful in organic synthesis (see Chap. 21.1). These complexes are clusters, and they react in acidic medium to give the famous Seyferth cluster:[10.14]

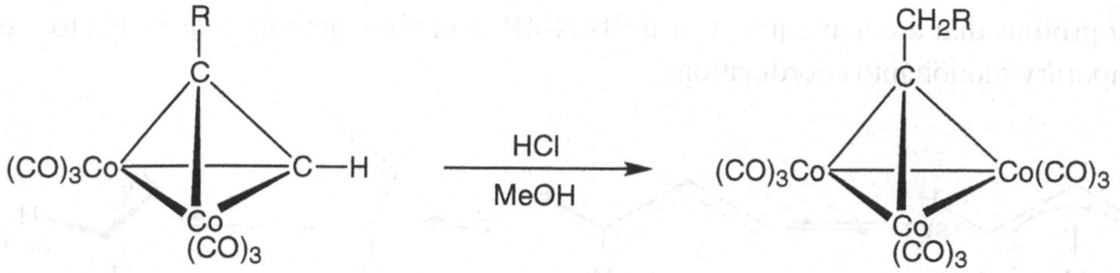

Transition metals induce the stoichiometric or catalytic dimerization, trimerization or tetramerization of alkynes.[10.15] When they are carried out with metal carbonyls, these reactions lead to various cyclic ketone derivatives. Catalytic versions of such reactions have been developed (see Chap. 21.4):

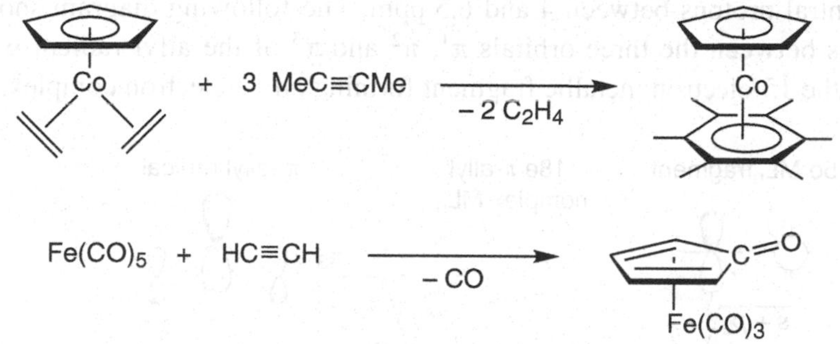

Phospha-alkynes RC≡P behave as alkynes, giving mono- or bimetallic π complexes, for instance with the fragments $Pt(PPh_3)_2$ and $Co_2(CO)_6$ respectively (the "side on" coordination by the heteroatom found with nitriles being very rare).

4. π-ALLYL COMPLEXES

π-Allyl palladium complexes play a n important role in Pd-catalyzed organic reactions (see Part V),[10.6] but π-allyl complexes are also well known with other transition metals.[10.16, 10.17]

4.1. STRUCTURE

The allyl ligand can bind to a metal in a monohapto fashion like an alkyl, in which case it is a 1-electron X ligand. Most of the time, however, it is bonded to the metal as a π 3-electron LX trihapto ligand.

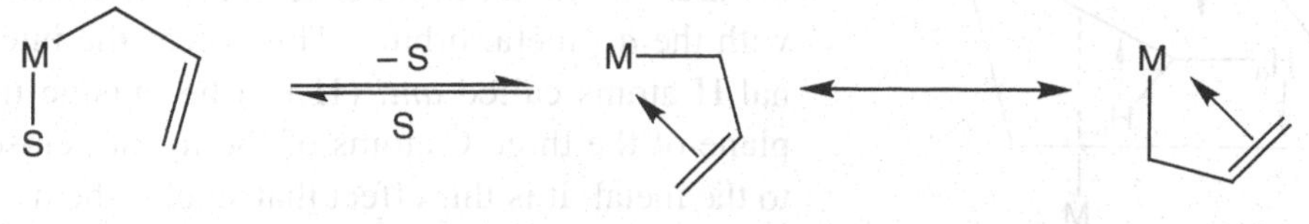

The two forms rapidly interconvert, which is facilitated by a weakly coordinated solvent. This interconversion probably explains the equilibrium between the *syn* and *anti* protons that are non-equivalent in ^{1}H NMR, and the reactivity that is due to the temporary monohapto coordination:

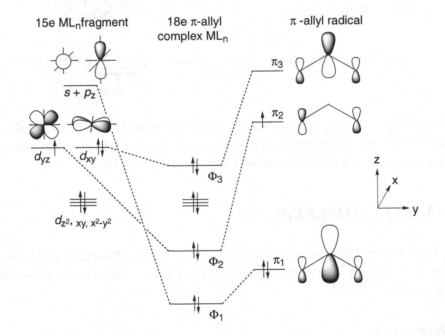

For static η^3-allyl complexes, the *anti* protons are closer to the metal than the *syn* protons, thus these *anti* protons are shielded and appear at high field (typically between 1 and 3 ppm) in ^{1}H NMR. The *syn* protons appear between 2 and 5 ppm, and the central protons between 4 and 6.5 ppm. The following diagram shows the interactions between the three orbitals π^1, π^2 and π^3 of the allyl radical with the orbitals of the 15-electron metallic fragment forming an 18-electron complex.

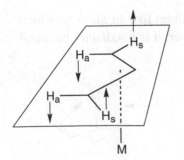

The π^1 orbital engages its doublet towards the vacant metal orbitals of adequate symmetry for the directional σ bond. The π^2 orbital must point towards the metal in order to share an electron with the d_{xy} metal orbital. This forces the internal H atoms called *anti* (H$_a$) to be outside the plane of the three C atoms of the ligand, closer to the metal. It is this effect that shields them.

4.2. SYNTHESIS

The various synthetic routes to allyl complexes are:

▸ allyl transfer using a source of anionic or cationic allyl (this method can lead to binary metal-allyl complexes, but the latter are very sensitive:

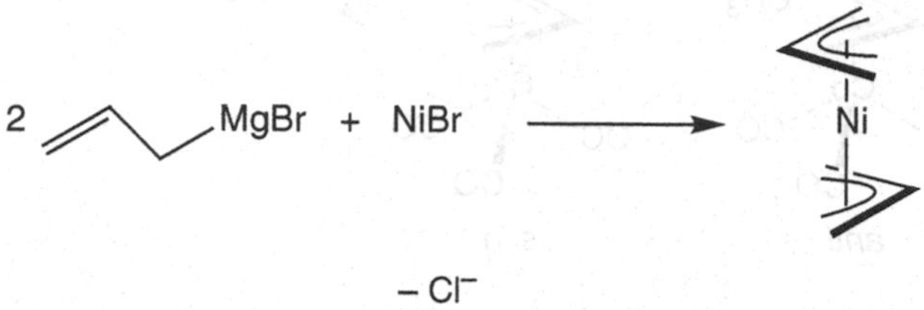

$$CH_2=CHCH_2Cl + Mn(CO)_5^- \xrightarrow{-Cl^-} [Mn(\eta^1\text{-}CH_2=CHCH_2)(CO)_5]$$

▸ chelation of a monohapto allyl ligand:

$$[Mn(\eta^1\text{-}CH_2=CHCH_2)(CO)_5] \xrightarrow[-CO]{\Delta} [Mn(\eta^3\text{-}CH_2CHCH_2)(CO)_4]$$

▸ use of oxidative addition, nucleophilic addition or insertion of a mono- or diene ligand. The oxidative addition of an allylic C-H bond of an olefin is convenient, easy to carry out and produces the classic reagents $[M(\eta^3\text{-allyl})X]_2$, M = Ni or Pd, X = halogen; it is very often used in organic synthesis, and has allowed the first synthesis of a π-allyl complex by Smidt and Hafner in 1959:

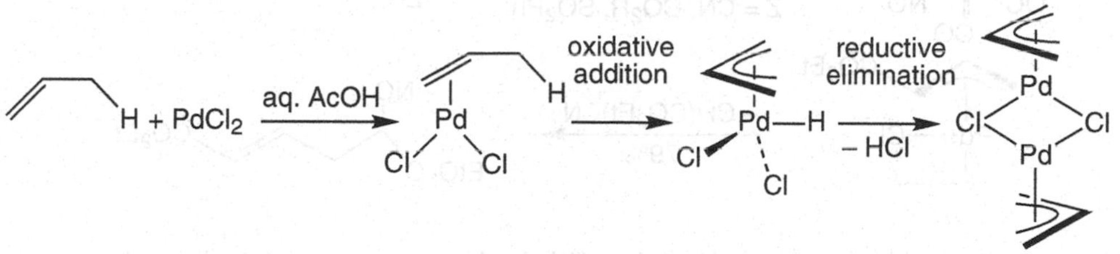

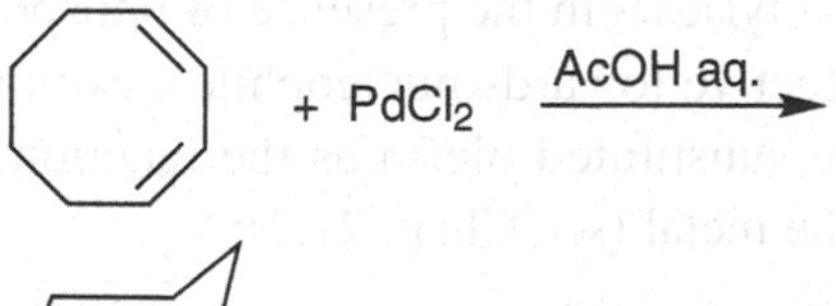

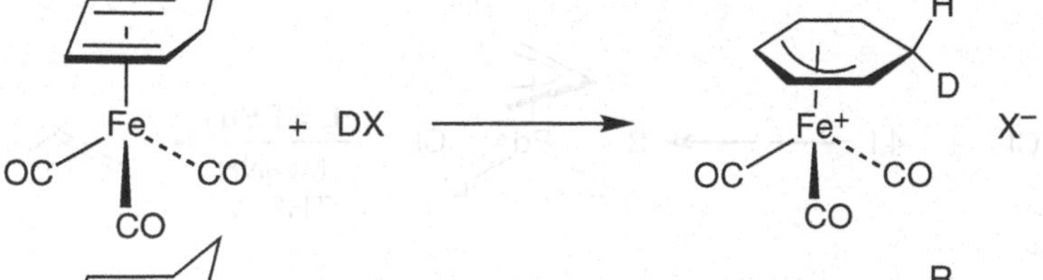

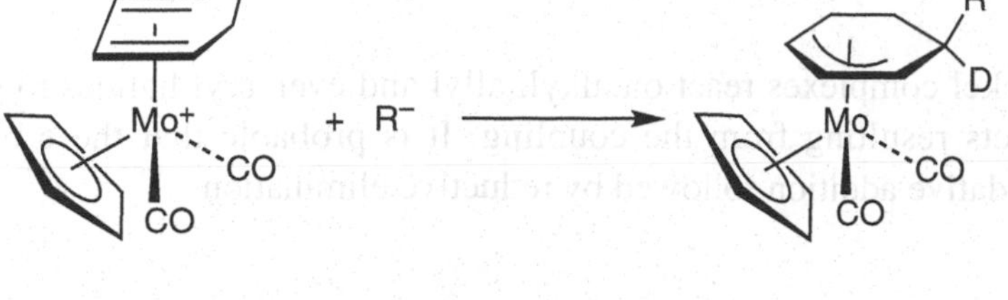

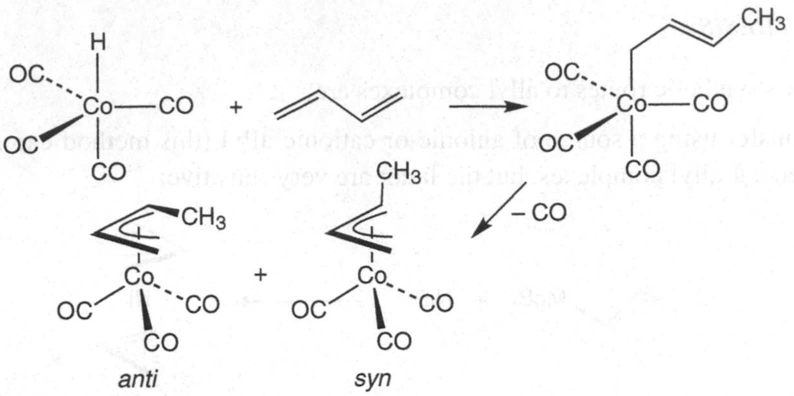

anti   syn

4.3. REACTIONS

The nucleophilic and electrophilic additions are the most common reactions and have a large number of applications in organic synthesis and palladium catalysis. The nucleophilic additions are more or less easy depending on whether the complex is cationic or neutral. In the latter case, the ancillary ligands must be electron withdrawing (CO, NO), or the allyl group must bear an electron-withdrawing substituent in order to allow the reaction:

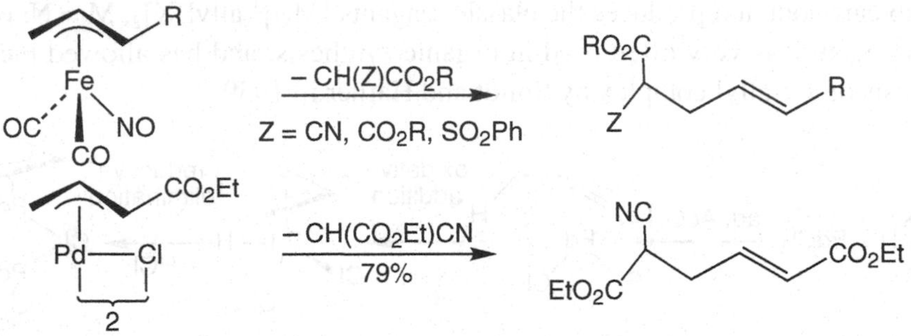

The reaction of the complex $[Pd(\eta^3\text{-allyl})Cl]_2$ in the presence of phosphine allows to produce a cationic complex, thus reactive towards nucleophiles, even neutral ones. The nucleophilic attack leads to the substituted olefin as the reaction product that spontaneously decoordinates from the metal (see Chap. 21.2):

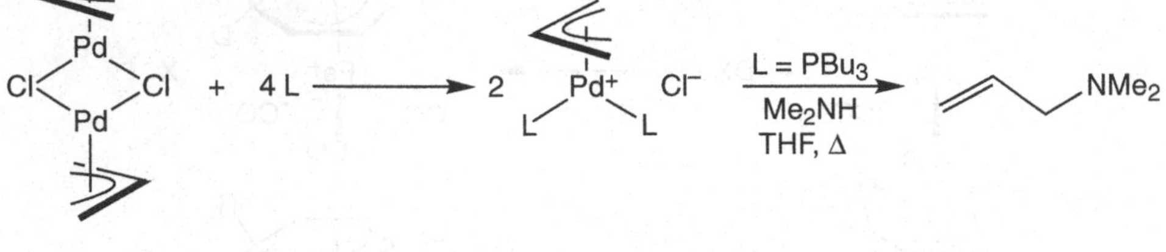

The π-allyl-nickel complexes react on alkyl, allyl and even aryl halides to give the organic products resulting from the coupling. It is probable that these reactions proceed by oxidative addition followed by reductive elimination.

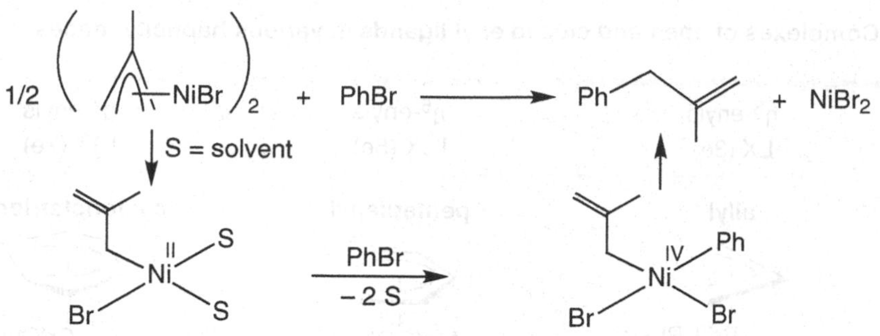

Finally, various insertion reactions can be carried out. It is possible, for instance, to obtain ketones from CO in this way.

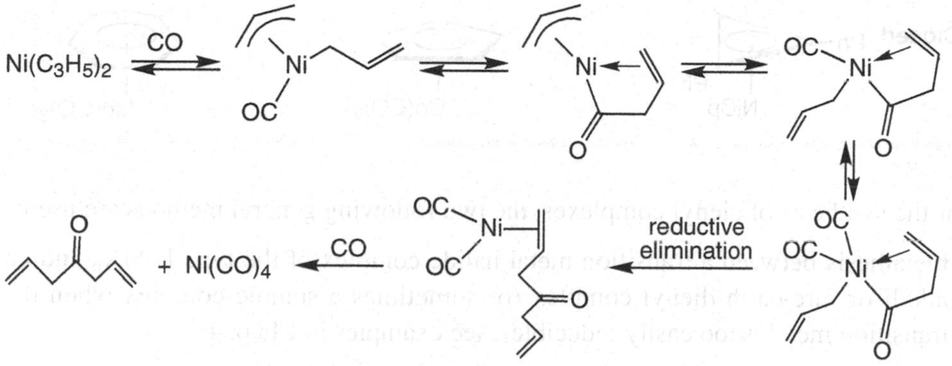

The catalytic aspects and the applications of nucleophilic attacks on Pd-allyl complexes in organic synthesis will be developed in Chap. 21.2.

5. POLYENYL COMPLEXES

5.1. VARIETY OF ENYL COMPLEXES

The open and closed enyl radicals are odd-electron ligands providing 3, 5 or 7 electrons to the metal, this number being the same as that of the hapticity. The cationic complexes are attacked by nucleophiles to give even-electron ligands whose hapticity and numbers of given electrons are one less than those of the starting complexes (see Chap. 4). The synthetic routes and the chemistry of these complexes are rather well connected within this family and resemble the above description given for the π-allyl complexes. Besides the latter, the most important ligands are the cyclopentadienyl, one of the most used ancillary ligands in organo-metallic chemistry (see Chap. 11), and the cyclohexadienyl for its applications in organic chemistry (see Chap. 4). Both are L_2X 5-electron ligands, as all the other dienyl ligands. M.L.H. Green has also extensively developed the chemistry of the 7-electron L_3X cycloheptatrienyl ligand.[10.18] Examples of the mono-, di- and trienyl ligands are represented in the table below.

Complexes of open and closed enyl ligands in various hapticity modes

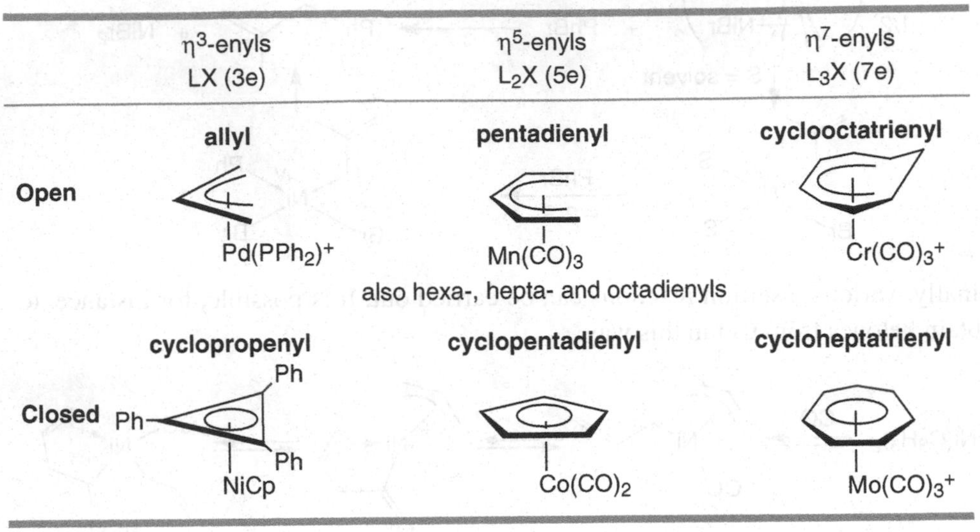

	η^3-enyls LX (3e)	η^5-enyls L$_2$X (5e)	η^7-enyls L$_3$X (7e)
Open	allyl Pd(PPh$_2$)$^+$	pentadienyl Mn(CO)$_3$ also hexa-, hepta- and octadienyls	cyclooctatrienyl Cr(CO)$_3$$^+$
Closed	cyclopropenyl Ph—, Ph, Ph NiCp	cyclopentadienyl Co(CO)$_2$	cycloheptatrienyl Mo(CO)$_3$$^+$

For the synthesis of dienyl complexes, the two following general methods are used:

▸ metathesis between a transition metal halide complex of the type L$_n$MX$_p$ and an alkali or rare-earth dienyl complex (or sometimes a stannic complex when the transition metal is too easily reducible); see examples in Chap. 4:

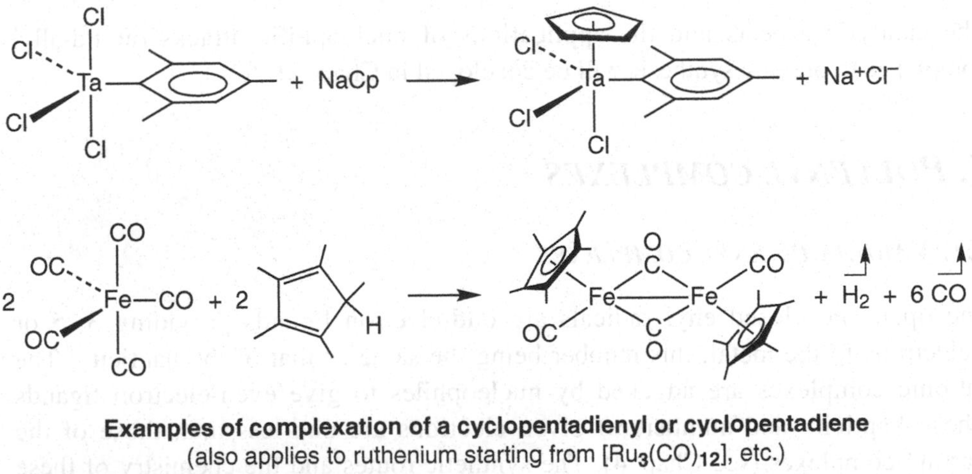

Examples of complexation of a cyclopentadienyl or cyclopentadiene
(also applies to ruthenium starting from [Ru$_3$(CO)$_{12}$], etc.)

▸ reaction of a metal (0), most often a metal carbonyl or halide complex that is reduced *in situ*, with the diene, followed by spontaneous intermolecular oxidative addition of a C-H bond, or abstraction or addition of a proton (or of another electrophile) or hydride (or of another nucleophile); see examples in Chap. 5.

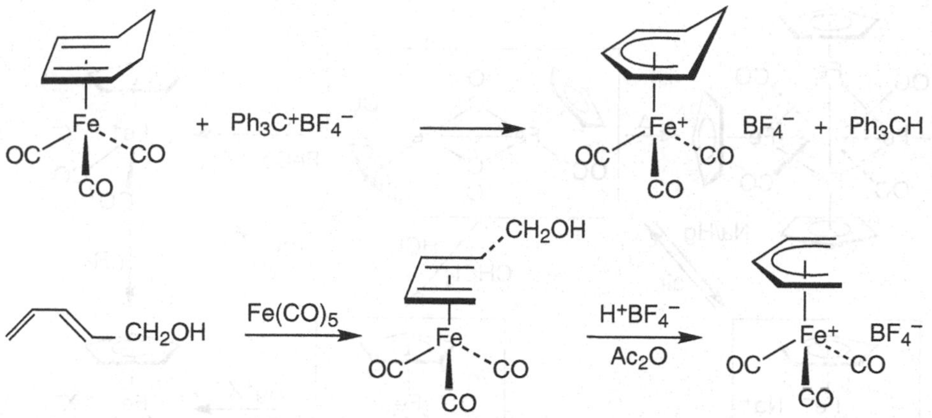

5.2. CYCLOPENTADIENYL COMPLEXES

The metals bonded to two cyclopentadienyls (Cp) in a pentahapto mode form the metallocenes (see Chap. 11 also describing the Cp-M orbitals). In many other families of compounds, the metal is bonded to only one Cp or derivative. Common examples of this category include ternary families containing, besides the Cp-type ligand, carbonyls, phosphines, nitrosyls, hydrides, oxo, imido, methyls, etc.[10.19]

Examples of complexes of Cp or Cp* (see the metallocenes, Chap. 11)

In the chemistry of these complexes, the Cp ligand usually plays an ancillary (spectator) role, whereas reactions occur on the other ligands. Virtually occupying three coordination sites and giving 5 electrons, the Cp ligand fills the metal coordination sphere (the NVE often is 18) and occupies bulk. An example of chemistry based on $[FeCp(CO)_2]_2$ is summarized in the following scheme:

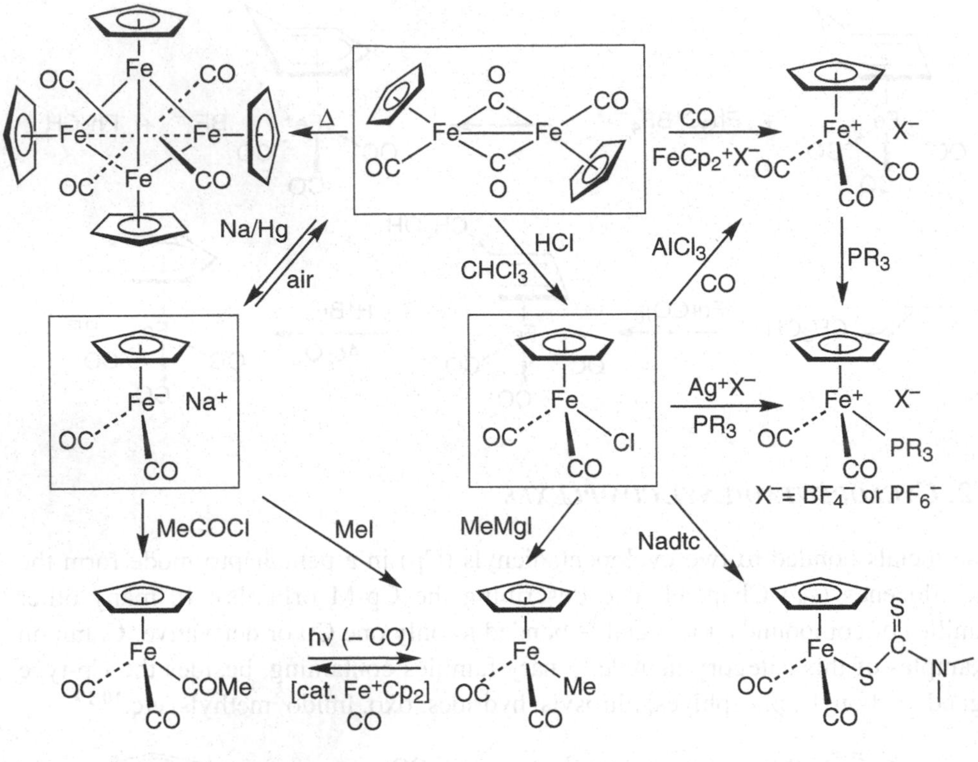

Reactions of Fp complexes {Fp = CpFe(CO)$_2$}

Other classic families are [MCp(CO)$_3$]$_2$, M = Cr, Mo or W, [NiCp(CO)]$_2$, mononuclear [VCp(CO)$_4$], [MCp(CO)$_3$], M = Mn, Tc or Re, [MCp(CO)$_2$], M = Co, Rh or Ir and ternary nitrosyl complexes. With neutral Cp complexes such as [VCp(CO)$_4$], [MnCp(CO)$_3$] and [ReCp(CO)$_3$], it is possible to achieve electrophilic substitutions on the Cp ring similar to those known with ferrocene (see Chap. 11), which allows to functionalize complexes in order to branch them on polymers, electrodes, dendrimers and derivatized silica for chromatography.

Although Cp is most of the time pentahapto, there are complexes of main groups and transition metals containing one or several Cp ligands that are monohapto coordinated (as for the allyl ligand). In [Re(η^1-Cp)(=NPh)$_3$], the monohapto coordination of Cp allows three imido ligands to adopt the π coordination of 4-electron donors.

The complex [Fe(η^5-Cp)(CO)$_2$(η^1-Cp)] is very stable with its d^6 18-electron configuration. It is also fluxional, however, i.e. has a dynamic structure in which the five carbon atoms of the monohapto Cp bind in turn the metal according to a 1,2-migration process of iron around the Cp ring. This process is faster than the NMR frequency, which leads to the observation of an average signal for the 5 H in ^{1}H NMR or the 5 C in ^{13}C NMR. Low-temperature NMR allows analyzing the migration mechanism (figure below).

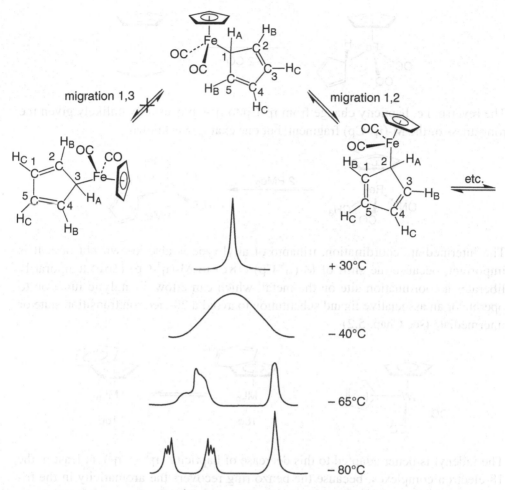

**Variable-temperature ^{1}H-NMR spectrum of the fluxional η^1-Cp ligand
of the complex [Fe(CO)$_2$(η^5-Cp)(η^1-Cp)]**

Indeed, this migration is frozen at –80°C in ^{1}H NMR, and the three signals H$_A$, H$_B$
and H$_C$ corresponding to the stable monohapto structure are observed. When the
temperature is increased to –65°C, one can see on the figure that the signal of the
H$_B$ protons broadens faster than that of the H$_C$ protons. In the 1,2-migration
mechanism, both H$_B$ protons leave the H$_B$ position, but only one H$_C$ proton leaves
the H$_C$ position. This is why the rate of the H$_C$ proton exchange is only half that of
the H$_B$ proton exchange, explaining the above rate order of signal broadening. If the
mechanism was 1,3-migration, only one H$_B$ would leave the H$_B$ position, whereas
both H$_C$ protons would leave the H$_C$ position. The opposite situation would be
observed, i.e. the signal of the H$_C$ protons would broaden faster than the signal of
the H$_B$ protons. It is thus possible, based on this ^{1}H-NMR experiment, to propose
the 1,2-migration mechanism and to exclude the 1,3-migration mechanism.

Heating the complex provokes the η^1-Cp to η^5-Cp hapticity change, i.e. the
formation of ferrocene, resulting from the loss of the two carbonyl ligands.

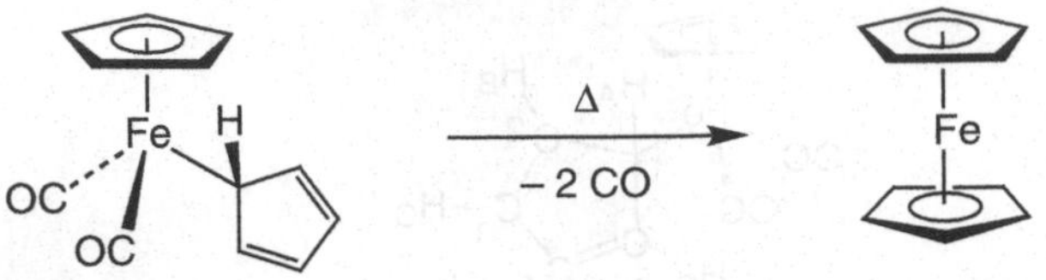

The reverse, i.e. hapticity change from η^5-Cp to η^1-Cp is usually unlikely given the robustness of the M-(η^5-Cp) fragment, but one example is known:

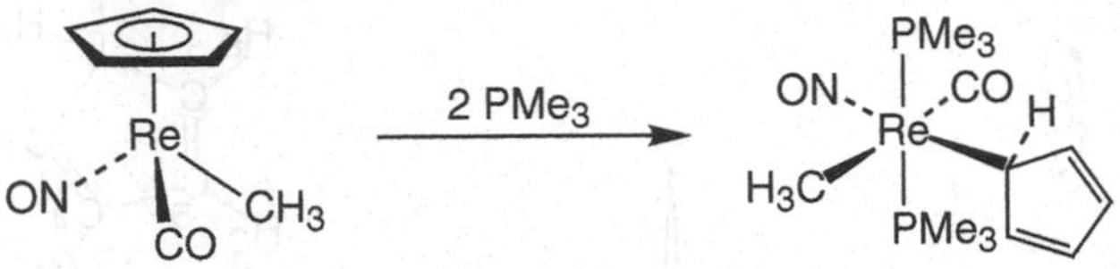

The intermediate coordination, trihapto of allyl type is also known but rare. It is important, because the shift of M-(η^5-Cp) (18 e) to M-(η^3-Cp) (16 e) temporarily liberates a coordination site on the metal, which can allow a catalytic function to operate or an associative ligand substitution to avoid a 20-electron transition state or intermediate (see Chap. 5.2).

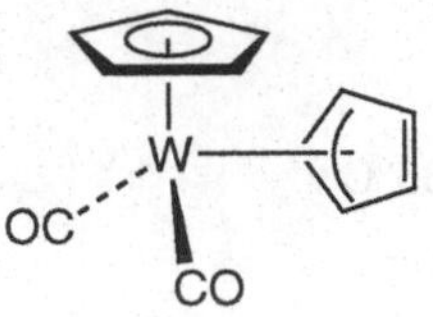

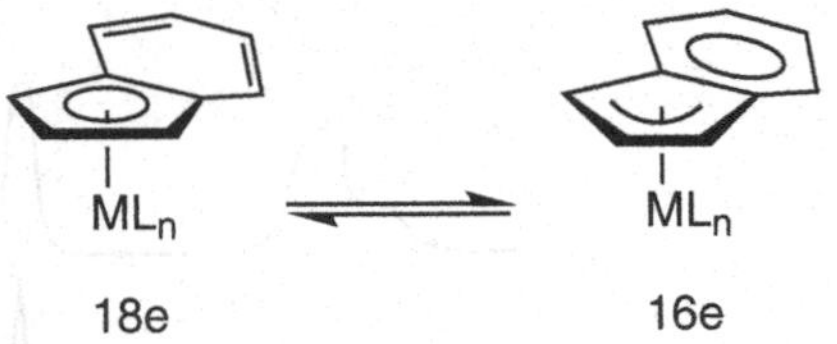

The indenyl is better adapted to this decrease of hapticity ($\eta^5 \rightarrow \eta^3$), at least in the 18-electron complexes, because the benzo ring recovers the aromaticity in the tri-hapto mode of the indenyl. Therefore, the indenyl ligand is often used in catalysis. On the other hand, upon monoelectronic reduction of an 18-electron complex, the Cp ligand shows a larger ability to decoordinate than the indenyl complex. The lower energy of the indenyl ligand-based LUMO in the latter complex avoids a genuine 19-electron electronic structure of the metal in the monoreduced complex. It is then better described as an 18-electron complex with the extra electron locali-zed on the indenyl ligand whose pentahapto structure remains unchanged.

The ancillary tendency is still increased with the pentamethylcyclopentadienyl (Cp*) bulkier than Cp. This ligand has been very much used, because it stabilizes the complexes and increase their ease to crystallize, increase the electronic density on the metal center and thus finally leads to an even richer chemistry than that of Cp.[10.20] The introduction of Cp* instead of Cp has indeed allowed *inter alia* to stabilize highly reactive ligands such as $=CH_2$ in $[FeCp^*L_2(=CH_2)]^+$ or ligands that are difficult to stabilize such as N_2 in $[TiCp^*_2(N_2)]$ and $[Cp^*_2TiNNTiCp^*_2]$, paramagnetic complexes such as $[Fe^{III}Cp^*(dtc)_2]$ or of high oxidation state such as $[Fe^{IV}Cp^*(dtc)_2]^+$ or $[W^VCp^*Cl_4]$. All these complexes do not exist or are instable with Cp. Other C_5R_5 ligands are also known for R = i-Pr, CHEt$_2$, CH(CH$_2$Ph)$_2$ (these ligands have a directionality that induces a plane of chirality at low temperature). Other R groups are: allyl, Ph, CO$_2$Me, CF$_3$, Cl, CH$_2$Ar, with various stereoelectronic effects, but their use is only occasional. Tridentate inorganic and organic ligands that are comparable to Cp (including some heterocycles)[10.22] are described in Chap. 1.

6. ARENE COMPLEXES

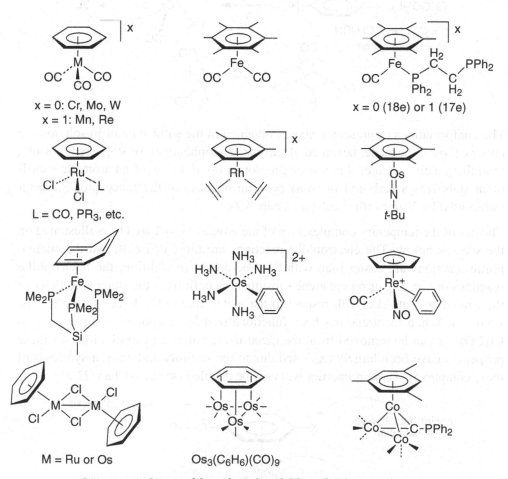

Arene complexes with various hapticities of the arene ligand

Most arene complexes are hexahapto, and ternary complexes usually have 18 valence electrons. The most known and used ones for their applications in arene synthesis are the arene-metal-tricarbonyls (columns of chromium and manganese), in particular the arene-chromium-tricarbonyl and the arene-manganese-tricarbonyl cations (see Chap. 21.2). Di- and tetrahapto complexes are also known, and this peculiar hapticity mode is most of the time encountered in order to respect the 18-electron rule. H. Taube and W.D. Harman have shown that Os^{II} has a tendency to form dihapto arene complexes in which the nucleophilicity of the arene is enhanced.[10.23] Such dihapto complexes are also known with Ag^I but usually not well characterized. Arene ligands are engaged in cluster structures with various metals. A key class of metal-arene complexes is that of sandwich complexes (see Chap. 11).

The synthesis of the arene-chromium-tricarbonyl complexes proceeds by heating the arene with $[Cr(CO)_6]$ in dibutyl ether or, if the arene is thermally fragile, by reaction with another labile complex such as $[Cr(naphthalene)(CO)_3]$ or $[Cr(CO)_3L_3]$, $L = NH_3$ or CH_3CN:

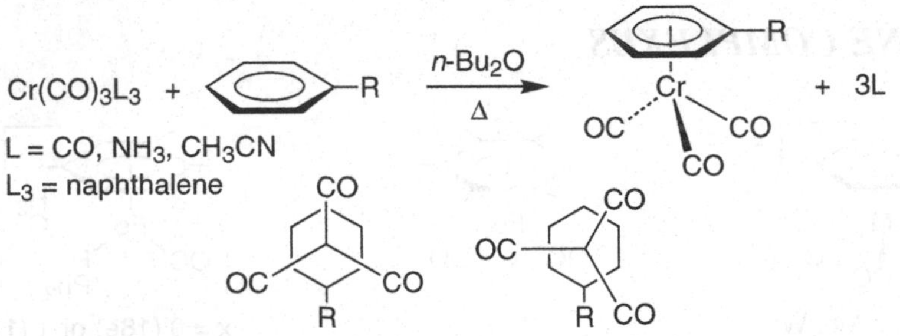

The conformations (represented above) obtained in the solid state or in solution are either eclipsed (benzene, hexamethylbenzene, acetophenone), or staggered (anisole, toluidine). This indicates that a strong mesomer donor effect of an aromatic substituent stabilizes a carbonyl in *trans* position because of the π acceptor character transmitted by "*trans* influence" (see Chap. 5.2).

The use of the temporary complexation of the aromatics by $Cr(CO)_3$ is illustrated on the scheme below. The electrophilic reactions are more difficult, and the nucleophilic reactions are easier than with the free arene. In addition, the nucleophilic reactions on the ring or on the arene substituent benefit from the stereospecificity of the *exo (trans)* attacks, with respect to the metal, due to the bulk of the $Cr(CO)_3$ fragment. When the arene has been functionalized by temporary complexation to $Cr(CO)_3$, it can be removed from the metal using solar-light irradiation. All these properties have been largely exploited due to the stability and ease of synthesis of these complexes. These properties will now be detailed (see also Chap. 21.2).

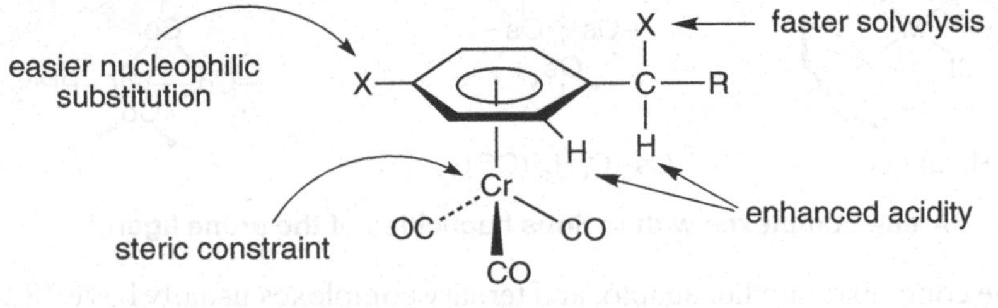

Semmelhack has shown that some very reactive carbanions can form C-C bonds with the benzene ring (stabilized enolates do not react, however, and LiMe and *n*-BuLi deprotonate the ring). A methoxy substituent on the ring orientates preferentially in *meta* (see Chap. 4 and ref 4.5-4.9). In halogenobenzene complexes, the nucleophilic substitution of the halide can be achieved by an even milder anionic nucleophile (oxoanion, stabilized carbanion) if X = Cl, and by an aliphatic amine (but not an aromatic amine) if X = F (but not X = Cl)

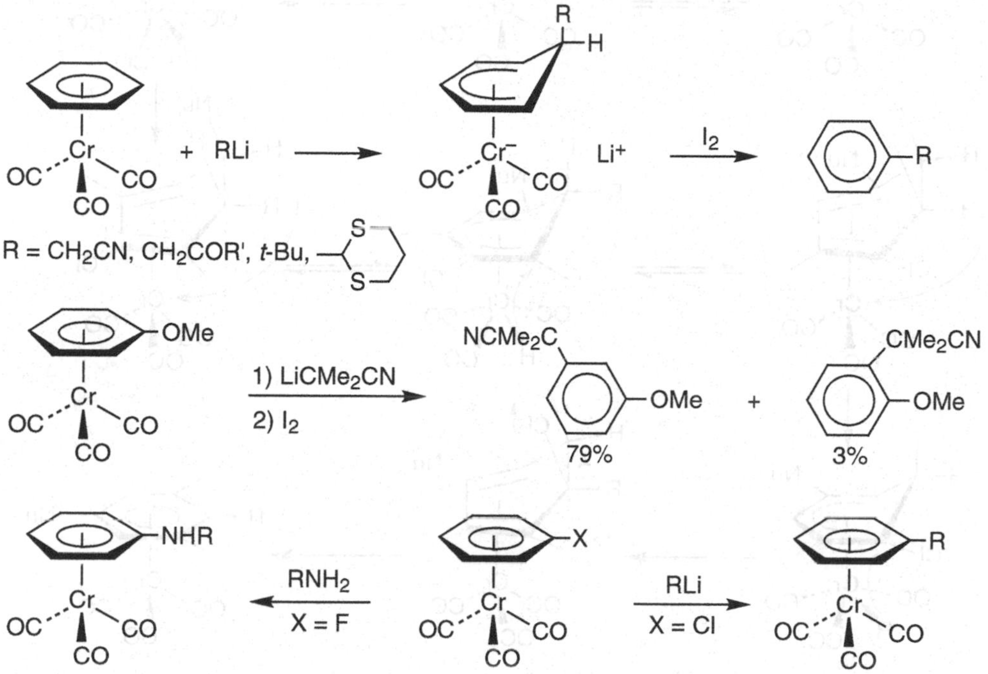
R
H
Cr
OC
CO
CO
+ RLi
Cr⁻
Li+
I2
R
R = CH2CN, CH2COR', t-Bu,
S
S
OMe
1) LiCMe2CN
2) I2
NCMe2C
OMe
+
CMe2CN
OMe
79%
3%
NHR
RNH2
X = F
X
RLi
X = Cl
R

Cr(CO)$_3$-induced nucleophilic reactions of arenes in the complexes [Cr(arene)(CO)$_3$]

F. and E. Rose have disclosed rich reactions when the nucleophilic addition is followed by a treatment with acid: these are new nucleophilic additions (SN_{Ar}) called "*cine*" and "*tele*" observed when the nucleophile attacks a ring position that is different from that bearing the leaving group (*ipso*).[10.24] These reactions are:

- the "*cine*" SN_{Ar} corresponding to the attack of the nucleophile in *ortho* position with respect to the leaving group;
- the "*tele*" *meta* SN_{Ar} corresponding to the attack of the nucleophile in *meta* position with respect to the leaving group;
- the "*tele*" *para* SN_{Ar} corresponding to the attack of the nucleophile in *para* position with respect to the leaving group.

Example of SN_{Ar} "*tele*" *para* mechanism:

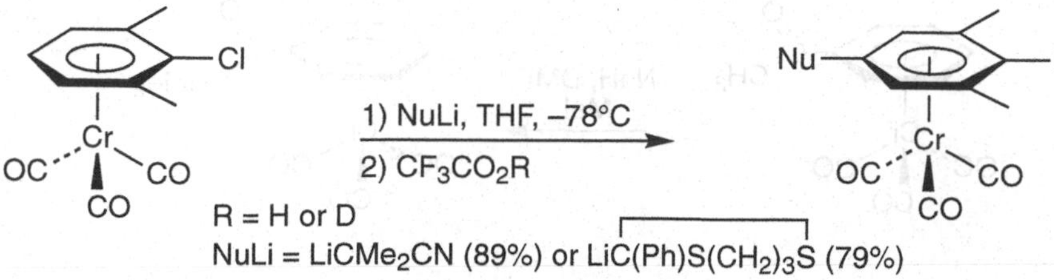
Cl
Cr
OC
CO
CO
1) NuLi, THF, –78°C
2) CF3CO2R
R = H or D
NuLi = LiCMe2CN (89%) or LiC(Ph)S(CH2)3S (79%)
Nu
R

***"Tele" para* SN_{AR} reaction**

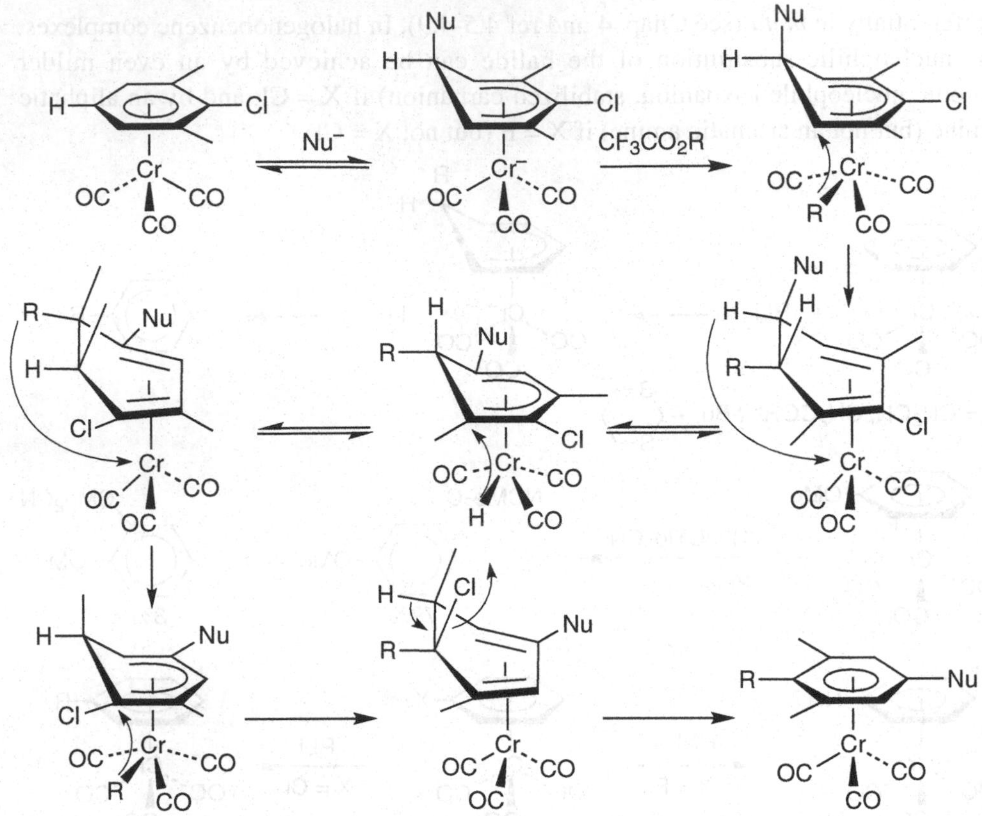

Proposed mechanism for the *"tele" para* SN$_{AR}$ reaction

The lithiation reaction is easier than on the free arene and, in the presence of a substituent, can be directed in *ortho*. The side chain can be protonated and deuterated in α position in the presence of *t*-BuOK in d$_6$ DMSO; it can be alkylated in α (or in β if the α carbon is a carbonyl as in the example below) in DMF in the presence of NaH and an organic halide.

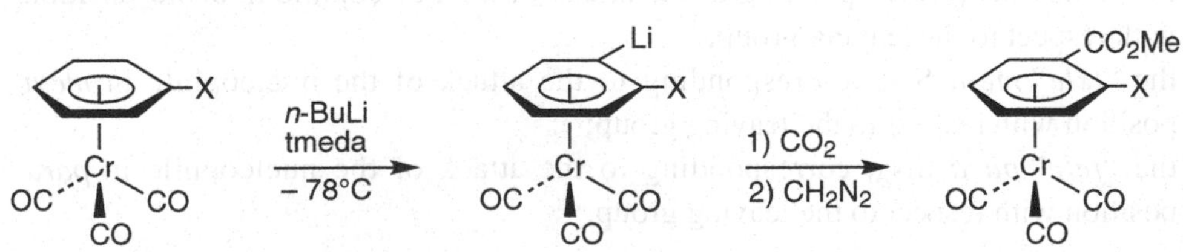

X = H, F, OMe

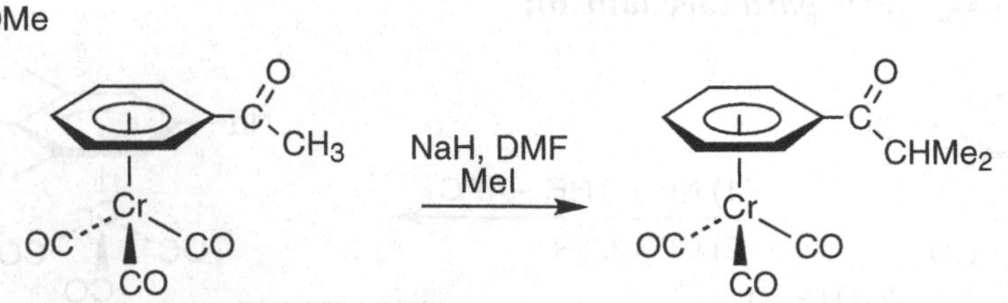

Finally, the Cr(CO)$_3$ group stabilizes the carbocation in exocyclic (α) position as in ferrocenes. This stabilization allows to achieve nucleophilic substitution in two reactions on this α carbon:

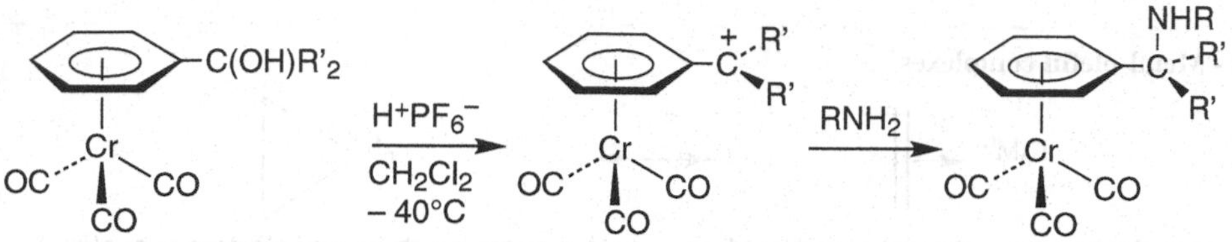

The complexation of arenes by the 12-electron cationic groups [Mn(CO)$_3$]$^+$ and [FeCp]$^+$ provides an even larger activation than with the Cr(CO)$_3$ group (see Chap. 4.1). The complexation with [Mn(CO)$_3$]$^+$, a 12-electron fragment isoelectronic to Cr(CO)$_3$, is carried out upon heating the arene with [Mn(CO)$_5$Br] in the presence de AlCl$_3$, which does not allow the presence of functional groups that are not compatible with AlCl$_3$ such as most heteroatom-containing groups. The use of a more sophisticated starting material such as [Mn(CO)$_3$(MeCN)$_3$]$^+$ BPh$_4^-$ can answer this problem. The addition of a stabilized carbanion, impossible with the Cr(CO)$_3$ moiety, now becomes feasible. Although the *endo* H$^-$ abstraction by Ph$_3$C$^+$ to recover the functionalized arene structure is not possible for stereoelectronic reasons, one could oxidize the cyclohexadienyl adduct using a sufficiently strong monoelectronic oxidant. Indeed, such an oxidation is thermodynamically much more difficult than in the chromium series, however, because of the one-unit charge difference.[10.25]

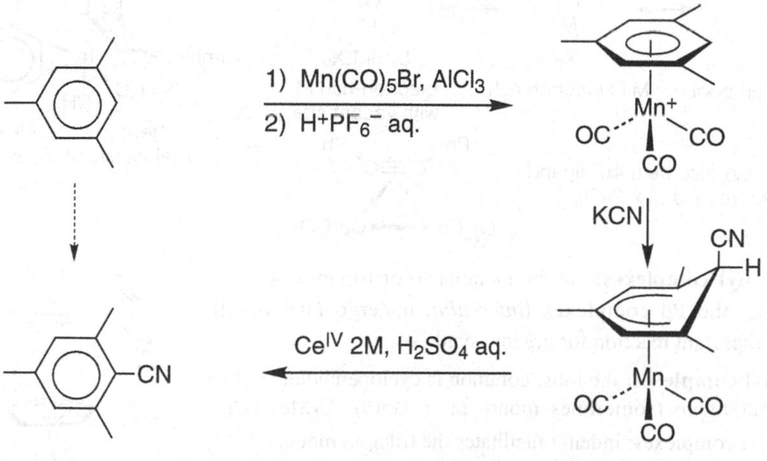

Many other reactions of arenes that are not possible using the Cr(CO)$_3$ fragment become possible with the much stronger [Mn(CO)$_3$]$^+$ fragment. The chemistry involving the activation by 12-electron fragments is far from being completely explored and *a fortiori* exploited (see Chap. 4). Such nucleophilic reactions are possible with the strong activator [FeCp]$^+$ whose multiple possibilities will be examined in the following chapter.

SUMMARY OF CHAPTER 10
π COMPLEXES OF MONO- AND POLYENES AND ENYLS

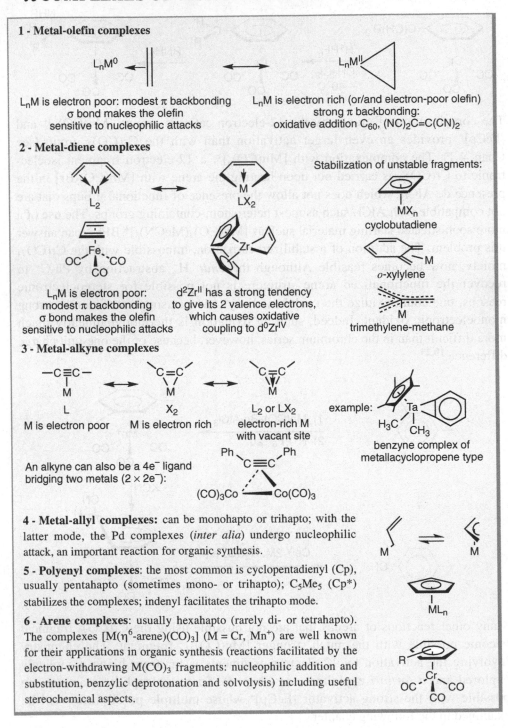

1 - Metal-olefin complexes

L_nM^0

L_nM is electron poor: modest π backbonding
σ bond makes the olefin
sensitive to nucleophilic attacks

L_nM^{II}

L_nM is electron rich (or/and electron-poor olefin)
strong π backbonding:
oxidative addition C_{60}, $(NC)_2C=C(CN)_2$

2 - Metal-diene complexes

M
L_2

M
LX_2

Stabilization of unstable fragments:

Fe
OC   CO
CO

Zr

MX_n
cyclobutadiene

M
o-xylylene

L_nM is electron poor:
modest π backbonding
σ bond makes the olefin
sensitive to nucleophilic attacks

d^2Zr^{II} has a strong tendency
to give its 2 valence electrons,
which causes oxidative
coupling to d^0Zr^{IV}

M
trimethylene-methane

3 - Metal-alkyne complexes

$-C\equiv C-$
M
L
M is electron poor

C=C
M
X_2
M is electron rich

C=C
M
L_2 or LX_2
electron-rich M
with vacant site

example:

Ta
H_3C CH_3

benzyne complex of
metallacyclopropene type

An alkyne can also be a $4e^-$ ligand
bridging two metals ($2 \times 2e^-$):

Ph   Ph
C≡C
$(CO)_3Co$ —— $Co(CO)_3$

4 - Metal-allyl complexes: can be monohapto or trihapto; with the latter mode, the Pd complexes (*inter alia*) undergo nucleophilic attack, an important reaction for organic synthesis.

M   ⇌   M

5 - Polyenyl complexes: the most common is cyclopentadienyl (Cp), usually pentahapto (sometimes mono- or trihapto); C_5Me_5 (Cp*) stabilizes the complexes; indenyl facilitates the trihapto mode.

ML_n

6 - Arene complexes: usually hexahapto (rarely di- or tetrahapto). The complexes $[M(\eta^6\text{-arene})(CO)_3]$ (M = Cr, Mn$^+$) are well known for their applications in organic synthesis (reactions facilitated by the electron-withdrawing $M(CO)_3$ fragments: nucleophilic addition and substitution, benzylic deprotonation and solvolysis) including useful stereochemical aspects.

R
Cr
OC   CO
CO

EXERCISES

10.1. Among the following olefin complexes, which ones are genuine olefin complexes and which one are rather metallacyclopropanes:
$[FeCp(CO)_2(C_2H_4)]^+$, $[Os(CO)_4(C_2H_4)]$, $[Pt(PMe_3)_2(C_2H_4)]$, $[Ni(C_6H_6)(C_2F_4)]$?

10.2. Give two modes of synthesis of $[WCp(CO)_3(C_2H_4)]^+X^-$ (X = BF_4 or PF_6).

10.3. Give the isomers of $[Fe(CO)_4(\eta^3\text{-2-butenyl})]^+$.

10.4. Describe the different steps of the reaction of cyclopentadiene with $[Fe(CO)_5]$ giving $[FeCp(CO)_2]_2$.

10.5. Why is the nucleophilic substitution of a halide X^- by aniline impossible in $[Cr(C_6H_5X)(CO)_3]$ and possible in $[Mn(C_6H_5X)(CO)_3]^+$?

METALLOCENES

AND SANDWICH COMPLEXES

Metal-sandwich complexes occupy a major place in organometallic chemistry for various reasons. First, from the historical point of view, the discovery of the ferrocene structure by Wilkinson et Fischer in 1952 has been a starting point for the boom of the chemistry of π complexes.[11.1] Then, ferrocene has provided a very rich chemistry as a superaromatic compound.[11.2] Its redox properties have led to multiple applications in the field of materials and molecular engineering:[11.3] molecular ferromagnets,[11.4] modified electrodes for redox catalysis (titration of glucose in blood), polymers and dendritic electrochemical sensors for molecular recognition, antitumor drugs and paintings. The family of metallocenes (MCp_2) and decamethylmetallocenes (MCp^*_2) now includes all the transition and main group metals. Metal-arene complexes also have such properties, especially the permethylated ones that are reservoirs of electrons, protons, hydrides and hydrogen atoms. They are efficient for transition-metal-mediated aromatic synthesis and for the synthesis of dendrimers.[11.5] Early transition-metal metallocenes, in particular, do not have the sandwich structure, and the Cp rings make an angle of about 130°. The metal bears up to three other ligands. They also have remarkable properties. For instance, $[TiCp_2Cl_2]$ is an antitumor agent and the zirconium and lanthanide derivatives are known as a new generation of initiators of olefin polymerization (see Chap. 15.1).

1. STRUCTURE OF THE METALLOCENES

Metallocenes, in a broad sense, have the composition $[MCp_2]$ and are known for all the transition metals and many main-group and rare-earth metals. The sandwich structure is reserved to the first row of transition metals from Ti to Ni and for the iron column for which the 18-electron rule brings a special robustness. It is also known for a few others metals including main-group ones.

Let us first examine these sandwich metallocenes. Ferrocene has two isoenergetic conformations: eclipsed (D_{5h} symmetry) and staggered (D_{5d} symmetry). X-ray crystallographic and electronic diffraction studies in the gas phase indicate an eclipsed conformation as well as for ruthenocene and osmocene. The rotation barrier about the iron-Cp ring axis is very weak (2 to 5 kcal·mol⁻¹; 8 to 21 kJ·mol⁻¹). Thus, this rotation is free and fast in fluid solution, i.e. there is only one 1,1'-disubstituted isomer (by 1 and 1', it is meant that the substituents are located on different rings) because this fast rotation makes all the positions on the first ring equivalent for the second ring. In decamethylferrocene, the solid-state conformation is staggered.

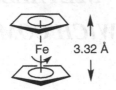

MCp₂: eclipsed D₅ₕ   **FeCp*₂: staggered D₅d**

Fe-C: 2.04 Å (FeCp₂)
Ru-C: 2.21 Å (RuCp₂)
Os-C: 2.22 Å (OsCp₂)

Ferrocene is an air-stable orange crystalline powder that sublimes at 184°C under 1 atm and is stable up to 400°C; it is soluble in hydrocarbons, has a nil dipole moment and thus covalent metal-ring bonds. The molecular orbital (MO) diagram below shows various metal-ring interactions: σ (Cp ⟶ Fe), π (Cp ⟶ Fe) and δ (Fe ⟶ Cp).

The other metallocenes are represented by the same diagram, although the energy levels of the orbitals may vary from one metal to the other. The occupation of the orbitals does not correspond to a low-spin structure except for ferrocene and cobaltocene (see the table below). Decamethylmetallocenes [MCp*₂] and the cations [MCp₂]⁺ and [MCp*₂]⁺ all are low spin.[11.6] Indeed, the Cp* ligand field is much stronger than the Cp ligand field because of the electron-releasing effect of the methyl groups. Manganocene has 5 d electrons, i.e. one per d orbital (semi-occupation of the 5 d orbitals), which means a high-spin structure. A spin cross-over transition 5/2 ⇌ 1/2 is observed for manganocene and 1,1-dimethyl-manganocene, the pink low-spin form being the stable one at low temperature, and the high-spin state observed above the transition temperature (see Chap. 1). In manganocene, cooperative intermolecular interactions are responsible for the antiferromagnetic coupling provoking the spin lowering. The energy difference between high spin and low spin is only 0.5 kcal·mol⁻¹ (2.1 kJ·mol⁻¹). Decamethyl-manganocene always stays low spin, as [FeCp₂]⁺ and [FeCp*₂]⁺. The electronic structure of the other metallocenes is indicated in the table following the MO diagram of ferrocene.

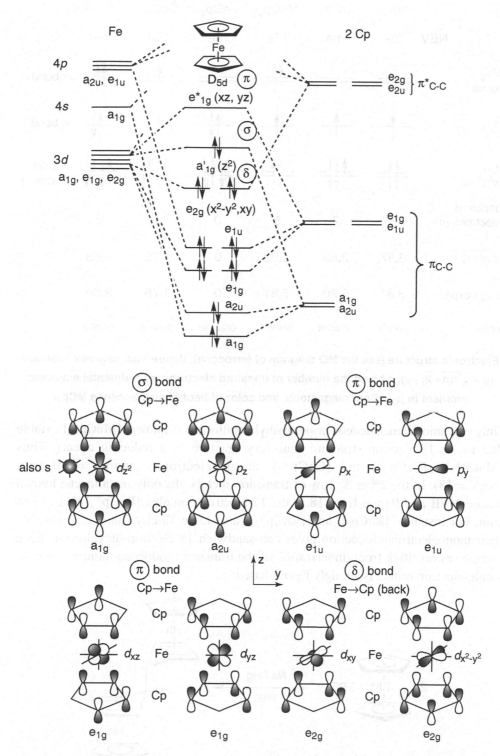

Molecular orbital diagram and interactions in ferrocene (staggered conformation).
In cobaltocene and nickelocene, the double antibonding e*$_{1g}$ level is occupied
respectively by one electron (Jahn-Teller distorsion) and two unpaired electrons.

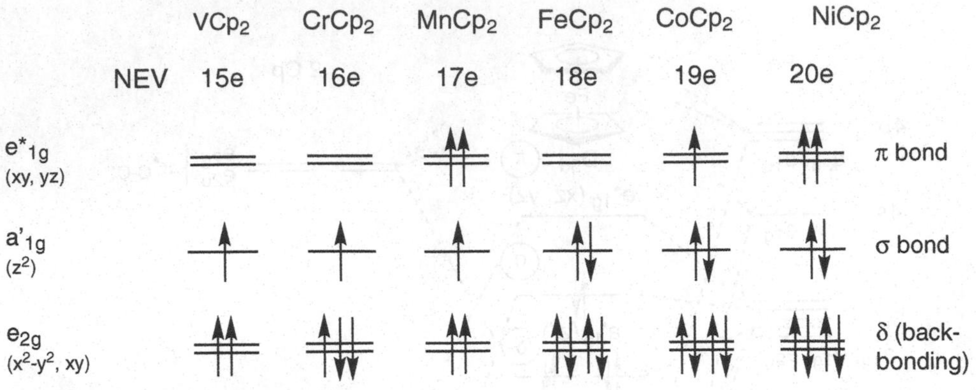

	VCp$_2$	CrCp$_2$	MnCp$_2$	FeCp$_2$	CoCp$_2$	NiCp$_2$	
NEV	15e	16e	17e	18e	19e	20e	
e*$_{1g}$ (xy, yz)			↑↑		↑	↑↑	π bond
a'$_{1g}$ (z^2)	↑	↑	↑	↑↓	↑↓	↑↓	σ bond
e$_{2g}$ (x^2-y^2, xy)	↑↑	↑↓ ↑	↑↑	↑↓ ↑↓	↑↓ ↑↓	↑↓ ↑↓	δ (back-bonding)
unpaired electrons (n)	3	2	5	0	1	2	
$\sqrt{n(n+2)}$	3.87	2.83	5.92	0	1.73	2.93	
μ(μ$_B$) exper.	3.84	3.20	5.81	0	1.76	2.86	
color	purple	scarlet	brown	orange	purple	green	

Electronic structure (see the MO diagram of ferrocene), theoretical magnetic moment ($\mu = \sqrt{n(n+2)}\ \mu_B$), n being the number of unpaired electrons, experimental magnetic moment in μ_B = Bohr magnetons, and color of neutral metallocenes MCp$_2$.

Only one anionic metallocene or permethylmetallocene is known, [MnCp*$_2$]$^-$, stable due to its 18-electron structure (but very reactive as a reducing agent). Thus [MnCp*$_2$]$^-$, [FeCp*$_2$] and [CoCp*$_2$]$^+$ are isoelectronic, having a nil spin (NEV = 18). In the 2nd et 3rd row of transition metals, the only metallocene known whose NVE is different from 18 is the 17-electron complex [ReCp*$_2$] that is low spin. For instance, [RhCp$_2$]$^+$ and [RhCp*$_2$]$^+$ are stable 18-electron complexes, but their monoelectronic reduction gives non-sandwich 18-electron dicyclopentadiene complexes resulting from dimerization of the transient rhodocene monomer whose single-electron orbital is strongly ligand-based:

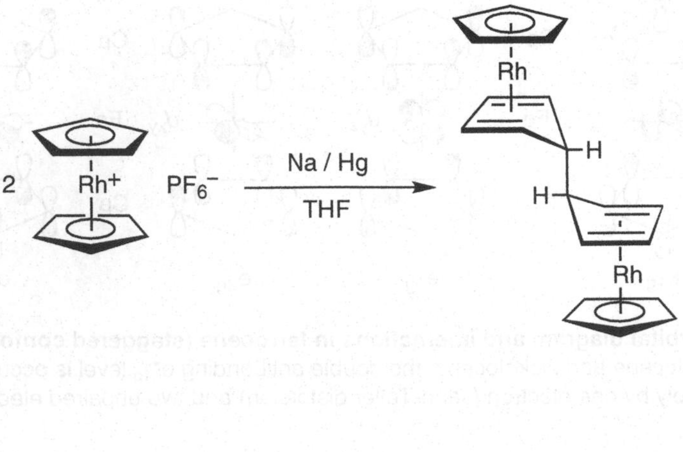

2. REDOX PROPERTIES
OF THE SANDWICH METALLOCENES

A key property of metallocene is their ability to exist in the form of various oxidation states with the sandwich structure and a variable number of d electrons (NVE between 14 and 20), even if the stability decreases as the NVE is further away from 18. Each metallocene exists in various oxidation states, which is usually not the case for the other families of inorganic or organometallic complexes. These various oxidation states are characterized by cyclic voltammetry, each wave corresponding to a change of oxidation state. The shape of the wave shows that, for metallocenes, this oxidation or reduction is reversible and occurs without structural change. For instance, for ferrocene, the groups of Laviron (reduction, DMF) and Bard (oxidation, SO_2) have reached extreme oxidation states (Fe^I and Fe^{IV}). The potentials below are given $vs.$ the saturated calomel reference electrode (SCE):

$$\underset{16e}{Cp_2Fe^{2+}} \overset{+3.5\,V}{\rightleftharpoons} \underset{17e}{Cp_2Fe^{+}} \overset{+0.4\,V}{\rightleftharpoons} \underset{18e}{Cp_2Fe} \overset{-3.0\,V}{\rightleftharpoons} \underset{19e}{Cp_2Fe^{-}}$$

With $[CoCp_2]$ and $[NiCp_2]$, Bard could observe even up to the dications and dianions. Since the metallocene structure does not tolerate more than 20 valence electrons, the most reduced forms $[CoCp_2]^{2-}$, $[NiCp_2]^{-}$ and $[NiCp_2]^{2-}$ presumably involve partial decoordination by slippage (for example $\eta^5 \longrightarrow \eta^3$) of at least one Cp ligand (see Chap. 10).

Monocationic metallocenes have been isolated and their stability is modest when their NVE is different from 18, but dications are only isolable for decamethylmetallocenes that are much more robust. The ferrocenium cations are stable as various salts, but they are slightly air- and light-sensitive in solution. Nevertheless, these salts are very often used as mild monoelectronic oxidants, and the reduction of ferrocenium salts can be easily achieved using an aqueous solution of dithionite or $TiCl_3$ or, in organic solution, using $[FeCp^*_2]$:

Ferrocene is thus a poor reductant. For instance, it cannot reduce TCNE, although it forms a charge-transfer complex. However, $[FeCp^*_2]$, with a Fe^{III}/Fe^{II} redox potential of -0.15 V $vs.$ SCE, can exergonically reduce TCNE. The product, $[FeCp^*_2]^{+},TCNE^{\bullet-}$, was reported by Miller who showed that it has a solid-state packing alternating the cation and the anion, which provided the first molecular ferromagnet.[11.4]

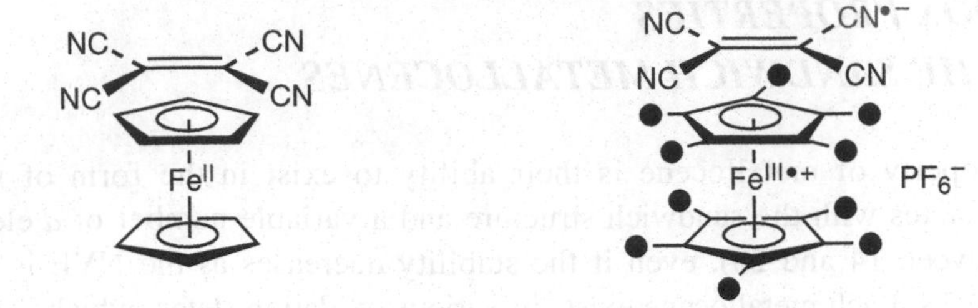

charge-transfer complex

ionic salt, ferromagnet
—● = CH₃

Cobaltocene and especially decamethylcobaltocene are "19-electron complexes" (or rather 18.5-electron complexes since the HOMO is only 50% metal-based) that are good single-electron reductants with redox potentials of –0.89 and –1.45 V *vs*. SCE respectively.

3. METALLOCENE SYNTHESIS

First, it is necessary to crack dicyclopentadiene at 180°C, which provides the monomer that must be used immediately as such or reduced to the anion Cp⁻ by deprotonation providing LiCp, NaCp, KCp, TlCp or [SnCp(n-Bu₃)]. The choice of the metal is often crucial if the transition metal used in the further reaction is reducible.

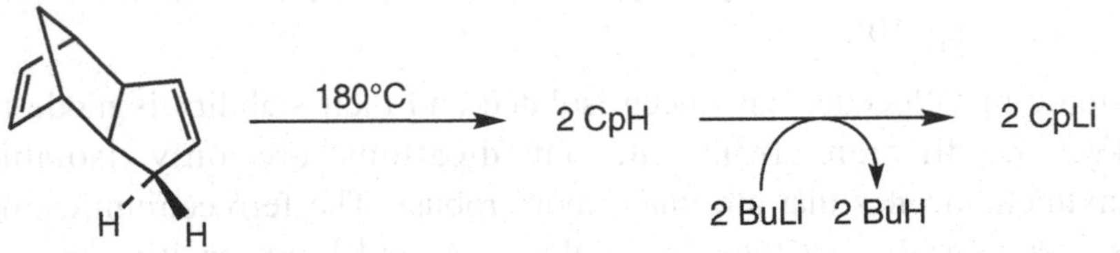

$$MX_2 + 2\,M'Cp \longrightarrow [MCp_2] + 2\,M'X^-$$

This method is used for various metallocenes. It led Pauson to achieve one of the first ferrocene syntheses, whereas he was trying to make fulvalene using the coupling of the Cp• radical by oxidation of Cp⁻ using the mild oxidant FeCl₃.[11.9] The erroneous structure of "bis-cyclopentadienyl-iron" proposed by Pauson resembled the hoped-for one:

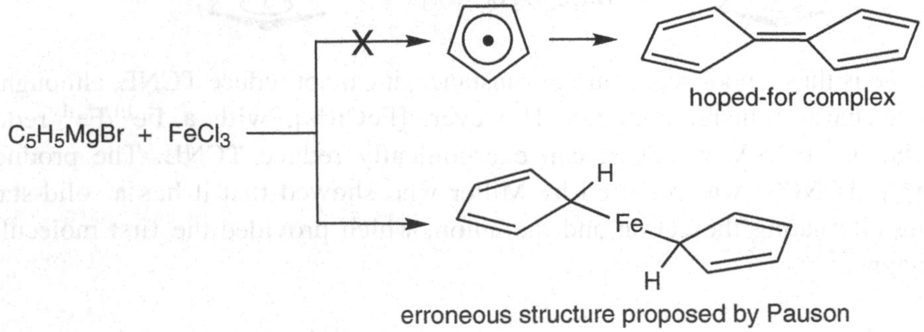

hoped-for complex

$C_5H_5MgBr + FeCl_3$

erroneous structure proposed by Pauson

On the other hand, this oxidation of the anion to the radical worked in the case of $C_5(i\text{-}Pr_5)$ for which Sitzmann [11.10a] was trying to make the elusive decaisopropyl ferrocene:

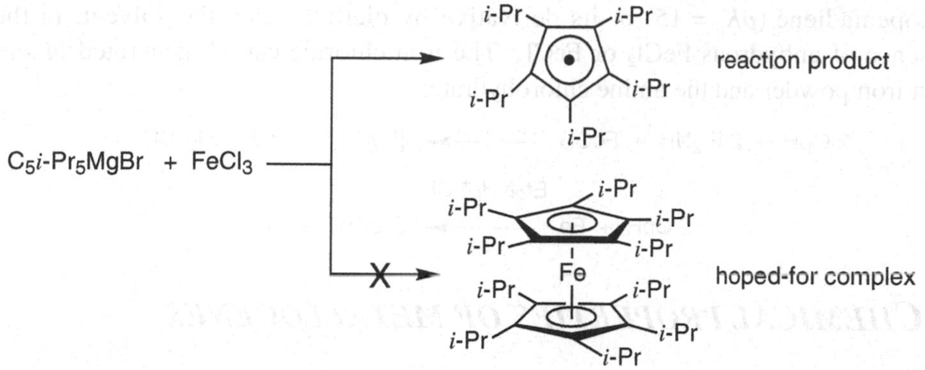

Sometimes, the reduction of the transition-metal halide is necessary *in situ*, either by the salt MCp itself or by Zn:

$$RuCl_3(H_2O)_x + 3\,C_5H_6 \xrightarrow{\text{Zn, EtOH}} [RuCp_2] + C_5H_8$$

This first classic method can also be used sequentially in order to introduce two different cyclopentadienyl ligands in a transition-metal sandwich complex, especially if one of the Cp derivatives is Cp*: [11.10b]

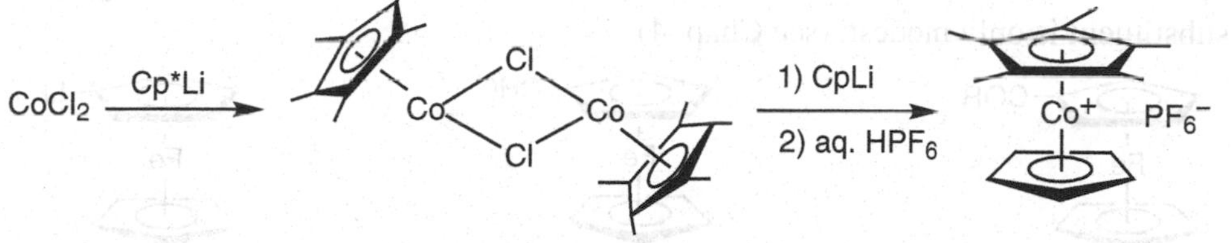

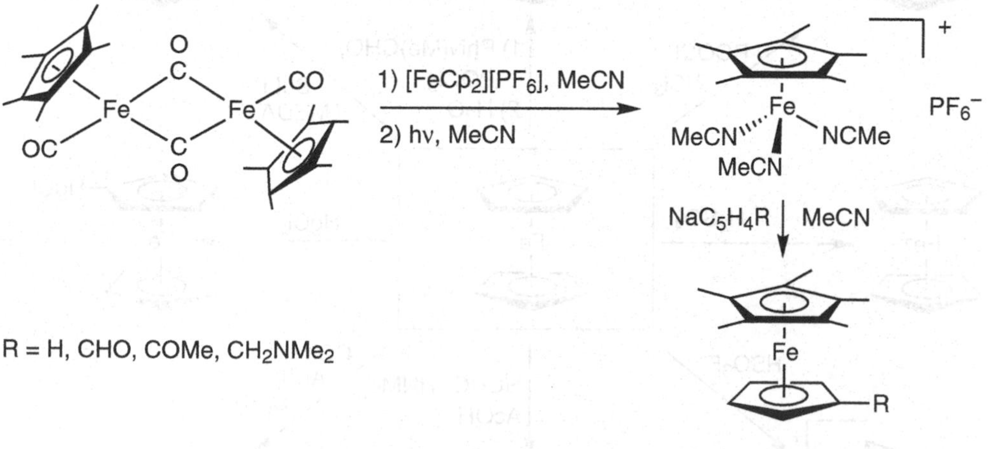

R = H, CHO, COMe, CH$_2$NMe$_2$

The second method, published by Miller, Tebboth and Tremaine [11.11] also in 1951, starts from cyclopentadiene and freshly reduced iron:

$$2\,CpH + Fe \xrightarrow{300°C} [FeCp_2] + H_2$$

It works well for instance with the metal-vapor technique that consists in vaporizing metal atoms at low temperature and pressure. In fact, the most practical method for the syntheses of ferrocene and its derivatives consists in the *in situ* deprotonation of cyclopentadiene ($pK_a = 15$) or its derivative by diethylamine, the solvent, in the presence of anhydrous $FeCl_2$ or $FeCl_3$. The iron chloride can be generated *in situ* from iron powder and the amine chlorohydrate:

$$2\ CpH\ +\ 2\ Et_2NH\ +\ FeCl_2\ \longrightarrow\ [FeCp_2]\ +\ 2\ Et_2NH_2^+Cl^-$$

$$2\ CpH\ +\ Fe\ \xrightarrow{\ Et_2NH_2^+\ Cl^-\ }\ [FeCp_2]\ +\ H_2$$

4. CHEMICAL PROPERTIES OF METALLOCENES

4.1. FERROCENE AND ITS DERIVATIVES [11.12]

Ferrocene undergoes many electrophilic reactions, more rapidly than benzene, although they are limited by oxidation reactions with electrophiles that are strong oxidants (H_2SO_4 or HNO_3). Formylation and carboxylation reactions give only monofunctionalization, because the functional group strongly deactivates the ferrocenyl group. Indeed, there is, to a certain extent, transmission of the electronic effect through the metal center. On the other hand, metallation and acylation reactions can be followed by an identical reaction on the other ring leading to 1,1'-disubstituted derivatives, because the deactivation of the second Cp ring by the substituent is only modest. (see Chap. 4).

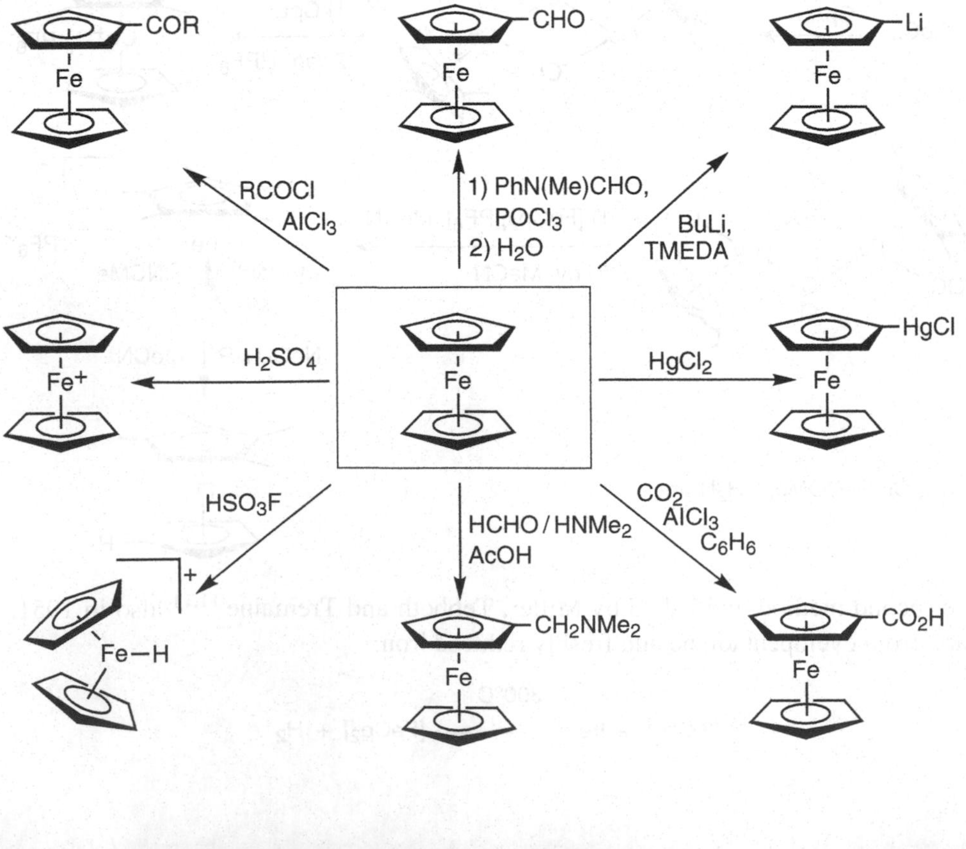

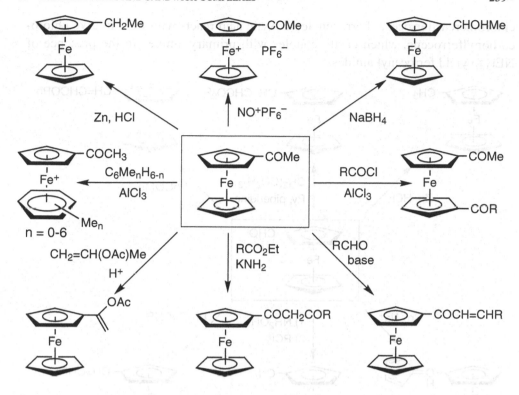

Lewis acids such as $AlCl_3$ form complexes leading to the cleavage of the Fe-Cp ring upon heating. The cleaved ring is protonated *in situ* and becomes an electrophile towards ferrocene. The reaction continues another time to give the cyclopentylene bridge. The *syn* and *anti* isomers of the di-bridged derivatives also form, and they have been separated, yielding cage structure in which the iron is encapsulated by four 5-carbon rings. In the presence of an aromatic or CO, these reactions are minimized, because the main reaction is substitution of a ferrocene ring by an arene or three CO ligands. This reaction is the classic mode of synthesis of the sandwich complexes $[FeCp(arene)]^+$.[11.13]

The most useful ferrocene derivatives for the development of functional ferrocene chemistry are acetylferrocene, ferrocenyl carbaldehyde, ferrocenyl lithium and chloromercuriferrocene. Ferrocenylcarboxylic acid reacts with PCl_5 to give chloro-carbonylferrocene, which easily couples with primary amines in the presence of NEt_3 to yield ferrocenyl amides.

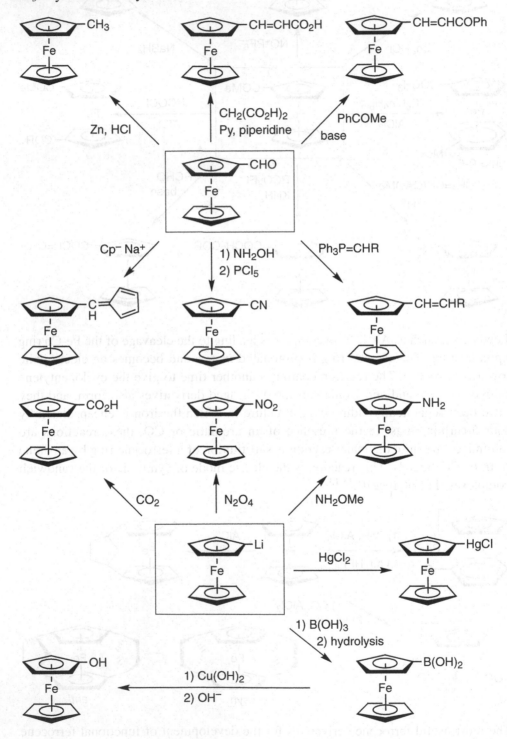

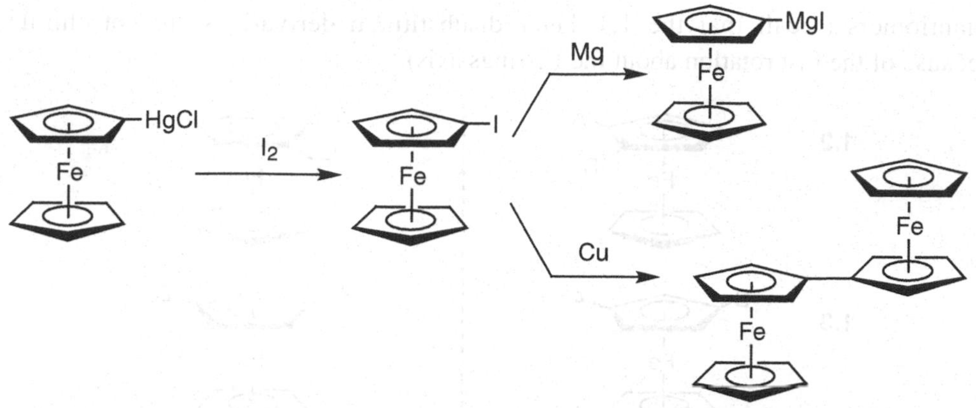

An essential characteristic of ferrocene chemistry is the stabilization of ferrocenyl carbonium ions. These carbocations are mesomers of the corresponding hexahapto fulvene complexes $[FeCp(\eta^6\text{-fulvene})]^+$. They are even more stable than the trityl cation Ph_3C^+. The stabilization of the α-ferrocenyl carbonium ions explains the acetolysis of vinylferrocene, the hydrolysis of the acetate formed, the ease of nucleophilic substitution in α position, and the OH^- abstraction from the α-ferrocenyl alcohols. This stabilization is still enhanced by going down in the iron column of the periodic table, because the size of the d orbital increases, which facilitates their insertion with the "carbocation" and accelerates the solvolysis of acetates $(Os > Ru > Fe)$.

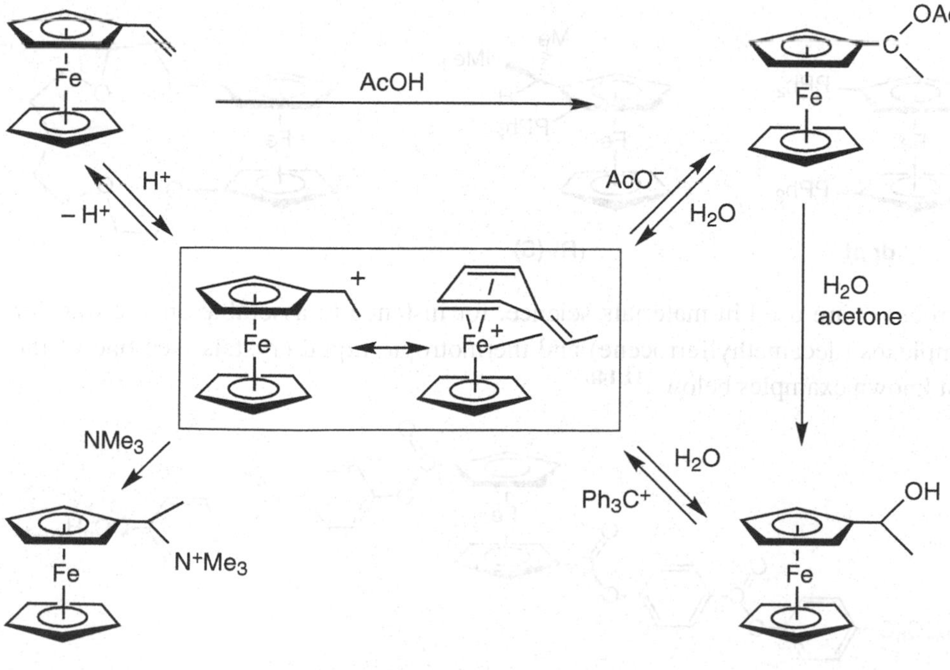

Finally note that the presence of two different substituents on the same ferrocene ring introduces a planar chirality. Many 1,2- or 1,3-heterodisubstituted ferrocene derivatives (amines, carboxylic acids, ketones, etc.) have indeed been split into enantiomers (recall that the 1,1'-heterodisubstituted derivatives are not chiral because of the fast rotation about the Fe-rings axis).

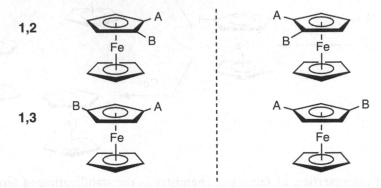

1,2

1,3

Planar metallocenic chirality for 1,2- or 1,3-heterodisubstituted ferrocenes

Applications of ferrocenes are numerous,[11.2] in particular with the use of ferrocenyl phosphines, including chiral ones (see for insance the aminophosphine below), as ligands for catalysis. 1,1'-bis(diphenylphosphino)ferrocene (dppf, below) is the best known ferrocenyl-based ligand in the catalysis of most classic reactions, and its chemistry and uses are prolific. The redox property of ferrocene has been used to attach it to macrocycles (below), cryptands, calixarenes and other *endo-receptors* for sensing, a chemistry beautifully developed by Paul Beer at Oxford.[11.14a]

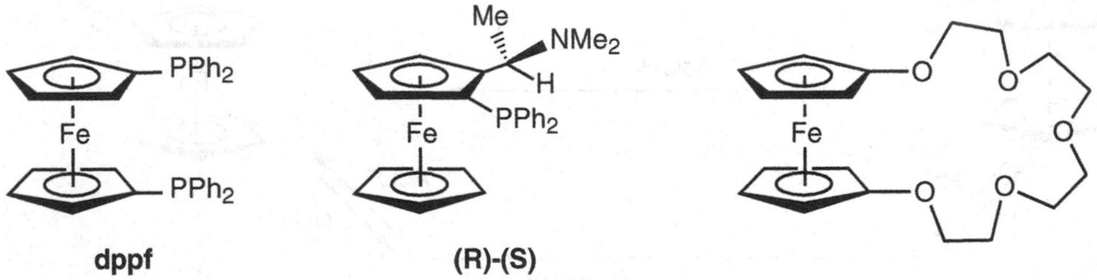

dppf   **(R)-(S)**

Ferrocenes are used in materials science, for instance to assemble charge-transfer complexes (decamethylferrocene) and thermotropic liquid crystals (see one of the best known examples below).[11.14b]

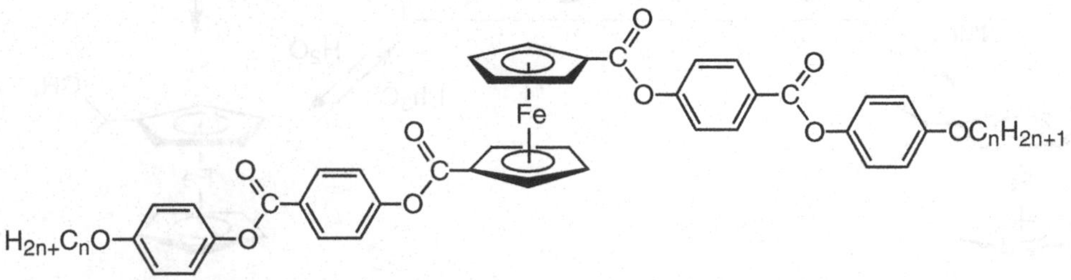

Ferrocenes are key components of polymers. For instance, Manners disclosed remarkable ring-opening polymerization of ferrocenophanes (below).[11.14c] Ferrocene-containing polymers have also been synthesized using classic routes by cationic, radical or ROMP polymerization (see Chap. 15).[11.14d] They are used *inter alia* to derivatize electrodes,[11.14e] which can provide excellent redox sensors.

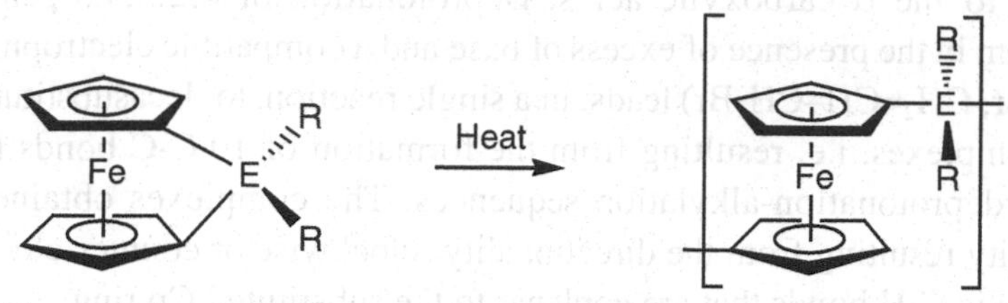

E = Si or Ge; R = Me, Et, etc.

Finally, ferrocenes, as many other cyclopentadienyl-metal complexes, have been widely studied as antitumor drugs or other therapeutic agents.[11.14f] The redox activity of the ferrocenyl group (below) may also add valuable synergistic effects on the cytotoxicity.[11.14g]

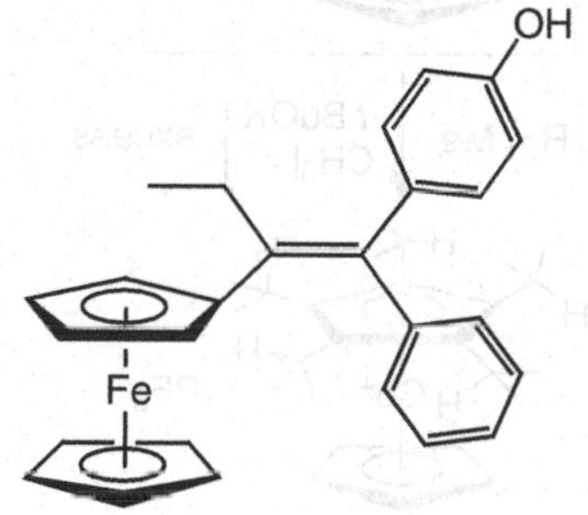

"Hydroxyferrocifen" related to the anticancer drug tamoxifen in which a ferrocenyl group has replaced a phenyl substituent with benefit on the cytotoxicity [11.14g]

4.2. OTHER METALLOCENE SANDWICHS

Chromocene reacts with 2-electron donors such as CO and $F_3CC \equiv CCF_3$ to give 18-electron adducts. Manganocene reacts with CO to give [MnCp(CO)$_3$]. Ruthenocene and osmocene have chemistries comparable to that of ferrocene, but have been much less studied. The 17-electron cations [MCp*$_2$]$^+$ (M = Ru, Os) are known, but not the parent complexes [MCp$_2$]$^+$. For instance, the oxidation of osmocene gives the binuclear Os-Os bonded complex.

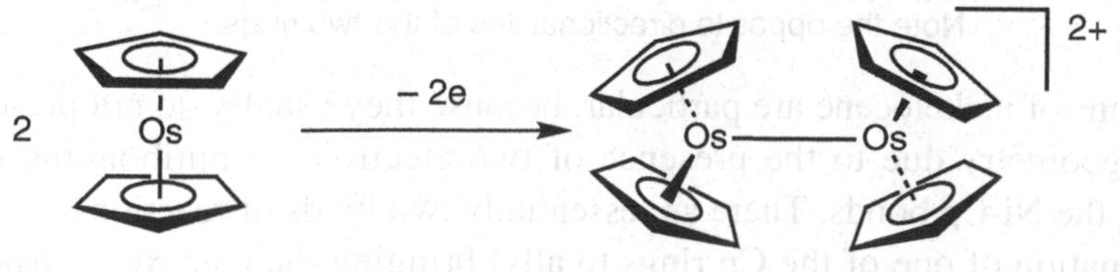

Cobaltocene reacts with oxygen and oxidants, tetrafluoroethylene and halides. The reaction of cobaltocene with dihalogenomethanes gives ring extension. For instance, borabenzene complexes are accessible in this way by reaction with alkyldihalogenoboranes.

The robust cobaltocenium cation $[CoCp_2]^+$ reacts (usually as a PF_6^- salt) with anionic nucleophiles, and alkylcobaltocenium cations can be deprotonated in α position or oxidized to the α carboxylic acids. Deprotonation of 1,2,3,4,5-pentamethylcobaltocenium in the presence of excess of base and a compatible electrophile (CH_3I, $PhCH_2CH_2Br$, $CH_2=CH–CH_2Br$) leads, in a single reaction, to decasubstituted cobalt sandwich complexes, i.e. resulting from the formation of 10 C-C bonds following 10 identical deprotonation-alkylation sequences. The complexes obtained have a planar chirality resulting from the directionality (clockwise or counter-clockwise) of the 5 *exo*-cyclic C-H bonds that are coplanar to the substituted Cp ring.

Pentamethyl- and decamethylrhodocenium analogously lead to respectively penta- and decaisopropylrhodocenium by reaction of excess CH_3I and *t*-BuOK. In $[Rh\{C_5(i-Pr_5)_2\}]^+$, the directionalities of the two rings are opposite, being self-organized in the course of the reaction: [11.15]

Synthesis of decaisopropylrhodocenium (the methyl groups are omitted for clarity).
Note the opposite directionalities of the two rings.

The reactions of nickelocene are particular, because they usually do not preserve the sandwich geometry due to the presence of two electrons in antibonding orbitals weakening the Ni-Cp bonds. There are essentially two kinds of reactions:

▸ hydrogenation of one of the Cp rings to allyl bringing the electronic structure to 18 valence electrons;

▸ facile thermal loss of a Cp ring, more frequent and interesting, because it yields a source of the very reactive 15-electron fragment that reacts with a large variety of unsaturated substrates to complete the 18-electron structure.

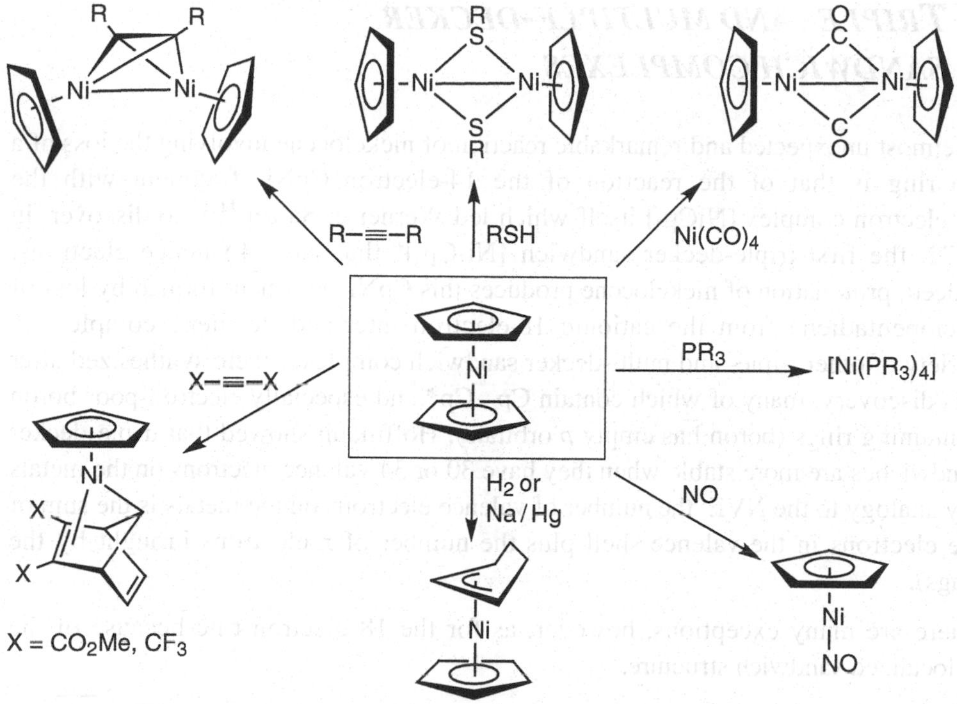

R–≡–R | RSH | Ni(CO)$_4$

X–≡–X

PR$_3$ ⟶ [Ni(PR$_3$)$_4$]

Ni

H$_2$ or
Na/Hg

NO

X = CO$_2$Me, CF$_3$

Ni
NO

Finally, the family of metallocenes includes all the main-group metals, and robust examples involve the permethylmetallocenes, a chemistry developed by Jutzi.[13.9a] One Cp ligand is often bonded in a monohapto mode (η^1-C$_5$H$_5$) for main-group metallocenes, but many compounds also have the bis-pentahapto-Cp-metal sandwich structure.[13.9a] For instance, in beryllocene, the energies of the η^5/η^5 and η^5/η^1 structures have been calculated to be about the same, only favoring the mixed structure by a few kcal.mol^{-1}. Beryllocene is highly fluxional, only one set of resonances being found for the two C$_5$H$_5$ groups in the ^{1}H- and ^{13}C-NMR spectra at $-135°$C. There are 1,5 sigmatropic shifts of the Be(η^5-Cp) unit around the periphery of the η^1-C$_5$H$_5$ ring and a molecular inversion that interchanges the η^1-C$_5$H$_5$ and η^5-C$_5$H$_5$ rings. The Carmona group reported remarkable examples representative of the monohapto/pentahapto dichotomy with the synthesis of tetra- and penta-methyl Cp sandwich Be complexes below. For these complexes, the X-ray crystal structures have been determined, illustrating the subtle difference between the η^5/η^5 and η^5/η^1 structures:

Be
H
K$^+$
toluene/ether
20°C, overnight

BeCl$_2$

K$^+$
toluene
115°C, 4 days

Be

–• = CH$_3$

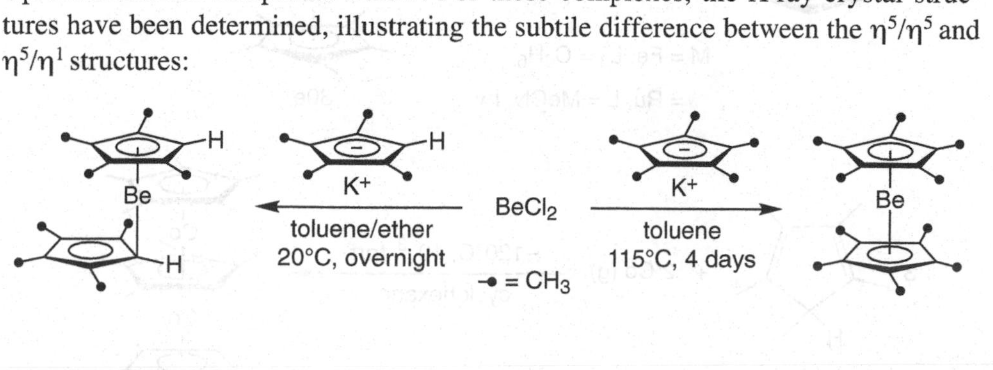

5. TRIPLE- AND MULTIPLE-DECKER SANDWICH COMPLEXES

The most unexpected and remarkable reaction of nickelocene involving the loss of a Cp ring is that of the reaction of the 14-electron CpNi$^+$ fragment with the 20-electron complex [NiCp$_2$] itself which led Werner et Salzer [11.16] to discover, in 1972, the first triple-decker sandwich [Ni$_2$Cp$_3$]$^+$ that has 34 valence electrons. Indeed, protonation of nickelocene produces this CpNi$^+$ fragment formed by loss of cyclopentadiene from the cationic 18-electron intermediate diene complex. A variety of other triple- and multi-decker sandwich complexes were synthesized after this discovery, many of which contain Cp , Cp* and especially electron-poor boron containing rings (boron has empty p orbitals). Hoffmann showed that triple-decker sandwiches are more stable when they have 30 or 34 valence electrons on the metals (by analogy to the NVE, the number of valence electrons on the metals is the sum of the electrons in the valence shell plus the number of π electrons brought by the rings).

There are many exceptions, however, as for the 18-electron rule because of the delocalized sandwich structure.

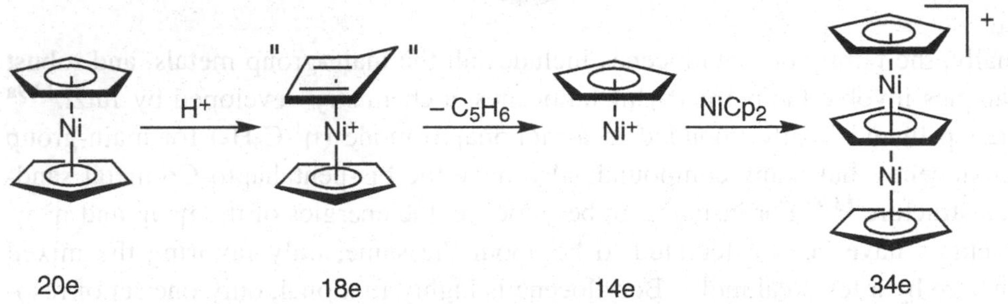

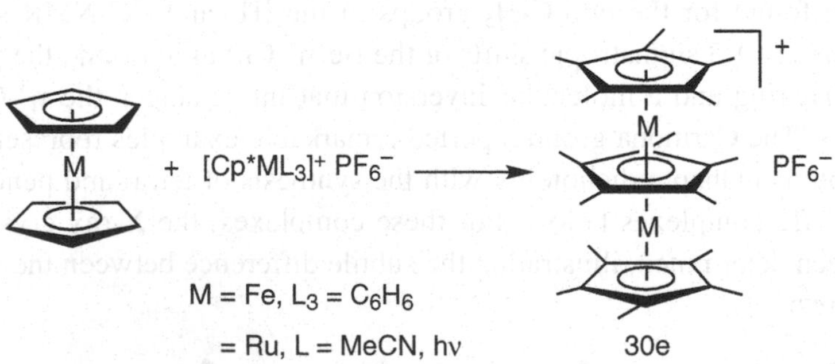

20e   18e   14e   34e

M = Fe, L$_3$ = C$_6$H$_6$

= Ru, L = MeCN, hν 30e

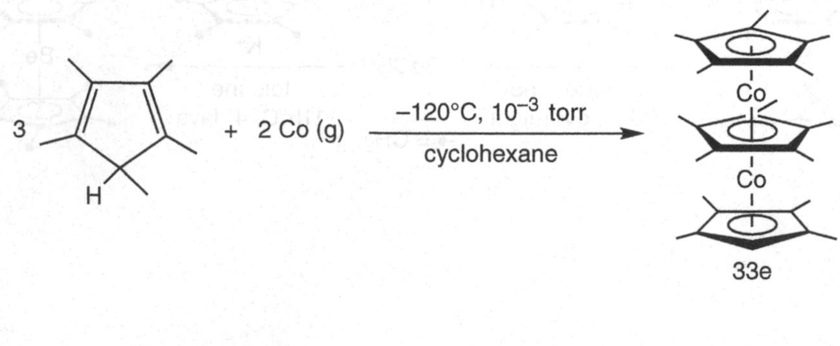

33e

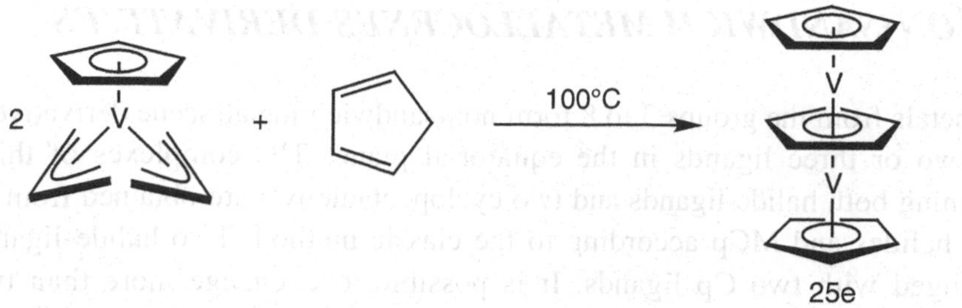

Siebert and Herberich have systematically synthesized boron-containing electron-deficient rings, especially because they found that these rings were excellent stabilizers of the triple-and multiple-decker structures. For instance, borole C_4H_4BH (4e) and 2,3-dihydro-1,3-diborole (3e) are bonded to the metal in a pentahapto fashion as Cp, yet they have and bring a reduced number of π electrons. The triple-decker sandwich complexes formed herewith can have different oxidation states as known in transition-metal sandwich chemistry.[11.16c]

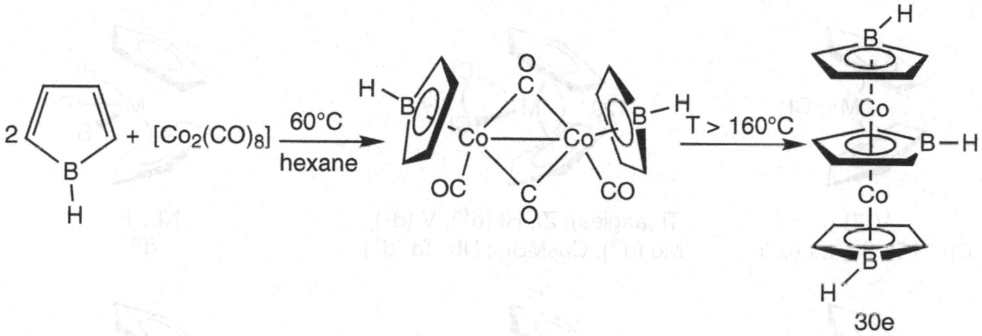

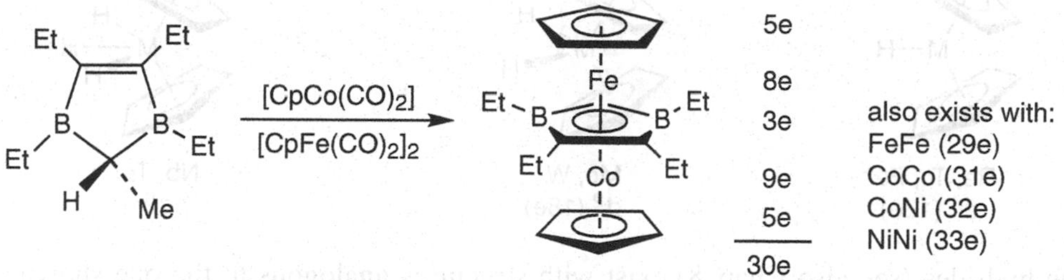

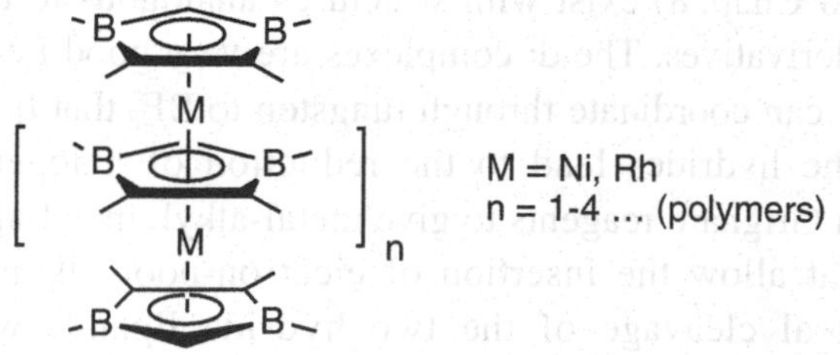

6. NON-SANDWICH METALLOCENES DERIVATIVES

The metals from the groups 3 to 8 form non-sandwich metallocene derivatives with one, two or three ligands in the equatorial plane. The complexes of this type containing both halide ligands and two cyclopentadienyls are obtained from binary metal halides and MCp according to the classic method. Two halide ligands are exchanged with two Cp ligands. It is possible to exchange more than two Cp ligands, but beyond two, the additional Cp ligands bind in a monohapto fashion except in lanthanides and actinides (see Chap. 12). With excess MCp, the halide substitution sometimes goes along with some reduction of the metal:

$$NbCl_5 \xrightarrow{\text{excess NaCp}} [Nb(\eta^5\text{-}Cp)_2 (\eta^1\text{-}Cp)_2]$$

The halide ligands can be substituted by various nucleophiles (hydride, carbanion, alkoxy, amido, imido).

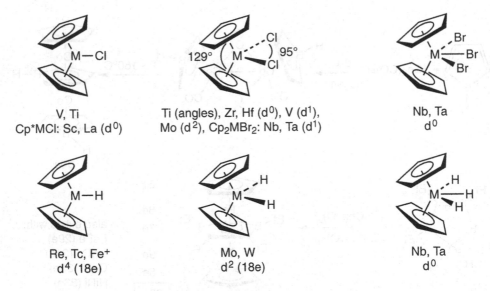

V, Ti	Ti (angles), Zr, Hf (d^0), V (d^1),	Nb, Ta
Cp*MCl: Sc, La (d^0)	Mo (d^2), Cp$_2$MBr$_2$: Nb, Ta (d^1)	d^0
Re, Tc, Fe$^+$	Mo, W	Nb, Ta
d^4 (18e)	d^2 (18e)	d^0

The hydrides (see also Chap. 8) exist with structures analogous to the one shown above for the halide derivatives. The d^2 complexes are very good Lewis bases. For instance, [WCp$_2$(H)$_2$], can coordinate through tungsten to BF$_3$ that becomes a zero-electron Z ligand. The hydrides lead to the reduction of halogenated organic derivatives, react with Grignard reagents to give metal-alkyl, metal-allyl and metal-benzyl complexes that allow the insertion of electron-poor alkynes. They also undergo photochemical cleavage of the two hydride ligands with reductive elimination of H$_2$ (especially with Mo and W), giving exchange with CO, C$_2$H$_4$, PR$_3$, C$_2$Ph$_2$. The transient reactive metallocenes (MCp$_2$, d^2) formed upon photolysis of the dihydride can also undergo oxidative addition of a C-H bond of benzene (but not alkane), activated olefins and a C-O bond of epoxides:

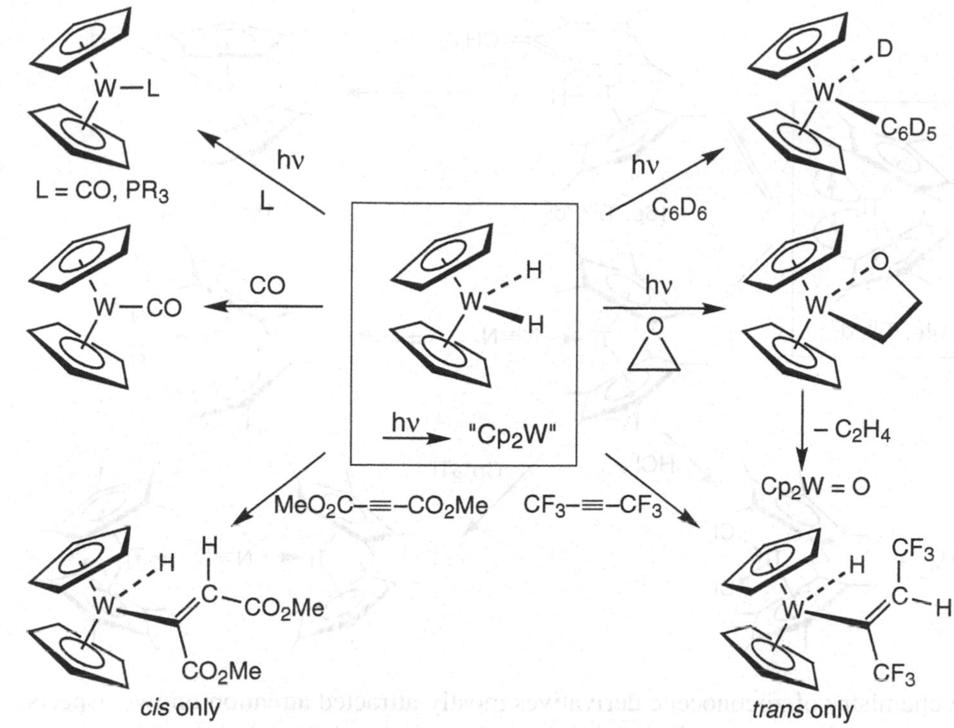

cis only   *trans* only

The instable titanocenes [TiCp$_2$] and [TiCp*$_2$] are formed upon reduction of [TiCp$_2$Cl$_2$] by Na or Na/Hg or by thermolysis of [TiCp*$_2$H$_2$], and rearrange in an original way. A binuclear complex is formed from [TiCp$_2$], and two mechanistic possibilities for the dimerization are shown below:

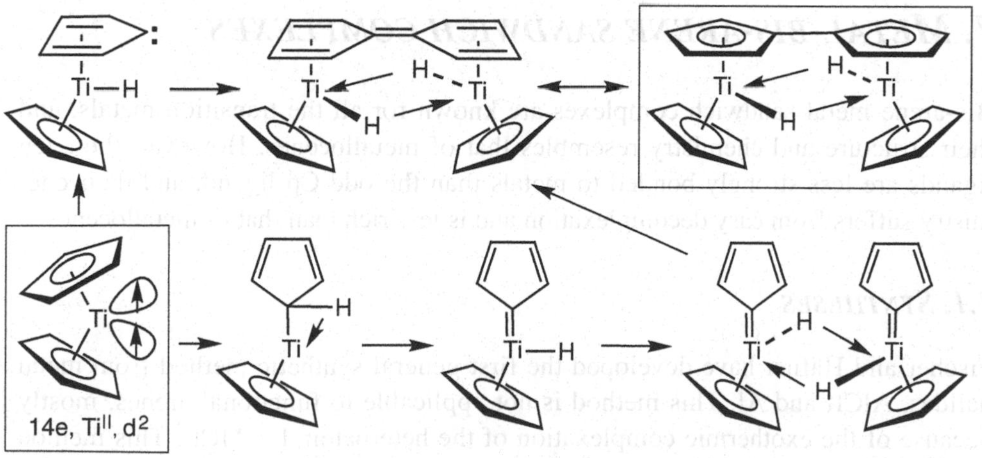

Two mechanistic possibilities for the dimerization of titanocene

[TiCp*$_2$], generated as indicated above, binds molecular nitrogen giving two forms, monohapto and dihapto of a mononuclear complex that can be protonated to hydrazine (a rare case of transition-metal-induced reduction of dinitrogen).

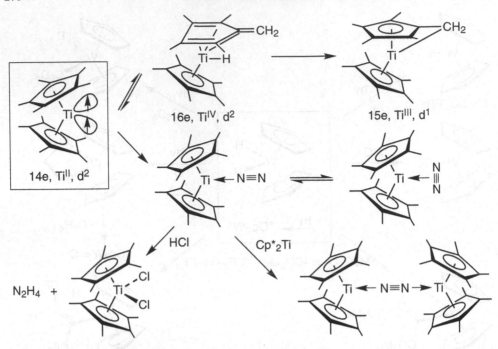

The chemistry of zirconocene derivatives mostly attracted attention on two aspects. First, Schwarz reagent [ZrCp$_2$(H)(Cl)] is able to regiospecifically insert olefins into the Zr-H bond (see Chaps 6 and 21). Then, a second more recent aspect is the zirconocene-based initiation of olefin polymerization with 14-electron precursor complexes of the type "[ZrCp$_2$(Me)]$^+$" (see Chap. 15.1).

7. METAL-BIS-ARENE SANDWICH COMPLEXES

Bis-arene metal sandwich complexes are known for all the transition metals, and their structure and chemistry resembles that of metallocenes. However, the even ligands are less strongly bonded to metals than the odd Cp ligand, and their chemistry suffers from easy decomplexation and is less rich than that of metallocenes.

7.1. SYNTHESES

Fischer and Hafner have developed the first general synthetic method from metal halides, AlCl$_3$ and Al. This method is not applicable to functional arenes, mostly because of the exothermic complexation of the heteroatom by AlCl$_3$. This method also leads to the isomerization of the methyl- and alkylaromatics by retro-Friedel-Crafts reaction, especially for the 2nd- and 3rd-row transition metals. It has been applied with success for V, Cr, Fe, Co and Ni with benzene and its polymethyl derivatives, in particular C$_6$Me$_6$ with Co and Ni. The monocations [Cr(η^6-arene)$_2$]$^+$ are produced after hydrolysis, and the dications are obtained with the 1st-row transition metals:

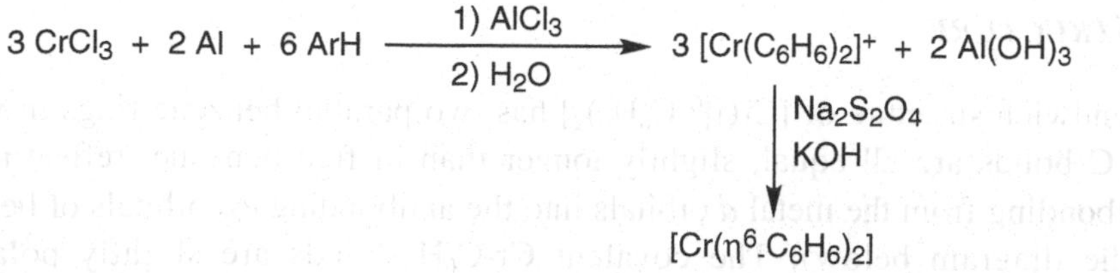

$$3 \, CrCl_3 + 2 \, Al + 6 \, ArH \xrightarrow[\text{2) } H_2O]{\text{1) } AlCl_3} 3 \, [Cr(C_6H_6)_2]^+ + 2 \, Al(OH)_3$$

$$\downarrow \begin{array}{c} Na_2S_2O_4 \\ KOH \end{array}$$

$$[Cr(\eta^6\text{-}C_6H_6)_2]$$

The second general method consists in condensing metal vapors with the aromatic at low temperature and pressure. It has been pioneered by the groups of Timms, Green and Cloke and applied to all the transition metals. Eschenbroich and Kundig have also applied it to condensed polyaromatics, and the metals bind to the rings located at the periphery, because these rings are more electron-rich than the other ones due to the reduced number of common bonds.[11.17a]

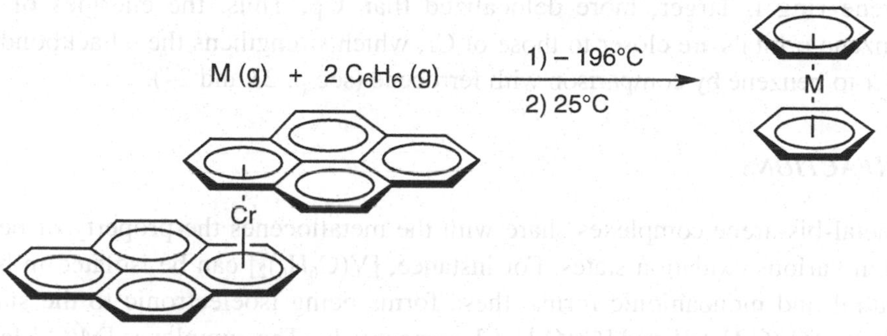

$$M \, (g) + 2 \, C_6H_6 \, (g) \xrightarrow[\text{2) } 25°C]{\text{1) } -196°C} M$$

There are other synthetic methods that are restricted to particular cases such as the catalyzed trimerization of alkynes and the historical reaction of PhMgBr with $CrCl_3$ leading to $[Cr(\eta^6\text{-}C_6H_6)(\eta^6\text{-diphenyl})]$. A useful method is the reaction of arenes with $[Ru(\eta^6\text{-}C_6H_6)(\text{acetone})_3]^{2+}$ pioneered by Bennett which allows to make Ru^{II} sandwich complexes containing two different arenes. This method has allowed Boekelheide [11.17c] to synthesize cyclophane complexes.

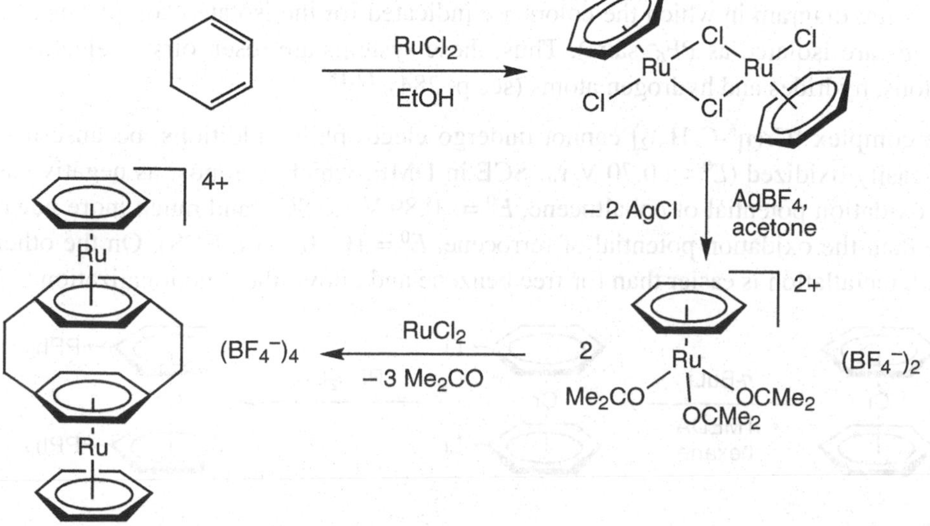

7.2. STRUCTURE

The sandwich structure of $[Cr(\eta^6\text{-}C_6H_6)_2]$ has two parallel benzene rings in which the C-C bonds are all equal, slightly longer than in free benzene, reflecting the δ backbonding from the metal d orbitals into the antibonding e_{2g} orbitals of benzene (see the diagram below). The covalent $Cr\text{-}C_6H_6$ bonds are slightly polarized ($\delta^+_{Cr} = +0.7$; $\delta^-_{C_6H_6} = -0.35$). The bond energy (40 kcal·mol^{-1}; 167 kJ·mol^{-1}) is smaller than in ferrocene (52 kcal·mol^{-1}; 217 kJ·mol^{-1}), and the rotation barrier about the Cr-ring axis is very weak (< 1 kcal·mol^{-1}; 4.2 kJ·mol^{-1}). The MO diagram including the orbital interactions in $[Cr(\eta^6\text{-}C_6H_6)_2]$ and the electronic structures of the other metal-bis-arene sandwich complexes of the first-row transition metals are represented on p. 283. This diagram resembles that of ferrocene, but the benzene orbitals have lower energies than the Cp orbitals, because the arene ring is larger, more delocalized than Cp. Thus, the energies of the e_{2g} benzene orbitals are closer to those of Cr, which strengthens the δ backbonding from Cr to benzene by comparison with ferrocene (see p. 23 and 24).

7.3. REACTIONS

The metal-bis-arene complexes share with the metallocenes the property of being found in various oxidation states. For instance, $[V(C_6H_6)_2]$ can be isolated in both its neutral and monoanionic forms, these forms being isoelectronic to the stable complexes $[Cr(C_6H_6)_2]^+$ and $[Cr(C_6H_6)_2]$, respectively. The complexes $[M(C_6Me_6)_2]$ (M = Fe, Co, Ni) exist as neutral, mono- and dicationic complexes (although the structures of complexes that would have more than 20 valence electrons in the fully sandwich form do not exist as such but partial decoordination reduces their valence electron count). In the iron-bis-arene series, the neutral, mono- and dicationic complexes are also isolable, and these three complexes are thermally stable in the case of C_6Me_6. Moreover, it is also possible to transfer one or two hydrides (see Chap. 4.1), one or two protons and one or two hydrogen atoms, which gives the following diagram in which the colors are indicated for the isolated complexes (the cations are isolated as PF_6^- salts). Thus, these systems are reservoirs of electrons, protons, hydrides and hydrogen atoms (see p. 284). [11.18]

The complex $[Cr(\eta^6\text{-}C_6H_6)_2]$ cannot undergo electrophilic additions, because it is too easily oxidized ($E^0 = -0.70$ V $vs.$ SCE in DME, which is almost as negative as the oxidation potential of cobaltocene, $E^0 = -0.89$ V $vs.$ SCE and much more negative than the oxidation potential of ferrocene, $E^0 = +0.40$ V $vs.$ ECS). On the other hand, metallation is easier than for free benzene and allows the functionalization.

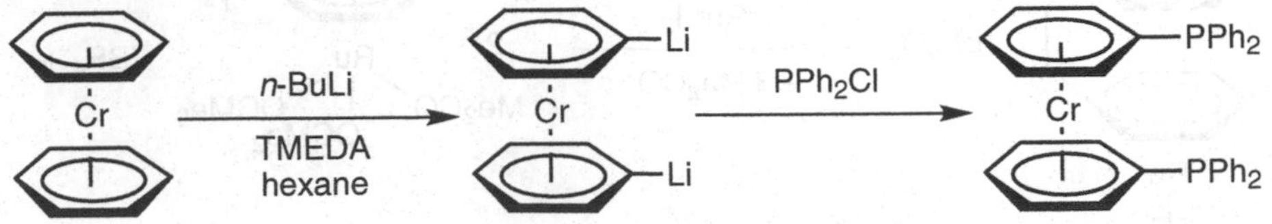

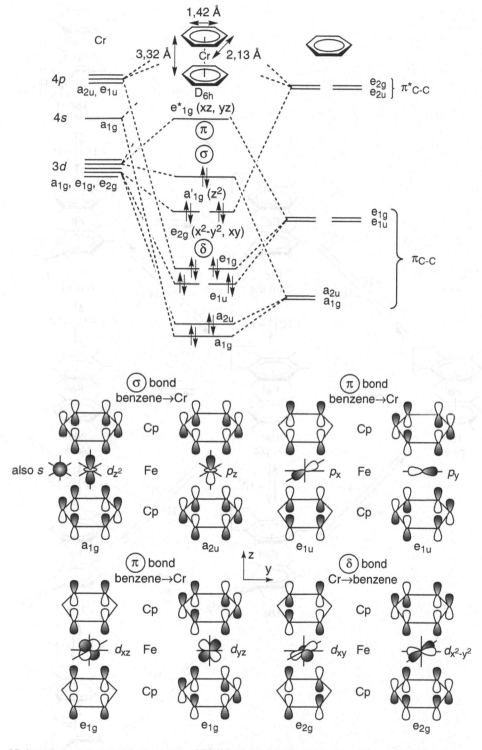

Molecular orbital (MO) diagram of [Cr(C₆H₆)₂] (top) and interactions of the benzene orbital with the Cr orbitals of appropriate symmetry to form [Cr(C₆H₆)₂] (bottom)

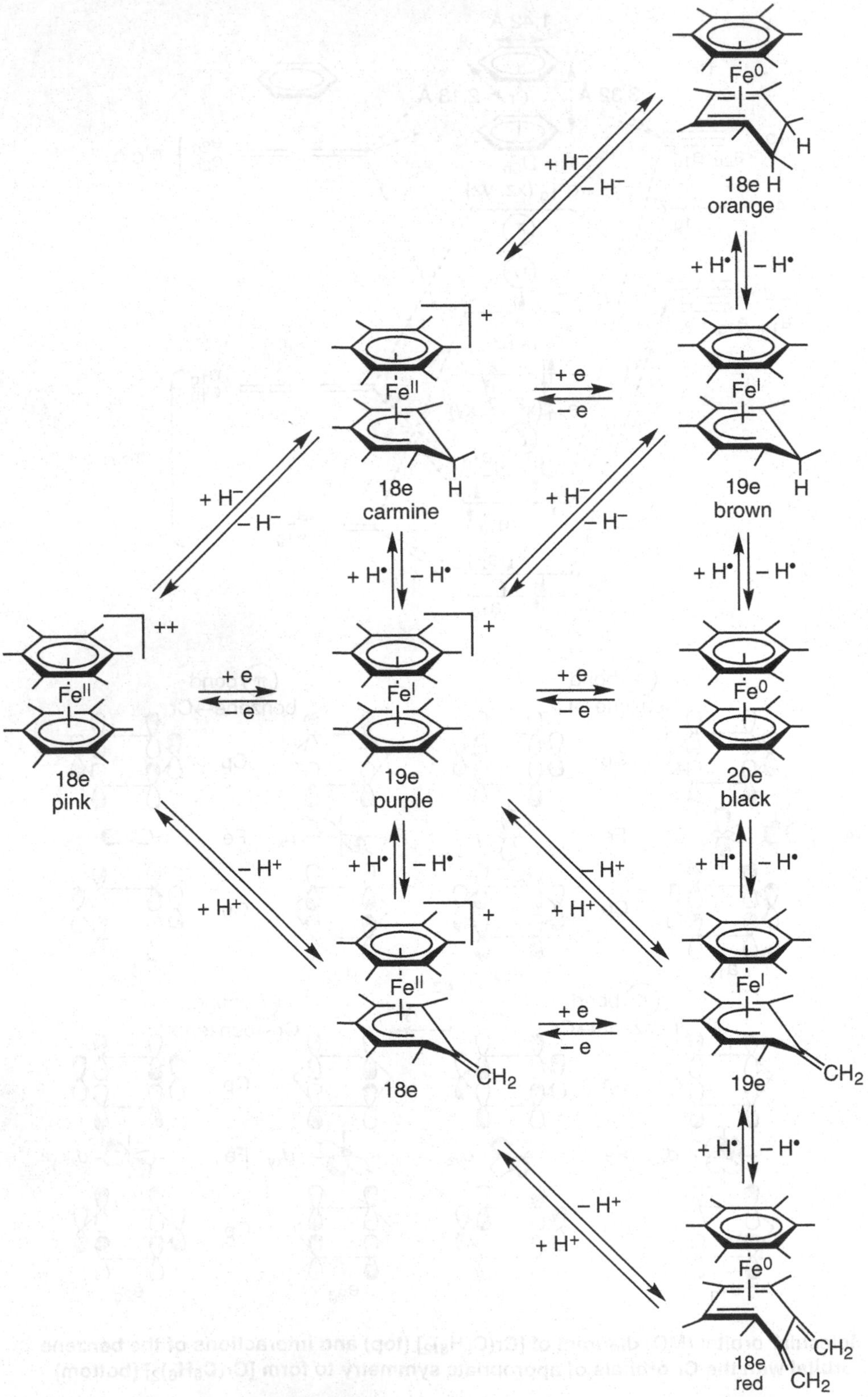

Network of iron-bis-arene sandwich complexes as reservoirs of electrons, hydrides, protons and hydrogen atoms

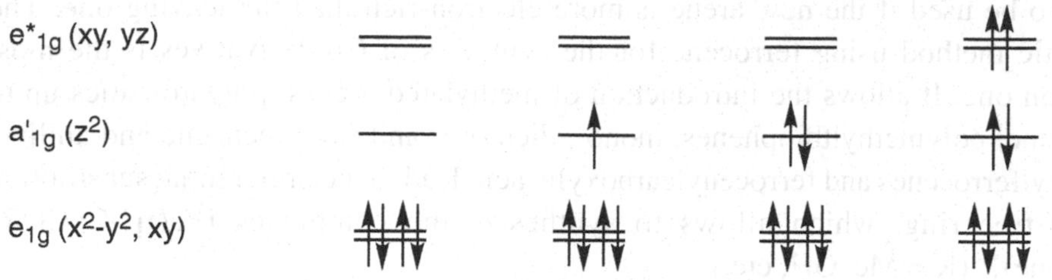

	$Ti(C_6H_6)_2$	$V(C_6H_6)_2$	$Cr(C_6Me_6)_2$	$Fe(C_6Me_6)_2$
NEV	16e	17e	18e	20e
e^*_{1g} (xy, yz)	═══	═══	═══	⥮⥮
a'_{1g} (z^2)	───	↑	⥮	⥮
e_{1g} (x^2-y^2, xy)	⥮⥮	⥮⥮	⥮⥮	⥮⥮
unpaired electrons (n)	0	1	0	2
$\sqrt{n(n+2)}$	0	1.73	0	2.83
exp. $\mu(\mu_B)$	0	1.68	0	3.08
color	red	dark red	brown	pink

Electronic structure (see the MO diagram of $[Cr(\eta^6\text{-}C_6H_6)_2]$ above), theoretical
($\mu = \sqrt{n(n+2)}$ μ_B) and experimental (μ in Bohr magnetons) magnetic moment
and color of the neutral metal-bis-arene sandwich complexes

8. CYCLOPENTADIENYL-METAL-ARENE COMPLEXES [11.19]

These complexes usually have the advantage of robustness, the same type of redox properties as metallocenes, and offer new possibilities of arene chemistry easily applicable in organic chemistry.[11.5,11.12,11.18] They are known for the groups 7, 8 and 9, but especially for iron and ruthenium. The 18-electron forms are represented below (some other redox forms of these complexes are known).

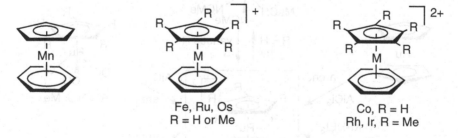

Fe, Ru, Os
R = H or Me

Co, R = H
Rh, Ir, R = Me

8.1. SYNTHESES

The cations $[Fe(\eta^5\text{-}C_5R_5)(\eta^6\text{-arene})]^+$ are synthesized by reactions between ferrocene (R = H), $[Fe(\eta^5\text{-}C_5R_5)(CO)_2X]$ (R = H or Me, X = Cl or Br) or $[Fe(\eta^5\text{-}Cp^*)(\eta^2\text{-acac})]$ and an arene in the presence of $AlCl_3$.

It is also possible to carry out arene exchange using visible-light photochemistry in dicholoromethane using a starting complex $[Fe(\eta^5\text{-}Cp)(\eta^6\text{-}arene)]^+$ that has a labile arene such as naphthalene, but benzene, toluene or a xylene ligand in the complex can also be used if the new arene is more electron-rich than the leaving one. The synthetic method using ferrocene for the synthesis of Cp derivatives is the most common one. It allows the introduction of methylated arenes, polyaromatics up to corronene, polymethylthiophenes, mono-, dichloro- and fluorobenzene and aniline, etc. Acylferrocenes and ferrocenylcarboxylic acid lead to the preferential substitution of the free ring, which allows to synthesize the complexes $[Fe(\eta^5\text{-}CpCOR)(\eta^6\text{-}arene)]^+$ (R = Me, OH, etc.).

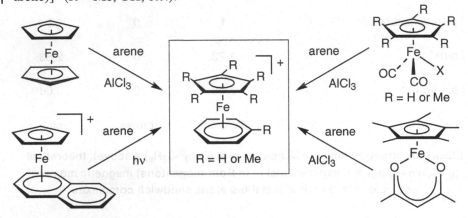

The analogs $[Ru(\eta^5\text{-}C_5R_5)(\eta^6\text{-}arene)]^+$ are accessible using comparable methods, but they are more common and rapidly accessible with the Cp* ligands than with the Cp one. The complexes $[Ru(\eta^5\text{-}C_5R_5)(\eta^1\text{-}MeCN)_3]^+$ (R = H or Me) react with aromatics to give the sandwich arene complexes, which is not the case in the iron series (Ru, unlike Fe, favors the temporary tetrahapto arene structure, a key intermediate).

8.2. NUCLEOPHILIC SUBSTITUTIONS AND ADDITIONS

The nucleophilic substitutions of the halide in halogenoarene ligands by alcohols, thiols and amines first proceed by attack of this nucleophile on the *ipso* carbon to give a cyclohexadienyl intermediate. This intermediate is very acidic and easily dehydrohalogenated by a weak base (the bases that are used in these reactions are too weak to deprotonate the starting nucleophile). The first step, i.e. attack by the nucleophile, is the rate-limiting step. For instance, the fluoroarene complexes react faster than the analogous chloroarene complexes, because the *ipso* site of attack is more positively charged in the fluoroarene complexes than in the chloroarene complexes. The fluoride is not as good a leaving group as the chloride, but this step does not influence the rate.

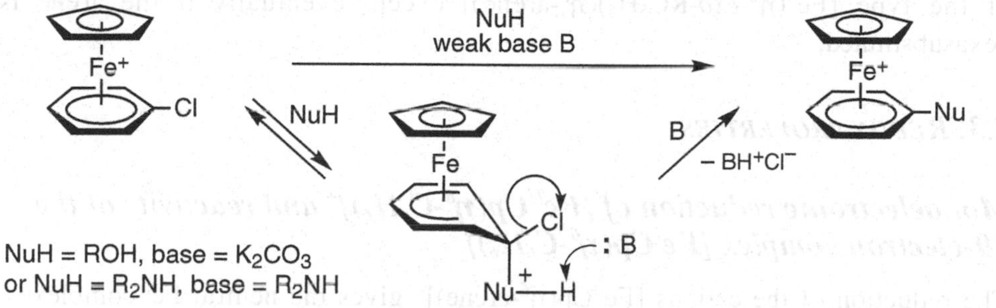

NuH = ROH, base = K_2CO_3
or NuH = R_2NH, base = R_2NH

Nevertheless, the more practical chloroarene complexes are commonly used as they react well, whereas the activation by $Cr(CO)_3$ is therefore not sufficient.

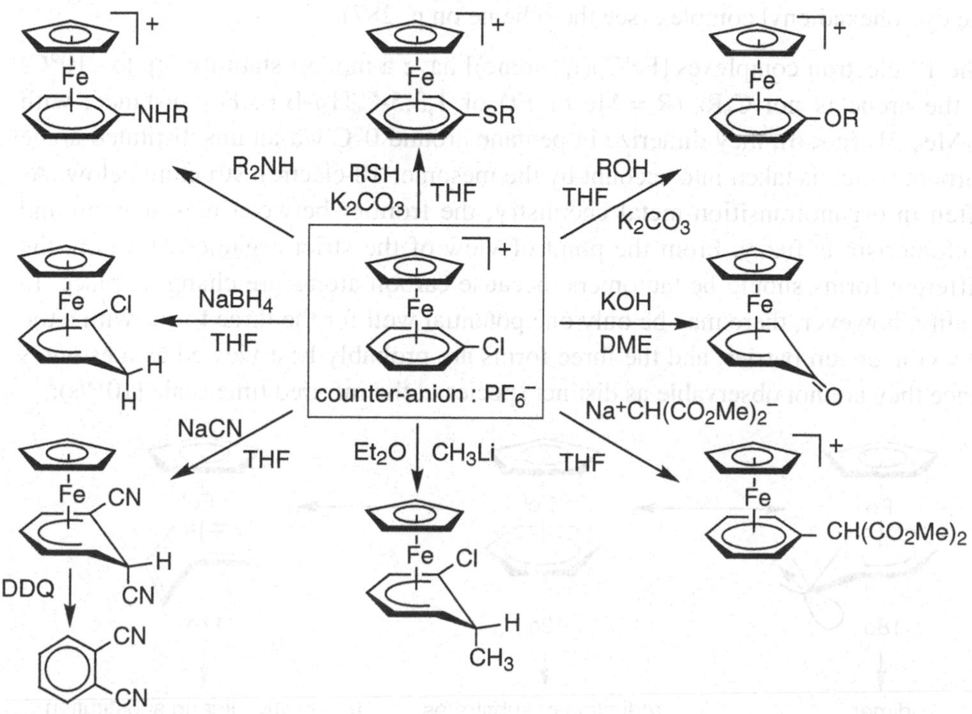

Non-stabilized carbanions such as CH_3Li attack the chlorobenzene ligand in *ortho* to give stable cyclohexadienyl complexes, the reaction being analogous to that of the benzene complex. The regioselectivity is the same with the methylbenzoate ligand, but attack occurs in *meta* on the anisole ligand. Substitution of Cl^- occurs by reaction with stabilized carbanions such as CN^-, but in this case the reaction continues with a second nucleophilic attack of CN^- in *ortho* to give a disubstituted cyclohexadienyl complex. In these nucleophilic reactions, substitution of H^- never occurs, because H^- is not a good leaving group. It can be obtained, however, upon monoelectronic oxidation to the 17-electron radical cation of the cyclohexadienyl complex that looses an H atom to recover the aromaticity (which means abstraction first of an electron, then an H atom instead of direct H^- abstraction). The Cp ring is not attacked by nucleophiles in the complexes $[FeCp(\eta^6\text{-arene})]^+$ to give complexes of the type $[Fe^0(\eta^4\text{-}exo\text{-}RC_5H_5)(\eta^6\text{-arene})]$ except eventually if the arene is hexasubstituted.

8.3. REDOX PROPERTIES

Monoelectronic reduction of $[Fe^{II}Cp(\eta^6\text{-}C_6H_6)]^+$ and reactivity of the 19-electron complex $[Fe^ICp(\eta^6\text{-}C_6H_6)]$

The reduction of the cations $[Fe^{II}Cp(\eta^6\text{-arene})]^+$ gives the neutral Fe^I complexes, and can be carried our either using a classic monoelectronic reductant such as Na/Hg or using a hydride. In the later case, the Fe^I complex is an intermediate, and the reaction usually continues by $H^\bullet$ transfer onto the arene ligand to finally yield the cyclohexadienyl complex (see the scheme on p. 287).

The 19-electron complexes $[Fe^ICp(\eta^6\text{-arene})]$ have a modest stability (up to $-10°C$) if the arene is not C_6R_6 (R = Me or Et) or $1,3,5\text{-}C_6H_3t\text{-}Bu_3$. For instance, with $C_6Me_{6-n}H_n$ (n < 6), they dimerize in pentane around $0°C$ *via* an unsubstituted arene carbon, which is taken into account by the mesomer 18-electron structure below. As often in organotransition-metal chemistry, the frontier between mesomerism and tautomerism is fuzzy. From the point of view of the strict organic definition, the different forms should be tautomers, because carbon atoms are changing place. In reality, however, there may be only one potential well for the three forms with a too low conversion barrier, and the three forms are probably best viewed as mesomers since they are not observable as distinct species at the infrared time scale (10^{13}/s):

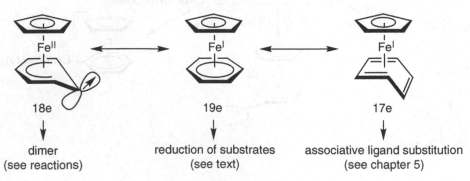

18e	19e	17e
dimer (see reactions)	reduction of substrates (see text)	associative ligand substitution (see chapter 5)

The complexes [$Fe^I(\eta^5$-Cp)(η^6-arene)] in which the arene is C_6Me_6, C_6Et_6 or 1,3,5-t-Bu$_3$C$_6$H$_3$ are stable and do not undergo dimerization or ligand substitution reactions, which allowed to isolate them, study their physical and chemical properties, and use them as electron-reservoir systems with arene = C_6Me_6, since they are strong reductants.

Electron and proton reservoirs

Cyclic voltammetry studies allowed access to four oxidation states of the complexes [$Fe(\eta^5$-Cp)(η^6-arene)]^x (x = –1 to +2) starting from the 18-electron monocations.

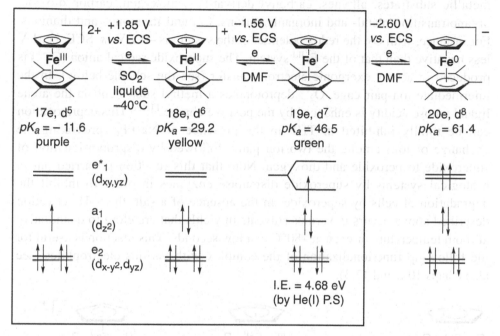

Electron reservoir complexes [FeCp(η^6-C$_6$Me$_6$)]^x:
the three first forms have been isolated and are isostructural

Note the extreme values of the potentials and the large pK_a variations from one form to the other (solvent: DMSO) calculated using the pK_a of [FeCp(η^6-C$_6$Me$_6$)]$^+$ (18e) determined by ^{1}H NMR and thermodynamic diagrams that also allow to calculate the C-H bond dissociation energies (BDE). The ionization energy (I.E.) of [Fe^{I}Cp(η^6-C$_6$Me$_6$)] was determined by He (I) photoelectron spectroscopy (P.S.).

With [$Fe(\eta^5$-C$_5$R$_5$)(η^6-C$_6$Me$_6$)]$^+$PF$_6^-$, R = H or Me, the three cyclic voltammetry waves connecting these four oxidation states are chemically and electrochemically reversible. This shows the stability of the sandwich structure along the redox processes between the 17-electron Fe^{III} dications and the 20-electron Fe^0 anion. Among these four oxidation states, three of them (Fe^{III}-Fe^I) have been isolated in crystalline form in the series [FeCp*(η^6-C$_6$Me$_6$)]^x, x = 0, 1 and 2.[11.19b] For the Fe^I complex, the extra electron is borne by the $e*_{1g}$ orbital that is essentially metal-based (80% metal character by Mössbauer spectroscopy). The Fe^I radical center is located at the heart of the electron reservoir and protected by the permethylated rings, which explains its stability. The ionization potentials determined by He(I) photoelectron spectroscopy are the lowest ever recorded for neutral molecular complexes, which means that these complexes are the most electron-rich molecules known to-date. They form, together with the isostructural Fe^{II} and Fe^{III} complexes, electron-reservoir systems whose Fe^{I} forms are excellent reductants and Fe^{III} forms excellent oxidants.

The electron-reservoir complexes have the following properties:

▸ **Stoichiometric reductions**

The Fe^{I} complexes can reduce a large number of organic, inorganic and organo-metallic substrates: alkynes, carbonyl derivatives, dioxygen, carbon dioxide, organotransition-metal- and inorganic cations, C_{60} and its mono- and dianions. For instance, with O_2, the redox potential of the system is -0.7 V vs. SCE, i.e. 1 V less negative than that of the $Fe^{II/I}$ system. The superoxide radical anion ($O_2^{\bullet-}$) is produced in a very exergonic electron transfer (see the scheme below). In the intermediate ion-pair cage, $O_2^{\bullet-}$ deprotonates a methyl subsistent on the arene ligand whose acidity is enhanced by the positive charge.[11.19c] This deprotonation can be totally inhibited in THF in the presence of $Na^+PF_6^-$ provoking the exchange of ions among the two ion pairs, followed by disproportionation of superoxide to peroxide and dioxygen. Note that this function is carried out in biological systems by superoxide dismutase enzymes in order to inhibit the degradation of cells by superoxide. In the absence of a salt, the C-H activation described above occurs in various solvents in yields that are close to quantitative at room temperature or even at $-80°C$ in a few seconds. This reaction is useful for the follow-up functionalization of the complex with various electrophiles (see also Chaps 10.2 and 17.3):

▸ **Initiation of electrocatalytic (ETC) reactions**

The complexes $[Fe^{I}(\eta^5-C_5R_5)(\eta^6-C_6R'_6)]$, R and R' = H or Me, can be used in catalytic amount to initiate electron-transfer-chain (ETC) reactions shown in Chap. 5 (sometimes called electrocatalytic). Depending on the redox potential required for the initiation step, the adequate initiator is chosen in the library of electron-reservoir reagents. In the example shown in the following scheme, the introduction of the first PMe_3 ligand can be carried out using $[Fe^ICp(\eta^6\text{-}C_6Me_6)]$ in THF, but not the second one. The introduction of the second PMe_3 ligand needs to be achieved using the more powerful reductant $[FeCp*(\eta^6\text{-}C_6Me_6)]$ (whose ionization potential is 4.1 eV, as low as that of K).[11.22a]

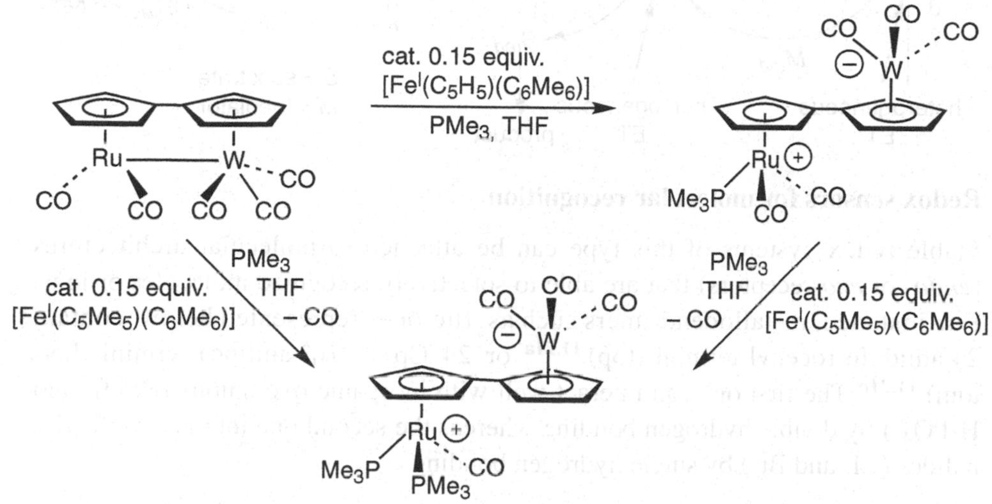

▸ Redox catalysis

Whereas electrocatalysis (ETC) most often deals with overall non-redox reactions that are catalyzed by a redox reagent, redox catalysis is the catalysis of an overall oxidation or reduction. For example, the cathodic reduction of nitrates and nitrites to ammonia cannot be carried out in aqueous medium on a mercury cathode, because NO_3^- or NO_2^- are not electroactive under these conditions (i.e. at negative potentials on the cathodic side, water is reduced and these anions are not). This cathodic reduction is possible under the same conditions, however, if a catalytic amount of the electrocatalyst $[Fe^{II}(\eta^5\text{-}Cp)(\eta^6\text{-}C_6Me_6)][PF_6]$ (or its water-soluble version $[Fe^{II}(\eta^5\text{-}CpCO_2)(\eta^6\text{-}C_6Me_6)]$), is introduced into the electrochemical cell and the potential set at the reduction potential of this Fe^{II} complex to Fe^I. The complex formed reduces NO_3^- or NO_2^- in the bulk of the solution producing ammonia and the Fe^{II} cation that is then reduced again to Fe^I at this working potential for another catalytic cycle. Thus, the $Fe^{II/I}$ system catalyzes the cathodic reduction of these inorganic anions in water. This property is due to the fact that the structure of the electrocatalyst does not change upon interconversion between Fe^{II} and Fe^I, which allows fast electron transfer with the electrode. The reduced Fe^I form can reduce the substrate more rapidly in the bulk than the electrode does at its surface, because the bulk is tridimensional, which allows optimal adjustment of the relative positions and geometry of the donor and acceptor for electron transfer (optimizing orbital interactions). On the other hand, there is a large overpotential (surtension) at the electrode surface, because this match is much more difficult on this two-dimensional space:[11.22b,c]

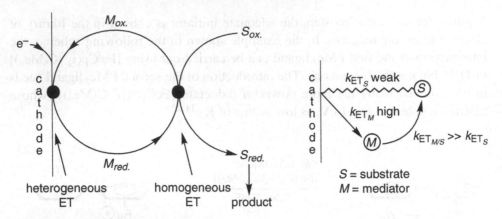

heterogeneous   homogeneous   product
ET   ET

S = substrate
M = mediator

▸ Redox sensors for molecular recognition

Stable redox systems of this type can be attached to molecular architectures (*endo-* or *exo-*receptors) that are able to selectively recognize molecules or ions. For instance, metallodendrimers such as the ones represented below contain 24 amidoferrocenyl termini (top)[11.21a] or 24 Cp*Fe$^+$(η^6-anilino) termini (bottom).[11.21b] The first one can interact well with inorganic oxo anions (HSO$_4^-$ and H$_2$PO$_4^-$) by double hydrogen bonding whereas the second one interacts well with halides (Cl$^-$ and Br$^-$) by single hydrogen bonding.

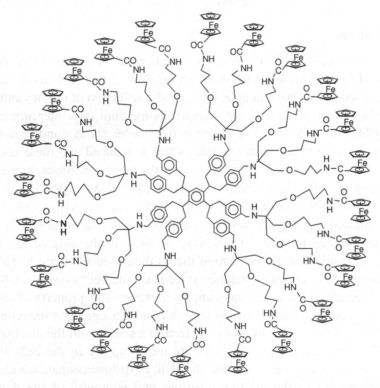

24-Fc - Recognition of H$_2$PO$_4^-$ and HSO$_4^-$ by variation of the redox potential $E°$(Fe$^{III/II}$)

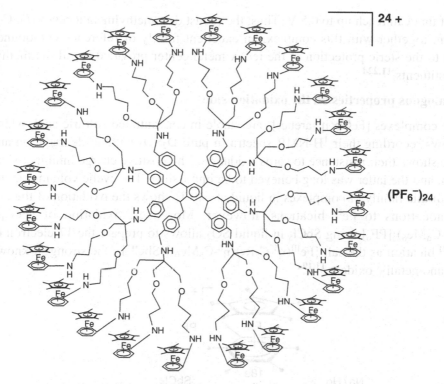

24-FeAr - Recognition of Cl⁻ and Br⁻ by variation of δ (NH, ¹H NMR)

Recognition is achieved by variation of a physico-chemical parameter when a *n*-Bu₄N⁺ salt of the anion is added to the solution of the metallodendrimer acting as a sensor at the dendritic periphery (*exo*-receptor). Encapsulation of the anion near the redox center perturbs this center, which provokes a variation of its redox potential. Likewise, the NH chemical shift in ¹H NMR varies. This variation is specific for each ion and results from the synergy between the electrostatic effect, hydogen bonding and topology factor. The dendritic effect, i.e. the better recognition by the dendrimer than by the monomer or even tripod, is important because of the topological factor. Indeed, microcavities form at the surface of these macromolecules approaching the globular shape and define narrow channels enhancing the specificity of the interaction (a phenomenon that resembles recognition by the fractal surface of viruses).[11.21]

▶ **Reference redox systems for the determination of redox potentials**

The determination of electrochemical redox potentials requires standard reference compounds whose potential, ideally, does not vary. The redox couple [FeCp* (η⁶-C₆Me₆)₂]⁺/⁰ has a redox potential $E° = -1.85$ V *vs*. SCE that is independent on the nature of the solvent and electrolytes, because solvent molecules and electrolyte ions cannot reach the iron center due to the steric bulk of the ring methyl groups. This provides an excellent reference system, much better than ferrocene, the IUPAC reference, because these interactions can occur between the two ferrocene rings provoking variations of the potential from one solvent to the next that can reach up to 0.5 V. Thus, the robust decamethylmetallocenes (Fe,Co) form, together with this complex, an excellent family of reference compounds due to the steric protection of the redox metal center provided by all the methyl substituents.[11.22d]

▸ **Analogous properties on the oxidation side**

The complexes $[FeCp(\eta^6\text{-arene})]^+$ are stable in concentrated sulfuric acid, which allows recording their ^{1}H-NMR spectra in pure D_2SO_4 without degradation and also shows their resistance towards oxidation. The positive charge inhibits oxidation, and the latter was long believed to be impossible. The cyclic voltammogram of all this family of complexes in liquid at –40°C shows the oxidation of the Fe^{II} monocations to Fe^{III} bications, however. Moreover, oxidation of $[FeCp^*(\eta^6\text{-}C_6Me_6)][PF_6]$ using $SbCl_5$ in liquid SO_2 allows to prepare the 17-electron d^5 Fe^{III} bication as the salt $[Fe^{III}(\eta^5\text{-}C_5R_5)(\eta^6\text{-}C_6Me_6)][SbCl_6]_2$, the strongest known organometallic oxidant.[11.19a]

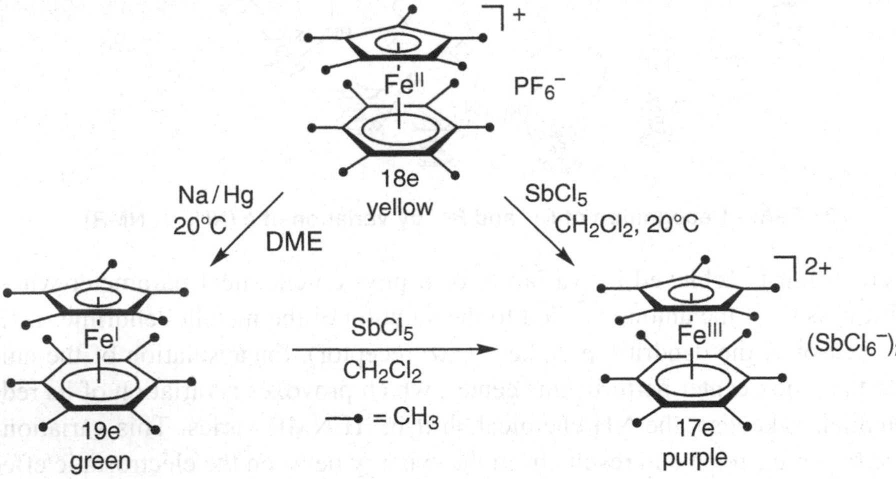

This Fe^{III} complex is stable even in dry air, has a reduction potential 1 V more positive than that of ferrocene/ferrocenium, and its electronic structure resembles that of ferrocenium. It can stoichiometrically oxidize many substrates, initiate electrocatalytic reactions that require a very positive potential and serve as redox catalyst for compounds that are difficult to oxidize such as furfural. It plays a symmetrical role in oxidation to the one described above in reduction with $[Fe^I(\eta^5\text{-}C_5R_5)(\eta^6\text{-}C_6Me_6)]$.[11.19a]

▸ **Proton reservoirs**

The deprotonation of aromatics that are coordinated to $CpFe^+$ and have at least one proton on the exocyclic atom such as toluene, aniline, phenol and thiophenol is possible due to the strong electron-withdrawing property of the cationic group. In addition, the neutral deprotonated species formed are described by a stabilized ferrocene-like bis-pentahapto structure with a zwitterionic mesomer form shown below. Thus, $[CpFe(\eta^6\text{-}C_6Me_6)]^+$ has a pK_a that is 14 units lower than that of the free arene. Likewise, the heterolytic cleavage of the aryl ether complexes is considerably activated by comparison with such a cleavage in the free arene (this property is useful for the dendrimer syntheses shown in Chap. 21.2.2).

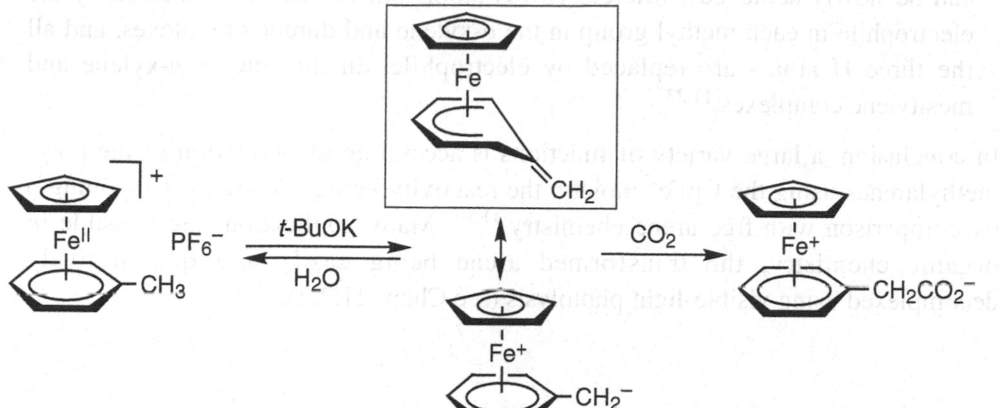

If deprotonation of a polymethylbenzene complex is carried out in the presence of an excess base and compatible electrophile such as an alkyl iodide or allyl- or benzyl bromide, the deprotonation-alkylation sequences spontaneously follow one another *in situ* until steric inhibition is reached. This reaction has been exploited with arene = C_6Me_6 for the synthesis of star-shaped molecules (examples below) and living polymers and with arene = durene or mesitylene for the synthesis of dendritic cores (see for instance Chap. 21.2.2).

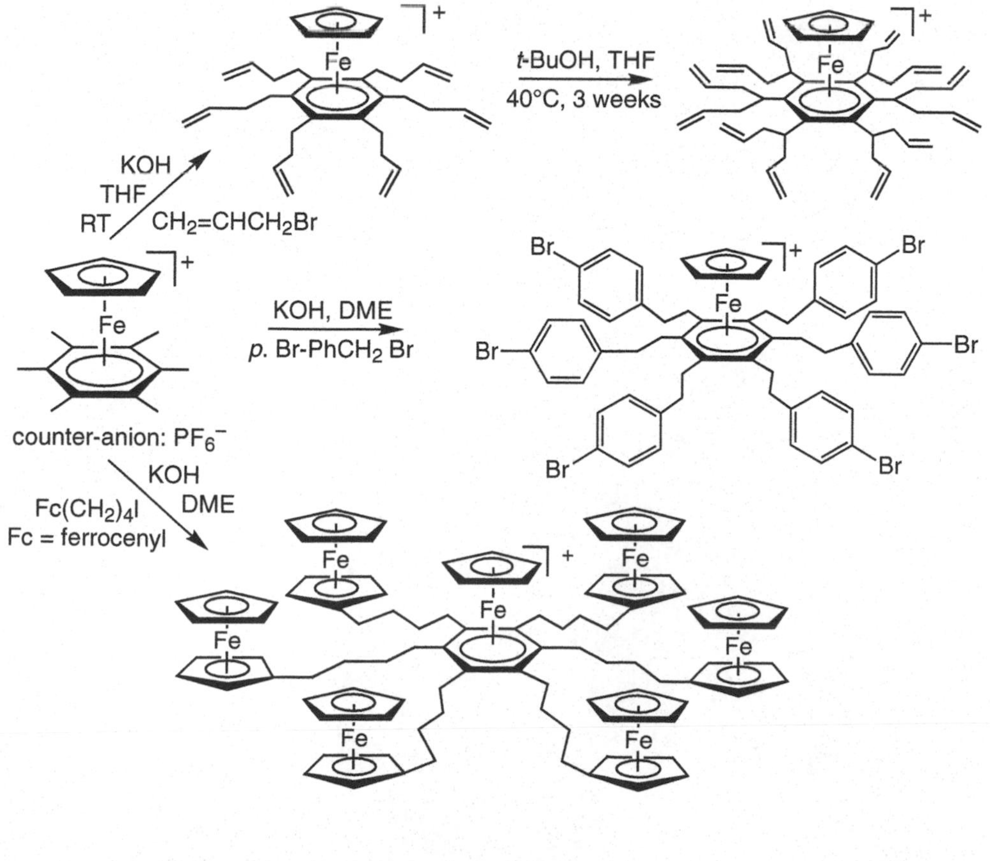

Indeed, with C_6Me_6, only one H out of three is usually replaced by the electrophilic group (with the reactive allyl bromide, a slow second substitution can be slowly achieved), whereas two H atoms out of three are replaced by the electrophile in each methyl group in the o-xylene and durene complexes, and all the three H atoms are replaced by electrophiles in the tolene, p-xylene and mesitylene complexes.[11.23]

In conclusion, a large variety of functions is accessible by activation of the poly-methylarenes using the CpFe$^+$ moiety, the reactivity being reversed ("Umpolung") by comparison with free arene chemistry.[11.19c] Many applications are possible in organic chemistry, the transformed arene being easily and quantitatively decomplexed using visible-light photolysis (see Chap. 21.2.2).

SUMMARY OF CHAPTER 11
METALLOCENES AND METAL-SANDWICH COMPLEXES

Metallocenes are known with most metals (discovery of ferrocene in 1951 and report of its sandwich structure in 1952), and have applications in organic, polymer and medicinal chemistry. With the MCp_2 structure, they are sandwich complexes in which the two Cp rings are parallel (ferrocene). The sandwich MCp_2 structure is very robust with NVE = 18 (Fe, Ru, Os). The neutral metallocenes of the 1st-row with 15 to 20 e$^-$ (V, Cr, Mn, Fe, Co, Ni) correspond to the filling of the $5d$ orbitals split in three MO levels (e_{1g}: bonding, double; a_{1g}: non-bonding, simple, e*$_{1g}$; antibonding, double) under the influence of the pseudo-octahedral field of the two Cp ligands. The metallocenes are all the more fragile as their NVE is further from 18, but they are stabilized by the Cp* (η^5-C_5Me_5) ligand.

Ferrocene is a diamagnetic (low spin) orange, crystalline solid stable up to 400°C, covalent (nil dipole moment), soluble in hydrocarbons, with free rotation about the Fe-Cp ring axis. It readily oxidizes to blue ferrocenium [FeCp$_2$]$^+$, (most often synthesized as the PF$_6^-$ salt), a useful single-electron mild oxidant. Ferrocene is a superaromatic, starting point of many syntheses, for instance of Friedel-Crafts type. The α-ferrocenylcarbonium ions FcCR$_2^+$ (Fc = ferrocenyl) are stabilized by conjugation with Fe.

Cobaltocene [CoCp$_2$] (19e) is electron rich and air sensitive; the 18e [CoCp$_2$]$^+$ cation is robust.

Nickelocene [NiCp$_2$] (20e) is a practical source of the fragment CpNi and its protonation led to the first triple-decker sandwich [Ni2Cp$_3$]$^+$.

Bent metallocenes derivatives with non-parallel rings (angle: about 130°), mostly known with early transition metals, are of the type MCp$_2$(R), MCp$_2$(R)$_2$ or MCp$_2$(R)$_2$ (M = Sc, Lu, Ti, V, Cr, etc.; R = H, Cl, alkyl, etc.), and often are stabilized with Cp* (in particular for Sc and the rare earths).

Metal-bis-arene sandwich complexes, accessible by reactions of the Fischer type (MCl$_3$ + arene + AlCl$_3$ + Al) or by the metal-vapor condensation technique, have electronic structures and redox properties close to those of the metallocenes.

Cyclopentadienyl-metal-arene complexes are mostly known with Fe and Ru. The CpM$^+$ group (M = Fe or Ru) activates many aromatic syntheses (nucleophilic addition and substitution, benzylic deprotonation, etc.). With Fe, they are stable in the 17, 18 and 19e forms with permethylated rings, which provides electron-reservoir properties (stoichiometric and catalytic) and proton-reservoir functions (perfunctionalization of polymethylbenzene ligands).

EXERCISES

11.1. What is the electronic structure and the theoretical magnetic moment of the following metal-sandwich complexes:
$[(FeCp_2]^+, [CoCp_2]^-$ and $[Co(\eta^6-C_6Me_6)_2]^+$?

11.2. Why do first-row transition-metal-sandwich complexes exist in various oxidation states and not the second- and third-row ones ?

11.3. Why does decaisopropylferrocene not form from LiC_5i-Pr_5 and $FeCl_2$? Given that 1,2,3,4,5-pentamethylcobalticinium reacts with excess KOH and CH_3I to yield 1,2,3,4,5-pentaisopropylcobalticinium, what does the same reaction of decamethylcobalticinium and decamethylrhodocenium produce? Compare these reactions of the two latter complexes.

11.4. Explain how ferrocene reacts with aromatics in the presence of $AlCl_3$ under CO_2 giving, at 80°C, the complexes $[Fe(\eta^5-C_5H_4CO_2H)(\eta^6-arene)]^+$.

11.5. Shifts of exocyclic double bonds from planarity towards the metal or away from the metal: why is the exocyclic carbon shifted towards the metal in ferro-cenylcarbonium $[FcCR_2]^+$ (mesomer form of $[FeCp(\eta^6-fulvene)]^+$), whereas it is shifted away from the metal in the sandwich complex $[FeCp(\eta^5-C_6H_5=CH_2)$ resulting from the deprotonation of $[FeCp(\eta^6-toluene)]^+$?

11.6. How many isomers are there for simple 1,2-heterodisubstituted ferrocenes?

11.7. Among the two d^6 Fe^{II} 18-electron complexes $[FeCp_2]$ and $[FeCp(\eta^6-C_6H_6)]^+$, which one is more difficult to oxidize to d^5 17-electron Fe^{III}? Why?

Chapter 12

IONIC AND POLAR METAL-CARBON BONDS: ALKALI AND RARE-EARTH COMPLEXES

1. INTRODUCTION

The electronegativity index of carbon is 2.5, whereas that of metals is between 0.8 for cesium and 2.5 for gold. As a consequence, the metal-carbon bonds are more or less polarized $M^{\delta+}-C^{\delta-}$. For late transition metals (groups 8 to 12), the polarity is very weak or nil. It is marked, however, for alkali metals, rare earths and early transition metals. The ionic character of the M-C bond increases with the polarity, and alkali metal- and lanthanide organometallic compounds can be considered as essentially ionic. This classification resembles that holding for hydrides, but it also depends on the hybridation of the carbon atom bonded to the metal. Indeed, the electronegativity of sp^2 carbon is 2.75, and that of sp carbon is 3.29, which leads to the acidity order: $C_2H_6 < C_2H_4 < C_2H_2$. The alkynyl metal complexes are thus the more polar organometallic derivatives. Finally, the polarity obviously depends on the nature of the ligands of the metal and substituents of the carbon atom bonded to the metal. It is more marked if the carbanion is stabilized. The type of organometallic chemistry described up to here concerns transition-metal complexes that are very slightly polar. Let us now examine, in this chapter, ionic or polar complexes of alkali metals and rare earths.[12.1] This ionic character makes the complexes very sensitive to air, water and a large number of reactants.

2. ORGANOMETALLIC COMPLEXES OF ALKALI METALS

2.1. STRUCTURE

The polarity of the metal-carbon bond increases upon going down in the periodic table: the lithium alkyls have some covalent character and form tetrameric clusters, whereas cesium alkyls are purely ionic. The degree of clusterification of lithium alkyls varies with the nature of the solvent between dimer (LiCH$_3$: TMEDA) and hexamer (Li-n-C$_4$H$_9$: cyclohexane), as can be checked by osmometry, NMR and EPR. ^{7}Li NMR (I = 3/2; abundance: 92.6%) and ^{6}Li NMR (I = 1/2; abundance: 7.4%) allow to also show the dynamic fluxionality phenomena around the Li$_4$ tetrahedron, reversible dissociation of tetramers to dimers and ion pairs (contact ion pairs, CIP, and solvent-separated ion pairs, SSIP).[12.2,12.3] For instance, with LiC_5H_5 (equation below), the chemical shifts δ measured *vs.* the external reference 0.1 M $LiCl_{aq}$ in ^7Li NMR for the Li^+ cation in the contact ion pair and the solvent-separated ion pair are very different: [12.3]

$$Li^+, C_5H_5^- + solvent \longrightarrow Li^+, solvent, C_5H_5^-$$
$$Et_2O: δ = -8.37 \text{ ppm} \qquad HMPA: δ = -0.88 \text{ ppm}$$

$$R = n\text{-butyl} \qquad S = THF$$

Dimer-tetramer equilibrium of *n*-butyllithium in THF

Lithium-aryls are dimers or tetramers depending on the solvent (ether resp. TMEDA). Lithium-benzyl is a monomer, and lithium-allyl has a zigzag chain structure with one coordinated ether molecule bonded to each lithium.

2.2. SYNTHESIS OF ALKALI-METAL-ALKYL AND -ARYL COMPLEXES AND REDUCING POWER OF ALKALI METALS

Lithium metal and *n*-butyllithium in hexane solution being commercial, the two following methods of syntheses of lithium-alkyl complexes are currently used:

$$CH_3Br + 2 Li \xrightarrow[20°C]{Et_2O} CH_3Li + LiBr$$

$$RCpH + n\text{-BuLi} \xrightarrow[-78°C]{THF} RCpLi + n\text{-BuH}\nearrow$$

The concentrations of lithium-alkyls can be easily followed by ^1H NMR using the signal of the protons α to Li whose δ values are found near –1 ppm.

The alkali metals all having a very negative redox potential (near –3 V *vs.* SCE) are well known as extremely strong reducing agents that can be dangerous (spontaneous inflammation in air). This energy is best used when it is tempered or canalized. Thus, the Na/Hg amalgam (with a 0.8% Na) is liquid at 20°C. It is a good reducing agent ($E° = -2$ V *vs.* SCE) that is very much used, because it is relatively stable in air and can be easily stored. The C_8K alloy is another useful and clean reducing agent that can be used with a variety of solvents. Its use is not reserved to the synthesis of organometallic potassium salts, but is quite general. For instance, the synthesis of polysilanes $[-Si(Me)(R)-]_n$, that are very useful materials (see Chap. 13), gives much better yields using C_8K (for example: THF, 0°C) than the Würtz method (Na, toluene, 110°C).[12.2c]

Aromatic hydrocarbons and organic molecules with delocalized π systems can be reduced, for instance, either by Na or Na/Hg depending on their reduction potential. The delocalization of the π system gives rise to degeneracies of the π orbitals and to the presence of low-lying π^* orbital levels, which corresponds to relatively accessible reduction potentials. Naphthalene gives a radical anion whose sodium salt is sometimes used in THF or DME as a powerful reductant, its redox potential being -2.5 V $vs.$ SCE.

$$Na + C_{10}H_8 \xrightarrow{\text{THF}} Na^+ C_{10}H_8^{\bullet -} \xrightarrow[-C_{10}H_8]{\text{MnCp}^*_2} Na^+ [MnCp^*_2]^-$$

$E° = -2.17$ V $vs.$ ECS

Benzophenone whose reduction potential is -1.88 V $vs.$ SCE, is reduced by Na to its purple radical anion Na^+ $PhCOPh^{\bullet -}$ in THF. This redox couple is used as a colorimetric indicator in the absence of water, since this radical anion rapidly reacts with water with discoloration.

The main reactions of alkali organometallics are metallation or deprotonation, reduction (*vide supra*), alkylation of transition-metal halides and addition on multiple bonds:

Deprotonation-metallation

n-Butyllithium is one of the most currently used bases in organic synthesis. The hydrocarbons and other organic compounds are considered as acids [12.4] and n-butyllithium as the base corresponding to butane ($pK_a = 44$). The deprotonation of the following hydrocarbons by n-butyllithium is thus thermodynamically favorable:

Hydrocarbon	cyclopentadiene	fluorene	acetylene	toluene	benzene
pK_a	15	21	24	35	37

In the case of benzene, the large kinetic barrier does not allow the reaction to proceed. The latter is possible, however, in the presence of TMEDA. The role of this chelate ligand is first to break the hexameric cluster in hexane by complexation of the Li^+ cation, which strongly polarizes the Li-C bond. The addition of benzene or ferrocene then rapidly leads to deprotonation-metallation of the aromatic derivative, because the n-C$_4$H$_9^-$ anion is rendered more basic once it is disengaged from the covalent bond with Li.

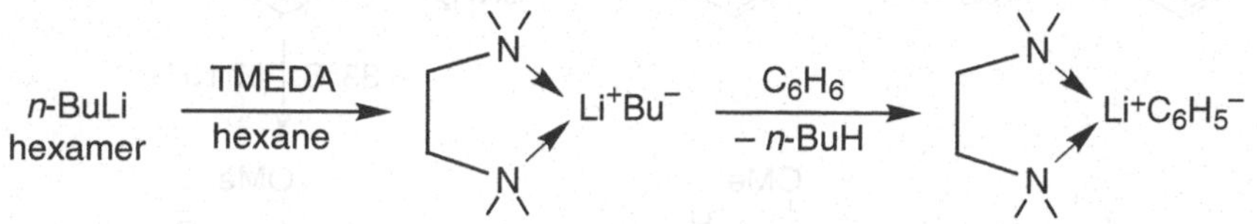

$$n\text{-BuLi hexamer} \xrightarrow[\text{hexane}]{\text{TMEDA}} \left[\begin{array}{c}N\\\\N\end{array}\right]Li^+Bu^- \xrightarrow[-\,n\text{-BuH}]{C_6H_6} \left[\begin{array}{c}N\\\\N\end{array}\right]Li^+C_6H_5^-$$

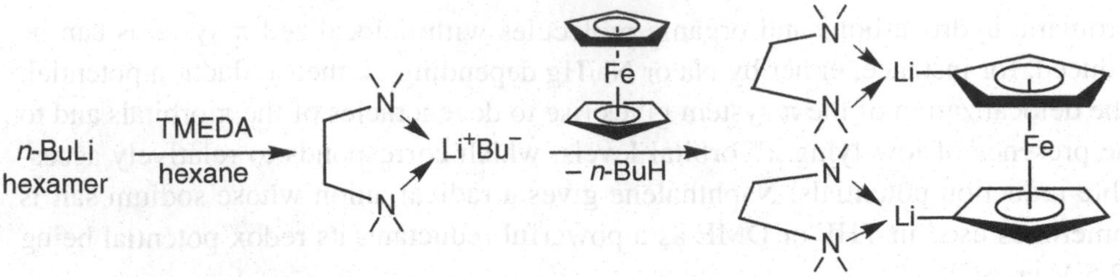

Another means to enhance the basicity of *n*-butyllithium consists in using it in a mixture with potassium *ter*-butoxide. Here again, coordination of lithium with the oxygen atom of *t*-BuOK breaks the tetrameric cluster (*n*-BuLi)$_4$, which leads to the formation of a very active monomer. Alternatively, it is possible to add a cosolvent such as hexamethylphosphoramide (HMPA) to a commercial solution of *n*-butyllithium in hexane, which has the same effect as TMEDA indicated above. Finally, one can also evaporate the hexane solvent and add a more polar solvent such as ether with or without cosolvent. Note that *n*-butyllithium is not a bulky base, and thus it also often reacts as a nucleophile. In these cases, it is necessary to prepare a bulky base by reaction of *n*-butyllithium with a bulky amine such as diisopropylamine, bis(trimethylsilyl)amine or tetramethylpiperidine.

$$R_2NH + \textit{n-BuLi} \longrightarrow R_2NLi + \textit{n-BuH}$$

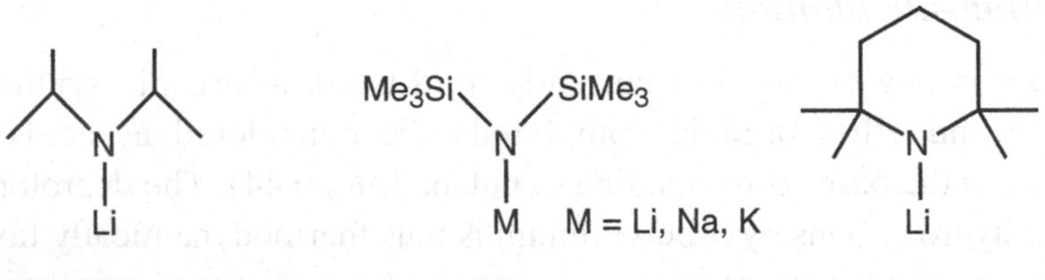

Alkali amides: strong non-nucleophilic bases

We have indicated in the previous section that alkali metals can reduce aromatics to yield alkali derivatives of aromatic radical anions. The basic properties of the latter have been used in organic synthesis: this is the famous Birch reduction of aromatics by Li or Na in NH$_3$ involving two sequences of electron transfer from the alkali metal to the aromatic followed by proton transfer from NH$_3$ to the organic anion. Example:

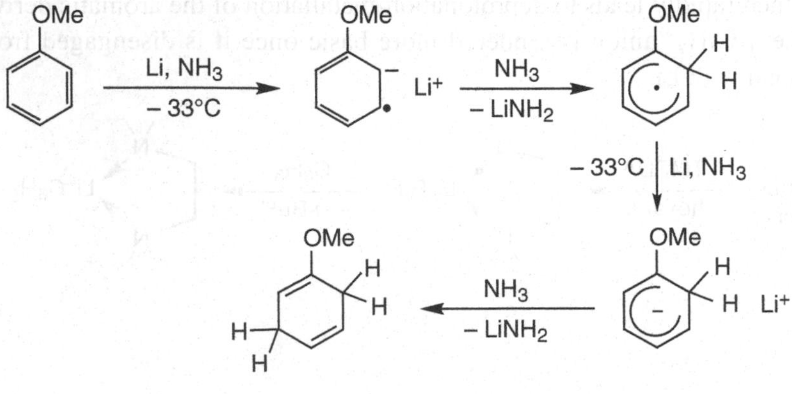

Alkali metal hydrides such as NaH and KH are also very much used as bases to deprotonate very weak acids. They often allow avoiding the side electron-transfer reactions observed with n-butyllithium in the case of the deprotonation of cations. Finally, t-BuOK ($pK_a = 28$ in DMSO) is a very common base that is used when the acid is not too weak.

Alkylation of element halides

Element halides are alkylated by alkali metal organometallic derivatives. Depending on the stoichiometry, one, several or all the halides are substituted by alkyl groups. The number of introduced alkyl groups is sometimes larger than the number of halides on the element, which yields "ate" (i.e. anionic) complexes. With transition-metal halides, the reaction sometimes goes along with some reduction of the oxidation state of the transition metal.

$$SiCl_4 + 4 \, CH_3Li \longrightarrow SiMe_4 + 4 \, LiCl$$

$$ZrCl_4 + 2 \, CpNa \longrightarrow ZrCp_2Cl_2 + 2 \, NaCl$$

$$FeCl_3 + 7 \, PhLi \longrightarrow [FePh_4]^{2-} \, [Li^+]_2 + 3/2 \, Ph\text{-}Ph + 3 \, LiCl$$

Addition on the multiple bonds

These additions are of synthetic interest. For instance, the addition of lithium compounds on nitriles is a synthetic route to ketones, whereas the addition of lithium compounds on N,N-dimethylformamide leads to aldehydes.

$$R\text{–}C\equiv N \xrightarrow[\text{2) } H_2O]{\text{1) } R'Li} \underset{R}{\overset{R'}{>}}C=NH \xrightarrow{R'Li} \underset{R}{\overset{R'}{>}}C=NLi \xrightarrow{H_2O} \underset{R}{\overset{R'}{>}}C=O$$

Finally, recall that we have indicated in the chapter devoted to the metal-carbene complexes (Chap. 9) that they are accessible by reactions of lithium derivatives on a carbonyl ligand.

3. GRIGNARD REAGENTS

3.1. INTRODUCTION

The electronegativity of divalent metals of the groups 2 (alkaline earth: Be, Mg, Ca, Sr, Ba) and 12 (Zn, Cd, Hg) is between 0.9 for the more electropositive one (Ba) and 2.0 for the less electropositive one (Hg), Mg being intermediate with 1.3. The electropositivity regularly decreases as follows: Ba > Sr > Ca > Mg > Be > Zn > Cd. The polarity of the metal-carbon bond follows the same trend, of course. The metals of group 12 are more electron-rich because of the presence of the full d electron shell (these electrons are not involved in the reactions).

The chemistry of these metals is largely dominated by the magnesium compounds, the Grignard reagents being of primary importance in organic synthesis.[12.5-12.11] The other group-12 metals are used to a lesser extent [12.7,12.8] (see Chap. 13.3, however).[12.8]

3.2. PREPARATION, STRUCTURE AND MODE OF FORMATION OF THE GRIGNARD REAGENTS

The discovery by Grignard, at the very beginning of the xx^{th} century, of the insertion of metallic magnesium into the alkyl-halogen bond of alkyl halides and the use of the RMgX reagent formed as masked carbanions to alkylate a large variety of substrates has had a considerable impact on the whole of organic chemistry. It is indeed one of the major modes of carbon-carbon and carbon-element bond formation.

$$Mg + RX \xrightarrow{\text{Et}_2\text{O}} RMgX(Et_2O)_n \xrightarrow{\text{R'R''CO}} RR'R''COH$$
$$X = \text{halogen}$$

The surface of magnesium can be activated by addition of a small amount of iodine, because MgI_2 that is formed removes the traces of water from the medium. Alternatively, a mixture of ether (Et_2O or THF) and hydrocarbon can be used as solvent.

Given the very electropositive character of Mg, close to that of Li, Mg is a strong reducing agent. The single-electron transfer from Mg to RX is very exergonic, and thus very favorable. It is possible to bring about experimental evidence for the formation of intermediate radicals R$^{\bullet}$ that are formed in solvent cages in the course of this reduction.[12.9-12.11]

$$Mg(s) + RX \longrightarrow [RX^{\bullet -}, Mg(s)^+] \longrightarrow (R^{\bullet}, XMg^{\bullet}) \xrightarrow{\text{solvent}} RMgX(\text{solv.})$$

After its formation, the Grignard reagent is most of the time in dynamic equilibrium between the monomer, the dimer, sometimes the polymer and the species resulting from dismutation, i.e. MgR_2 and MgX_2.

$$2\ RMgX(S)_n \xrightleftharpoons{K} MgR_2(S)_m + MgX_2(S)_{n-m}$$

The nature of the solvent (S), of the group R and the halide X and the concentration have, of course, a large influence on equilibria. For instance, in ethers, $K = 0.2$ for EtMgBr. The Grignard agents exist as monomers in this solvent only in dilute solution (< 0.1 M). These ether solutions are weakly conducting, which indicates a weak dissociation.

$$2\ RMgX \underset{Et_2O}{\rightleftharpoons} RMg^+ + RMgX_2^-$$

Oligomers, in which the solvated monomeric units are bridged by halogens, are formed in ethers at higher concentrations. On the other hand, the Grignard reagents exist only as monomers $RMgX(THF)_2$ in a high range of concentrations.

The choice of solvent also allows to control the required conditions for the formation of the organomagnesium derivatives MgR_2. Indeed, dioxane, while precipitating $MgX_2(dioxane)_2$, displaces the dismutation equilibrium in this direction.

$$\underset{solution}{2\ RMgX} + 2\ dioxane \longrightarrow \underset{solution}{R_2Mg} + \underset{precipitate}{MgX_2(dioxane)_2}$$

The alternative synthesis of these magnesium derivatives, i.e. for instance from Mg and HgR_2, is less elegant.

$$HgR_2 + Mg \longrightarrow MgR_2 + Hg$$

3.3. REACTIONS OF THE MAGNESIUM REAGENTS

These reactions are varied in organic- and even inorganic synthesis. The main classes of reactions are summarized in the following scheme (the reaction products are obtained after hydrolysis): [12.11]

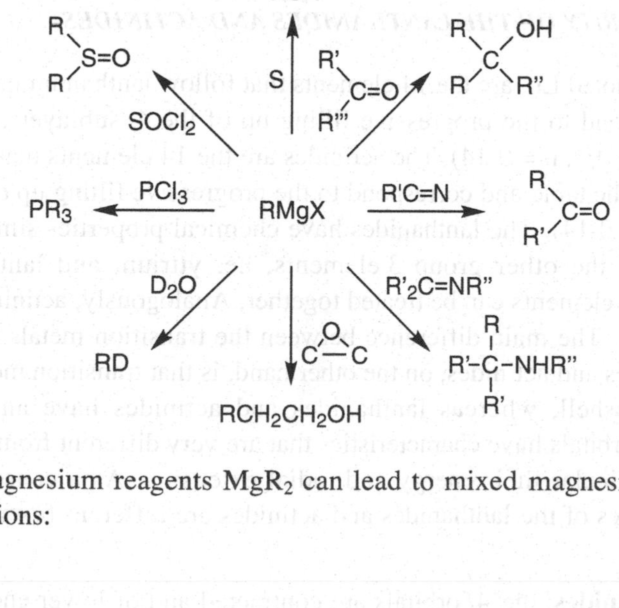

The organomagnesium reagents MgR_2 can lead to mixed magnesium reagents or magnesiate anions:

4. THE SCANDIUM, YTTRIUM, LANTHANIDES AND ACTINIDES COMPLEXES [12.12-12.16]

The lanthanides (table I) are well known for their applications as materials for high technology. The uses of these elements as metals or oxides is due to their optical, magnetic, electric and mechanical properties. Derivatives of these metals have also been found to be useful as catalysts for a variety of reactions. Organometallic chemists became interested in these elements only in the late 70s and especially in the 80s.[12.12-12.16] The applications of this chemistry are now bursting, as exemplified by the reactions of SmI_2 initiated by Kagan such as the reduction of carbonyl derivatives. The organometallic chemistry of the actinides (table II), previously only known for their radioactivity and radiochemistry, is also developing in parallel to that of the lanthanides.

4.1. PARTICULARITY OF THE LANTHANIDES AND ACTINIDES

The lanthanides, noted Ln, are the 14 elements that follow lanthanum in the periodic table and correspond to the progressive filling up of the $4f$ sublayer that contains seven orbitals (Xe $4f^n$, n = 0-14). The actinides are the 14 elements that follow actinium in the periodic table and correspond to the progressive filling up of the $5f$ sublayer (Kr $5f^n$, n = 0-14). The lanthanides have chemical properties similar to those of scandium and the other group 3 elements, i.e. yttrium and lanthanum, and therefore these 17 elements can be treated together. Analogously, actinium is treated with the actinides. The main difference between the transition metals on one hand and the lanthanides and actinides, on the other hand, is that transition metals have an incomplete d subshell, whereas lanthanides and actinides have an incomplete f subshell. The f orbitals have characteristics that are very different from those of the d orbitals, in particular their energy and radial extension. As a consequence, the chemical properties of the lanthanides and actinides are different from those of the transition metals:

▸ **For the lanthanides**, the $4f$ orbitals are contracted and of lower energy that the $5d^1$ and $6s^2$ valence orbitals. Thus, the lanthanide complexes are ionic, with domination of the oxidation state III resulting from the loss of these three valence electrons, the very stable Ln^{3+} ions having the electronic structure of xenon. For the electronic structures corresponding to an empty, semi-filled or filled f subshell (or getting close to such electronic structures), other oxidation states can be stabilized such as Ce^{IV} (f^0), Eu^{II} (f^7) and Yb^{II} (f^{14}) and, under more extreme conditions, Pr^{IV} (f^0), Tb^{IV} (f^7) and Sm^{II} (f^6).

Table I - Lanthanides

Atomic number	Name	Symbol	Electronic configuration
21	Scandium	Sc	$3d^1 4s^2$
39	Yttrium	Y	$4d^1 5s^2$
57	Lanthanum	La	$5d^1 6s^2$
58	Cerium	Ce	$4f^5 5d^1 6s^2$
59	Praseodymium	Pr	$4f^3 6s^2$
60	Neodymium	Nd	$4f^5 6s^2$
61	Promethium	Pm	$4f^5 6s^2$
62	Samarium	Sm	$4f^6 6s^2$
63	Europium	Eu	$4f^7 6s^2$
64	Gadolinium	Gd	$4f^7 5d^1 6s^2$
65	Terbium	Tb	$4f^9 6s^2$
66	Dysprosium	Dp	$4f^{10} 6s^2$
67	Holmium	Ho	$4f^{11} 6s^2$
68	Erbium	Er	$4f^{12} 6s^2$
69	Thulium	Tm	$4f^{13} 6s^2$
70	Ytterbium	Yb	$4f^{14} 6s^{12}$
71	Lutetium	Lu	$4f^{15} 5d^1 6s^2$

The common electronic configuration of the ions is d^0 Ln^{3+}, i.e. $4f^n 5d^0 6s^0$.

▸ **For the actinides**, the situation is significantly different. Indeed, the shielding of their $5f$ orbitals by the partially filled $6s$ and $6p$ orbitals is not as complete for the first actinides. The result is that the f orbitals have sufficient radial extension to cover the ligand orbital up to a certain extent. As a consequence, various oxidation states and partially covalent metal-ligand bonds are found for the first actinides. This covalent character remains minor, however, and the ionic character dominates. The accumulation of nuclear charges provokes the contraction of the $5f$ orbitals as one moves forward in the actinide series. Again, the oxidation state III dominates and the complexes are completely ionic for the later actinides as for the lanthanides.

Table II - Actinides

Atomic number	Name	Symbol	Electronic configuration	Common oxidation states
89	Actinium	Ac	$6d^1 7s^2$	<u>3</u>
90	Thorium	Th	$6d^2 7s^2$	(3)<u>4</u>
91	Protactinium	Pa	$5f^e 6d^1 7s^2$	(3)4,<u>5</u>
92	Uranium	U	$5f^3 6d^1 7s^2$	3,4,5,<u>6</u>
93	Neptunium	Np	$5f^5 7s^2$	3,4,<u>5</u>,6,7
94	Plutonium	Pu	$5f^6 7s^2$	3,<u>4</u>,5,6,7
95	Americium	Am	$5f^7 7s^2$	(2),<u>3</u>,4,5,6
96	Curium	Cm	$5f^7 6d^1 7s^2$	<u>3</u>,4
97	Berkelium	Bk	$5f^8 6d^1 7s^2$	(2),<u>3</u>,4
98	Californium	Cf	$5f^{10} 7s^2$	(2),<u>3</u>
99	Einsteinium	Es	$5f^{11} 7s^1$	(2),3
100	Fermium	Fm	$5f^{12} 7s^2$	(2),<u>3</u>
101	Mendelevium	Md	$5f^{13} 7s^2$	2,<u>3</u>
102	Nobelium	No	$5f^{14} 7s^2$	<u>2</u>,3
103	Lawrencium	Ln	$5f^{14} 6d^1 7s^2$	<u>3</u>

The oxidation states noted in parentheses are not known in aqueous solution.
The most stable oxidation state of each element is underlined.

The electronic structure of the lanthanide and actinide complexes is high spin, because the contraction of the f orbitals leads to a crystal-field induced splitting that always remains weak. For this reason, the $f \longrightarrow f$ transitions are very narrow and of weak intensity and are forbidden by the Laporte's rules (these rules, that are not strictly respected, state that the transitions between terms of different multiplicities are forbidden and that pure $d \longrightarrow d$ or $f \longrightarrow f$ transitions are forbidden [12.17]), which explains why the colors of the d electron-free Ln^{III} complexes are pale. Moreover, these transitions are insensitive to changes of ligands given the absence of covalence. This contrasts with the transition-metal complexes that are very colored because of the large and intense bands corresponding to the $d \longrightarrow d$ transitions, the d orbital levels being largely split by the ligand field. Only the Ln complexes involving d electrons in their optical transitions, such as Ln^{II} and Ce^{IV}, have intense colors. By the way, the Sm^{II} and Ce^{IV} complexes have become the most important lanthanide derivatives in organic chemistry because of their reductant and oxidant properties, respectively. For the first actinides, the extension of the f orbitals renders them sensitive to the ligand field; $f \longrightarrow f$ transitions that are an order of magnitude more intense than those of lanthanides are observed as well as more intense and higher-energy $f \longrightarrow d$ transitions.

Concerning the steric aspect, the lanthanide and actinide ions are larger than those of the transition metals. For instance, the trivalent ions have a radius between 1.17 Å (La^{3+}) and 1.00 Å (Lu^{3+}) for the lanthanides and between 1.26 Å (Ac^{3+}) and 1,06 Å (Cf^{3+}) for the actinides. For comparison, the radius of Fe^{3+} is only 0.53 Å. As a result, the coordinance C is markedly higher for the lanthanide and actinide complexes than for the transition-metal complexes. It is usually of the order of 8 to 10 and can reach 12 (but it can also decrease to 3 or 4 with very bulky ligands). The well-known lanthanide contraction is increasing as *f* electrons are introduced. It is explained by the fact that the *f* orbitals do not shield one another, i.e. they do not protect one another from the attraction by the nucleus. The above numbers show that the sizes of the ions decrease as one moves forward in each of the 4*f* and 5*f* series.

4.2. THE ORGANOLANTHANIDES [12.14-12.18]

This chemistry is dominated by the trivalent cyclopentadienyl and alkyl derivatives with various structures that are prepared from the trichloride $LnCl_3$. These compounds are very air- and moisture sensitive because of their ionic character. They are mild Lewis acids that coordinate the solvent used for their synthesis (most often THF). In the case of scandium (III), the Lewis acidity is much stronger (even larger than that of $AlMe_3$), because of the small ion size.

$$LnCl_3 + 3\ MCp \xrightarrow{\text{THF}} [LnCp_3] \text{ or } [LnCp_3(THF)] + 3\ MCl$$

(M = Li, Na, K, MgBr)

$$LnCl_3 + 2\ MCp \xrightarrow{\text{THF}} 1/2\ [LnCp_2Cl]_2 + 2\ MCl$$

$$LnCl_3 + MCp \xrightarrow{\text{THF}} [LnCpCl_2(THF)_3] + MCl$$

The binary compounds $[MCp_3]$ are not monomeric, except $[YbCp_3]$, but oligomers or polymers, one of the Cp ligands bridging two metals with a pentahapto or monohapto coordination. Likewise, the compounds $[LnCp_2Cl]_2$ are bridged by the two common chloride ligands, each monomer unit bringing one Cl. This latter class of compounds is extremely useful to reach the complexes $[MCp_2R]$ with various R groups that are X ligands: OR', O_2CR', NH_2, PR_2, BH_4, acac, alkyl, allyl. The hydride complexes are accessible by hydrogenolysis of the Ln-alkyl bond according to the σ-bond metathesis. Many of these complexes are dimers, with two H or R bridging ligands. In other cases, the complexes are monomers with R = alkyl, but then THF is coordinated.

$$[LnCp_2Cl]_2 + 2\ MR \xrightarrow{\text{THF}} [LnCp_2(R)]_2 + 2\ MCl$$

(M = Li, Na, etc.)

$$[Er(MeC_5H_4)_2(CH_2)(SiMe_3)]_2 + 2\ H_2 \xrightarrow[80°C]{toluene} [Er(MeC_5H_4)_2H]_2 + 2\ SiMe_4$$

R = H, Cl, Me, OR, NH₂

Despite the above common representation with bond lines, one must not forget that the metal-ligand bonds are in fact ionic. The complexes of the [LnCp₂R] type have been synthesized with a large number of variations of the Cp structure, in particular with the Cp* ligand because of its stabilizing stereoelectronic properties. The complexes [LnCp*₂(R)] show remarkable catalytic properties for the oligo-merization and polymerization of olefins and other unsaturated hydrocarbons and for various additions onto these substrates (see Chap. 15.1). Particularly useful pre-catalysts are those for which R = CH(SiMe₃)₂. These complexes are formed as monomers without solvent coordination because of the steric bulk of this alkyl ligand. These complexes also are excellent sources of hydride complexes upon hydrogenolysis: [12.16]

$$+ LiCH(SiMe_3)_2 \longrightarrow \qquad + 2\ Li^+Cl^- + 2\ H_2O$$

Et₂O | Cp*Li

LnCl₃

Ln = Y, Ce, Nd, Sm, Lu

Although it had been believed for a long time that lanthanides could not form com-plexes [LnCp*₃] for steric reasons, Evans' group succeeded in the synthesis of these complexes starting from Sm^{II} or Ln^{III} (scheme below). Evans also showed that these complexes, the first of which was [SmCp*₃], disclose a remarkable reactivity due to the large ionic metal-ligand bonds. Indeed, the complex [SmCp*₃] sometimes behaves as the ion pair [SmCp*₂]⁺ Cp*⁻. The anion Cp*⁻ of this ion pair is so little solvated, due to the large size of the counter-cation [SmCp*₂]⁺, that it is a strong reducing agent, able to reduce even Se=PPh₃ that cannot be reduced by KCp*. Another original aspect of the reactivity of [SmCp*₃] is the insertion of unsaturated substrates such as CO, PhCN (scheme below), PhN=C=O and ethylene (leading to polymerization in the latter case). These insertions occur in the Sm-η¹-Cp* bond. One must thus consider that there is an equilibrium between two species: one (characterized by X-ray diffraction) in which the three Cp* ligands are bonded to samarium in a pentahapto fashion, and the other one in which one Cp* is monohapto: [12b,c]

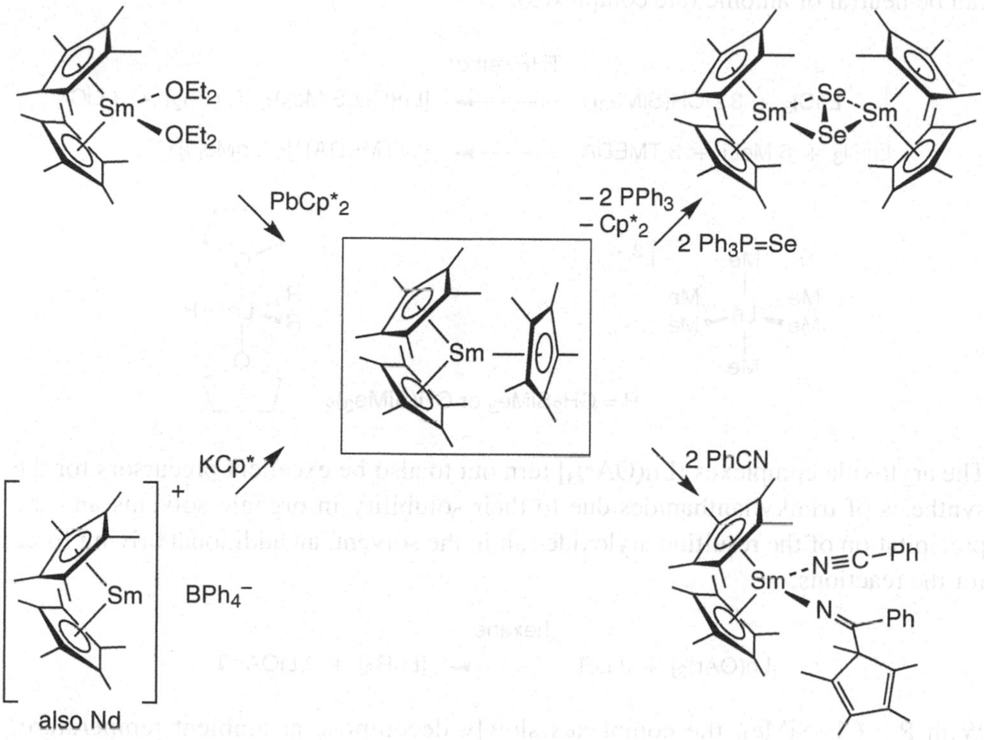

There also is a whole variety of ansa-metallocenes, i.e. complexes in which the two Cp ligands are bridged to each other, which stabilize the complexes due to the chelate (entropic) effect. These complexes were first designed by Brintzinger from $[Zr(C_5H_4-X-C_5H_4)R_2]$ in order to make isotactic polypropylene (see Chap. 6.4). The active form in the polymerization is the cation $[Zr(C_5H_4-X-C_5H_4)R]^+$ isoelectronic to $[Y(C_5H_4-X-C_5H_4)R]$ and to the lanthanide (III) complexes of identical formula. On this principle, other variations (below) are known; they contain a nitrogen ligand (then, the coordination sphere of the metal is completed with another ligand such as PPh_3) or other chelating ligands.

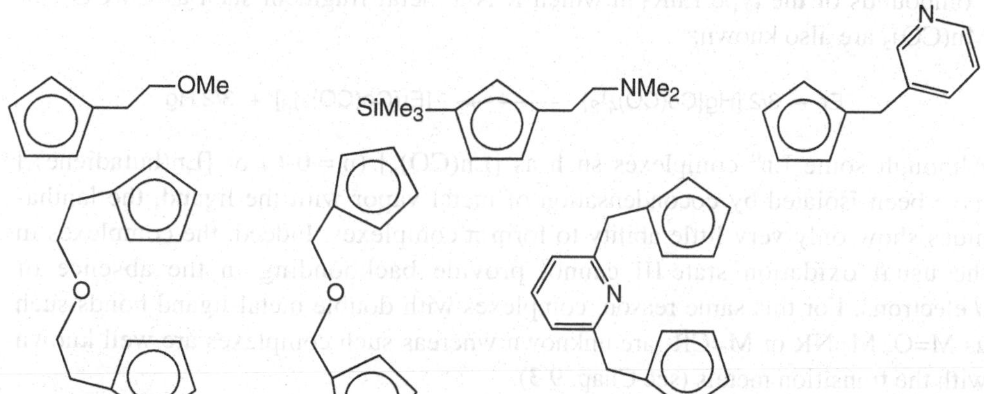

There are many binary complexes [LnR$_3$] with the alkyl, aryl and alkynyl ligands that do not contain β-H atoms, which avoids β-elimination. The complexes formed can be neutral or anionic (ate complexes).

$$LnCl_3 + 3\ LiCH(SiMe_3)_2 \xrightarrow{\text{THF / ether}} [Ln\{CH(SiMe_3)_2\}_3(THF)_2] + 3\ LiCl$$

$$LnCl_3 + 6\ MeLi + 3\ TMEDA \longrightarrow [Li(TMEDA)^+]_3[(LnMe_6)]^{3-}$$

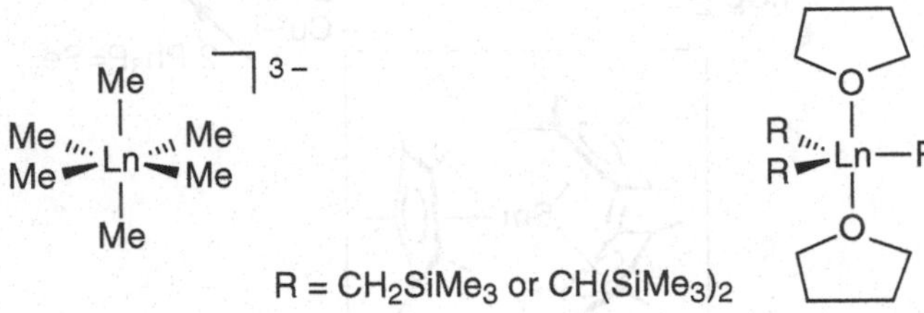

R = CH$_2$SiMe$_3$ or CH(SiMe$_3$)$_2$

The aryloxide complexes [Ln(OAr)$_3$] turn out to also be excellent precursors for the synthesis of trialkyllanthanides due to their solubility in organic solvents and the precipitation of the resulting aryloxide salt in the solvent, an additional driving force for the reactions:

$$[Ln(OAr)_3] + 3\ LiR \xrightarrow{\text{hexane}} [LnR_3] + 3\ LiOAr \downarrow$$

With R = CH$_2$SiMe$_3$, the complexes slowly decompose at ambient temperature, with abstraction of an H atom in α to one of the ligands, giving a bridging alkylidene complex:

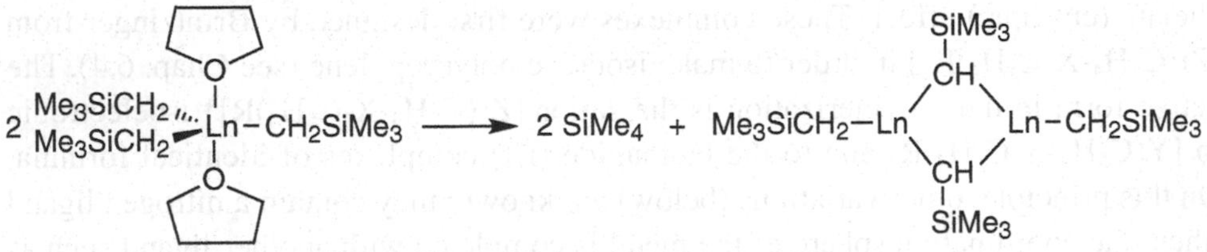

Compounds of the type LnR$_3$ in which R is a metal fragment such as Co(CO)$_4$ or Mn(CO)$_5$ are also known:

$$Er + 3/2\ [Hg\{Co(CO)_4\}_2] \longrightarrow [Er\{Co(CO)_4\}_3] + 3/2\ Hg$$

Although some Ln^0 complexes such as [Ln(CO)$_n$] (n = 0-6) or [Ln(butadiene)$_3$] have been isolated by cocondensation of metal vapor with the ligand, the lanthanides show only very little ability to form π complexes. Indeed, the complexes in the usual oxidation state III cannot provide backbonding in the absence of d electrons. For this same reason, complexes with double metal-ligand bonds such as M=O, M=NR or M=CR$_2$ are unknown whereas such complexes are well known with the transition metals (see Chap. 9.3).

The inorganic lanthanide triflate complexes Ln(OTf)$_3$ (made in aqueous solution) have been shown by Kobayashi to be efficient Lewis-acid catalysts for hydroxymethylation (using commercial aqueous formaldehyde solutions) of silicon enolates in aqueous medium (water + THF) or even in water alone in the presence of a surfactant. In these reactions, activation proceeds by coordination of the aldehyde oxygen atom by the Ln center that is a strong Lewis acid due to its hard character. Among the lanthanide triflates, ytterbium triflate was found to be the most active catalyst, but scandium triflate can sometimes also be efficiently used. Enantioselective versions are also known in the presence of chiral macrocyclic ligands. The water-soluble catalyst is recovered in water after extraction of the organic products.

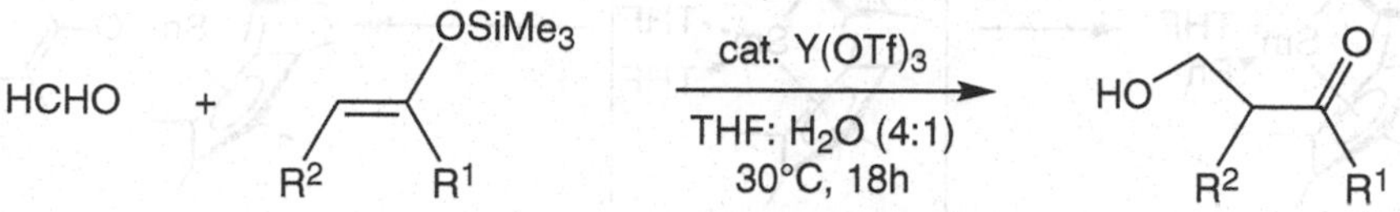

chiral ligand used
for enantioselective
versions of the reaction

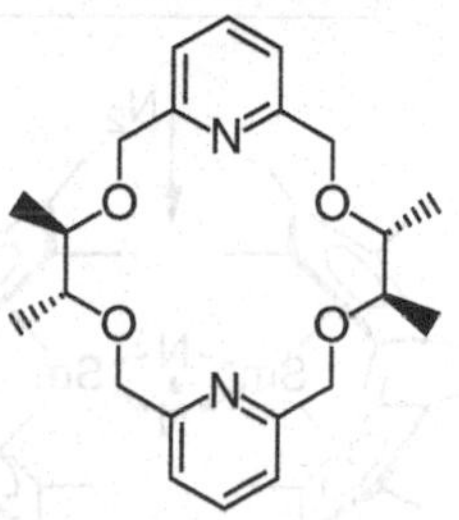

Mannich-type reactions yielding β-amino ketones from aldehydes, amines and silyl enolates can also be performed in water alone with a surfactant according to this principle. Aldehydes react with amines in a hydrophobic reaction field created in water in the presence of a catalytic amount of a metal triflate to produce imines which can then react with hydrophobic silyl enolates:

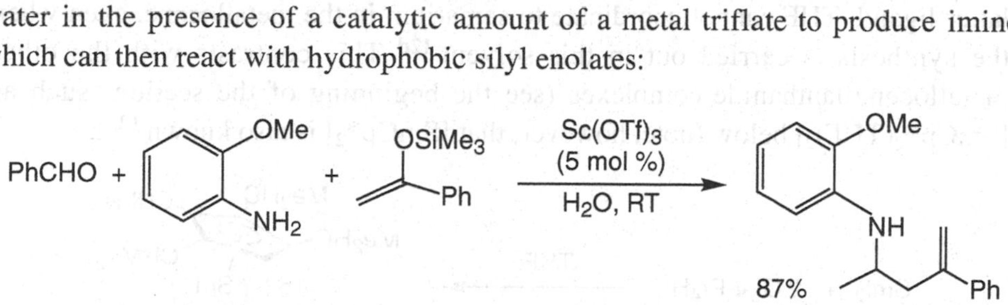

Finally, other oxidation states are known:

▸ Sm^{II} especially with the ligands Cp in [SmCp$_2$(THF)] or Cp* in [SmCp*$_2$] (with and without coordinated THF).

$$[SmCp_3] + K^+(C_{10}H_8)^- \xrightarrow{\text{THF / ether}} [SmCp_2(THF)] + KCp + C_{10}H_8$$

[SmCp*$_2$(THF)$_2$], obtained by reaction of Cp*Li with SmI$_2$ in THF, shows an especially rich chemistry: [12a]

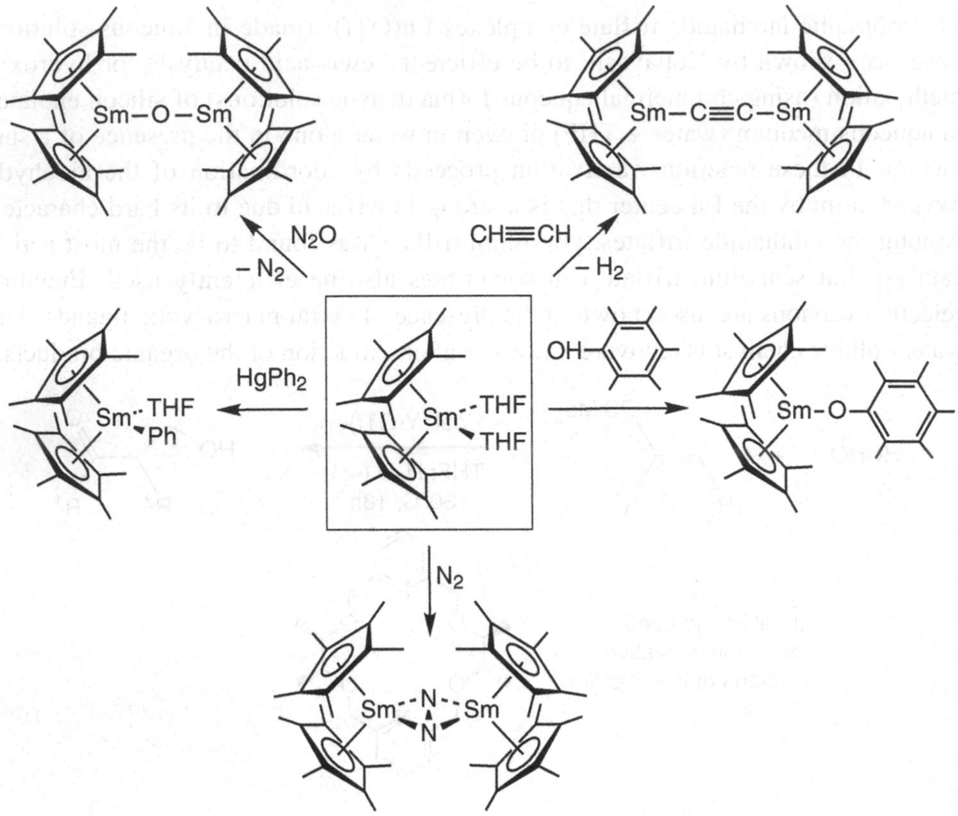

With the ligand $C_5i\text{-}Pr_4H$, whose electron richness is comparable to that of Cp*,[11.15] but whose steric constraints are much more demanding than that of the latter ligand, THF cannot coordinate to samarium in the metallocene, even when the synthesis is carried out in this solvent.[12d] This contrasts with the other metallocene lanthanide complexes (see the beginning of the section) such as [SmCp*$_2$(THF)$_2$] below (note, however, that [SmCp*$_2$] is also known [12a]).

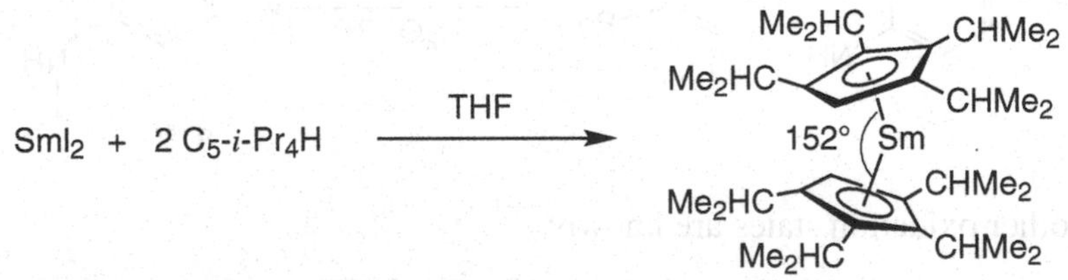

Kagan has carried out a simple preparation of SmI_2 from metallic samarium and diiodoethane to perform a number of useful reduction reactions in organic synthesis such as the reduction of aldehydes and ketones to alcohols and the Barbier reaction (below). Subsequent to this pioneering work, SmI_2 became "Kagan's reagent", and has been used in the Reformatsky reaction, the deoxygenation of epoxides and sulfoxides, the cleavage of C-C bond in α position from a carbonyl and the coupling between carbonyl derivatives and activated olefins.[12.18]

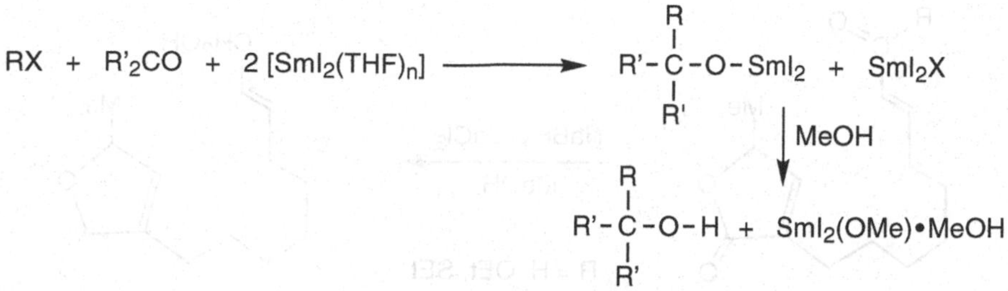

$$RX + R'_2CO + 2\,[SmI_2(THF)_n] \longrightarrow R'-\overset{\overset{\displaystyle R}{|}}{\underset{\underset{\displaystyle R'}{|}}{C}}-O-SmI_2 + SmI_2X$$

$$\downarrow \text{MeOH}$$

$$R'-\overset{\overset{\displaystyle R}{|}}{\underset{\underset{\displaystyle R'}{|}}{C}}-O-H + SmI_2(OMe)\bullet MeOH$$

SmI_2 (solvated by THF ligands) is of course a strong reducing agent, with a $Sm^{III/II}$ potential of -1.5 V *vs.* SCE. It reduces RX to the radical R$^\bullet$ that reacts on the carbonyl group of a ketone, for instance. The mechanism can thus proceed according to a combination of inner-sphere and outer-sphere electron transfers or only by inner sphere. Samarium is finally cleaved from the organic ligand by easy protonation of this ionic ligand using methanol (see scheme above).

▶ **Ce^{IV}**, with a variety of organic ligands (such as Cp or cyclotetraenyl COT) or inorganic ones (such as O*i*-Pr), the syntheses being carried out from pyridinium hexachlorocerate or [Ce(O-*i*Pr)$_4$. *i*-PrOH].

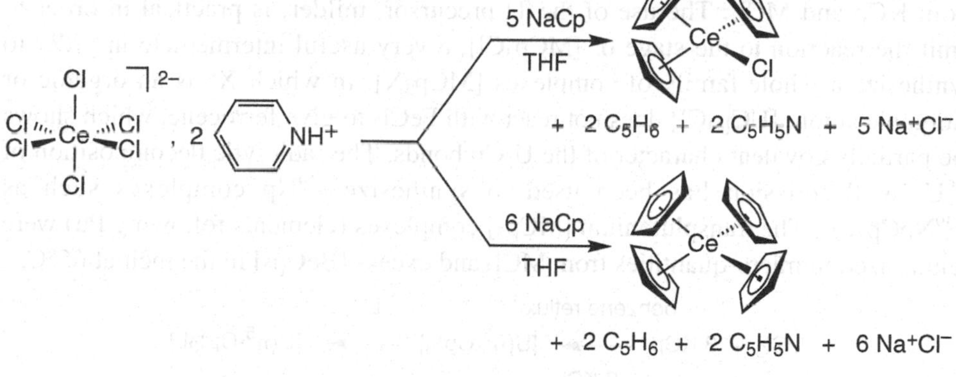

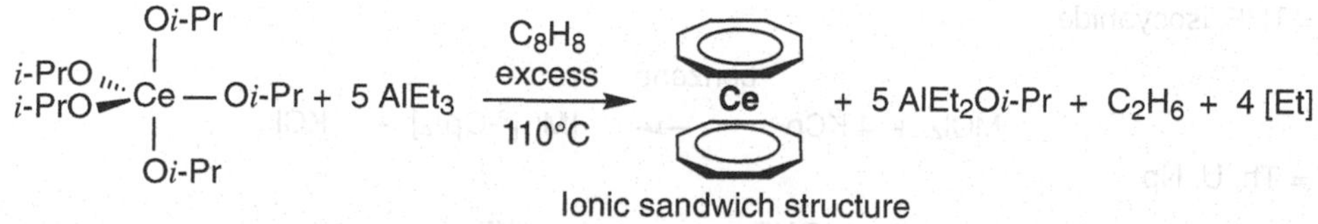

Reaction scheme: $[CeCl_6]^{2-}$, 2 pyridinium with 5 NaCp in THF giving $CeCp_3Cl$ + $2\,C_5H_6$ + $2\,C_5H_5N$ + $5\,Na^+Cl^-$; or with 6 NaCp in THF giving $CeCp_4$ + $2\,C_5H_6$ + $2\,C_5H_5N$ + $6\,Na^+Cl^-$

$$(i\text{-PrO})_4Ce + 5\,AlEt_3 \xrightarrow[110°C]{\substack{C_8H_8 \\ \text{excess}}} Ce(C_8H_8)_2 + 5\,AlEt_2O i\text{-Pr} + C_2H_6 + 4\,[Et]$$

Ionic sandwich structure

[Ce^{IV}(NO$_3$)$_6$(NH$_4$)$_2$] is very often used to oxidatively remove the organic ligand from organometallic complexes. Robust 18-electron complexes are oxidized in this way by Ce^{IV} to labile 17-electron species. The latter decompose, because the metal-ligand bonds have become too weak, the process being eventually facilitated upon nucleophilic attack by the coordinating solvent. Luche's reagent, a combination of NaBH$_4$ and CeCl$_3$, is very selective and largely used (typically in methanol) for the reduction of carbonyl groups of α-enols or ketones.[12.19]

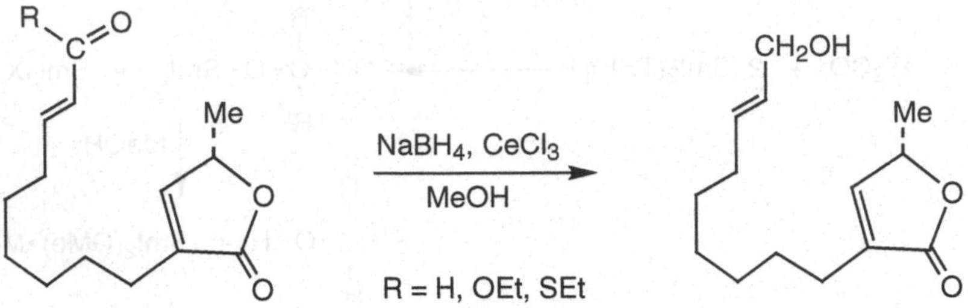

R = H, OEt, SEt

4.3. THE ORGANOACTINIDES [12.12, 12.14, 12.16]

The early organoactinides have been the most studied ones (especially Th and U). The latter show some covalent character of the metal-ligand bonds, contrary to the lanthanides and late organoactinides. This chemistry is dominated by the oxidation states III and IV, by the ligands Cp, Cp*, COT, alkyls, and hydrides, by the oxophilicity of these metals and by insertion and β-elimination reactions.

The most common ancillary ligands are Cp and Cp*. It is possible to introduce up to 4 Cp ligands, but only up to 2 Cp* ligands onto Th or U. The complexes [MCp₃] are strong Lewis acids, contrary to [MCp₄]. The latter, of symmetry S_4, are accessible from KCp and MCl₄. The use of the Tl precursor, milder, is practical in order to limit the reaction to the stage of [MCp₃Cl], a very useful intermediate in order to synthesize a whole family of complexes [MCp₃X], in which X⁻ is an organic or inorganic anion. [UCp₃Cl] does not react with FeCl₂ to give ferrocene, which shows the partially covalent character of the U-Cp bonds. The radiolytic decomposition of ^{239}U by β emission has been used to synthesize ^{239}Np complexes such as [^{239}NpCp₃Cl]. The transplutonium [MCp₃] complexes (elements following Pu) were synthesized in micro-quantities from MCl₃ and excess [BeCp₂] in the melt at 65°C.

$$UCl_3 + 3\ KCp \xrightarrow[-\ 3\ KCl]{\text{benzene reflux}} [U(\eta^5\text{-Cp})_3] \xrightarrow{L} [U(\eta^5\text{-Cp})_3L]$$

L = THF, isocyanide

$$MCl_4 + 4\ KCp \xrightarrow{\text{benzene}} [M(\eta^5\text{-Cp})_4] + 4\ KCl$$

M = Th, U, Np

$$MCl_4 + 3\ TlCp \xrightarrow[-\ 3\ TlCl]{\text{DME}} [MCp_3Cl] \xrightarrow[-\ Cl^-]{X^-} [MCp_3X]$$

M = Th, U, Np; X⁻ = variety of anionic ligands

$$^{239}UCp_3X \xrightarrow{-\ \beta^-} [^{239}NpCp_3X]$$

$$2\ MCl_3 + 3\ [BeCp_2] \xrightarrow{65°C} 2\ [MCp_3] + 3\ BeCl_2$$

M = Pu, Am, Cm, Bk, Cf

The complexes are very labile with a number of Cp ligands lower than 3. On the other hand, their Cp* homologs are much more stable and are easily made, especially [MCp*$_2$Cl$_2$] (M = Th or U).

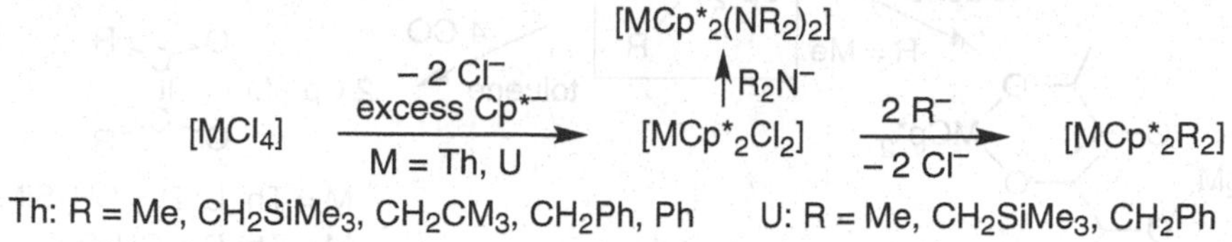

$$[MCl_4] \xrightarrow[\substack{M = Th, U}]{\substack{-2\ Cl^- \\ excess\ Cp^{*-}}} [MCp^*_2Cl_2] \xrightarrow[-2\ Cl^-]{2\ R^-} [MCp^*_2R_2]$$

$$[MCp^*_2(NR_2)_2] \uparrow R_2N^-$$

Th: R = Me, CH$_2$SiMe$_3$, CH$_2$CM$_3$, CH$_2$Ph, Ph   U: R = Me, CH$_2$SiMe$_3$, CH$_2$Ph

The alkyl complexes [MCp*$_2$(Cl)R] and [MCp*$_2$R$_2$] (M = U or Th) are stable only in absence of β-H atom on the alkyl ligand (in contrast with the [MCp$_3$R] series for which the β-elimination is inhibited by the steric constraints). They show a remarkable reactivity. First, their hydrogenolysis leads to the metal-hydrides by σ-bond metathesis. Then, their CO insertion reactions are driven by the metal oxophilicity. With R = Me, a dinuclear complex resulting from intermolecular C-C coupling is obtained, whereas with larger alkyl ligands, C-C coupling occurs intramolecularly. An especially spectacular reaction proceeds in the presence of only one bulky alkyl ligand: it is possible to couple 4 CO molecules between the two metals by continuing the reaction with one more CO molecule! The most probable reaction (scheme below) consists in involving the formation of an intermediate ketene that would be responsible for the coupling of two organometallic fragments.

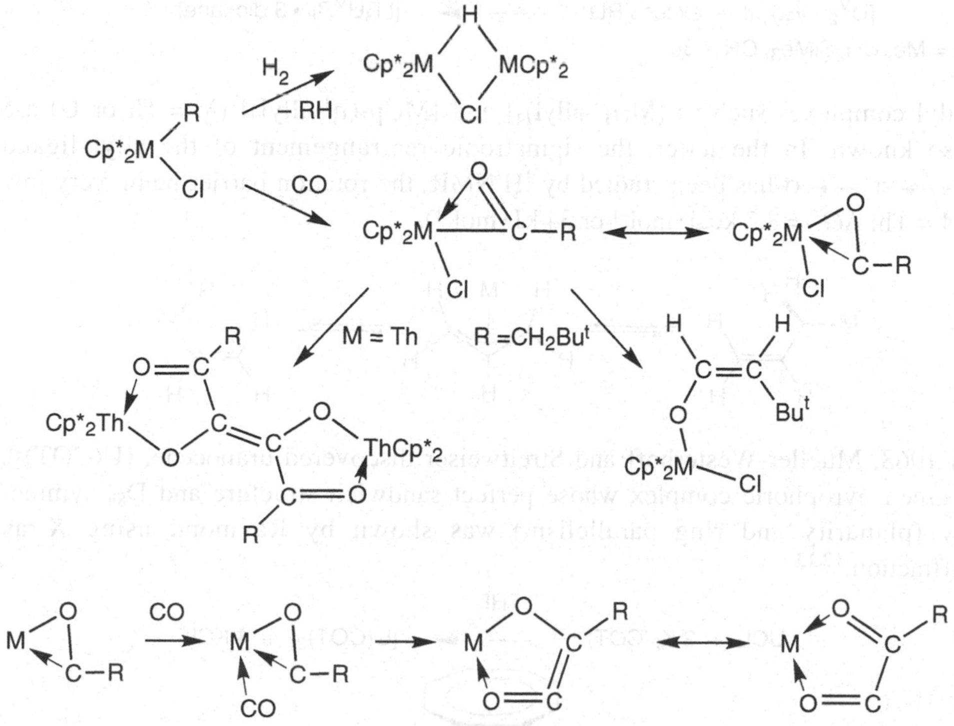

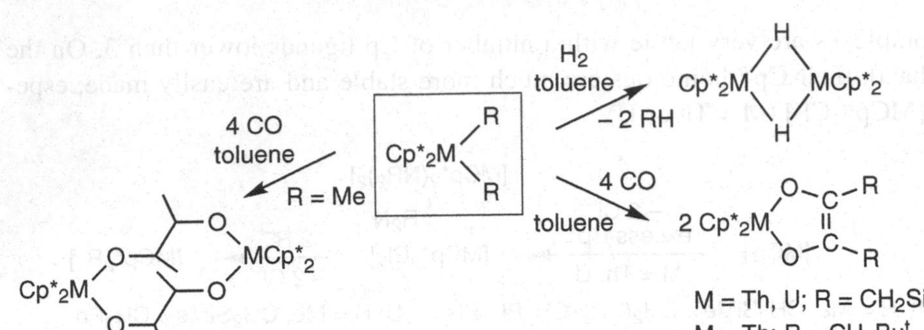

M = Th, U; R = CH$_2$SiMe$_3$
M = Th; R = CH$_2$Bu^t

The binary alkyl complexes are in principle accessible from UCl$_4$ and a lithium reagent in ether or THF at low temperature, but are not usually well characterized because of their instability at room temperature. Neutral complexes [MR$_4$] or anionic (ate) complexes [Li$_2$MR$_6$] are obtained depending on the stoichiometry in lithium reagent. All these complexes are accessible at low temperature except if the alkyl group has a β-hydrogen atom, in which case β-H elimination is very fast. This prevents isolation, even at low temperature. The most stable complex of [MR$_4$] type is [Th(CH$_2$Ph)$_4$] whose structure is close to those of the Ti, Zr and Hf analogs. On the other hand, the octa-alkyl uranates [U(V)R$_8$]$^{3-}$, crystallized with three molecules of dioxane, the solvent used to precipitate them from the reaction medium, have shown an excellent stability at ambient temperature:

$$[U^V_2(OEt)_{10}] + \text{excess RLi} \longrightarrow [Li_3U^VR_8 \cdot 3 \text{ dioxane}]$$
R = Me, CH$_2$SiMe$_3$, CH$_2$t-Bu

Allyl complexes such as [M(η^3-allyl)$_4$] and [MCp$_3$(η^1-allyl)] (M = Th or U) are also known. In the latter, the sigmatropic rearrangement of the allyl ligand σ $\longrightarrow$ π $\longrightarrow$ σ has been studied by ^{1}H NMR, the rotation barrier being very low (M = Th: $\Delta G^* = 8.2$ kcal·mol^{-1} or 34 kJ·mol^{-1}).

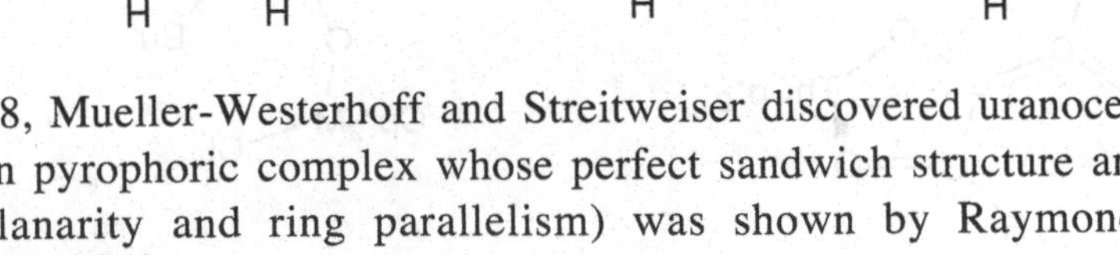

In 1968, Mueller-Westerhoff and Streitweiser discovered uranocene, [U(COT)$_2$], a green pyrophoric complex whose perfect sandwich structure and D$_{8h}$ symmetry (planarity and ring parallelism) was shown by Raymond using X-ray diffraction.[12,13]

UCl$_4$ + 2 K$_2$(COT) $\xrightarrow{\text{THF}}$ [U(COT)$_2$] + 4 KCl

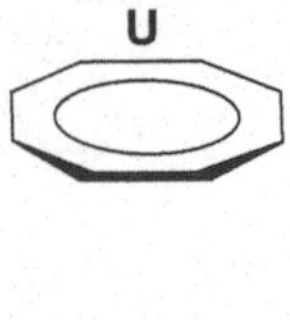

Uranocene   **U**

The dianion COT^{2-} is very stable because of its aromaticity with its $10\,\pi$ electrons (Hückel's rule, $4n + 2$ with $n = 2$), but pyrophoric (very sensitive to oxygen). It is decomposed only slowly by water, acetic acid or electrophiles. The reactions of the latter cannot lead to ring functionalization, contrary to the case of ferrocene. Given the marked ionic character of the actinide complexes and the stability of the oxidation state IV, it is not surprising that uranocene is stable in its structure with aromatic planar rings. One may observe that, remarkably, uranocene is a homolog of ferrocene, but with f orbitals.

$2\ C_8H_8{}^{2-}$ $\qquad$ U^{4+}

$U(C_8H_8)_2$

$NEV = 22\,(para)$

	Th(COT)$_2$	Pa(COT)$_2$		Np(COT)$_2$	Pu(COT)$_2$
e_{3u}	— —	↑ —		↑↓ ↑	↑↓ ↑↓
e_{2u}	↑↓ ↑↓	↑↓ ↑↓		↑↓ ↑↓	↑↓ ↑↓
NEV:	20	21		23	24
Magnetism:	dia	para		para	dia

The interaction and overlap of the HOMO of COT (e_{2u}) with the orbitals of appropriate symmetry f_{xyz} and $f_{z(x^2-y^2)}$ of uranium recalls the interaction of the HOMO of the Cp (e_{1g}) with the d_{xz} and d_{yz} orbitals of iron in ferrocene.

The other actinide complexes [M(COT)$_2$] with Th, Np, Pu, Pa have also been synthesized according to the same equation. [Th(COT)$_2$] is even more ionic than [U(COT)$_2$]. Some substituted derivatives such as [U(1,3,5,7-R$_4$COT)$_2$] with R = Me or Ph and Th complexes with only one COT ligand are also known.

$$[Th(COT)_2] + ThCl_4 \xrightarrow{\ \ THF\ \ } 2\,[Th(COT)Cl_2 \bullet 2\,THF]$$

Finally, actinide complexes such as [MCp*$_2$R$_2$] (M = Th or U, R = alkyl or H) are very active catalysts for the hydrogenation and polymerization of olefins. The complexes [U(allyl)$_3$X] (X = Cl, Br, I) are excellent initiators for the stereospecific polymerization of butadiene, which produces rubbers that have remarkable mechanical properties. Some other complexes are active for the heterogeneous CO reduction and alkene metathesis. The field of catalysis using organoactinide complexes should considerably expand in the near future.

SUMMARY OF CHAPTER 12
IONIC AND POLAR METAL-CARBON BONDS: ALKALINE, ALKALINE EARTH, LANTHANIDE AND ACTINIDE COMPLEXES

1 - Organoalkali

Polarity: Li-C < Na-C < K-C< Cs-C

The aggregation of M-R in clusters depends on the solvent (solvent heteroatoms are coordinated).

Naphthylsodium in THF is the strongest common homogeneous reductant.

Butyllithium (pK_a = 44) is one of the most used bases in organic synthesis.

Bulky alkali amides such as i-Pr$_2$NLi are strong, non-nucleophilic bases.

NaH and t-BuOK are other good, weakly-nucleophilic bases.

Organoalkali reagents are powerful alkylating agents in organic and organometallic chemistry.

2 - Organomagnesium

Famous since Grignard, they alkylate numerous substrates including carbonyl- and nitrile derivatives:

$$Mg + RX \xrightarrow[\text{X = halogen}]{Et_2O} RMgX(Et_2O)_n \xrightarrow{R'R''CO} RR'R''COH$$

3 - Organoscandium, organoyttrium and organolanthanides

These organometallic compounds are ionic, essentially trivalent, i.e. without valence d electrons.

The lanthanides are the 14 elements following lanthanum. Their properties are very similar, because they only differ by filling from 1 to 14 e$^-$ the internal f sublayer. [MCp*$_2$(CH$_3$)] activates CH$_4$ by σ-bond metathesis with M = Sc or Lu. The most common compound is Kagan's reagent, SmI$_2$, a very good reductant leading to Sm^{III}. The most well-known organometallic compounds are [Cp*$_2$M(R)] and [SmCp*$_3$], a strong reducing agent, source of Cp*$^-$.

4 - Organoactinides

Mainly Th and U have been studied. The complexes [MCp*$_2$R$_2$] are the most common ones, when R lacks β-H atoms (otherwise, they decompose by β-elimination). They undergo hydrogenolysis by σ-bond metathesis and CO insertion driven by the oxophilicity of the metal. They are efficient olefin hydrogenation and polymerization catalysts.

Uranocene [U(COT)$_2$] is a green pyrophoric complex, with a perfect sandwich structure, whose stability is provided by the aromaticity of the dianionic cyclo-octatetraenyl ligands. Contrary to its ferrocene homolog, it does not give ring functionalization because of its strong ionic character. The analogs [M(COT)$_2$] with M = Th, Np, Pu and Pa are also known.

Uranocene

EXERCISES

12.1. Classify the following compounds according to the order of decreasing polarity of their metal-methyl bonds: LiMe, NaMe, KMe, $ZnMe_2$, $MgMe_2$, $CdMe_2$, $TaMe_5$, $LaMe_6^{3-}$, UMe_8^{3-}.

12.2. Among the following metal-cyclopentadienyl complexes, which ones react with $FeCl_2$ to give ferrocene: $LuCp_3$, $ScCp_3$, YCp_3, KCp, UCp_4, $ThCp_4$, $TiCp_4$, CuCp?

12.3. a. Has butane a higher or lower pK_a value than acetylene and benzene? Why?
b. As a consequence, is n-butyllithium able to deprotonate acetylene? Why?
c. Same question with benzene.

12.4. Does the metal-hydride complex $[SmCp*_2(H)]$ exist as a monomer or dimer? Why?

Chapter 13

COVALENT CHEMISTRY OF THE ORGANOELEMENTS OF FRONTIER (11, 12) AND MAIN (13-16) GROUPS

1. INTRODUCTION

The elements of groups 11 to 16 have in common a Pauling electronegativity between 1.7 and 2.6 (except O and N that are not treated here). Thus, they form, with carbon whose Pauling electronegativity is 2.5, apolar or very weakly polar bonds. Here is a rich chemistry of covalent complexes containing metal-alkyl σ bonds, π complexes being rarer. The energy of the σ M—C is very variable, from 87 kcal·mol^{-1} (365 kJ·mol^{-1}) for BMe$_3$ to 34 kcal·mol^{-1} (141 kJ·mol^{-1}) for BiMe$_3$. It decreases upon going down in the periodic table because of the radial extension of the metallic orbitals that makes the orbital overlap progressively less favorable. The metal-alkyl complexes of the heavy elements thus thermally decompose in a homolytic way giving free alkyl radicals (thermodynamic instability). For the complexes in which the metal has at least one vacant d orbital (group 13) and in which the alkyl group has at least onehydrogen atom in β position, the β-H-elimination process occurs as for transition-metal complexes (kinetic instability) even if the metal-alkyl bond is strong.

The particular character of each of these groups requires a treatment for each group rather than the treatment by ligand type that was used for transition-metal complexes. The elements of the groups 11 (Cu, Ag, Au) and 12 (Zn, Cd, Hg) are at the frontier between the transition groups and the main groups. They also are marginal because of their specific properties that make them difficult to classify in a clear way in either category; a specific treatment is thus appropriate.

The groups 12 to 16 are usually classified as main groups whose elements form organo-element or organometallic compounds with carbon. The latter word could logically be reserved to metals, the metallic character increasing upon going down in each column of the periodic table. Thus, the elements located below B, Si, As, and Se are usually classified as metals (these four elements together with P and S being sometimes called "*metalloids*"). These considerations, however, are not rigidly fixed in order not to cleave scientific communities. The distinction is here more relevant to use than to dogma.

Table distinguishing between the metallic (bold) and non-metallic elements (italic), at least in the context of organometallic chemistry

11	12	13	14	15	16
		B	*C*	*N*	*O*
		Al	*Si*	*P*	*S*
Cu	**Zn**	**Ga**	**Ge**	**As**	**Se**
Ag	**Cd**	**In**	**Sn**	**Sb**	**Te**
Au	**Hg**	**Tl**	**Pd**	**Bi**	**Po**

2. GROUP 11 (Cu, Ag, Au)

The organometallic chemistry of group 11 is dominated by the considerable applications of the organocuprate complexes [LiCu^{I}R$_2$] in organic synthesis.[13.1-13.4] Copper and silver[13.5] exclusively give organometallic complexes in the oxidation state I whereas, for gold, the oxidation state III is also known.[13.6, 13.7] However, the complexes [Cu^{I}R] and [Ag^{I}R] are insoluble, very sensitive to air and hydrolysis, their stoichiometry is ill-defined in the presence of solvent or halide, and they are sometimes thermally sensitive (for instance, CuMe explodes above –15°C, and AgMe decomposes above –50°C). They are stable when R is aryl, alkynyl, perfluoroalkyl or a bulky ligand that does not contain an H atom in β position. The organocuprates show a pronounced covalent character in spite of the "ate" nomenclature.

$$CuX + LiR \xrightarrow[-\,LiX]{\text{Et}_2\text{O}} CuR \xrightarrow{LiR} \underset{\text{cuprate}}{LiCuR_2}$$

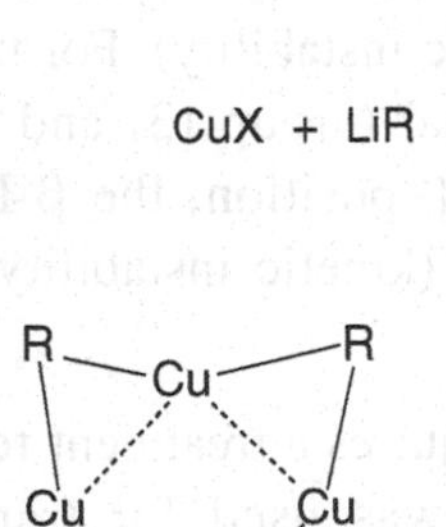

[CuR]$_4$, R = CH$_2$SiMe$_3$
The Cu-R bond can be considered as a
2-electron donnor for the neighboring Cu

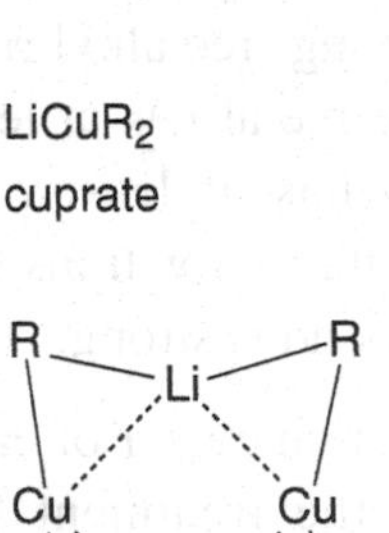

[LiCuR$_2$]$_2$, R = Me
The ^{7}Li NMR shows an equilibrium:
[LiCuR$_2$]$_2$ ⇌ [LiMe] + [LiCu$_2$Me$_3$]

Cuprates have been used in organic synthesis since Kharash, in 1941, observed the catalysis by cuprous ions in the 1,4 addition of Grignard reagents on enones. Cuprates allow the alkylation of alkyl halides, carbonyl groups, nitriles, epoxides and enones (see Chap. 21.3.1).

$$[LiCuR_2] + R'X \xrightarrow[-20°C]{Et_2O} R-R' + LiX + [CuR]$$

Copper also exists in the oxidation state II (inorganic cupric reagents such as $CuCl_2$ and $Cu(OAc)_2$). These compounds allow carrying out oxidative coupling with aryl lithium and terminal alkynes. These coupling reactions can also be achieved using $Cu^I + O_2$.

$$R-C\equiv CH \xrightarrow[\text{pyridine}]{Cu(OAc)_2} R-C\equiv C-C\equiv C-R$$

$$ArLi \xrightarrow[Et_2O]{CuCl_2} Ar-Ar$$

Aurates are also known with both oxidation states I and III, but the neutral compounds are unknown (AuR) or very instable. The complexes [AuMe$_3$] decompose above 40°C if they are not coordinated to an L ligand. The 14-electron (linear aurate (I)) and 16-electron (trigonal aurate (I) and square-planar aurate (III)) structures, however, are usually found rather than 18-electron tetrahedral [LiAu^{I}R$_2$L$_2$] that is rare. All these aurates are much more stable if the cation is complexed to a polyamine:

$$[Au^I(PEt_3)Cl] + MeLi \xrightarrow[-LiCl]{} [Au^I(PEt_3)Me] \xrightarrow[-PEt_3]{MeLi} [Au^IMe_2]^-Li^+$$
$$\text{14e} \qquad\qquad\qquad\qquad \text{14e} \qquad\qquad\qquad \text{14e}$$

MeI, PPh$_3$ | – LiI

$$\begin{bmatrix} Me & \overset{III}{\underset{Au}{\diagup}} & Me \\ Me & & Me \end{bmatrix}^- Li^+ \xleftarrow[-PPh_3]{MeLi} [Au^{III}(PPh_3)Me_3]$$
$$\text{16e} \qquad\qquad\qquad\qquad\qquad \text{16e}$$

The binary gold carbonyl complexes are instable, but the family of inorganic 14-electron complexes [Au^I(Cl)L] is known for its stability with a variety of L ligands (CO, PR$_3$, RNC, pyridine, etc.). These complexes are also prepared from the precursor complex [Au(Cl)(Me$_2$S)] or [Au(Cl)(tmt)], tmt = tetramethylthiophene. The isonitrile complexes undergo the addition of nucleophiles (alcohols, amines) to yield carbene complexes:

$$[Au(R_2S)Cl] + L \longrightarrow [Au(L)Cl] + R_2S$$
$$R_2S = Me_2S \text{ or } (CH_2)_4S; \ L = CO, PR_3, RNC, py, \text{ etc.}$$

$$[Au^I(CNR)Cl] + MeOH \longrightarrow \begin{matrix} RHN \\ \diagdown \\ C=Au^{III}-Cl \\ \diagup \\ MeO \end{matrix}$$

With the nitriles, Au^{III} gives square structures. Finally, for gold and silver, the formation of metal-metal bonds is easy (see the oxidative addition onto Au^I, Chap. 3) as well as that of giant clusters (see Chap. 2). ^{197}Au Mössbauer spectroscopy allows to distinguish the various geometries around gold.

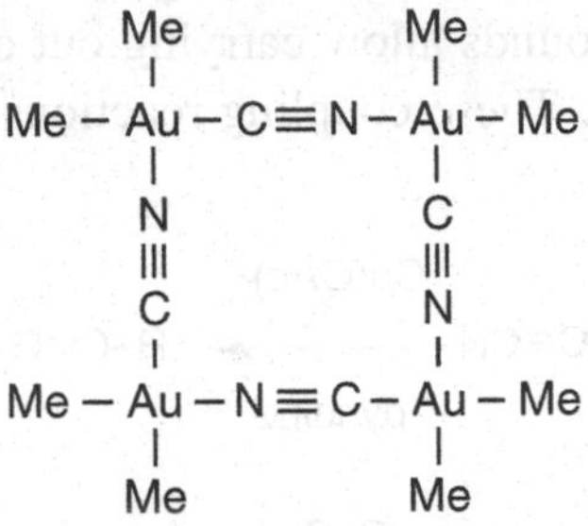

$$
\begin{array}{ccc}
Me & & Me \\
| & & | \\
Me - Au - C \equiv N - Au - Me \\
| & & | \\
N & & C \\
\text{III} & & \text{III} \\
C & & N \\
| & & | \\
Me - Au - N \equiv C - Au - Me \\
| & & | \\
Me & & Me
\end{array}
$$

3. Group 12 (Zn, Cd, Hg)

The organozinc complexes ZnR_2 and $RZnX$ are weakly polar and mild alkylating agents that are very useful in organic synthesis.[12.8,13.8,13.9] They are complementary to organomagnesium reagents that are very polar, and to cuprates. The organozinc reagents ZnR_2 are prepared by reaction of $ZnEt_2$ with RI (R being a primary or secondary alkyl group, reaction catalyzed by CuCl), with BR_3 or with a terminal olefin (reaction catalyzed by $NiCl_2$).

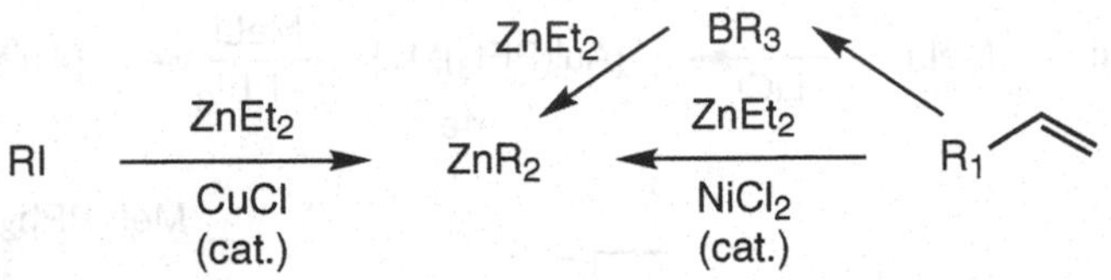

$$
RI \xrightarrow[\text{CuCl (cat.)}]{ZnEt_2} ZnR_2 \xleftarrow[\text{NiCl}_2 \text{ (cat.)}]{ZnEt_2} R_1 \diagup\!\!\diagdown
$$

with $ZnEt_2$ / BR_3 pathways converging at ZnR_2.

The organozinc halides $RZnX$, easily prepared *in situ,* are less air sensitive than the binary compounds ZnR_2.

$$
RLi + ZnCl_2 \xrightarrow{\text{THF, 20°C}} RZnCl + LiCl
$$

$$
RI + Zn \text{ (powder)} \xrightarrow{\text{THF, 20°C}} RZnI
$$

This second method of preparation is particularly simple and practical. Zinc powder can be made more reactive if necessary, using various means including sonication or preparation by reduction of $ZnCl_2$ (1.5h, 20°C) using lithium and naphthalene as electron-transfer mediator in THF (Riecke Zn noted Zn*).[12.8,13.9]

$$
ZnCl_2 + 2 Li \xrightarrow[\text{THF, 1.5h}]{\text{naphthalene}} Zn^* + 2 LiCl
$$

Organozinc complexes have been used for a long time as carbene transfer reagents on double bonds (Simmons-Smith cyclopropanation) and arene rings which leads to their extension to cyclohexatrienes.

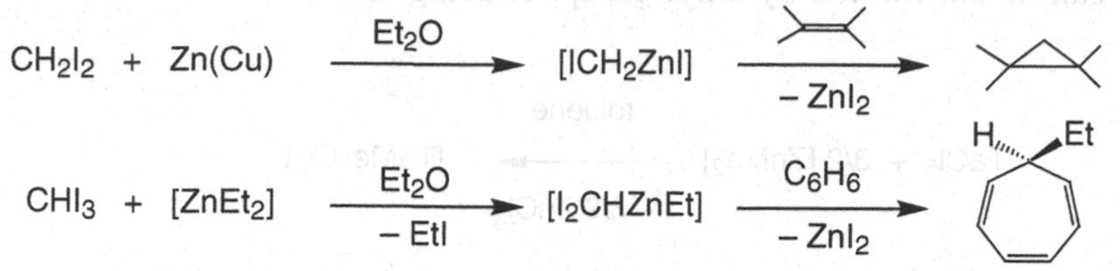

They are also very useful for the alkylation of alkyl halides and enones forming C-C bonds.

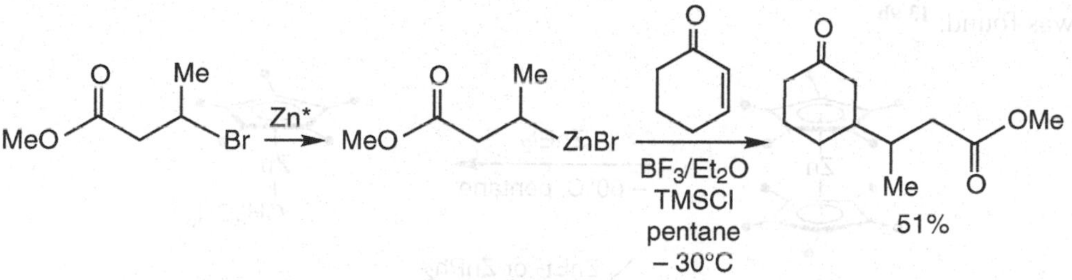

The use of zinc reagents in organic synthesis is now very much expanding.[12.8,13.8b] A large variety of alkylation of functional halides with C-C bond formations of this type are now possible, many of them being catalyzed by Cu, Ni, Co, Fe and especially Pd complexes.

Organozinc reagents are insufficiently reactive to react with carbonyl derivatives and double bonds, except allyl and propargylzinc reagents that react with carbonyl derivatives, nitriles and alkynes.

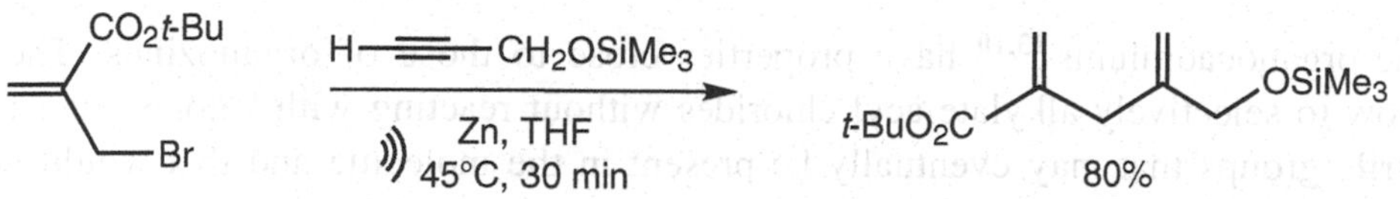

Organozinc reagents ($RZnX$ or ZnR_2) are sufficiently reactive to alkylate chlorophosphines.[13.9a]

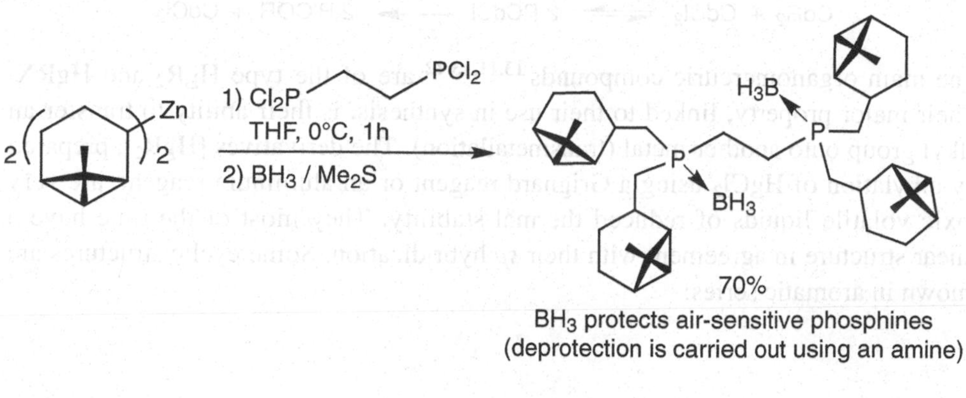

BH$_3$ protects air-sensitive phosphines
(deprotection is carried out using an amine)

Analogously, zinc reagents react with halides of various elements. They are particularly useful for the alkylation of halides of transition metals such as $NbCl_5$ and $TaCl_5$ that are too easily reducible by alkali metal alkyl reagents. Up to three halides can be substituted by alkyl groups R using ZnR_2 in these transition-metal halides.

$$TaCl_5 + 3/2\ [ZnMe_2] \xrightarrow[- 3/2\ ZnCl_2]{\text{toluene}} [TaMe_3Cl_2]$$

Metallocenes (MCp_2 and MCp^*_2 complexes) are known with Zn as with the other main-group metals.[13.9a] More unexpected was the finding by the Carmona group in 2004 of sandwich complexes in which a relatively strong Zn-Zn bond (60 kcal·mol^{-1}) was found: [13.9b]

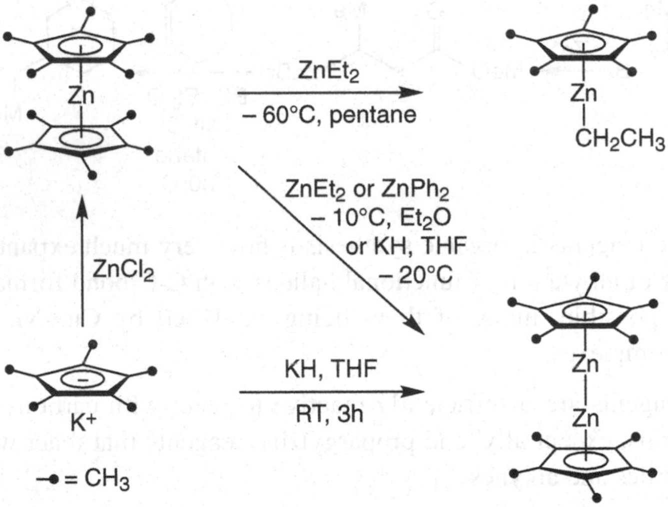

$-\bullet = CH_3$

The organocadmiums [13.10] have properties close to those of organozincs. They allow to selectively alkylate acyl chlorides without reacting with ketone, ester or nitrile groups that may eventually be present in the molecule and that would be attacked by the Grignard reagents which are more reactive.

$$CdR_2 + CdCl_2 \rightleftharpoons 2\ RCdCl \xrightarrow{R'COCl} 2\ R'COR + CdCl_2$$

The main organomercuric compounds [13.11,13.12] are of the type HgR_2 and $HgRX$. Their major property, linked to their use in synthesis, is their ability to transfer an alkyl group onto another metal (transmetallation). The derivatives [HgR_2], prepared by alkylation of $HgCl_2$ using a Grignard reagent or an aluminum reagent, are very toxic volatile liquids of reduced thermal stability. They most of the time have a linear structure in agreement with their sp hybridization. Some cyclic structures are known in aromatic series:

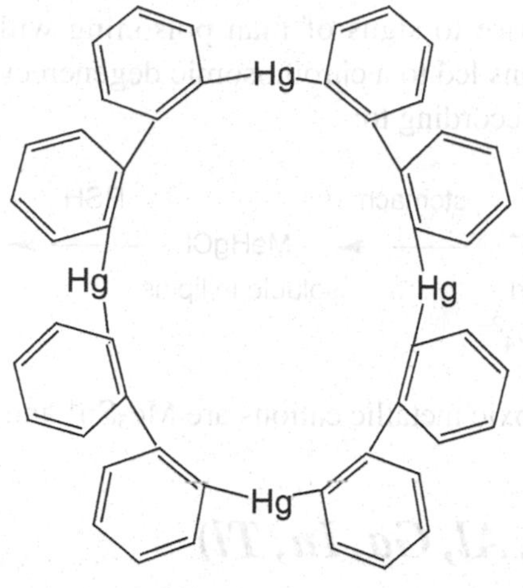

The mixed derivatives RHgX can be obtained especially upon photochemical irradiation of Hg in the presence of RI, by selective alkylation using a Grignard reagent
or of another organometallic compound (Sn, Pb, Bi, Cd, Tl, etc.), by reaction of the
Na/Hg amalgam on an alkyl halide or by redistribution of a mixture of HgR$_2$ and
HgX$_2$. The mercuration of aromatics using Hg(OAc)$_2$ in acetic acid is a well-known
electrophilic substitution that is favored by electron-releasing substituents as the
other reactions of this type. It applies to π ligands of neutral complexes that are able
to undergo electrophilic substitutions. It is essentially the case of Cp in [FeCp$_2$] (see
Chap. 11), [RuCp$_2$] and [MnCp(CO)$_3$]. The compound [FeCp(C$_5$H$_4$HgCl)], formed by
reaction with [FeCp$_2$], is a useful intermediate in ferrocene chemistry (see Chap. 11).
Two reactions that are largely used up to now are the additions of HgCl$_2$ to olefins
in the presence of alcohols (oxymercuration) or amine (aminomercuration):

$$H_2C=CH_2 + HgCl_2 + ROH \longrightarrow ROCH_2CH_2HgCl$$

$$H_2C=CH_2 + HgCl_2 + R_2NH \longrightarrow R_2NCH_2CH_2HgCl$$

The mixed derivatives [RHgX] are crystalline solids that are sometimes sublimable.
They are solubilized in water for $X^- = F^-$, NO_3^-, 1/2 SO_4^{2-}, ClO_4^- while forming the
RHg$^+$ cations.

The traditional uses of organomercury compounds (fungicides, bactericides and
antiseptic) tend to decrease given the very high toxicity of the ions R-Hg^{II+}, that
appeared especially after the disaster of Mimamata in Japan between 1953 and
1960. The industrial water containing the Hg^{2+} ions were discharged into the sea,
contaminating the aquatic environment (microorganisms, fishes), then men. The
Hg^{2+} ions are alkylated by methylcobalamine, a vitamin B$_{12}$ derivative.

$$CH_3(Co) + aq.\ Hg_2^+ \xrightarrow{\quad H_2O \quad} [H_2O(Co)]^+ + aq.\ [CH_3Hg]^+$$

The CH_3Hg^+ ions give rise to signs of man poisoning with oral doses as low as 0.3 mg per day. These ions led to a chromosomic degenerecy due to transferring the thiol group of enzymes according to:

$$\text{aq. MeHg}^+, \text{X}^- \xrightarrow{\text{stomach}} \text{MeHgCl} \xrightarrow{\text{RSH}} \text{MeHgSR}$$

soft Lewis acid   soluble in lipids

X^-: NO_3^-, 1/2 SO_4^{2-}

The other homologous toxic metallic cations are Me_3Sn^+ and Me_2As^+.

4. GROUP 13 (B, Al, Ga, In, Tl)

4.1. INTRODUCTION

The chemistry of boron and aluminum dominates this group although the organo-metallic Ga and In compounds are presently the subject of scrutiny as precursors of ultrapure Ga-In alloys for electronic industry. Boranes have become extremely important in organic synthesis because of the efficiency of the hydroboration reactions that were discovered and exploited by Brown.[13.13] The organoaluminum compounds are excellent and cheap alkylating agents. The MR_3 derivatives of this trivalent group do not follow the octet rule; they only have 6 electrons on the central atom, BR_3 being isoelectronic to the R_3C^+ carbocations. They are well-known Lewis acids, which guides their structure and chemistry. This Lewis-acid property peaks for Al^{III}, then decreases upon going down in the column. Only thallium shows, in its chemistry, current interconversion between its oxidation states I and III (only a few In^I compounds are known). The usefulness of thallium in organic synthesis is severely tempered by its toxicity. The π complexes (a few pentahapto Cp derivatives) are not common in this group, as well as in the preceding one.

4.2. ORGANOBORANES

Boranes and polyboranes have already been introduced in the context of clusters (see Chap. 2), and organoborane chemistry will now be presented.[13.13-13.15] Boron has two isotopes ^{10}B (20%) and ^{11}B (80%). It is the ^{11}B isotope that is useful in NMR. The chemical shifts δ of ^{11}B spread over 250 ppm and depend on the coordination, charge and substituents of boron. The ^1J couplings (^{11}B-^1H) are between 100 and 200 Hz for terminal BH and 30 to 60 Hz for bridging BHB.[13.15]

The binary organoboranes are prepared from organomagnesium reagents and are monomeric, contrary to BH_3, BH_2R and BHR_2 that are bridged dimers.

$$Et_2OBF_3 + 3\,RMgX \longrightarrow BR_3 + 3\,MgXF + Et_2O$$

R = alkyl or aryl

The halogenoboranes are prepared in particular from BCl_3 or BR_3. They are precursors of a large variety of derivatives BR_2Nu that are obtained by reaction with a nucleophile NuH or Nu^- (alcohol, thiol, amine, carbanion, hydride).

$$BCl_3 + SnPh_4 \longrightarrow PhBCl_2 + Ph_3SnCl$$

The trialkoxyboranes also are accessible from trialkylboranes by reaction with trimethylamine oxide.

$$BR_3 + I_2 \longrightarrow R_2BI + RI$$

$$BR_2X + NuH \longrightarrow BR_2Nu + HX$$

(NuH = ROH or R_2NH)

$$BR_2X + NuM \longrightarrow BR_2Nu + MX$$

(NuM = HNa, R'Li, RSNa)

$$BR_3 + 3\,Me_3NO \longrightarrow B(OR)_3 + 3\,Me_3N$$

All these "6-electron" compounds are Lewis acid, thus they react with Lewis bases such as NR'_3. The carbanions lead to borates such as BPh_4^- that are sometimes used to precipitate organometallic cations. BH_3 is currently used to protect air-sensitive phosphines (see section 3) and to crystallize them. The complexes formed are air-stable, and the free phosphine can easily be regenerated by reaction with an amine giving, with BH_3, a more robust complex.[13.16]

$$BR_2X + NR'_3 \longrightarrow (X)R_2BNR'_3$$

$$NaBF_4 + 4\,PhMgBr \longrightarrow NaBPh_4 + 4\,MgBrF$$

$$PR_3 + BH_3 \longrightarrow H_3BPR_3 \xrightarrow{NR'_3} H_3BNR'_3 + PR_3$$

The aminoboranes are prepared from halogenodialkylboranes or lithiodialkylboranes and secondary amines. In aminoboranes, the nitrogen atom gives its free electron pair to boron intramolecularly, which increases the order of the B-N bond and allows to isolate *cis-trans* isomers.

$$R_2BCl + R_2NH \longrightarrow R_2BNR_2 + HCl$$

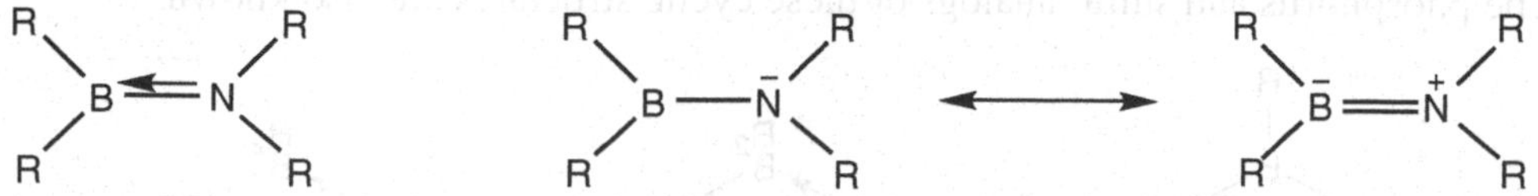

With small substituents, the nitrogen doublet can be given intermolecularly, which leads to the formation of cyclic structures (see bottom of p. 322).

The great application of boranes in organic synthesis is the regio- and stereoselective hydroboration (*cis* addition) of terminal alkenes (anti-Markovnikov) that is selective when the R groups are large enough. Boranes currently used are disiamylborane and 9-BBN.[13.13]

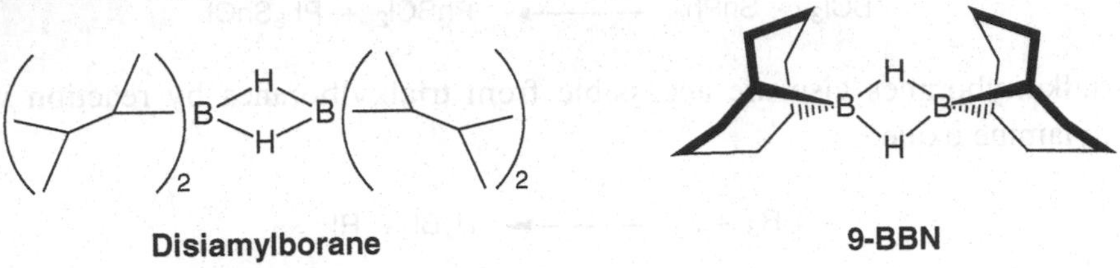

Disiamylborane
9-BBN

Hydroboration is mainly currently used to prepare primary alcohols from terminal olefins, but other functional groups are also accessible. The hydroboration reaction can also chemoselectively lead to the reduction of the carbonyl group of aldehydes and α, β-unsaturated ketones.

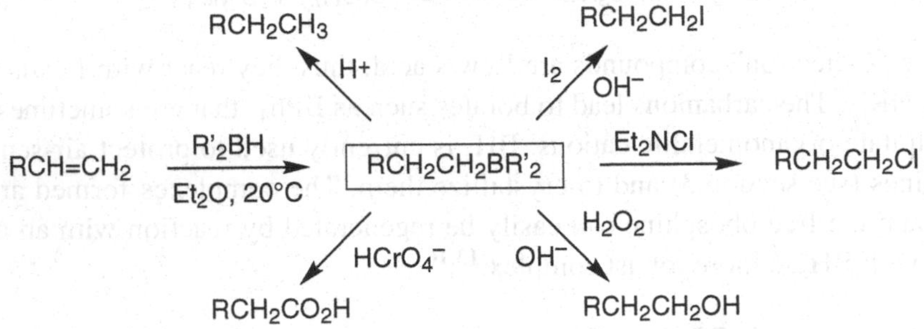

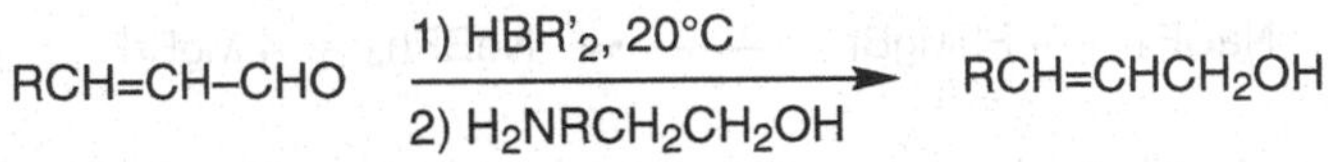

RCH2CH3
H+
RCH2CH2I
I2
OH−
RCH=CH2
R'2BH
Et2O, 20°C
RCH2CH2BR'2
Et2NCl
RCH2CH2Cl
HCrO4−
H2O2
OH−
RCH2CO2H
RCH2CH2OH

RCH=CH–CHO
1) HBR'2, 20°C
2) H2NRCH2CH2OH
RCH=CHCH2OH

The boronic and borinic acids, $RB(OH)_2$ and $R_2B(OH)$ respectively, are prepared by hydrolysis of halides or corresponding esters. They are weakly acidic and readily converted to the corresponding anhydrides, sometimes spontaneously. The esters are liquids that are moisture sensitive and useful synthetic reagents. They are obtained from trialkoxyboranes and one or two equivalents of a magnesium or lithium reagent.

Heterocyclic boron chemistry is very rich. The boronic acids $RB(OH)_2$ dehydrate to cyclic trimers, boroxines, and aminoboranes R_2BNR_2 also form cyclic trimers.[13.17] The phosphorus and sulfur analogs of these cyclic structures are also known.

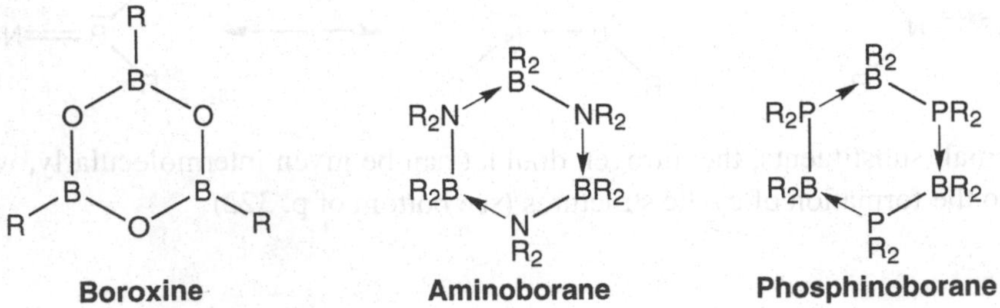

Boroxine
Aminoborane
Phosphinoborane

A classic route to boron heterocycles involves the hydroboration of a diene:

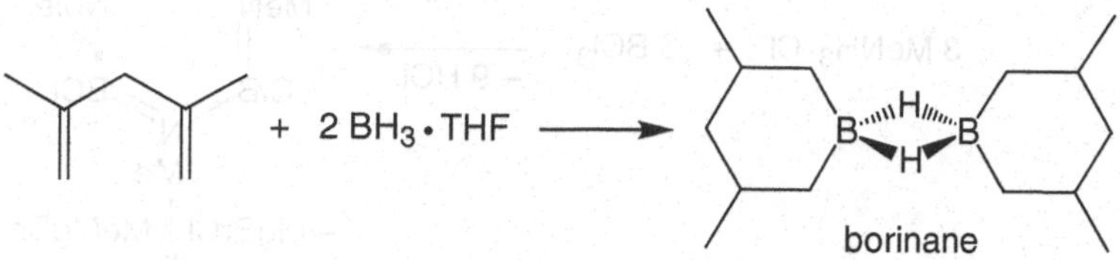

The hydrostannation of a diyne followed by transmetallation using a borane leads to the phenylboratabenzene anion, isoelectronic to biphenyl (the prefix *borata* means that a benzene CH group is replaced by the isoelectronic BH^- group). It is equivalent to the cyclopentadienyl anion and gives transition-metal sandwich complexes that are equivalent to metallocenes with a variety of metals (V, Cr, Fe, Co, Ru, Os).[13.17]

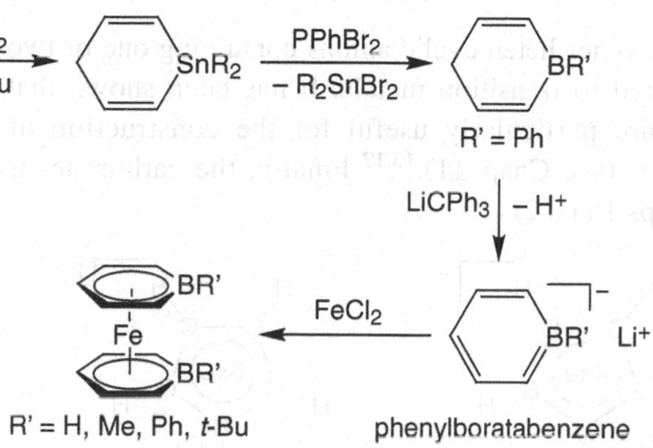

An alternative synthetic route to boratabenzenes has been disclosed from cobaltocene.

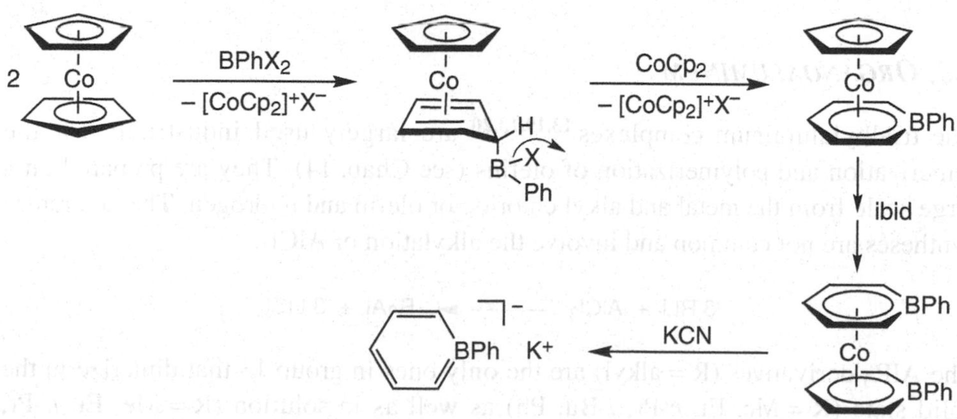

Another series isoelectronic to benzene derivatives is that of borazines or borazoles in which each C-C couple is replaced by the isoelectronic B-N couple:

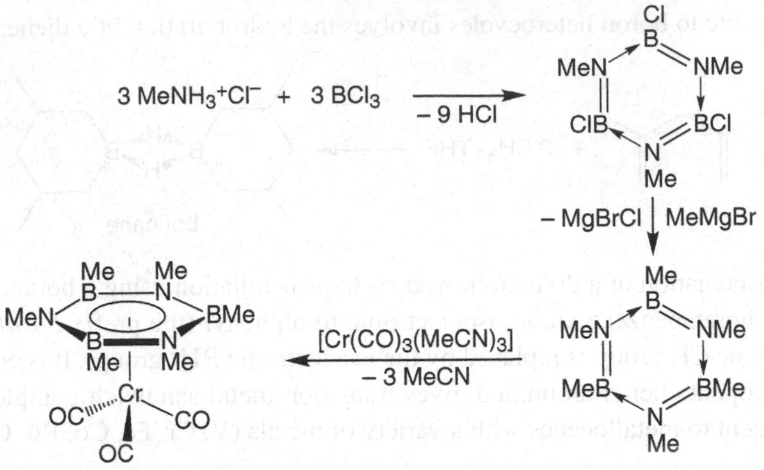

There are other heterocyclic anions containing one or two boron atoms that can be coordinated to transition metals. It has been shown that these electron-deficient ligands are particularly useful for the construction of multilayered sandwich complexes (see Chap. 11).[13.17] Finally, the carboranes also are excellent ligands (see Chaps 1 and 2).

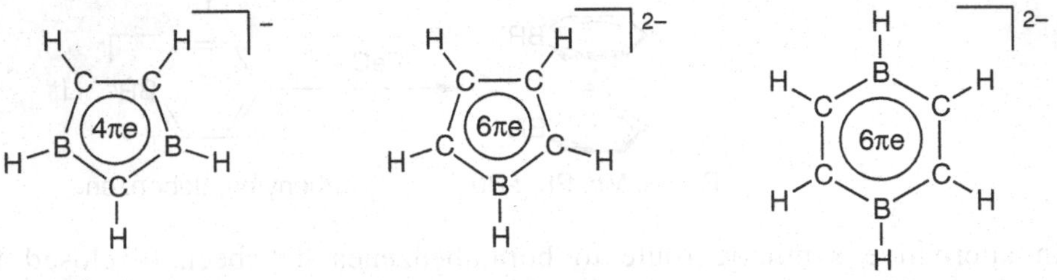

Dihydro-2,3-diborol-1,3-yle   **Borole dianion**   **Diborata-1,4-benzene**

4.3. ORGANOALUMINUM

The trialkylaluminum complexes [13.19,13.20] are largely used industrially for the dimerization and polymerization of olefins (see Chap. 14). They are prepared on a large scale from the metal and alkyl chloride or olefin and hydrogen. The laboratory syntheses are not common and involve the alkylation of $AlCl_3$:

$$3\ RLi\ +\ AlCl_3\ \longrightarrow\ R_3Al\ +\ 3\ LiCl$$

The AlR_3 derivatives (R = alkyl) are the only ones in group 13 that dimerize in the solid state (R = Me, Et, n-Pr, i-Bu, Ph) as well as in solution (R = Me, Et, n-Pr, equilibrium with R = Ph). This dimerization is slowed down or inhibited by large substituents. It corresponds, as for bridging hydrides, to the bridging 3-center 2-electron bond that was already met with boranes. For $[Al_2Me_6]$, it may be noted that the Al-Al bond is very short (260 pm), of the order of that found in the metal (256 pm).

$$\begin{array}{c}\text{Me}\diagdown\quad\diagup\text{Me}\\ \text{Me}\text{''''''}\text{Al}\rightleftharpoons\text{Al}\text{''''''}\text{Me}\\ \diagup\quad\diagdown\\ \text{Me}\qquad\text{Me}\end{array}$$

Ethylene insertion into the Al-R bond is facile because of the possible pre-coordination of the olefin on the vacant site in the monomer that is in equilibrium with the dimer (the existence of this pre-equilibrium is confirmed by kinetic studies). This insertion allows the polymerization to proceed.[13.21] The polymers can be thermally disengaged from the metal by β-elimination that is in competition with chain growth, or by hydrolysis. This latter method produces unbranched primary alcohols whose sulfates $ROSO_3H$ are biodegradable detergents.

$$AlEt_3 \xrightarrow[110°C,\ 100\ bar]{CH_2=CH_2} Al{\Big\langle}\begin{array}{l}(C_2H_4)_xEt\\(C_2H_4)_yEt\\(C_2H_4)_zEt\end{array} \xrightarrow{3/2\ O_2} Al{\Big\langle}\begin{array}{l}O(C_2H_4)_xEt\\O(C_2H_4)_yEt\\O(C_2H_4)_zEt\end{array}$$

$$\Big\downarrow CH_2=CH_2 \qquad\qquad\qquad \Big\downarrow 3\ H_2O$$

$$3\ CH_2=CH(C_2H_4)_{x,y,z}H \qquad 3\ Et(C_2H_4)_{x,y,z}OH\ +\ Al(OH)_3$$

The AlR_3 derivatives are extremely reactive. The reactions are explosive with $CHCl_3$ or CCl_4. The numerous other reactions must be carried out in the rigorous absence of air.[13.19,13.20]

$$\begin{array}{ccc}
& (RAINR')_4 & \\
& \Big\uparrow R'NH_2 & \\
(R'C\!\equiv\!CR')\ \nwarrow & \boxed{AlR_3} & \nearrow\ AIR_n(OR')_{3-n}\ (R'OH)\\
R'Cl\ \swarrow & & \searrow\ LiR\\
R-R'\ +\ R_2AlCl & \Big\downarrow H_2 & LiAlR_4\\
& R_2AlH &
\end{array}$$

The hydroalumination reactions, as hydroborations, are stereospecifically *cis*. An advantage, compared to hydroboration, is the ease of cleavage of the Al-C bond that does not require a peroxide. The regioselectivity, however, is not as good as that of the hydroboration reactions, which clearly is a major disadvantage.

The AlR_3 complexes are better Lewis acids than BR_3. They react with hard Lewis bases such as Me_3N, which shows their properties as hard Lewis acids. They react with carbanions to give tetraalkylaluminates of polymeric structure, also accessible upon reduction by Na.

$$AlEt_3 \xrightarrow{LiEt\ or\ Na} MAlEt_4$$
$$(M = Li\ or\ Na)$$

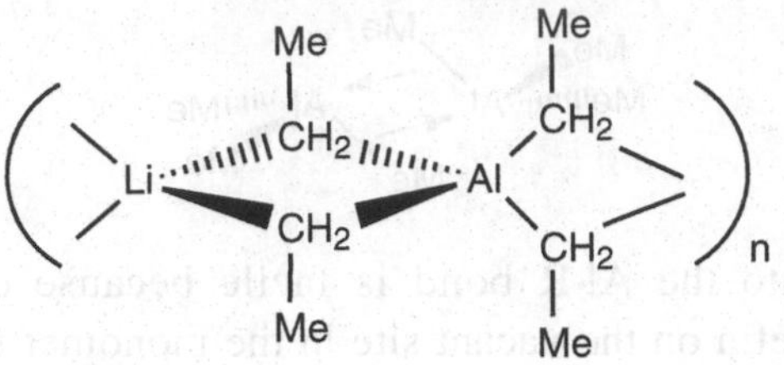

The reaction of Al_2Me_6 with Al_2Cl_6 gives the ternary sesquichloride dimers in equilibrium with the binary dimers resulting from disproportionation. This mixture, industrially produced from Al and MeCl, is also used for the synthesis of Al_2Me_6. In Al_2Me_6 and in partially chlorinated dimers, it is the Cl atoms that bridge the two aluminum metals. The bridging Cl forms, as with transition metals, bonds in which each Cl brings 3 electrons as an LX ligand. These dimers are thus not electron deficient, contrary to the complexes in which the bridges are hydride or alkyl ligands.

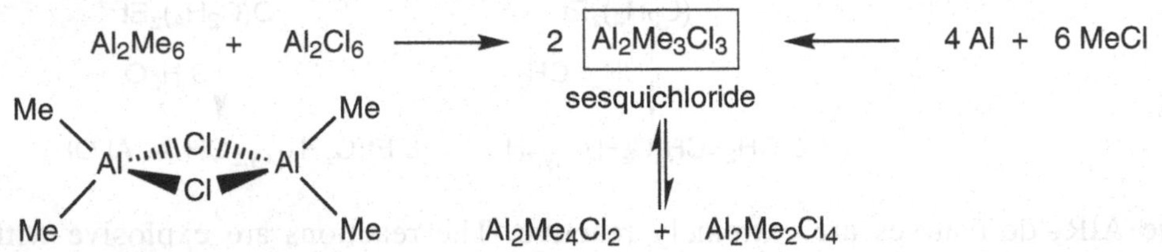

^{27}Al NMR allows, using chemical shifts, to determine the coordination number C of Al (110 to 130 ppm for C = 3, 140 to 180 ppm for C = 4 and 220 to 280 ppm for C = 5).

4.4. ORGANOGALLIUM, INDIUM AND THALLIUM [13.22]

The chemistry of these elements is much less developed than that of B, and even than that of Al. Since Ga-As layers serve as components for semiconductors, the corresponding organometallic precursors can lead to the formation of materials of quality by deposition of metal vapors. This relatively new technique should allow to avoid the traditional method that requires very high temperatures:

$$GaMe_3(g) + AsH_3(g) \xrightarrow{\;700\text{-}900°C\;} GaAs\,(s) + 3\,CH_4(g)$$

The binary trialkyl derivatives are prepared from metals or by alkylation of trihalides:

$$2\,M + 3\,HgMe_2 \longrightarrow 2\,MMe_3 + 3\,Hg$$
M = Ga, In

$$GaBr_3 + 3\,MeMgBr \longrightarrow GaMe_3.OEt_2 + 3\,MgBr_2$$

The compounds GeR_3 et InR_3 are monomeric and very air sensitive. They react with Lewis bases. The dialkylated derivatives are obtained by mild alkylation of MCl_3 or by action of an acid with a trialkylated derivative:

$$InCl_3 + 2\ MeLi \longrightarrow InMe_2Cl + 2\ LiCl$$

$$MMe_3 + HCN \longrightarrow MMe_2CN + CH_4$$
$$M = Ga,\ In$$

The di- and trialkylgallanes react with Lewis bases. Cations resistant to the hydrolysis of the metal-carbon bond can be obtained, which is not common:

$$GaMe_2Cl + 2\ NH_3 \longrightarrow [GaMe_2(NH_3)_2]^+\ Cl^-$$

$$GaMe_3 + 2\ H_2O \xrightarrow{\ aq.\ HCl\ } [GaMe_2(H_2O)_2]^+\ Cl^- + CH_4$$

On the other hand, the alkylthalliums are very air sensitive and can even decompose in an explosive way. They can be obtained by oxidative addition on Tl^I complexes and are strong oxidants.

$$Tl^IMe + MeI \longrightarrow Tl^{III}Me_2I$$

$$2\ Tl^{III}R_3 + 3\ Hg \longrightarrow 3\ HgR_2 + 2\ Tl$$

The cations TlR_2^+, obtained by hydrolysis, are very stable and adopt a linear geometry. They are isoelectronic to HgR_2 and SnR_2^{2+}. TlR_2OH is a strong base in aqueous solution.

$$TlR_2Cl \xrightarrow{\ H_2O\ } aq.\ TlR_2^+,\ Cl^-$$

The Tl^{III} derivatives are excellent electrophiles that are used in organic synthesis for reactions with olefins and aromatics. In particular, the trifluoroacetate $Tl^{III}(TFA)_3$ allows to carry out highly regioselective aromatic electrophilic substitution reactions in trifluoroacetic acid (see Chap. 17.3.2). It is probable that thallium toxicity will limit this use in the future.

4.5. THE π COMPLEXES

The pentahapto π Cp and Cp* complexes are known for Ga^{III}, In^I, In^{III} and Tl^I, with variable geometries. The geometry of $[InCp*]$ is an indium octahedron recalling $[(NiCp)_6]$ whereas $[AlCp*]$ is an Al_4 tetrahedron recalling $[Fe_4Cp_4(CO)_4]$.

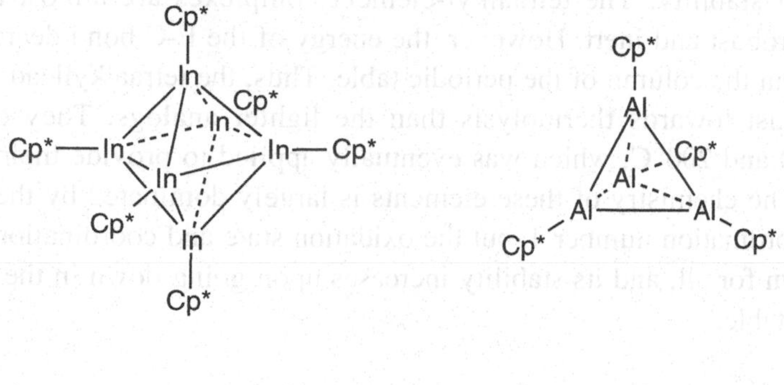

This stable tetrameric Al(I) complex $(AlCp^*)_4$ is prepared from Cp^*_2Mg and $AlCl$ or $[AlCp^*X(\mu\text{-}X)]_2$ and Na/K alloy. It reacts with Se, Te, Ph_2SiF_2 and Me_3SiN_3 to give cubes, polyhedrons and squares: [13.20]

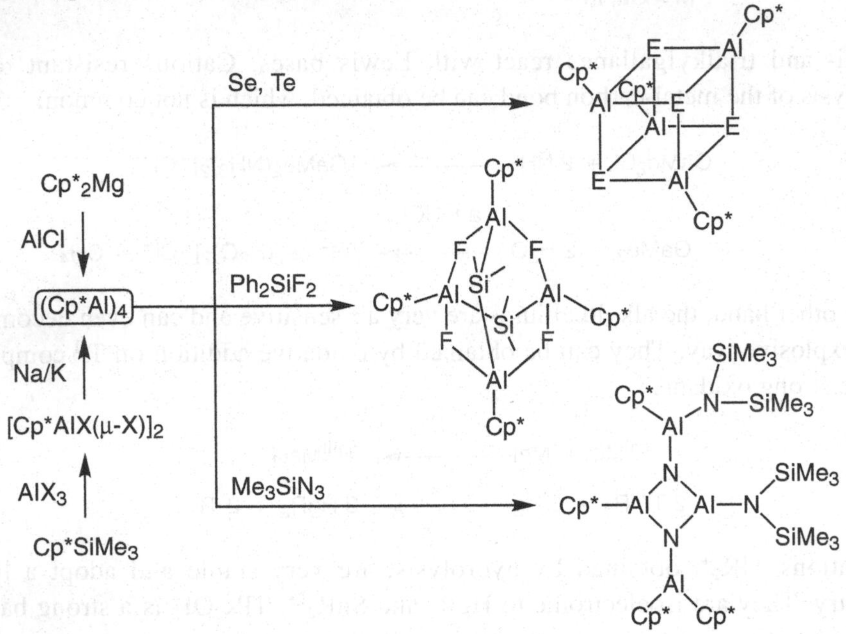

[TlCp] is very covalent and can be prepared in water, and sublimed and handled in air. It is currently used to transfer Cp to transition metals.

$$Tl_2SO_4(aq.) \;+\; 2\,C_5H_6 \;+\; 2\,NaOH \longrightarrow 2\,[TlCp] \;+\; Na_2SO_4 \;+\; 2\,H_2O$$

π Complexes involving weak interactions between $GaCl_3$ or $InBr_3$ and benzene or its methylated derivatives could be crystallized with various structures some of which are polymeric.

5. GROUP 14 (Si, Ge, Sn, Pb)

5.1. INTRODUCTION

These elements (E) have four valence electrons and are tetravalent as carbon below which they are located. The compounds ER_4 thus follow the octet rule which confers them a great stability. The tetraalkyl-element complexes are almost apolar and particularly robust and inert. However, the energy of the E-C bond decreases upon going down in the column of the periodic table. Thus, the tetraalkyllead complexes are less robust towards thermolysis than the lighter analogs. They decompose between 100 and 200°C, which was eventually applied to provide their antiknock properties. The chemistry of these elements is largely dominated by the oxidation state and coordination number 4, but the oxidation state and coordination number 2 is also known for all, and its stability increases upon going down in the column of the periodic table.

The industrial applications of the elements of group 14 are very important: 700 000 tons of silicones and 35 000 tons of organotin (stabilizer for plastics and fungicides) are produced every year, whereas the production of tetraalkylleads is dropping due to their toxicity.

5.2. ORGANOSILICON [13.23-13.32]

5.2.1. Formation and cleavage of the Si-C bond

The great industrial process for the production of alkylchlorosilanes is the heterogeneous Rochow-Müller process whose mechanism remains somewhat obscure. Essentially dialkyldichlorosilanes are produced in the mixture R_nSiCl_{4-n} (n = 1-4):

$$2\ RCl\ +\ Si\,/\,Cu\ \xrightarrow{\Delta T}\ R_2SiCl_2\ +\ ...$$
$$R = alkyl,\ aryl$$

Other classical methods (alkylation, hydrosilylation) can be used for special cases:

$$R_3SiCl\ +\ R'MgX\ \longrightarrow\ R_3R'Si\ +\ MgXCl$$

$$2\ R_2SiCl_2\ +\ LiAlH_4\ \longrightarrow\ 2\ R_2SiH_2\ +\ LiCl\ +\ AlCl_3$$

Tetramethylsilane is stable up to 700°C because of the very large dissociation energy of the Si-C bond (80 kcal·mol^{-1}, 334 kJ·mol^{-1}) that is comparable to that of the C-C bond (76 kcal·mol^{-1}, 318 kJ·mol^{-1}). It undergoes electrophilic attacks on carbon by strong acids whereas nucleophilic attacks occur on silicon, but the latter are much slower and require a very polar solvent.

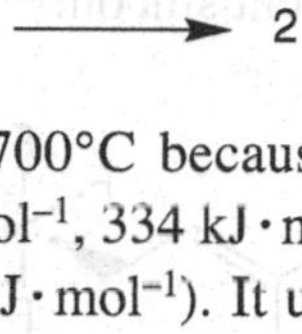

$$\underset{Si\text{---}CR_3}{\overset{Nu\quad El}{\underset{\delta+\quad\ \delta-}{}}}$$

$$SiMe_4\ +\ HCl\ \xrightarrow{AlCl_3}\ SiMe_3Cl\ +\ CH_4$$

$$SiMe_4\ \xrightarrow[-\ CH_4]{RO^-}\ SiMe_3OR$$

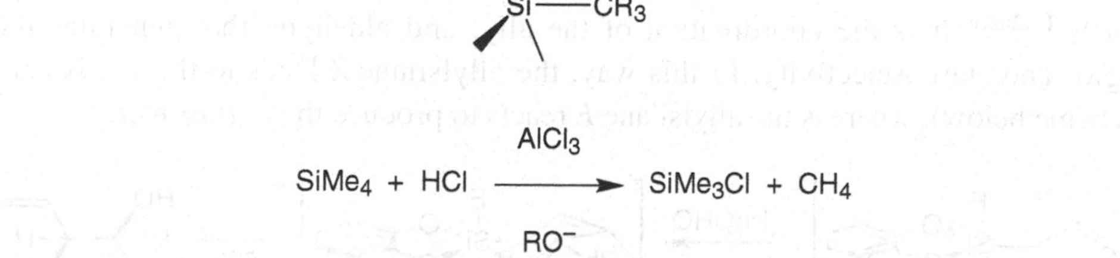

Nucleophilic reactions (F⁻ on Si)

In the above reaction, the attack of F⁻ on Si generates the allyl anion that reacts with the carbonyl. The alcoholate formed then displaces F⁻ in SiMe₃F. Since F⁻ is recycled, it can be used in catalytic quantities. The covalent radius of silicon (1.17 Å) is much larger than that of carbon (0.77 Å), and the bonds with the other atoms are much longer than with carbon (Si-C: 1.89 Å, C-C: 1.54 Å). This explains the steric capacities of silicon to reach high coordination numbers such as 5, 6 and even 7.[13.24] Many organosilicon reactions involve penta- or hexacoordinate silicon. Indeed, the presence of *d* orbitals of relatively low energy facilitates the access to pentavalent silicon (above example), which makes a clear distinction with carbon. Thus, nucleophilic substitutions at silicon always proceed according to the associative mechanism, the pentacoordinate intermediate in which the NVE of silicon is 10 being not of high energy.

Silicon is also more electropositive (Pauling electronegativity index: 1.8) than carbon (2.4), and the silicon-carbon bond is polarized $Si^{\delta+}$-$C^{\delta-}$. Tetravalent silicon, a weak Lewis acid, is thus essentially attacked by anions (F⁻, etc.) to form "hypervalent" silicon derivatives. This phenomenon is used for the steric control of reactions that proceed in the coordination sphere of silicon. For instance, fluorides and catecholate anions, among others, catalyze the stereoselective reaction between an allylsilane and an aldehyde leading to the corresponding homoallylic alcohols *via* an anionic intermediate of pentavalent silicon:[13.24c]

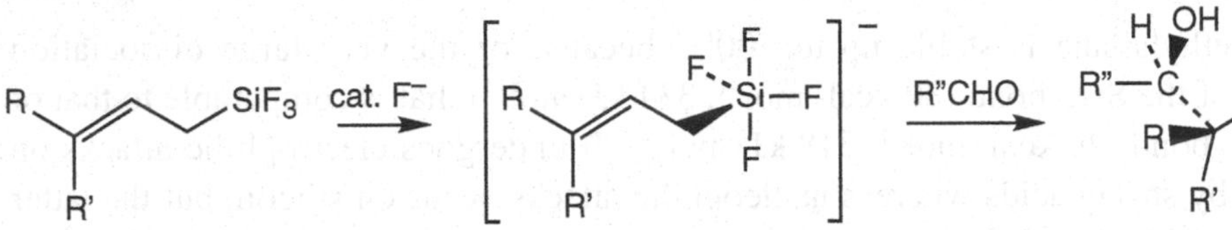

Corriu and his group have synthesized the first stable pentavalent silicon derivative (below). This compound also reacts with aldehydes leading to homoallylic alcohols regio- and stereoselectively *via* a transition state in which silicon is hexacoordinate.[13.24a,b] It is the coordination of the allyl and aldehyde that generates the regio- and stereoselectivity. In this way, the allylsilane Z leads to the *syn* isomer (scheme below), whereas the allylsilane E reacts to produce the *anti* isomer.

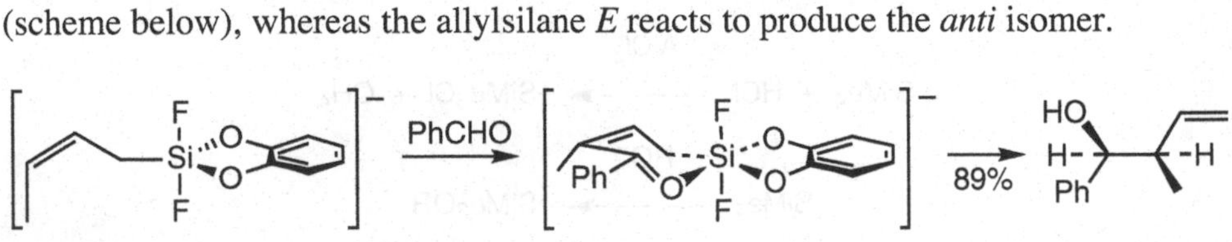

counter-cation: Et₄N⁺   *syn / anti*: 95 / 5

It sometimes happens that the hypervalent silicon intermediate lacks stereochemical rigidity and thus can racemize by pseudo-rotation. Thus, nucleophilic reactions do not always proceed with retention.

The polarity of the Si-C bond can be more or less marked depending on the substituents. For instance, electrophilic and nucleophilic attacks on aryl- and vinylsilanes are much more facile than with alkylsilanes, and they give applications in organic synthesis.[13.23,13.35]

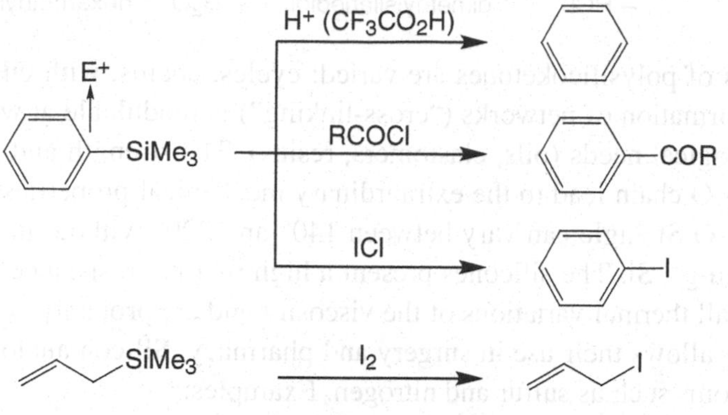

Electrophilic reactions (C on E⁺)

The SiMe$_3$ group can thus be considered as a "super-proton". Another property that is used in synthesis is the substitution of the trialkylsilyl groups by OH upon oxidation. Thus, in the course of his rapid synthesis of estrone, Vollhardt synthesized aryltrimethylsilanes by condensation of a diyne on bis-trimethylsilylacetylene, then he used one of the SiMe$_3$ groups for a regioselective electrophilic aromatic substitution as in the above scheme. The second SiMe$_3$ group was used to introduce the OH group by oxidation, i.e. to generate a phenol from a trimethylarylsilane (see the overall synthetic scheme Chap. 19). The oxidation of a trialkyl group is stereospecific, which brings an additional interest of this reaction in organic synthesis.[13.25]

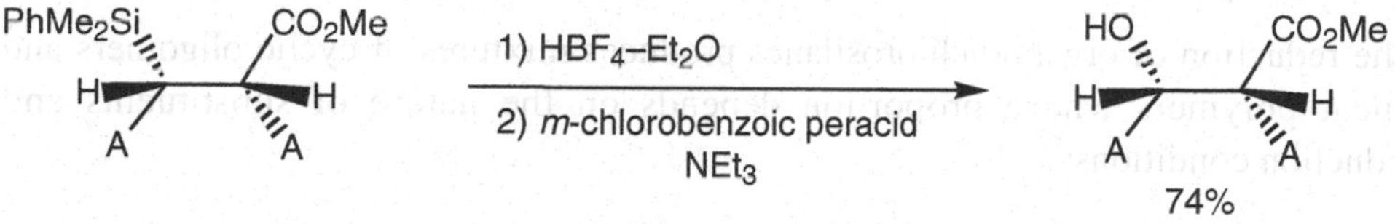

In general, electrophilic reactions of β-hydroxyalkyl-, vinyl-, allyl-, crotyl-, homoallyl- and dienylsilanes proceed with control of the stereochemistry of the olefinic systems formed, which is a great advantage for the synthesis of precursors of natural products such as pheromones and vitamins.[13.25]

5.2.2. Silicones

The hydrolysis of chlorosilane is thermodynamically favorable, because the energy of the Si-Cl bond (91 kcal·mol⁻¹, 381 kJ·mol⁻¹) cleaved by hydrolysis is smaller than that of the Si-O bond formed (108 kcal·mol⁻¹, 452 kJ·mol⁻¹). As for the formation of silica (-O-Si-O-)$_n$, it is this large energy of the Si-O bond that is the driving force for the dehydratation of silanols to siloxanes and of silanediols to polysilicoketones (silicones).[13.26]

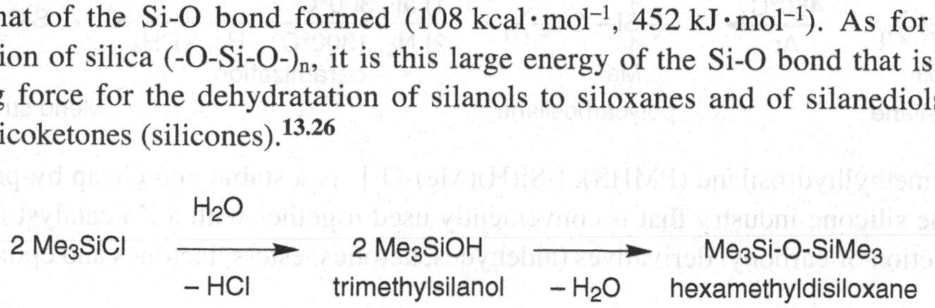

$$n\ Me_2SiCl \xrightarrow[-\ HCl]{H_2O} n\ \underset{\text{dimethylsilanediol}}{Me_2Si(OH)_2} \xrightarrow[-\ H_2O]{} \underset{\text{hexamethyldisiloxane}}{(Me_2Si\text{-}O\text{-})_n}$$

The structures of polysilicoketones are varied: cycles, chains, with OH or $OSiMe_3$ ending. The formation of networks ("cross-linking") is modulable at will according to the technological needs (oils, elastomers, resins). The strength and flexibility of the $O\text{-}Si(Me)_2\text{-}O$ chain lead to the extraordinary mechanical properties of silicones. Indeed, the Si-O-Si angle can vary between 140° and 220° without involving more than $1\ kJ \cdot atom\text{-}g^{-1}$ Si. The silicones present a high thermal resistance to corrosion, show only small thermal variations of the viscosity and are probably non-toxic. This latter property allows their use in surgery and pharmacy. Silicon analogs exist with other heteroatoms such as sulfur and nitrogen. Examples:

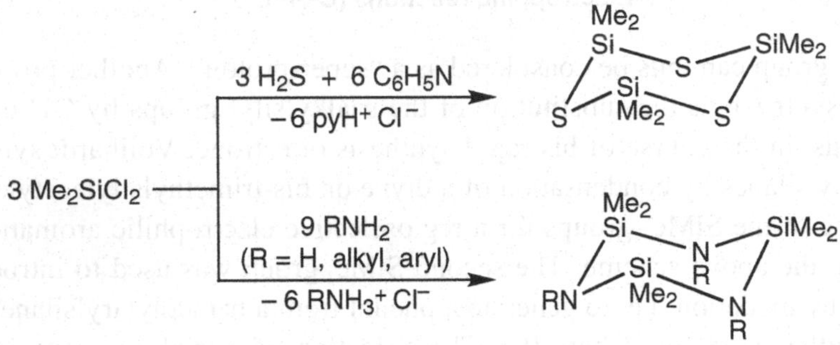

5.2.3. Polyorganosilanes and polycarbosilanes

The reduction of organodichlorosilanes produces mixtures of cyclic oligomers and linear polymers whose proportion depends on the nature of substituents and reduction conditions:

$$Me_2SiCl_2 \xrightarrow[-\ NaCl]{2\ Na/K} \underset{\text{rings: n = 5-7}}{(Me_2Si)_n} + \underset{\text{chains: m <100}}{Me(Me_2Si)_mMe}$$

The polyorganosilanes serve as precursors for the fabrication of ceramic fibers made of β silicon carbide. In a first step, low-molecular-weight (around 8000) carbosilanes are thermolyzed, then "ceramization" is achieved (Yajima process):

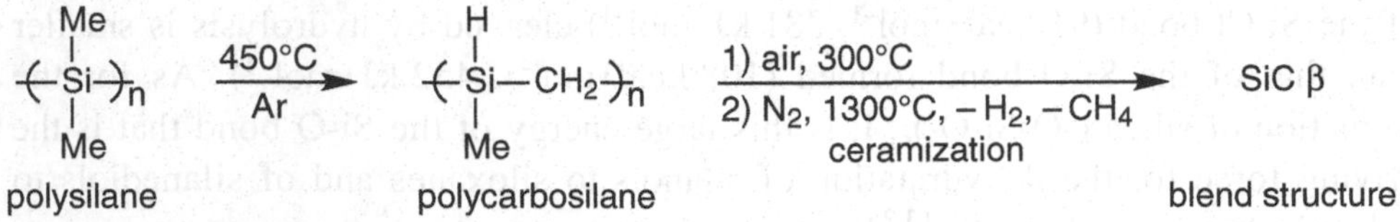

Polymethylhydrosilane (PMHS), $[\text{-}Si(H)(Me)\text{-}O\text{-}]_n$ is a stable and cheap by-product of the silicone industry that is conveniently used together with a Zn catalyst for the reduction of carbonyl derivatives (aldehydes, ketones, esters, lactones and epoxides).

It has been developed by Firmenich with the Venpure™ process. For instance, tri-glycerides are reduced to fatty alcohols, using this hydride in ton scale for the cosmetic industry.[13.27]

5.2.4. Silylenes

The silylenes SiR_2, analogous to carbenes CR_2, are supposed to be intermediates in the reduction of organodichlorosilanes to polycarbosilanes($-SiR_2-)_n$ (above).[13.28] They are also obtained by irradiation in the ultraviolet region. They can add to olefinic double bonds or insert into a Si-OR' bond. They spontaneously dimerize (West: Si=Si bond [13.29]), trimerize or polymerize depending on the bulk of the substituents.[13.28]

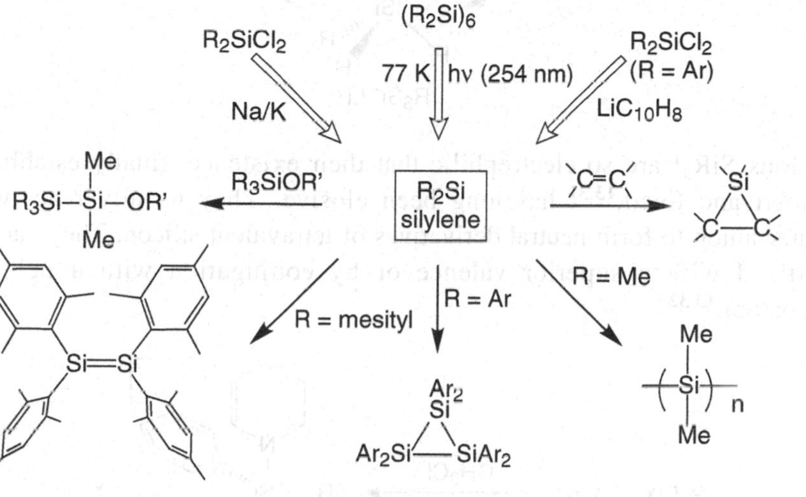

The only stable silylene known is $SiCp*_2$.[13.30] Silicocene is diamagnetic and shows a perfect sandwich structure with axial symmetry.

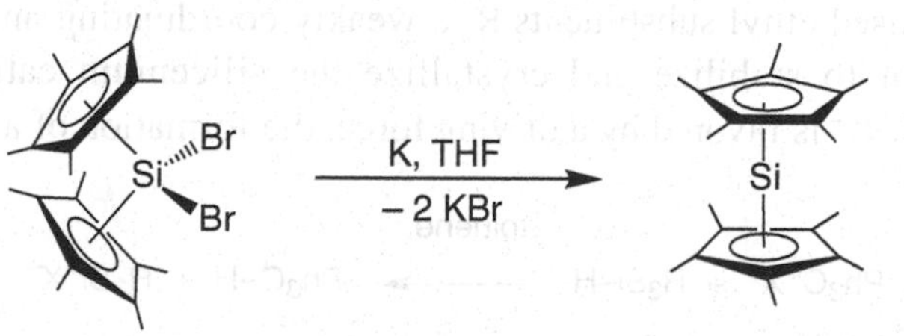

As the Si=Si bond, the Si=C bond is not stable if it is not sterically protected. The electronic stabilization can also be introduced by a 3-center delocalization analo-gous to that found in Fisher-type metal-carbene complexes.

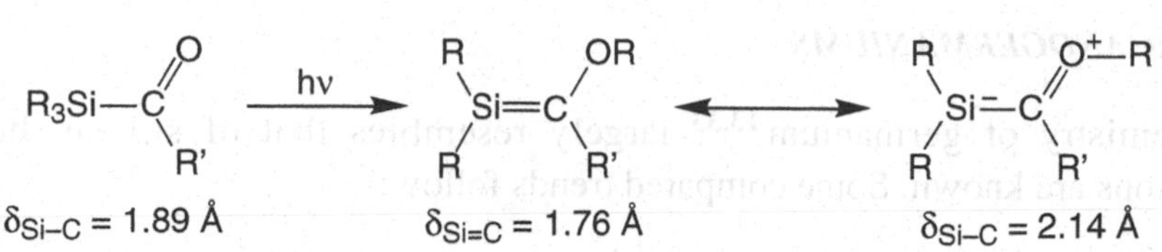

$\delta_{Si-C} = 1.89$ Å $\delta_{Si=C} = 1.76$ Å $\delta_{Si-C} = 2.14$ Å

5.2.5. The anions SiR₃⁻, the radicals SiR₃˙ and the cations SiR₃⁺

▸ The anion SiMe₃⁻ and the radical SiMe₃˙ both have a pyramidal structure, and they are sources of ligands. The radicals $R_3Si˙$ dimerize more easily than their carbon homologs $R_3C˙$ because the Si-Si bond formed is much longer than the C-C bond. For instance, with R = t-Bu, $R_3C˙$ remains monomeric whereas $R_3Si˙$ dimerizes, forming a Si-Si bond whose length is 2.70 Å.

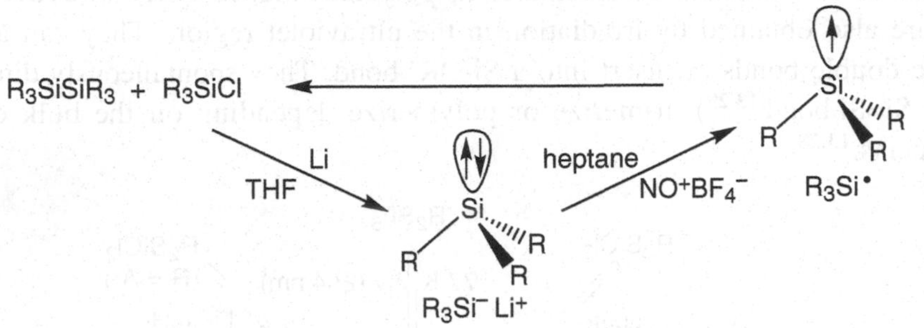

▸ The ions SiR₃⁺ are so electrophilic that their existence, finally established by Lambert and Reed,[13.31] has long been elusive. They usually react with the counter-anion to form neutral derivatives of tetravalent silicon. They can also be stabilized with a superior valence or by conjugation with a neighboring heteroatom.[13.32]

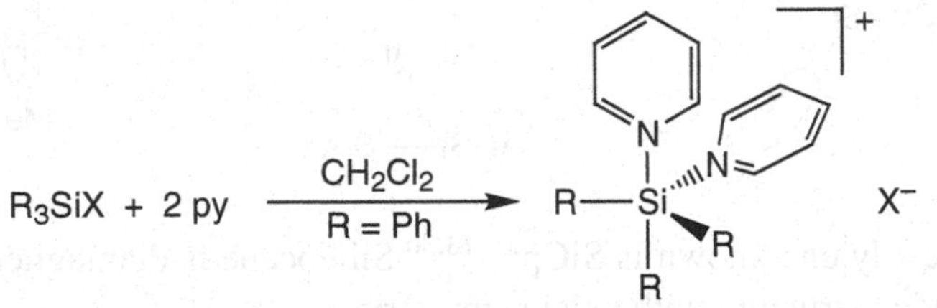

Lambert has used ethyl substituents R, a weakly coordinating anion and toluene as the solvent to stabilize and crystallize the silicenium cation SiR₃⁺. The synthesis of SiR₃⁺ is favored by a driving force, the formation of a C-H bond.[13.31]

$$Ph_3C^+X^- + R_3Si\text{-}H \xrightarrow{\text{toluene}} Ph_3C\text{-}H + R_3Si^+X^-$$

In the solid state, the X-ray diffraction structure shows that Et_3Si^+ is shifted from planarity by only 15°, that the anion is not coordinated and that only toluene is weakly coordinated, but without charge transfer.[13.31]

5.3. ORGANOGERMANIUMS

The chemistry of germanium[13.33] largely resembles that of silicon, but few applications are known. Some compared trends follow:

▸ The [GeR₃X] derivatives are much less reactive than their silicon analogues because of the reduced ability of germanium to use its d orbitals for penta-coordination.

▸ The germanes [Ge(R)H₃] are so weakly reactive, because of the completely apolar Ge-H bond, that they can be prepared in water:

$$GeMeBr_3 \xrightarrow[H_2O]{NaBH_4} GeMeH_3$$

▸ The hydrogermanation reactions are easier than the hydrosilylation reactions:

$$GePh_3H + CH_2{=}CHPh \xrightarrow{120°C} GePh_3(CH_2CH_2Ph)$$

$$HC{\equiv}CH + GeHCl_3GeHCl_3 \longrightarrow Cl_3GeCH_2{-}CH_2GeCl_3$$

▸ It has been shown that the polygermylenes are accessible from alkoxygermanes by α-elimination reaction of the alcohol:

$$GeRH_2OMe \xrightarrow[- MeOH]{\Delta T} \text{"RGeH"} \longrightarrow (-RGe^-)_n$$

Contrary to [SiCp*₂], [GeCp*₂] shows a bent structure as the Sn and Pb analogs. One of the Cp* ligands can be removed by protonation to give [GeCp*]⁺ (this same reaction is also known for Sn and Pb). These cations [MCp*]⁺ (M = Ge, Sn, Pb) can also be considered as clusters following Wade's theory (see Chap. 2) with the adequate number of electron pairs corresponding to the nido structure:

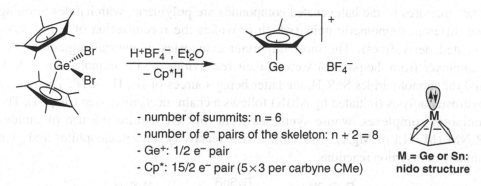

- number of summits: n = 6
- number of e⁻ pairs of the skeleton: n + 2 = 8
- Ge⁺: 1/2 e⁻ pair
- Cp*: 15/2 e⁻ pair (5×3 per carbyne CMe)

M = Ge or Sn:
nido structure

▸ EPR allows to determine the geometry of the GeR₃• radicals using the hyperfine coupling constant of the ⁷³Ge nucleus (7.6%, I = 9/2) that is proportional to the s character of the orbital containing the unpaired electron (SUMO). This s character shows the sp^3 hybridization, thus the pyramidal structure; the latter is confirmed by the retention observed in the homolytic substitution reactions of germanes provoked by radicals.

5.4. ORGANOTINS

The applications of organotins are multiple.[13,34] The compounds R_2SnX_2 are used as stabilizers of polyvinylchloride (PVC). They act by substituting the chlorine atom in the reactive regions of the polymer, which suppresses the release of HCl during the thermal process at 180-200°C. In the continuation of this process, the fixed organotin derivatives also are UV stabilizers because of their ability to quench radical reactions. The compounds of the type SnR_3X are also used as biocides, fungicides, antibactericide disinfectants and as additive protecting paints. The toxicity of organotin reagents increases with the degree of alkylation and decreases with the alkyl chain length. The molecular chemistry of Sn benefits from two useful tools related to the ^{119}Sn nucleus: NMR (in solution) and Mössbauer spectroscopy (in the solid state). Both techniques have parameters that are very sensitive to the tin coordination varying between 2 and 6.

The main mode of synthesis of organotin compounds are the following:

$$4\ RMgX + SnCl_4 \longrightarrow SnR_4 + 4\ MgXCl$$

$$SnR_4 \xrightarrow[-\ RBr]{Br_2} SnR_3Br \xrightarrow[-\ RBr]{Br_2} SnR_2Br_2$$

$$SnR_4 + HX \longrightarrow SnR_3X + RH$$

$$Sn/Cu + 2\ MeCl \xrightarrow{200\text{-}300°C} SnMe_2Cl_2$$

$$SnCl_2 + RCl \xrightarrow{cat.\ SbCl_3} SnRCl_3$$

The structures of the halogenated compounds are polymeric, with halides bridging the different monomeric units (which provokes the racemization of chiral halogenated derivatives). The most important categories of neutral organotin (IV) complexes from the point of view of their reactivities are the monohalides SnR_3X and the monohydrides SnR_3H, the latter being sources of H⁻, H⁺ or H⁻. The radical hydrostannolysis (initiated by AIBN) follows a chain mechanism (see Chap. 4). The aminotin complexes, whose syntheses from SnR_3Cl require the use of amides R_3NM (M = Li or MgX), also show a rich reactivity with nucleophiles and give numerous insertion reactions.

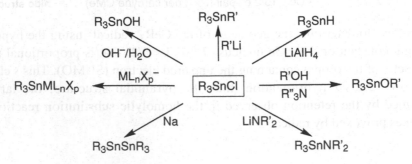

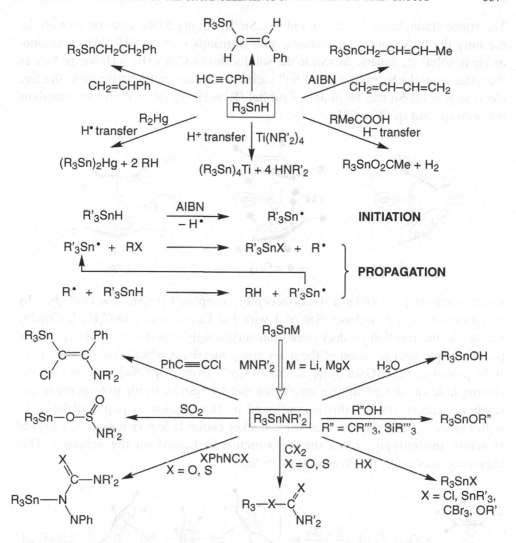

The anions SnR₃⁻ are excellent sources of SnR₃ ligands for transition metals. The radicals SnR₃• have, as GeR₃• and PbR₃•, a pyramidal structure. In the case of SnR₃•, this structure is deduced by EPR from the coupling constant with ^{119}Sn that indicates a strong 5s character of the orbital containing the single electron.

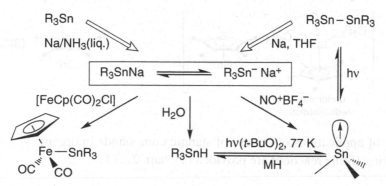

The stable stannylenes, i.e. the complexes $Sn^{II}R_2$ having a free electron pair on Sn, are only the bulky ones, for instance with R groups such as $CH(SiMe_3)_2$ (monomeric in solution, dimeric Sn=Sn in the solid state) or $C_5R'_5$ (R' = H, Me or Ph). In the latter case, the hybridization of Sn^{II} varies depending on the steric bulk: the free electron pair on Sn can be in an sp^2 orbital (R' = H), sp (R' = Ph) or intermediate between sp^2 and sp (R' = CH_3).

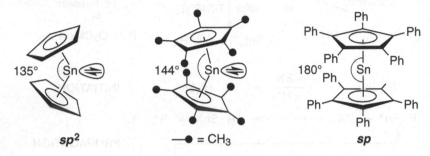

Upon attempting to make a donor-acceptor complex $Cp*_2Sn \longrightarrow Ga(C_6F_5)_3$ by treatment of the Lewis base $[SnCp*_2]$ with the Lewis acid $[Ga(C_6F_5)_3]$, Cowley found that the reaction product somewhat surprisingly was the $[Ga(C_6F_5)_4]^-$ salt of the first triple-decker cation of the main-group metals, $[Cp*Sn(\mu^2,\eta^5-Cp*SnCp*]^+$. It is probable that $[Ga(C_6F_5)_3]$ abstracts $Cp*^-$ from $[Cp*_2Sn]$ to generate the electrophilic cation $Cp*Sn^+$ that attacks another $Cp*_2Sn$ molecule to form the triple-sandwich cation. Indeed, this is confirmed by the reaction of $[Cp*Sn][B(C_6F_5)_4]$ with $Cp*_2Sn$ giving the expected triple-decker cation below (whose X-ray crystal structure intriguingly shows the *cis* structure indicated on the scheme). This chemistry also works for Pb as well as for Sn: [13.36]

Examples of applications of the use of stannic compounds in organic synthesis for C-C coupling (Stille reaction) are provided in Chap. 21.3.5.

5.5. ORGANOLEADS

This chemistry is dominated by the tetraalkyllead compounds $[PbR_4]$ (for instance, the derivatives $[PbRX_3]$ are rare). Tetraethyllead has been used since 1922 in gasoline as an anti-Knock reagent at a concentration of 0.01%. The action of lead consists in deactivating hydroperoxides by PbO and to end-chain radical reactions of the combustion process using radicals produced by thermolysis of $[PbEt_4]$:

$$PbEt_4 \longrightarrow PbEt_3^{\bullet} + Et^{\bullet}$$

$$PbEt_4 + Et^{\bullet} \longrightarrow PbEt_3CH_2CH_2^{\bullet} + C_2H_6 \text{ etc.}$$

Lead oxide is eliminated from the combustion chamber using $BrCH_2CH_2Br$ that transforms PbO into volatile lead derivatives. Lead-loaded gasoline has disappeared, being replaced because of its toxicity. Indeed, as for Hg and Sn, the high toxicity is due to the cations $[PbR_2]^{2+}$ and $[PbR_3]^{+}$ that are generated by biomethylation of inorganic lead. The $[PbR_3]^{+}$ cation inhibits the oxidative phosphorylation and the action of glutathione transferase. $[PbR_2]^{2+}$ blocks the enzymes containing thiol groups. Pb^{2+} accumulates in bones, in which it takes the place of calcium and cannot be removed by using chelating therapies.

For the identification of organolead derivatives, ^{207}Pb NMR is used (23%, $I = 1/2$), with chemical shifts covering a range of 1300 ppm. The coupling constants ^2J (^{207}Pb, ^1H) deduced from the ^{207}Pb or ^1H spectra (satellite lines due to ^{207}Pb) vary between 60 and 160 Hz depending on the s character of the lead orbital that is engaged with R. The oxidation state II dominates for inorganic lead compounds, but not for organolead compounds. The organolead (II) derivatives $[PbR_2]$ disproportionate to Pb^{IV} and Pb. For instance, the loss of metallic lead is avoided by addition of MeI or other alkylating agents during the syntheses of tetramethyllead below:

Laboratory: $2\,PbCl_2 + 4\,MeMgI \longrightarrow 2\,PbMe_2 + 4\,MgICl$

$$\downarrow$$

$$PbMe_4 + Pb$$

Industry: $4\,MeCl + 4\,NaPb \xrightarrow{\text{HMPA}} PbMe_4 + 3\,Pb + 4\,Na^+Cl^-$

In the same way, the diplumbanes, accessible at low temperature, dismutate at 20°C:

$$12\,RMgX + 6\,PbCl_2 \xrightarrow[-20°C]{Et_2O} \boxed{2\,R_3Pb\text{–}PbR_3} + 2\,Pb + 12\,MgXCl$$

$$\downarrow 20°C$$

$$3\,PbR_4 + Pb$$

The chemistry of organolead compounds [13.34b] is distinct from those of other group 13 elements by the following characteristics:

▸ Coordination numbers very often vary between 2 (for $[PbR_2]$, for instance R = $CH(SiMe_3)_2$) and 8 in $[PbPh_2(acac)_3]^-$.

▸ $[PbR_2(OH)_2]$ is amphoteric, giving $[PbR_2(OH)_3]^-$ at a pH higher than 10 and $[PbR_2(aq)]^{2+}$ at a pH lower than 5.

▸ Contrary to tin hydrides that are much used, lead hydrides $[PbR_3H]$ are instable and also very air- and light sensitive, decomposing to H_2 and diplumbanes $[R_3Pb\text{-}PbR_3]$. Their thermal stability increases with the length of the alkyl chain; they are very reactive (for instance, they reduce alkyl halides and carbonylated derivatives).

▸ Diplumbanes, in spite of a relatively weak Pb-Pb bond, do not dissociate to radicals $PbR_3{}^{\bullet}$, even with R = mesityl, because the Pb-Pb bond is very long. On the other hand, it is possible to achieve the oxidative addition of the Pb-Pb bond using an electron-rich metal center such as Pt^0:

$$[Pt(PPh_3)_4] + Ph_3Pb\text{–}PbPh_3 \longrightarrow [trans\text{-}Pt(PPh_3)_2(PbPh_3)_2] + 2\,PPh_3$$

▸ Pb^{II} does not always give Pb^{IV}: whereas $[SnCp_2]$ reacts with MeI by oxidative addition to give $[Sn^{IV}Cp_2(Me)I]$, the substitution of a Cp ligand by I occurs in $[Pb(\eta^5\text{-}Cp)_2]$:

$$Pb(OR)_2 + 2\,CpH \xleftarrow{\;ROH\;} \boxed{[Pb(\eta^5\text{-}Cp)_2]} \xrightarrow{\;MeI\;} [Pb(\eta^5\text{-}Cp)I] + CpMe$$

6. GROUP 15 (As, Sb, Bi)

6.1. INTRODUCTION

With 5 valence electrons, the elements E of this group show a duality of oxidation states (III and V) as well as a facile access to charged complexes. The result is a rather rich chemistry with various coordination numbers.[13.35,13.36] There is no important application of this chemistry presently, however, due to the toxicity of these elements. The arsines, stilbines and bismutines ER_3 are good σ donors and modest π acceptors, but this ligand property decreases upon going down in this column of the periodic table. The energy of the E-C, E-H, E-metal and E-E bonds also concomitantly decreases, as for the groups 13 and 14, with the usual consequences on the bond energies of the compounds. The E-C and E-H are weakly polar. The NMR of ^{75}As (100%, I = 1/2) and especially Mössbauer spectroscopy of ^{121}Sb are useful for the investigation of structures and stereoelectronic effects. The main structural types that do not have E-E bonds are the following:

Binary organoelements of group 15 (E = As, Sb, Bi)

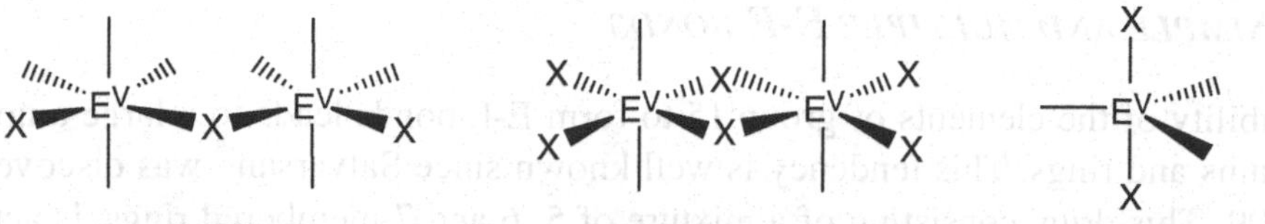

R_4EX R_2EX_3 (except Bi) R_3EX_2
(ex.: Me_4SbF)   (ex.: Ph_2SbCl_3)   (ex.: Me_3AsCl_2)

Ternary organoelements of group 15 (E = As, Sb, Bi)

Given the similarities between the two metalloids As and Sb and the metal Bi, this family is now being treated together rather than by element.

6.2. POLYALKYL COMPOUNDS [ER₅] AND [ERₙX₅₋ₙ]

The pentaalkyl derivatives $[ER_5]$ are thermally stable.[13.35] They cannot really be obtained from the pentahalides $[EX_5]$ that are too strong oxidants, but only *via* the trialkyls ER_3 (for instance, by oxidative addition of X_2 or RX followed by addition of a carbanion). The stability of the derivatives $[R_nEX_{5-n}]$ decreases with n, the decomposition products being E^{III} derivatives and most often RX (the reverse of synthesis reactions).

6.3. OXYGENATED ORGANOELEMENT COMPOUNDS

The hydrolysis of $[AsR_4Cl]$ leads to strong bases $[AsR_4]^+OH^-$. In this family, $[AsMe_3CH_2COO]^-$ has been detected in the abdominal muscle of lobsters. The arsonic and antimonic acids $[ERO(OH)_2]$ are among the most well-known compounds of this family since the discovery, in 1905, of the therapeutic effects of $[As(p-H_2NC_6H_4O)(OH)O]Na$ (Atoxyl) against sleeping sickness. Some of these acids (R = Ar) are still used as herbicides, fungicides and bactericides. Being of average strength, they are resistant to oxidation and form polymeric anhydrides $[(AsRO_2)_n]$. The arsinic acids $[AsR_2O(OH)]$, prepared by hydrolysis and/or oxidation of $[AsR_2Cl]$, form dimeric structures analogous to those of carboxylic acids and are amphoteric; the analogs $[SbR_2O(OH)]$, polymeric, are obtained by hydrolysis of $[SbR_2Cl_3]$.

6.4. HYDRIDES [ERₙH₃₋ₙ]

Of increasing stability upon going down in the group 15, they are very air sensitive (producing oxides), but not water sensitive given the absence of polarity of the E-H bonds.

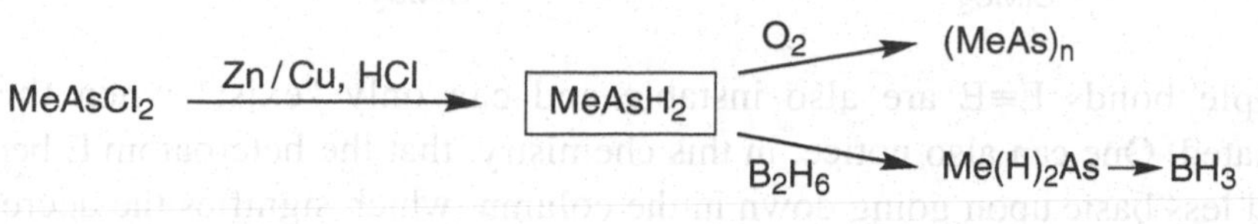

6.5. SIMPLE AND MULTIPLE E-E BONDS

The ability of the elements of group 15 to form E-E bonds leads to a large amount
of chains and rings. This tendency is well known since Salvarsan® was discovered
in 1909. This drug, consisting of a mixture of 5, 6 and 7-membered rings, is active
against the microorganism causing syphilis and was used till the advent of the
antibiotics.

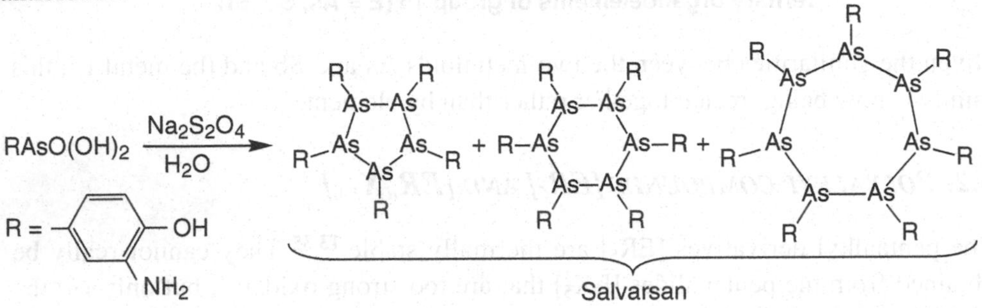

The analogs of hydrazines R_2N-NR_2 and diphosphines R_2P-PR_2 are the diarsanes,
distilbanes et dibismutanes. Tetramethyldiarsane is one of the main components of
the smoking liquid prepared by Cadet in 1760 by reaction of CH_3COOK with
As_2O_3. The weakness of the As-As bond makes it reactive, for instance towards
insertion reactions, which provides an ideal source of the ligand $AsMe_2$ for main-
group and transition metals. Tetramethyldistilbane is pyrophoric, thermochromic,
and forms chains in the solid state.

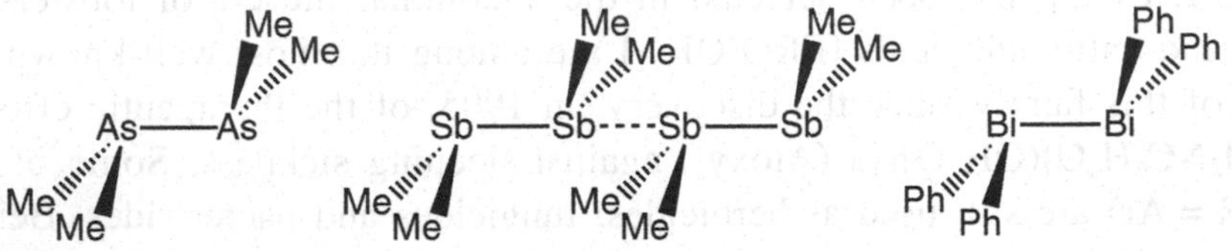

The analogs of the azo RN=NR and diphosphene RP=PR derivatives that contain a
double E=E bond are all the more instable as one goes down in the periodic table.
Yet, Cowley and his group were able to synthesize diarsene and distilbene by
reduction of $(SiMe_3)_2RCECl_2$ (R = $SiMe_3$ or H, respectively). The latter requires the
protection of the π bond by complexation: [13.36]

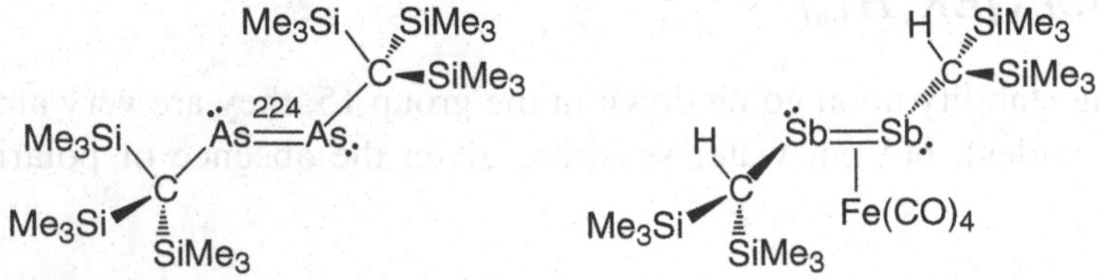

The triple bonds E≡E are also instable and can only "exist" when they are
coordinated. One can also notice, in this chemistry, that the heteroatom E becomes
less and less basic upon going down in the column, which signifies the decrease of
the *p* character of its free electron pair.

Other exotic structures are accessible by reduction of [AsMeI$_2$] (ladder) or upon demethylation of [cyclo-(AsMe)$_5$] (triple sandwich) by reaction with [MoCp(CO)$_3$]$_2$.

--- : bond order = 0.5
semi-conductor

Mo=Mo distance = 276 pm
analoguous to that of [Cp*Mo(cyclo-P$_6$)MoCp*]

6.6. HETEROCYCLES

The pyridine and phosphabenzene analogs are known for all the elements of group 15. Their stability decreases, however, upon going down in the periodic table, and stilba- and bismutabenzene are in equilibria with their dimers.

elementabenzene
E = P, As, Sb, Bi

E = Sb, Bi

In the same way, arsoles, analogous to pyrroles and phospholes are known. Here again, the aromaticity decreases together with the p_p-p_p interaction, i.e. with the stability of the double bond E=C that also decreases upon going down in the column. In arsole, As shows a pyramidal structure and its free electron pair is localized. Whereas pyrrole has an aromatic character (6 π-electron system), phosphole and arsole do not, and are diolefins. Stable enantiomers are not isolable for phosphole, however, whereas they are with arsole.

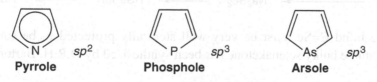

Pyrrole   Phosphole   Arsole

7. GROUP 16 (Se, Te)

Below oxygen and sulfur, two elements of group 16, selenium and tellurium, are usually considered in the context of organoelement chemistry (polonium, below tellurium in this column, is a famous powerful emitter of α radioactivity). The importance of selenium in biological processes is recognized; it is a toxic element if one absorbs more than 1 μg/g food (selenite SeO_3^{2-}, selenate SeO_4^{2-} and the element Se are more toxic than $SeMe_2$). At the same time, it is also indispensable for life, however, playing the role of an antioxidant in the glutathione peroxidase enzyme; selenium deficiency creates a syndrome. The analysis of organoselenate compounds is favored by ^{77}Se NMR (7.5%, I = 1/2), the chemical shifts covering a range of 2500 ppm.

Selenium and tellurium are also well known for the physical properties of materials that contain them. The perchlorate salt of the bis(tetramethyltetraselenafulvalene) radical cation becomes a superconductor at 1.5 K, and the analogous derivative with tellurium is also known [NOTA BENE: *do not use perchlorate salts of any kind since they dangerously and randomly explode; a number of other inorganic anions are safe and available*].

tetraselenafulvalene (E = Se)
tetratellurafulvalene (E = Te)

Finally, thin layers of CdTe, obtained by decomposition of [CdMe$_2$] and [TeEt$_2$] using the codeposition of metal vapors, have promising properties as semi-conductors for micro-electronics. Selenium has become important in organic synthesis because of the *syn*-elimination reactions of SeROH (synthesis of enones from carbonyl derivatives, conversion of epoxides and alkenes to allylic alcohols, see Chap. 19). Selenium chemistry very much resembles sulfur chemistry; it is dominated by the oxidation state II, the degree IV being present in inorganic compounds and oxygenated organoseleniate reaction intermediates.

The double bond C=Se must be very well sterically protected to be stable. For instance, a stable bulky selenaketone has been synthesized by D.R.H. Barton: [13.37]

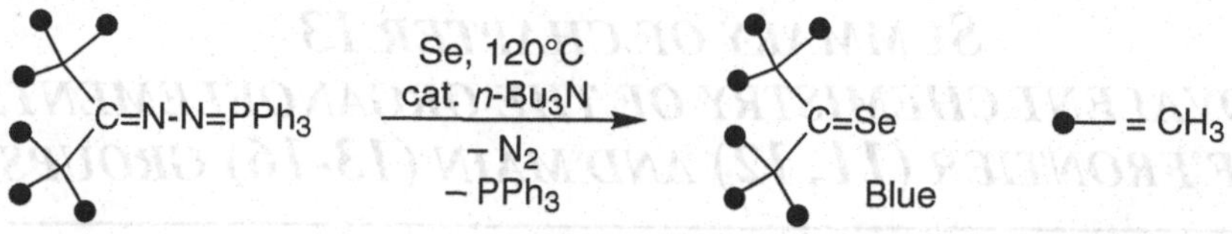

Although C=Se polymerizes at –160°C, it is stabilized as a ligand in complexes:

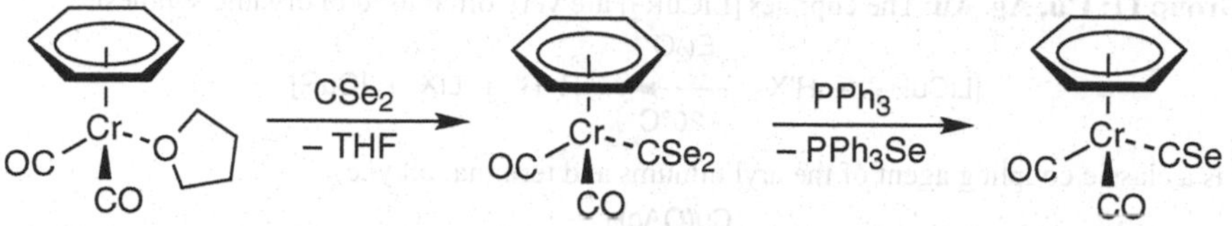

The analogs of furan and thiophene are easily accessible and stable (examples below). The order of aromaticity is:

benzene > thiophene > selenophene > tellurophene > furan.

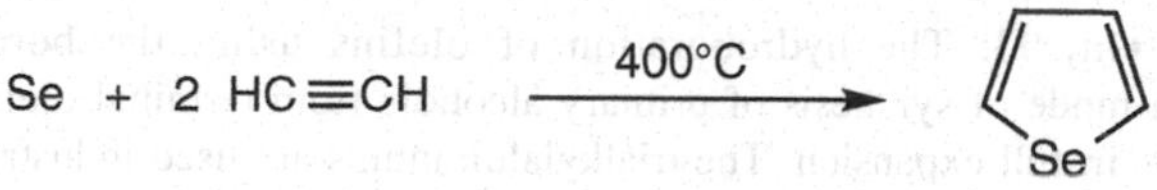

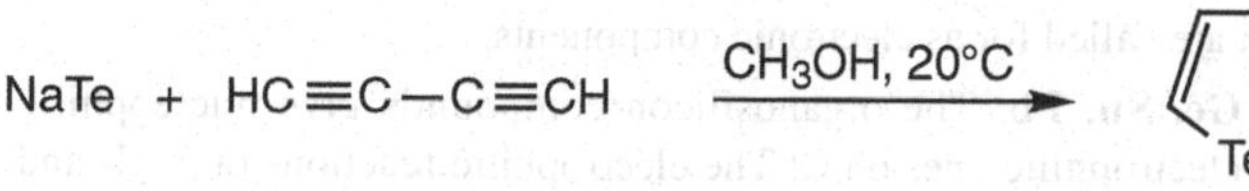

$$Se \; + \; 2 \; HC\equiv CH \xrightarrow{\;400°C\;}$$

$$NaTe \; + \; HC\equiv C-C\equiv CH \xrightarrow{\;CH_3OH,\; 20°C\;}$$

SUMMARY OF CHAPTER 13
COVALENT CHEMISTRY OF THE ORGANOELEMENTS OF FRONTIER (11, 12) AND MAIN (13-16) GROUPS

The general property of the groups 11 to 16 is the covalency (very weak polarity) of the E-C bonds.

1 - Group 11: Cu, Ag, Au: The cuprates $[LiCuR_2]$ are very often used in organic synthesis:

$$[LiCuR_2] + R'X \xrightarrow[-20°C]{Et_2O} R-R' + LiX + [CuR]$$

Cu^{II} is a classic coupling agent of the aryl lithiums and terminal alkynes.

$$2\ RC{\equiv}CH \xrightarrow[pyridine]{Cu(OAc)_2} RC{\equiv}C-C{\equiv}CR$$

The derivatives $[Au^{I}L(Cl)]$ and the complexes with M-M bonds (Ag, Au) are numerous.

2 - Group 12: Zn, Cd, Hg: The dialkylmetals ZnR_2, CdR_2 and HgR_2 are mild alkylating agents, but they are toxic, especially HgR_2.

3 - Group 13: B, Al, Ga, Tl: The hydroboration of olefins using the boranes BHR_2 (R = siamyl) is the main mode of synthesis of primary alcohols from terminal olefins, and the organoboron chemistry is in full expansion. The trialkylaluminums are used industrially for the dimerization and the polymerization of olefins. Ga and In organometallics are precursors to ultrapure metals that are called for as electronic components.

4 - Group 14: Si, Ge, Sn, Pb: The organosilicon compounds give nucleophilic additions on Si and electrophilic ones on C: The electrophilic reactions on aryl- and vinylsilanes lead to applications in synthesis. The high energy of the Si-O bond (108 kcal·mol^{-1}; 452 kJ·mol^{-1}) and the variety of polymeric structures $(R_2Si-O-)_n$, the polysilicoketones or silicones, are at the origin of the development of these materials that are biocompatible at the industrial level (surgery, pharmacy). The Si-O bond is generated by hydrolysis of the Si-Cl bond of the mono-, di-, or trichlorosilanes. The reduction of $SiMe_2Cl_2$ leads to polysilanes $(SiMe^{2-})_n$ that produce polycarbosilanes $(-SiMeHCH^{2-})_n$ upon heating (450°C) and finally the blend structure SiC β (1300°C), a ceramic. The silylenes $=SiR_2$, homologs of the carbenes $=CR_2$, have a rich chemistry (dimerization to $R_2Si=SiR_2$ with R = mesityl, etc.) but they are stable only for R = Cp*. The silicenium cations SiR_3^+, whose existence was controversal, have finally been isolated by Lambert and Reed; they are powerful electrophiles stabilized by coordination (for instance with 2 pyridines) leading to "hypervalent" silicon. The germanes GeR_3H are very stable and hydrogermanation of olefins, contary to the hydrosilylation, is possible without catalyst. The toxicity of organotins increases with the degree of alkylation and decreases with the length of the substituents. The isotope ^{119}Sn is active in NMR (solution) and in Mössbauer spectroscopy (solids). The chemistry of organotins such as R_3SnCl, R_3SnH and R_3SnNa is well developed. $PbEt_4$, used for a long time as an anti-Knock reagent in gasoline because of its radical-type decomposition, is now banned due to its toxicity.

5 - Group 15: As, Sb, Bi: These elements have the oxidation states III and V; their chemistry is little developed because of toxicity. The compounds ER_3 are good ligands ($AsPh_3$ is well known). The structures with the valences 4 (cationic), 5 (neutral) and 6 (anionic), the mono- or bis-bridged dimeric structures and multiple architectures with E-E bonds are known.

6 - Group 16: Se, Te: Se plays an important role in biochemistry: it is indispensable to life, but toxic at higher concentration. The thin layers of CdTe are promising semiconductors in microelectronics. Some bis(tetraselenafulvalene) salts are supraconductors at 1.5 K and analogs with Te are known.

tetraselenafulvalene (E = Se)

tetratellurafulvalene (E = Te)

EXERCISES

13.1. Do you know areas in which the redox chemistry Cu^I/Cu^{II} is important?

13.2. Why are dialkylzincs able to mono- or dialkylate tantalum (V) by reaction with $TaCl_5$ whereas lithium and magnesium alkyl derivatives, for instance, cannot do so?

13.3. Why do trialkylalanes $[AlR_3]$ dimerize, whereas trialkylboranes BR_3 do not?

13.4. Why are silicenium cations SiR_3^+ much more reactive cations than carbonium ions CR_3^+? Compare your answer to that of the preceding exercise 13.3.

13.5. Indicate an easy spectroscopic means to distinguish Sn^{II} from Sn^{IV} (for example $[Sn^{IV}(\eta^5-C_5R_5)_2Cl_2]$ from $[Sn^{II}(\eta^5-C_5R_5)_2])$.

PART IV

CATALYSIS

Introduction to catalysis

Chapter 14
Hydrogenation and hydroelementation of alkenes

Chapter 15
Transformations of alkenes and alkynes

Chapter 16
Oxidation of olefins

Chapter 17
C-H activation and functionalization of alkanes and arenes

Chapter 18
Carbonylation and carboxylation reactions

Chapter 19
Bio-organometallic chemistry: enzymatic catalysis

Chapter 20
Heterogeneous catalysis

*See also Chapter 21
for catalysis in organic synthesis*

Part IV

Catalysis

See also Chapter 21
for catalysis in organic synthesis

INTRODUCTION TO CATALYSIS

A catalyst is a compound that allows a reaction that does not proceed in its absence. It is added to the reaction mixture in quantities that are much lower than stoichiometric ones (between 10^{-6} and 10^{-1}) and, in principle, it is found unchanged at the end of the reaction. Thus, it does not appear in the reaction balance, and is usually written on the reaction arrow in order to emphasize this feature:

$$A + B \xrightarrow{\text{[cat.]}} C + D$$

The catalyst does not influence the thermodynamics of a reaction. For instance, it does not allow a thermodynamically impossible reaction to proceed. On the other hand, it changes the reaction pathways, i.e. the kinetics; in particular it lowers the energy of transition states.

The characteristic data of catalyst are the turnover frequency (TOF, i.e. the number of turnovers per mol. catalyst per unit of time) and the turnover number (TON, i.e. the total number of turnovers per mol. catalyst until it is no longer alive).

There are different types of catalysis:

▶ **acid and base catalysis**, usually encountered in organic chemistry [IV.1] (example: hydrolysis of esters), but also sometimes in homogeneous transition-metal catalysis (e.g., the Monsanto process,[IV.2] see Chap. 18.1).

▶ **electrocatalysis** (electron-transfer-chain catalysis or atom-transfer-chain catalysis):[IV.3] see Chap. 3.

▶ **photo-catalysis**: it can be the induction of a reaction by light or by a photo-catalyst called a photosensitizer. In both cases, a stoichiometric quantity of photons is being used. For instance, in water photo-splitting, the excitation by visible light for a colored photo-catalyst such as $[Ru(bpy)_3]^{2+}$ allows to store the energy necessary for water photo-splitting by reaction between the photo-excited state of the photo-catalyst and water.[IV.4]

▶ **redox catalysis**: catalysis of a reduction or oxidation. It can be achieved in an homogeneous phase (chemical reagents, photochemistry, biology) or heterogeneously (electrochemistry) using an organic, inorganic or organometallic redox catalyst. The mechanism can proceed by outer sphere (mediation without coordination to the catalytic center) or by inner sphere (catalysis with coordination to the catalytic center).[IV.5,6]

▶ **homogeneous catalysis**:[IV.7] everything is soluble in the liquid phase including the catalyst, a molecular complex of a transition metal, main-group metal, lanthanide or actinide. The advantage is that one is able to carry out spectroscopic and kinetic studies most often leading to the knowledge of the mechanism. This methodology allows to improve the efficiency and the chemio-, regio- and stereo-selectivity of the catalyst. The inconvenient is that the catalyst is very difficult or impossible to separate at the end of the reaction, which poses economical problems (cost and eventually rarity of catalyst) as well as ecological ones (pollution of the reaction product by the catalyst or its by-products).

▸ **heterogeneous catalysis**: [IV.8] the catalyst is insoluble and thus easily removed by filtration at the end of the reaction. Another notable advantage is that one can heat at high temperatures, far beyond temperatures that can be supported by homogeneous catalysts. Thus, it is possible to carry out reactions that are kinetically difficult, without perturbing the catalyst (synthesis of NH_3, selective oxidation of CH_4 to CH_3OH, reduction of CO). These advantages bring heterogeneous catalysis at the forefront of efficient industrial processes in *petrochemistry* (cracking, isomerization, increase of the octane index of gasoline), *synthesis of fine chemicals* (CO reduction by H_2, i.e. the *Fischer-Tropsch* process, water-gas-shift reaction, synthesis of H_2SO_4, HNO_3, NH_3, phosphate and nitrogen-containing fertilizers, polymers) and *environmental chemistry* (car exhaust pipes). The major problem is the control of selectivity. Modern approaches involve single-site catalysts by grafting organometallic catalysts on oxide support and control of nanoparticle catalysis (see Chap. 20).

▸ **supported catalysis**: [IV.9,IV.10] the homogeneous catalyst is immobilized on a support in order to benefit from the mildness, efficiency and selectivity of the reaction and of the easy separation of the catalyst at the end of the reaction. The support can be organic (example: polystyrene) or inorganic (example: silica). It is clear that the reaction conditions are deeply modified, however, and the supported catalyst must be specifically studied. The major drawback is leaching of the metal catalyst species from the support into the solution.

▸ **metallodendritic catalysis**: [IV.11,IV.12] the catalyst is fixed inside or at the periphery of dendrimers, which allows homogeneous catalysis, but the metallodendritic catalysts are easily removed by precipitation, ultrafiltration using membranes or ultracentrifugation as large biomolecules. This area has recently exploded, and practically all the catalyzed reactions have been developed using this technique, including asymmetric syntheses.

▸ **biphasic catalysis**: [IV.13] the homogeneous catalyst is solubilized in *water* by introduction of *water-solubilizing functional groups* onto a ligand in order to separate the catalyst in the water phase by decantation of the organic products at the end of the reaction. For instance, triphenylphosphine was sulfonated in *meta* on the phenyl substituents by Kunz at Rhône-Poulenc, and the rhodium complexes of the hydrosoluble ligand (TPPTS) have been used by Rhône-Poulenc for the hydroformylation of propene to butyraldehyde (see Chap. 18.2).

A *fluorinated solvent* can also be a phase. This *"fluorous"* chemistry, initiated by Horvath,[IV.14,IV.15] is presently in expansion. Indeed, organic solvents are usually insoluble in fluorous solvents at ambient temperature, but they become so upon heating, a reversible phenomenon that allows the recovery and re-use of the fluorous catalyst. This principle has also been used for the hydroformylation of α-olefins, for instance using the catalyst $[Rh(H)(CO)[P\{CH_2CH_2(CF_2)_5(CF_3)_3$.

Phase-transfer catalysis is a particular case of biphasic catalysis, the two phases being

▸▸ an organic phase containing the reactants and products and

▸▸ an aqueous phase with a transfer agent, typically a tetraalkylammonium salt that carries the anions from the aqueous phase to the organic one for reaction. This cation increases the nucleophilicity of anions resulting from the weakness of the electrostatic attraction due to its large size (see Chap. 17.6).[IV.13]

▸ **enzymatic catalysis**: [IV.16] catalysis by enzymes is highly efficient, but also very complex rendering its understanding often difficult. Some mechanisms are known, such as that of cytochrome P450. The modeling of enzyme reactions by molecular metal-complex models is a domain belonging to bioinorganic and bio-organometallic chemistry (see also Chap. 19). Biological engineering is more and more used to catalyze reactions *in vitro*.

In conclusion to this brief summary, let us emphasize that the development of catalysts that are well-defined, bring about efficient and selective transformations and can be separated from products and recycled-remains a key challenge for the XXIth century. Indeed, the use of the above catalytic techniques along with that of ionic liquids [IV.17] (because of their lack of vapor tension and their ease of separation from apolar products; the most used are imidazolium salts) and supercritical fluids [IV.18] (mostly sc. CO_2 because it is a non-polluting easily removed solvent, also compatible with fluorous chemistry) reflects efforts to improve the *"Green Chemistry"* aspects of catalytic processes.

Let us recall here the twelve principles of *"Green Chemistry"*: [IV.19]

1. Prevent waste

2. Atom economy

3. Use and generate no (less) toxic chemicals

4. Minimize product toxicity during function

5. Safer solvents and auxiliaries

6. Energy economy (carry out processes at ambient temperature and pressure)

7. Use renewable feedstocks

8. Reduce derivatives and steps

9. Selective catalytic processes preferred to stoichiometric ones

10. Design for degradation

11. Real-time analysis for pollution prevention

12. Safer chemistry for accident prevention

In this Part IV, we will examine the main homogeneous catalytic processes that have given rise to industrial applications or are currently used in the laboratory as well as the most promising types. The principles derived therefrom are essential to understand other catalytic reactions involving metals. Finally, Chap. 20 brings an insight into heterogeneous catalysis, and multiple use of catalysts in applications to organic synthesis appear in Chap. 21.

In homogeneous catalysis, a few general constraints can be defined:

▸ The catalytically active metal species must have a vacant coordination site, i.e. NVE = 16 at most or even 14, in order to allow substrate molecules to coordinate. Sometimes, weak ligands or solvent ligands can be present and are easily displaced by substrate molecules. Bulky phosphines such as triphenylphosphine are easily dissociatively displaced and thus constitute a reservoir of vacant coordination sites that can be filled or emptied at will. Noble metals (2nd and 3rd lines of transition metals of groups 8, 9 and 10) easily forming 16-electron species are privileged catalysts.

▸ In catalytic cycles, the various entities successively involved have most of the time NVEs alternating between 16 and 18.

▸ The role of the other ligands ("ancillary") is to avoid the precipitation of the metal and to insure a correct stereoelectronic balance (electronic density on the metal center, steric effect, trans effect) allowing all the individual reactions along the catalytic cycle to proceed at a good rate and with a good selectivity.

▸ The catalytic rings are represented in the following way. All the catalytic steps proceed at the same rate within a given cycle, but one of them is rate determining (it can change even in the course of the same reaction):

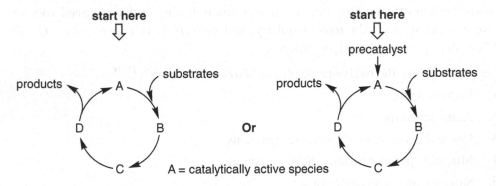

▸ Some schemes and reactions of simple and multiple metal-carbon bonds involving metallosquares (Chauvin-type) mechanisms are recalled and represented on the next page.[IV.20]

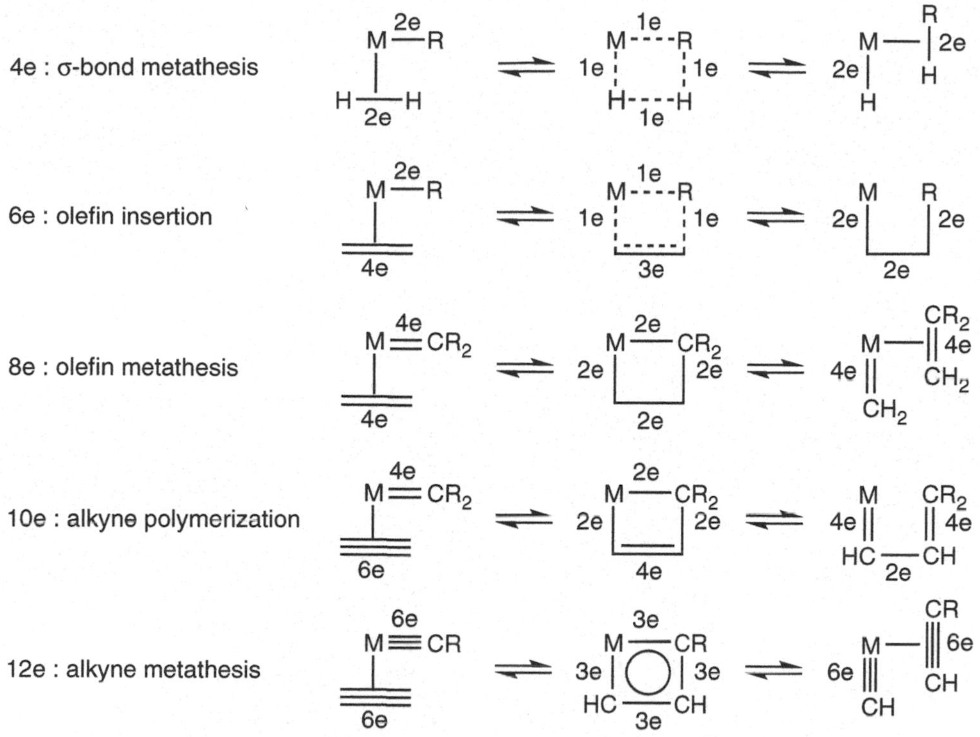

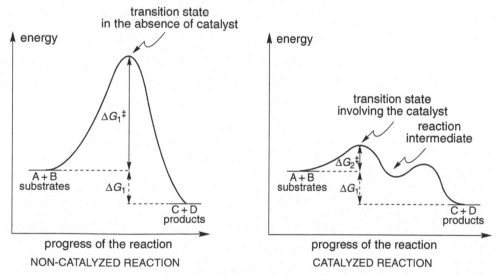

Schemes involving Chauvin-type metallosquares for the transformations of simple, double and triple metal-carbon bonds (for σ-bond metathesis of Ln-R with H_2, calculations lead to prefer linear R···H···H shape in a triangular transition state) [IV.21]

Comparison of the profiles of the uncatalyzed and catalyzed reaction A + B ⟶ C + D

Remarks:

▸ *The energy levels of the starting substrates on one hand and reaction products on the other hand are the same with or without catalyst ($\Delta G°$ constant), but the activation energy $\Delta G^{\ddagger}$ is much lower when the reaction is catalyzed ($\Delta G_1^{\ddagger} >> \Delta G_2^{\ddagger}$).*

▸ *A catalyzed reaction can eventually involve one or several reaction intermediates (for instance, one intermediate in the right figure above).*

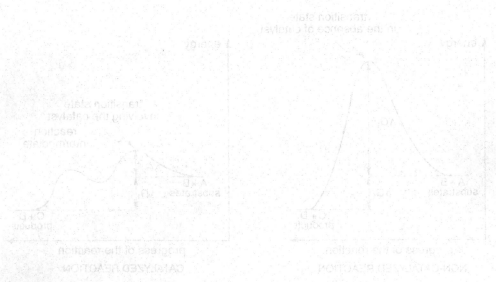

Chapter 14

HYDROGENATION AND
HYDROELEMENTATION OF ALKENES

The insertion of alkenes into M-H bonds has been examined in Chap. 4. This reaction is very important because, it leads to the dimerization, oligomerization and polymerization of alkenes. It is broad and concerns not only transition metals, but also main-group metals (group 13 Lewis acids), lanthanides and actinides. For instance, $AlEt_3$ is an excellent initiator of olefin polymerization. This reaction can also be considered as the hydrometallation or the hydroelementation of an olefin, and stoichiometric examples have been shown. If the element E does not have the property of a Lewis acid allowing olefin pre-coordination onto a vacant site and thus facilitating insertion, the insertion reaction is not possible without a catalyst.

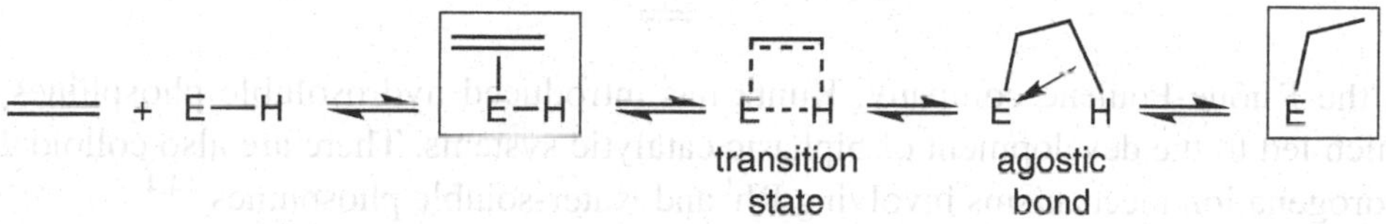

E : transition metal (16e), lanthanide, actinide, group 13 element

1. HYDROGENATION OF OLEFINS

The hydrogenation of olefins, impossible without catalyst, can be catalyzed by numerous transition-metal complexes. All the known mechanisms involve the formation of metal-hydride species. Four modes of formation of hydrides corresponding to four mechanisms of catalytic hydrogenation of olefins are known. Two mechanisms involve mono- or binuclear oxidative addition, and the two others involve a H_2 complex and homo- or heterolytic σ-bond metathesis.

1.1. MECHANISMS INVOLVING OXIDATIVE ADDITION

Classic mechanism with oxidative addition onto a metal center

This mechanism is the most common and the best known. It often uses the Wilkinson-Osborn catalyst, $[RhCl(PPh_3)_3]$, whose action was discovered in 1964. It involves three reversible steps: oxidative addition of H_2, substitution by the olefin of

PPh$_3$ that is labilized in *trans* position *vs.* the hydride ligand (*trans* effect). The last step, reductive elimination of the alkyl and hydride ligands to yield the alkane, is totally irreversible [14.1-14.3], and has allowed to successfully use this system for asymmetric catalysis (*vide infra*)

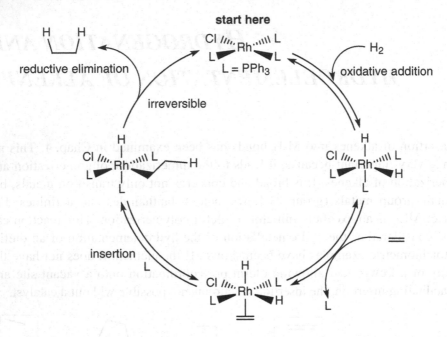

At the Rhône-Poulenc company, Kuntz has introduced hydrosoluble phosphines, which led to the development of biphasic catalytic systems. There are also colloidal hydrogenation mechanisms involving Rh^I and water-soluble phosphines.[14.4]

Oxidative addition involving two metal centers with a radical mechanism

This system was disclosed in 1942 by Igushi, and it involves H$_2$ addition onto the 17-electron cobalt center of $[Co(CN)_5]^{3-}$ that serves as a reservoir of hydrogen atoms. The latter are transferred, one by one, by the 18-electron cobalt hydride complex $[HCo(CN)_5]^{3-}$ onto the olefin. The intermediate organic radical must be sufficiently stable, which is the case for the benzyl radical. Only activated olefins can react in this way; thus, this mechanism is rather rare.[14.5]

$$[(CN)_5Co]^{\bullet 3-} + H\!-\!H + [(CN)_5Co]^{\bullet 3-} \longrightarrow (CN)_5Co^{\bullet 3-}\text{-}\,\text{-}\,\text{-}H\text{-}\,\text{-}\,\text{-}H\text{-}\,\text{-}\,\text{-}^{\bullet}Co(CN)_5^{3-}$$

$$\downarrow$$

$$2\,[HCo(CN)_5]^{3-}$$

$$[HCo(CN)_5]^{3-} + Ph\diagup\!\!\!\diagdown\!\!\!\diagup CO_2^- \longrightarrow [(CN)_5Co]^{\bullet 3-} + Ph\diagup\!\!\!\diagdown\!\!\!\diagup\!\!\!\diagdown CO_2^-$$

$$Ph\diagup\!\!\!\diagdown\!\!\!\diagup\!\!\!\diagdown CO_2^- + [HCo(CN)_5]^{3-} \longrightarrow Ph\diagup\!\!\!\diagdown\!\!\!\diagup\!\!\!\diagdown CO_2^- + [(CN)_5Co]^{\bullet 3-}$$

1.2. MECHANISM INVOLVING σ-BOND METATHESIS

Classic mechanism

Lanthanide complexes such as [LuCp*$_2$(H)] are extremely active catalysts for the hydrogenation of terminal olefins. First, the olefin coordinates onto the metal center, then insert into the Lu-H bond to give a Lu-alkyl intermediate. Then, the reversible addition of H$_2$ onto the d^0 metal center gives a transitory dihydrogen complex that cannot undergo oxidative addition of H$_2$ in the absence of d electrons on Lu. Thus, this formation of the Lu-H$_2$ species is followed by σ-bond metathesis irreversibly giving the alkane and a Lu-H species.[14.6]

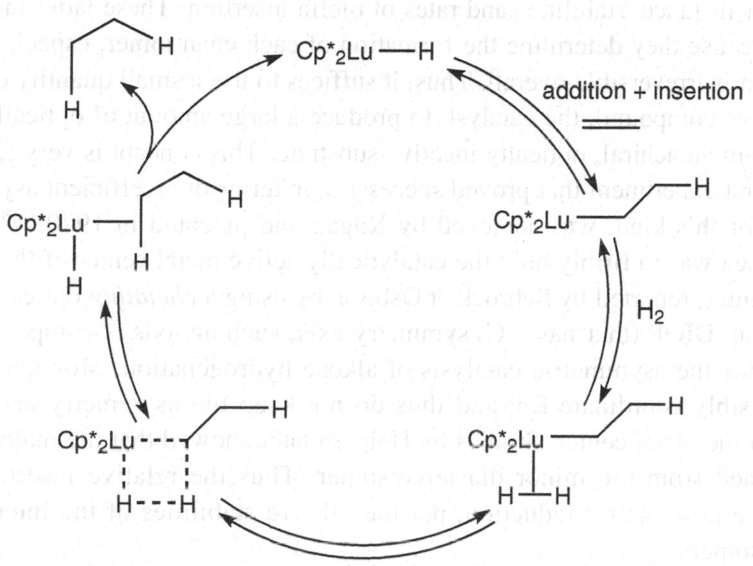

Generation of the catalytically active species by heterolytic H$_2$ activation

The complex [RuCl$_2$(PPh$_3$)$_3$] is a hydrogenation catalyst for which the catalytic cycle is the same as the above one, the catalytically active species being [RuCl(H)(PPh$_3$)$_3$].[14.7] This species is generated by heterolytic H$_2$ activation of an acidic H$_2$ complex that is deprotonated either intramolecularly by the basic Cl ligand or intermolecularly by an amine such as NEt$_3$.

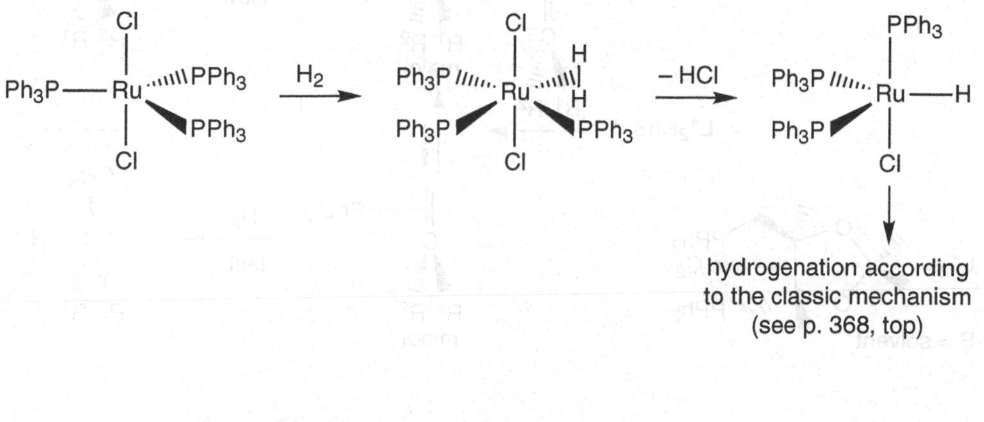

hydrogenation according to the classic mechanism (see p. 368, top)

2. ASYMMETRIC HYDROGENATION

An olefin having two different substituents on the same carbon of the double bond
is prochiral (except if it has a symmetry plane in the middle of the double bond). An
addition reaction on such an olefin gives a chiral product in the form of two enantiomers (the chirality is centered on carbon). If the hydrogenation catalyst is optically
inactive, the two enantiomers are obtained in equal quantities, i.e. the optically inactive racemic is obtained. On the other hand, if the catalyst is chiral and, in addition, optically active, the coordination of the olefin onto the metal center produces
two diastereoisomeric intermediates having different chemical and physical properties, i.e. for instance, stabilities and rates of olefin insertion. These latter factors are
crucial, because they determine the formation of each enantiomer, especially since
the reaction is irreversible overall. Thus, it suffices to use a small quantity of an optically active compound, the catalyst, to produce a large amount of optically active
product from an achiral, optically inactive substrate. This concept is very far reaching. The first experiment that proved successful, in terms of an efficient asymmetric
induction of this kind, was achieved by Kagan and patented in 1970.[14.8] Indeed,
Kagan's idea was to firmly hold the catalytically active metal center of the catalyst
$[RhL_2]^+$ center, reported by Schrock et Osborn, by using a *chelating* optically active
diphosphine, DIOP (that has a C_2 symmetry axis; such an axis is compatible with
chirality) for the asymmetric catalysis of alkene hydrogenation. Monophosphines
only reversibly coordinate Rh^I and thus do not keep the asymmetry and optical
activity on the metal center. Studies by Halpern later showed that the major enantiomer formed from the minor diastereoisomer. Thus the relative insertion rates
directed the asymmetric induction, not the relative stabilities of the intermediate
diastereoisomers.

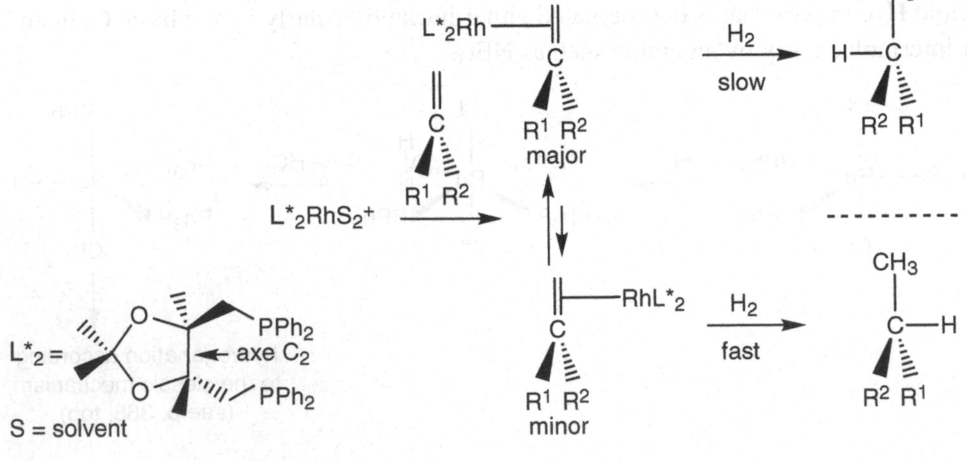

A very important use of Kagan's concept with an optically active *chelating* diphosphine has been developed in 1975 by Knowles' group at the Monsanto company: the synthesis of *L*-DOPA, a drug against Parkinson disease.[14.9] Subsequently, a large number of chiral *chelating* diphosphines have been shown to be extremely efficient in asymmetric catalysis and largely applied to the synthesis of optically active organic compounds including drugs. The efficiency of optically active asymmetric catalysts is measured by the enantiomeric excess, e.e. $= (r - s) / (r + s)$, r and s being the molecular quantities of each enantiomer R and S formed in the asymmetric reaction.

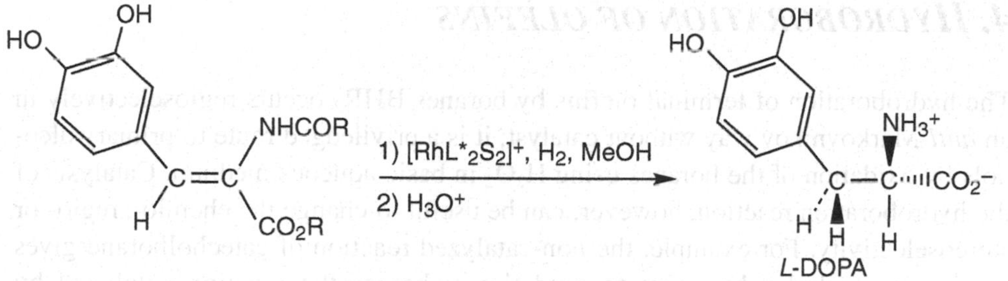

$$\text{1) [RhL*}_2\text{S}_2\text{]}^+, \text{H}_2, \text{MeOH}$$
$$\text{2) H}_3\text{O}^+$$

L-DOPA

3. *HYDROSILYLATION OF OLEFINS*

This reaction is useful in organic synthesis [14.10] and for the fabrication of materials, in particular silicones (especially with R' = Cl):

$$\text{RCH=CH}_2 + \text{R'}_3\text{SiH} \longrightarrow \text{RCH}_2\text{CH}_2\text{SiR'}_3$$

The results of mechanistic studies are much less secure than in the case of the hydrogenation reaction. A mechanism similar to that of the classic oxidative addition onto a metal center has been proposed with olefin insertion into the M-H bond or more hypothetically into the M-SiR'$_3$ bond. The Wilkinson-Osborn catalyst, among others, follows this mechanism.

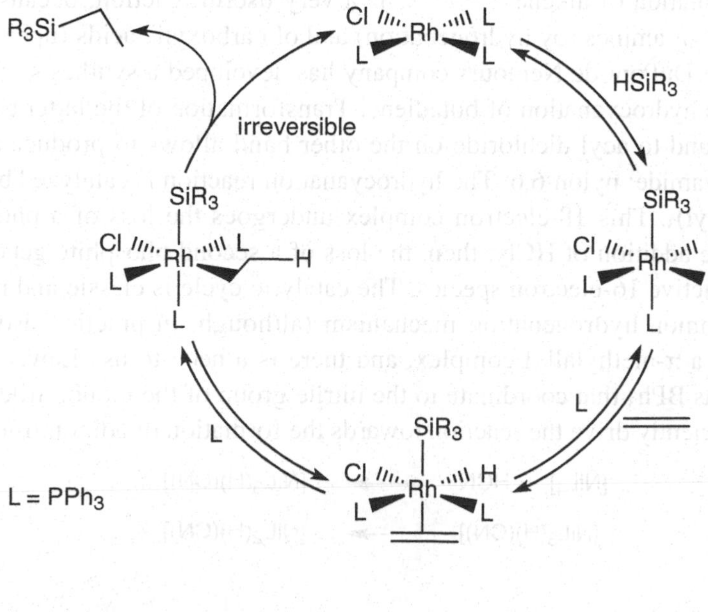

irreversible

L = PPh$_3$

It has been shown, however, that in some cases, catalysts working efficiently even if they are used in very small quantities (0.1 ppm) such as the Speier catalyst H_2PtCl_6 or the Karstedt catalyst $[Pt(CH_2=CHSiMe_2OSiMe_2CH=CH_2)_2]$, are indeed functioning in a colloidal nanoparticle form. Such nanoparticles are giant metal clusters containing several hundred atoms that are readily produced for instance by reduction of noble metal salts or complexes (see Chap. 2.2.1 and 20.7.3).

It is also possible to catalyze the hydrosilylation of C=O and C=N bonds.

4. HYDROBORATION OF OLEFINS

The hydroboration of terminal olefins by boranes BHR_2 occurs regioselectively in an *anti*-Markovnikov way without catalyst; it is a priviledged route to primary alcohols by oxidation of the boranes using H_2O_2 in basic aqueous medium. Catalysis of the hydroboration reaction, however, can be useful to change the chemio-, regio- or stereoselectivity. For example, the non-catalyzed reaction of catecholborane gives the linear alcohol subsequent to oxidation, whereas the reaction catalyzed by $[RhCl(PPh_3)_3]$ yields 99% of branched alcohol: [14.12]

5. HYDROCYANATION OF OLEFINS

The hydrocyanation of alkenes [14.13,14.14] is a very useful reaction, because nitriles are precursors of amines (by hydrogenation) and of carboxylic acids (upon hydrolysis). Thus, the DuPont de Nemours company has developed a synthesis of adiponitrile using the hydrocyanation of butadiene. Transformation of the latter to diamine on one hand and to acyl dichloride on the other hand allows to produce a famous synthetic polyamide: nylon-6,6. The hydrocyanation reaction is catalyzed by $[NiL_4]$, L = P(O-*o*-tolyl)$_3$. This 18-electron complex undergoes the loss of a phosphite to give oxidative addition of HCN; then, the loss of a second phosphite generates the catalytically active 16-electron species. The catalytic cycle is classic and resembles the most common hydrogenation mechanism (although, in practice, isomers are produced via a π-methylallyl complex, and there is a need to use Lewis acid promoters such as BPh_3 that coordinate to the nitrile group of the mononitrile intermediate and efficiently drive the reaction towards the formation of adiponitrile):

$$[NiL_4] + HCN \longrightarrow [NiL_3(H)(CN)] + L$$

$$[NiL_3(H)(CN)] \longrightarrow [NiL_2(H)(CN)] + L$$

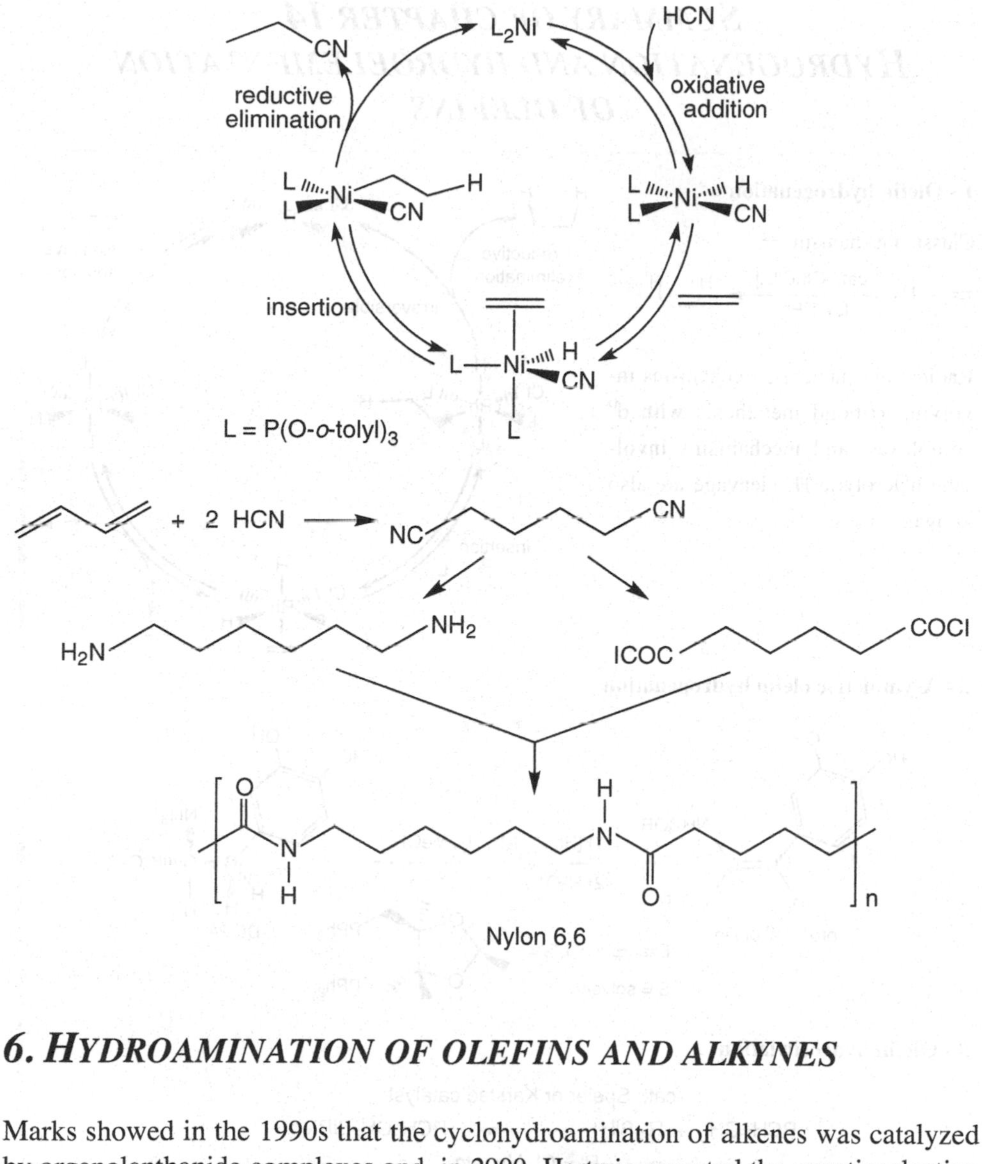

Nylon 6,6

6. HYDROAMINATION OF OLEFINS AND ALKYNES

Marks showed in the 1990s that the cyclohydroamination of alkenes was catalyzed by organolanthanide complexes and, in 2000, Hartwig reported the enantioselective hydroamination of *p*-trifluromethylstyrene by aniline using Noyori's catalyst [R-BINAP-Pd(OSO$_2$CF$_3$)$_2$] (toluene, 45°C, 81% yield, 81% ee).

Inter- and intramolecular hydroamination of internal alkynes by primary amines giving imines was shown by Doye in 1999 to be catalyzed by Cp$_2$TiMe$_2$. Thus, amines, ketones and amino acids are subsequently accessible from imines using this route. The hydroamination mechanism first involves the formation of the pre-catalyst Cp$_2$Ti(NHR)$_2$. This is followed by α-H elimination giving an imido complex [Cp$_2$Ti=NHR], the catalytically active species. The latter then forms a titana-aza-cyclobutene by reaction with the alkyne, which is followed by reaction with the amine leading to ring opening and formation of a second Ti-amino bond. Finally α H-transfer to free the imine closes the cycle by reforming the Ti-imido catalyst.

SUMMARY OF CHAPTER 14
HYDROGENATION AND HYDROELEMENTATION OF OLEFINS

1 - Olefin hydrogenation

Classic mechanism ⇒

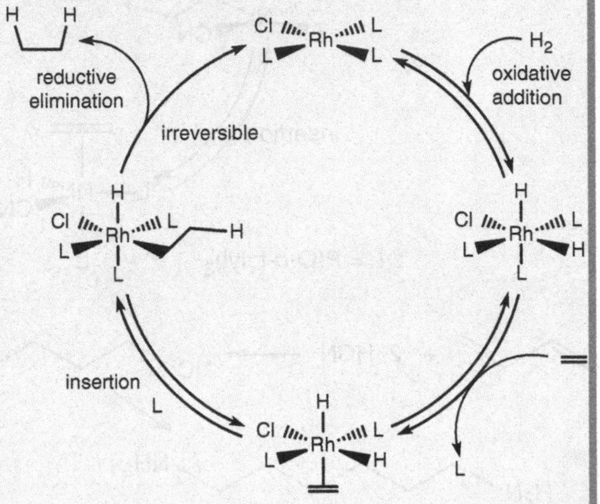
reductive
elimination
irreversible
H_2
oxidative
addition
insertion

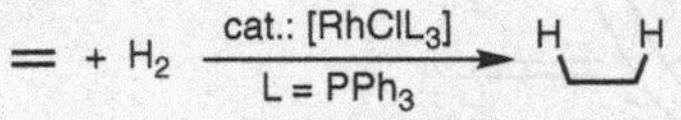
cat.: [RhClL3]
L = PPh3

Radical mechanisms, mechanisms involving σ-bond metathesis with d^0 complexes, and mechanisms involving heterolytic H_2 cleavage are also known.

2 - Asymmetric olefin hydrogenation

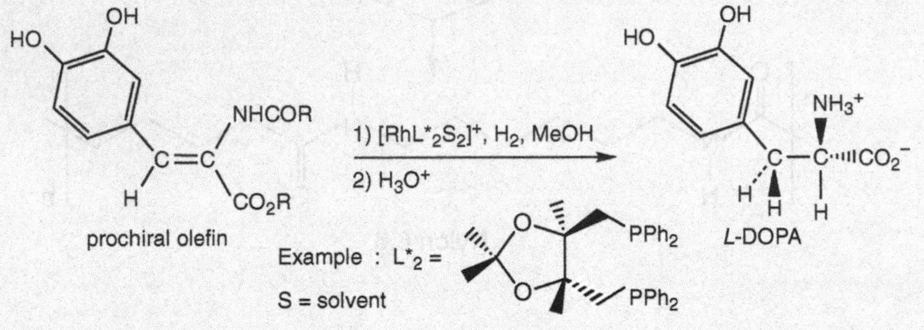
NHCOR
CO2R
prochiral olefin
1) [RhL*2S2]+, H2, MeOH
2) H3O+
Example : L*2 =
S = solvent
PPh2
PPh2
NH3+
CO2−
L-DOPA

3 - Olefin hydrosilylation

$$RCH{=}CH_2 + R'_3SiH \xrightarrow[\text{R' = Cl, Me, etc.}]{\text{cat.: Speier or Karsted catalyst}} RCH_2CH_2SiR'_3$$

4 - Olefin hydroboration

Hydroboration is carried out without catalyst, but the regioselectivity can be reversed by using a catalyst.

5 - Olefin hydrocyanation

Butadiene hydrocynation leads to precursors of nylon 6,6.

$$RCH{=}CH_2 + HCN \xrightarrow{\text{cat.: NiL}_4} RCH_2CH_2CN \qquad L = P(O\text{-}o\text{-}tol)_3$$

6 - Olefin and alkyne hydroamination

Enantioselective olefin hydroamination with aniline is catalyzed by Noyori's catalyst, and hydroamination of internal alkynes with primary amines is catalyzed by Cp_2TiMe_2.

EXERCISES

14.1. In the asymmetric hydrogenation, does the produced major enantiomer result from insertion in the most stable diastereoisomeric intermediate formed by coordination of the prochiral olefin with the optically active catalyst?

14.2. Is it possible to catalyze the dehydrogenation of alkanes, the reverse reaction of olefin hydrogenation?

14.3. Why did the DuPont de Nemours company choose the ligand $P(O\text{-}o\text{-tolyl})_3$ for the catalysis of butadiene hydrocyanation by Ni^0?

14.4. Suggest a mechanism for the hydrogenation of benzene to cyclohexane by H_2 catalyzed by a transition-metal complex.

14.5. Can a 20-electron complex of the type $[Fe(\eta^6\text{-arene})_2]$ be a good catalyst?

Chapter 15

TRANSFORMATIONS OF ALKENES
AND ALKYNES

The catalyzed transformations of olefins without other substrates are *isomerization, dimerization, oligomerization, polymerization and metathesis*. Alkynes are also concerned with most of these reactions. Note that the Ziegler-Natta type polymerization, that consists in a series of olefin (or alkyne) insertions into metal-carbon bonds, usually leaves the metal moiety at the polymer chain termini. Thus, it does not fit with the definition of a catalytic system whereby the catalytically active metal species is recovered at the end of the catalytic cycle. This system should be defined as using an *"initiator"*. Also note, however, that the polymer chain can eventually release the initiator either thermally by β-hydride elimination of by hydrogenolysis (*vide infra*) to form a metal-hydride species that again serves as an initiator in the medium. In this case, such an *initiator* also corresponds to the definition of a *catalyst*. Likewise, the metathesis of cyclic olefins (and eventually non-cyclic alkynes) produces polymers to which the initiator remain attached by a M=C bond and thus does not strictly correspond to the definition of a catalyst. Both catalysis and initiation, however, are treated in this chapter.

1. ZIEGLER-NATTA-TYPE OLEFIN POLYMERIZATION

Industry produces 15 millions tons of polyethylene and polypropylene per year with the initiators discovered by Ziegler and Natta, among which the best known is the $TiCl_3/Et_2AlCl$ mixture, active at 25°C and 1 atm. The mixture being heterogeneous and very active, its mechanism of action is not known but only speculated from studies of initiators working homogeneously. It is probable that the Ti-bond, formed at the surface of the $TiCl_3$ crystals upon alkylation of $TiCl_3$ by Et_2AlCl, undergoes the insertion of the pre-coordinated olefin: [6,9]

$$Ti\text{——}Et + \text{==} \longrightarrow Ti\text{——}Et \longrightarrow Ti\text{-}CH_2\text{-}CH_2\text{-}Et \longrightarrow etc.$$

This mechanism has been proposed by Cosse in 1975. In 1982, Patricia Watson at Du Pont discovered a soluble initiator, $[LuCp*_2CH_3]$, active in the absence of a co-initiator, for the polymerization of ethylene and propylene. She could demonstrate the different steps of this mechanism.[6.10] In particular, successive olefin insertion was demonstrated. The reverse reaction, β-elimination is favored for β-H elimination as in transition-metal chemistry, but β-alkyl elimination is also encountered, contrary to most transition-metal alkyl complexes (competition between β-H and β-alkyl elimination is also known only for some *early* transition-metal alkyl complexes). Since then, numerous systems have been shown to be active Ziegler-Natta-type polymerization catalysts, in particular those of the type [ZrCp$_2$R$_2$] (*vide infra*).

The agostic C-H-M bond sometimes plays a keyrole in polymerization according to this mechanism. Indeed, in the complexes that are responsible for the polymerization, the metal coordination sphere is unsaturated (most often: NVE = 14) and then traps the σ C-H bond of the α carbon (i.e. the agostic bond). If the complex is d^0, the oxidative addition of this C-H bond cannot occur, and the agostic bond that is involved all along the insertion mechanism can influence the stereochemistry of this insertion: [6.11,6.12]

It is possible to show, using kinetic isotope effects, how the agostic C-H bond is involved. For instance, the use of a prochiral olefin leads, after two well-controlled successive insertions and hydrogenolysis, to two diastereoisomers, *threo* and *erythro*. In this case, the second olefin molecule can coordinate to Zr in agostic interaction with the C-H or the C-D bond. This implies that the CH$_2$(*n*-Bu) substituent of the α carbon is located above or below the plane of the sheet respectively, and that the *n*-Bu olefin substituent coordinates on the opposite side in order to minimize the steric bulk. Thus, the agostic C-H bond must lead to the *erythro* isomer whereas the agostic C-D bond must lead to the *threo* isomer. If the agostic bond was not involved, both stereoisomers would be formed in equal amounts. The experiment shows that the *erythro* isomer is favored in the ratio 3:1, which is in agreement with the fact that the agostic C-H bond is, according to calculations, more favorable than the agostic C-D bond: [6.11]

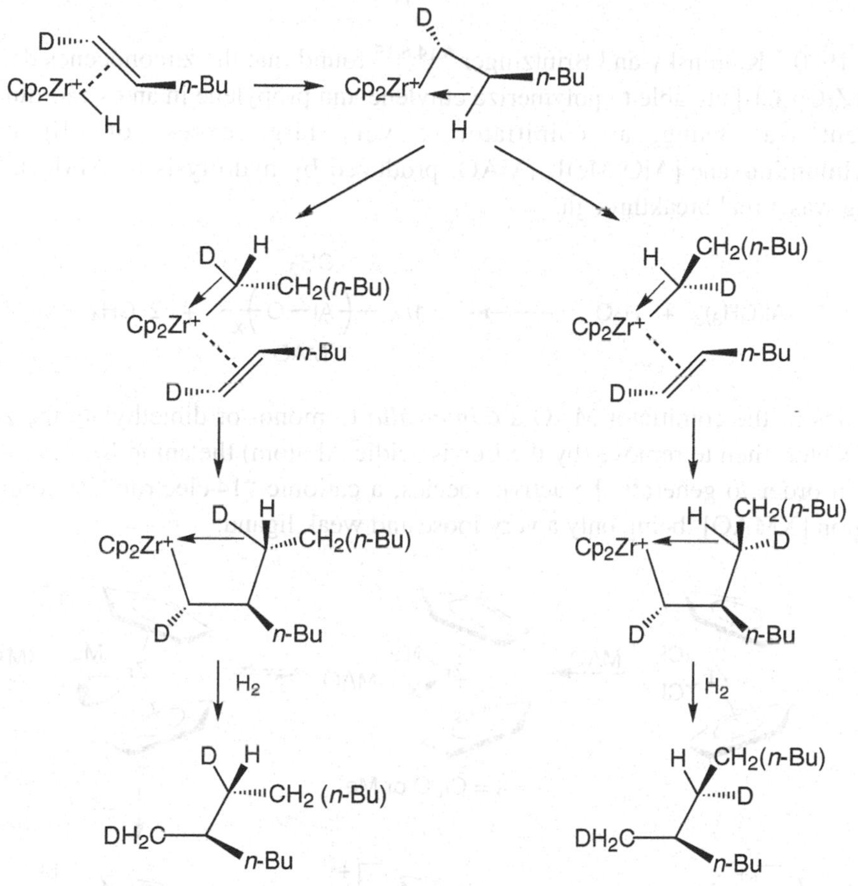

threo   erythro

The polymerization is not interrupted (when the olefin is in excess *vs.* the initiator) with d^0 complexes, because extrusion is disfavored by the instability of the intermediate olefin complex due to the absence of π backbonding. This is not the case for the metals located on the right of the periodic table, rich in d electrons, for which the insertion of an olefin onto a metal-alkyl bond is less frequent and extrusion very favorable in the presence of a vacant coordination site. If even a late metal is sufficiently electron poor, however, as for instance in cationic complexes, extrusion can be unfavorable and insertion leads to rapid polymerization. Brookhart [6.13] has shown that the key to this polymerization with $[CoCp^*\{P(OMe)_3\}(C_2H_5)]^+$ is in the agostic β C-H bond with the metal, a witness of the low metal basicity. The coordination site that is occupied by the agostic bond must become vacant in order to let the olefin enter the coordination sphere:

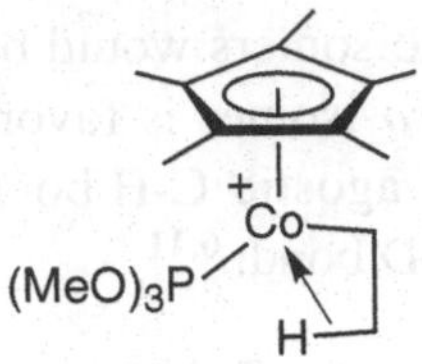

In the 1980s, Kaminsky and Brintzinger [6.14,6.15] found that the zirconocenes derived from [ZrCp$_2$Cl$_2$] are able to polymerize ethylene and propylene in an extraordinarily efficient way using, as coinitiator, a very large excess of oligomeric methylaluminoxane [AlO(Me)]$_x$ (MAO, produced by hydrolysis of AlMe$_3$). This finding was a real breakthrough.

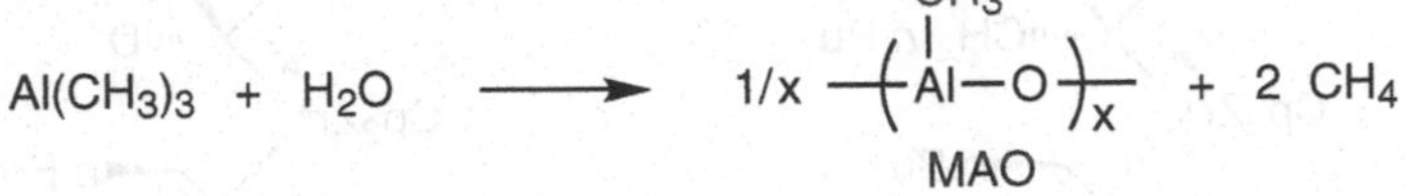

The roles of the coinitiator MAO are *inter alia* to mono- or dimethylate the zirconium center, then to remove (by the Lewis acidic Al atom) the anion X$^-$ (X = Cl$^-$ or Me$^-$) in order to generate the active species, a cationic "14-electron" Zr complex, the anion [XMAO]$^-$ being only a very loose and weak ligand:

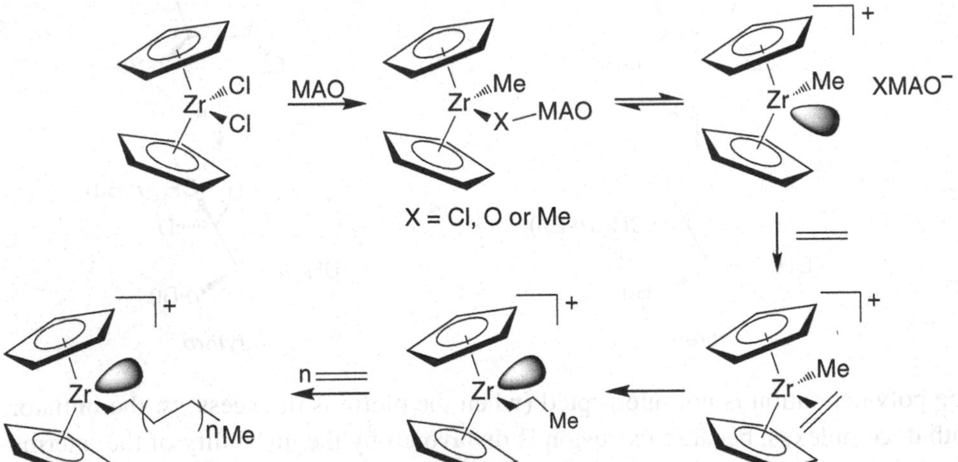

In 1984, Brintzinger [6.14] synthesized the first chiral derivative of this family, an *ansazirconocene* containing two indenyl ligands bridged by an ethylene chain (represented on the Scheme below). This initiator, of C_2 symmetry, is able, in its racemic form, to produce a stereoregular polypropylene, isotactic polypropylene (see the mechanism on the following page). Thus, Ziegler's hypothesis according to which the stereoregularity of a polypropylene was specifically due to a surface phenomenon of an heterogeneous catalyst was demolished.

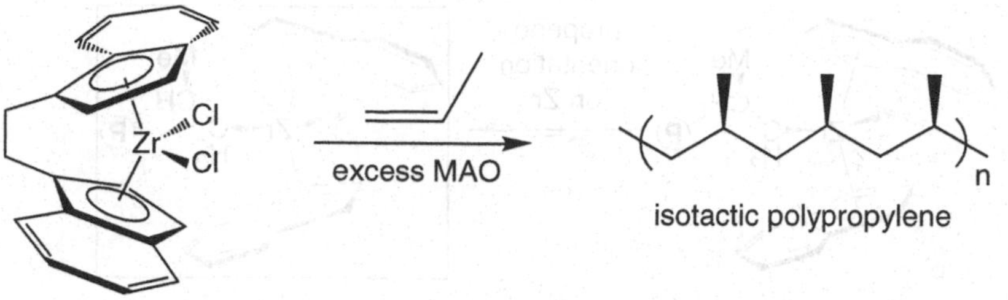

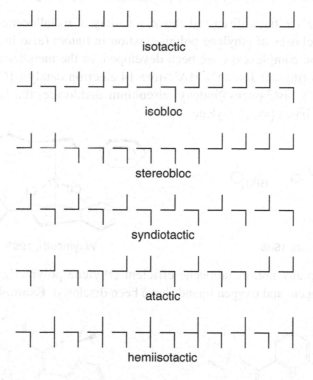

ansa structure (bridged)
C_2 symmetry

isotactic polypropylene

MAO (methylaluminoxane) = $-(Al-O)_x$, with Me group

The complex [ZrCp$_2$Cl$_2$] alone produces an atactic polypropylene. Other types of tacticities (below) can be obtained depending on the nature of the initiator. More than hundred initiators with formula related to Brintzinger's one have been synthesized given the enormous market of these polymers, in order to improve the performances of the initiator regarding the stereoregularity and the molecular mass of the polypropylene.

isotactic

isobloc

stereobloc

syndiotactic

atactic

hemiisotactic

Various possible tacticities for polypropylene polymers

The origin of the tacticity resides in the steric constraints that the ligands impose to the olefin coordination. [6.15] For example, with the initiators bis-indenyl zirconocenes of C_2 symmetry described above, the stereoselectivity is due to the privileged orientation of the propene ligand in the framed molecule of the Scheme below resulting from the non-bonding interactions with the other ligands:

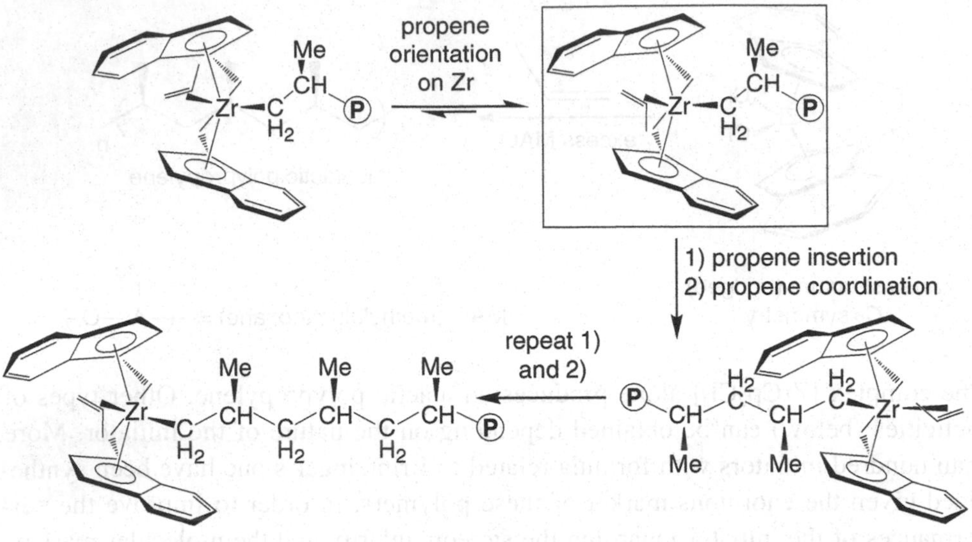

Origin of the stereospecificity and mechanism of isotactic propene polymerization with the initiator [Zr(η⁵-indenyl)(Me)] + [MeMAO]⁻ of C₂ symmetry

After Brinzinger's initiator discussed above, a variety of metallocene- and non-metallocene-based classes of ethylene polymerization initiators (also including lanthanide metallocene complexes) have been developed. In the metallocene series, particularly noteworthy are Jordan's MAO-free 14-electron catalyst $[Cp_2ZrCl][BPh_4]$ and Waymouth's bis(3-phenyl)indenyl zirconium dichloride, the latter providing isotactic-atactic block polypropylene:

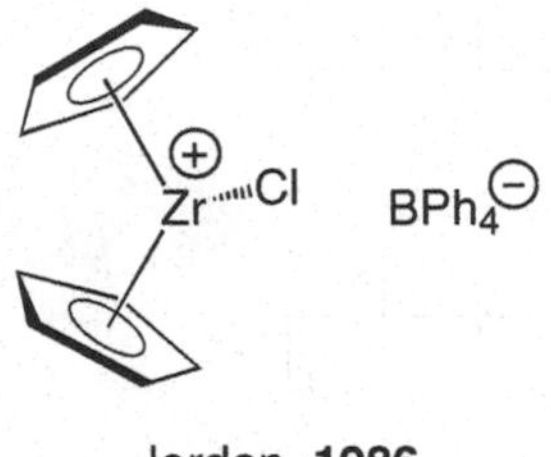

Jordan, **1986**

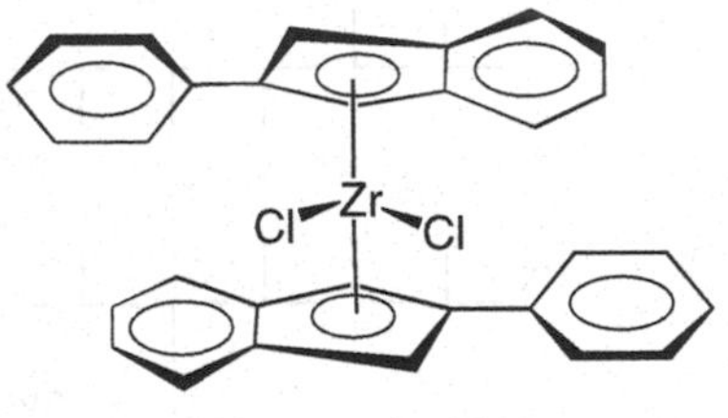

Waymouth, **1995**

Subsequently, many non-metallocene efficient ethylene polymerization initiators containing nitrogen- and oxygen ligands have been disclosed. **Examples:**

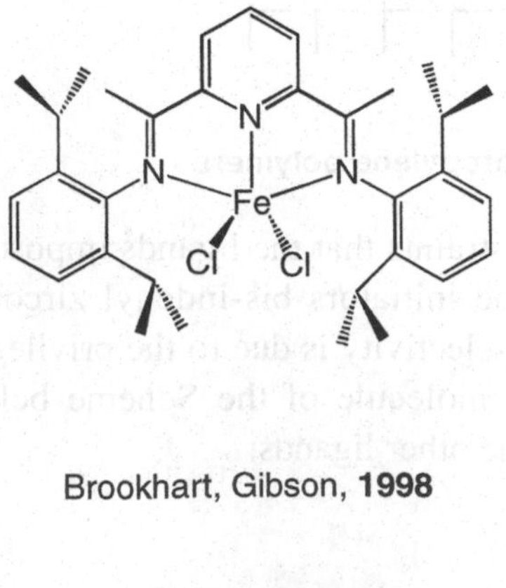

Brookhart, Gibson, **1998**

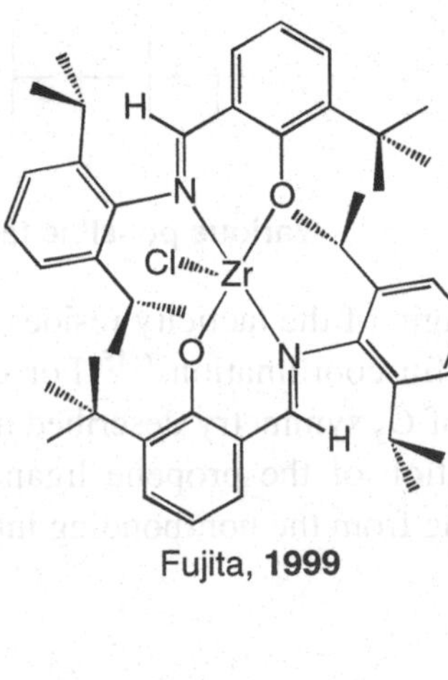

Fujita, **1999**

Although the symmetry-related control of the stereoselectivity becomes highly challenging, once the metallocene-based structure is abandoned, a variety of stereoselective single-site polymerization initiators are known for various α-olefins.

Examples:

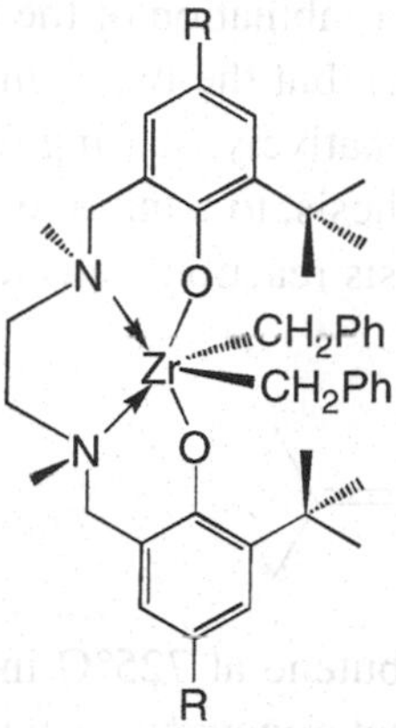

Isotactic poly(1-hexene)
Tshuva, Goldberg, Kol, **2000**

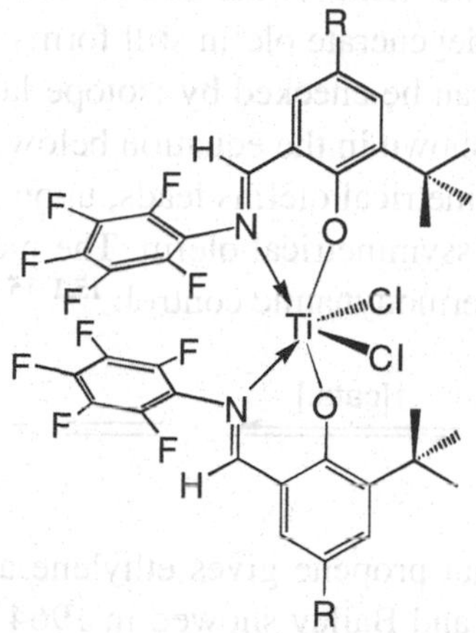

Syndiotactic poly(propylene)
Coates, Fujita, **2001**

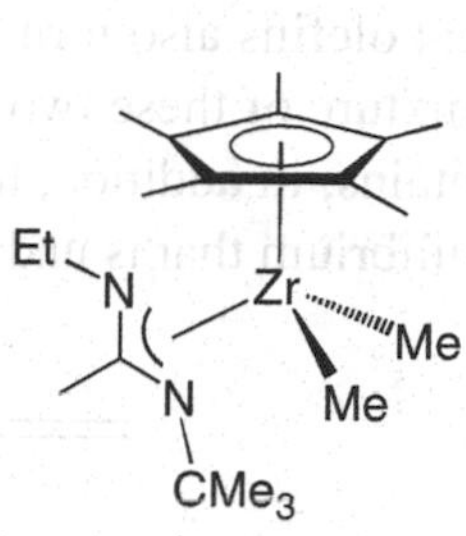

Isotactic poly(1-hexene)
Sita, **2001**

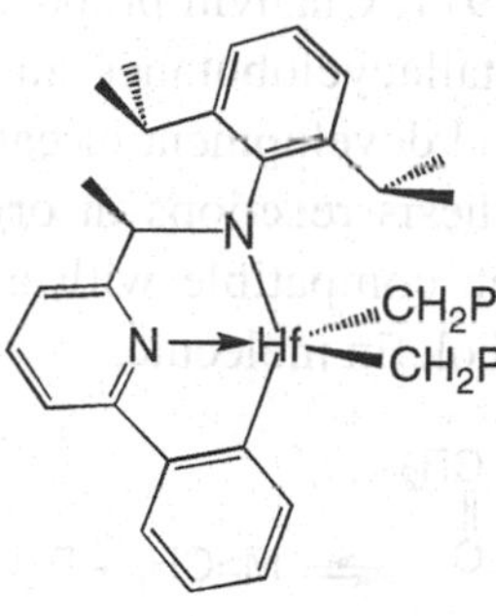

Isotactic poly(propylene)
Stevens, Vanderlende (Dow), **2003**

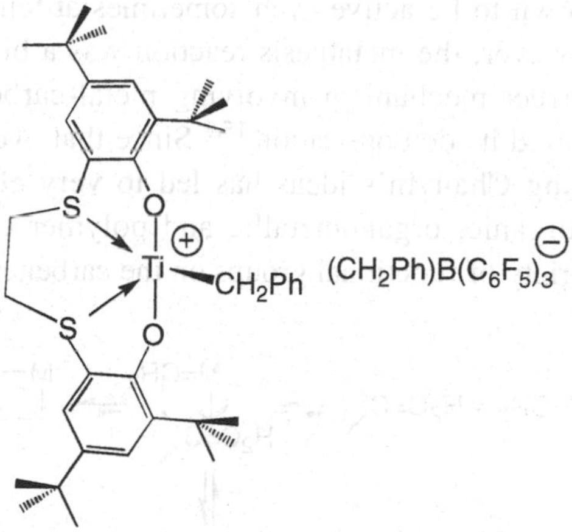

Isotactic polystyrene
Okuda, **2003**

2. METATHESIS OF ALKENES, ALKYNES AND CYCLOALKENES

2.1. ALKENE METATHESIS

The etymology of the word *metathesis* comes from the Greek μεταθεσις (metathesis) that means *transposition*. Thus, metathesis is invoked when, for instance, ions are exchanged in a solution containing two ion pairs in order to produce the most stable ion pairs. The metathesis of σ bonds and π bonds has already been examined in Chaps 3 and 9, respectively. Let us recall that the metathesis of an alkene is the redistribution of the two carbene fragments of the olefin to form other olefins or the same one. If the two carbenes of an olefin are identical with identical substituents (i.e. CH_2, CMe_2, etc.), the redistribution is degenerate, i.e. forms the olefin identical to the initial one. On the other hand, if the two carbenes of the olefin are different, as in isobutene below, the degenerate olefin still forms by recombination of the two different carbenes (which can be checked by isotope labeling), but the two symmetrical olefins also form, as shown in the equation below. Alternatively, starting from a mixture of these two symmetrical olefins leads, upon metathesis, to a mixture that contains, in addition, the dissymmetrical olefin. The metathesis reaction leads to an equilibrium that is under thermodynamic control: [15.1-15.5]

Since 1931, it is known that propene gives ethylene and 2-butene at 725°C in the absence of catalyst. Banks and Bailey showed in 1964 that heterogeneous catalysts such as [Mo(CO)$_6$] or [W(CO)$_6$] on alumina or Re_2O_7 on silica facilitate the reaction (150 to 500°C).[15.6] Later, homogeneous catalysts such as WCl_6 + $SnMe_4$ were shown to be active even sometimes at temperatures down to –77°C.[15.4] Till 1971, however, the metathesis reaction was a black box. In 1971, Chauvin proposed the correct mechanism involving metal-carbenes and metallacyclobutanes and also started its demonstration.[15.7] Since that event, the rational development of catalysts using Chauvin's ideas has led to very efficient metathesis reactions in organic, inorganic, organometallic and polymer chemistry even compatible with a large variety of functional groups on the carbenes of the initial olefin molecule.

Chauvin's mechanism for alkene metathesis

Metallacyclobutanes, with topological symmetry, can undergo a horizontal shift (giving back the initial olefin) or a vertical one (metathesis). The new metal-carbene species formed are the catalytically active species of the metathesis reaction: they react on the starting olefin in the same way as the initial metal-carbene complex to give olefins ultimately found in the overall metathesis stoichiometry. This mechanism now is universally accepted.[15.1] In particular, Schrock showed in 1980 that metal-carbene complexes are active in the absence of cocatalyst. This finding was made possible by the key discovery by Schrock's group at that time that introduction of alkoxide ligands in Ta alkylidene complexes made these complexes olefin metathesis catalysts.[15.8] Tungsten alkylidene complexes reported in 1980 also follow this principle and, importantly, such species were spectroscopically characterized as intermediates in the metathesis cycle, a definitive proof of Chauvin's mechanism.[15.2,15.8,15.9] Osborn also showed that well-defined complexes, formed from a Lewis acid catalyst and a tungsten-carbene complex, were stable and active.[15.10]

With the knowledge of the metathesis mechanism, it is retrospectively possible to understand how the homogeneous systems composed of a metal halide of the groups 4 to 9 and an alkylating agent (for instance, alkylaluminum or alkyltin) work. The catalytically active entities in these systems are metal-carbene complexes that are formed by α-elimination:

$$MCl_n + M'(CH_3)_m \longrightarrow Cl_pM(CH_3)_2 \longrightarrow CH_4 + Cl_pM{=}CH_2$$
(M = Mo, W, Re; M' = Sn, Al)

This α-elimination reaction, involving d^0 complexes (M-CH$_3$ $\longrightarrow$ M(H)(=CH$_2$)), also is a metathesis reaction, but of σ bonds (see Chap. 3).

In other systems including heterogeneous catalysts working at high temperatures, it has been noticed that the presence of dioxygen accelerated the metathesis reactions. It is probable that a M=O bond formed by double oxidative addition of O$_2$ to M forming M(=O)$_2$ plays the role of a metal-carbene M=CR$_2$ to which it is isoelectronic. Alternatively, a metal-alkoxide species resulting from oxo insertion into metal-alkyl bond could β-eliminate, then double oxidative addition to M(H)(=O)(=CH$_2$) might proceed: [15.2,15.4]

$$(M({=}O)(CH_3)) \longrightarrow MOCH_3 \longrightarrow M(H)(CH_2{=}O) \longrightarrow M(H)({=}CH_2)({=}O)$$

Note, however, that from an olefin R$_2$C=CR$_2$, the analogous double oxidative addition to a bis-carbene M(=CR$_2$)$_2$ has never been shown to occur, because olefins are far less electron-withdrawing ligands than O$_2$ and therefore cannot overtake the stage of a single oxidative addition to a metallacyclopropane. Thus, it is the above α-elimination mechanism from a metal-alkyl intermediate M-CH$_2$R $\longrightarrow$ M=CHR(H) that best accounts for the initiation of alkene metathesis in the above Ziegler-type systems.

Numerous applications of alkene metathesis have been developed.

Examples:
▸ Propene is industrially transformed into ethylene and n-butenes in Texas (PHILIPS process exploited by ARCO).
▸ C$_8$ and C$_{16}$ internal olefins are metathesized in the course of a multi-step process (ethylene oligomerization, isomerization, metathesis, hydroformylation) leading to C$_{11}$-C$_{15}$ primary alcohols that can be used as detergents (SHOP process at Shell).
▸ Neohexene (musk scent for detergent perfume industry) is synthesized by metathesis from ethylene and 2,4,4-trimethyl-2-pentene.

In the case of functional olefins, the difficulty resides in the fact that the functional group coordinates to the vacant site of the metal center of the catalyst, which inhibits its reactivity. This major drawback has considerably slowed down the development of applications. These problems, however, have finally been overtaken by the discovery of the unimolecular olefin metathesis catalysts that were designed and optimized, such as Schrock and Grubbs catalysts, for being compatible with most functional groups.[15.1-15.3] These two groups of catalysts (below) are complementary in terms of tolerance to functional groups, the ruthenium catalysts being somewhat more air and moisture stable than the group 6 catalysts.

The family of stable, very efficient, high-oxidation-state Mo and W alkylidene complexes working as unimolecular alkene metathesis catalysts has been prepared since 1980 [15.8] and optimized by the Schrock group. These catalysts (included in the Chart below) are up to date the most active unimolecular metathesis catalysts.[15.1,15.2,15.8]

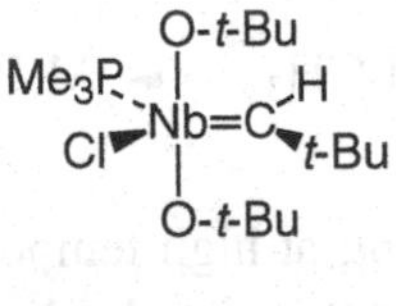

1st unimolecular metal-alkylidene
metathesis catalyst (*cis*-pentene)
Schrock R.R., *J. Mol. Catal.* **8**, 73, **1980**

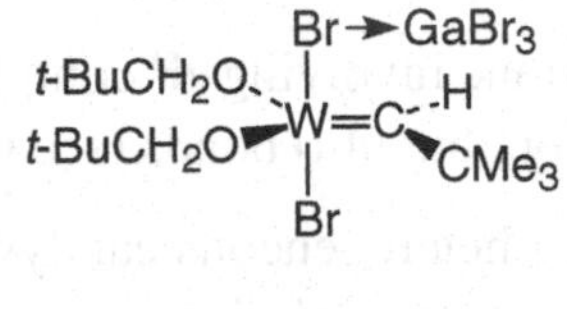

Kress-Welosek-Osborn
catalyst, **1983**

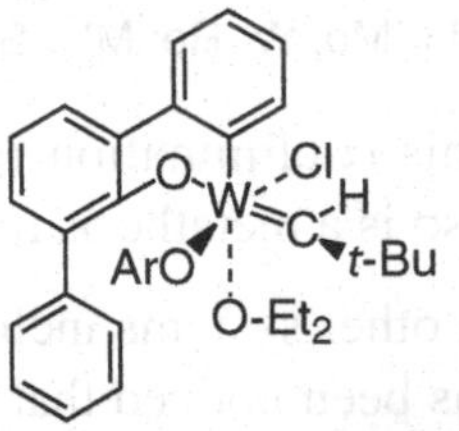

Basset's catalyst, **1985**

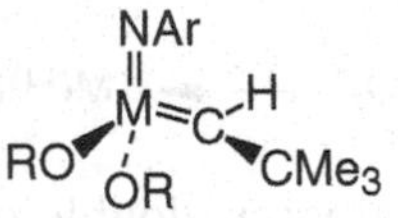

General formula of the family of Schrock's
metathesis catalysts (M = Mo or W; R and Ar
are bulky substituents, **1990**) For an example of
more recent variations, see formula on the right

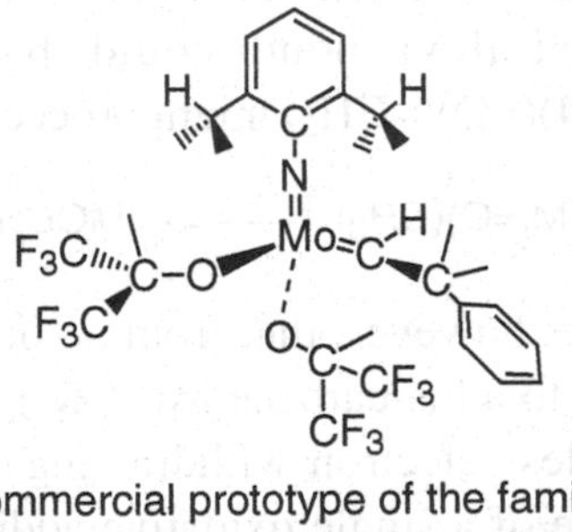

Commercial prototype of the family
of Schrock's metathesis catalysts
(can achieve RCM of tri-and
tetrasubstituted olefins)

Below are summarized the members of the other family, i.e. ruthenium metathesis catalysts from $RuCl_3$ to the most recent variations of Grubbs catalysts. The most universally used of these commercial ruthenium catalysts presently are $[RuCl_2(=CHPh)(PCy_3)_2]$ (1st generation Grubbs catalyst) and $[RuCl_2(=CHPh)(PCy_3)$ L], L = saturated or unsaturated N-heterocyclic carbene (NHC) in which the carbene ligand is substituted on the two nitrogen atoms by mesityl groups, also called 2nd generation Grubbs catalyst (see the Chart p. 377).[15.3,15.5,15.11]

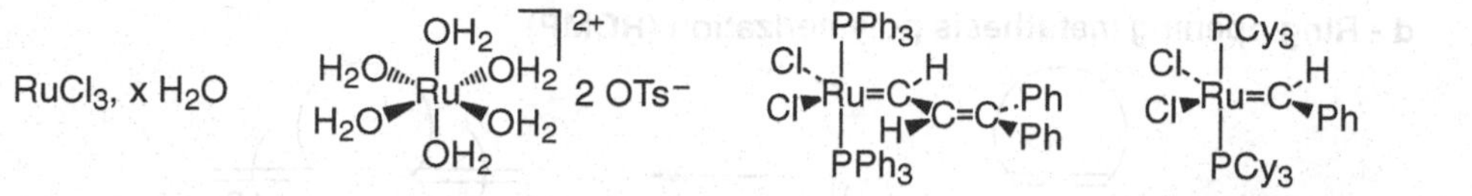

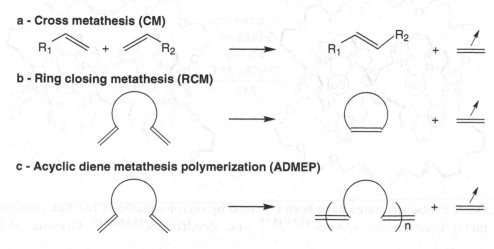

RuCl₃, x H₂O

Natta, **1965**
Grubbs, **1988**

Grubbs, **1988**

Grubbs, **1992**

Grubbs, **1995**
commercial, 1ˢᵗ generation
most used catalyst

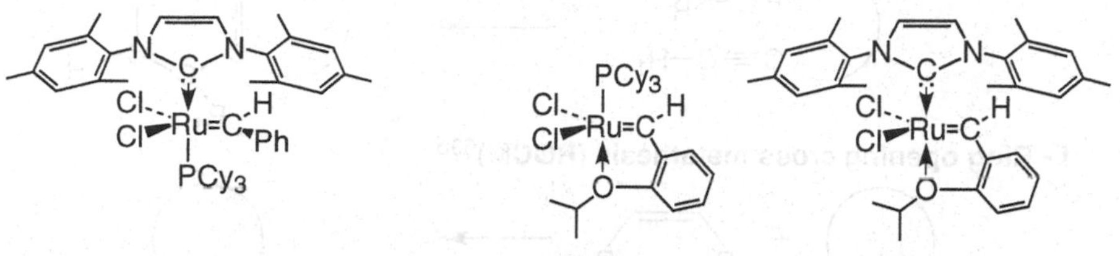

Grubbs, Nolan, Hermann, Fürstner, **1999**
2ⁿᵈ generation, commercial,
second most used catalyst

Hoveyda, **1999**
commercial

Hoveyda, **1999**
commercial

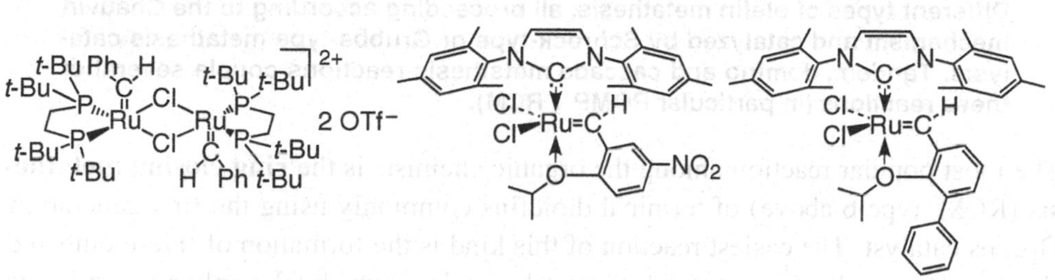

Hofmann, **1999**

Grela, **2000**
(also meta nitro isomer)

Blechert, **2002**
(also binaphthyl derivative)

**Historic progression of the family of unimolecular ruthenium metathesis
catalysts derived from Grubbs commercial 1ˢᵗ generation benzylidene
catalyst [RuCl₂(=CHPh)(PCy₃)₂]. These benzylidene catalysts are tolerant
to a large variety of organic functional groups in metathesis. Analogs
are known with the saturated heterocyclic carbene ligand.**

Six types of metathesis reactions are known, and all of them can be catalyzed by
Schrock and Grubbs metathesis catalysts. They can lead to extensive organic transformations including the synthesis of low-dispersity polymers.[15.1]

a - Cross metathesis (CM)

R_1 + R_2 ⟶ R_1 R_2 +

b - Ring closing metathesis (RCM)

⟶ +

c - Acyclic diene metathesis polymerization (ADMEP)

⟶ +

d - Ring opening metathesis polymerization (ROMP)

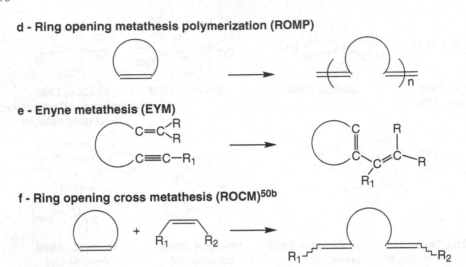

e - Enyne metathesis (EYM)

f - Ring opening cross metathesis (ROCM)[50b]

Different types of olefin metathesis, all proceeding according to the Chauvin mechanism and catalyzed by Schrock-type or Grubbs-type metathesis catalysts. Tandem, domino and cascade metathesis reactions couple several of these reactions (in particular ROMP + RCM).

The most popular reaction among the organic chemists is the **ring closing metathesis** (RCM, type **b** above) of terminal diolefins commonly using the first generation Grubbs catalyst. The easiest reaction of this kind is the formation of five-membered rings such as cyclopentenes and heterocyclic analogs which take only a few minutes at room temperature using, for instance, $[Ru(=CHPh)(PCy_3)_2Cl_2]$, but it is also relatively easy to form large rings if terminal diolefins are used as precursors [15.12] as for instance, for the synthesis of Sauvage's molecular knots and catenanes (below). Yields obtained for these compounds using RCM are much higher than those obtained with traditional methods.[15.13] The problem of the formation of a mixture of *cis* and *trans* isomers, a general one in RCM forming such large rings, is circumvented by subsequent hydrogenation of the double bond (compare alkyne metathesis that also avoids this problem, section 2.2).[15.12]

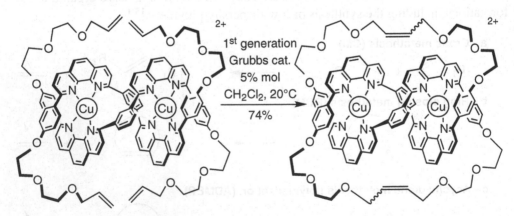

Other exotic structures have been obtained by olefin metathesis reactions around metal coordination spheres [15.15-15.17] and dendrimers.[15.18,15.19] Organic and organometallic chemists can now synthesize large macrocycles some of which, *inter alia*, have important biological and medical properties. Although the first generation Grubbs catalyst is the most popular one in organic chemistry, more active catalysts are sometimes necessary, for instance to carry out **cross metathesis** (CM) reactions.[15.14] Schrock-type catalysts or the second-generation Grubbs catalyst must then be used. Such CM reactions are useful, for instance, for the synthesis of insect pheromones and in the chemistry of oils and perfumes.[15.1,15.5,15.12]

A recent application for using the second-generation Grubbs catalyst is the synthesis of organo-bimetallic and organic cyclophane capsules in two steps from the simple mesitylene sandwich complex [FeCp(η^6-mesitylene)][PF$_6$]. First, the CpFe$^+$-induced nona-allylation of mesitylene under ambient conditions using KOH and allyl bromide gives the nona-allylated compound [FeCp{η^6-1,3,5-C$_6$H$_3$ (C(allyl)$_3$)$_3$}] [PF$_6$] (Scheme below). Metathesis of this complex or its free arene ligand yields the capsules in one pot resulting from nine metathesis reactions including six RCM and three CM reactions.[15.19a] On the other hand, if the metathesis reaction is carried out in the presence of acrylic acid or methylacrylate, the formation of this cyclophane capsule is inhibited because heterobifunctionalization by RCM + CM is much faster. The CM reaction in the presence of olefins bearing an electron-withdrawing substituent is very stereoselective, giving only the *trans* olefin. Even large polyolefin dendrimers could be cleanly functionalized to yield carboxylic-acid terminated dendrimers in this way.[15.19b] This example, *inter alia*, shows that cross methathesis is an extremely efficient, selective and useful method of olefin functionalization.[15.14]

Formation of organic and organometallic cyclophane capsules by triple RCM + triple CM and its inhibition to yield functional dendrimers by CM in the presence of CH$_2$=CHCO$_2$R (R = H or CH$_3$)

Enantioselective metathesis catalysis is presently another major challenge. The first example of a chiral metathesis catalyst was reported by Schrock in 1993 reaction and the first example of very efficient enantioselective ROMP was published by the Schrock and Hoveyda groups using a chiral molybdenum catalyst in 1998; [15.20] it was later followed by another example reported by Grubbs' group.[15.1,15.3] Examples of enantioselective RCM reactions are becoming numerous, including in particular cascade or domino enantioselective reactions by ROMP followed by RCM (see Chap. 21). One advantage of Schrock-type catalysts, also developed by Basset's group on oxide supports,[15.21] over their ruthenium analogs, is the extremely high stereoselectivity (up to 99%) of these metathesis catalysts that is of considerable importance in enantioselective synthesis.[15.20]

2.2. ALKYNE METATHESIS

Metathesis has been applied to dissymmetrical alkynes using MoO_3 or WO_3 as heterogeneous catalysts:

$$2\ R_1C\equiv CR_2 \xrightleftharpoons[M = Mo\ or\ W]{\substack{[MoO_3] \\ 300-400°C}} R_1C\equiv CR_1 + R_2C\equiv CR_2$$

Alkyne metathesis was also shown to proceed under homogeneous conditions using $[Mo(CO)_6]$ + phenol in high boiling solvents,[15.22,15.23] then under mild (ambient) conditions using metallocarbyne complexes of the family $[M(CR)(OR')_3]$ (M = Mo or W; OR' = bulky and electron withdrawing) in the absence of a cocatalyst. For instance, the prototype $[W(C-t-Bu)(O-t-Bu)_3]$, synthesized as shown below by the metathesis reaction between the triply metal-metal bonded complex $[W_2(t-BuO)_6]$ and neoheptyne in large scale, can achieve several hundred turnovers per minute, and alkyne metathesis reactions often proceed at room temperature:[15.24]

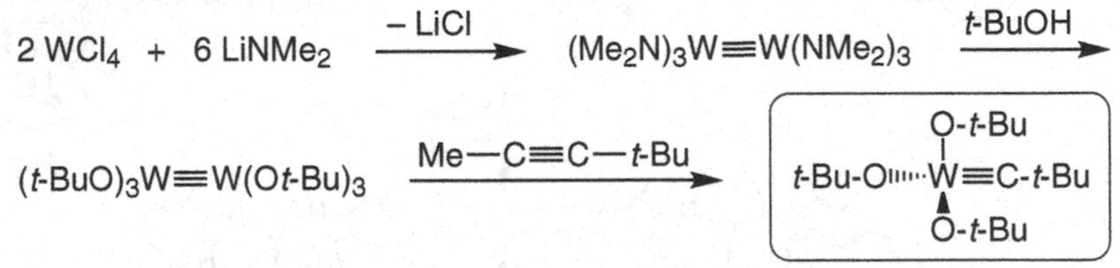

**Synthesis of Schrock's W^{VI} tungsten-carbyne complex,
the prototype of active alkyne metathesis catalysts**

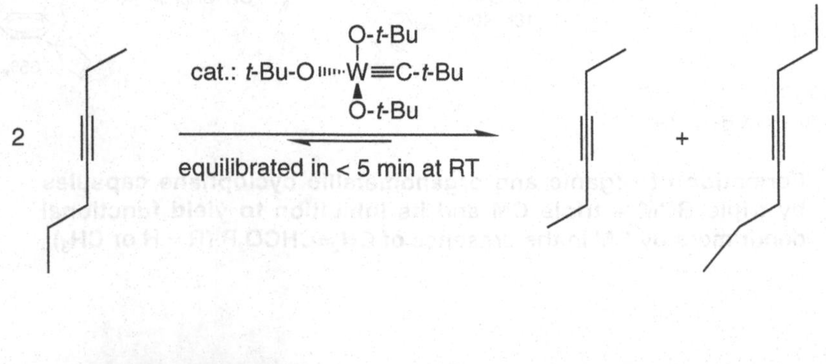

The metallo-square Chauvin's mechanism is still operating,[15.4b,15.7] as demonstrated by the isolation of metallacyclobutadiene complexes formed by [2 + 2] cycloaddition of alkynes to alkylidyne complexes. The metallacyclobutadiene complexes themselves can also serve as alkyne metathesis catalysts confirming their intermediacy in the catalytic reactions starting from the alkylidyne complexes. Interestingly, they do not react readily with alkenes, rendering alkyne metathesis selective in the presence of olefinic bonds.[15.24]

$$M - \boxed{\bigcirc} - R_x$$

metallacyclobutadienes intermediates

⇓

Mechanism:

$$R_1C{\equiv}CR_2 + {\underset{R_1C{\equiv}CR_2}{M{\equiv}CR}} \rightleftharpoons {\underset{\underset{R_1}{C}={\underset{R_2}{C}}}{M={\overset{R}{C}}}} \leftrightarrow {\underset{\underset{R_1}{C}-{\underset{R_2}{C}}}{M-{\overset{R}{C}}}} \rightleftharpoons {\underset{\underset{R_2}{C}}{M={\overset{\overset{R}{C}}{C}}}} \rightleftharpoons M{\equiv}CR_1 + RC{\equiv}CR_2 \quad etc.$$

⇕

$$R_2C{\equiv}CR_1 + {\overset{}{M{\equiv}CR}} \rightleftharpoons {\underset{\underset{R_2}{C}={\underset{R_1}{C}}}{M={\overset{R}{C}}}} \rightleftharpoons {\underset{\underset{R_2}{C}-{\underset{R_1}{C}}}{M-{\overset{R}{C}}}} \rightleftharpoons {\underset{\underset{R_2}{C}}{M={\overset{\overset{R}{C}}{C}}}} \rightleftharpoons M{\equiv}CR_2 + R_1C{\equiv}CR \quad etc.$$

It is believed that the Mortreux system, $[Mo(CO)_6]$ + a phenol (best phenols: p-ClPhOH, p-F$_3$CPhOH and o-FPhOH) [15.25] involves some form of $[Mo^{VI}(CR)(OAr)_3]$ as well as catalytically active species with the alkylidyne ligand coming from CO or the alkyne.[15.1,15.24] Improvement of such a system with a polyether (best efficiency: 1,2-diphenoxyethane) over a bed of molecular sieves leads to metathesis of phenylpropyne at 50°C.[15.26] Another improvement of this system is pre-activation with heptyne and makes it work at only 30 to 60°C higher than the unimolecular tungsten neopentylidyne catalyst.[15.27] It has produced polymer materials with various interesting physical properties by *Acyclic Diyne Metathesis* (ADIMET).

A very efficient bulky triamido Mo pre-catalyst $[Mo\{NR(Ar)\}_3Cl]$ [15.28] is generated by addition of CH_2Cl_2 to Cummins complex $[Mo\{NR(Ar)\}_3]$,[15.29] and is probably also active via a Mo^{VI} alkylidyne intermediate. The methylidyne complex $[Mo\{NR(Ar)\}_3(CH)]$, however, also formed along with the pre-catalyst (Scheme below). It does not sustain catalytic turnover,[15.28] although higher alkylidyne complexes, synthesized from higher gem-dichlorides, do. For instance, $EtCMo[NAr(t-Bu)]_3$ catalyzes the remarkable synthesis of arylene ethynylene macrocycles driven by the precipitation of the metathesis product.[15.30]

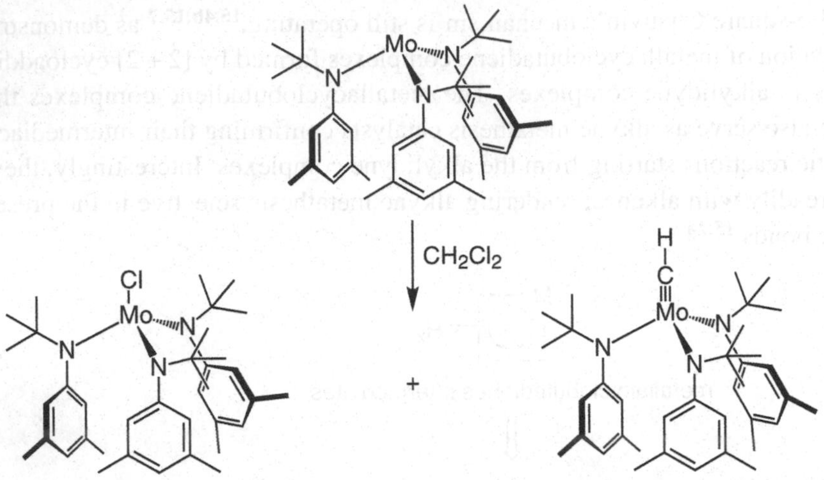

CH₂Cl₂

+

**Powerful alkyne metathesis
precatalyst tolerant to
polar functional groups**

**This methylidyne complex reacts with
alkynes but does not sustain catalytic
turnover, whereas analogous (higher)
alkylidyne complexes do**

The pre-catalyst [Mo{NR(Ar)}₃Cl], as [W(C-*t*-Bu)(O-*t*-Bu)₃], exhibits tolerance towards many polar functional groups. Impressive synthetic applications of alkyne metathesis in organic chemistry using this Mo^{IV} catalyst precursor disclosed by Fürstner have been developed. For instance, homometathesis led to the synthesis of dehydrohomoancepsenolide, cross metathesis (ACM) led to the total synthesis of prostaglandin E₂ (PEG₂), whereas ring closing alkyne metathesis (RCAM) led to the stereoselective synthesis of olfactory lactones and prostaglandin lactones. Indeed, an important advantage of RCAM over RCM is that the RCM reaction is not stereoselective when applied to large rings, giving mixtures of *E* and *Z* olefins that are difficult to separate. This problem does not intervene in RCAM followed by regioselective Birch reduction to the E olefin or Lindnar reduction to the Z olefin: [15.28a]

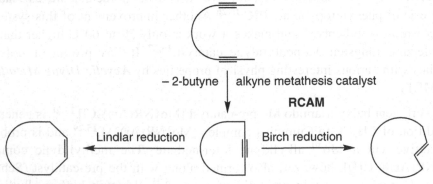

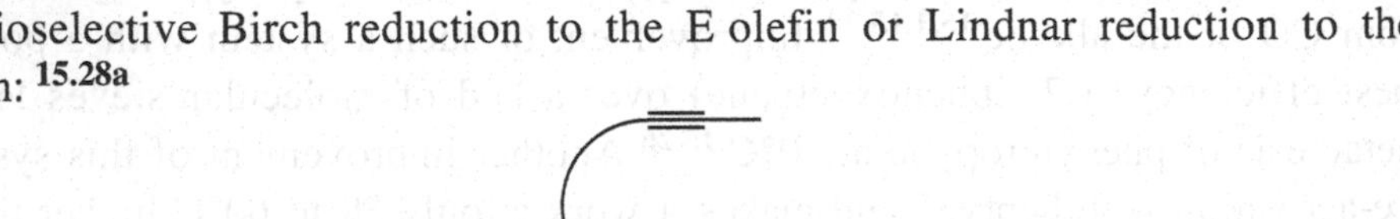

– 2-butyne | alkyne metathesis catalyst

RCAM

Lindlar reduction ← | → Birch reduction

Although the metathesis of terminal alkynes also proceeds rapidly, polymerization of the terminal alkynes displaces the carbyne exchange reactions towards the "wrong side" (towards the starting terminal alkyne that polymerizes) except if suitable parameters are adjusted such as temperature (80°C), solvent (diethyl ether) and especially the presence of the potential ligand quinuclidine together with the

Schrock alkylidyne complex [W(CCMe$_3$)(OCMe$_3$)] (80% carbyne exchange metathesis compounds from 1-heptyne were obtained at 80°C within 1 min using 4 mol% catalyst).[15.28b] Thus, the carbyne exchange metathesis of alkynes terminated by either a methyl group or a hydrogen atom can proceed efficiently, under specific conditions, due to the formation of a volatile sacrificed alkyne (2-butyne resp. acetylene) that displaces the metathesis equilibrium toward the formation of the symmetrically substituted alkyne products. This feature renders these reactions all the more useful in synthesis as no stereoisomers are formed contrary to the situation usually encountered in olefin metathesis.

Discrete alkylidene W and Mo complexes on silica prepared by Basset's group such as the silica-supported alkylidyne-alkylidene complex [(SiO)(Re(C-t-Bu)(=CH-t-Bu)(CH$_2$-t-Bu)] also catalytically metathesize 2-pentyne. They are intermediate between homogeneous and heterogeneous with the advantages of both types.[15.21]

Metallacyclobutadienes formed by addition of an alkyne on a metal-alkylidyne complex are isolable (as also are some metallacyclobutane complexes), but very reactive. Indeed, their formation is reversible and, in the presence of a ligand, these metallacyclobutadienes give back the alkylidyne complex *via* a dissociative mechanism, including if the incoming ligand is an alkyne. In the presence of a chelating ligand, however, they yield complexes of cyclopropenyl, a fragment that is a tautomer of the metallocycle. They can also continue to insert alkyne molecules (see section 4). For instance, the insertion of one more alkyne molecule leads to metallabenzenes (whose isolable examples are known with iridium) that irreversibly rearrange in the present case of Schrock's tungsten complexes to W-η^5-cyclopentadienyl, a much more stable tautomeric fragment.

Metathesis also stoichiometrically applies to nitriles, yielding nitrido complexes *via* the metallaazacyclobutadiene in which the nitrogen atom is bound to the metal.[15.24]

2.3. RING OPENING METATHESIS POLYMERIZATION OF CYCLIC ALKENES (ROMP)

The same metathesis mechanism, described above for linear olefins, can be applied to cyclic olefins and then leads to polymers, the metallocarbenic species remaining linked to the polymer chain ends. This type of polymerization is called Ring Opening Metathesis Polymerization, in short: ROMP): [15.31,15.32]

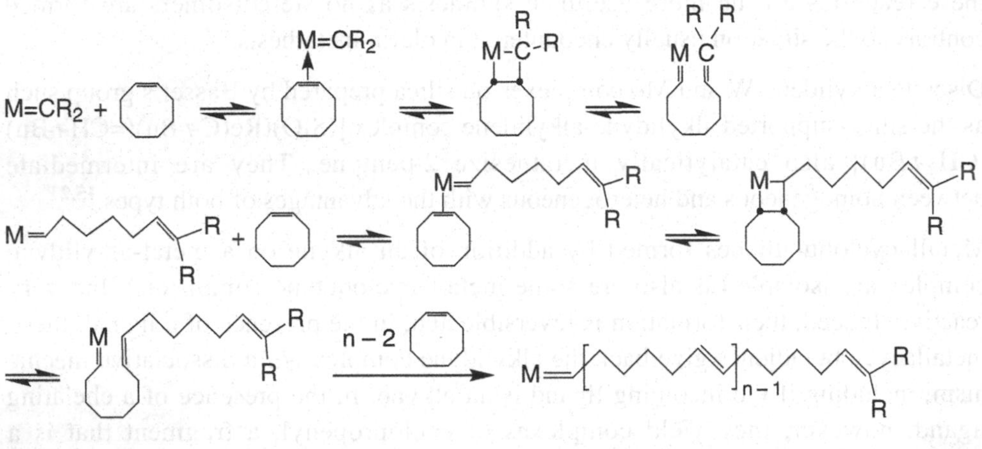

Grubbs has shown that the metathesis of cyclooctatetraene derivatives yielded conducting polyacetylenes although the polymerization system was not "living" (i.e. the polydispersity was not close to 1, and side-reactions have destroyed the initiating chain after the polymer was formed): [15.32]

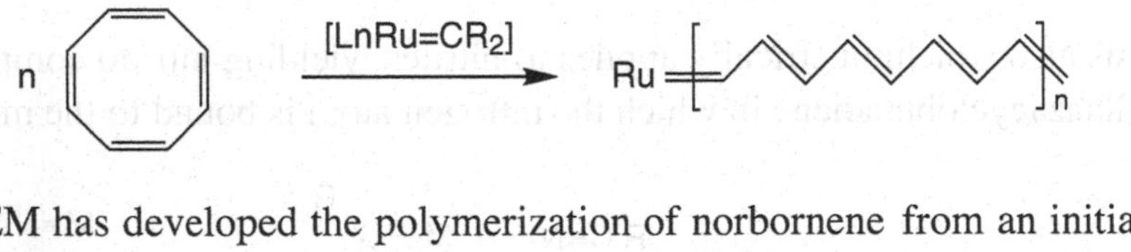

ATOCHEM has developed the polymerization of norbornene from an initiator that was simply $RuCl_3 \cdot n\, H_2O$, butanol and HCl, producing an active metallocarbene species. Numerous applications and outlets of this type of products are foreseen.

It is possible to combine ROMP with cross metathesis. With a diacrylate and a cyclic olefin, the same factors that give selective cross metathesis between an acrylate and a terminal olefin (*vide supra*, lack of homodimerization of acrylates) lead to the formation of alternating polymers if the monomers are present in precisely a 1:1 ratio.[15.3]

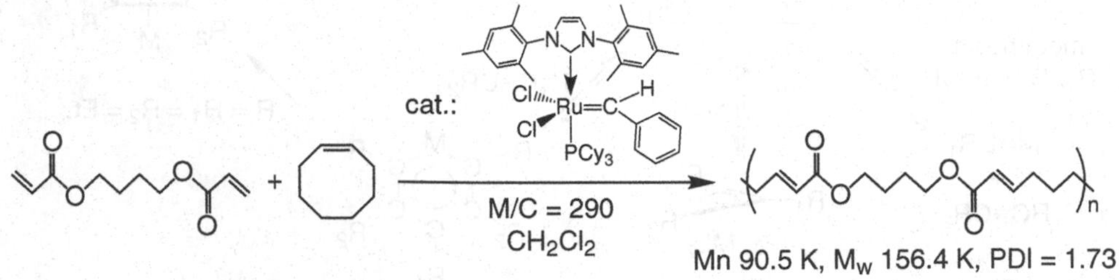

Mn 90.5 K, M_w 156.4 K, PDI = 1.73

2.4. ALKYNE POLYMERIZATION

Alkyne polymerization can be initiated by metal-alkylidenes or metal-alkylidyne complexes. The mechanism involves metallacyclobutene or metallacyclobutadiene key intermediates, in the same way as alkene metathesis and ROMP involve metal-lacyclobutane intermediates (Katz mechanism, bottom of this page, an extension of the Chauvin mechanism for alkyne metathesis). For instance, Schrock's catalyst [W(Ct-Bu)(Ot-Bu)$_3$], shown in section 2.2 as catalyst of disymmetrical alkyne metathesis, also initiates the polymerization of acetylene and terminal alkynes rather than metathesizing them. Indeed, the molecules of terminal alkynes successively insert into the W-C bond of the metallacyclobutadiene intermediate: [15.24]

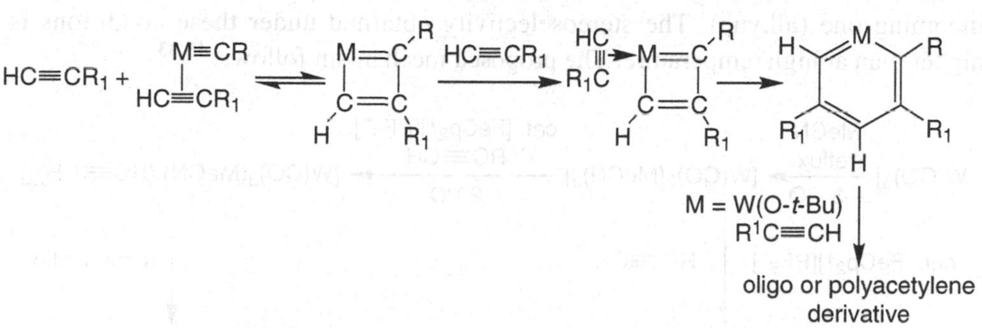

With a carbyne complex of related structure, [W(Ct-Bu)(dme)(Cl)$_3$], the first inser-tion complex, a metallabenzene derivative, irreversibly rearranges to the more stable η^5-cyclopentadienyl complex that has been isolated (see section 2.2). [15.24]

Not only metal-carbyne complexes, but also metal-carbene complexes initiate the polymerization of terminal alkynes. For instance, [W(CO)$_6$] initiates the polymeri-zation of terminal alkynes upon heating several days at 100°C. The thermal decoor-dination of at least two CO ligands opens vacant coordination sites for the coordi-nation of alkyne molecules. The W-terminal alkyne species formed isomerizes to a vinylidene complex W=C=CHR, a common process that has been shown with dis-crete complexes. This vinylidene complex forms, together with another alkyne ligand, a metallacyclobutene rearranging to a metal-carbene according to the metal-losquare mechanism. [15.4b,15.8] Successive iterations yield the polyacetylene derivative.

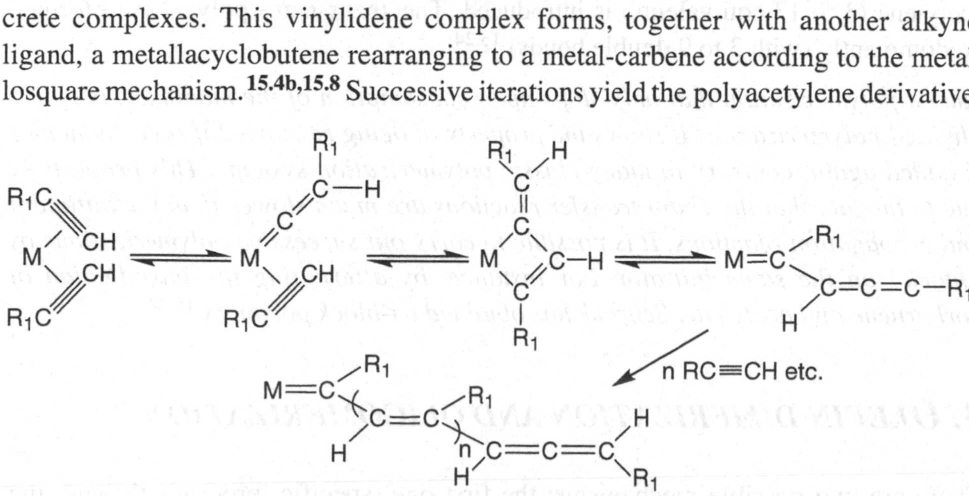

Katz mechanism of terminal alkyne polymerization

It is possible to considerably accelerate this reaction in order to provide cleaner polymers, however. For this purpose, the polymerization process can be coupled with an electron-chain-catalyzed ligand substitution reaction that considerably facilitates introduction of the alkyne onto the W center. Thus, the complex [W(CO)$_3$ (MeCN)$_3$], easily prepared by reaction of [W(CO)$_6$] with MeCN very rapidly initiates this reaction under ambient conditions if a monoelectronic oxidant such as [FeCp$_2^+$][PF$_6^-$] is added in catalytic amount with respect to the tungsten complex (i.e., in very small amount). The reaction then rapidly proceeds at room temperature. The adequate molecular engineering leads to the use of an oxidizing initiator (a 17-electron complex) in order to obtain an exergonic cross electron-transfer step in the propagation cycle, the leaving ligand (MeCN) being a better donor than the incoming one (alkyne). The stereoselectivity obtained under these conditions is higher than at high temperature. The proposed mechanism follows: [15.33]

The Schrock complex [W(=CHt-Bu)(=NAr)(O-t-Bu)$_2$], previously mentioned (section 2.2) as an alkene metathesis catalyst, also initiates the polymerization of acetylene in toluene in the presence of quinuclidine with metallacyclobutene formation followed by opening of the metallacyclobutene, which again confirms the metallo-square mechanism. This polymerization can be controlled if a limited amount of acetylene (3 to 13 equivalents) is introduced. The *trans-trans* polyenes are formed predominantly (with 3 to 9 double bonds).[15.24]

Such a polymerization that only stops upon consumption of the monomer, is called "living" polymerization. It shows the property of being re-started if some monomer is added again, contrary to many classic polymerization systems. This property is due to the fact that the chain transfer reactions are much slower than the initiation and propagation reactions. It is possible to carry out successive polymerizations by "block" on the same initiator. For instance, by alternating the introduction of norbornene and acetylene, Schrock has obtained tri-block polymers.[15.24]

3. OLEFIN DIMERIZATION AND OLIGOMERIZATION

There are two possible mechanisms: the first one, specific, proceeds through the intermediacy of a metallacyclopentane, whereas the second one, non-specific, consists in the insertion of two olefin molecules into a M-H bond as for polymerization. This second method thus also yields oligomers.

3.1. ETHYLENE DIMERIZATION VIA THE INTERMEDIACY OF A METALLACYCLOPENTANE

It consists in coordinating two ethylene molecules to a metal center that then undergo oxidative coupling, then β-elimination followed by reductive elimination of 1-butene. It has been developed by Chauvin at the Institut Français du Pétrole using Ti(O-n-Bu)$_4$ associated with AlEt$_3$ (ALPHABUTOL process).[15.34-15.36] In this extraordinarily active catalyst (20 ppm suffice), Ti has a low number of valence electrons and at least two non-bonding electrons, which is indispensable for an oxidative addition such as oxidative coupling.

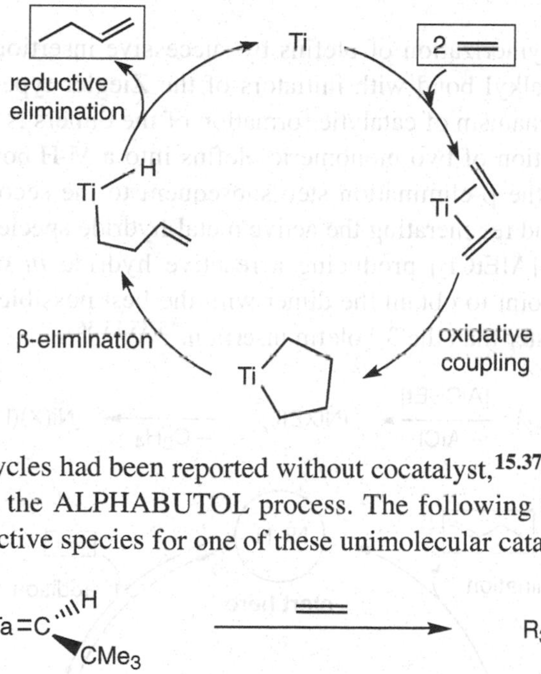

Similar catalytic cycles had been reported without cocatalyst,[15.37,15.38] but they were less active than in the ALPHABUTOL process. The following Scheme shows the generation of the active species for one of these unimolecular catalysts.[15.37]

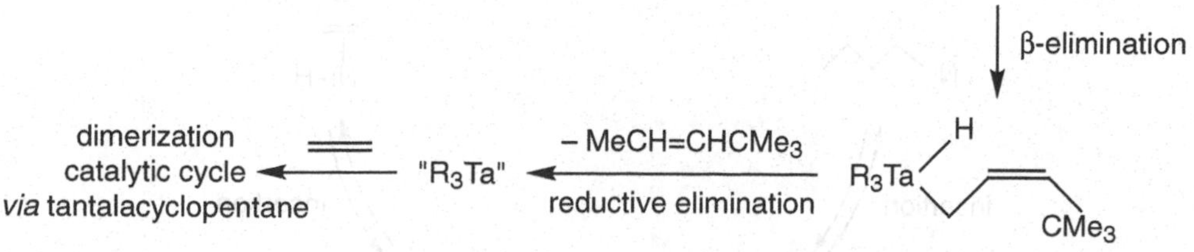

Although the dimerization of olefins proceeds by oxidative coupling to metallacyclopentanes, Schrock has shown using deuterium labeling, that in the case of tantalacyclopentanes, propene dimerization does not involve direct decomposition of this metallocycle by the above mechanism. Ring contraction from tantalacyclopentane to tantalacyclobutane (see exercise **15.5**) intermediates indeed occurs before β-elimination, followed by reductive elimination producing the olefin dimers:

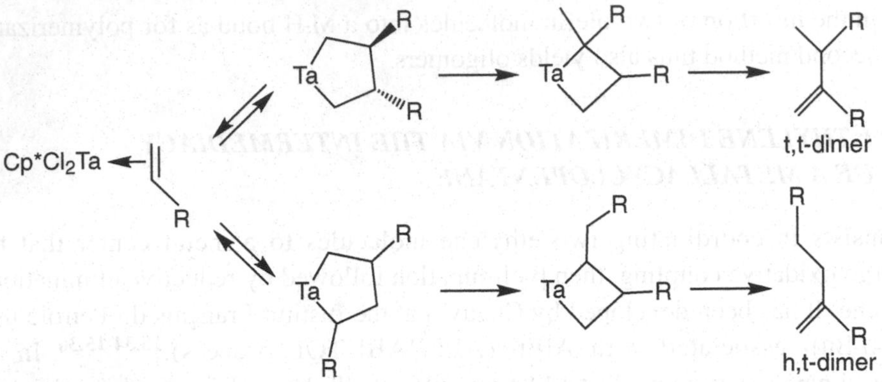

3.2. DIMERIZATION AND OLIGOMERIZATION
BY SUCCESSIVE INSERTIONS

In Chap. 4, the polymerization of olefins by successive insertions of olefin molecules into a metal-alkyl bond with initiators of the Ziegler type has already been examined. The mechanism of catalytic formation of the dimers is the same, but it is limited to the insertion of two monomeric olefins into a M-H bond. The system is made catalytic by the β-elimination step subsequent to the second insertion, producing the dimer and regenerating the active metal-hydride species. A mixture such as $[Ni(PR_3)_2Cl_2] + [AlEtCl_2]$ producing a reactive hydride *in situ* is used as the catalyst. The key point to obtain the dimer with the best possible selectivity is that β-elimination be faster than the 3rd olefin insertion.[15.35,15.36]

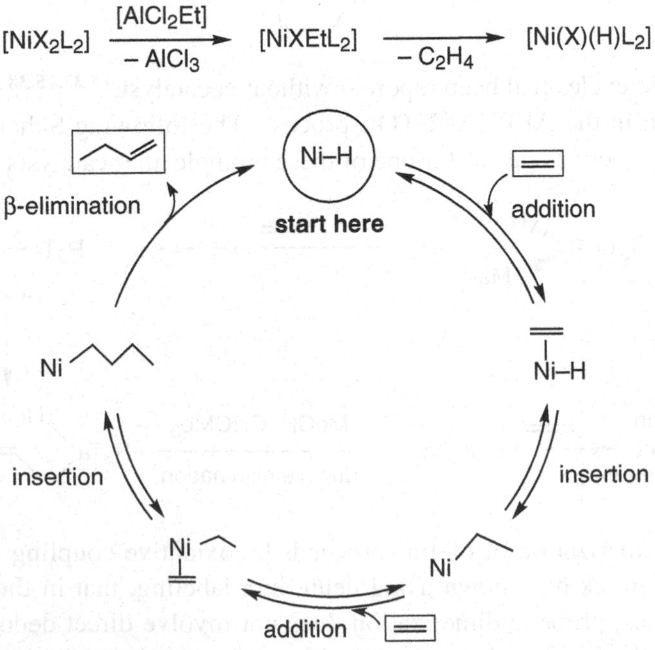

$$[NiX_2L_2] \xrightarrow[- \ AlCl_3]{[AlCl_2Et]} [NiXEtL_2] \xrightarrow[- \ C_2H_4]{} [Ni(X)(H)L_2]$$

Such a selectivity can be obtained with hemilabile bidentate ligand, a concept that allows optimum control in various types of catalysis.[15.39] For instance, the

Ni^{II} complexes below afforded mostly dimers and trimers in the catalytic conversion of ethylene. In the presence of only 6 equiv. of $AlEtCl_2$ as a cocatalyst, a turnover frequency (TOF) of 45,900 mol C_2H_4 / mol. hemilabile Ni complex / h was obtained compared to 27,200 mol C_2H_4 / mol. Ni / h for $[NiCl_2(PCy_3)_2]$:[15.39]

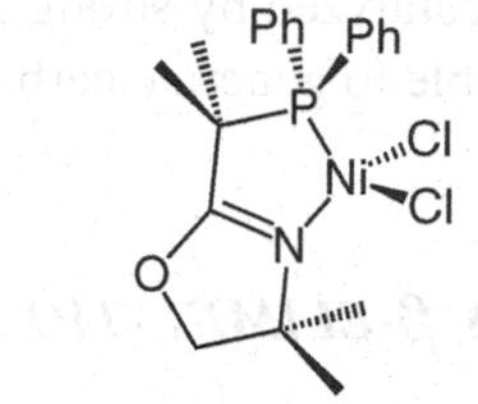

Hemilabile
phosphino-oxazoline Ni^{II}

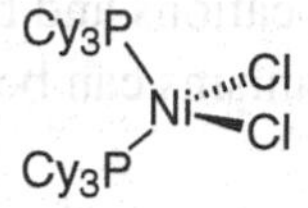

Non-hemilabile
tricyclohexylphosphine Ni^{II}

On the other hand, if insertion of additional olefin molecules is faster than β-elimination, the reaction yields polymers. *Oligomers* are intermediates between the dimer and the polymers. The term *oligomer* is usually reserved to a chain containing between 3 and 100 monomer units, and *the term polymer is used above 100 monomer units* in the chain. In the case of the formation of oligomers, the same mode of termination, β-H elimination, is found to close the catalytic cycle. The nature of the ligands is a key parameter influencing the length of the oligomer chain. In the SHOP process from Shell (section 1), a group of oligomers between C6 and C18 is targeted. Square planar Ni^{II} complexes containing various chelating ligands show the optimal selectivity,[15.35,15.36] in particular those with other hemilabile ligands.[15.39]

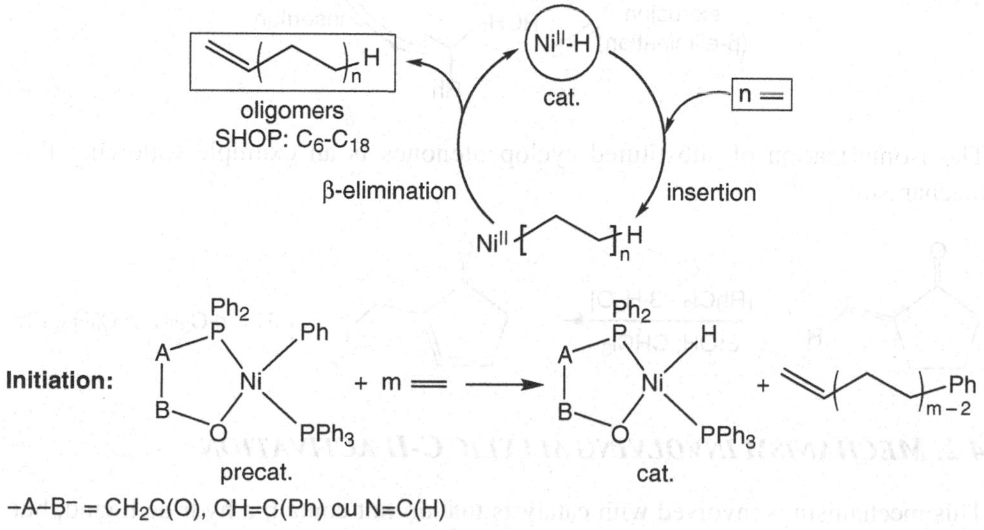

Initiation:

precat.

cat.

$-A-B^- = CH_2C(O), CH=C(Ph)$ ou $N=C(H)$

Finally, it is possible to cyclo-dimerize, trimerize, cyclo-trimerize, oligomerize and co-oligomerize 1,3-butadiene (with other alkenes and alkynes) using catalysts based on Ni^0, Ti^{IV} and Rh^I. A variety of large rings can be obtained in this way with high turnover numbers.[15.36]

4. OLEFIN ISOMERIZATION

Isomerization can occur via migration of the double bond. Typically, this is case of the isomerization of a terminal olefin to the thermodynamically more stable internal olefin, also known in organic chemistry for being catalyzed by strong acids that are able to generate carbocations and by strong bases able to generate carbanions. Three organometallic mechanisms can be examined here.

4.1. MIGRATION OF THE DOUBLE BOND BY β-ELIMINATION

Numerous 16-electron metal-hydride complexes or complexes of noble metals iso-merize terminal olefins according to a mechanism involving reversible insertion of the olefin into the M-H bond followed by reversible extrusion (β-elimination), i.e. overall isomerization under thermodynamic control. A mixture of *cis* and *trans* olefins is obtained, the more stable *trans* form being the major one.[15.36]

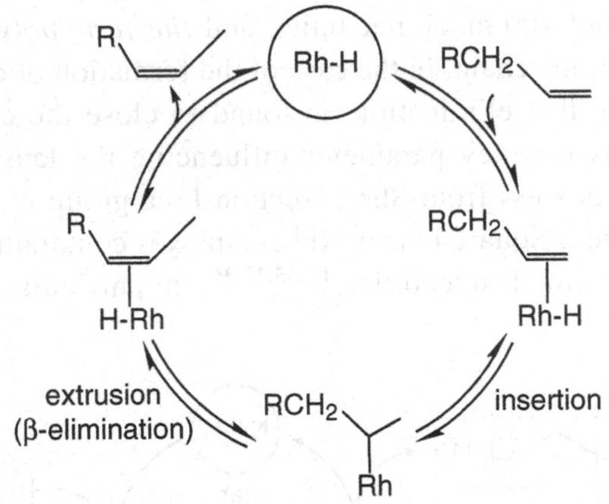

The isomerization of substituted cyclopentenones is an example following this mechanism:

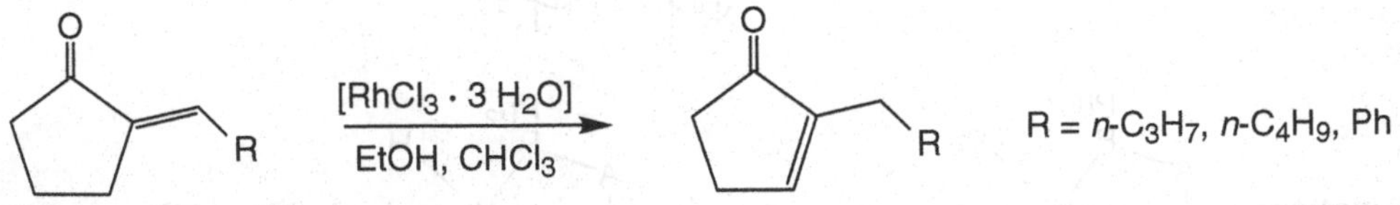

4.2. MECHANISM INVOLVING ALLYLIC C-H ACTIVATION

This mechanism is involved with catalysts that do not contain a hydride ligand, but potentially have at least two vacant coordination sites. Such a situation is rare. Indeed, only one case is known, with the 14-electron species $Fe(CO)_3$ generated by fragmentation of the cluster $[Fe_3(CO)_{12}]$ or upon photolysis of $[Fe(CO)_5]$.

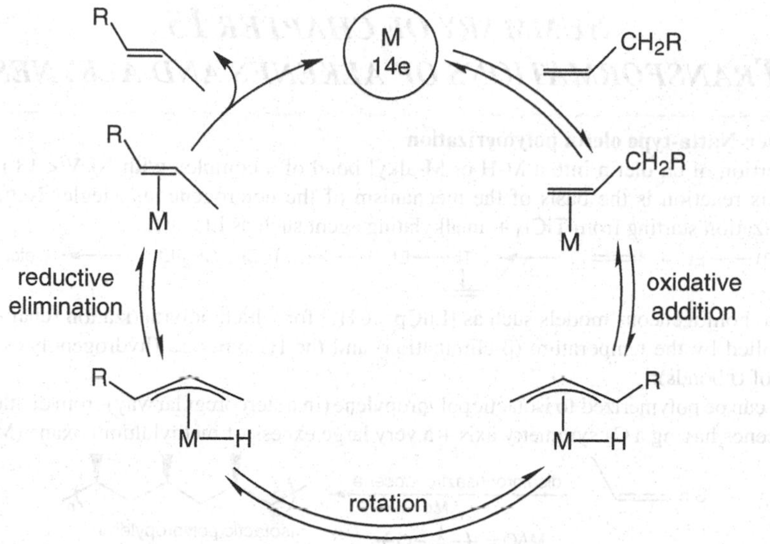

Example:

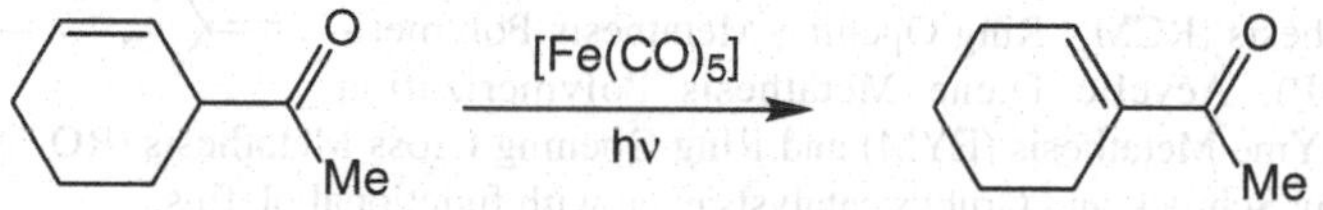

4.3. CIS-TRANS OR Z-E ISOMERIZATION VIA METALLOCARBENES

Metathesis allows the recombination of two carbene fragments of the olefin. When an olefin is subjected to *cis-trans* or *Z-E* isomerization, one of these recombinations proceeds according to the opposite geometry:

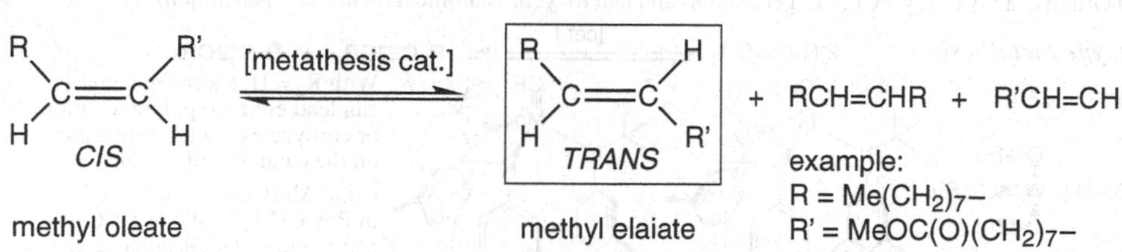

CIS

methyl oleate

TRANS

methyl elaiate

+ RCH=CHR + R'CH=CHR'

example:
R = Me(CH$_2$)$_7$–
R' = MeOC(O)(CH$_2$)$_7$–

SUMMARY OF CHAPTER 15
TRANSFORMATIONS OF ALKENES AND ALKYNES

1 - Ziegler-Natta-type olefin polymerization

The insertion of an olefin into a M-H or M-alkyl bond of a complex with NEV < 18 is favorable. This reaction is the basis of the mechanism of the heterogeneous Ziegler-Natta olefin polymerization starting from $TiCl_4$ + an alkylating agent such as Et_2AlCl.

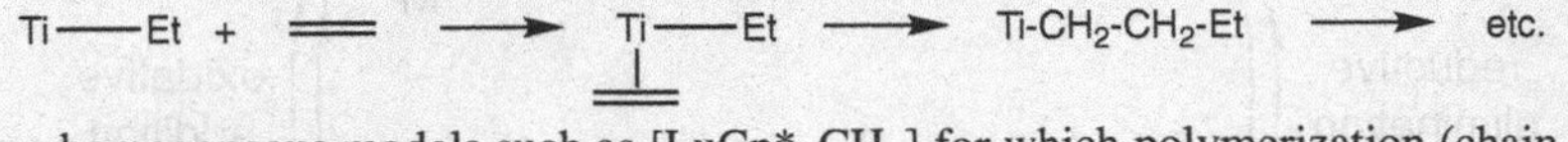

Ti——Et + ==== ——→ Ti——Et ——→ Ti-CH$_2$-CH$_2$-Et ——→ etc.

There are homogeneous models such as $[LuCp*_2CH_3]$ for which polymerization (chain length) is controlled by the temperature (β-elimination) and the H_2 pressure (hydrogenolysis by metathesis of σ bonds).

Propene can be polymerized to isotactic polypropylene (in a stereoregular way) from dichloroansazirconocenes having a C_2 symmetry axis + a very large excess of methylaluminoxane (MAO).

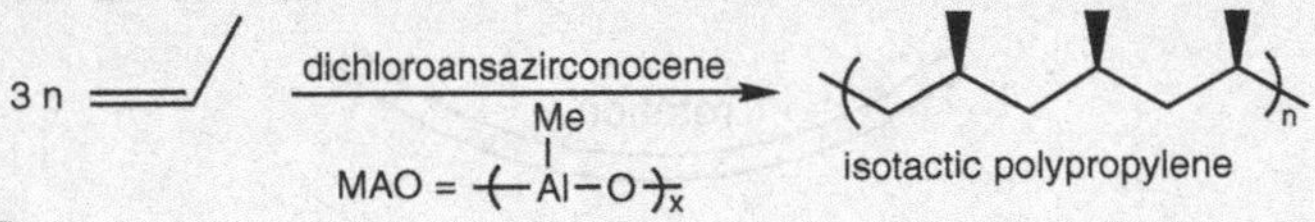

$$3\,n \quad \diagup \quad \xrightarrow[\substack{\text{Me} \\ | \\ \text{MAO} = (\text{Al}-\text{O})_x}]{\text{dichloroansazirconocene}} \quad \text{isotactic polypropylene}$$

2 - Metathesis

Alkene metathesis - Main types: Cross Metathesis (CM), Ring Closing Metathesis (RCM), Ring Opening Metathesis Polymerization (ROMP), Acyclic Diene Metathesis Polymerization (ADMEP), EnYme Metathesis (EYM) and Ring Opening Cross Metathesis (ROCM). All types are catalyzed by Schrock and Grubbs catalysts even with functional olefins.

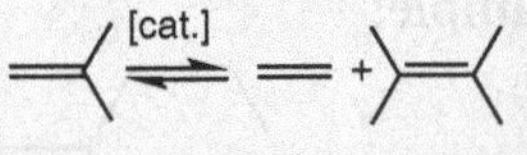

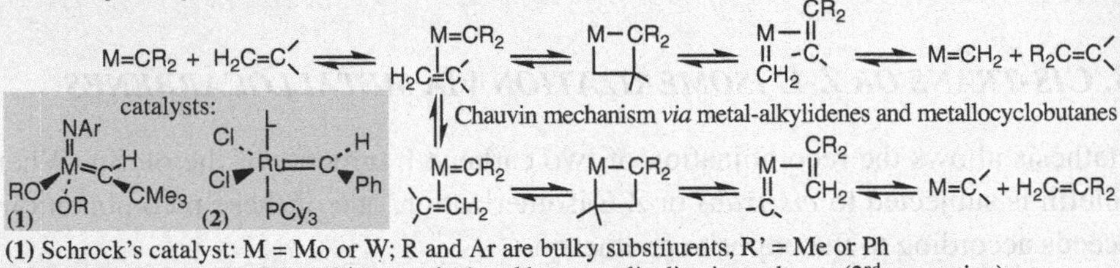

$$M=CR_2 + H_2C=C \rightleftharpoons H_2C=C \rightleftharpoons \rightleftharpoons -C \rightleftharpoons M=CH_2 + R_2C=C$$

catalysts:

Chauvin mechanism *via* metal-alkylidenes and metallocyclobutanes

(1) Schrock's catalyst: M = Mo or W; R and Ar are bulky substituents, R' = Me or Ph
(2) Grubbs catalyst: L = PCy_3 (1st generation) and heterocyclic diaminocarbenes (2nd generation)

Alkyne metathesis $2\ R_1C\equiv CR_2 \xrightarrow{\text{[cat.]}} R_1C\equiv CR_1 + R_2C\equiv CR_2$

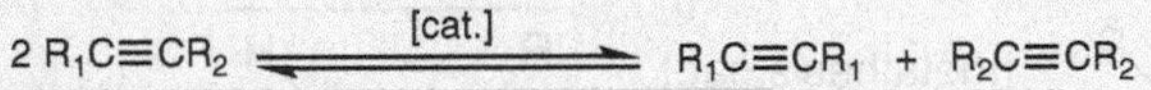

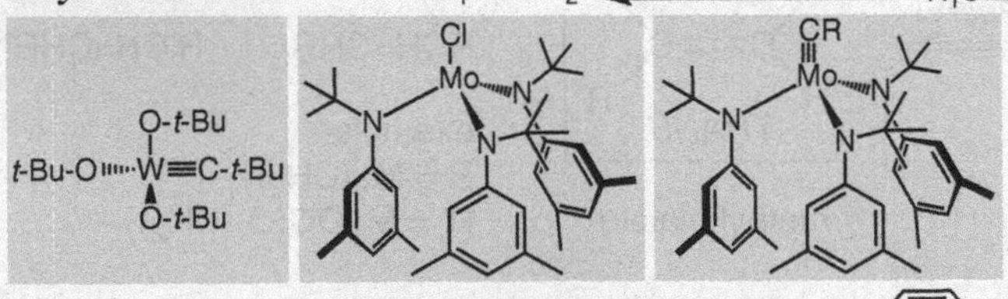

With R_2 = H, alkyne metathesis can lead either to polymerization or carbyne exchange depending on the catalyst and solvent.
cat. = $Mo(CO)_6$ + p.ClC_6H_4Cl in $PhOCH_2CH_2OPh$ at 50°C or one of the 3 unimolecular catalyst

Ring Opening Metathesis Polymerization (ROMP)

$$n \quad \bigcirc \quad \xrightarrow{M=CR_2} \quad M = \left(\quad \right)_n = CR_2$$

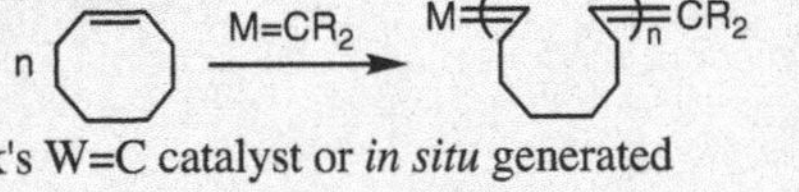

Alkyne polymerisation can be carried out using Schrock's W=C catalyst or *in situ* generated W-vinylidene:

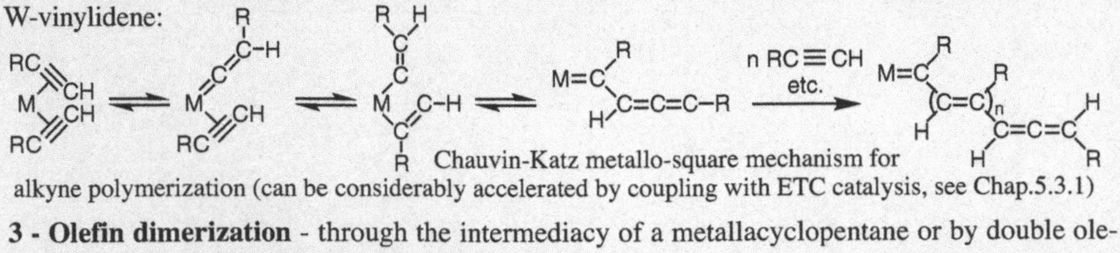

Chauvin-Katz metallo-square mechanism for alkyne polymerization (can be considerably accelerated by coupling with ETC catalysis, see Chap.5.3.1)

3 - Olefin dimerization
through the intermediacy of a metallacyclopentane or by double olefin insertion into a M-H bond + β-elimination.

4 - Olefin isomerization
• double-bond migration by insertion-β-elimination or by allylic C-H activation.
• *E-Z* isomerization by alkene metathesis

EXERCISES

15.1. During the early work on catalyzed olefin metathesis, chemists observed that the metathesis reaction worked better if some air was admitted in the reaction mixture. How do you explain this phenomenon?

15.2. Olefin metathesis usually is under thermodynamic equilibrium, which could provide a limitation of applications. For which type of olefin is the metathesis reaction totally shifted towards the formation of the products?

15.3. What is the result of cyclooctene metathesis? Give the mechanism of product formation.

15.4. Should an olefin dimerization catalyst have an electron-rich or electron-poor metal center. Why? Example?

15.5. Suggest a mechanism for the ring contraction from tantalacyclopentane to tantalacyclobutane accounting for olefin dimerization catalysis of section 3.1.

Chapter 16

OXIDATION OF OLEFINS

1. HISTORIC INTRODUCTION
AND RADICAL-TYPE OXIDATION

The oxidation reactions of organic compounds are numerous and varied, include homogeneous and heterogeneous processes as well as biological and industrial ones and involve a variety of mechanisms.[16.1-16.3]

The first synthetic example, oxidation of ethanol to acetic acid by air catalyzed by platinum, was discovered by Dany in 1820. It is only one century later that catalytic oxidation reactions were developed with radical autoxidation processes. Indeed, in 1939, Criegee discovered autoxidation of cyclohexene to allyl hydroperoxide:

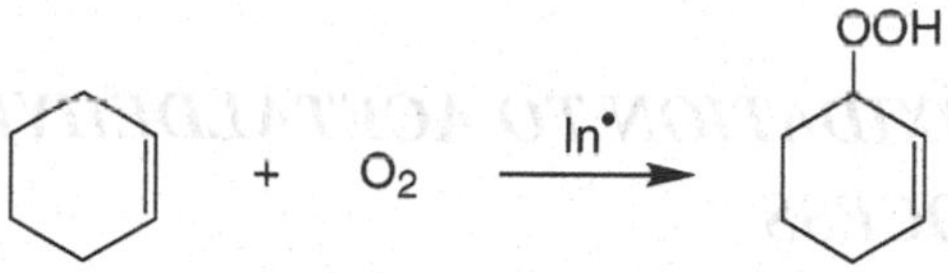

$$\text{(cyclohexene)} + O_2 \xrightarrow{In^\bullet} \text{(cyclohexenyl hydroperoxide, OOH)}$$

$$RH + O_2 \longrightarrow ROOH$$

initiation: $RH + In^\bullet \longrightarrow R^\bullet + InH$ ($In^\bullet$ = radical initiator)

propagation: $R^\bullet + O_2 \longrightarrow ROO^\bullet$ (fast, diffusion-controlled step)

     $ROO^\bullet + RH \longrightarrow ROOH + R^\bullet$ (rate-determining step)

termination: $R^\bullet + ROO^\bullet \longrightarrow ROOR$

     $2\,ROO^\bullet \longrightarrow RO_4R \longrightarrow O_2 + \text{non-radical products}$

(Russell fragmentation: $RO_4R \longrightarrow 2RO^\bullet + O_2$ is a minor reaction sometimes observed, followed by: $RO^\bullet \longrightarrow ROH + R^\bullet$)

The stabilization of the radical $R^\bullet$ by conjugation with a double bond (allyl) or an aromatic ring (benzyl) is necessary in order to facilitate the reaction. Indeed, the energy of an allylic or benzylic C-H bond is only 85 kcal·mol⁻¹ (355 kJ·mol⁻¹), lower than that of a O-H bond in hydroperoxides (90 kcal·mol⁻¹, 376 kJ·mol⁻¹) or C-H bonds in saturated hydrocarbons (around 100 kcal·mol⁻¹, 418 kJ·mol⁻¹). This stabilization can also be steric (t-Bu⁻ radical). The process is always industrially exploited, in particular for the synthesis of cumyl hydroperoxide $PhC(CH_3)_2OOH$ (precursor of phenol), $PhCH(CH_3)OOH$ (precursor of styrene) and t-BuOOH [16.1,16.2] (industrial agent for olefin epoxidation, *vide infra*).

Important industrial processes such as oxidation of butane to acetic acid, of *p*-xylene to terephthalic acid and ethylene to acetaldehyde have appeared at the end of the 1950s. They require a catalytic system based on transition metals. At the laboratory scale, the catalytic oxidation reactions using *ter*-butyl hydroperoxide are numerous:

$$t\text{-BuOOH} + R\text{–CH}_2\text{–CH=CH}_2 \quad \xrightarrow[-\,t\text{-BuOH}]{}$$

$$\xrightarrow{[\text{Mo}^{VI}]} R\text{–CH}_2\text{–CH}\overset{O}{\diagup}\text{CH}_2$$

$$\xrightarrow{[\text{Os}^{VIII}]} R\text{–CH}_2\text{–}\underset{\overset{|}{OH}}{CH}\text{–CH}_2$$

$$\xrightarrow{[\text{Se}^{IV}]} R\text{–CH}_2\text{–}\underset{\overset{|}{OH}}{CH}\text{=CH}_2$$

More recently, research concerning catalytic oxidation reactions has emerged on one side from bioinorganic chemistry with cytochrome P450 models and non-porphyrinic methane mono-oxygenase models, and on the other side, from organic chemistry with asymmetric epoxidation and dihydroxylation.

2. ETHYLENE OXIDATION TO ACETALDEHYDE: WACKER PROCESS

The Wacker process (1953) presently leads to the production of 4 million tons of acetaldehyde *per* year:

$$H_2C\text{=CH}_2 + 1/2\,O_2 \xrightarrow{\text{PdCl}_2,\ \text{CuCl}_2,\ \text{H}_2\text{O}} CH_3CHO \qquad \Delta H° = -243\ \text{kJ}\cdot\text{mol}^{-1}$$

The stoichiometric ethylene oxidation reaction has been discovered by F.C. Phillips in 1894:

$$H_2C\text{=CH}_2 + H_2O + PdCl_2 \longrightarrow CH_3CHO + 2\,HCl + Pd^0$$

In aqueous medium, $PdCl_2$ is actually in the form $[PdCl_4]^{2-}$. The detailed mechanism has only been proposed in 1979 by Bäckwall and Stille [16.3-16.5,16.6a] (see the catalytic cycle below). The isomerization of the hydroxyethyl ligand by β-elimination and re-insertion before decomposition to acetaldehyde has been demonstrated by the fact that addition of D_2O, known to deuterate an enol, does not lead to the incorporation of deuterium in acetaldehyde. The rate law is:

$$v = d(CH_3CHO)/dt = k([PdCl_4]^{2-})(C_2H_4)/(Cl^-)^2(H^+)$$

It can be deduced that the rate-limiting step involves, in its transition state, a palladium complex containing an ethylene molecule and having lost two chloride ligands and a proton.

In this stoichiometric reaction, the palladium metal precipitates. In the presence of oxygen, the thermodynamics is favorable to the re-oxidation of Pd^0 to Pd^{II}. The structural transformation required for Pd oxidation slows down this re-oxidation, however. The Pd^0 colloid formation is thus faster and the kinetics is unfavorable for catalysis. It is the introduction of $CuCl_2$ as a cocatalyst that allowed to make this process catalytic. Indeed, $CuCl_2$ can rapidly re-oxidize Pd^0 to $PdCl_2$ because of fast inner-sphere Cl transfer *via* bridging Cu and Pd. CuCl formed can be oxidized by O_2. The redox Cu^I/Cu^{II} system works as a redox catalyst in a way very similar to that of biological systems. The coupling between coordination catalysis and redox catalysis is thus a biomimetic concept. The overall catalytic cycle follows:

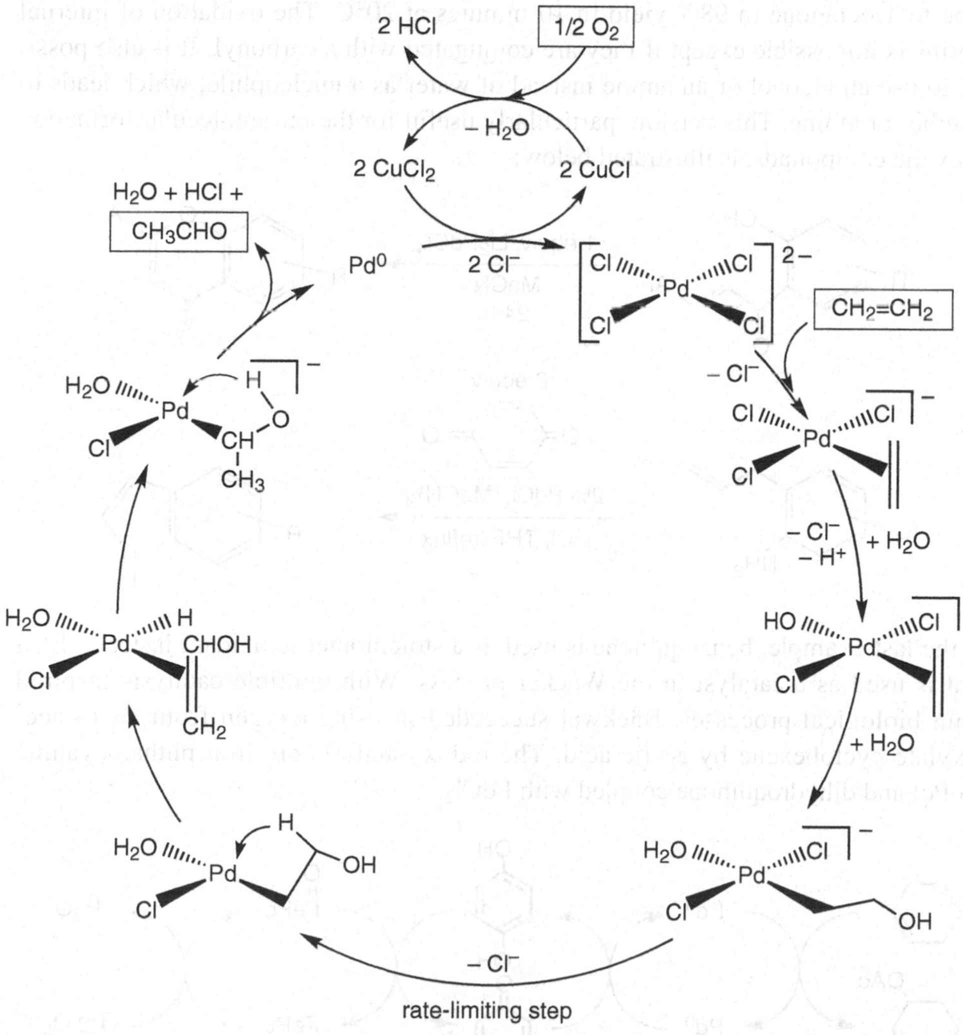

Accepted mechanism for the Wacker process

or, in summary:

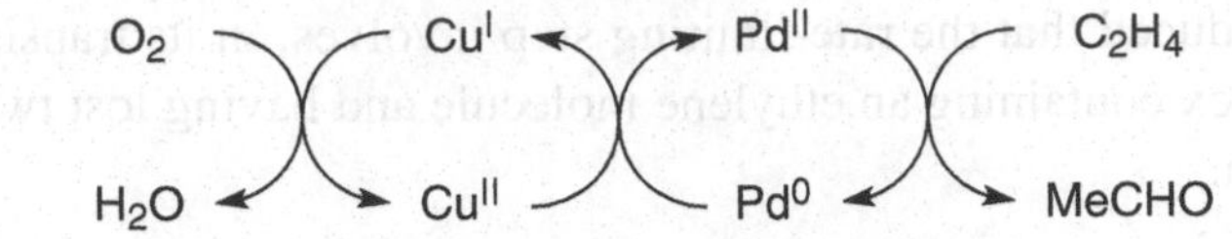

The Wacker process has also been applied to the ketonization of terminal olefins. Although these applications are complicated by the isomerization of the olefins, good selectivities are now obtained in particular if DMF is added. The reaction is currently used in organic synthesis.

$$RCH=CH_2 + 1/2\ O_2 \xrightarrow{\ Pd^{II}/Cu^{II}/H_2O/DMF\ } RCOCH_3$$

For this ketonization reaction, t-BuOOH is a more selective oxidant than O_2, however. For instance, 1-octene is oxidized under these conditions by t-BuOOH in benzene to 1-octanone in 98% yield in 10 minutes at 20°C. The oxidation of internal olefins is impossible except if they are conjugated with a carbonyl. It is also possible to use an alcohol or an amine instead of water as a nucleophile, which leads to an ether or amine. This version, particularly useful for the intramolecular formation of cyclic compounds, is illustrated below:

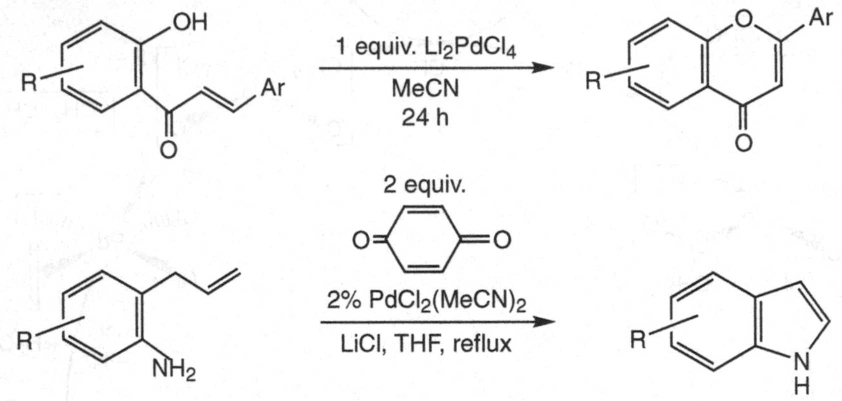

In the last example, benzoquinone is used as a stoichiometric oxidant instead of Cu that is used as a catalyst in the Wacker process. With multiple catalysis inspired from biological processes, Bäckwal succeeded in using oxygen from air to acetoxylate cyclohexene by acetic acid. The redox catalysts are iron phthalocyanine (FePc) and dihydroquinone coupled with $PdCl_2$:

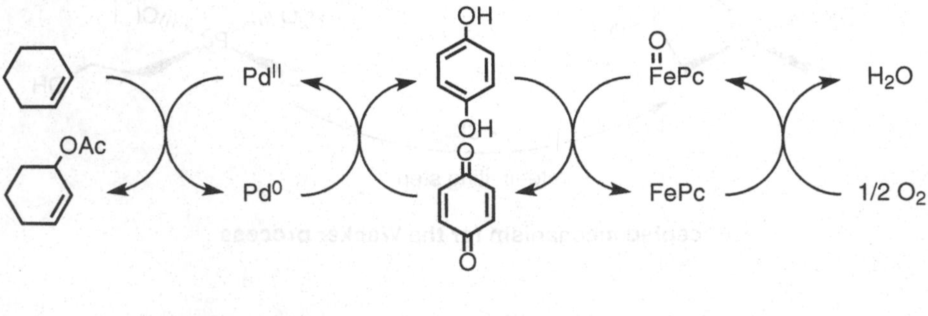

There are many variations of catalytic systems for the ketonization of terminal olefins, for instance those using the Wilkinson-Osborn catalyst [RhCl(PPh$_3$)$_3$]. It is also often possible to avoid Cu by using this rhodium catalyst with PPh$_3$ as a co-substrate or with [Pd(OAc)$_2$] using H$_2$O$_2$ as the oxidant. These catalytic systems have a mechanism that is far from that of the Wacker process, however. They involve metal-oxo or metal-peroxo species that have remarkable properties (see section 5 and Chap. 18.5).

Multiply bonded O ligands in the model organometallic complex [Pt$_4$(1,5-COD)$_4$ (μ_3-O)$_2$Cl$_2$][BF$_4$]$_2$ have been shown to oxidize ethylene to acetaldehyde and norbornene to a platinaoxetane complex. Such a reaction might suggest an unexpected role of oxo complexes in oxidation chemistry (e.g. Wacker chemistry).[16.6b]

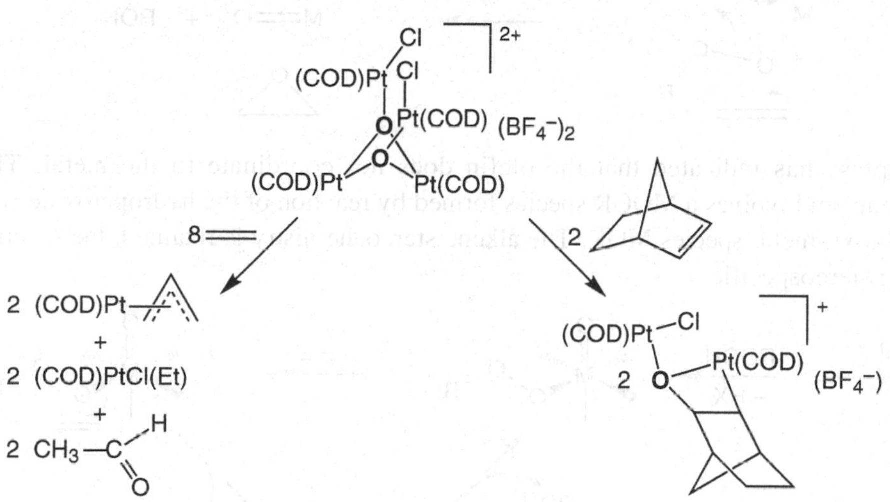

4. EPOXIDATION OF OLEFINS

The epoxidation of olefins was discovered by a Russian chemist, N. Prileschajew, in 1909 using peracids RCO$_3$H, and this reaction has been used for a long time. Per-acids often are dangerous (explosive), however, and must therefore imperatively be avoided. Transition-metal catalysis is now used together with hydroperoxides ROOH (most often R = t-Bu, in short TBHP) whose first example was discovered by Hawkins in 1950 with [V$_2$O$_5$] as catalyst. The catalysts most often are d^0 complexes with an oxophilic Lewis acid able to bind an oxygen atom of hydroperoxide.

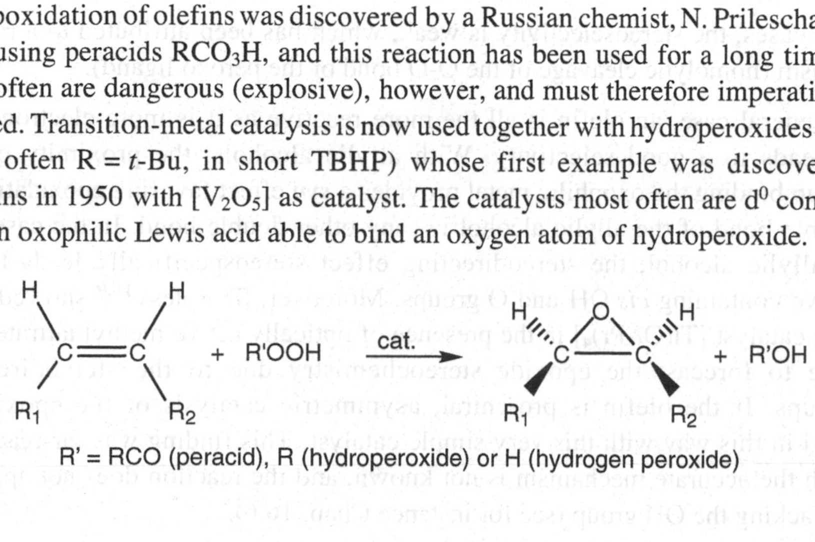

R' = RCO (peracid), R (hydroperoxide) or H (hydrogen peroxide)

The mechanistic analogy between the reaction with a peracid and that using a transition-metal catalyst and a hydroperoxide has been pointed out by Sharpless.[16.7]

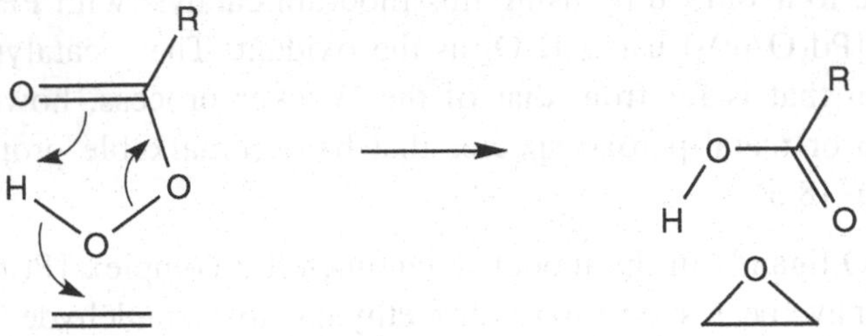

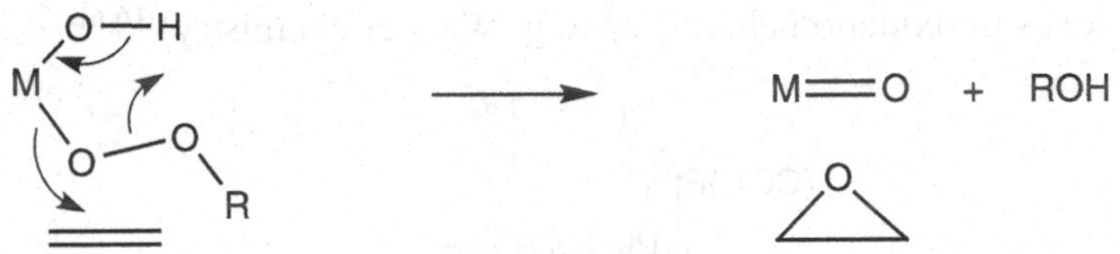

Sharpless has indicated that the olefin does not coordinate to the metal. This mechanism involves a MOOR species formed by reaction of the hydroperoxide with an alkoxy-metal species MOR. The alkene stereochemistry is retained, the reaction being stereospecific.

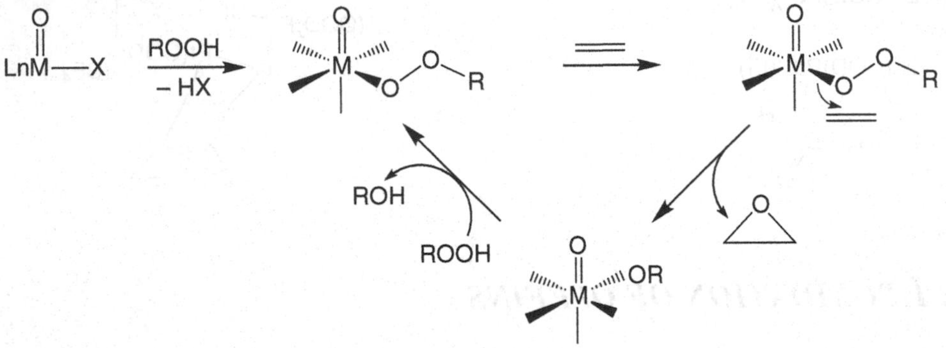

In some cases, the stereoselectivity is weak, which has been attributed to a radical mechanism (homolytic cleavage of the O-O bond of the peroxo ligand).

In the general case, an olefin is all the more reactive as it is more electron rich, which leads to a good selectivity. With allylic alcohols, the proximity of the OH group binding the oxophilic metal provides a *syn* effect favoring epoxidation of the double bond of the allylic alcohol *vs.* any other double bond. In the case of a cyclic allylic alcohol, the stereodirecting effect stereospecifically leads to the derivative containing *cis* OH and O groups. Moreover, Sharpless[16.9] showed that, with the catalyst [Ti(Oi-Pr)$_4$] in the presence of optically active methyl tartrate, it is possible to forecast the epoxide stereochemistry due to the stereodirecting OH groups. If the olefin is pro-chiral, asymmetric catalysis of the epoxide is achieved in this way with this very simple catalyst. This finding was far-reaching, although the accurate mechanism is not known, and the reaction does not apply to olefins lacking the OH group (see for instance Chap. 16.6).

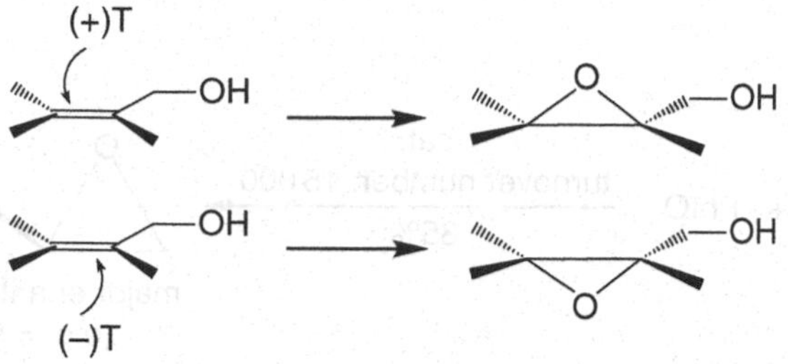

In 1990, Jacobsen found a catalyst able to achieve asymmetric epoxidation of simple alkenes using iodosylarenes ArIO or NaOCl as the oxidant.

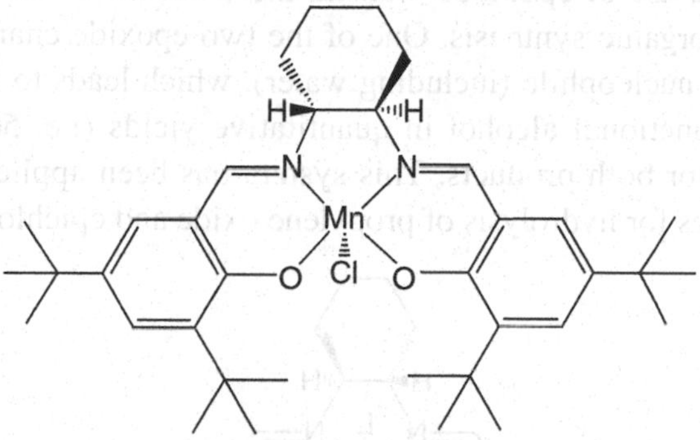

The modest performances of this catalyst in terms of stability and turnover have been largely improved by the Collman-Rose catalyst using a chiral porphyrin.

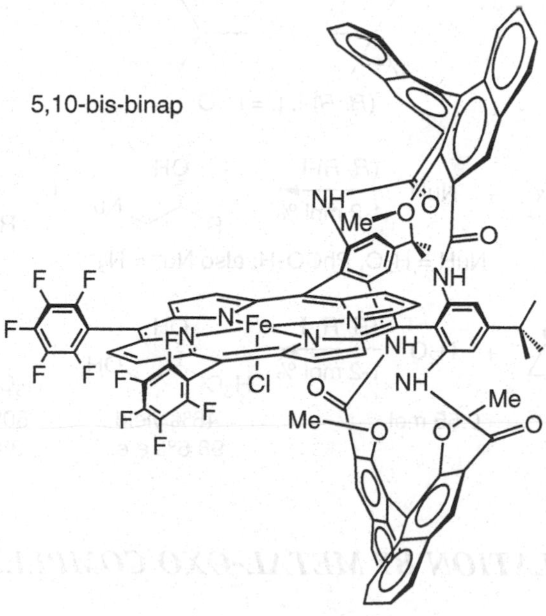

Collman-Rose catalyst

Numerous prochiral olefins can be asymmetrically epoxidized by this catalyst.[16.10]

Example:

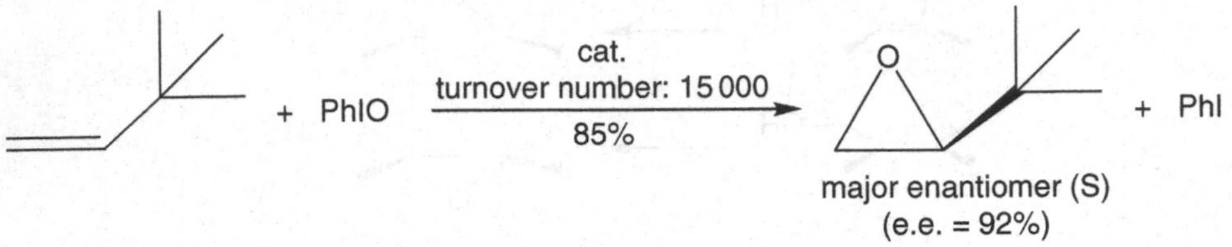

major enantiomer (S)
(e.e. = 92%)

Jacobsen's hydrolytic kinetic resolution of epoxides catalyzed by a Co(salen) catalyst analogous to the one used for asymmetric epoxidation has brought a considerable advance to the use of epoxides. Indeed, these substrates are among the most useful reagents in organic synthesis. One of the two epoxide enantiomers is selectively opened by a nucleophile (including water), which leads to both the terminal epoxide and the functional alcohol in quantitative yields (i.e. 50% of each) and more than 98 e.e. for both products. This system has been applied industrially by Rhodia on ton-scales for hydrolysis of propylene oxide and epichlorhydrin.[16.11]

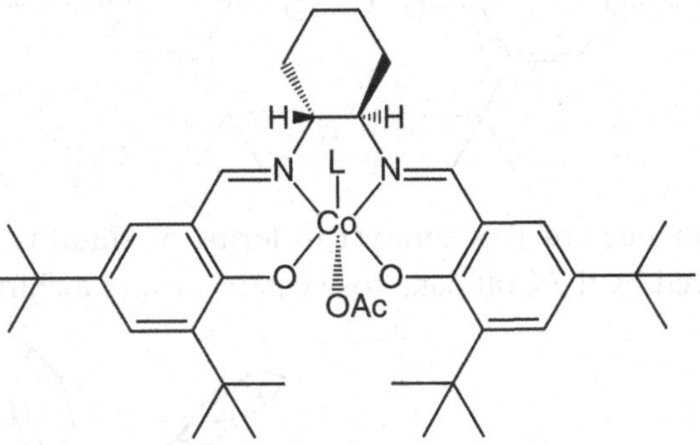

(R, R)-I, L = H_2O

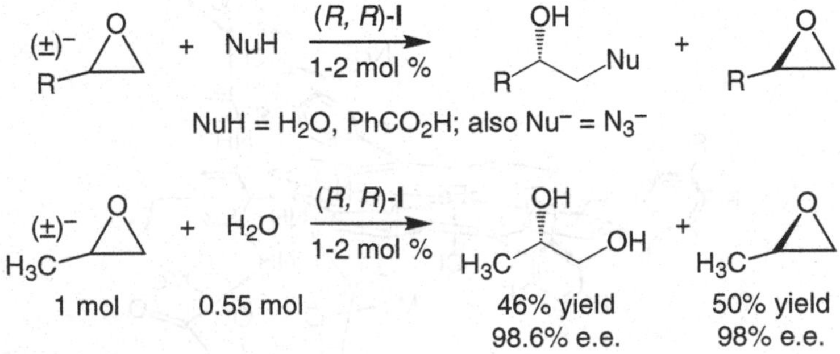

NuH = H_2O, $PhCO_2H$; also Nu^- = N_3^-

1 mol	0.55 mol	46% yield	50% yield
		98.6% e.e.	98% e.e.

5. HYDROXYLATION BY METAL-OXO COMPLEXES

5.1. METAL-OXO COMPLEXES IN OXIDATION CATALYSIS

The complexes with M=O bonds have very different reactivities depending on the nature of the transition metal M. The early transition metals are very oxophilic and form M=O bonds that are not very reactive. These compounds are called **oxides**. On the other hand, late transition metals form labile M=O bonds because of the repulsion between the filled d metal orbitals and p oxygen orbitals. They are called **metal-oxo** complexes. The metal-oxo complexes can form and regenerate by transfer of an oxygen atom onto a transition metal using an oxygen atom donor such as H_2O_2 or from O_2 by double oxidative addition giving a metal-dioxo complex. They play an essential role in oxidation catalysis. They can also, as oxidants, remove one electron from an oxidizable substrate (for instance, in the case of $[MnO_4]^-$ for alkylated aromatics). There are many binary mono- and polymetallic complexes, i.e. containing only one type of metal and the oxo ligands. There are also many compounds containing one or several oxo ligands in addition to other ligands (see Chaps 2 and 9). There are many oxidation reactions that are catalyzed by metal-oxo complexes as illustrated by the non-exhaustive following table.

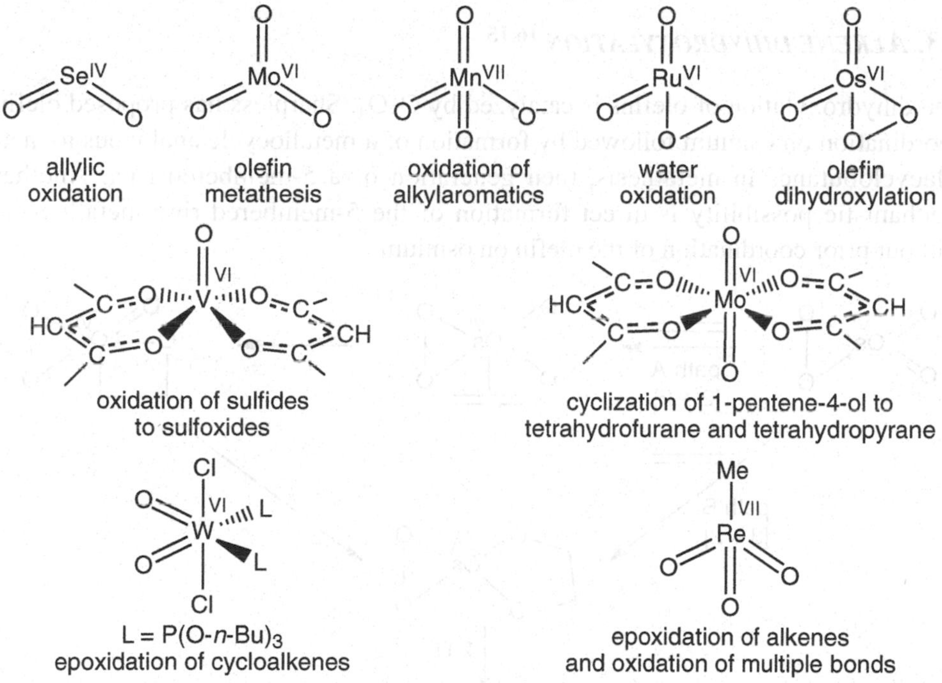

| allylic oxidation | olefin metathesis | oxidation of alkylaromatics | water oxidation | olefin dihydroxylation |

oxidation of sulfides to sulfoxides

cyclization of 1-pentene-4-ol to tetrahydrofurane and tetrahydropyrane

L = P(O-n-Bu)$_3$
epoxidation of cycloalkenes

epoxidation of alkenes and oxidation of multiple bonds

Finally, the heteropolymetallates, shown in Chap. 2, are especially active in oxidation catalysis because of their marked redox and acido-basic properties.[2.6a] They have been used especially in the Wacker process instead of corrosive cuprous chloride.

5.2. ALLYLIC OXIDATION

SeO_2 catalyzes the oxidation of olefins to allylic alcohols in the presence of an oxygen donor such as t-BuOOH as a co-substrate.[16.12] The mechanism is probably concerted (**a**) as in the Prins reaction of aldehydes with alkenes (**b**).

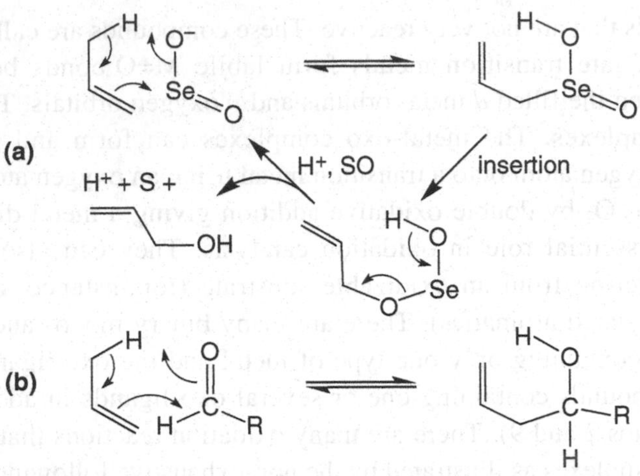

(a)

(b)

5.3. ALKENE DIHYDROXYLATION [16.15]

The dihydroxylation of olefins is catalyzed by OsO_4. Sharpless has proposed olefin coordination on osmium followed by formation of a metallocycle analogous to metallacyclobutanes in metathesis, then generation of a 5-membered ring. Another mechanistic possibility is direct formation of the 5-membered ring metallocycle without prior coordination of the olefin on osmium.

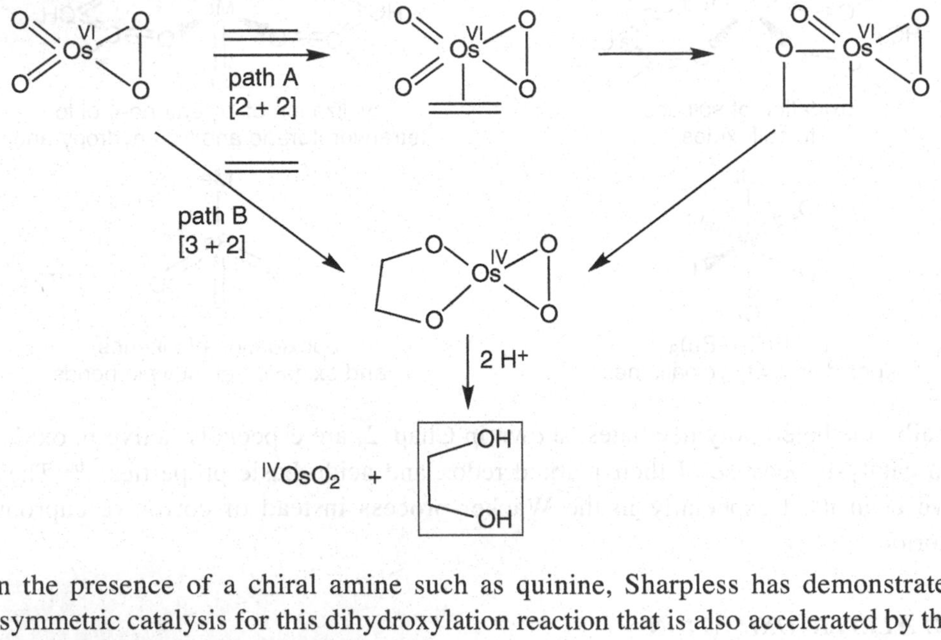

In the presence of a chiral amine such as quinine, Sharpless has demonstrated asymmetric catalysis for this dihydroxylation reaction that is also accelerated by this type of ligand. The oxidizing agent (oxygen donor) is then amine oxide. This system compares with Jacobsen's efficient hydrolytic kinetic resolution shown in section 4, but extension to the industrial scale is more problematic with the Sharpless system.

6. PHASE-TRANSFER CATALYSIS IN OXIDATION CHEMISTRY [16.16-16.18]

Although transition-metal oxide anions are used in stoichiometric amounts in phase-transfer catalysis, this technique should be noted, because it is practical. The insolubility in water of substrates combined with the insolubility in common organic solvents of sodium and potassium salts of transition-metal oxide anions led to low oxidation yields. Under these conditions, it was necessary to utilize large quantities of oxidant, well superior to stoichiometric amounts. The discovery of phase-transfer catalysis allows to use these metal oxide anions in nonpolar solvents such as toluene and methylene chloride. Good selectivities are obtained under mild conditions with this technique. It consists in involving two phases, the aqueous one and the organic one and a phase-transfer reagent, most often in catalytic amounts (which justifies the catalytic nomenclature). This latter reagent is a tetraalkylammonium- or tetraalkylphosphonium salt containing long alkyl chains or a crown ether. The principle consists in obtaining a large cation that allows:

▶ to transport the inorganic anion into the organic phase in which it is solubilized due to the lipophilicity of the counter cation;

▶ to render this anion very reactive in the organic phase. This is due to the decreased strength of electrostatic binding between the anion and the counter cation resulting from the large size of the latter.

Currently used anion oxides as their sodium or potassium salts include MnO_4^-, CrO_4^-, $Cr_2O_7^{2-}$, ClO^-, IO_4^- and FeO_4^{2-}.

The most classic examples of application are Meunier's epoxidation [16.18] (also using a transition-metal catalyst), Sharpless' dihydroxylation and oxidative cleavage of double C=C bonds (which requires catalysis by RuO_2).

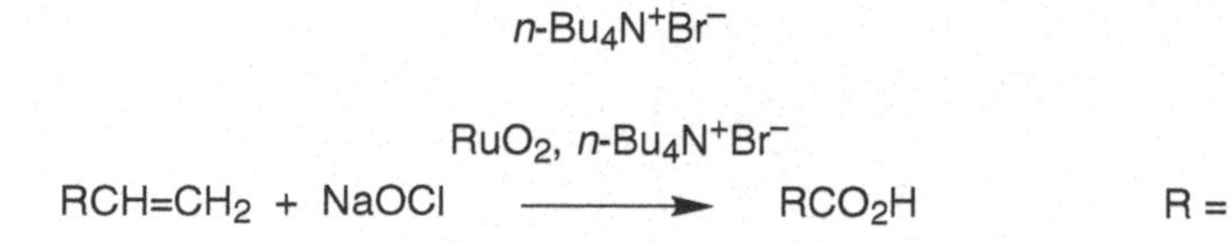

$$Ph\text{-}CH=CH_2 + NaOCl \xrightarrow[R_4N^+Cl^-]{MnTPP} Ph\text{-epoxide}$$

$$RCH=CHR + KMnO_4 \xrightarrow[n\text{-}Bu_4N^+Br^-]{NaOH} RCHOH\text{-}CHOHR$$

$$RCH=CH_2 + NaOCl \xrightarrow[NaOH,\ CH_2Cl_2\ \ 100\%]{RuO_2,\ n\text{-}Bu_4N^+Br^-} RCO_2H \qquad R = CH_3(CH_2)_{12}$$

$$RCH=CH_2 + KMnO_4 \xrightarrow{R_4N^+Cl^-} RCO_2H + HCO_2H \qquad R = n\text{-}C_8H_{17}$$

91% yield from 1-decene

For the latter reaction, a mechanism involving a thermally allowed (2+4) cycload-dition followed by electron transfer (oxidation of Mn^V to Mn^{VI}) and also thermally allowed chelotropic (2+2+2) elimination has been suggested:

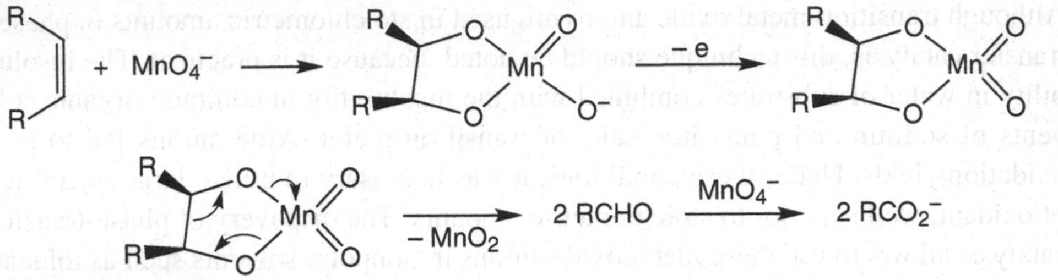

In fact, the interaction of olefins with transition-metal oxides gives five-membered metallocycles. This can result from (3+2) cycloaddition as shown above (inorganic mechanism) or (2+2) olefin addition on the metal followed by insertion of the two oxygen atoms (organometallic mechanism) as proposed by Sharpless in the case of OsO_4. Indeed, MnO_4^-, RuO_4 and OsO_4 can be viewed as 16-electron complexes with a vacant site available on the metal center for the attack of the olefin. On the other hand, in the presence of another ligand such as pyridine, the metal center is electronically saturated, and the (3+2) cycloaddition occurs.

SUMMARY OF CHAPTER 16
OXIDATION OF OLEFINS

1 - History: Radical autoxidation of alkenes (Criege, 1939):

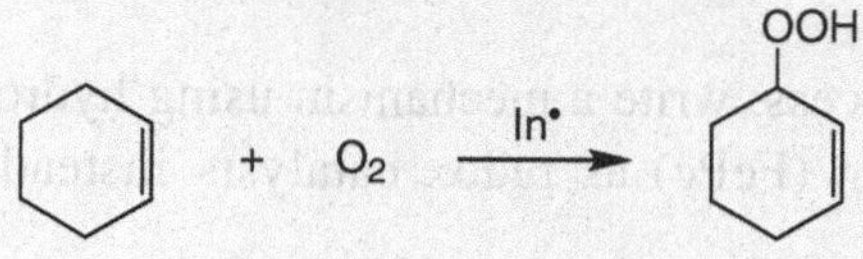

Radical-type mechanism. No metal is involved, but in more modern reactions,:transition-metal complexes catalyze alkene oxidation reactions in various ways:

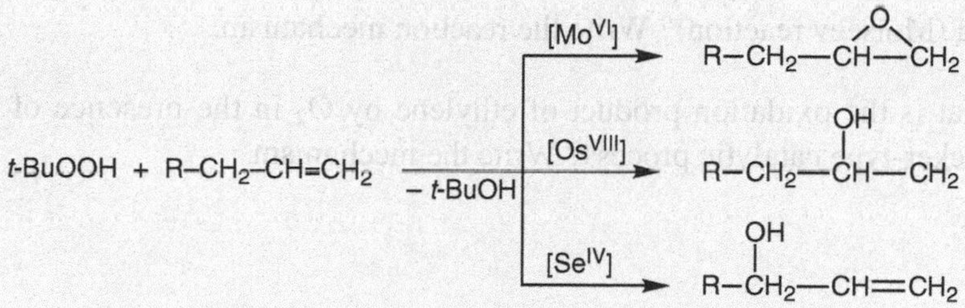

2 - Oxidation of ethylene to acetaldehyde (Wacker process)

$$H_2C=CH_2 + 1/2\ O_2 \xrightarrow{PdCl_2,\ CuCl_2,\ H_2O} CH_3CHO$$

3 - Extension to the oxidation of terminal olefins to ketones

$$RCH=CH_2 + 1/2\ O_2 \xrightarrow{PdCl_2,\ CuCl_2,\ H_2O,\ DMF} RCOCH_3$$

5 - Epoxidation of olefins

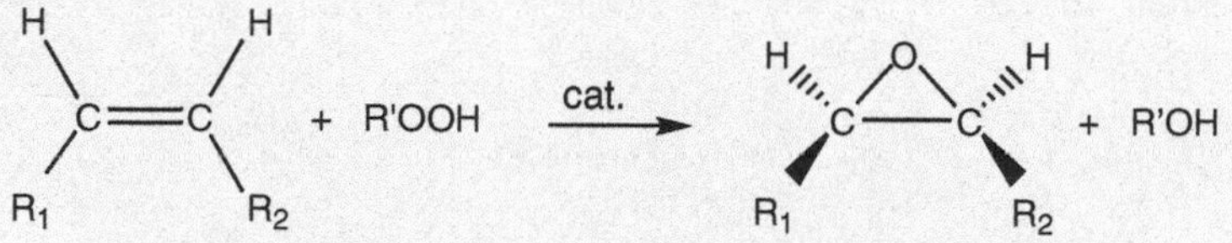

R' = RCO (peracid), R (hydroperoxide) or H (hydrogen peroxide)

6 - Sharpless epoxidation of prochiral allylic alcohols. Asymmetric cat.:
$$[Ti(O\text{-}iPr)_4] + \text{diethyl tartrate}$$

7 - Epoxidation by asymmetric cat. of simple prochiral olefins by PhIO or other O-atom donors is also possible using other specific catalysts (such as some chiral metalloporphyrins).

8 - The various metal-oxo complexes catalyze numerous reactions: allylic oxidation (SeO_2), olefin metathesis (MoO_3), aromatic oxidation (MnO_4^-), water oxidation (RuO_4), alkene dihydroxylation (OsO_4), oxidation of sulfides to sulfoxides ($[VO(acac)_2]$), epoxidation of alkenes (WO_2ClL_2 or ReO_3Me) and cyclization of 1-pentene-4-ol to THF and THP ($[MoO_2(acac)_2]$).

These oxidation reactions are considerably improved by using **phase-transfer catalysis (PTC)** when the catalyst is an anionic oxo complex. PTC enhances the reactivity of transition-metal oxide anions by the introduction of a large organic cation such as R_4N^+ (with long alkyl arms for R). This organic cation, the phase-transfer catalyst, carries the oxo-anion from the aqueous phase into the organic phase and renders it very reactive by decreasing the electrostatic binding within the ion pair due to its large size.

EXERCISES

16.2. Explain the difference of function among MO_4 (Ru *vs.* Os). Why does RuO_4 oxidize water whereas OsO_4 dihydroxylates olefins?

16.2. In the Wacker process, write a mechanism using hydroquinone (QH_2) and an iron phthalocyanin (FePc) as redox catalysts instead of corrosive cuprous chloride.

16.3. What is the oxidation product of ethylene by O_2 (Wacker process) in acetic acid (Moiseev reaction)? Write the reaction mechanism.

16.4. What is the oxidation product of ethylene by O_2 in the presence of CO in a Wacker-type catalytic process? Write the mechanism.

Chapter 17

C-H ACTIVATION AND FUNCTIONALIZATION
OF ALKANES AND ARENES

1. INTRODUCTION

Although alkanes are considered to be inert, some classical reactions have been known for a long time, for instance with halogens, reactive radicals and superacids. The C-H bond that is cleaved in such reactions is the weakest one, i.e. reaction occurs at the most substituted carbon. Thus, the radicals R$^•$ or carbocations R$^+$ formed react at their most substituted carbon atom giving branched products. The selectivity in the above reactions always is: tertiary > secondary > primary.

In Chap. 3, we have examined the addition of an alkane C-H bond to a 16-electron metal center M giving a transient 18-electron metal-alkane intermediate M(RH) that ultimately gives *oxidative addition* of the C-H bond to M(R)(H). In this organometallic reaction, activation of the C-H bond occurs at the less substituted carbon for steric reasons. For instance, a linear alkane yields an *n*-alkyl-metal species *via* C-H activation at a terminal methyl group. The selectivity is thus primary > secondary > tertiary. This opens the valuable possibility of generating linear alkyl-functionalized products if such an oxidative addition mechanism can be made catalytic:

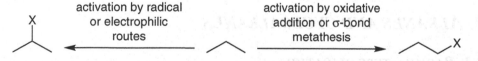

However, in addition to facing the functionalization of the strongest C-H bond (the terminal one) of the alkane molecule, it is necessary to overcome the fact that the monofunctional products are usually more reactive than the alkane starting material. Alcohols tend to form ketones, aldehyde and CO_2, and monohalogenated alkanes may react further to give gem-dihalogenated products. Finally, most reactions of alkanes such as carbonylation and dehydrogenation are endergonic. Thus, in view of these problems, the goal of selective terminal alkane functionalization is very challenging. In fact, no industrial alkane functionalization process is based on alkane oxidative addition.

Yet, it is of great interest to now examine what these remarkable organometallic concepts of *oxidative addition* and *σ-bond metathesis* elaborated in the early 1980s have brought in terms of practical catalytic functionalization of alkanes and arenes.[17.1] It is also essential to include all the types of systems that can activate C-H bonds by metals by extending our scope to include non-classical organometallic systems, which allows comparison, mutual enrichment and confrontation of hypotheses.[17.1]

Thus, as in the case of alkenes in the preceding chapter, we start with the radical type of activation that is much older. Transition-metal compounds play a key role in radical activation, because they provide very strong oxidants that can oxidize hydrocarbons either by (reversible) electron transfer or H-atom transfer (more rarely by hydride transfer). Biological oxidation of hydrocarbons involves reactive metal-oxo species in methane mono-oxygenases and many related synthetic models, and a number of simple metal-oxo complexes also work. The clear criterion of distinction between an organometallic C-H activation and a radical activation is the above selectivity in activated C-H bonds.

Finally, as we move to arenes, it is found that C-H activation is more facile than with alkanes, because arenes are more electron rich than the latter, i.e. for instance they are easier to oxidize by single electron transfer (their oxidation potential is lower than that of alkanes), and they react more readily than alkanes with electrophiles. They can also coordinate more strongly than alkanes to transition metals, because of their ability to accept much more backbonding from transition metals than alkanes do, which facilitates further activation. Electrophilic activation of C-H bonds known from Friedel-Crafts reactions with organic electrophiles, has also been known with metal electrophiles since the end of the XIXth century and now provide really efficient catalytic C-H activation reactions. Finally, given the richness and variety of the C-H activation field, it is not surprising that organic chemists have focused on the possibility of functionalizing alkane segments in complex organic substrates, in particular *via* coordination-directed C-H bond activation.[17.1b,c]

2. ALKANES AND CYCLOALKANES

2.1. RADICAL-TYPE OXIDATION

The direct oxidation of methane and alkanes by very strong oxidants according to a single electron-transfer pathway is followed by deprotonation of the extremely acidic radical cationic species $RH^{\bullet+}$:

$$RH \longrightarrow RH^{\bullet+} \longrightarrow R^{\bullet} \longrightarrow products$$

For instance, peroxodisulfate oxidizes methane to alcohols in the absence of a metal at 105-115°C. Metal complexes can be used as mediators for such reactions which permits carrying out the reactions at lower reaction temperatures.[17.1] In the case of cycloalkanes, strongly electrophilic and oxidizing transition-metal ions play an irreplaceable role in the initiation step.

Butane can be oxidized with O_2 to acetic acid at 180°C under pressure using Co^{3+} as catalyst, but other acids are also obtained in the reaction medium. This industrial process has been abandoned in favor of the Monsanto process starting from methanol.

$$C_4H_{10} + 5/2\ O_2 \xrightarrow{\text{[Co(OAc)}_3\text{]}} 2\ CH_3COOH + H_2O$$

Likewise, cyclohexane is oxidized by O_2 to adipic acid (*via* cyclohexyl hydroperoxide, cyclohexanol and cyclohexanone). Adipic acid, which was used for the synthesis of nylon-6,6 by DuPont de Nemours, is now produced from butadiene (see Chap. 15.5).

$$\text{cyclohexane} + O_2 \xrightarrow[\text{80°C, AcOH}]{\text{[Co(OAc)}_3\text{]}} \text{(CH}_2\text{)}_4(\text{CO}_2\text{H)}_2$$

The accurate mechanism of this type of reaction is still uncertain, however. For instance, in the first step, the formation of the radical R• from the hydrocarbon RH, it is difficult to distinguish between a single-electron oxidation (outer-sphere mechanism without formation of bond) and an electrophilic attack (inner-sphere mechanism involving intermediate coordination). In both hypotheses, the cationic species formed is extremely acidic and immediately looses a proton to form the neutral radical:

$$Co^{III} + RH \longrightarrow Co^{II} + RH^{•+} \longrightarrow Co^{II} + R^{•} + H^{+}$$

$$RH + X_2Co^{III\ ]^+} \longrightarrow \left[\begin{array}{c} R \\ \diagdown \\ H \\ X_2Co^{III} \end{array} \right]^{+} \longrightarrow [X_2Co^{III}\text{–R}] + H^{+} \downarrow X_2Co^{II} + R^{•}$$

In the second hypothesis, a C-H bond of RH is a 2-electron L ligand for the metal, analogous to the dihydrogen ligand H_2 (although weak and distorted because of the steric bulk of R). The H_2 complexes are acidic, and so are the alkane complexes.

An example of typical H-atom abstraction from alkanes following the selectivity trends observed for radical reactions was discovered by Crabtree using photo-excited Hg atoms in the gas phase. The reaction applies not only to alkanes, but also to alcohols, ketones, amines, etc.

$$RH \xrightarrow{Hg^*} R^{•} + H^{•} \longrightarrow 1/2\ R\text{-}R + 1/2\ H_2 \nearrow$$

with RH = alkane, alcohol, ketone, amine, etc.

Let us end this section by considering the Fenton reaction, discovered in the 1890s, whereby Fe^{II} salts catalyze the hydroxylation of alkanes and other substrates. In these systems, iron (II) salts reduce H_2O_2 to OH^- and supposedly "OH•" that is assumed to subsequently abstract a hydrogen atom from the substrates: [17.3]

$$Fe^{II} \longrightarrow H_2O_2 \longrightarrow Fe^{III} + OH^- + \text{"OH}^{•}\text{"}$$

This system is of industrial importance and biological interest because of the cytotoxicity of the generated oxygen species.[17.3,17.4] Since OH° is of extremely high energy, however, it is unlikely for this very thermodynamic reason, that it be liberated in the free state (especially in the presence of iron salts). Indeed it has been proposed that inner-sphere OH transfer may intervene *via* an Fe-OH species.[17.3b] If we refer to the following section and to Chap. 19 which is devoted to Fe-based methane monoxygenase sites and their models, one realizes that Fe-OH is a protonated form of the ferryl species Fe=O, and therefore it is probable that Fenton chemistry is relevant to the biological alkane activation systems.[17.4a]

2.2. HYDROXYLATION
USING METHANE MONOXYGENASE MODELS AS CATALYSTS

Metal-porphyrin complexes (noted N_4M in the scheme below, see Chap. 1) are able to catalyze the transfer of an oxygen atom to an alkane RH giving an alcohol ROH, the donor substrate being PhIO (iodosylbenzene), H_2O_2, NaOCl, $KHSO_5$, etc.[17.4-17.6] Such systems mimic the monoxygenase enzymes whose function is examined in Chap. 19. The metal can be Fe^{III}, Ru^{II} or Mn^{III}. The simple porphyrin ligands are themselves attacked, but perchlorinated porphyrins are not. Other substrates S are also mono-oxygenated in the same way to SO: these are olefins (to epoxides), phosphines, amines and sulfides to the corresponding oxides. The intermediate M=O species are very reactive. In the case or Fe, the mesomeric iron-oxyl form Fe^{IV}-O° has a slightly greater energy (3 to 5 kcal higher) than the Fe^V=O form.

This mechanism represents a shortcut in biological mono-oxygenase enzyme systems that use O_2 as oxygen atom sources and whose mechanism is much more complex because of the requirement to cleave O_2. Non-porphyrinic binuclear methane mono-oxygenase model complexes are also able to activate methane in the same way (see Chap. 19).[17.7] When S = RH, Groves originally proposed the well-known "rebound" mechanism in which the Fe^V=O species removes an H atom from RH, then transfers OH° to produce the alcohol ROH: [17.4]

$$M=O + HR \longrightarrow [M–OH + R^°] \longrightarrow M + HOR$$

This mechanism has been the subject of considerate debate,[17.4-17.6] and the first step, which consists in an overall H-atom transfer, may in fact, be an electron transfer followed by a stepwise proton transfer or an inner-sphere H-atom transfer in the cage or any intermediate along the continuum between the two situations.[17.8]

2.3. FUNCTIONALIZATION VIA ACTIVATION BY Pt^{II}: SHILOV CHEMISTRY

In Chap. 3.3.7.4, it was indicated that a mixture of $PtCl_2$ and H_2PtCl_6 activates methane and alkanes, *via* the intermediacy of a Pt-alkyl species which is formed either by direct deprotonation of the transient Pt^{II}-alkane complex or by oxidative addition of the alkane ligand within this complex giving a Pt^{IV}-alkyl-hydride intermediate.

$$RH + Pt^{II} \longrightarrow Pt^{II}\text{-}R + H^+$$

This remarkable activation system, which was discovered in 1969 by Shilov,[17.9] was ignored for more than a decade. The reaction is catalytic in $PtCl_2$ but only stoichiometric in $H_2Pt^{IV}Cl_6$. A revival of interest starting in the 1980s has contributed to gaining insight into the details of the mechanism using model complexes.[17.10-17.12] The role of $H_2Pt^{IV}Cl_6$ is to re-oxidize Pt^{II}-R to Pt^{IV}-R in each catalytic cycle, as demonstrated by Bercaw and Labinger.[17.10] Nucleophilic attack on the carbon atom of the Pt^{II}-R bond by H_2O or Cl^- (shown by stereochemical inversion at carbon) finally produces methanol or methyl chloride respectively (equations below, $R = CH_3$ or alkyl).[17.10] In model complexes, strong evidence has been provided by the Bercaw-Labinger[17.10] and Goldberg[17.11] groups that this nucleophilic attack occurs at five-coordinate Pt^{IV} rather than six-coordinate Pt^{IV}.

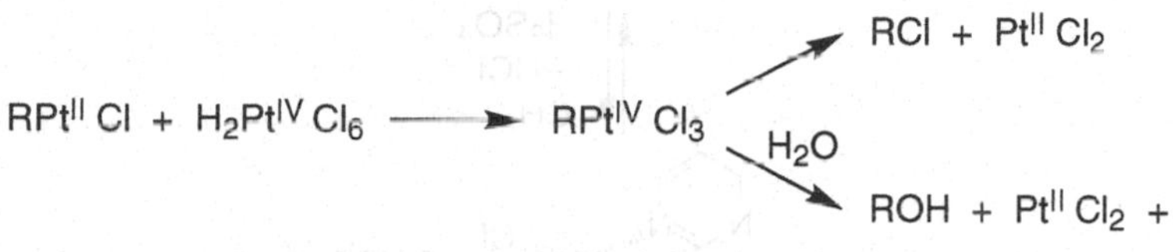

$$RPt^{II}\,Cl + H_2Pt^{IV}\,Cl_6 \longrightarrow RPt^{IV}\,Cl_3 \begin{cases} \nearrow RCl + Pt^{II}\,Cl_2 \\ \searrow\!{}_{H_2O}\; ROH + Pt^{II}\,Cl_2 + H^+\,Cl^- \end{cases}$$

The overall stoichiometries and mechanism proposed by Shilov are as follows: [17.9]

$$RH + H_2\,Pt^{IV}\,Cl_6 + H_2O \xrightarrow{\text{cat. } Pt^{II}} ROH + H_2\,Pt^{II}\,Cl_4 + 2\,H^+\,Cl^-$$

$$RH + H_2\,Pt^{IV}\,Cl_6 \xrightarrow{\text{cat. } Pt^{II}} RCl + H_2\,Pt^{II}\,Cl_4 + H^+\,Cl^-$$

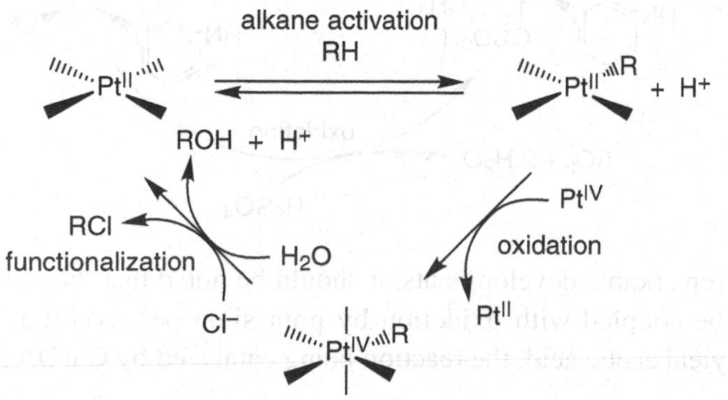

Related systems using less expensive alternative stoichiometric oxidants have been disclosed.[17.13-17.16] One of the most significant developments was found at Catalytica by Periana using SO_3 in fuming H_2SO_4, which acts as both the solvent and the stoichiometric oxidant. The Pt^{II} complex $[PtCl_2(bpym)]$ is used as the catalyst for this process (see below). Methylbisulfate is obtained in 72% yield with 90% conversion of methane and 81% selectivity, and the turnover numbers are greater than 300. The catalyst is inhibited by H_2O or methanol, causing low reaction rates below 90% H_2SO_4. This process is thus remarkably efficient, although not economically competitive compared with existing heterogeneous industrial processes. The mechanism is similar or very closely related to that of the Shilov system (i.e., it is not clear wether oxidative oxidation or σ-bond metathesis is the methane C-H activation step): [17.13]

$$CH_4 + H_2SO_4 \xrightarrow{\quad Pt^{II} \quad} CH_3OSO_3H + 2\,H_2O + SO_2$$

Proposed mechanism for the Catalytica methane oxidation system:

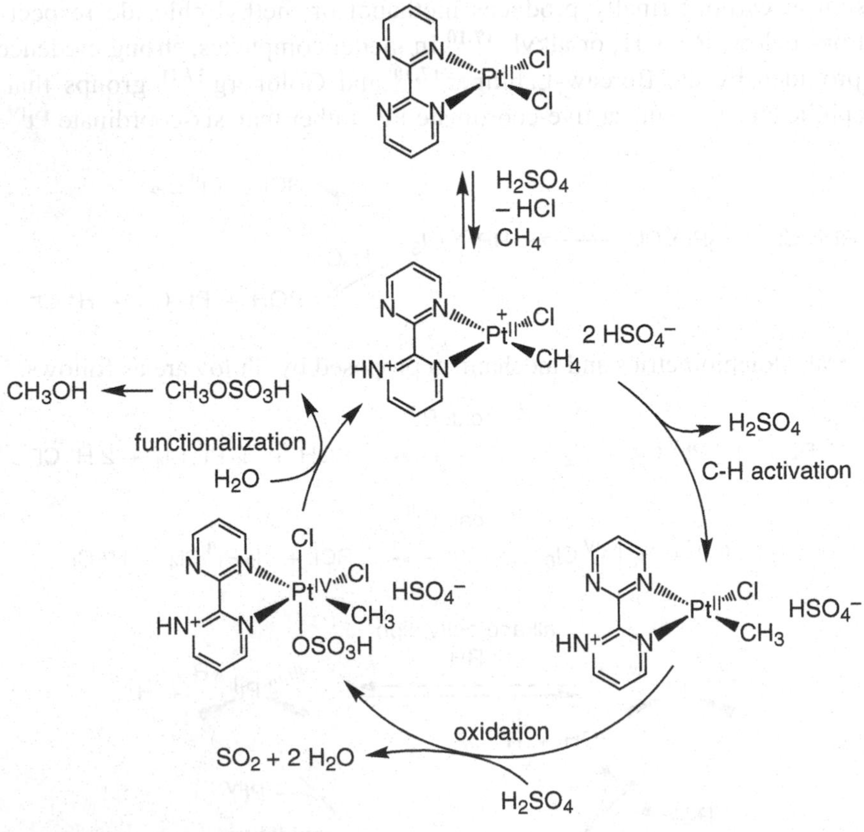

Among other remarkable developments, it should be noted that the carboxylation of methane can be coupled with oxidation by potassium peroxysulfate in trifluoro-acetic acid to yield acetic acid, the reaction being catalyzed by $Cu(OAc)_2$: [17.15]

$$CH_4 + CO + K_2S_2O_8 \xrightarrow{\quad Cu^{II} \quad} CH_3CO_2H + K_2SO_4$$

Interestingly, carbonylation can also proceed in the absence of CO, although in only 10% yield, using $PdSO_4$ as the catalyst in concentrated sulfuric acid as solvent and oxidant: [17.16]

$$2\,CH_4 + 4\,H_2SO_4 \xrightarrow{\quad Pd^{II} \quad} CH_3CO_2H + 4\,SO_2 + 6\,H_2O$$

2.4. DEHYDROGENATION INVOLVING OXIDATIVE ADDITION

Oxidative addition is probably the most powerful concept of organometallic chemistry, and it has been applied in the preceding chapters to hydrogenation and hydroelementation of olefins and in the last chapter to catalysis of organic halide functionalization. We know that oxidative addition of alkanes is much more difficult than that of many substrates, however, and only known in few cases. Thus, its implication in catalysis is extremely challenging, and therefore the discovery by Crabtree in 1979 that cyclopentane and cyclooctane can be dehydrogenated by $[Ir(Me_2CO)_2H_2(PPh_3)_2]^+$ to the corresponding alkadiene was a real breakthrough.[17.17] This was reversing the transition-metal catalyzed hydrogenation of alkenes to alkanes, and indeed the mechanism was proposed to start by oxidative addition of the cycloalkane to a cycloalkyl-metal-hydride species. The reaction was carried out in the presence of a good hydrogen acceptor, ideally t-butylethylene, in order to compensate the endergonicity ($\Delta H = 30$ kcal·mol^{-1}) of the dehydrogenation of the alkane to the free olefin and H_2.

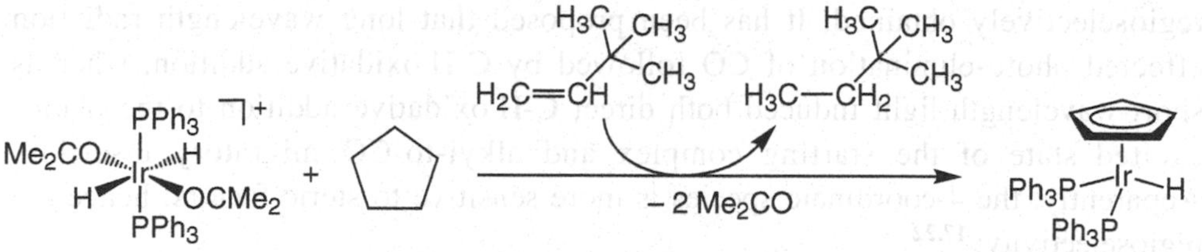

A year later, the dehydrogenation of cyclooctane to cyclooctene was made catalytic by Felkin's group with the complex $[ReH_7(PPh_3)_2]$,[17.17b] (equation below) and Crabtree showed that alkanes can also be catalytically dehydrogenated using the complex $[Ir\{(CF_3)_2CO)_2\}H_2(PPh_3)_2]^+$.

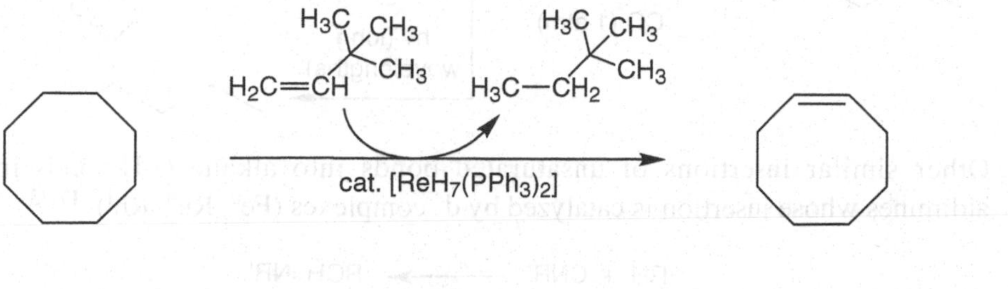

Futhermore, Goldman has found a complex that features a pincer ligand and is the most efficient catalyst for alkane dehydrogenation.[17.19a] Double bonds can be introduced into polymers and dehydrogenation of tertiary amines leads to enamines.[17.19b]

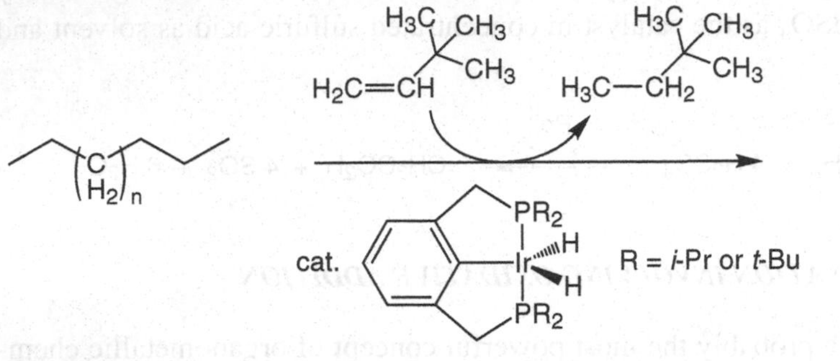

2.5. CARBONYLATION INVOLVING OXIDATIVE ADDITION

Tanaka reported that alkanes can be catalytically carbonylated to aldehydes under 1 atm of CO using the catalyst [*trans*-Rh(PR$_3$)$_2$(CO)Cl].[17.20] This disclosure (with R = Me) followed the report by Eisenberg of the carbonylation of benzene to benzaldehyde under the same conditions but using the related catalyst in which R = Ph (see section 3.4.1)[17.21] The electron-rich phosphine PMe$_3$ facilitates oxidative addition of the alkane by increasing the electron density on the rhodium center, compared with PPh$_3$. These reactions would normally be endergonic, but when carried out under photochemical conditions, they are exergonic, since light drives them in each catalytic cycle. A puzzling feature of alkane carbonylation is that, with short wavelength radiation, *n*-alkanes are carbonylated regioselectively at the terminal position whereas with longer wavelengths, branched aldehydes are regioselectively obtained. It has been proposed that long wavelength radiation effected photo-elimination of CO followed by C-H oxidative addition, whereas short-wavelength light induced both direct C-H oxidative addition to the photo-excited state of the starting complex and alkyl-to-CO migratory insertion. Apparently, the 4-coordinate species is more sensitive to steric factors; hence the regioselectivity:[17.22]

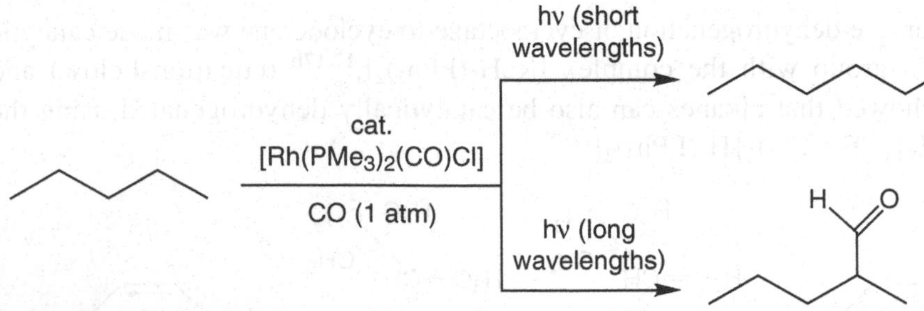

Other similar insertions of unsaturated bonds into alkane C-H bonds include aldimines whose insertion is catalyzed by d^8 complexes (Fe^0, Ru^0, Rh^I):[17.23]

$$RH + CNR' \longrightarrow RCH=NR'$$

2.6. REGIOSELECTIVE BORYLATION INVOLVING σ-BOND METATHESIS

After having shown that various transition-metal-boryl complexes react with alkanes under photochemical or thermal conditions to produce alkyl-boron compounds (first equation below),[17.24a] Hartwig discovered that diboron compounds also react regioselectively with alkanes to afford alkyl-boron compounds in good to excellent yields, these reactions being catalyzed by organotransition-metal complexes.[17.24b-e] The catalytic reactions proceed by ligand loss (upon irradiation [17.24b] or heating [17.24c]) from an 18e complex to produce a transient 16e intermediate. The latter effects the oxidative addition of the diboron compound such as bis-pinacoldiborane to generate a transition-metal-boryl bond that resembles the one shown below in the stoichiometric reaction (with a different metal). Alternatively, starting from HB (pinacolate) also provides this bond. The catalyst used under photochemical conditions is [Cp*Re(CO)$_3$],[17.24b] whereas those used under thermal conditions are [Cp*IrH$_4$], [Cp*Rh(C$_2$H$_4$)$_2$], [Cp*Rh(C$_6$Me$_6$)] and [(Cp*RhCl)$_2$].[17.24c,e]

The reactions typically proceed at 150°C with n-octane and di-pinacolboronate for 5-24 h, and the amount of catalyst typically is 1-5%. The yields of 1-octylboronate ester are good and the reaction is regioselective in the terminal alkyl position (second equation below, top of p. 418). The proposed mechanism involves oxidative addition of the B-B or B-H bond, followed by σ-bond metathesis between the M-B and R-H bonds, which is driven towards formation of the B-R bond by the Lewis-acid property of boron. Note that 16e species (see Scheme p. 418) could also be involved in oxidative addition of the alkane to give an 18e intermediate M(Bpin)$_2$(H)(R) or M(Bpin)(H)$_2$(R) that would provide R-Bpin as well by reductive elimination. Calculations showed, however, that the σ-bond metathesis path is preferred by about 10 kcal·mol^{-1} over this alkane oxidative-addition path.

The strength of the B-C bond (111-113 kcal·mol^{-1}),[17.24d] which is larger than that of a terminal alkane C-H bond, is responsible for the driving force of the overall reaction, and the lack of a β-hydrogen atom in the pinacolboryl group inhibits side reactions of the metal-boryl intermediates. Altogether, this remarkable catalytic regioselective borylation of alkanes is probably the most efficient and useful process so far for regioselective alkane functionalization.[17.24e]

Example of stoichiometric alkane borylation:

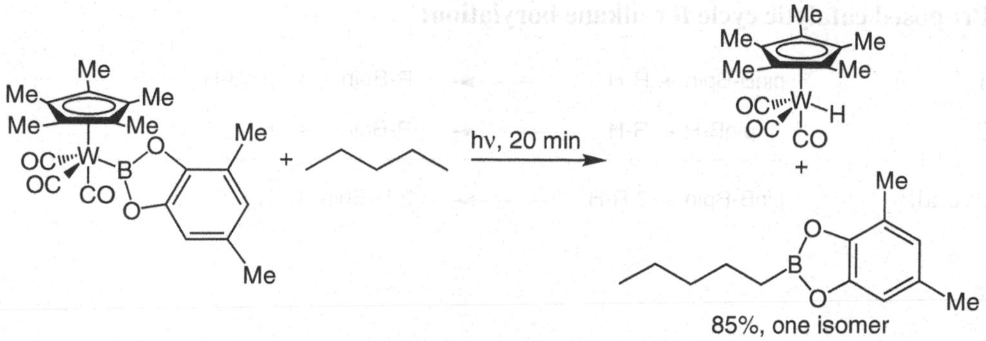

85%, one isomer

Example of catalytic alkane borylation:

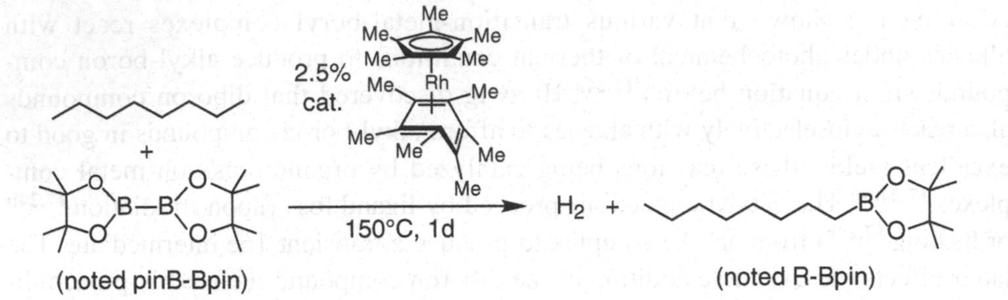

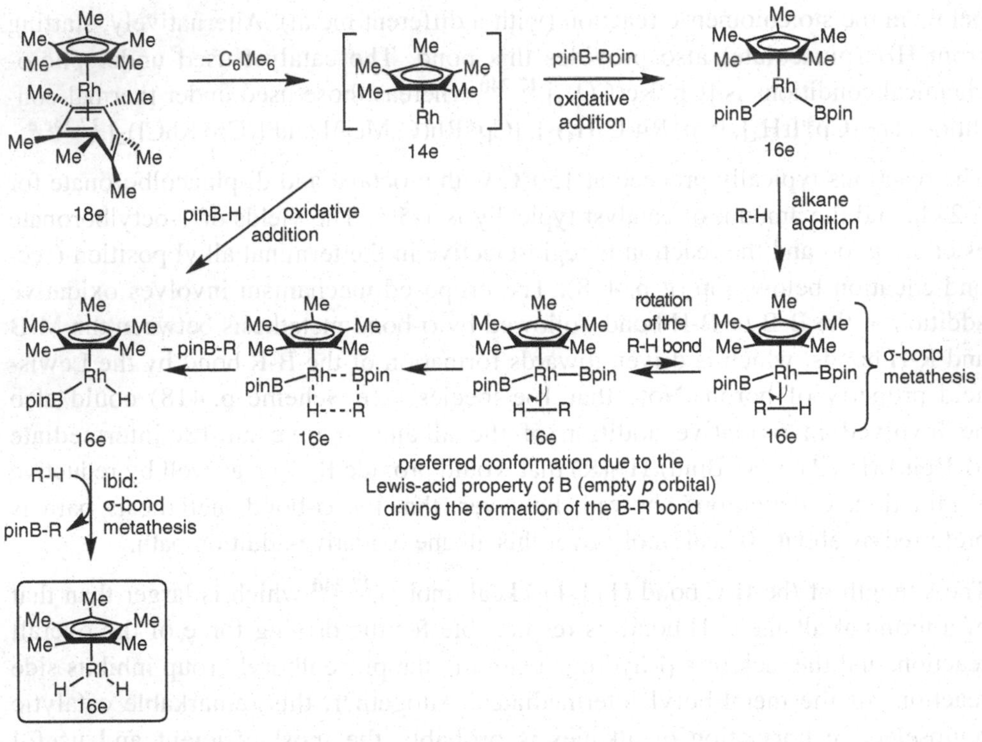

(noted pinB-Bpin)

150°C, 1d

H_2 +

(noted R-Bpin)

Generation of the catalytically active Cp*Rh(H)$_2$ species from the precatalyst Cp*RhC$_6$Me$_6$ by oxidative addition of either pinB-Bpin or H-Bpin (pin = pinacolate, see above) followed by σ-bond metathesis of the Rh-B and R-H bonds

Proposed catalytic cycle for alkane borylation:

1.   pinB-Bpin + R-H ⟶ R-Bpin + pinB-H

2.   pinB-H + R-H ⟶ R-Bpin + H$_2$

overall: pinB-Bpin + 2 R-H ⟶ 2 R-Bpin + H$_2$

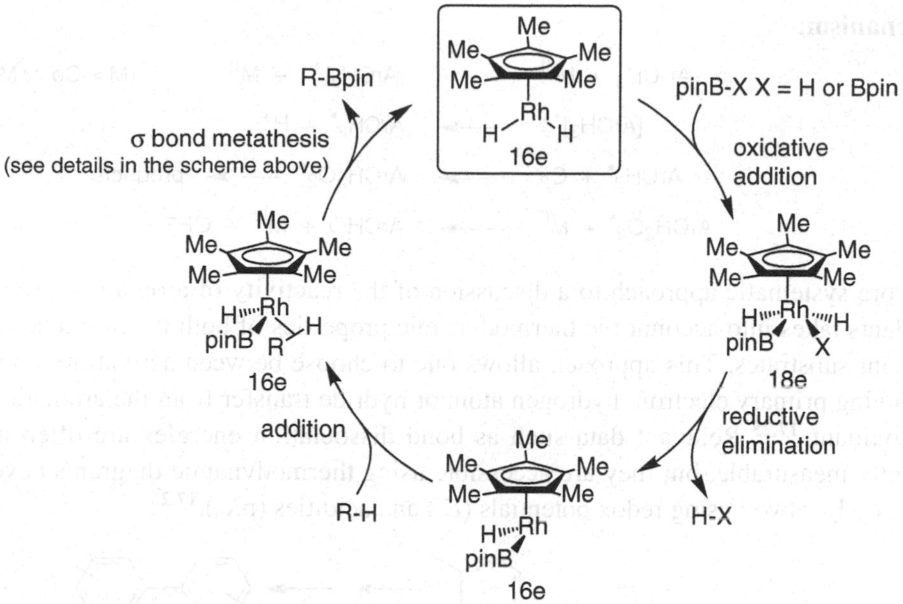

3. AROMATICS

3.1. RADICAL-TYPE REACTIONS

Molecular oxygen oxidizes toluene to benzoic acid and *p*-xylene to terephthalic acid *via* *p*-toluic acid. These reactions are catalyzed by [Co(OAc)$_3$] or [Mn(OAc)$_3$]. The Mn^{III} salts, less oxidizing than those of Co^{III}, catalyze the oxidation of these alkylaromatics, but not the oxidation of alkanes whose redox potentials are higher than those of alkylaromatics. It is probable that the M^{III} ions are regenerated from M^{II} and the ArCH$_2$OO$^•$ radicals (mechanism below).

Benzoic acid is a precursor of phenol (but the process from cumene is more competitive). Terephthalic acid is used as a precursor to polymers.

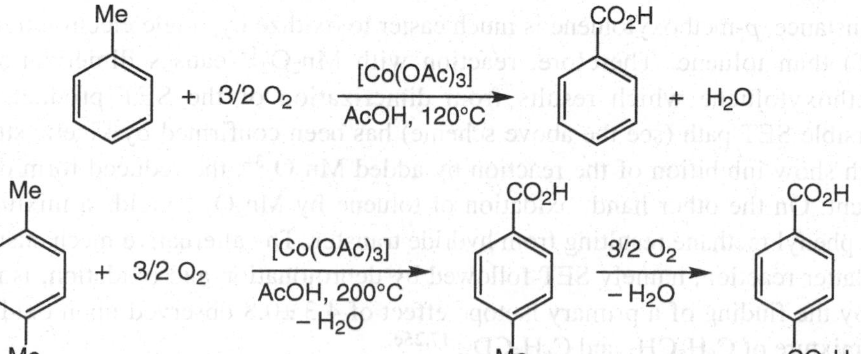

Mechanism:

$$Ar\text{-}CH_3 + M^{III} \longrightarrow [ArCH_3]^{\bullet+} + M^{II} \qquad (M = Co\ or\ Mn)$$

$$[ArCH_3]^{\bullet+} \longrightarrow ArCH_2^{\bullet} + H^+$$

$$ArCH_2^{\bullet} + O_2 \longrightarrow ArCH_2O_2^{\bullet} \longrightarrow products$$

$$ArCH_2O_2^{\bullet} + M^{II} \longrightarrow ArCHO + M^{III} + OH^-$$

A more systematic approach to a discussion of the reactivity of aromatics towards oxidants takes into account the thermodynamic properties of both the aromatic and oxidant substrates. This approach allows one to choose between activation modes involving primary electron, hydrogen atom or hydride transfer from the aromatic to the oxidant. [17.25] Relevant data such as bond dissociation energies are often not directly measurable, but they are accessible using thermodynamic diagrams developed by Bordwell using redox potentials (E^0) and acidities (pK_a). [17.26]

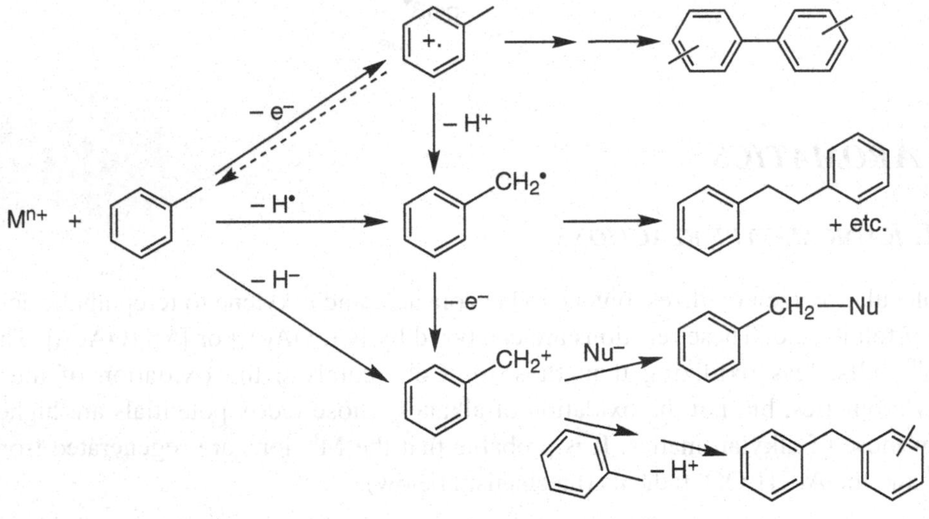

Mayer's reactivity scheme accounting for the products resulting from the reaction of toluene with electrophilic complexes [17.25c]

For instance, p-methoxytoluene is much easier to oxidize by single electron transfer (SET) than toluene. Therefore, reaction with $Mn_2O_2^{4+}$ causes dimerization of p-methoxytoluene which results from dimerization of the SET product. The reversible SET path (see the above scheme) has been confirmed by kinetic studies which show inhibition of the reaction by added $Mn_2O_2^{3+}$, the reduced form of the oxidant. On the other hand, oxidation of toluene by $Mn_2O_2^{4+}$ yields a mixture of tolyl phenyl methane resulting from hydride transfer. The alternative mechanism for this latter reaction, namely SET followed by deprotonation and oxidation, is ruled out by the finding of a primary isotope effect of 4.3 ±0.8 observed upon oxidation of a mixture of $C_6H_5CH_3$ and $C_6H_5CD_3$: [17.25c]

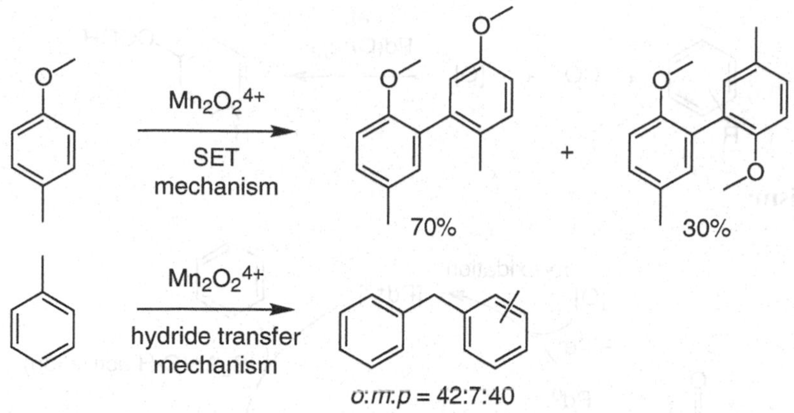

70%   +   30%

$o:m:p = 42:7:40$

3.2. ELECTROPHILIC REACTIONS

This first example of C-H activation with a metal compound is the electrophilic aromatic mercuration of benzene using $Hg(OAc)_2$ which was reported by Dimroth in 1898. This reaction is the analog of Friedel-Crafts reactions that use organic electrophiles.[17.27a]

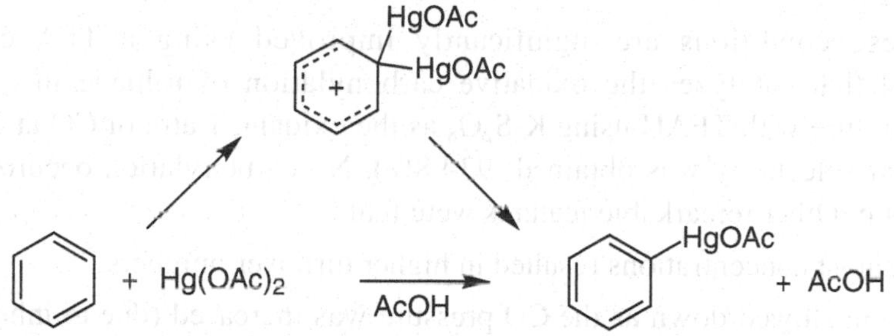

The electrophilicity of Hg(II) is drastically enhanced by using mercury trifluoroacetate $Hg(TFA)_2$ instead of $Hg(OAc)_2$, and CF_3CO_2H as the solvent which is more ionizing than CH_3CO_2H in order to favor the formation of the Wheland intermediate (above, σ complex). This results in mercuration that is 7×10^5 times faster with $Hg(TFA)_2$ than with $Hg(OAc)_2$,[17.27b] so that five mercurations of benzene become possible whereas only one is possible, with $Hg(OAc)_2$:[17.27c]

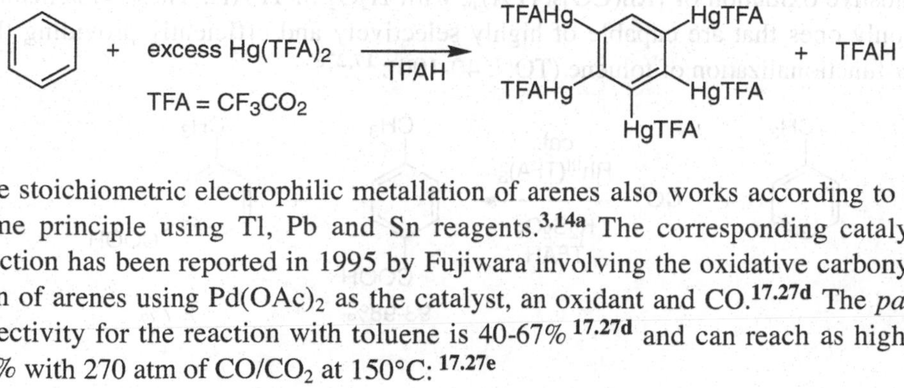

The stoichiometric electrophilic metallation of arenes also works according to the same principle using Tl, Pb and Sn reagents.[3.14a] The corresponding catalytic reaction has been reported in 1995 by Fujiwara involving the oxidative carbonyla-tion of arenes using $Pd(OAc)_2$ as the catalyst, an oxidant and CO.[17.27d] The para-selectivity for the reaction with toluene is 40-67%[17.27d] and can reach as high as 90% with 270 atm of CO/CO_2 at 150°C:[17.27e]

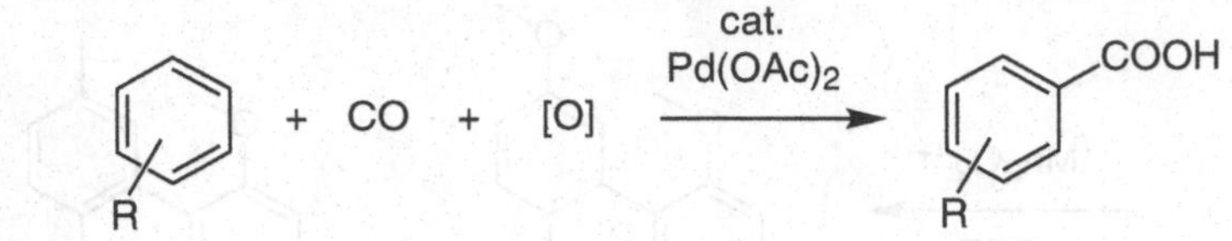

Mechanism:

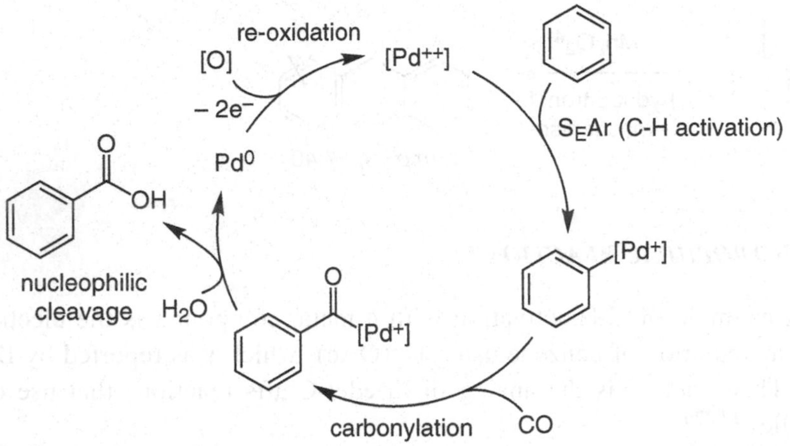

Again, these conditions are significantly improved using a TFA complex, [Rh(TFA)$_3$] that catalyzes the oxidative carbonylation of toluene in a 1:1 (by volume) mixture with TFAH using K$_2$S$_2$O$_8$ as the oxidant, 1 atm of CO at 20-65°C. A high *para*-selectivity was obtained (93-98%). No carboxylation occurred at the methyl group. Other remarkable features were that:

▸ lower catalyst concentrations resulted in higher turnover numbers,

▸ the reaction slowed down as the CO pressure was increased (due to inhibition of the catalytic activity by formation of rhodium carbonyl complexes),

▸ the kinetic isotope effect of 3 was high, probably due to the presence of a bulky Rh^{III} electrophile (5-coordinate in the reactive electrophilic form),

▸ the reaction is inhibited by Cl$^-$ (a ligand that quenches the electrophilic properties of Rh^{III}).

The chloride-free [Rh(TFA)$_3$] that is used for this reaction can be prepared by exhaustive oxidation of [Rh(CO)$_2$(TFA)]$_n$ with H$_2$O$_2$ in TFAH. These systems are the only ones that are capable of highly selectively and efficiently providing the *para*-functionalization of toluene (TON: 40-100): [17.27e]

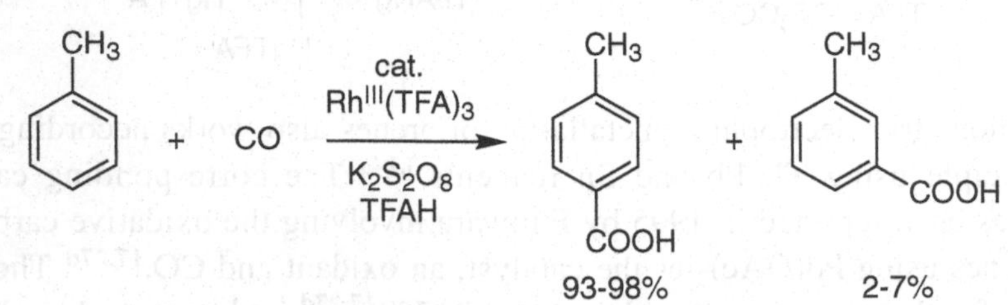

In addition to benzoic acid and its *para*-methyl derivative described above, other simple arylcarboxylic acids can also be synthesized using this method: [17,27e]

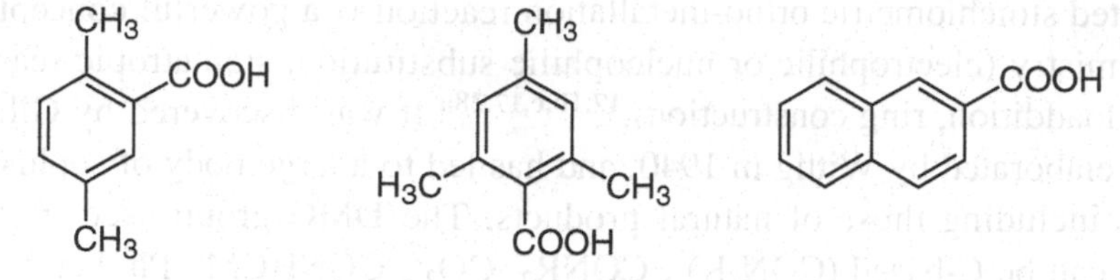

3.3. OXIDATIVE ADDITIONS COUPLED WITH INSERTIONS

3.3.1. Carbonylation: insertion of CO in the Ph-H bond

Note that the above electrophilic carbonylation reaction catalyzed by very electron-deficient Rh^{III} has its counterpart in the chemistry of electron-rich Rh^{I} catalysis of the carbonylation which takes place subsequent to the oxidative addition of the Ph-H bond. This oxidative addition process with a Rh^{I} complex bearing PPh_3 ligands is more facile than the alkane activation described in section 2.5 that requires the more electron-releasing PMe_3 ligands.[17,21]

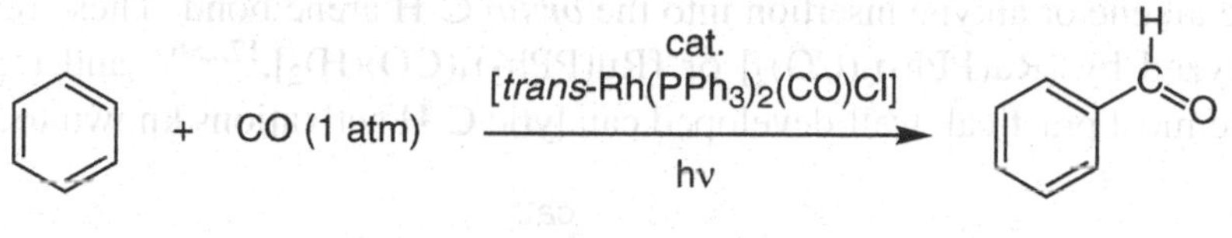

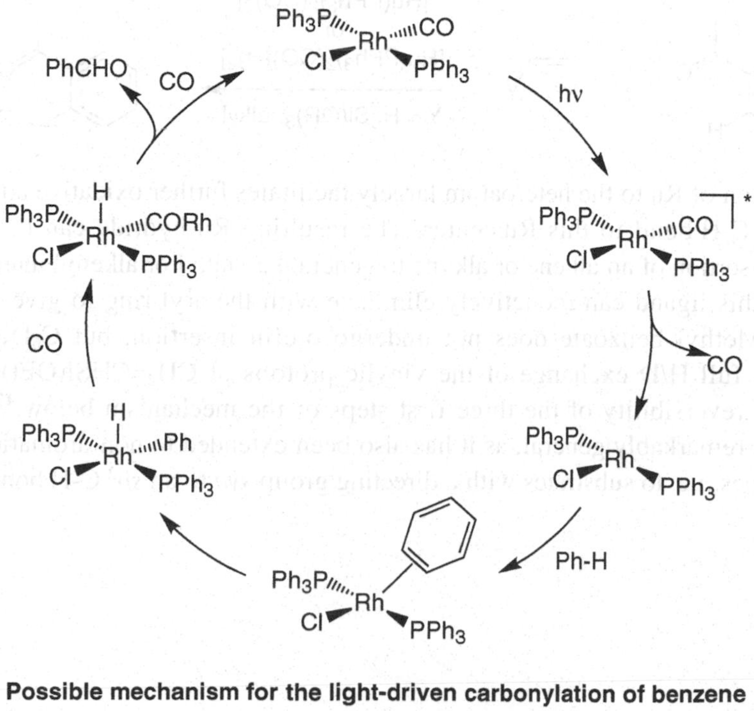

Possible mechanism for the light-driven carbonylation of benzene

3.3.2. Insertion of alkenes and alkynes in the Ar-H bond directed by an ortho substituent

The directed stoichiometric ortho-metallation reaction is a powerful concept in organic chemistry (electrophilic or nucleophilic substitution, sigmatropic rearrangement, cycloaddition, ring construction).[17.1b,c,17.28a] It was discovered by Gilman in 1939 and elaborated by Wittig in 1940, and has led to a large body of sophisticated syntheses including those of natural products. The DMG group used in organic syntheses can be C-based (CON(R)⁻, CONR₂, CO₂⁻, CONHCMe₂Ph, heterocyclic) or heteroatom based (N-Boc, N⁻COt-Bu, OCH₂Me, OCONEt₂, OCO(NMe)CMe₂Ph, SO₂⁻NR, SO₂NR₂, SO₂N⁻CMe₂Ph or P(O)(t-Bu)₂).[17.1c,17.28a]

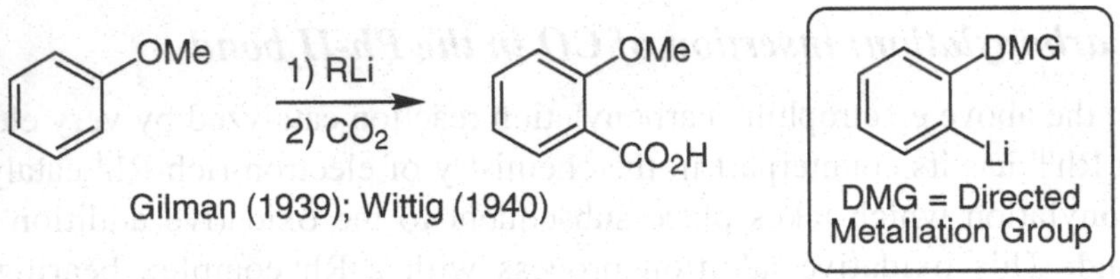

Gilman (1939); Wittig (1940)

DMG = Directed Metallation Group

The extension to catalytic reactions was discovered by Murai in 1993.[17.28b] When an arene bears an acyl or an imino substituent, the heteroatom of this group can direct the alkene or alkyne insertion into the *ortho* C-H arene bond. These reactions are catalyzed by [Ru(PPh₃)₃(CO)₂] or [Ru(PPh₃)₃(CO)(H)₂],[17.28b,c] and represent one of the most practical, well-developed catalytic C-H activations known today.

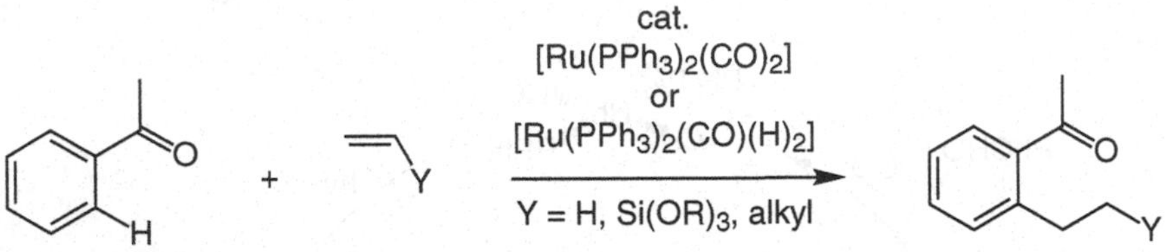

Coordination of Ru to the heteroatom largely facilitates further oxidative addition of the *ortho* C-H bond to this Ru center. The resulting Ru-hydride can then easily undergo insertion of an alkene or alkyne to generate an alkyl or alkenyl intermediate in which this ligand can reductively eliminate with the aryl ring to give the final product. Methyl benzoate does not undergo olefin insertion, but $C_6D_5CO_2CH_3$ undergoes full H/D exchange of the vinylic protons of $CH_2=CHSi(OEt)_3$, which shows the reversibility of the three first steps of the mechanism below.[17.28c] The reaction is remarkably general, as it has also been extended to non-aromatic enones and enimines, i.e. to substrates with a directing group *syn* to an sp^2 C-H bond.[17.28d]

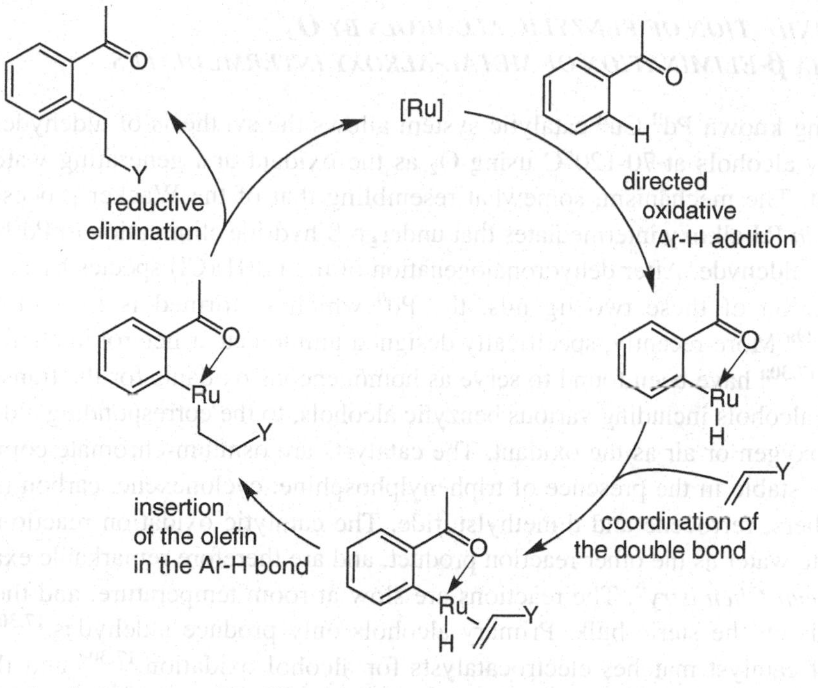

[Ru]

reductive
elimination

directed
oxidative
Ar-H addition

insertion
of the olefin
in the Ar-H bond

coordination of
the double bond

Mechanism proposed by Murai for the ortho-directed olefination of arenes

3.4. BORYLATION INVOLVING σ-BOND METATHESIS

The catalytic alkane borylation whose mechanistic scheme is detailed in section 2.6 also works according to the same mechanism with benzene in the sense that it can be borylated by pinB-Bpin to form the phenylboronate ester in 82% yield using only 0.25 mol% of the same catalyst, $[Cp*Rh(\eta^4\text{-}C_6Me_6)]$. This efficient reaction is extremely useful in view of the Suzuki coupling leading to aryl functionalization using such boron derivatives (see Chap. 19.3.8): [17.24c]

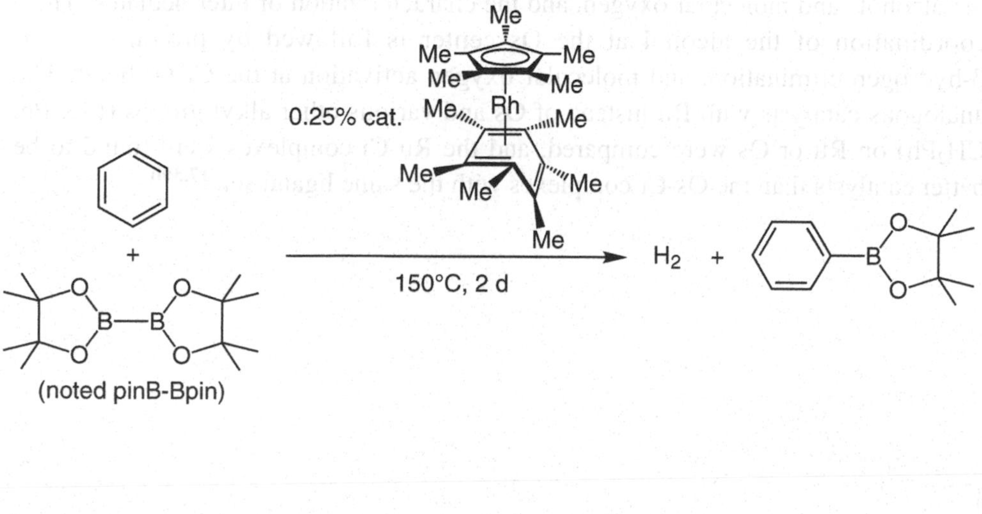

0.25% cat.

150°C, 2 d

H_2 +

(noted pinB-Bpin)

3.5. OXIDATION OF BENZYLIC ALCOHOLS BY O_2
VIA β-ELIMINATION OF METAL-ALKOXY INTERMEDIATES

The long-known Pd^{II}-Cu^{II} catalytic system allows the synthesis of aldehydes from primary alcohols at 70-120°C using O_2 as the oxidant and generating water as a product. The mechanism, somewhat resembling that of the Wacker process, proceeds *via* Pd-alkoxy intermediates that undergo β-hydride elimination to Pd-hydride and the aldehyde. After dehydrohalogenation of the Pd(H)(Cl) species by reductive elimination of these two ligands, the Pd^0 which is formed is re-oxidized by Cu^{II}.[17.29a] More recently, specifically designed unimolecular heterobimetallic complexes [17.30a] have been found to serve as homogeneous oxidants for the transformation of alcohols including various benzylic alcohols, to the corresponding aldehydes using oxygen or air as the oxidant. The catalysts are osmium-chromate complexes that are stable in the presence of triphenylphosphine, cyclohexene, carbon monoxide, ethers, ferrocene and dimethylsulfide. The catalytic oxidation reactions only generate water as the other reaction product, and are therefore remarkable examples of *"Green Chemistry"*. The reactions are slow at room temperature, and their rate depends on the steric bulk. Primary alcohols only produce aldehydes.[17.30b] This type of catalyst matches electrocatalysts for alcohol oxidation,[17.30c] and there is promise of asymmetric oxidation with recently prepared chiral analogs and related strategies with sulfido-bridged heterotrinuclear complexes.[17.30d]

R = CH₃, CF₃, OCH₃

The mechanism was determined using reaction kinetics, the isotopic distribution of the alcohols and molecular oxygen, and the characterization of intermediates. Thus, coordination of the alcohol at the Os center is followed by proton transfer, β-hydrogen elimination, and molecular oxygen activation at the Cr-Os bond. The analogous catalysts with Ru instead of Os and various other alkyl groups (Me, Ph, CH_2Ph) on Ru or Os were compared, and the Ru-Cr complexes were found to be better catalysts than the Os-Cr complexes with the same ligand set.[17.30b]

Mechanism:

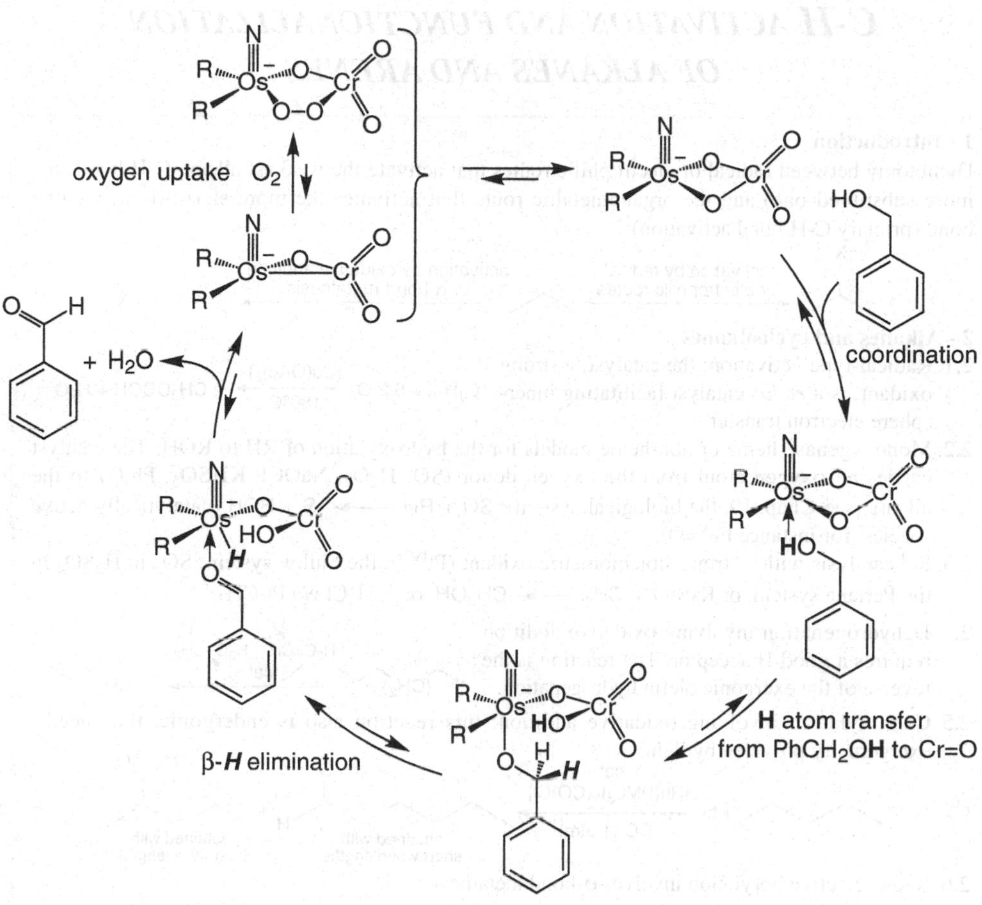

Mechanism proposed by Patricia Shapley for the Cr-Os-catalyzed alcohol oxidation by O_2

SUMMARY OF CHAPTER 17
C-H ACTIVATION AND FUNCTIONALIZATION
OF ALKANES AND ARENES

1 - Introduction

Dichotomy between radical or electrophilic routes that activate the weakest alkane C-H bond (the more substitued one) and the organometallic route that activates the more sterically accessible bond (primary C-H bond activation):

activation by radical or electrophilic routes

activation by oxidative addition or σ-bond metathesis

2 - Alkanes and cycloalkanes

2.1. Radical-type activation: the catalyst, a strong oxidant, is a *redox* catalyst facilitating inner-sphere electron transfer

$$C_4H_{10} + 5/2\ O_2 \xrightarrow[180°C]{[Co(OAc)_3]} 2\ CH_3COOH + H_2O$$

2.2. Monooxygenase heme or non-heme models for the hydroxylation of RH to ROH. The catalyst carries an oxygen atom from the oxygen donor (SO: H_2O_2, NaOCl, $KHSO_5$, PhIO) to the alkane (see Chap. 19, the biological system): SO + RH $\longrightarrow$ S + ROH (catalytically active species: for instance $Fe^V=O$).

2.3. Pt^{II} catalysis with a strong stoichiometric oxidant (Pt^{IV} in the Shilov system, SO_3 in H_2SO_4 in the Periana system, or $K_2S_2O_8$): $CH_4 \longrightarrow CH_3OH$ or CH_3Cl *via* $Pt-CH_3$.

2.4. Dehydrogenation involving oxidative addition requires a good H acceptor. The reaction is the reverse of the exergonic olefin hydrogenation.

2.5. Carbonylation involving oxidative addition: this reaction also is endergonic, thus needs exergonicity provided by light:

obtained with short wavelengths

obtained with long wavelengths

2.6. Regioselective borylation involves σ-bond metathesis:

3 - Aromatics

3.1. Radical reactions are more facile than with alkanes, because arenes are easier to oxidize (lower redox potentials) than alkanes: $PhCH_3 + 3/2\ O_2 \longrightarrow PhCO_2H + H_2O$ catalyzed by $Co(OAc)_3$ at 120°C. Thermodynamic data permit to investigate the mechanisms.

3.2. Electrophilic Fujiwara reaction:

The para selectivity is largely improved by using $Rh(CF_3CO_2)_3$, a powerful cat.

3.3. Oxidative addition coupled with insertion:

3.3.1. Oxidative addition + CO insertion

3.3.2. Oxidative addition + ortho-directed olefin or alkyne insertion: the very useful and well-developed Murai reaction:

3.4. Borylation of arenes works with high efficiency (see section 2.5), which provides useful substrates for the Suzuki coupling reaction (Chap. 21).

3.5. Oxidation of benzylic alcohols by O_2 with bimetallic catalysis yields aldehydes and H_2O. Unimolecular Os-Cr complex, a well-defined catalyst.

Exercises

17.1. Why is $Pd(CF_3CO_2)_2$ a much more efficient catalyst for arene carbonylation than $Pd(OAc)_2$?

17.2. What is the difference between the σ-bond metathesis and the electrophilic reaction in methane activation?

17.3. Why does the σ-bond metathesis between a metal-boryl species give the alkyl-borane and the metal-hydride and not the borane and the metal-alkyl, whereas the σ-bond metathesis between a lutetium-methyl and methane (Chap. 3) does not also give a lutetium hydride and ethane, but only degenerate metathesis?

17.4. Why is it possible to oxidize both alkanes and arenes using $Co(OAc)_3$ as a catalyst, whereas only arenes, not alkanes can be oxidized by $Mn(OAc)_3$?

17.5. Why are Mn and Fe porphyrin complexes good oxo transfer catalysts for the oxidation of alkanes, whereas V and Cr analogs are not?

Chapter 18

CARBONYLATION AND CARBOXYLATION REACTIONS

The catalytic incorporation of carbon monoxide CO into organic molecules is a major part of industrial catalytic processes. Carbon monoxide is accessible by combustion of coal (T > 200°C, cat.: Fe/Cu), natural gas or petroleum residues in the presence of water vapor, which generates a mixture of CO and H_2, "syngas"("Water-Gas-Shift Reaction", WGSR, see Chap. 20).

The main industrial processes involving the use of CO are:
▶ the water-gas-shift reaction,
▶ the synthesis of methanol,
▶ the Fischer-Tropsch process (reduction of CO by H_2 to alkanes and ethylene glycol),
▶ the Roelen hydroformylation process (oxo process) and related processes,
▶ the synthesis of acetic acid,
▶ the production of formic acid,
▶ the synthesis of phosgene and its applications,
▶ the synthesis of dimethylcarbonate,
▶ the synthesis of polyketones.

The catalytic transformations of CO and CO_2 involve both homogeneous and heterogeneous processes. In this Chapter, we are dealing with homogeneous processes whereas heterogeneous processes are discussed in Chap. 20.

The catalytic carbonylation reaction involves CO insertion into a metal-alkyl bond. This metal-alkyl species is provided either by the oxidative addition of an organic halide RX (Monsanto process) or the insertion of an olefin into a M-H bond (oxo process, Reppe – and related processes). We will successively examine these three industrial processes. The carbonylation reactions make a rich and varied chemistry, several examples of which are provided concerning applications to Pd(0)-catalyzed organic synthesis in Chaps 16.4 and 21.

1. CARBONYLATION OF METHANOL: MONSANTO PROCESS

The transformation of methanol into acetic acid produces more than one million tons of the latter *per* year. It is the most economical synthesis of acetic acid [18.1,18.2] (the alternative routes are the oxidation of acetaldehyde and the oxidation of C4 petroleum fractions). Acetic acid is used as a solvent for the synthesis of esters (isopropyl-, butyl- and especially vinyl acetates, the latter being a polymer precursor), for the synthesis of cellulose acetate and as an acylating agent (aspirin). Methanol is produced from "syngas" CO + H$_2$ (heterogeneous catalysis). In the Monsanto process, the catalyst is [RhI$_3$ · n H$_2$O] or [RhCl(CO)(PPh$_3$)$_2$], and the cocatalyst is HI.

$$CH_3OH + CO \xrightarrow{\text{180°C, 30 atm, } CH_3I, \text{ [RhI}_3 \cdot \text{n H}_2\text{O] } (10^{-3}\ M)} CH_3COOH\ (99\%)$$

The catalytically active metal species is the anion [Rh(CO)$_2$I$_2$]$^-$ that is formed according to:

$$[RhI_3] + 3\,CO + H_2O \longrightarrow [RhI_2(CO)_2]^- + CO_2 + 2H^+ + I^-$$

HI is necessary to the prior transformation of methanol to methyl iodide, because methanol does not undergo oxidative addition onto rhodium. Methyl iodide does undergo oxidative addition onto rhodium (I) leading to the formation of the Rh-CH$_3$ bond in which CO can insert. Finally, reductive elimination of acetyl iodide is followed by its hydrolysis to acetic acid, which regenerates the cocatalyst HI. The mechanism thus consists in a coupling between organometallic catalysis and acid catalysis.

This double catalytic cycle proceeds in the following way (Monsanto process):

$$CH_3OH + HI \longrightarrow CH_3I + H_2O$$

$$CH_3I + CO \xrightarrow{\text{cata Rh}^I} CH_3COI$$

$$CH_3COI + H_2O \longrightarrow CH_3COOH + HI$$

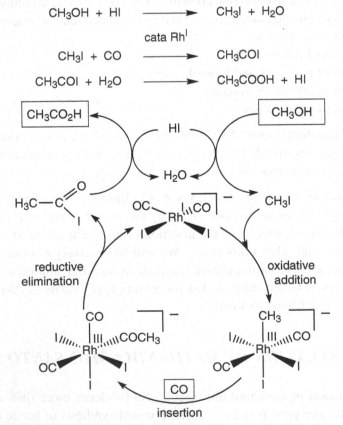

Mechanism of methanol carbonylation involving double catalysis: Rh^I and HI

2. OLEFIN HYDROFORMYLATION: OXO PROCESS

The catalysis of alkene hydroformylation, discovered by Roelen in 1938 is, with 4 million tons per year, one of the most important industrial processes [18.3,18.4] (especially the synthesis of n-butanal from propene)

$$R–CH=CH_2 + H_2 + CO \longrightarrow R–CH_2–CH_2–CHO$$

The minor iso-butanal isomer R-CH-(CHO)-CH$_3$ also forms. The aldehydes are intermediates of choice for the synthesis of alcohols (by hydrogenation), leading to detergents and plastifiers, and for the synthesis of carboxylic acids (by oxidation using O$_2$). For instance, propionic acid is synthesized from ethylene via propanal.

The olefin hydroformylation catalyst is still that of Roelen, [Co$_2$(CO)$_8$] (120 to 170°C, 200 to 300 atm) or especially [RhCl(CO)$_2$(PPh$_3$)$_2$] in molten PPh$_3$ (100°C, 50 atm), a catalyst discovered by Union Carbide in 1976. The role of PPh$_3$ is to stabilize the catalytically active species and to block the metal coordination sphere favoring propene insertion, which leads to the linear isomer. PBu$_3$ has also been used with cobalt, which increases its reactivity and the selectivity for the linear aldehyde product. With PBu$_3$, however, the cobalt catalyst is so reactive that it also catalyzes the complete hydrogenation of the aldehyde to the alcohol. This method with PBu$_3$ has not met a great success because of the catalyst recycling problems.

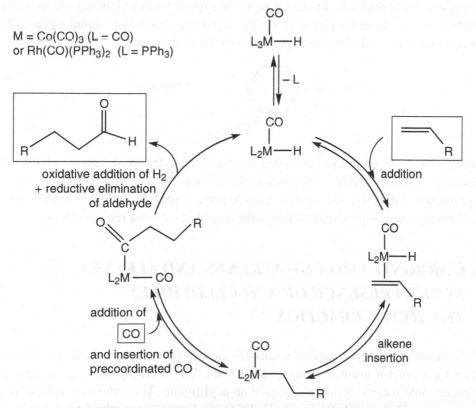

**Mechanism of propene hydroformylation to butyraldehyde (R = CH$_3$):
Roelen process (cobalt) and Union Carbide process (rhodium)**

The irreversibility of the overall hydroformylation cycle is due to the irreversibility of the last step leading to the reductive elimination of the aldehyde from the intermediate metal-hydride-formyl species generated by oxidative addition of H_2.

In the cobalt carbonyl catalysis, the catalytically active 16-electron species $[HCo(CO)_3]$ is generated by binuclear oxidative addition of H_2 leading to $[HCo(CO)_4]$:

$$[Co_2(CO)_8] \longrightarrow 2\,[Co(CO)_4]^{\bullet} \xrightarrow{\Delta,\ H_2} [(CO)_4Co\!-\!H\!-\!H\!-\!Co(CO)_4] \longrightarrow 2\,[HCo(CO)_4]$$

followed by the reversible loss of a carbonyl ligand upon heating. In the rhodium catalysis, the stable rhodium complexes $[Rh_4(CO)_{12}]$ and $[Rh_6(CO)_{16}]$ are converted to the hydride $[HRh(CO)_4]$, then $[HRh(CO)_2(PPh_3)_2]$, under the $H_2 + CO$ pressure. The catalytically active 16-electron species is formed from the latter by CO loss.

Rhodium is of the order of 1000 times more expensive than cobalt, but it is rhodium that is mostly used, because this ratio is about the same in favor of rhodium concerning the respective activities of these two metals in hydroformylation catalysis.

In order to address the problems of cost and rarity of rhodium, Kunz,[18.4b] at Rhône-Poulenc, adjusted a water-soluble phosphine, $P(m\text{-}C_6H_4SO_3^-Na^+)_3$ (TPPTS), obtained by sulfonation of $[HPPh_3]^+$ in *meta*. This phosphine solubilizes the rhodium catalyst in water without loss of activity for the hydroformylation of propene to butyraldehyde. In such a biphasic system (under vigorous stirring), the catalyst in the aqueous phase is easily separated from the organic phase by decantation at the end of the reaction, recovered and recycled.

TPPTS

Finally, Union Carbide has developed the hydroformylation of internal olefins by rhodium catalysis using very bulky phosphites. Given the present burst of asymmetric catalysis, the search of adequate chiral ligands for asymmetric hydroformylation of prochiral olefins achieving high e.e.s is a real challenge.

3. CARBONYLATION OF ALKENES AND ALKYNES
IN THE PRESENCE OF A NUCLEOPHILE:
THE REPPE REACTION [18.5]

With olefins, the Reppe reaction resembles the oxo process, because it starts as the latter by an olefin insertion into an M-H bond. The resulting metal-alkyl species migrates onto a carbonyl ligand to give an acyl ligand. The latter is captured by a nucleophile (H_2O, ROH, RNH_2, RSH, RCO_2H). For instance, $[Fe(CO)_5]$ catalyzes the hydroformylation of olefins using $CO + H_2O$:

$$\text{RCH=CH}_2 + \text{CO} + \text{H}_2\text{O} \xrightarrow[\text{OH}^-,\ 90°\text{C}]{[\text{Fe(CO)}_5]} \text{CO}_2 + \text{RCH(CHO)CH}_3 + \text{RCH}_2\text{CH}_2\text{CHO}$$

The key intermediate is the metal-hydride generated by reaction of water on [Fe(CO)$_5$] without requiring a vacant coordination site on the metal. As shown by Pettit, the reaction depends on the *pH*, because it is OH$^-$ that attacks the carbon atom of a CO ligand. Contrary to previous belief, however, the formation of the iron hydride [Fe(H)(CO)$_4$]$^-$ does not result from β-H elimination from [(CO)$_4$Fe(COOH)]$^-$. Des Abbayes has shown that deprotonation of this latter species by OH$^-$ generates transient [(CO)$_4$Fe(CO$_2$)]$^{2-}$ giving Collman's reagent [Fe(CO)$_4$]$^{2-}$, subsequent to CO$_2$ loss, and that it is protonation of the latter that gives [Fe(H)(CO)$_4$]$^-$:

$$[\text{Fe(CO)}_5] + \text{OH}^- \longrightarrow [(\text{CO})_4\text{Fe(COOH)}]^-$$

$$[(\text{CO})_4\text{Fe(COOH)}]^- + \text{OH}^- \longrightarrow [(\text{CO})_4\text{Fe(CO}_2)]^{2-} + \text{H}_2\text{O}$$

$$[(\text{CO})_4\text{Fe(CO}_2)]^{2-} \longrightarrow [\text{Fe(CO)}_4]^{2-} + \text{CO}_2$$

$$[\text{Fe(CO)}_4]^{2-} + \text{H}_2\text{O} \longrightarrow [\text{HFe(CO)}_4]^- + \text{OH}^-$$

This monohydride complex reversibly reacts with water to be protonated, giving the neutral dihydride that is supposed to be the active catalyst.

$$[\text{HFe(CO)}_4]^- + \text{H}_2\text{O} \rightleftharpoons [\text{H}_2\text{Fe(CO)}_4] + \text{OH}^-$$

$$\text{M–H(CO)} + \text{olefin} \longrightarrow \text{M–R(CO)} \longrightarrow \text{M–C(O)R}$$

$$\text{M-C(O)R} + \text{NuH} \longrightarrow \text{RC(O)Nu} + \text{MH}$$

NuH = R'NH$_2$, R'OH, H$_2$O

Hydroesterification of ethylene to methyl propionate using methanol (instead of water, above) with a Pd catalyst is an industrially valuable method developed by INEOS to produce methyl methacrylate (MMA), an important intermediate in industry (2 million tons/year) for polymer synthesis (mainly "Plexiglass" or "Perpex", a crystal-clear artificial glass with high hardness, resistance to fracture, and chemical stability).[18.5b]

In the absence of methanol, a perfectly alternating copolymer ethylene/CO is formed with high selectivity and activity of the Pd-bis-(diphenylphosphino)propane catalyst, which led to the Carillon™ process: [18.5c]

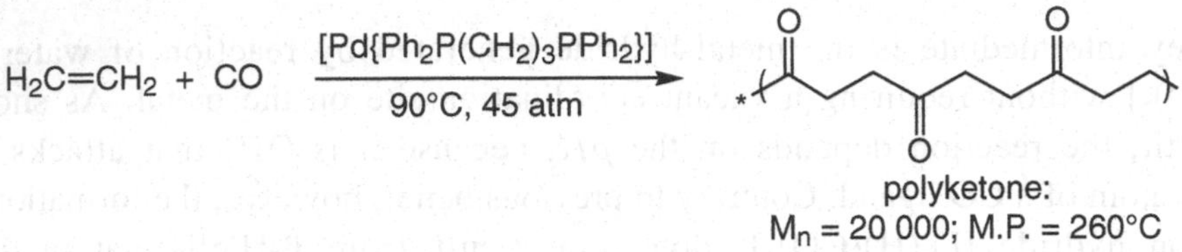

$$H_2C=CH_2 + CO \xrightarrow[90°C,\ 45\ atm]{[Pd\{Ph_2P(CH_2)_3PPh_2\}]}$$

polyketone:
M_n = 20 000; M.P. = 260°C

A noteworthy industrial application of the Reppe reaction is the hydrocarboxylation of acetylene catalyzed by $[Ni(CO)_4]$ leading to acrylic acid:

$$HC≡CH + CO + H_2O \xrightarrow[120°-220°C,\ 30\ atm]{[Ni(CO)_4]} H_2C=CHCO_2H$$

Mechanism:

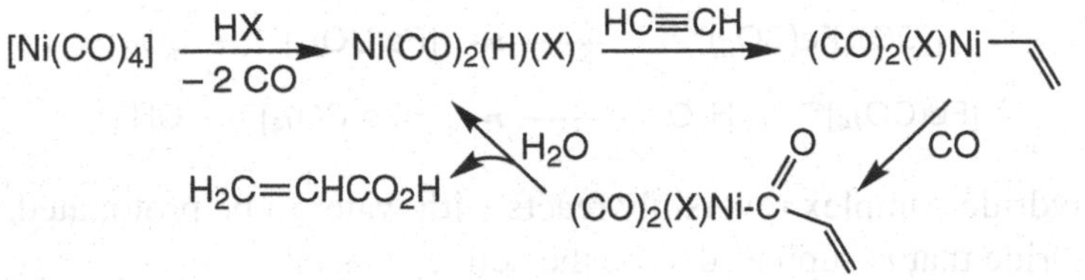

$$[Ni(CO)_4] \xrightarrow[-2\ CO]{HX} Ni(CO)_2(H)(X) \xrightarrow{HC≡CH} (CO)_2(X)Ni$$

$$H_2C=CHCO_2H \xleftarrow{H_2O} (CO)_2(X)Ni-C \xleftarrow{CO}$$

Drent, at the Shell company, has also proposed another excellent synthesis of MMA from propyne (propyne is present in 0.2 to 1% in vapocracking of naphthas) according to this principle.[18.5d]

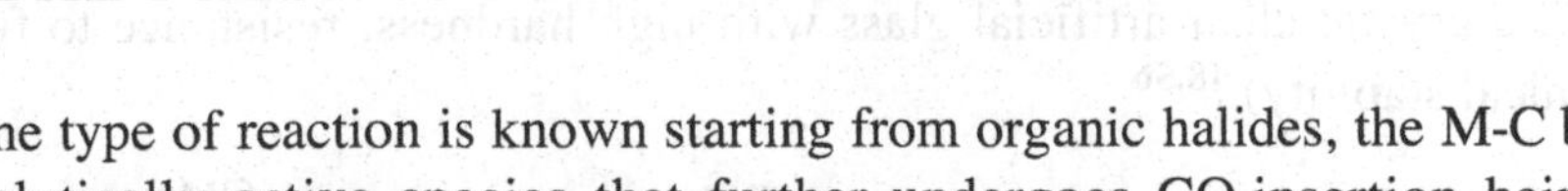

$$H_3C—C≡CH + CO + CH_3OH \xrightarrow[L\ =\ Ph_2P(2-pyridyl)]{[Pd(OAc)_2 + L]}$$

4. CARBONYLATION OF ARYL HALIDES IN THE PRESENCE OF A NUCLEOPHILE

The same type of reaction is known starting from organic halides, the M-C bond of the catalytically active species that further undergoes CO insertion being now generated by oxidative addition of the RX substrate. The Pd^0 complexes catalyze the carbonylation of vinyl, aryl, benzyl and allyl halides (but not alkyl halides that have one or several β hydrogen atoms) in the presence of a nucleophile NuH and a base (NEt$_3$) or NuM. Various nucleophiles can be used: alcohols, amines, H_2 (NuH) and carbanions (NuM): [18.5]

$$RX + CO + NuH + NEt_3 \xrightarrow{Pd\ cat.} RCONu + NHEt_3^+ X^-$$

These reactions are very useful in organic synthesis and thus will be detailed in the 5th part of the book together with the other Pd^0 catalyzed reactions.

5. CATALYSIS OF CO_2 TRANSFORMATION

Carbon dioxide is present in very large quantities in the biosphere, resulting from an equilibrium between its production through respiration (combustion product) and its consumption by plants during the photosynthetic process. It is thus a cheap raw material, but of low energy, thus difficult to activate.[18.6] Carbon dioxide can sometimes bind transition metals according to the classic mode $M(\eta^2\text{-}CO_2)$ or, more frequently, be reduced by the metal to give a metallacarboxylate $M^+\text{-}CO_2^-$, a mode stabilized by the presence of an alkali cation bonded to an oxygen atom. The metal coordination *via* an oxygen atom $M\leftarrow O=C=O$ is also known, especially if a nucleophile interacts with the electrophilic carbon atom.

Among the numerous reactions of CO_2 with the complexes, the most important one, implied in the catalytic processes, is the insertion into M-X bonds (X = H, OH, alkyl, NR_2) leading to the species $M\text{-}CO_2X$, $M\text{-}O\text{-}C(O)X$ or $M(\eta^2\text{-}O_2CX)$.[18.7]

5.1. FORMATION OF CO FROM CO_2

These processes relevant to heterogeneous catalysis are discussed in Chap. 20.

5.2. SYNTHESIS OF ALKYL- AND DMF FORMIATES

The metal-hydride species M-H generated by reaction of a catalyst and H_2 can insert CO_2; the formed intermediates M-O-C(O)H are trapped by alcohols and amines acting as nucleophiles, and these processes are catalytic:

$$CO_2 + H_2 + NuH \xrightarrow{\text{cat.}} H-\overset{\overset{\displaystyle O}{\|}}{C}-Nu + H_2O$$

Nu = RO : 160°C, 100-200 bar, cat. : $[Ni(dppe)_2]$

Nu = NMe_2 : 100-125°C, 60-120 bar, cat. : $[Co(dppe)_2H]$

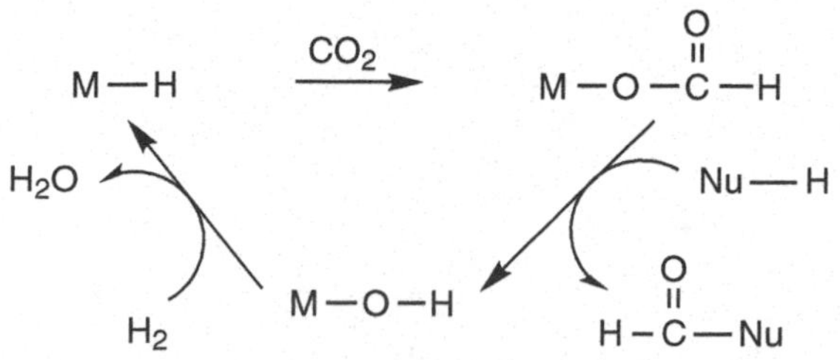

Proposed mechanism for the catalytic carboxylation of alcohols and dialkylamines

SUMMARY OF CHAPTER 18
CARBONYLATION AND CARBOXYLATION REACTIONS

---

1 - Carbonylation of methanol: Monsanto process

$$CH_3OH + CO \xrightarrow[180°C, 30\ atm]{cat.\ [RhI_2(CO)_2]^- + HI} CH_3COOH$$

2 - Hydroformylation of olefins: oxo process

$$RCH=CH_2 + CO + H_2 \xrightarrow{cat.\ [Co_2(CO)_8]\ or\ [RhCl(CO)_2(PPh_3)_2]} RCH_2CH_2CHO$$

3 - Carbonylation of alkenes and alkynes in the presence of a nucleophile: the Reppe reaction

$$RCH=CH_2 + CO + H_2O \xrightarrow[OH^-,\ 90°C]{[Fe(CO)_5]} CO_2 + RCH(CHO)CH_3 + RCH_2CH_2CHO$$

$$HC≡CH + CO + H_2O \xrightarrow[30\ atm,\ 120\text{-}220°C]{[Ni(CO)_4]} H_2C=CHCO_2H$$

4 - Carbonylation of organic halides in the presence of a nucleophile

$$RX + CO + NuH + NEt_3 \xrightarrow[\substack{NuH\ (alcohol,\ amine,\ H_2) \\ or\ NuM\ (M = metal)}]{[Pd(PPh_3)_4]} RCONu + NHEt_3^+ X^-$$

5 - Catalysis of CO₂ transformation

Water-gas-shift reaction (WGSR) and other heterogeneous reactions (see Chap. 20).

Synthesis of formiates

$$CO_2 + H_2 + NuH \xrightarrow{cat.} HC(O)Nu + H_2O \begin{cases} NuH = ROH : cat. = [Ni(dppe)_2] \\ NuH = NMe_2H : cat. = [Co(dppe)_2H] \end{cases}$$

EXERCISES

18.1. What is (are) the product(s) of 1-pentene hydroformylation?

18.2. The anionic complexes of the type $[Rh(CO)(PPh_3)_2]^-$ are active catalysts for olefin hydroformylation. How can they be obtained from $[Rh(CO)(PPh_3)Cl]$?

18.3. The cluster $[HRu_3(CO)_{11}]^-$ catalyzes the hydroformylation of olefins, but this catalysis is suppressed by the presence of PPh_3. Do you think that PPh_3 leads to a simple exchange of a CO ligand by PPh_3?

18.4. What does the reaction of 1-hexyne, catalyzed by $[Rh_4(CO)_{12}]$ or $[Co_2Rh_2(CO)_{12}]$ at 25°C, give with hydrosilanes $HSiR_3$ and CO? Write the stoichiometric reaction (by analogy with hydrosilylation), then imagine a molecular mechanism taking into account the initial formation of a disilyl derivative by oxidative addition of two molecules of silane and reductive elimination.

18.5. The reductive carbonylation $(CO + H_2)$ of methanol to acetaldehyde at 140°C is catalyzed by $[Rh(CO)_2(acac)] + PPh_2(CH_2)_3PPh_2$. Write a mechanism.

18.6. In the above reaction, homologation of methanol to ethanol is obtained if the cocatalyst $[NMe_4]_3[Ru(CO)_3I_3]$ is added. Explain.

Chapter 19

BIO-ORGANOMETALLIC CHEMISTRY: ENZYMATIC CATALYSIS

1. INTRODUCTION

The biological catalysts are called enzymes, and biological catalysis is named enzymatic catalysis.[19.1,19.2] The enzymes are proteins made of amino-acid chains. There are more than 20 different amino acids, all with the L configuration. More than half the enzymes are metallo-enzymes in which the metal plays the very important role of the active site. These metals, present in biological systems as traces because their role is catalytic, are Mg, V, Cr, Mn, Fe, Co, Ni, Cu, Zn and Mo. The other elements, also present as traces but essential, are B, Si, Se, F, I, and possibly Br. Together with all the elements that are present in large quantities: Na, K, Ca, P, S, Cl, all these elements constitute the field of bioinorganic chemistry, an area that presently is rapidly expanding.[19.3,19.4] The main biological ligands and the metals to which they most frequently bind are the following:

CO_2^-	CO_2^-		OH		$S-CH_3$
CH_2	CH_2	(imidazole)		SH	CH_2
	CH_2	CH_2	CH_2	CH_2	CH_2
$H-C-NH_3^+$	$H-C-NH_3^+$	$H-C-NH_3^+$	$H-C-NH_3^+$	$H-C-NH_3^+$	$H-C-NH_3^+$
CO_2^-	CO_2^-	CO_2^-	CO_2^-	CO_2^-	CO_2^-
aspartate	glutamate	histidine	tyrosine	cysteine	methionine
Zn^{II}, Mg^{II}, Ca^{II}, Mn^{III}, Fe^{II}, Fe^{III}		Zn^{II}, Cu^{II}, Cu^{I}, Fe^{II}	Fe^{III}	Zn^{II}, $Cu^{I/II}$, $Ni^{I/III}$, $Fe^{II/III}$, $Cu^{I/II}$, $Mo^{IV/V/VI}$, $Fe^{II/III}$	

Amino acids of the proteins forming the most stable complexes in biological media

Bioinorganic chemistry also involves the synthesis and study of low-molecular-weight models allowing to approach and understand biological systems. A restricted part of this field is bio-organometallic chemistry that, in principle, only concerns systems containing metal-carbon bonds.[19.5] Let us note, however, that mechanisms are most often uncertain, and that one can variously consider several mechanisms, as for instance in oxidation catalysis, as for instance those involving metal-element bonds and others involving metal-carbon bonds. In any case, several examples of metalloenzyme catalysis will be presented in this chapter because of their great importance, even if their inclusion in organometallic chemistry is arbitrary.

2. COBALAMIN: COENZYME VITAMIN B₁₂

A coenzyme such as coenzyme B_{12} is a required common element for the function of a group of enzymes to which it binds.[19.6] The human body contains 2 to 5 mg of coenzyme B_{12}, and a normal person needs less than 10 µg per day. It became challenging in chemistry and medicine when George Minot and William Murphy discovered in 1926 that pernicious anemia could be cured by feeding the patient with liver. Its X-ray crystal structure (as cyanocobalamine, *vide infra*) was solved by D. Hodgins in 1956, and its total synthesis was achieved by R.B. Woodward and published in 1966. Coenzyme B_{12} is an octahedral d^5 cobalt (III) 18-electron complex of the type $[Co^{III}L_4X_2]^+$. The central planar unit (core), containing the Co atom at the center bonded to four nitrogen atoms of pyrrole rings (like a porphyrin), is called corrin (see below).

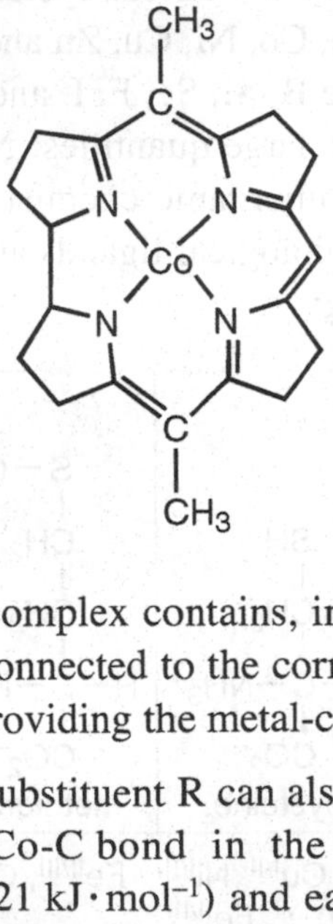

Corrin core of cobalamin (coenzyme B₁₂)
(substituents on the four pyrroles and the two apical ligands are not shown here, see p. 453)

This complex contains, in addition to the corrin, an apical benzimidazole L ligand also connected to the corrin and another apical X ligand, adenosyl (noted R, above Co) providing the metal-carbon bond.

This substituent R can also be CN^-, (most common commercial form), CH_3 or OH^-. The Co-C bond in the adenosyl cobalamine is weak (29 ±5 kcal·mol⁻¹ or 123 ±21 kJ·mol⁻¹) and easily undergoes reversible homolytic cleavage at ambient temperature to give the adenosyl radical $RCH_2^\bullet$ and the 17-electron Co^{II} radical.[198.7] This cleavage is the basis of the enzymatic mechanisms of vitamin B_{12} coenzyme.

$$[(L_4X)Co^{III}-CH_2R]^+ \longrightarrow [(L_4X)Co^{II}]^+ + RCH_2^\bullet$$

The weakly bonded adenosyl ligand is thus easily substituted *in vivo* by other ligands in axial position such as H_2O (aquacobalamine or B_{12a}) or CH_3 (methyl-cobalamine). The 18-electron Co^{III} complex aquacobalamine B_{12a} is easily reduced first to the 17-electron Co^{II} radical complex that has lost H_2O, noted B_{12r} (r for reduced), then to tetra- or pentacoordinated Co^I, noted B_{12s} (s for superreduced). The latter is a strong nucleophile that is readily alkylated by alkylating reagents such as alkyl iodide. For instance, the natural methylating agent, N^5-methyl tetrahydrofolate (noted MeTHF), gives, with B_{12s}, methylcobalamine. It is the substitution of adenosyl by cyanide that produced cyanocobalamine whose X-ray crystal structure was determined.

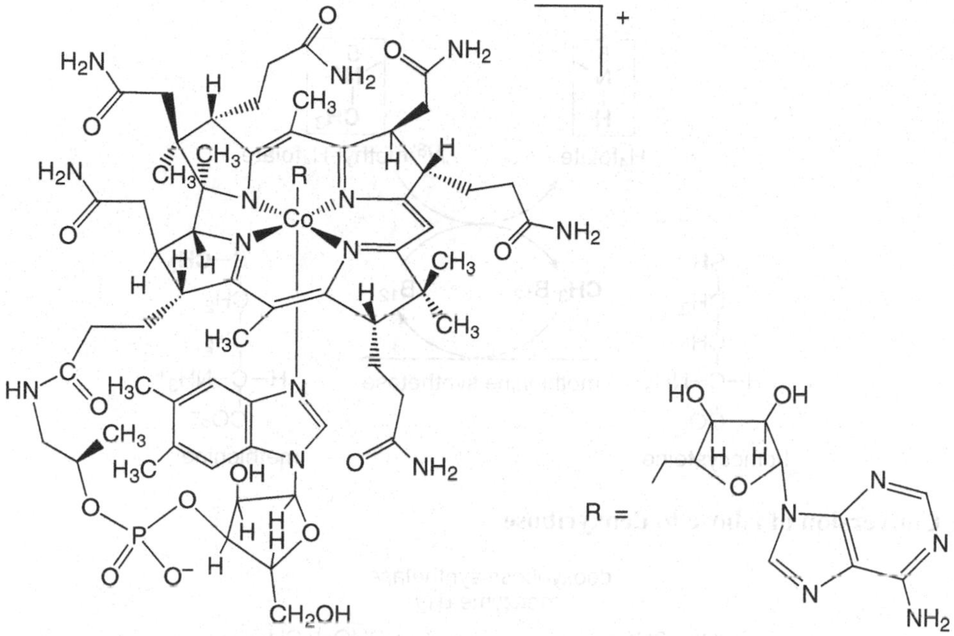

Cobalamin: coenzyme vitamin B_{12}

The coenzyme B_{12} reacts with several enzymes to catalyze three types of reactions: [19.8]

1. **Isomerase** - the H and X atoms bonded to two neighboring carbon atoms are exchanged (for instance, in coenzyme A).

$$R'CHX{-}CH_2R'' \xrightarrow{\text{enzyme, coenzyme } B_{12}} R'CH_2{-}CHXR''$$

Mechanism:

$$[(L_4X)Co^{III}{-}CH_2R] \longrightarrow [(L_4X)Co^{II}] + RCH_2^\bullet$$

$$RCH_2^\bullet + R'CHX{-}CH_2R'' \longrightarrow RCH_3 + R'CHX{-}C^\bullet HR''$$

$$R'CHX{-}C^\bullet HR'' \longrightarrow R'C^\bullet H{-}CHXR''$$

$$R'C^\bullet H{-}CHXR'' \longrightarrow R'CH_2{-}CHXR'' + RCH_2^\bullet$$

2. **Methylation of a substrate** - vitamin B_{12} catalyzes the transfer of a methyl group onto several biological substrates. In particular, it can catalyze the methylation of homocystein by N^5-tetrahydrofolate to methionine and tetrahydrofolate:

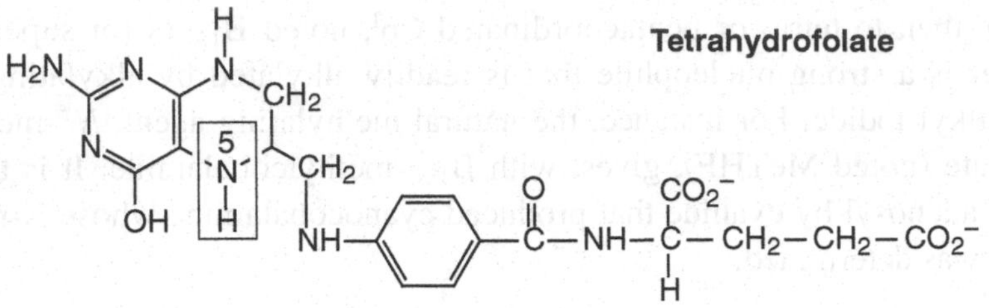

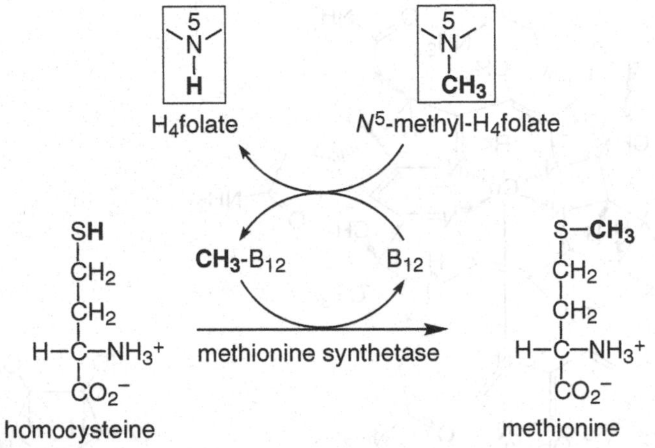

Tetrahydrofolate

H$_4$folate N^5-methyl-H$_4$folate

homocysteine   methionine

3. Conversion of ribose to deoxyribose

deoxyribose synthetase
coenzyme B$_{12}$

–CHOH–CHOH– $\longrightarrow$ –CHOH–CH$_2$–

Ribose leads to the formation of DNA (deoxyribonucleic acid) whereas deoxyribose is the basis of the formation of RNA (ribonucleic acid).

Since 1960, the synthesis of vitamin B$_{12}$ models, the cobaloximes, could be achieved, in particular with dimethylglyoxime (dmgH) forming with itself hydrogen bonds around cobalt. DmgH can be written below as an L$_2$ anionic ligand of a positively charged Co^{III} center (left) or as a neutral LX ligand of Co^{III} (right).

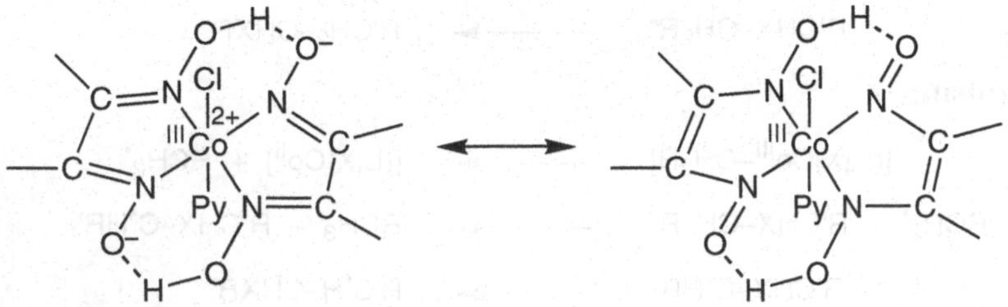

The complex [Co(dmg)$_2$(py)(X)] can be reduced in two steps (X = Cl) like aquacobalamine (*vide supra*) to finally yield [Co(dmg)$_2$(py)]$^-$, a powerful nucleophile that can be methylated by CH$_3$I. The methylated derivative obtained, [Co(dmg)(py)(CH$_3$)] is a model of methylcobalamine.

3. BIOLOGICAL REDOX MEDIATORS

There are many *oxidoreductase enzymes*, i.e. enzymes able to catalyze redox reactions: oxidases, dehydrogenases, peroxidases, superoxide dismutases and finally the mono- and dioxygenases. All of them contain active sites that are able to transfer electrons using mediators or redox catalysts that can be classified into four categories: [19.9]

1. **The cytochromes**, Fe-porphyrin complexes containing, for instance, as apical ligands, the imidazole of an histidine or the thioether of a cysteine (cytochrome c). The redox reactions occur because of the ease of interconversion between the oxidation states Fe^{II} and Fe^{III}.

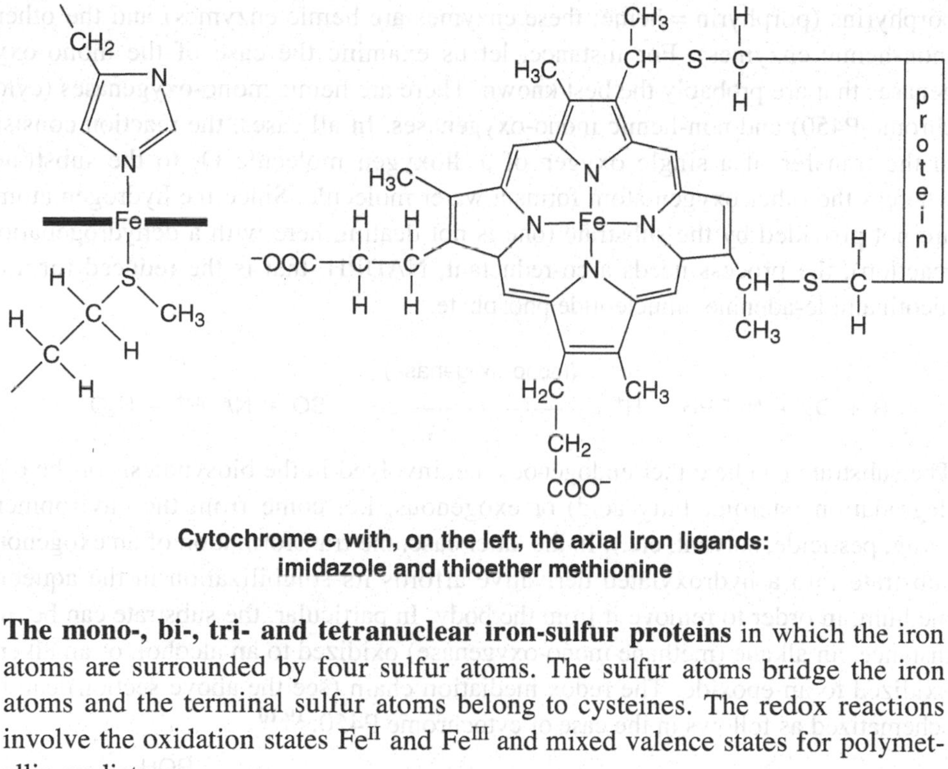

Cytochrome c with, on the left, the axial iron ligands:
imidazole and thioether methionine

2. **The mono-, bi-, tri- and tetranuclear iron-sulfur proteins** in which the iron atoms are surrounded by four sulfur atoms. The sulfur atoms bridge the iron atoms and the terminal sulfur atoms belong to cysteines. The redox reactions involve the oxidation states Fe^{II} and Fe^{III} and mixed valence states for polymetallic mediators.

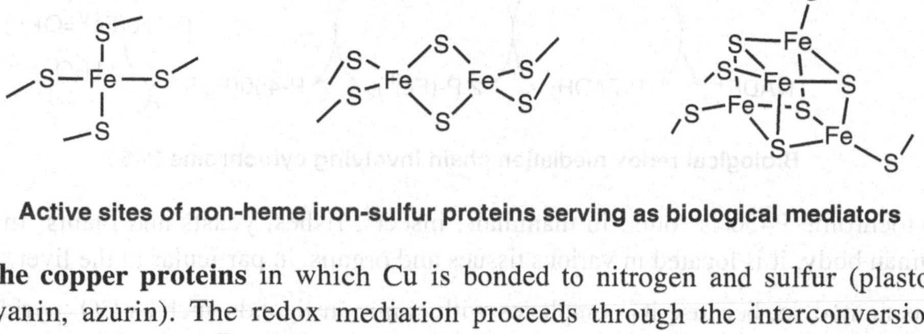

Active sites of non-heme iron-sulfur proteins serving as biological mediators

3. **The copper proteins** in which Cu is bonded to nitrogen and sulfur (plastocyanin, azurin). The redox mediation proceeds through the interconversion between Cu^I and Cu^{II}.

4. The flavodoxines and quinones whose structure is based on flavin groups. Whereas the three first categories above involve monoelectronic inorganic mediators, flavins do not contain metals and transfer electrons between NADPH and oxygenase.

An example of the redox mediation chain is involved in the next section.

4. EXAMPLES OF OXIDOREDUCTASE ENZYMES:
THE MONO-OXYGENASES

The oxidoreductase enzymes can be divided into two large categories: the metallo-porphyrins (porphyrin = heme; these enzymes are hemic enzymes) and the others (non-hemic enzymes). For instance, let us examine the case of the mono-oxy-genases that are probably the best known. There are hemic mono-oxygenases (cyto-chrome P450) and non-hemic mono-oxygenases. In all cases, the reaction consists in the transfer of a single oxygen of a dioxygen molecule O_2 to the substrate, whereas the other oxygen atom forms a water molecule. Since the hydrogen atoms are not provided by the substrate (one is not dealing here with a dehydrogenation reaction), the process needs a co-reductant, NADPH, that is the reduced form of nicotinamide-adenine dinucleotide phosphate.

$$\text{S} + \text{O}_2 + \text{NADPH} + \text{H}^+ \xrightarrow{\text{(mono-oxygenase)}} \text{SO} + \text{NADP}^+ + \text{H}_2\text{O}$$

The substrate can be either endogenous, i.e. involved in the biosynthesis or the bio-degradation (steroid, fatty acid) or exogenous, i.e. come from the environment (drug, pesticide, solvent, etc.). In the latter case, the transformation of an exogenous substrate into a hydroxylated derivative affords its solubilization in the aqueous medium, in order to remove it from the body. In particular, the substrate can be, for instance, an alkane (methane mono-oxygenase) oxidized to an alcohol, or an alkene oxidized to an epoxide. The redox mediation chain (see the above section) can be schematized as follows in the case of cytochrome P450: [19.10]

Biological redox mediation chain involving cytochrome P450

Cytochrome P450 is found in mammals, insects, fishes, yeasts and plants. In the human body, it is located in various tissues and organs, in particular in the liver.

It is named so, because its complexes with common ligands such as CO, pyridine, etc. show an absorption maximum around 450 nm. There are more than hundred varieties, all containing an iron-porphyrin complex with a cysteinate axial ligand bonded to the metal by the sulfur atom. The other axial site is that of dioxygen activation. Hemoglobin and myoglobin have the same iron-porphyrin structure of the active site as cytochrome P450, but with a histidine bound to iron by a nitrogen atom of an imidazole. These two hemoproteins only transport dioxygen to the blood and tissues by reversibly forming a bent Fe-O-O bond. This latter bond is formed by reaction between the Fe^{II} form and O_2 without activating the latter, i.e. without electron transfer onto the terminal oxygen atom. The extraordinary difference of reactivity towards O_2 between cytochrome P450 and these two hemoproteins is thus solely due to the nature of the axial ligand, i.e. to its trans effect on the reactivity of bound O_2. The mechanistic cycle of activation in the case of cytochrome P450 is represented below:

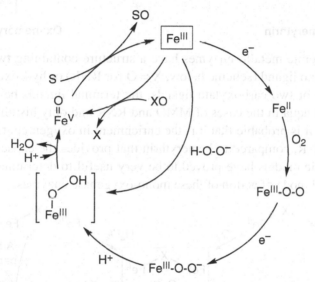

**Oxidation mechanism of a substrate S by cytochrome P450,
an iron-protoporphyrin IX complex containing an axial cysteine thiolate ligand**

This scheme contains shortcuts. Only these shortcuts could be mimicked catalytically and in an efficient way by many model compounds having a closely related structure. Indeed, O_2 activation is difficult and complex, as one can notice on the mechanistic scheme. In particular, iron and manganese porphyrin complexes have been synthesized in order to mimic cytochrome P450 using single oxygen-atom donors such as H_2O_2, etc. (XO on the scheme).

The active species resulting from O_2 activation is an $Fe^V=O$ or $Fe^{IV}-O^{\bullet}$ complex, and the intimate mechanism that has been proposed for alkane C-H activation is shown below:

$$PorFe^V=O + H-C\begin{matrix} R_1 \\ -R_2 \\ R_3 \end{matrix} \longrightarrow \left[PorFe^{IV}-OH \; ^{\bullet}C\begin{matrix} R_1 \\ -R_2 \\ R_3 \end{matrix} \right] \longrightarrow PorFe^{III} + OH-C\begin{matrix} R_1 \\ -R_2 \\ R_3 \end{matrix}$$

**Proposed mechanism for the hydroxylation of alkane C-H bonds
catalyzed by cytochrome P450**

Among the non-hemic mono-oxygenase enzymes, two most important ones are methane mono-oxygenase (MMO) and ribonucleotide reductase (RNR). The same parallel can be made between these two metallo-enzymes and hemerythrin whose structure resembles those of MMO and RNR, but whose function, as for hemoglobin and myoglobin, only consists in transporting dioxygen, not in activating it.[19.11]

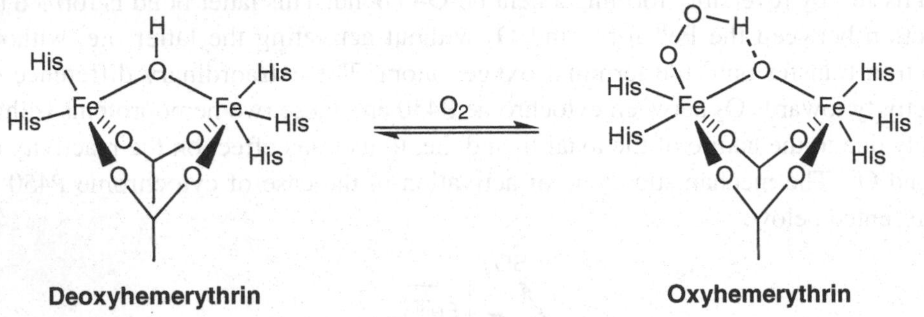

Deoxyhemerythrin   **Oxyhemerythrin**

All these non-hemic metallo-enzymes have a structure containing two iron atoms bridged by an oxo ligand (scheme below, X = O for RNR) or hydroxo (X = OH for MMO) and one or two carboxylato ligands, the terminal ligands being histidines, carboxylato and aqua in the cases of MMO and RNR and only histidine in the case of hemerythrin. It is probable that it is the enrichment in oxygen-containing ligands in MMR and RNR, compared to hemerythrin that provides the capacity to activate O_2. The synthetic models have proved to be very useful to determine the structure and hypothetical mode of action of these mono-oxygenase enzymes.

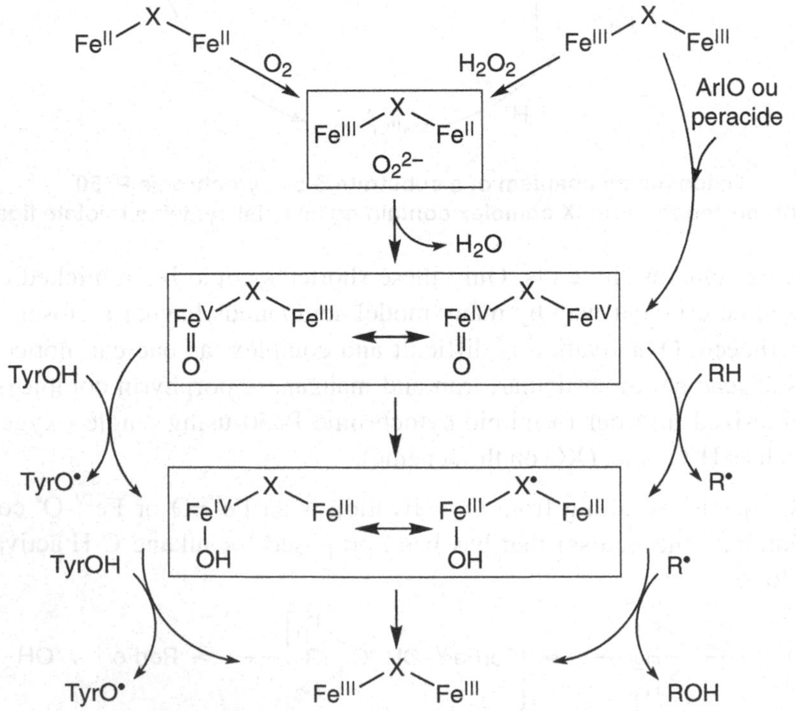

Common mechanisms for dioxygen activation proposed for ribonucleotide reductase (RNR) and methane mono-oxygenase (MMO)

The probable structure of one of the components of RNR and its mode of action leading to the formation of the tyrosinyl radical 122 (TyO•) bound to the protein are represented on the next page (left part). The right part of the scheme concerns the hydroxylation of methane (RH = CH_4). Note that the active species has the $Fe^V=O$ structure that is very similar to the one of cyctochrome P450 in the hemic series.

There are many other heme and non-heme oxidoreductase enzymes and the understanding of these systems using synthetic models is in full expansion.

5. *NITROGEN FIXATION BY NITROGENASE ENZYME* [19.12-19.14]

The conversion of dinitrogen to ammonia by nitrogenase enzyme in bacteria (sometimes in symbiosis with plants) is called nitrogen fixation. This process is essential to life, because it is the starting point of the synthesis of amino acids that are the components of proteins. Three kinds of nitrogenase enzymes are known, each of them containing two air-sensitive metalloproteins that can be easily separated:

▸ A 60.000-Dalton iron metalloprotein containing a single Fe_4S_4 cluster that has a distorted cubic structure and serves as redox mediator (see section 3), the reducing agent being ATP. This metalloprotein is the same in the three sorts of nitrogenase enzymes.

▸ A second iron metalloprotein that is larger (about 220.000 Daltons) and also eventually contains another transition metal that is different for the three kinds of nitrogenase enzymes: Mo (MoFe protein), V (VFe protein) or Fe (FeFe protein). It is the MoFe protein that is the best known. It contains:

▸▸ two MoFe clusters called cofactors (MoFe-co) containing the active site responsible for the coordination and activation of dinitrogen. The X-ray crystal structure of the cofactor has been solved in 1990. It is not yet known, however, whether N_2 coordinates to Mo or Fe in this cluster, although the coordination to Fe is more probable because of its low valency (the four central Fe atoms are trivalent whereas Mo is hexa-coordinated). The resolution of the structure indicated the presence of an unknown ligand Y at the center, a light atom (such as for instance nitrogen) that was too difficult to identify near heavier atoms in such a large structure.

**Cofactor (MoFe-co) of a nitrogenase enzyme
containing the active site of nitrogen fixation**

▶▶ an iron-sulfur cluster called P cluster and made of two distorted Fe_4S_4 cubes connected by two SR bridges (cysteinic radicals). It has the structure of a redox mediator.

"P" cluster serving as a redox mediator for a nitrogenase enzyme

The reduction of N_2 to $2 NH_3$ requires $6H^+ + 6e^-$. In the biological nitrogen fixation system, however, the stoichiometry of dinitrogen reduction concomitantly involves the reduction of two protons to H_2:

$$N_2 + 10H^+ + 8e^- \xrightarrow{\text{[N}_2\text{ase]}} 2NH_4^+ + H_2$$

In the absence of N_2, the nitrogenase enzyme only reduces the protons to H_2. Dinitrogen fixation by the nitrogenase enzyme is inhibited by CO because of the irreversible coordination of the latter onto the active site of the enzyme. The nitrogenase enzyme can catalyze the reduction of substrates such as, eventually, acetylene and terminal alkynes (to olefins), cyanide (to methane, ammonia and methylamine), isocyanide (to alkane and ammonia), nitride (to ammonia and hydrazine) and cyclopropene (to propene and cyclopropane). The purely organic nature of some of these substrates and the design of organometallic model catalysts (in particular by Schrock) justify the introduction of this area of chemistry in the organometallic context. The study of the biochemical reduction of all these substrates as well as that of N_2 reduction by inorganic and organometallic model catalysts have allowed to approach the understanding of the biological N_2 fixation mechanism, although little is still certain. It is known that electrons provided by ATP are conveyed towards the active metal-N_2 site (for instance MoFe-co) through the intermediacy of a redox mediator (ferredoxin or flavodoxin; see section 3). The transformation of coordinated N_2 into NH_3 proceeds through a series of single electron-transfer/proton-transfer sequences (the proton source is the aqueous medium). The probable overall mechanism can be depicted as in the sheme, next page.

Since the first N_2 complex isolated by Allen and Senof in 1965, N_2 complexes have been disclosed with most transition metals, and these complexes are discussed in Chap. 7.5.3. Many inorganic models of nitrogenase enzymes and some organometallic ones able to stoichiometrically reduce N_2 have been synthesized, but only two of them are catalytic. One of the most impressive model systems is that developed by Schrock's group using an extremely bulky tris(amido)amine ligand for Mo complexes, so that N_2-bridged bimetallic complexes cannot form.

$$\text{overall}: \quad N_2 + 8\,H^+ + 8\,e^- \longrightarrow 2\,NH_3 + H_2$$

Currently accepted mechanism for nitrogen fixation by nitrogenase enzymes

In this way, it was shown that Mo can change its oxidation states from Mo^{III} to Mo^{VI} in order to accommodate the various intermediates in the above nitrogen fixation cycle resulting from successive electron-transfer/proton-transfer sequences. In the scheme of p. 452, all the framed complexes could be isolated and characterized and the dashed framed complexes were observed in solution. Thus, evidence for 9 out of the 12 possible intermediates were obtained.[19.14a] Moreover, catalytic N_2 reduction was achieved in heptane solvent using $[Cr(\eta^5\text{-}C_5Me_5)_2]$ as the electron-reservoir complex (progressively added by syringe over 6 hours in order to prevent predominantly H_2 formation) and $\{LutH\}\{Ar_{f4}\}$ as the proton source, which yielded ≈ 7 to 8 equivalents of ammonia with 63-66% yield. Only $^{15}NH_3$ was produced upon reduction of $^{15}N_2$. In this model, N_2 is reduced around -1 V vs. SCE (E^0 of $CrCp^*_2{}^{+/0}$) whereas non-catalytic N_2 reduction occurs around -3 V (E^0 of $Li^{+/0}$) by Li metal to LiN_3.[19.14b]

Many questions remain, however. For instance, are other metals viable sites? How is hydrazine formed? There is only one other truly catalytic system yielding mainly hydrazine (N_2H_4 and NH_3, $\approx 10:1$), disclosed by the Shilov group. It uses methanol as the protic solvent, contains Mo^{III}, $Mg(OH)_2$, a reducing agent ($Ti(OH)_3$, Na amalgam or Hg electrode), phosphines and a surfactant at the Hg surface (phosphatidylcholine). This catalytic system works at ambient temperature, 1 atm and yields up to 170 equivalents of ammonia per Mo at 110°C and 1 atm. The turnover rate can be close to that of nitrogenase itself ($\approx 1s^{-1} \cdot mol^{-1}$). Shilov's mechanistic proposal involves at least two Mo centers binding N_2 during the course of its reduction to hydrazine and ammonia.[19.12a,19.14c]

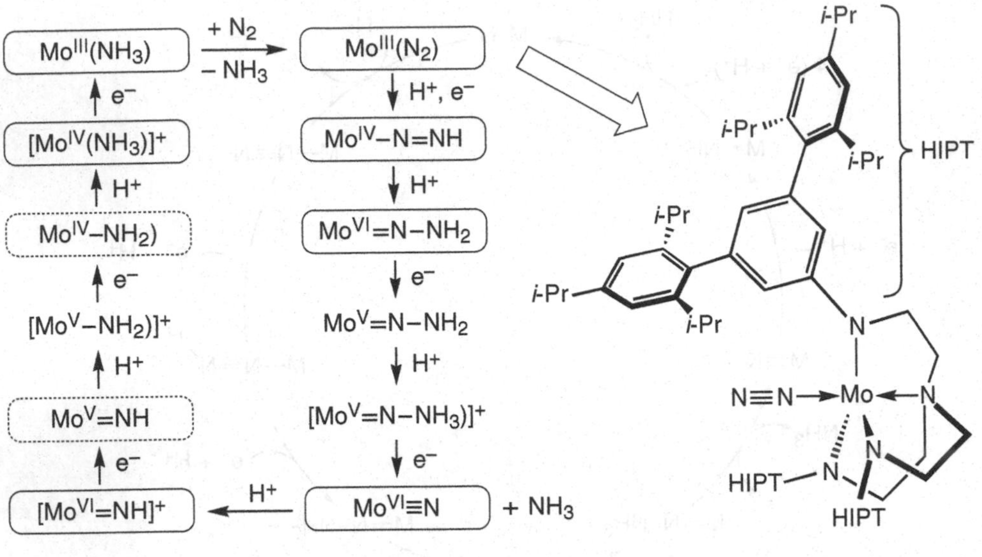

Proposed intermediates in the reduction of N$_2$ at a LMo center (see L on the right) through the stepwise addition of protons and electrons in the Schrock system
(framed species were isolated or characterized)

Despite these remarkable model systems, chemists have not yet been able to prepare a catalytic system that can compete with the XIXth century Haber-Bosch process (1895) that is still used in industry to synthesize NH$_3$ (see the mechanism in Chap. 20.2.2).

$$N_2 + 3H_2 \xrightarrow[\text{350°C, 90 atm}]{\text{[Fe]}} 2\,NH_3$$

This remains a formidable challenge for which the models, structural studies and biological information bring about a significant advance (despite the mechanistic difference).

6. NICKEL ENZYMES [19.5,19.15-19.17]

These enzymes, whose study is relatively recent, are not yet very well known:

CO dehydrogenase enzymes [19.15] catalyze two organometallic reactions. The first one resembles the water-gas-shift reaction (see Chap. 20) except that 2 H$^+$ and 2 e$^-$ are produced instead of H$_2$:

$$CO + H_2O \longrightarrow CO_2 + 2\,H^+ + 2\,e^-$$

In the second one, methyltetrahydrofolate (MeTHF) is carbonylated, the product being a thioester, acetyl coenzyme A (CoA being a thiol). This reaction resembles methanol carbonylation to acetic acid (Monsanto process, see Chap. 18):

$$MeTHF + CoA + CO \longrightarrow MeCOCoA + THF$$

The enzyme contains two nickel-centered fragments: the first one is a cluster containing Fe and Ni, whereas the second one, remote from the first one, is supposed to be responsible for CO oxidation. EXAFS shows that the Ni atoms of both sites have a nitrogen, oxygen and sulfur environment. Complexes based on nickel that serve as models for this enzyme have been synthesized by the groups of Crabtree [19.18] in Yale and Holm [19.19] in Harvard.

Methanogene enzymes reduce CO_2 to CH_4 and recover the resulting energy. In the last step, methyl-coenzyme M is hydrogenated to methane by a thiol cofactor HS-HTP, the reaction being catalyzed by the nickel enzyme.

$$\text{CH}_3\text{SCH}_2\text{CH}_2\text{SO}_3^- + \text{R'SH} \xrightarrow{\text{[Ni enzyme]}} \text{CH}_4 + \text{R'S-SCH}_2\text{CH}_2\text{SO}_3^-$$

methyl-coenzyme M   HS-HTP

A coenzyme containing a Ni-methanocorphin complex might be bound to the nickel enzyme and would contain the active site allowing the transformation of the methyl group.

This methyl radical transfer onto the nickel-methanocorphin center would proceed from the methyl-coenzyme M and would be followed by protonation to form CH_4. Protonation on nickel of a Ni-CH_3 complex indeed leads to methane through reductive elimination in the intermediate complex $L_n\text{Ni(H)(CH}_3)$. The structure of the enzyme is still unknown, however.

The hydrogenase enzymes allow some bacteria to feed with H_2 and some others to eliminate H_2. They catalyze the following reaction in the forward or backward direction:

$$\text{H}_2 \underset{\text{[H}_2\text{ase]}}{\rightleftharpoons} 2\,\text{H}^+ + 2\,\text{e}^-$$

The majority of hydrogenase enzymes contain Ni and Fe that are seemingly bridged by a carbonyl and a dithiolate ligand. The active site is a nickel atom, probably Ni^{III}. A minority of hydrogenase enzymes only contain Fe in a 6Fe-6S cluster recalling the structure of the FeMo-co cofactor of nitrogenase that also works like a hydrogenase enzyme in the absence of N_2 (see section 5). We know that an electron-poor transition-metal center binds H_2 without oxidative addition, and that the complexes M-(H_2) are consequently acidic. The activation process is completed by the transfer of electrons from nickel, which considerably increases the acidity of the nickel-hydride species.

$$\text{Ni}^{II} \xrightarrow{\text{H}_2} \text{Ni}^{II}(\text{H}_2) \xrightarrow{-\,\text{H}^+ -\,\text{e}^-} \text{Ni}^{III}(\text{H}) \xrightarrow{-\,\text{H}^+ -\,\text{e}^-} \text{Ni}^{II}$$

Nickel bacteria, the archaebacteria, are primary bio-organisms in the evolution, being able to survive with gases as elementary as H_2, N_2, CO and CO_2 as just indicated above. Nickel bio-organometallic research is presently expanding rapidly.[19.5]

These few examples show both the importance of bio-inorganic chemistry and its overlap with organometallic chemistry. One can be sure that the XXIth century will see a considerable development of the understanding of these metallo-enzymes. Finally, let us note other important branches of *bio-organometallic chemistry*: [19.20-19.22]

▸ *anticancer therapy* (for example, [TiCp$_2$(Cl)$_2$], probably working in a way related to that of [*cis*-PtCl$_2$(NH$_3$)$_2$], and ferrocenium are active), as well as other therapies;

▸ *toxicology* of some organometallic compounds (lead, mercury, tin: cf. Chap. 12);

▸ *medical imaging* (for instance, the chemistry of radioactive labels such as technetium; NMR contrast agents such as gadolinium-rich dendrimers);

▸ *supramolecular interactions* of organometallic compounds with amino acids, DNA, RNA involving molecular recognition and application thereof to specific drug transport;

▸ beyond bio-organometallic chemistry, the considerably increasing impact of *organometallic catalysts* on drug synthesis (see Chap. 21).

SUMMARY OF CHAPTER 19
BIO-ORGANOMETALLIC CHEMISTRY:
ENZYMATIC CATALYSIS

The biological catalysts are called enzymes. Examples:

1 - Cobalamin: vitamin B_{12} coenzyme

Octahedral d^5, 18 e^- Co^{III} complex containing an L_3X corrin ligand, and in apical position, an L benzimidazole ligand and an X ligand bonded to Co by a carbon atom [weak bond: 28.6 kcal·mol^{-1} (119.5 kJ·mol^{-1}) whose facile homolytic cleavage leads to the adenosyl radical]. Vitamin B_{12} coenzyme catalyzes radical reactions such as the isomerization of halogenated derivatives (for instance coenzyme A), the methylation of a substrate such as homocysteine and the conversion of ribose to deoxyribose.

2 - Biological redox mediators

There are four categories:
- cytochromes (Fe-porphyrin complexes);
- Fe-S proteins (mono-, bi- and tetranuclear);
- Cu proteins;
- flavodoxines and quinones (purely organic) transporting 2 H^+ + 2 e^-.

3 - The oxidoreductases

For example, the **mono-oxygenases** catalyze the transfer of an O atom

$$S + O_2 + NADPH + H^+ \xrightarrow{\text{[mono-oxygenase]}} SO + NADP^+ + H_2O$$

S = alkane ($\longrightarrow$ alcohol), alkene ($\longrightarrow$ epoxide), phosphine ($\longrightarrow$ phosphine oxide), etc.

There are two categories:
- the iron-porphyrin complexes
- the non-hemic mono-oxygenases that are bridged by an O atom between two Fe atoms (ribonucleotide reductase or RNR) or OH between two Fe (methane mono-oxygenase or MMO).

4 - The nitrogenases catalyzing nitrogen fixation
(see the catalytic Schrock and Shilov model systems)

$$8\,H^+ + 8\,e^- + N_2 \xrightarrow{\text{[N}_2\text{ase]}} 2\,NH_3 + H_2$$

5 - The nickel enzymes

- CO dehydrogenases:

$$CO + H_2O \xrightarrow{\text{[CO dehydrogenase]}} CO_2 + 2\,H^+ + 2\,e^-$$

$$MeTHF + CoA + CO \xrightarrow{\text{[CO dehydrogenase]}} MeCOCoA + THF$$

- Methanogenes:

$$\underset{\text{methylcoenzyme M}}{CH_3SCH_2CH_2SO_3^-} + RSH \xrightarrow{\text{[Ni]}} CH_4 + RSSCH_2CH_2SO_3^-$$

- Hydrogenases:

$$H_2 \underset{\text{[H}_2\text{ase]}}{\rightleftharpoons} 2\,H^+ + e^-$$

EXERCISES

19.1. Why do the cytochromes, hemic redox mediators, have lower Fe^{II}/Fe^{III} redox potentials than many organometallic complexes with the same oxidation states?

19.2. Why does vitamin B_{12} have low Co-C bond energy allowing the release of an alkyl radical?

19.3. Why do cytochrome P450 and its synthetic models easily transfer an oxygen atom to substrates, even to the most inert ones such as alkanes, which makes them excellent oxidation catalysts for these substrates?

19.4. Why do the large majority of M-(N_2) complexes release dinitrogen upon attempting to protonate them to yield ammonia, only a few of them being reduced to NH_3 or NH_2-NH_2. Suggest a simple experimental method allowing to select those in which the N_2 ligand is so reducible.

Chapter 20

HETEROGENEOUS CATALYSIS

1. INTRODUCTION

At first sight, heterogeneous catalysis looks to be a far cry from organometallic chemistry and homogeneous catalysis. It is indispensable to introduce this aspect, however, because it is complementary to homogeneous catalysis, and the majority of industrial processes are carried out with heterogeneous catalysts. Moreover, the molecular approach is now common in heterogeneous catalysis, and a continuity of disciplines is being established which runs from monometallic activation to solid-state activation *via* organometallic clusters, giant clusters, then to mono- or polymetallic nanoparticles of various sizes.

Homogeneous and heterogeneous catalytic processes were discovered at about the same time two centuries ago. In 1817, Davy observed that hydrogen was oxidized by air on platinum, and various technologies based on heterogeneous catalysis were introduced at the end of the XIXth century and at the turn of the century such as the oxidation of SO_2 to SO_3 on platinum (Messel, 1875), the oxidation of methane by water vapor ($CH_4 + H_2O \longrightarrow CO + H_2$) catalyzed by Ni (Mond, 1890), then the synthesis of NH_3 catalyzed by Fe (Haber-Bosch, 1895), its oxidation ($2 NH_3 + 5/2 O_2 \longrightarrow 2 NO + 3 H_2O$) catalyzed by Pt (Oswald, 1900), the reduction of ethene to ethane by H_2 catalyzed by Ni (Sabatier and Senderens, 1902) and the use of clays to carry out dehydrogenation, isomerization, hydrogenation and polymerization (Ipatief, 1905). The Fischer-Tropsch process arrived in 1923 and the second quarter of the century saw the development of gas production for engines.

The heterogeneous processes can be classified into three large groups:
1. Oil transformation to make gas
2. Synthesis of fine chemicals
3. Environmental aspects, including *inter alia* treatment of gases from automobile exhaust systems.

Main heterogeneous catalytic processes and catalysts [20.1]

Reactions	Catalysts
oxidation of CO and HC from car exhaust pipes	Pt, Pd on oxides (Si, Al, Zn, Cr)
reduction of NO_x from car exhaust pipes	Rh on Al_2O_3
oil cracking	zeolites
hydrotreatment of oil	Co-Mo, Ni-Mo, W-Mo
oil reforming-production of aromatics	Pt, Pt-Re and other alloys on Al_2O_3
hydrocracking	metals on zeolite or Al_2O_3
alkylation	H_2SO_4, HF, solid acids
vapor reforming	Ni on support
watergas shift reaction	Fe-Cr, CuO, Al_2O_3
methanation	Ni on support
synthesis of NH_3	Fe
oxidation of ethylene	Ag on support
synthesis of HNO_3 from NH_3	Pt, Rh, Pd
synthesis of H_2SO_4	V_2O_5
synthesis of acrylonitrile from propene	Bi, Mo oxides
synthesis of vinyl chloride from ethene	CuCl
hydrogenation of oils	Ni
polyethylene	Cr, CrO_3 on SiO_2

The use of catalysts, the respective markets (millions of dollars per year), the modes of production and use of syngas, olefins and aromatics are shown in the following tables.[20.2b]

Production and use of aromatic hydrocarbons [20.2b]

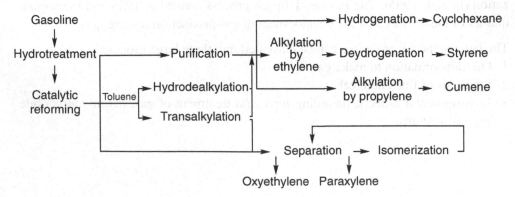

Various uses of catalysts and world markets per year in millions of dollars (M\$) [20.2b]

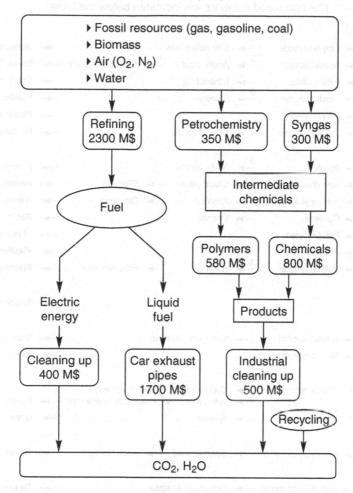

Production and use of syngas [20.2b]

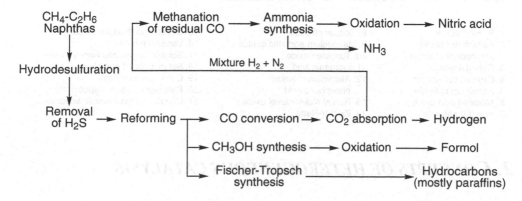

Main uses of olefins and aromatics in petrochemistry 20.2b
(the numbered catalysts are indicated below the table)

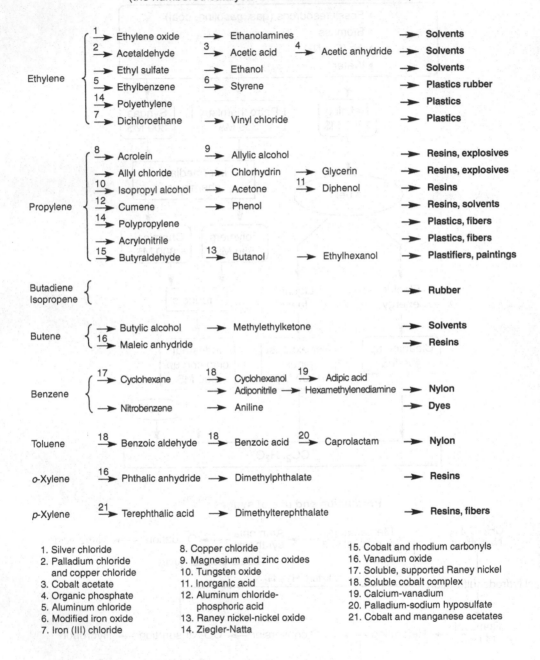

1. Silver chloride
2. Palladium chloride
   and copper chloride
3. Cobalt acetate
4. Organic phosphate
5. Aluminum chloride
6. Modified iron oxide
7. Iron (III) chloride

8. Copper chloride
9. Magnesium and zinc oxides
10. Tungsten oxide
11. Inorganic acid
12. Aluminum chloride-
    phosphoric acid
13. Raney nickel-nickel oxide
14. Ziegler-Natta

15. Cobalt and rhodium carbonyls
16. Vanadium oxide
17. Soluble, supported Raney nickel
18. Soluble cobalt complex
19. Calcium-vanadium
20. Palladium-sodium hyposulfate
21. Cobalt and manganese acetates

2. CONCEPTS OF HETEROGENEOUS CATALYSIS

2.1. THE ELEMENTARY STEPS

Heterogeneous catalysis is based on *surface phenomena*: atoms that are available at the surface of a solid are those which activate substrates. The activity of a heterogeneous catalyst does not depend on its mass *per se*, but on its contact surface with the fluid reagents. It becomes larger as the surface/mass ratio (specific area, $m^2 \cdot g^{-1}$) increases, i.e. as the solid is more porous and divided. For instance, Pt black has a specific area of 20 to 30 $m^2 \cdot g^{-1}$ and that of activated alumina is 100 to 300 $m^2 \cdot g^{-1}$.

A surface has *various types of elementary domains*, each one being able to bind a molecule or atom in a different way: therefore, usually, the *surface sites are not all identical*. This distribution of active surface sites with different properties represents an important distinction between heterogeneous and homogeneous catalysis.

The steps involved in molecular activation on a surface are:

▸ *diffusion* of substrates near the surface or into the cavities of porous materials,

▸ *physical adsorption*, an activation step,

▸ *chemical reaction* at the surface between adsorbed molecules or atoms: bond breaking, forming and rearranging, and

▸ *desorption* of the reaction products and diffusion away from the surface

For instance, in the formation of H_2O from H_2 and O_2, how does the Pt catalyst work? In spite of the very large free energy of the reaction ($\Delta G^\circ = -232$ kJ $\cdot$ mol^{-1}), the mixture of H_2 and O_2 in a glass container is indefinitely stable, because both H_2 and O_2 have large dissociation energies (412 kJ $\cdot$ mol^{-1} and 418 kJ $\cdot$ mol^{-1}). However, if a Pt gauze is introduced, an explosive reaction takes place immediately. This observation is explained by the fact that, in the presence of the large Pt surface, the reactions $H_2 + 2$ Pt $\longrightarrow$ 2 Pt-H and $O_2 + 2$ Pt $\longrightarrow$ 2 Pt-O occur without activation energy. The follow-up reactions on the surface then occur with small activation energies.[20.1] Thus, the overall result is that the Pt surface is able to dissociate molecules that have large bond energies and provide significantly lower activation energies for surface reactions.

2.2. PHYSISORPTION AND CHEMISORPTION

When a molecule diffuses near a metal surface, it is first physisorbed, i.e. it binds weakly to the surface without dissociation. The forces involved are of the Van der Waals type, i.e. weaker than 20 kJ $\cdot$ mol^{-1}. This step is reversible, and desorption can rapidly occur. With H_2, for instance, one can compare this with the formation of an H_2 complex at the vacant metal site of a 16e metal complex. Oxidative addition may eventually occur at the surface (depending on the nature of the metal). In the case of the monometallic complex, the H_2 complex can be stable and isolable (i.e. more strongly bonded to the metal than when physisorbed on a metal surface), or alternatively, it can be transient and rapidly form the metal-dihydride by oxidative addition. The difference is that, on the surface, the oxidative addition is bimetallic whereas, in organometallic chemistry, it is usually (but not always) monometallic:

Surface chemistry:

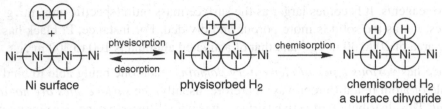

Molecular chemistry:

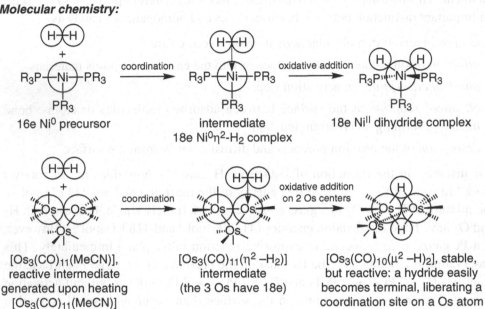

The synthesis of NH$_3$ from N$_2$ and H$_2$ catalyzed by Fe (with small proportions of Al$_2$O$_3$ and K$_2$O additives) around 300°C is one of the great classical examples of heterogeneous catalysis. The very strong bond between the two nitrogen atoms of the N$_2$ molecule ($\Delta G° = -1120$ kJ·mol^{-1}) precludes any dissociation in the gas phase. Reversible physisorption occurs at room temperature, but not dissociation. The dissociative chemisorption of N$_2$ occurs at 300°C, however. Heating to overcome this activation energy is against the thermodynamics of the system, because the reaction of N$_2$ with H$_2$ to form NH$_3$ is slightly exergonic ($\Delta G° = -11$ kcal·mol^{-1} at 300°C). Once again, one may make a comparison with molecular chemistry: Cummins N$_2$ complex forms and even dissociates the two N atoms of the μ_2-coordinated bimetallic N$_2$-centered molecule at room temperature to yields two nitrido-metal species, but it does not catalyze N$_2$ reduction.[7.12c]

If the Fe surface of the heterogeneous catalyst has the appropriate structure, the chemisorbed N atoms can then bind to the chemisorbed H atoms with a small activation energy (12 kJ·mol^{-1}). Detailed kinetic studies have shown that the mechanism involves a series of reversible surface-bonded H-atom transfer steps to the surface-bonded N-atoms: [20.3]

$$N_{ads} \; \underset{}{\overset{H_{ads}}{\rightleftarrows}} \; NH_{ads} \; \underset{}{\overset{H_{ads}}{\rightleftarrows}} \; NH_{2\,ads} \; \underset{}{\overset{H_{ads}}{\rightleftarrows}} \; NH_{3\,ads} \; \rightleftarrows \; NH_3 \nearrow$$

In molecular chemistry (including biology), the reduction of metal-bonded N_2 and metal-bonded nitrido complexes involves protonation-single electron transfer sequences.

Heterogeneous catalysts are extended transition-metal inorganic solids (metals, oxides, sulfides, carbides, etc.) in which the d orbitals play a key role in the chemisorption of the substrates because of their flexibility to donate and accept electron density to and from the substrates. This is the case in particular for the degenerate states in band structures. The electronic flexibility provided by the d electrons of the surface must be such that the bond with the substrate atoms be intermediate between weak and strong. The surface must be able to bind the substrate atoms strongly enough to provoke their dissociation in the chemisorption process. For instance, in the synthesis of NH_3, the layer of N atoms on the Fe surface reaches a coverage of 90%. The surface-atom bond created in this chemisorption process must not be too strong, however, so that the bonded substrate atom is able to react further with other surface-bonded atoms and form the products that can rapidly desorb. If the surface-atom bond is too strong, further reaction will be precluded. The catalytic activity as a function of transition metals shows a strong periodic effect with a maximum of reactivity for group-8 transition metals.[20.2] Indeed, Fe is the best catalyst for the synthesis of NH_3, which is confirmed by calculations involving the number of d electrons of the transition metals.

A classification of metals according to their abilities in chemisorption

Metals	O_2	C_2H_2	CO	H_2	CO_2	N_2
Ti, Zr, Hf, V, Nb, Ta,						
Cr, Mo, W, Fe, Ru, Os	+	+	+	+	+	+
Ni, CO	+	+	+	+	+	−
Rh, Pd, Pt, Ir	+	+	+	+	−	−
Mn, Cu	+	+	+	±	−	−
Al, Cu	+	+	+	−	−	−
Li, Na, K	+	+	−	−	−	−
Mg, Ag, Zn, Cd, In, Si,						
Ge, Sn, Pb, As, Sb, Bi	+	−	−	−	−	−

+ means strong chemisorption, ± means weak chemisorption, − means unobservable chemisorption

It is very difficult to connect the catalyst structure with reaction kinetics, however, and this remains a challenge for the XXIst century. There are two main reactivity models: the *Langmuir-Hinselwood* mechanism that involves reaction by interaction between two surface-bonded substrate atoms in the rate-determining step, and the *Rideal-Eley* mechanism that involves interaction between a surface-bonded substrate atom and a molecule in the gas phase in the rate-determining step. Most reactions have kinetics that are consistent with the first model.[20.1]

2.3. KINETICS: THE ROTATION FREQUENCY $\Re$

The rotation frequency F is defined as the number of product molecules formed per second. Its inverse, 1/F, is the rotation time, i.e. the time necessary to form a product molecule. Dividing the rotation frequency F by the catalyst surface area $\wp$ gives the specific rotation rate $(mol \cdot cm^{-2} \cdot s^{-1})$: $\Re = F/\wp$. $\Re$ is often refered to as the rotation frequency. For instance, for hydrocarbon transformations, $\Re$ varies between 10^{-4} and 100. Multiplying $\Re$ by the total reaction time t gives the number of rotations, i.e. the number of molecules formed per surface site (which must be larger than 100 to allow the use of the term catalysis). $\Re$ can be formulated as the product of the rate constant by a term related to the pressure P: $\Re = k \cdot f(P_i)$, in which P_i is the partial pressure of the reactants. The rate constant, k, can be expressed using an Arrhenius equation: $k = A \exp(-\Delta E^*/RT)$ in which A is a pre-exponential factor and ΔE^* the apparent activation energy of the reaction under the catalytic conditions. The transformations of hydrocarbons (isomerization, cyclization, dehydrocyclization and hydrogenolysis) often have ΔE^* values of the order of 140 to 180 $kJ \cdot mol^{-1}$, but the hydrogenation reactions only have ΔE^* values of 24 - 48 $kJ \cdot mol^{-1}$ at ambient temperature. Hydrogenation thus occurs under much milder reaction conditions than the above hydrocarbon transformations.

2.4. SELECTIVITY

Selectivity is a major problem in heterogeneous catalysis, because many heterogeneous processes are not selective and therefore provide mixtures of products. Considerable progress has been made in providing selectivities of catalysts, especially by variation of catalytic materials and the development of zeolites (*vide infra*). A broad range of products can be selectively obtained by changing the nature of the catalyst, as shown by the following example of propene oxidation:

BiPO$_4$	CuCo$_2$O$_4$	Th$_2$O$_3$	Bi$_2$(MoO$_4$)

NiMoO$_4$ + MoO$_3$	CoMoO$_4$	Mo(CO)$_6$	Bi$_2$O$_3$

Variation of the products obtained in heterogeneous catalysis that depend on the nature of the catalyst (noted below the reaction product) - Example of propene oxidation

Even with the same metal, various reactions can occur, and the selectivity must then be directed by other parameters such as the collected kinetic data for each specific reaction. The following example of the catalysis of n-hexane transformation using Pt as the catalyst illustrate the variety of products and the need for a effective optimization of selectivity: [20.3a]

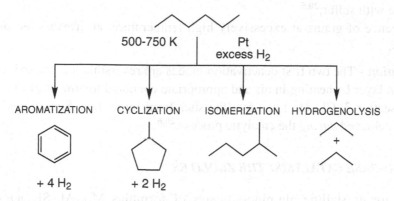

It is also known that the selectivity depends on the particular face of the crystal. For instance, with MoO_3, the {010} face is not selective for propene oxidation, but the {100} face gives acrolein. With methanol, the face {010} of MoO_3 selectively gives dehydrogenation to formaldehyde HCHO, the {100} face leads to $H_2C(OCH_3)_2$. [20.3b]

2.5. PREPARATION, DEACTIVATION AND REGENERATION OF THE CATALYST

Preparation - Since the activity of a heterogeneous catalyst is directly related to its surface area, it is important to use metal catalysts that are as finely divided as possible. For this purpose, the catalysts are prepared from alumina, silica, other supports (oxides of Mg, Zr, Ti, V), carbon (graphite or nanotubes), phosphate, sulfide or carbonate. Usually, the surface area reach 100 to 400 $m^2.g^{-1}$ on such supports. The microporous support is impregnated in the micropores with one or several metal salts which are reduced to metal nanoparticles (1-10 nm) upon heating. With such dimensions, most of the metal atoms of the nanoparticles are on the nanoparticle surface, and the polydispersity is supposed to be close to 1. Another oxide such as TiO_2 is frequently dispersed at the surface of the primary oxide to gain selectivity. Additives may include alkali metals acting as electron donors or halogens acting as electron acceptors. [20.4] The heterogeneous catalysts are produced in the form of pellets, rings, spheres, tablets, granules and extrudates. The sizes vary from a fraction of mm to several cm. The largest ones are used in fixed beds crossed by the fluid, liquid or gas, whereas the smallest ones are suspended in the fluid from which they have to be separated by decantation or filtration. The required properties are the *activity*, *selectivity* and *longevity*. Such systems can work at high rates up to several thousand hours with several millions rotations.

Deactivation - The catalysts that are used in the oil industry for reforming usually produce around 10^5 liters *per* pound of catalyst; then the catalyst must be replaced or regenerated. The reasons for deactivation are those that decrease the catalyst surface area:

▶ formation of a carbon layer upon decomposition of the hydrocarbon at the reforming temperature (see section 3),

▶ poisoning of the active sites, a chemical process that changes their reactivity, for instance with sulfur,[20.5]

▶ coalescence of grains at excessively high temperature, an irreversible physical process.

Regeneration - The two first deactivation modes are reversible, i.e. the oxidation of the carbon layer by heating in air and appropriate chemical treatment of the surface (H_2, see section 3; Cl_2, etc.) can free the sites from impurities. Sometimes, regeneration is achieved *during* the catalytic process.[20.6]

2.6. *ACID-BASE CATALYSIS: THE ZEOLITES*

Zeolites are crystalline alumino-silicates of formulas $M_{x/n}(Al_{x+}Si_{1-x})O_2 \cdot yH_2O$ ($M^+ = H^+$, NH_4^+, Na^+, Fe^{2+}, La^{3+}...). They are among the most common minerals in nature, and have been extensively used in heterogeneous catalysis since 1970. They present many holes, channels and cavities. Since oxides and zeolites are used extensively in heterogeneous catalysis, it is essential to consider their acid-base properties as defined by their Lewis acid or base and Brønsted acid or base properties. Indeed, these oxides not only play a role as support, but they also take a very active part in the activation processes.

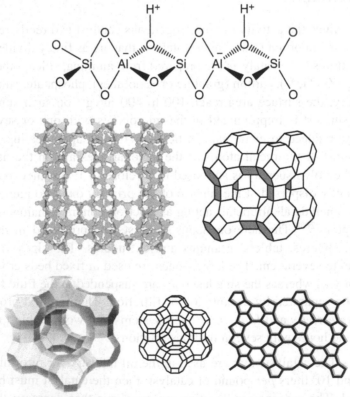

Some representations of zeolite structures

Lewis acids accept an electron pair from the absorbed substrate whereas Lewis bases give an electron pair to the adsorbed substrate. Brønsted acids can transfer a proton to a bound substrate whereas Brønsted bases can abstract a proton from the surface-bonded substrate. Lewis and Brønsted acidities usually increase upon increasing the cationic charge of a metal ion. Thus Na_2O, CaO, Al_2O_3 are basic, whereas SiO_2, TiO_2 and iron oxides, are acidic. The acidity of zeolites depends on the Si/Al ratio. Aluminosilicates are anionic and, since Al has one less electron than Si, it has a superior electronic affinity and is the site of the negative charge. Thus, protons bind the basic oxygen atoms of the AlO_4^- tetrahedral anion. Upon heating, water is removed, coordinatively unsaturated Al^{3+} ions form, and they are strong Lewis acids.[20.7]

Upon increasing the Si/Al ratio, the stability, hydrophobicity, strength of acid sites and resistance to acids increase, whereas the hydrophilicity and capacity for ion exchange are diminished. Zeolites with very large superficial internal surface area can be prepared (100 $m^2 . g^{-1}$), and their pore sizes can be controlled (0.8 to 2 nm). They protonate and cleave hydrocarbons at high temperatures, as superacids (composed of a mixture of strong Lewis and Brønsted acids such as HF and SbF_5 yielding H_2F^+, SbF_6^-) do in solution chemistry at ambient temperature. Thus, these solid acids are used in oil reforming:

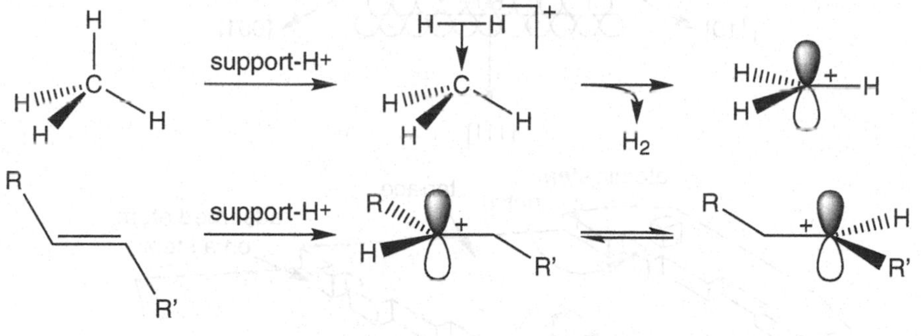

Zeolites have well-defined channels and cavities that govern the shape selectivity involved in the substrates (that are admitted in the catalytic area), transition states and final products. There are also several natural structures such as faujasite, mordenite, etc. that can be modified by cation exchange or by the introduction of various nanoparticles. Clays (montmorillite, beidellite, kaolin, hydrotalcite) are also used, having related structures.

2.7. STUDY OF THE CATALYST SURFACE

Surface scientists study the surface of the catalyst in order to obtain information that can be correlated with the kinetics and mechanisms of the catalytic reactions. In this way, they hope to be able to improve the activity, selectivity and lifetimes of the catalysts. Real catalysts with porous supports with large surfaces have very complex structures that can hide the metal sites, however. The more simple and rational approach uses crystals of the catalyst.[20.8] In section 2.4, the dramatic influence of the crystal face on the selectivity has already been noted. Various spectroscopic and electronic microscopy techniques of nanoscience are used under vacuum to study the nature of the surface (about 1 cm^2). The cell is then charged with substrates and connected to a gas chromatograph in order to study the product distribution. Theoretical models may help making the difficult correlation between the structure and reactivity. Crystal growth induces the formation of large terraces at their surface. STM analysis together with ionic microscopy with field effect (FIM) and slow electron diffraction (LEED) show the presence of steps and notches as well as punctual defects. The following view is static, but rearrangements rapidly occur on the surface (atoms "walk" on the surface and leave it, as carbonyl ligands in clusters and fluxional organometallic molecules).

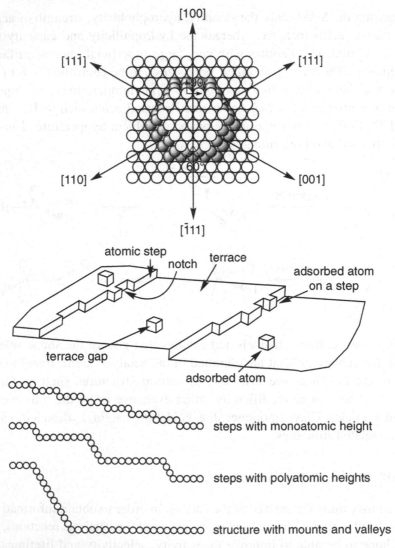

Top: Crystal of catalyst composed of well-defined atomic planes, steps and notches. Middle: Model of the heterogeneous surface of a solid representing various surface sites. Bottom: Structures of steps induced by rearrangements resulting from adsorption (adapted from Somorjai's book, ref. 20.1)

3. *CO* AND *CO₂* HYDROGENATION AND *FISCHER-TROPSCH* CHEMISTRY

The production of CH_3OH using CO, CO_2 and H_2 as feedstocks is about 3×10^6 kg per day. The catalyst is Cu/ZnO with Al or Cr additive: [20.9]

$$CO + 2 H_2 \xrightleftharpoons{\text{cat. Cu/ZnO}} CH_3OH \qquad \begin{array}{l} \Delta H°_{600} = 100.5 \text{ kJ} \cdot \text{mol}^{-1} \\ \Delta G°_{600} = 45.4 \text{ kJ} \cdot \text{mol}^{-1} \end{array}$$

Since CO_2 is an abundant, low-price feedstock, its use must be optimized. Thus, if the mixture of CO, CO_2 and H_2 is available, their ratio can be adjusted using the water-gas-shift reaction (WGSR) in the presence of a Cu-based catalyst:

$$CO_2 + H_2 \xrightleftharpoons{\text{cat. Cu}} CO + H_2O \qquad \Delta G°_{298} = 28 \text{ kJ} \cdot \text{mol}^{-1}$$

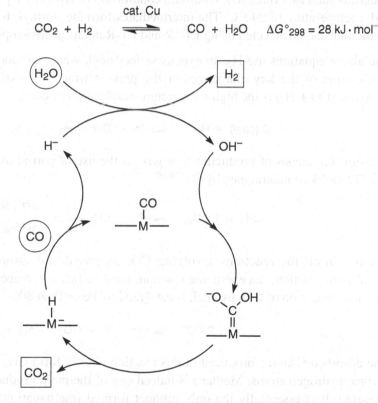

Possible mechanism for the water-gas-shift reaction (WGSR)

This reaction is industrially exploited to make pure H_2 or to remove CO from gas streams. For instance, in the Texaco Hy Tex process, the first stage involves conversion at 280-350°C with Fe oxide/Cr oxide catalyst and, in the second stage, the temperature is lowered to 180-260°C, in the presence of Cu-containing catalysts such as or CuO-ZnO in order to reduce the CO content to < 0.5%.

Coal also is an excellent reductant of CO_2 to usable CO, this heterogeneous reaction being catalyzed by Fe or Pt. Transition metals and main-group metals also stoichiometrically reduce CO_2 to give metallic oxides and CO between 500°C and 1200°C.

Analogously, methane reduces CO_2 to CO and H_2 between 600°C and 1000°C using Ni, Fe or Cu as catalyst. Finally, the electrochemical reduction of CO_2 also leads to CO and CO_3^{2-} under certain conditions (for instance in DMSO and/or on certain cathodes: Hg, Au, Pt).

The net result of the CO_2 hydrogenation to methanol can be formulated as follows:

$$CO_2 + 3\,H_2 \longrightarrow CH_3OH + H_2O$$

Its represents a considerable challenge, because concentration in air is quite high (320 ppm). Catalysts contain Cu that is involved as a redox Cu^I/Cu^{II} system in the course of CO_2 hydrogenation. Pratically, the heterogeneous CO_2 hydrogenation catalysts are $Cu/CeO_2/\gamma\text{-}Al_2O_3$ (eventually doped with yttria) or superconducting materials such as $YBaCu_3O_7$. Optimum conditions are typically a pressure of 3 MPa and a temperature of 240°C. The intermediates formate, formyl, formaldehyde and methoxide can be detected using FT-IR and FT-Raman spectroscopies.

The above equations use H_2, an expensive feedstock whose production and storage is also one of the key challenges at the present time. A classical synthesis of "*syngas*" (CO + H_2) is the high-temperature gasification of coal:

$$C\,(\text{coal}) + H_2O \rightleftharpoons CO + H_2 \qquad \Delta G°_{298} = 88\ \text{kJ} \cdot \text{mol}^{-1}$$

A promising means of producing "*syn gas*" is the use of partial oxidation of methane (72 mol% of natural gas) by O_2:[20.10]

$$CH_4 + 1/2\,O_2 \rightleftharpoons CO + 2\,H_2 \qquad \begin{aligned} \Delta H°_{298} &= -86.5\ \text{kJ} \cdot \text{mol}^{-1} \\ \Delta G°_{298} &= -35.5\ \text{kJ} \cdot \text{mol}^{-1} \end{aligned}$$

Finally, in all the reactions involving CO, its possible disproportionation (the *Boudouard* reaction), an exergonic reaction, must be taken into account (the reverse reaction, reduction of CO_2 by coal, is catalyzed by Fe or Pt at 800-1000°C):

$$2\,CO \rightleftharpoons C + CO_2 \qquad \Delta G°_{298} = -116\ \text{kJ} \cdot \text{mol}^{-1}$$

The adsorbed C atoms produced in this reaction readily form CH_4 by reaction with surface hydrogen atoms. Methane is indeed one of the main products of CO hydrogenation. It is essentially the only product formed (methanation) when Ni is the catalyst. The mechanism of this methanation reaction probably is similar for all the late transition metal catalysts (the activation energy being always between 96 and 105 kJ·mol^{-1}). It follows:

$$C_{ads} + 4\,H_{ads} \rightleftharpoons CH_{ads} + 3\,H_{ads} \rightleftharpoons CH_{2\,ads} + 2\,H_{ads} \rightleftharpoons CH_{3\,ads} + H_{ads} \rightleftharpoons CH_4 \nearrow$$

In organometallic chemistry for comparison, a single Fe center suffices to account for all the reduction steps from CO to CH_4 at ambient or subambient temperature:[20.11]

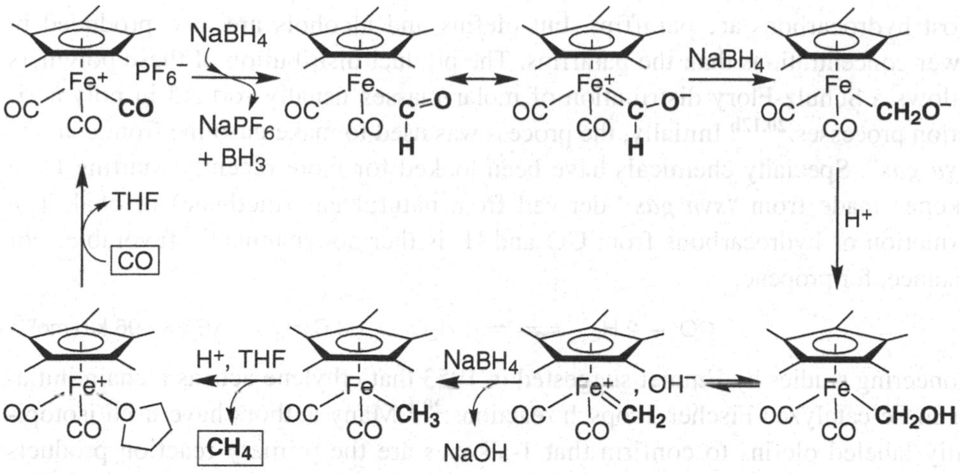

The mechanism of CO hydrogenation to form CH_3OH (also compare the above organoiron model with the [Fe]CH_2OH complex that can be protonated to methanol) in the presence of Cu/ZnO catalyst does not involve the dissociation of CO, but its hydrogenation by dissociated hydrogen atoms:

$$CO_{ads} + 4\,H_{ads} \rightleftharpoons CHOH_{ads} + 2\,H_{ads} \rightleftharpoons CH_2OH_{ads} + H_{ads} \rightleftharpoons CH_3OH \nearrow$$

On the other hand, CO_2 can dissociate to CO_{Ads} and O_{Ads} or be also hydrogenated to produce formate:

$$CO_{2\,ads} + H_{ads} \rightleftharpoons HCOO_{ads}$$

When CH_4 and higher alkanes are formed during the hydrogenation of CO, the cleavage of the C-O bond occurs at the interface between the oxide support and the metal (for instance, Ni supported on TiO_2 is several orders of magnitude more active than Ni alone):

$$CHOH_{ads} + H_{ads} \rightleftharpoons CO_{ads} + H_2O \nearrow$$

The *K additives* in CO hydrogenation result in strong reduction of the metal catalyst, increasing the density of electronic states (Fermi level) available to form bonds with adsorbed CO by back bonding into the antibonding π^* CO orbitals. This effect increases the adsorption energy in the same manner as it strengthens the bond between the metal and carbonyls in metal-carbonyl complexes (see Chap. 9). For instance, on a rhodium (111) surface, CO is adsorbed but does not dissociate at room temperature. When K is added in an optimal amount, CO dissociation occurs at the rhodium surface.[20.12a]

The Fischer-Tropsch process, catalyzed by Fe, Co or Ru, involves the polymerization of reduced CO in addition to CO reduction. A broad range of hydrocarbons are formed, and the term includes "gas to liquid" technologies:

cat. Fe

$$n\,CO + 2n\,H_2 \rightleftharpoons (-CH_2-)_n + n\,H_2O$$

Most hydrocarbons are paraffins, but olefins and alcohols are also produced in lower concentrations than the paraffins. The product distribution of these polymers follows a Schulz-Flory distribution of molar masses usually formed in polymerization processes.[20.12b] Initially, the process was used to make gasoline from coal *via* "*syn gas*". Specialty chemicals have been looked for more recently, starting from alkenes made from "*syn gas*" derived from natural gas (methane) or coal. The formation of hydrocarbons from CO and H_2 is thermodynamically favorable. For instance, for propene:

$$CO + 2\,H_2 \rightleftharpoons H_2O + 1/3\,C_3H_6 \qquad \Delta G° = -96\,kJ \cdot mol^{-1}$$

Pioneering studies by Emmet suggested in 1953 that ethylene acts as a chain initiator in Fe-catalyzed Fischer-Tropsch reactions.[20.1] Many authors have used isotopically labeled olefins to confirm that 1-alkenes are the primary reaction products (although they are thermodynamically instable under the reaction conditions). For instance, Somorjai's group has shown, using a polycrystalline iron sheet (1 cm^2), that the addition of ethylene or propene in the "*syn gas*" increases the length of the polymer chain. These olefins readsorb and readily form polymers by insertion into adsorbed C fragments. Thus, the Fischer-Tropsch mechanism involves two steps, namelly the formation of olefins and their polymerization: [20.14]

$$2\,CO + 4\,H_2 \underset{600\,K}{\overset{Fe,\ 6 \times 10^5\,Pa}{\rightleftharpoons}} C_2H_4(C_3H_6) + 2\,H_2O$$

$$CO + H_2 \underset{600\,K}{\overset{Fe,\ C_2H_4(C_3H_6)}{\rightleftharpoons}} C_{5\text{-}9}H_{12\text{-}20} + 2\,H_2O$$

Many mechanisms have been proposed for the Fischer-Tropsch reaction since the discovery of the process, but the surface and organometallic chemistry approaches have certainly brought new insights. Let us start from the original mechanism proposed by Fischer and Tropsch. This mechanism correctly takes into account the chemisorption of H_2 and CO giving H_{ads}, C_{ads} and O_{ads}, and the formation of CH_4 from the two former species (*vide supra*): [20.15]

In this mechanism, however, the primary C-C bond formation arises from two sp^3 carbon atoms, if (as highly probable) the methylenes are bridging two surface metal atoms. Modern theoretical studies indicate that such bond formation on the catalyst surface is unlikely, being prohibitively high-energy processes. Fischer-Tropsch reactions carried out under mild conditions including added labeled olefins showed that the added olefin accelerated the Fischer-Tropsch synthesis and that the C2 olefin fragments were incorporated and preserved in the alkane and alkene products, i.e. no cleavage of the C=C bond occurred under mild conditions. This clearly indicates that the primary C-C coupling process is a rate-limiting step. It is possible, for instance, that this step involves coupling between two adsorbed carbides, a carbide and a carbyne, two carbynes, a carbyne and a methylene or a carbide and a methylene.[20.16-20.18]

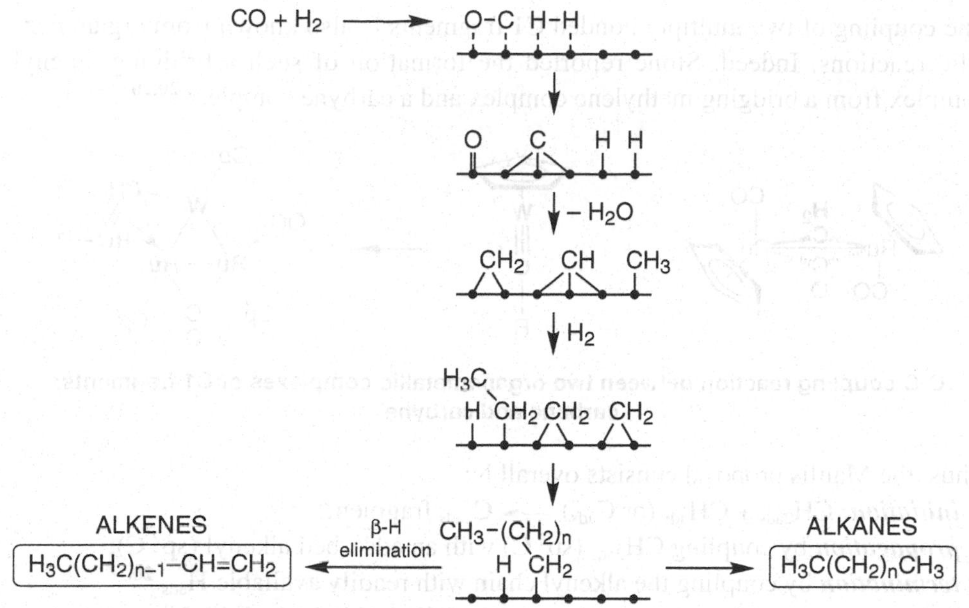

$$CO + H_2 \longrightarrow O\text{-}C \quad H\text{-}H$$

$$\downarrow$$

$$O \quad C \quad H \quad H$$

$$\downarrow - H_2O$$

$$CH_2 \quad CH \quad CH_3$$

$$\downarrow H_2$$

$$H_3C$$
$$H \; CH_2 \; CH_2 \; CH_2$$

ALKENES β-H   **ALKANES**
                           elimination $CH_3\text{-}(CH_2)_n$
$$\boxed{H_3C(CH_2)_{n-1}\text{-}CH=CH_2} \longleftarrow \quad H \; CH_2 \longrightarrow \boxed{H_3C(CH_2)_nCH_3}$$

Mechanism originally proposed by Fischer and Tropsch for catalysis of the heterogeneous Fischer-Tropsch reaction: CO + H$_2$ $\longrightarrow$ CH$_3$(CH$_2$)$_n$CH$_3$ + ...

Maitlis proposed that the initiator is a vinyl (CH$_2$=CH$_{ads}$) or a vinylidene (CH$_2$=C=$_{ads}$), these intermediates being produced by coupling on the surface, subsequently followed by hydrogenation by surface-bonded H atoms (see p. 474).[20.16]

In organometallic chemistry, stable methylene-bridged bimetallic complexes and methyne-capped trimetallic clusters are common, and rare carbide complexes have occasionally been formed by reaction of coordinated CO in clusters (for instance, Hübel's nido cluster shown below):[20.19]

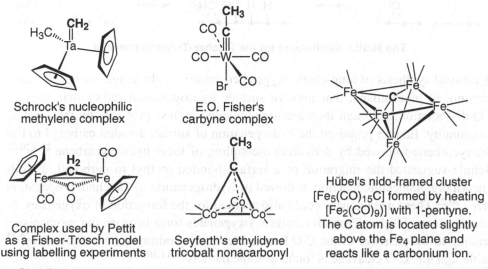

Schrock's nucleophilic
methylene complex

E.O. Fisher's
carbyne complex

Complex used by Pettit
as a Fisher-Trosch model
using labelling experiments

Seyferth's ethylidyne
tricobalt nonacarbonyl

Hübel's nido-framed cluster
[Fe$_5$(CO)$_{15}$C] formed by heating
[Fe$_2$(CO)$_9$)] with 1-pentyne.
The C atom is located slightly
above the Fe$_4$ plane and
reacts like a carbonium ion.

Classic examples of organometallic complexes of carbene, carbyne and carbide

The coupling of two multiply bonded C1 fragments is also known from organometallic reactions. Indeed, Stone reported the formation of such a bridging alkenyl complex from a bridging methylene complex and a carbyne complex: [20.20]

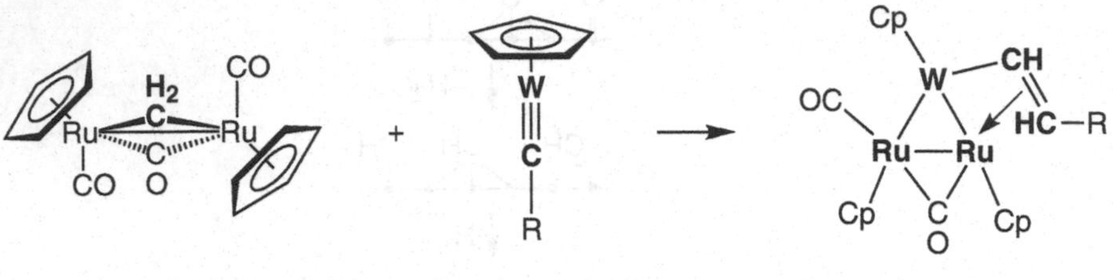

C-C coupling reaction between two organometallic complexes of C1 fragments: carbene and carbyne

Thus, the Maitlis proposal consists overall in:

▸ *initiation:* $CH_{2ads} + CH_{ads}$ (or C_{ads}) $\longrightarrow C_{2ads}$ fragment;

▸ *propagation* by coupling CH_{2ads} (sp^3 C) with an adsorbed alkenyl (sp^2 C);

▸ *termination* by coupling the alkenyl chain with readily available H_{ads}.[20.16]

This last step represents a bimetallic reductive elimination which is also well known in organometallic chemistry. These modern mechanistic views (scheme below) represent significant improvements with respect to the mechanism proposed originally by Fischer and Tropsch shown p. 473.

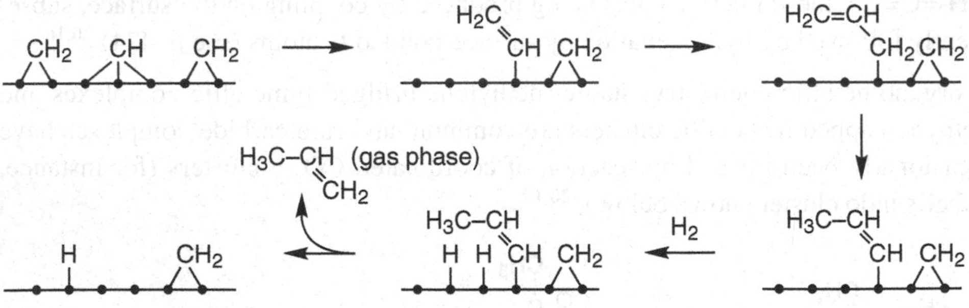

The Maitlis mechanism for the Fischer-Tropsch process

A rational synthesis of long-chain oxygenates remains a challenge, however. There are indeed mechanisms that involve surface hydrogenation rather than surface CO dissociation, although they are now regarded as less probable by the catalyst community. Emmet proposed the hydrogenation of surface-bonded carbonyl to hydroxycarbene followed by dehydration-coupling of these hydroxycarbene.[20.3,20.21] Schulz suggested the migration of a surface-bonded methyl to carbonyl, a well-known organometallic reaction, followed by hydrogenation as the chain propagation steps.[20.22] These mechanisms could also account for the formation of oxygenates. In homogeneous Fischer-Tropsch catalysis, oxygenates form because the mechanisms leading to the cleavage of the C-O bond under such conditions are not favored [20.23] (although this C-O cleavage is found in some models [20.11,20.24]).

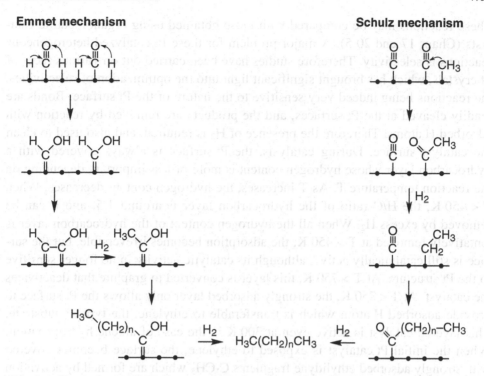

Fischer-Tropsch mechanistic proposals that do not involve initial CO$_{ads}$ cleavage

4. TRANSFORMATION OF HYDROCARBONS

Hydrocarbons are transformed into a variety of products under hydrogenation conditions. The heterogeneous catalyst is Pt which is a very reactive metal (also used as CO oxidation catalyst for automobile exhaust systems and in various other reactions). Linear aliphatic hydrocarbons are transformed into aromatics (dehydrocyclization) and branched hydrocarbons (isomerization).

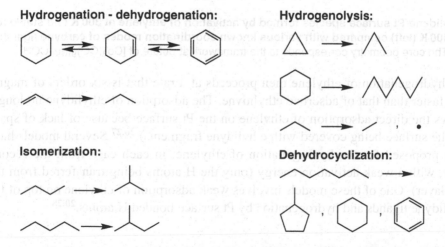

Various competitive reactions of hydrocarbons catalyzed by Pt

These reactions may be compared with those obtained using organometallic catalysts (Chaps 17 and 20.5). A major problem for these Pt-catalyzed heterogeneous reactions is selectivity. Therefore studies have been carried out on various faces of Pt crystals, which has brought significant light into the optimization of selectivities, the reactions being indeed very sensitive to the nature of the Pt surface. Bonds are readily cleaved at the Pt surfaces, and the products are removed by reaction with adsorbed H atoms. Therefore the presence of H_2 is required, and also used to clean the catalyst surface. During catalysis, the Pt surface is always covered with a hydrocarbon layer whose hydrogen content is more or less important depending on the reaction temperature T. As T increases, the hydrogen content decreases. When T < 450 K, the H/C ratio of the hydrocarbon layer is around 1.5, and it can be removed by excess H_2. When all the hydrogen content of the hydrocarbon layer is tentatively removed at T > 450 K, the adsorption becomes irreversible, but the surface is still catalytically active, although its catalytic activity in no longer sensitive to the Pt structure. At T > 750 K, this layer is converted to graphite that deactivates the catalyst. At T < 750 K, the strongly adsorbed layer only allows the Pt surface to provide adsorbed H atom which is transferable to ethylene, the typical substrate. This layered catalyst is active even at 300 K in the case of alkene hydrogenation. When the initial Pt catalyst is exposed to ethylene, the surface becomes covered with strongly adsorbed ethylidyne fragments $C-CH_3$ which are formed by activation of ethylene on Pt (scheme below).[20.25,20.26]

 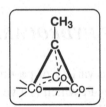

Very stable ethylidyne
formed at RT from
ethylene on the Pt surface

Examples of carbyne-metal bonds
in organometallic complexes, the adequate
model being the framed one, a robust cluster

Ethylidene-Pt surface species formed by activation of ethylene at 300 K and stable up to 400 K (left) compared with various known coordination modes of carbyne ligands
The core geometry corresponds to the framework structure of $[Co_3(CO)_9](\mu_3-CCH_3)$.

The hydrogenation of ethylene then proceeds at a rate that is six orders of magnitude faster than that of adsorbed ethylidyne. The adsorption of ethylidyne no longer allows the direct adsorption of ethylene on the Pt surface because of lack of space (all the surface being covered with ethylidyne fragments).[20.27] Several models have been proposed for the hydrogenation of ethylene, in each case from the second layer, with a weak activation energy (only the H atoms being transferred from the first layer). One of these models involves weak adsorption of ethylene on top of the ethylidyne ligands and hydrogenation by Pt surface bonded H atoms.[20.28]

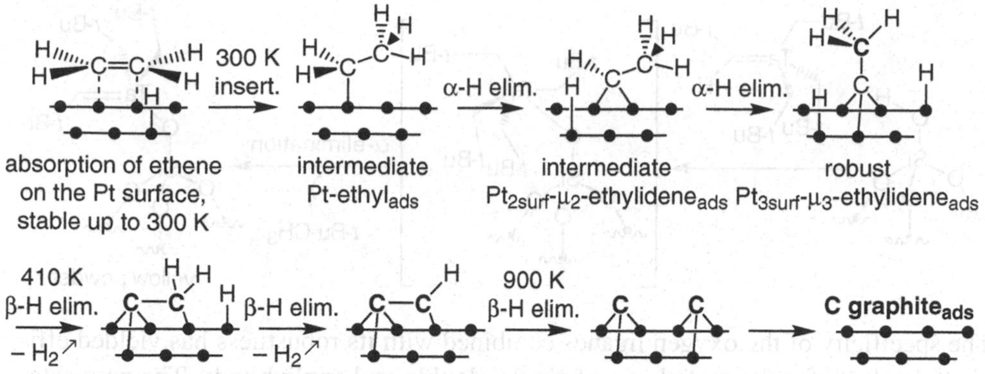

absorption of ethene
on the Pt surface,
stable up to 300 K

intermediate
Pt-ethyl$_{ads}$

intermediate
Pt$_{2surf}$-μ_2-ethylidene$_{ads}$ Pt$_{3surf}$-μ_3-ethylidene$_{ads}$

robust

Thermal transformation of ethylene on a Pt surface

The ethylidyne fragment is formed at 300 K by ethylene insertion into Pt-H$_{ads}$
and two α-H eliminations. This is followed by successive loss of 3 H atoms by
β-H eliminations as the temperature is increased. At high temperature, irreversible formation of graphite inhibits the catalytic reactions.

5. METATHESIS OF ALKANES, ALKENES AND ALKYNES

Metathesis is a particular case of the hydrocarbon transformations described in section 4. The metathesis of alkanes was discovered at the Chevron Company in the
1970s.[20.45] The strategy consisted of dehydrogenating alkanes by a Pt-based catalyst
followed by metathesis of the alkenes, formed in this way, using a W-based catalyst, and finally hydrogenation of the olefins, resulting from alkene metathesis, to
alkanes by the Pt-based catalyst. In this way, butane could be metathesized overall
to propane and pentane as the main products together with minor quantities of various other alkanes and alkenes, at 430°C using Pt/Re/Li on Al$_2$O$_3$ and WO$_3$ supported on silica as the heterogeneous catalysts:[20.45]

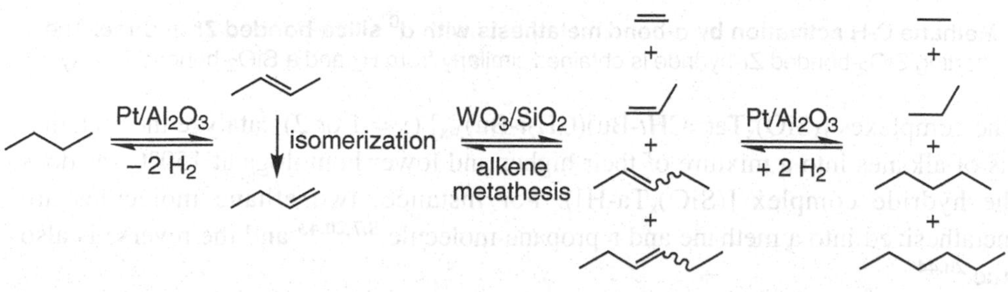

Basset and his group have observed that propane and propene metathesis give
similar C$_{n+1}$/C$_{n+2}$ ratios of cross-metathesis products on silica-supported tantalum-
neopentylidene catalyst at 150°C. The olefin-metathesis activity of these Schrock-
type supported complexes results from the presence of the silyloxy ligand (*vide
infra*).[8.7] Organometallic complexes are bound to silica or alumina by reaction of
soluble complexes and involve the formation of one or several bonds between the
central metal and the oxygen atom of the oxide support.[20.42]

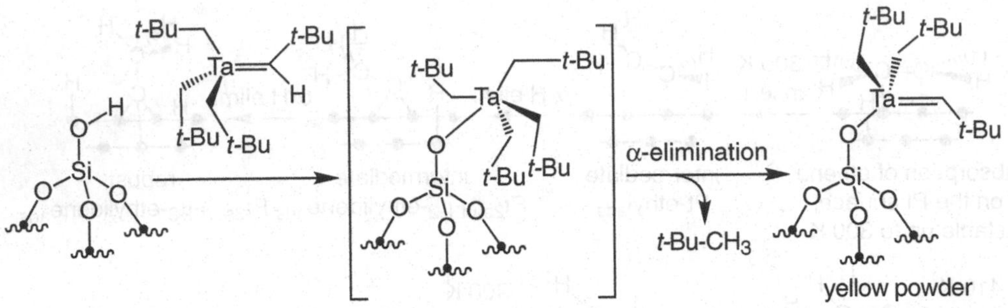

yellow powder

The specificity of the oxygen ligands combined with its robustness has yielded efficient catalysts for the metathesis of single, double and triple bonds. The new catalysts have been characterized using spectroscopic techniques:

▸ EXAFS gives access to the bond distances to the neighbors of the metal center and its average coordination number,

▸ IR allows to follow modification of the support upon grafting or further treatment of the surface complex,

▸ ESR indicates if the complex is paramagnetic and gives related information, and

▸ XANES gives valuable information about the oxidation state and geometry of the metal complex.

Complexes such as (-Si-O-)$_3$ZrMe are involved for C-H activation by σ-bond metathesis:[8.7,15.21]

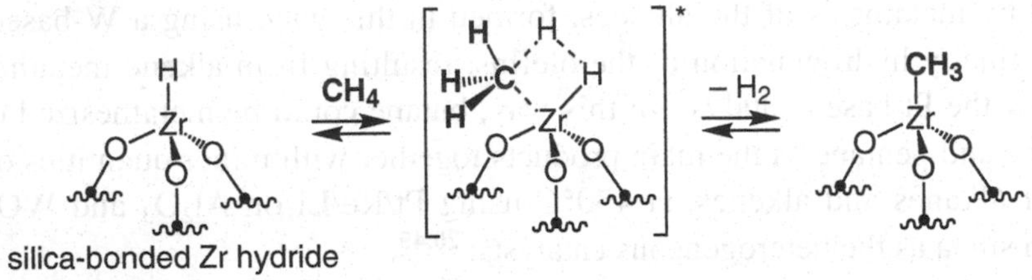

silica-bonded Zr hydride

Methane C-H activation by σ-bond metathesis with d^0 silica-bonded Zr hydride. The starting SiO$_2$-bonded Zr-hydride is obtained similarly from H$_2$ and a SiO$_2$-bonded Zr-alkyl.

The complexes [(SiO)$_x$Ta(=CHt-Bu)(CH$_2t$-Bu)$_{3-x}$] (x = 1 or 2) catalyze the metathesis of alkanes into a mixture of their higher and lower homologs at 150°C (as does the hydride complex [(SiO)$_x$Ta-H]). For instance, two ethane molecules are metathesized into a methane and a propane molecule,[8.7,20.43] and the reverse is also true.[20.44]

The mechanism is suggested to proceed by a composite series of σ-bond metatheses of C-H bonds and α- and β-eliminations (rather than direct σ-bond metathesis of C-C bonds). The α-elimination from a d^2 metal-methyl or metal-alkyl species generates respectively HTa=CH$_2$ or HTa=CHR, and the mechanism is proposed to then follow an alkene metathesis pathway with the olefins being generated by β-elimination (including metallacyclobutane intermediates as in the Chauvin mechanism).

Thus, only σ-bond metathesis of two kinds of σ bonds are observed in these systems: H-H, and C-H involving a Ta…C…H interaction but not a Ta…H…C interaction, because a negatively polarized pentacoordinated carbon intermediate is not favored:

| H-H bond activation giving a Ta-hydride species | C-H bond activation of alkanes giving a Ta-alkyl and H$_2$ | reverse C-H bond activation **not observed** due to a negatively polarized penta-coordinated β carbon |

Hydrogenolysis of Ta-alkyl and methanolysis of Ta-H by σ-bond metathesis for alkane disproportionation disclosed by Basset's group with the silica-supported catalyst [(SiO)$_x$Ta^V(=CH-t-Bu)(CH$_2$-t-Bu)$_{3-x}$]

Metal-alkylidene and alkylidyne complexes bonded to silica *via* silyloxy ligands are also well-defined active alkene and alkyne metathesis catalysts. Recall that Schrock had turned metathesis-inactive alkylidene complexes into active ones by the introduction of alkoxy ligands. In Basset's catalysts, this beneficial role is played by a silyloxy ligand from silica. [15.2,15.31a] Thus, the catalysts [(SiO)M(=CH-t-Bu)(CH$_2$-t-Bu)$_2$], M = Mo or W and [(SiO)Mo(=NH)(=CH-t-Bu)(CH$_2$-t-Bu)] are active at 25°C, unlike previously reported ill-defined heterogeneous catalysts and the early Mo and W oxides on silica or alumina. [15.21]

Active, well-defined silica-supported Mo, W and Re alkene and alkyne metathesis catalysts

The only oxide that has been used for catalyzed olefin metathesis at 25°C is Re$_2$O$_7$/Al$_2$O$_3$ (in the middle of the 1960s by British Petroleum), [20.45] but it suffered from a low number of active sites, side reactions caused by the acid support and deactivation of the catalyst. On the other hand, the silica-supported rhenium catalyst [(SiO)(Re(C-t-Bu)(=CH-t-Bu)(CH$_2$-t-Bu)] catalyzes the metathesis of propene at 25°C with an initial rate of 0.25 mol/(mol Re × s). The formation of 3,3-dimethylbutene and 4,4-dimethylpentene in a 3:1 ratio results from cross metathesis between propene and the neopentylidene ligand, and the ratio of cross-metathesis products matches the relative stability of the metallacyclobutane intermediates. Cross metathesis of propene and isobutene and self-metathesis of methyl oleate can also be achieved, and TONs reach 900 for the latter reaction, which is unprecedented for both heterogeneous and most homogeneous catalysts.[20.46] Highest performances, in terms of stability, reactivity and selectivity, are reached in olefin metathesis with the silica-supported catalyst $[Mo(S_{SiO_2}-O)(=NAr)(=CHCMe_2Ph)(NPh)]$ with $Ar = 2,6-iPr_2C_6H_3$.[20.46b] Finally, 1-pentyne is also metathesized owing to the presence of the carbyne ligand in this catalyst.[20.47]

6. OXIDATION OF HYDROCARBONS

Parallel to the homogeneous oxidation processes described in Chaps 16 and 19, the heterogeneous oxidation processes mainly concern hydrocarbons, but also include the production of H_2SO_4 from SO_2 and of HNO_3 from NO. Heterogeneous oxidation of hydrocarbons gives products such as acids and their derivatives that are more oxidized than in homogeneous oxidation chemistry. In particular, heterogeneous oxidative functionalization of alkanes is particularly challenging. The most important heterogeneous oxidation processes that have been studied are shown in the following table: [20.2.c]

Type of reaction	Reactant	Product	Nb e⁻	Catalyst	N. B.
Oxidative coupling	Methane	Ethane +	2	Li/MgO	N. I.
		Ethylene	4		
Oxidative dehydrogenation	Ethane	Ethylene	2	Pt, oxides	N. I.
	Propane	Propene	2	Pt, oxides	R
	n-Butane	Butene	2	Metal	N. I.
		Butadiene	4	Molybdates	
	Ethylbenzene	Styrene	2	FeO, AlO_4	P
	Methanol	Formaldehyde	4	FeMoO	I
	Isobutyric acid	Methacrylic acid	2	Fe phosphate	I
Oxichloration	Ethylene + Cl_2	Vinyl chloride	2	CuPdCl	I
	Ethane + Cl_2		4	AgMnCoO	N. I.
Oxidation	Methane	CO + H_2	2	Pt, Ni	R
		Formaldehyde	4	MoSnPO	R
	Ethylene	Ethyl oxide	2	Ag/Al_2O_3	I
		Acetaldehyde	2	V_2O_5 + $PdCl_2$	I
		Acetic acid	4		
	Ethane	Acetic acid	6	MoVNbO	N. I.
	Propene	Acrolein	4	BiCoFeMoO	I
	Propane		6	varied	R
	n-Butane	Maleic anhydride	14	$(VO)_2P_2O_7$	I
	i-Butene	Methacrolein	6	SnSbO	I
	i-Butane		8	Oxides, POM	R
	o-Xylene	Phthalic anhydride	12	V_2O_5/TiO_2	I
Ammoxidation	Propene + NH_3	Acrylonitrile	6	MoBFeCoNiO	I
	Propane + NH_3		8	VSbO, MoVO	N. I.

N.B.: I: industrialized; N. I: not (yet) industrialized; P: pilot plant; R: research; POM: polyoxometallates

The kinetics of most of these reactions correspond to a redox mechanism represented by:

$$RCH_2R' + 2\ CatO \longrightarrow RCOR' + H_2O + 2\ Cat$$

The catalyst (Cat) is re-oxidized by atmospheric O_2: $2\ Cat + O_2 \longrightarrow 2\ CatO$.

Thus, in most cases the catalyst is a naturally occurring oxide, with flexible redox and acido-basic properties, and capable of being doped. Polyoxometallates are promising examples of materials in this context. Sometimes, the oxidation sites can be isolated such as a V=O center in zeolites. The oxidation of *n*-butane to maleic acid catalyzed by $(VO)_2P_2O_7$ is commercially successful, and the ammoxidation of propene to acrylonitrile catalyzed *inter alia* by Bi_2MoO_6 or $Bi_2(MoO_4)_3$ is making good progress. Other reactions that are actively sought are the dehydrogenative oxidation of ethane, propene and isobutene and their fuctionalization to the corresponding carboxylic acid as well as the direct hydroxylation of benzene (with the aim of replacing the three-step cumene process).[20.2.c]

As shown in section 2.3, selectivity is a major problem in heterogeneous catalysis, especially in the case of oxidation reactions. Modern experimental and theoretical methods have brought significant advances in the field.[20.2,20.29-20.36] For instance, molecular routes to isolated catalytic sites on various forms of silica have been proposed by Tilley using a *Thermolytic Molecular Precursor* (TMP) synthesis.[20.36] This TMP method works well for precursors of the type $M[OSi(O^tBu)_3]_4$ (M = Ti, Zr, Hf) which are converted to $MO_2.4SiO_2$ at 373-473 K. For example, the following synthesis produces a transparent gel, which upon drying yields a high-surface-area xerogel (520 $m^2 \cdot g^{-1}$):[20.36,20.37]

$$Zr[OSi(O^tBu)_3]_4 \xrightarrow[\text{toluene}]{413\ K} ZrO_2 \bullet 4\ SiO_2 + 12\ CH_2{=}CMe_2 + 6\ H_2O$$

The chemical nature of the precursor has a profound influence on the properties of the resulting solid, and this correlation is useful for the design and synthesis of tailored catalysts. The TMP method is particularly useful for generating homogeneously dispersed metal sites in an amorphous SiO_2 support matrix. Co-thermolysis can produce high-performance catalysts:

The TMP method for the production of high-performance SiO$_2$-bonded catalysts

For instance, use of the molecular precursors $OV[OSi(O^tBu)_3]_3$ and commercially available $[Zr(OCMe_2Et)_4]$ yielded V/Si/Zr/O catalysts with V contents from 2 to 34 wt% V_2O_5 by co-thermolysis. Calcination at 773 K produced high-surface-area catalysts (from 300 to 465 $m^2 \cdot g^{-1}$) for oxidative propane dehydrogenation,[20.38] a highly valuable reaction given the many synthetic uses of propene:[20.39,20.40]

$$n\ OV[OSi(O^tBu)_3]_3 \xrightarrow{\Delta} VnSi_{3n}Zr_mO_x(OH)_y + 9n\ C_4H_8$$
$$+\ m\ Zr[OSi(O^tBu)_3]_4 \qquad\qquad +\ 4m\ C_5H_{10} + 1/2(9n + 4m)\ H_2O$$

$$\text{propane} + 1/2\ O_2 \xrightarrow{\text{cat.}} \text{propene} + H_2O \qquad \Delta H° = -25\ \text{kcal} \cdot \text{mol}^{-1}$$

The intrinsic selectivities for propene formation were 95.5% at 673 K for 23 wt% vanadia, and activities ranged between 28 and 100 mmol propene produced (g cat)$^{-1} \cdot h^{-1}$ from 673 K to 773 K (14% vanadia). These catalysts are more active and selective than materials of similar composition but prepared by conventional impregnation methods using NH_4VO_3-oxalic acid as the V source.[20.39]

Another example is the use of $Ti[OSi(O^tBu)_3]_4$ that yielded $TiO_2 \cdot 4\ SiO_2$, a good catalyst for cyclohexene epoxidation by cumylhydroperoxide (ROOH). Treatment of $Ti[OSi(O^tBu)_3]_4$ with aerosol silica, however, provided a material with 1.01 wt% Ti that, after calcination at 573 K, was more active (5.28 mmol $\cdot g^{-1} \cdot min^{-1}$) and selective for this reaction (94.4% yield after 2 h at 338 K). This catalyst was twice as active as the Shell catalyst which consists of silica treated with $[Ti(O^tPr)_4]$.[20.36]

$$\text{cyclohexene} + ROOH \xrightarrow{\text{Cat. Ti/Si/O}} \text{cyclohexene oxide} + ROH$$

The molecular precursor $Fe[OSi(O^tBu)_3]_3 \cdot THF$ has been employed for the generation of: single-site Fe^{III} catalysts on SBA-15 as the mesoporous support (after calcination at 573 K, 2 h, O_2). Oxidation of benzene at 60°C by a 0.50% Fe catalyst resulted in the production of phenol with 100% selectivity after 96 h at 42% H_2O_2 conversion and 7.5% benzene conversion. The turnover frequency was 2.5×10^{-3} mol product (mol Fe)$^{-1} \cdot s^{-1}$, which is comparable to those of other heterogeneous Fe-based catalysts that are less selective.[20.36]

$$\text{benzene} + H_2O_2 \xrightarrow{\text{cat. Fe}^{III}/\text{SB-A15}} \text{phenol} + H_2O$$

7. NANOPARTICLE CATALYSIS:
THE FAST-GROWING FRONTIER BETWEEN HOMOGENEOUS AND HETEROGENEOUS CATALYSIS

7.1. INTRODUCTION

In Chap. 2, metal nanoparticles (abbreviated MNPs) are introduced and defined as giant clusters of nanometer size with precise geometries resulting from the packing of atoms.[20.48] Ideally, their size and dispersity can be controlled during their synthesis. They can be covered with ligands, anionic stabilizers such as halides (see below), polyoxometallates, etc. or various polymer stabilizers that will define their solubility in organic solvent, water or fluorous solvent.

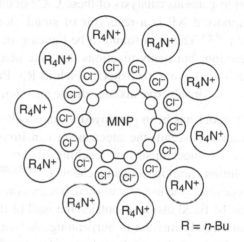

Stabilization of nanoparticles (NPs) by chloride anions subsequent to formation by reduction of the transition-metal chloride in the presence of a R_4N^+ salt as stabilizer.[20.49]
Synthesis can be carried out analogously in the presence of a supported oxide that is impregnated.

R = n-Bu

Alternatively, the ligands can be metal oxides, zeolites, etc. in which case they are insoluble and used as heterogeneous catalysts. Thus, significant interest is emerging in comparing their catalytic behavior in homogeneous solution and that observed when they are combined with an oxide in a heterogeneous catalyst.[20.50] As indicated earlier in this chapter, they currently take part in all the classical heterogeneous reactions discussed therein and in many others.

7.2. OXIDATION OF CO TO CO_2 BY O_2 UNDER AMBIENT OR SUBAMBIENT CONDITIONS BY OXIDE-SUPPORTED AuNPs

Although gold has the reputation of being an inert metal, Haruta discovered in the 1980s that AuNPs supported on metal oxides (best oxide support: TiO_2) catalyzed CO oxidation by O_2 under ambient or subambient conditions (down to 200 K). The reaction is believed to occur at the interface between the AuNP surface that activates O_2 forming superoxide at the surface: O_2^-/AuNP and TiO_2 that acts as a Lewis acid. Synergistic activation of CO favors nucleophilic attack by O_2^- on the polarized C atom of CO. The mechanism is still presently unknown and debated. Since the composite AuNP-Fe_2O_3 supported on a zeolite-coated paper honeycomb has been used commercially as an odor eater in modern Japanese restrooms since 1992, many oxidation reactions have been catalyzed by $AuNP-TiO_2$ for environmental purposes and fine organic chemical synthesis.[20.51,20.52]

7.3. ALKENE HYDROGENATION AND C-C BOND FORMATION BY OXIDE-OR CARBON-SUPPORTED PdNPs

NPs have been supported on a large variety of oxide supports (oxides of Si, Al, Ti, Ca, Mg, Zn) in the form of aerogels or sol-gels such as Gomasil G-200, high-surface silica, M41S silicates, silica spheres, micro-emulsion, hydroxyapatite, hydrotalcite, zeolites, molecular sieves, alumina membranes. Classical reactions that have been studied include hydrogenation and various C-C coupling reactions (Heck, Sonogashira, Suzuki, Stille, Corriu-Kumada, etc.).[20.50] It has been suggested that heterogeneous catalysis of these C-C bond formations proceeds by leaching, i.e. the supported NP is a reservoir of small atoms that reacts homogeneously in solution.[20.53] This was found to be the case in particular for the PdNP-catalyzed Heck reaction, but many reactions that take place around or above 100°C might follow such a course. Other metals such as Rh, Pt and Au are very active as well, in particular for selective alkene, alkyne and arene hydrogenation and oxidation reactions.

Charcoal has been employed as an excellent support for MNPs that catalyze these reactions. Again the mechanism can involve the adsorbed MNP (heterogeneous surface mechanism) or leaching of atoms or very small atom fragments in the solution (homogeneous mechanism).[20.54,20.55] Modern techniques are being applied to develop the use of carbon nanotubes on a support (SWNT, etc.) where the MNPs can be fixed inside or outside the wall of the nanotubes. This field is in its infancy, but already interest is burgeoning. Activity and selectivity are the most desirable qualities of the catalyst, but surface studies and mechanistic investigations are needed (such as comparisons between NPs and metal crystal surfaces).[20.56] Heterobimetallic NPs (such as cored system, one cored metal influencing the reactivity of the NP surface metal) on supports often are more active than monometallic supported catalysts. Various types are illustrated below:[20.57]

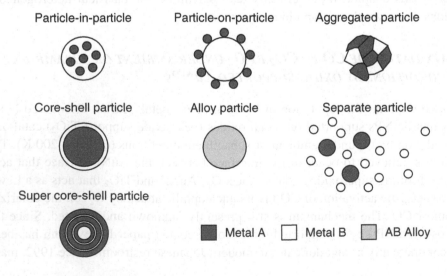

Particle-in-particle   Particle-on-particle   Aggregated particle

Core-shell particle   Alloy particle   Separate particle

Super core-shell particle

■ Metal A   □ Metal B   ■ AB Alloy

SUMMARY OF CHAPTER 20
HETEROGENEOUS CATALYSIS

1 - Most catalytic industrial processes are heterogeneous.
3 main areas: oil transformation to gasoline, synthesis of fine chemicals and environment.

2 - Concepts:
• physisorption (weak binding of substrates on surfaces)
• chemisorption (substrate dissociation on the surface)
The selectivity varies with catalysts (improved with zeolites, that are acid-base aluminosilicate catalysts) and surfaces (faces of crystals).

3 - CO and CO_2 hydrogenation and Fischer-Tropsch chemistry
CO, available by reaction of CO_2 with H_2 (watergas-shift reaction), is hydrogenated to CH_3OH (3×10^6 kg per day, cat: Cu/ZnO).
Further CO hydrogenation (Fischer-Tropsch process) gives $(-CH_2-)_n$ and water, but olefins and alcohols are also produced in lower amounts than paraffins.
The Fischer-Tropsch mechanism involves two steps: production of olefins and their polymerization. Mechanistic hypotheses involve prior dissociation of H_2 and CO to surface-bound atoms, but differ for their recombination.

4 - Transformation of hydrocarbons
Competitive reactions catalyzed by Pt involve hydrogenation-dehydrogenation, hydrogenolysis, isomerization and dehydrocyclization. Stable ethylidyne forms on Pt surfaces by chemisorption of ethylene.

5 - Alkane metathesis
Early transition-metal akyl and hydride complexes grafted on silica catalyze alkane disproportionation by σ-bond metathesis. For instance, a metal-hydride reacts with an alkane to yield a metal-alkyl on silica, then combination of C-H σ-bond metathesis with α and β-H eliminations leads to alkane metathesis (disproportionation), the mechanism involving metal-carbenes and metallocyclobutanes as for olefin metathesis.

6 - Oxidation of hydrocarbons is achieved selectively by using *Thermolytic Molecular Precursors* of the type $M[OSi(Ot-Bu)_3]_4$ (M = Ti, Zr, Hf). For instance, propane, cylohexene and benzene are oxidized to propene, cylohexene epoxide and phenol, respectively.

7 - Nanoparticle catalysis
Metal nanoparticles (MNPs) are excellent catalysts working under mild conditions either homogeneously or heterogenized on a solid support (zeolite, charcoal, nanotubes). For instance, Haruta discovered that AuNPs on TiO_2 catalyze CO oxidation by O_2 to CO_2 under subambient conditions, with many applications. Alkene and arenes hydrogenation under ambient conditions and cross C-C bond formation (Suzuki, Sonogashira, Heck, Stille, Corriu-Kumada) with aryl halides are catalyzed by noble metal NPs (Ru, Rh, Pd or Pt).

EXERCISES

20.1. Why do late transition-metal surfaces better chemisorb substrates than early transition-metal surfaces?

20.2. Why is Al_2O_3 basic whereas SiO_2 is acidic and why is a zeolite more acidic upon increasing the Si/Al ratio?

20.3. What is, in terms of raw materials, the most economical way to produce syngas?

20.4. What are the common points and differences between Ziegler-Natta olefin polymerization and the Fischer-Tropsch formation of paraffins?

20.5. Compare the Schrock and Basset catalysts (section 6) for alkene metathesis: what is the common point that makes both series of complexes active in alkene metathesis?

PART V

APPLICATIONS IN ORGANIC SYNTHESIS

Chapter 21
Organometallics in organic synthesis
Examples of applications

PART V

APPLICATIONS IN ORGANIC SYNTHESIS

Chapter 21
Organometallics in organic synthesis
Examples of applications

ORGANOMETALLIC COMPLEXES IN ORGANIC SYNTHESIS EXAMPLES OF APPLICATIONS

The potential applications of organometallic complexes in organic synthesis, whether they are stoichiometric or catalytic, are apparent in all the preceding chapters. In the present chapter, we will provide precise examples of applications of organometallic complexes in the synthesis of mainly natural products, i.e. we will view organometallic chemistry from the perspective of organic chemists.[21.1-21.4] This field is now expanding towards asymmetric synthesis using chiral and optically active ligands.

1. PROTECTION AND STABILIZATION OF UNSATURATED ORGANIC DERIVATIVES AND FRAGMENTS

It is possible to temporarily protect reactive organic functional groups by complexing them to appropriate metal-centered fragments forming stable 18-electron complexes. This process makes selective reactions on other parts of the molecule possible, thus alkenes and alkynes are protected respectively by the 14-electron species $Fe(CO)_3$ [21.6,21.7] and the fragment $Co_2(CO)_6$ [21.8]. In the latter, both cobalt atoms have 16 valence electrons, each of the two perpendicular π alkyne orbitals serving as an L ligand for a Co atom. Thus, in the complex $[Co_2(CO)_6$ $(\mu_2,\eta^2$-alkyne)]$, the Co_2C_2 species is a tetrahedron in which both Co atoms have 18 valence electrons.

In the synthesis of cyclocolorenone,[21.9] the complexation of the two conjugated tropone double bonds serving as a starting point allows to make the diazo-2-propane to react with the noncoordinated C=C bond. The synthesis continues through the use of other various organometallic compounds: alkylation with a cuprate, then silylation of the carbonyl and alkylation by a carbocation located in the α position with respect to a triple bond (propargyl) that is protected by $Co_2(CO)_6$, followed by the oxidative decomplexation of this dicobalt carbonyl fragment using Fe^{III} or Ce^{IV} derivatives. This fragment not only protects the triple bond, but it also stabilizes the propargyl carbocation. This type of carbocation stabilization is also general with

18-electron complexes because of the interaction with the vacant *p* orbital of the carbocation with the filled *d* orbital with adequate symmetry of the metal (the prototype example is the ferrocenyl carbocation).

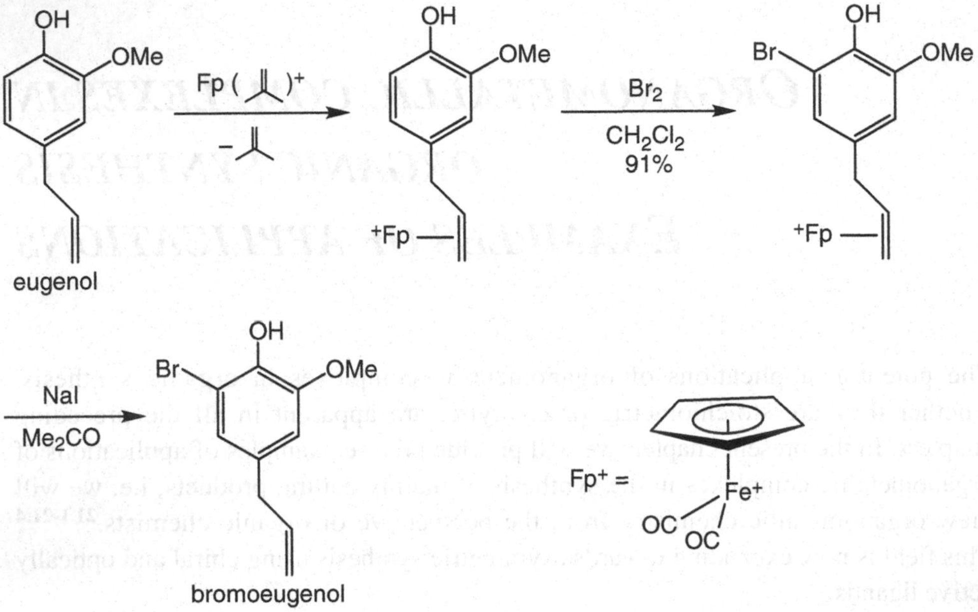

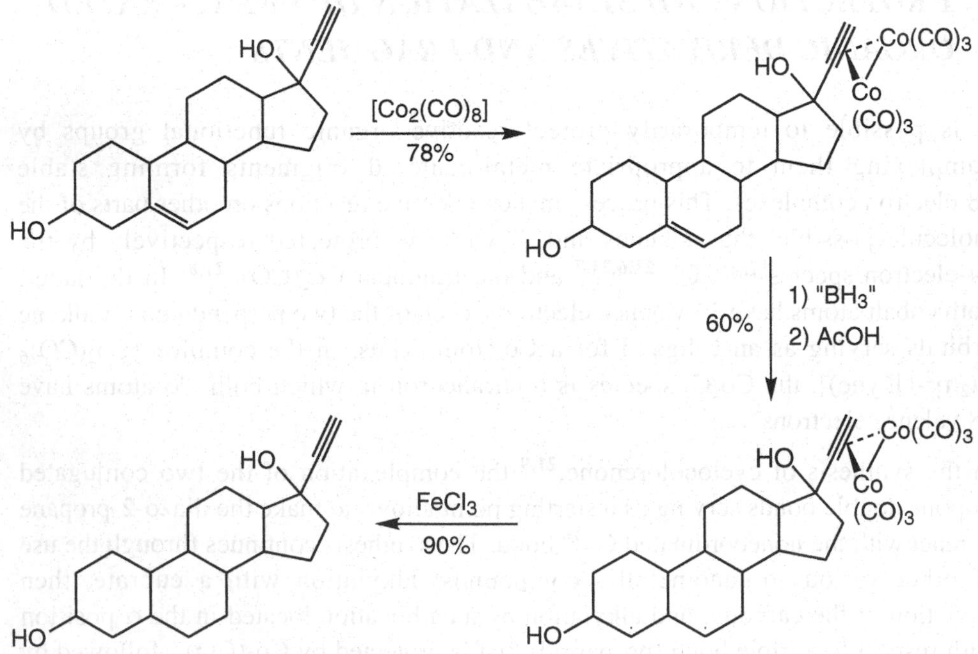

eugenol

bromoeugenol

$Fp^+ =$

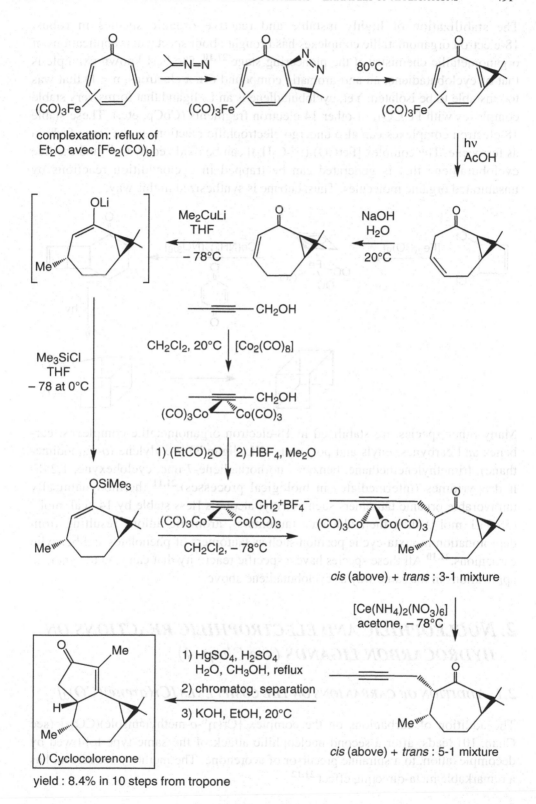

complexation: reflux of
Et₂O avec [Fe₂(CO)₉]

Me₂CuLi
THF
– 78°C

NaOH
H₂O
20°C

hν
AcOH

━━━━CH₂OH

CH₂Cl₂, 20°C | [Co₂(CO)₈]

━━━━CH₂OH
(CO)₃Co━━Co(CO)₃

Me₃SiCl
THF
– 78 at 0°C

1) (EtCO)₂O   2) HBF₄, Me₂O

━━━CH₂⁺BF₄⁻
(CO)₃Co━━Co(CO)₃

CH₂Cl₂, – 78°C

(CO)₃Co━━Co(CO)₃

cis (above) + trans : 3-1 mixture

[Ce(NH₄)₂(NO₃)₆]
acetone, – 78°C

1) HgSO₄, H₂SO₄
   H₂O, CH₃OH, reflux

2) chromatog. separation

3) KOH, EtOH, 20°C

cis (above) + trans : 5-1 mixture

() Cyclocolorenone

yield : 8.4% in 10 steps from tropone

The stabilization of highly instable and reactive organic species in robust 18-electron organometallic complexes has brought about spectacular applications of organometallic chemistry at the pioneering stage.[21.10] The best known example is that of cyclobutadiene, an anti-aromatic compound (4n π electrons, n = 1) that was too instable to be isolated. Yet, cyclobutadiene is an L_2 ligand that forms very stable complexes with $Fe(CO)_3$ or other 14-electron fragments (CoCp, etc.). These stable 18-electron complexes can also undergo electrophilic reactions in the same fashion as ferrocene. The complex $[Fe(CO)_3(\eta^4-C_4H_4)]$ can be oxidized by Ce^{IV} and the free cyclobutadiene that is generated can be trapped in cycloaddition reactions by unsaturated organic molecules. Thus, cubane is synthesized in this way:

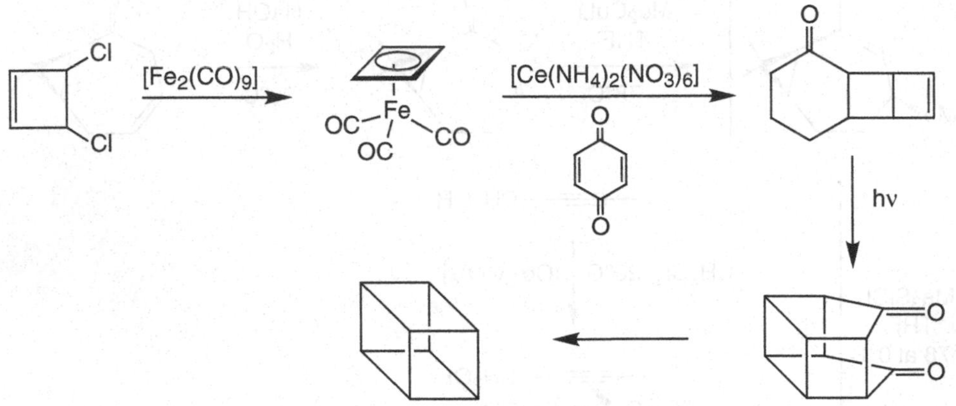

Many other species are stabilized in 18-electron organometallic complexes: carbenes and carbynes, enyls and polyenyls (XL_n ligands), o-xylylene (o-quinodimethane), trimethylenemethane, benzyne, norbornadiene-7-one, cyclohexyne, 1,2-dihydropyridines (intermediates in biological processes),[21.11] thermodynamically unfavorable organic tautomers such as vinyl alcohols [less stable by 14 kcal·mol⁻¹ $(58.5 kJ \cdot mol^{-1})$ than their aldehyde tautomers], aromatic anions resulting from deprotonation in juxta-cyclic position such as tautomers of phenolates and benzylic carbanions.[21.10] All these species have a specific reactivity that can lead to synthetic applications in the same way as cyclobutadiene above.

2. NUCLEOPHILIC AND ELECTROPHILIC REACTIONS ON HYDROCARBON LIGANDS (see Chap. 4)

2.1. ADDITION OF CARBANIONS ON THE COMPLEXES [Cr(arene)(CO)₃]

The addition of carbanions on the complex $[Cr(\eta^6$-o-methylanisole)$(CO)_3]$ (see Chap. 10) leads, after a second nucleophilic attack of the same type followed by decomplexation, to a spiranne precursor of acorenone. The methoxy substituent has a remarkable meta-directing effect.[21.12]

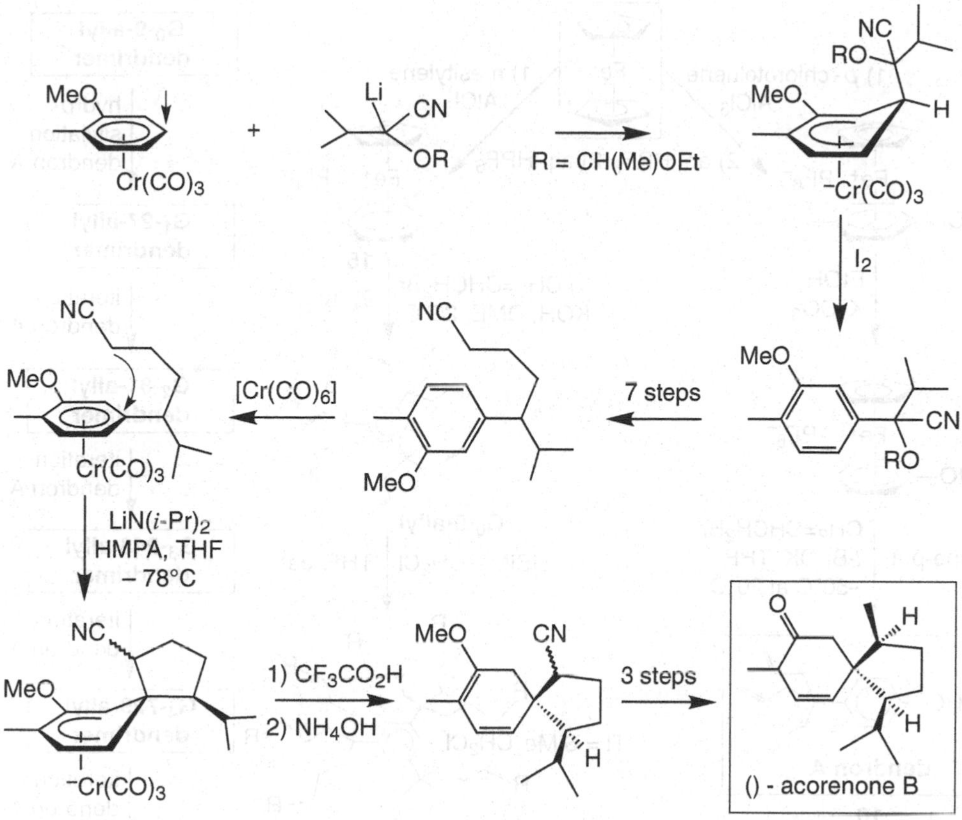

Yield : 5.3% from *o*-methylanisole in 15 steps

Intermediate in terpenic biosynthesis, antifungal agent and stress metabolite

2.2. APPLICATIONS OF THE CpFe⁺-INDUCED ACTIVATION OF NUCLEOPHILIC AROMATIC REACTIONS (see Chap. 11) TO THE SYNTHESIS OF DENDRONS AND DENDRIMERS

Nucleophilic aromatic substitution by various nucleophiles, heterolytic cleavage of C-O bonds of aryl ethers and benzylic deprotonation reactions are considerably facilitated by temporary π complexation by the 12-electron fragment CpFe⁺.[21.10] These reactions are used in the scheme below to synthesize dendrons and dendrimers.[21.13] Dendrimers, whose syntheses proceed by successive iterations of a series of reactions recalling the principle of replication in biology (DNA, cells), are nanosized well-defined macromolecules with a polydispersity equal to 1.0.[21.14] Nowadays, the key technique to characterize them is MALDI TOF mass spectroscopy, especially because it allows the observation of the molecular peak and impurities. Their shapes become globular after a few generations and close to those of biological molecules such as neurons and viruses that have a fractal surface. They have many potential applications such as catalysis, DNA transfection, gene therapy, drug transport and molecular imaging.

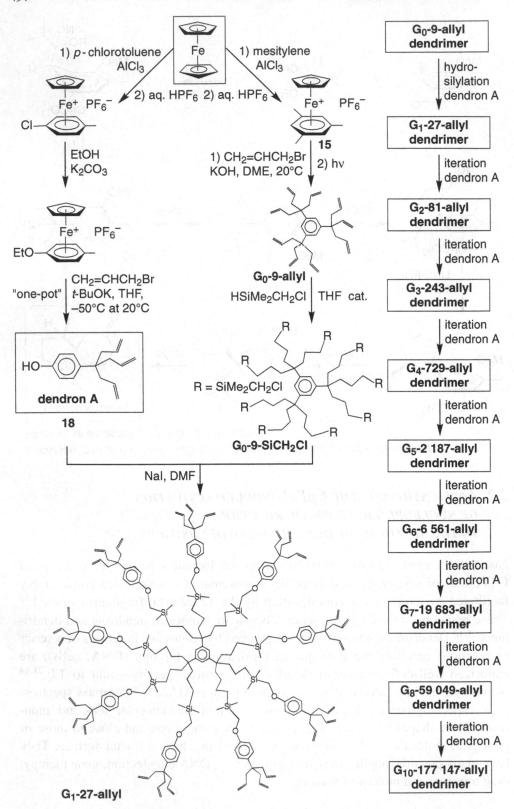

Synthesis of giant dendrimers with 3^n allyl branches using arene activation by CpFe$^+$

A very high number of termini can be reached (see the theoretical numbers on the scheme) if these termini are small and can fold back towards the center of the dendrimer. The construction is then no longer limited by the periphery (de Gennes' "dense-packing" limit), but by the volume.

2.3. REGIOSELECTIVE ADDITION OF A CARBANION ON THE COMPLEX [Fe(CO)₃(η⁵-METHOXYCYCLOHEXADIENYL)]

Regioselective hydride abstraction using the directing effect of a methoxy substituent in a 1,4-disubstituted diene coordinated to the $Fe(CO)_3$ group yields an organo-iron cation $[Fe(\eta^5\text{-cyclohexadienyl})(CO)_3]^+$. This cation undergoes a nucleophilic attack leading to the formation of a quaternary carbon. These key steps are involved in the total synthesis of (±) limaspermine starting from para-disubstituted anisole derivatives.[21.15]

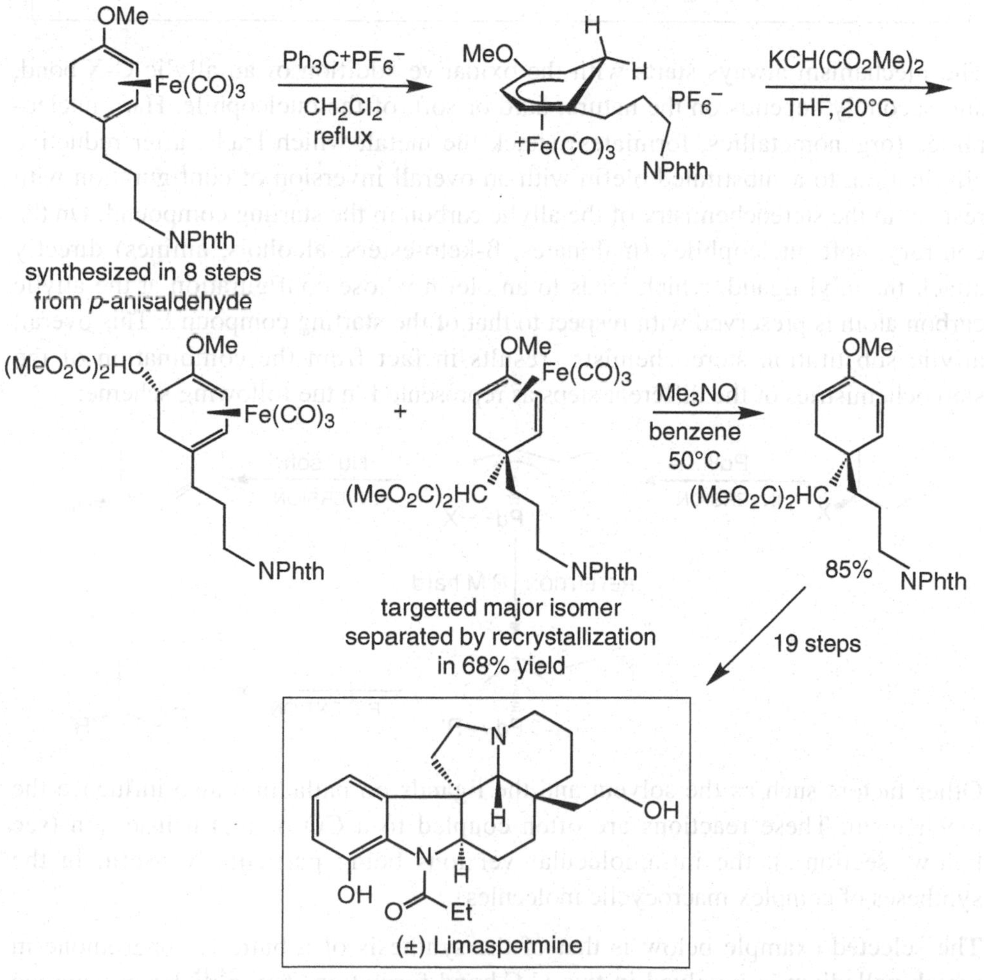

Synthesis with an overall 0.45% yield in 30 steps from *p*-anisaldehyde
Drug used against Hodgkins desease and child leukemia

2.4. *ADDITION OF A CARBANION ON A* $Pd^{II}(\eta^3$*-ALLYL) COMPLEX: TSUJI-TROST ALLYLIC SUBSTITUTION*

The applications of palladium in organic syntheses are numerous,[21.16,21.17] this metal possibly being the most important one in the field. Numerous reactions are known: oxidations of the Wacker type, C-C coupling by transmetallation and/or insertion of CO or olefin (see following section). A category of reactions that is also very common concerns the catalytic use of palladium to carry out the substitution of a nucleofuge (typically acetate) in allylic position by a carbanion or any other nucleophile. This is the Tsuji-Trost allylic substitution:

R X + NuH (or NuM) Pd0 R Nu + HX (or MX)
X = OAc, OCO2R, OH, OP(O)(OR)2, Cl, NO2, SO2R, NR2, NR3+
NuH = ROH, RNH2, etc. (or Nu− = RO−, RNH−, stabilized carbanion, etc.)

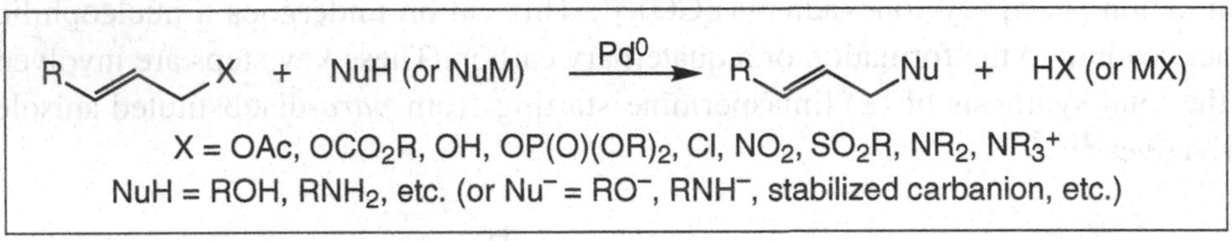

The mechanism always starts with the oxidative addition of an allylic C-X bond, and secondly depends on the nature, hard or soft, of the nucleophile. Hard nucleophiles (organometallics, formiates) attack the metal, which leads, after reductive elimination, to a substituted olefin with an overall inversion of configuration with respect to the stereochemistry of the allylic carbon in the starting compound. On the contrary, soft nucleophiles (malonates, β-keto-esters, alcohols, amines) directly attack the allyl ligand, which leads to an olefin whose configuration at the allylic carbon atom is preserved with respect to that of the starting compound. This overall allylic substitution stereochemistry results in fact from the combination of the stereochemistries of the different steps as represented in the following scheme:

X
Pd0
INVERSION
Pd—X
Nu− soft
INVERSION
Nu
RETENTION
R'M hard
Pd—R'
RETENTION
R'

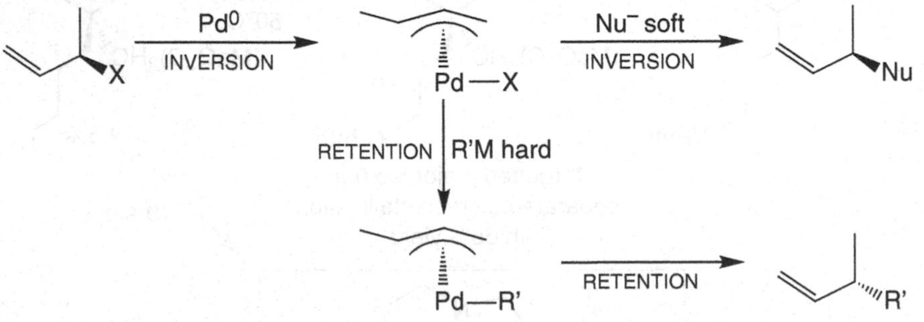

Other factors such as the solvent and the ligands on palladium also influence the mechanism. These reactions are often coupled to a CO or olefin insertion (see below, section 3), the intramolecular versions being particularly useful in the syntheses of complex macrocyclic molecules.

The selected example below is that of the synthesis of a butterfly pheromone in which palladium is involved in two C-C bond formation steps.[21.18] From isoprene epoxide, the action of palladium leads to the opening of the epoxide, and the attack of the first carbanion is not stereoselective. This is not inconvenient, however.

Indeed, in the second step, not only the formation of the C-C bond is stereoselective, but also the formation of the intermediate Pd-allyl complex proceeds along with a thermodynamic re-equilibration of the *syn-anti* mixture leading exclusively to the *E* stereoisomer coming from the *syn* complex.

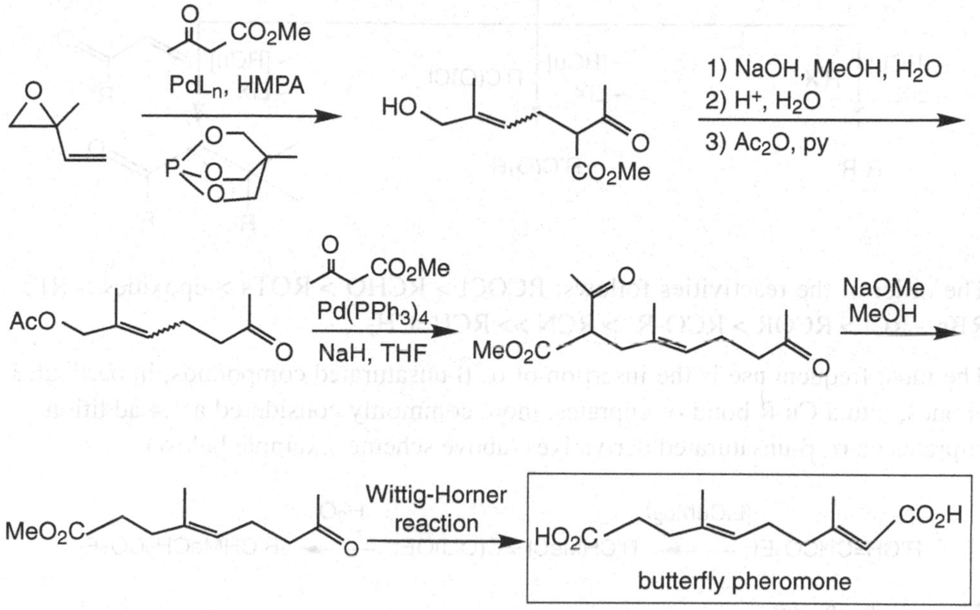

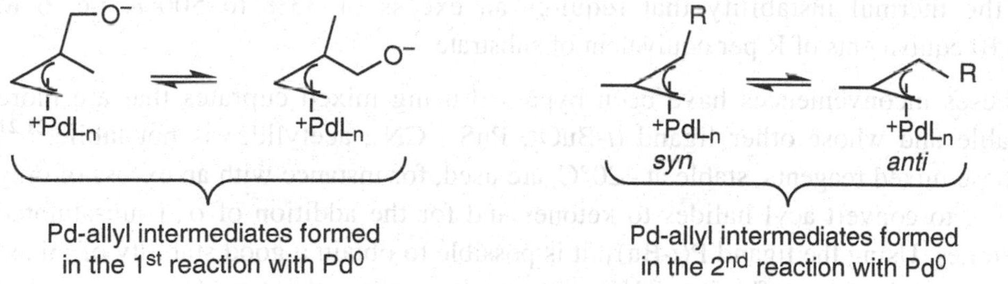

butterfly pheromone

Pd-allyl intermediates formed
in the 1st reaction with Pd⁰

Pd-allyl intermediates formed
in the 2nd reaction with Pd⁰

3. GENERAL METHODS OF C-C BOND FORMATION USING THE OXIDATIVE ADDITION OF AN ORGANIC HALIDE OR A RELATED ELECTROPHILE

3.1. ORGANOCUPRATES

Cuprates, obtained by reaction of a lithium reagent on cuprous iodide, do not undergo β-elimination and can alkylate organic halides (primary > secondary > tertiary) and tosylates, including allyl, benzyl, propargyl and acyl halides and various other electrophiles. [21.19,21.20]

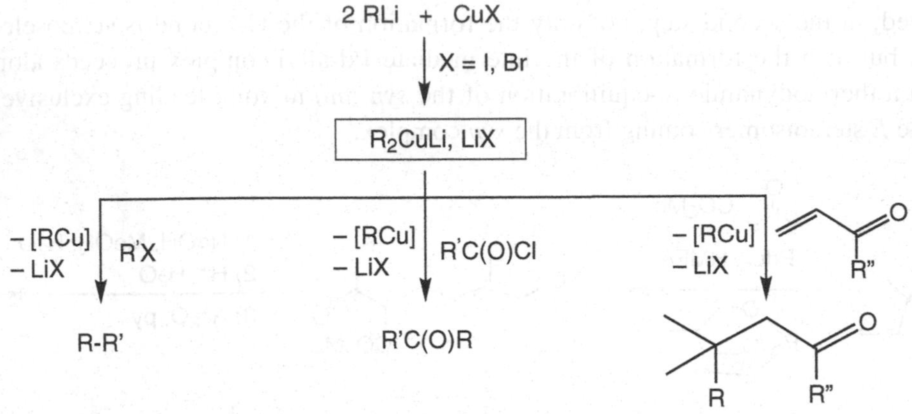

The order of the reactivities follows: RCOCl > RCHO > ROTs > epoxides > RI > RBr > RCl > RCOR > RCO$_2$R' > RCN >> RCH=CH$_2$.

The most frequent use is the insertion of α, β-unsaturated compounds, in particular enones, into a Cu-R bond of cuprates, more commonly considered a 1,4 addition of cuprates on α, β-unsaturated derivatives (above scheme, example below).

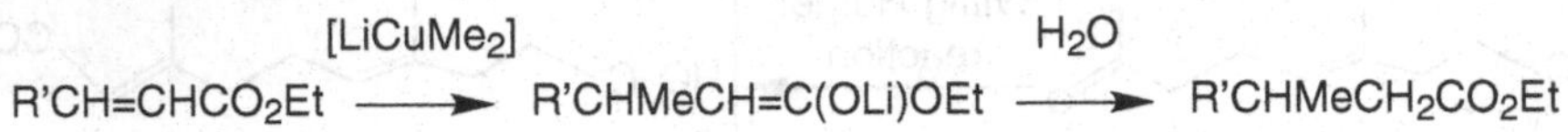

$$\text{R'CH=CHCO}_2\text{Et} \xrightarrow{\text{[LiCuMe}_2\text{]}} \text{R'CHMeCH=C(OLi)OEt} \xrightarrow{\text{H}_2\text{O}} \text{R'CHMeCH}_2\text{CO}_2\text{Et}$$

The drawbacks are
▸ the use of only one of the two R groups in the stoichiometry,
▸ the thermal instability that requires an excess of 33% to 500%, i.e. 6 to 10 equivalents of R per equivalent of substrate.

Theses inconveniences have been bypassed using mixed cuprates that are more stable and whose other ligand (t-BuO$^-$, PhS$^-$, CN$^-$, acetylide) is not labile.[21.21] These mixed reagents, stable at –20°C, are used, for instance with an excess of only 10%, to convert acyl halides to ketones and for the addition of α, β-unsaturated ketones. Using the ligand P(t-Bu)$_3$, it is possible to obtain a good stability of mixed cuprates even in refluxing THF without preventing the transfer of the alkyl group.[21.22] The low-temperature synthesis of the cuprate complexes [R$_2$Cu(CN)Li$_2$] allows access to a whole series of mixed Ar-Ar' derivatives subsequent to reaction with O$_2$ at –125°C. Normant and Alexakis have shown that the addition of BF$_3$ to organo-cuprous complexes RCu increases their reactivity in a crucial way, which allows monoalkylating ketals with removal of an alkoxy group. Thus allylic acetals can be cleaved with an excellent diastereoselectivity.

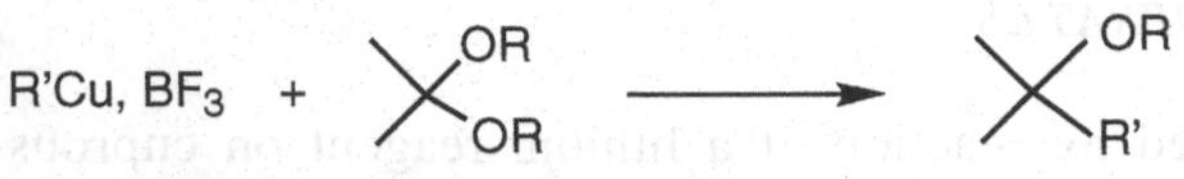

This technique opens the route to various applications. For example, the reaction has been used for the synthesis of a pheromone as shown below.[21.19,21.20]

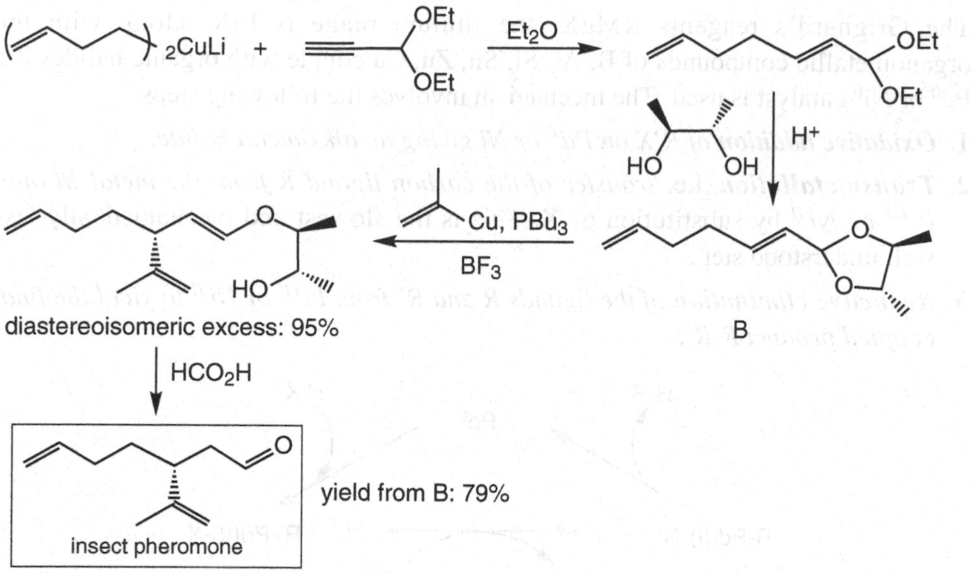

diastereoisomeric excess: 95%

yield from B: 79%

insect pheromone

The use of stannyl cuprates $[Cu(SnR_3)_2(CN)Li_2]$ and $[Cu(SnR_3)(R')(CN)Li_2]$ has also been developed by showing their regioselective *syn* addition to alkynes, the intermediate vinyl cuprate being eventually trapped with a variety of electrophiles.[21.23, 21.24]

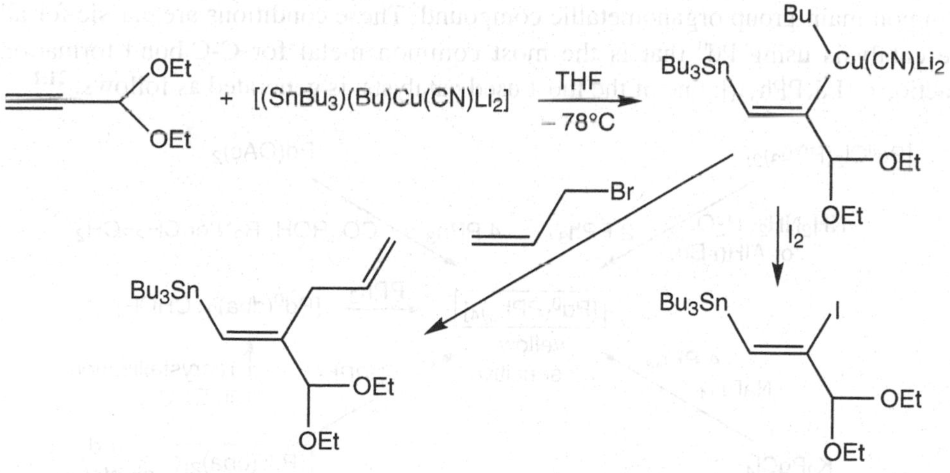

3.2. PRINCIPLE OF THE C-C COUPLING BY TRANSMETALLATION CATALYZED BY Pd⁰

The C-C coupling reaction of an organometallic compound with an organic halide usually catalyzed by Pd⁰:

$$RM + R'X \xrightarrow{Pd^0} R-R' + MX$$

is synthetically highly valuable, and it has consequently been highly scrutinized, especially because many metals, including main-group ones, are concerned.[21.1, 21.17]

The Grignard's reagents RMgX, the lithium reagents LiR, along with the organometallic compounds of B, Al, Si, Sn, Zn, Cu couple with organic halides if a Pd^0 or Ni^0 catalyst is used. The mechanism involves the following steps:

1. **Oxidative addition** *of R'X on Pd^0 or Ni giving an alkylmetal halide.*

2. **Transmetallation**, *i.e. transfer of the carbon ligand R from the metal M onto Pd^{II} or Ni^{II}* by substitution of X. This is the slowest and mechanistically less well understood step.

3. **Reductive elimination** *of the ligands R and R' from Pd^{II} or Ni^{II} to yield the final coupled product R-R'.*

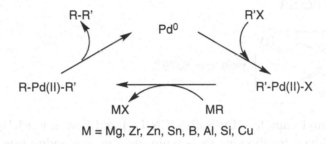

$$M = Mg, Zr, Zn, Sn, B, Al, Si, Cu$$

The Pd^0 catalysts are generated *in situ* from a Pd^{II} salt such as $PdCl_2$ or $Pd(OAc)_2$ using a reducing agent that can be an amine, phosphine, alcohol, olefin, CO or common main-group organometallic compound. These conditions are classic for all the catalysis using Pd^0 that is the most common metal for C-C bond formation reactions. [$Pd(PPh_3)_4$], one of the most used catalysts, is generated as follows: [21.1]

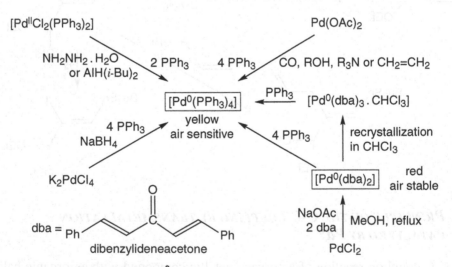

Synthesis of Pd^0 catalysts for C-C coupling reactions

Phosphines are well-known as being perhaps the most active ligand for Pd^0-catalyzed C-C bond formation. As shown by the remarkable results obtained *inter alia* by Hartwig, Fu and Buchwald, they indeed even allow the activation of the less reactive aryl chlorides. Alternatives are being actively sought after, however, in order to avoid phosphines for toxicity reasons.

Two lines are promising: Pd^0 nanoparticles and *N*-heterocyclic carbenes (NHC). The uses of NHCs have been successfully introduced into catalysis by Herrmann with various systems in the late 1990s, particularly in Pd^0 catalysis. Nolan (*inter alia*) has also brought important developments with efficient and practical uses of NHCs for all the reactions described thereafter. NHCs are strong ligands that are conveniently introduced in the reaction medium, as their air-stable imidazolium hydochloride salt, together with the base Cs_2CO_3 (instead of the amine) for deprotonation *in situ* during an activation period at 80°C before cooling, introducing substrates, and heating again. The Pd source is $Pd(OAc)_2$, $Pd(dba)_2$, or $Pd_2(dba)_3$. A typical solvent is dioxane, possibly mixed with THF. The active species presumably is, as with bulkyl active phosphines, an electron-rich low-coordinate NHC-Pd^0 species, stabilized by solvent coordination as a four-coordinate species with labile solvent ligands. This low coordination is favored by the bulkyness of the phosphine or NHC ligand, and facilitates the oxidative addition of the halide that is the rate-limiting step, and a crucial one given the difficulty in activating aryl chlorides. Therefore, the NHC ligand must bear very bulky nitrogen substituents, the most common ones being mesityl and 2,6-diisopropylphenyl (especially the latter).

Di-*N*, *N'*-mesityl-NHC
Heck

Di-*N*, *N'*-2,6-di-*iso*-propylphenyl-NHC
Suzuki (B), Kumada-Corriu (Mg),
Stille (Sn), Hiyama (Si), amination

N-Heterocyclic carbenes (NHCs) used in Pd⁰-catalyzed formation of C-C and C-N bonds

Simplified mechanism indicating how a single strongly bonded, electron-releasing bulky ligand L (bulky electron-rich phosphine or NHC) together with weakly bonded solvent ligands (S) favor the rate-limiting oxidative addition step RX + Pd⁰ → R-Pdᴵᴵ-X

3.3. ZINC COMPLEXES: EXCELLENT TRANSMETALLATION REAGENTS NEGISHI COUPLING

For Pd^0 catalysis, the organozinc derivatives are among the best coupling agents RM. The mixed derivatives $RZnI$[12.8] are easily generated from an iodide RI by sonication with zinc powder or by reaction with $ZnEt_2$, sometimes catalyzed by [Pd^0(dppf)]. The applications in synthesis are quite general, because the reaction tolerates a large variety of functional groups. The RZnX compounds are non-isolable and used without purification, however, which is a drawback compared to the more often utilized Suzuki and Stille couplings.

Example: [21.25]

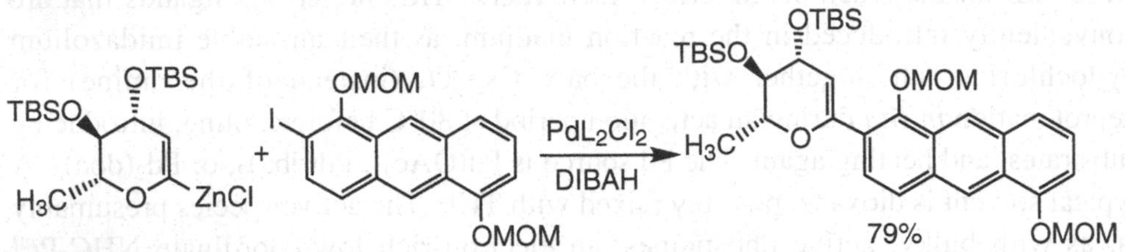

Homoleptic organozinc compounds ZnR$_2$, that are isolable but very reactive, can be efficiently used. A spectacular example of Negishi-type coupling is Vollhardt's synthesis of hexaferrocenylbenzene, an extremely bulky molecule, from hexaiodo-benzene and diferrocenylzinc (ZnFc$_2$). The major reaction product, however, is pentaferrocenylbenzene, substitution of the last iodine atom by hydrogen probably resulting from side-electron transfer from ZnFc$_2$ to the aryliodide due to the very slow, sterically hindered formation of the sixth C-C bond.

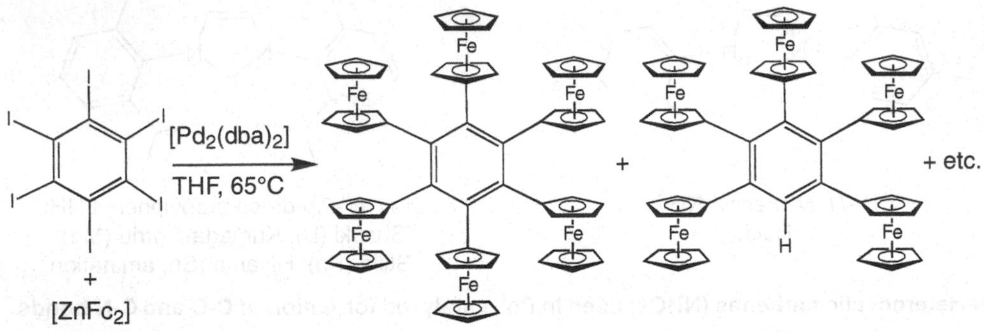

With the other organometallic derivatives RM that are less efficient than the Zn reagents such as the Li reagents, the reaction can proceed in the same way as with the Zn reagents, if one equivalent of ZnCl$_2$ in THF is added to the reaction medium. In this way, the mixed organozinc reagent is generated *in situ* by transmetallation as in the following section 3.4: [12,18]

$$MR + ZnCl_2 \longrightarrow Zn(R)Cl + MCl$$

3.4. SCHWARZ REAGENT [ZrCp$_2$(H)Cl]:
TRANSMETALLATION OF σ-VINYLZIRCONIUM

This 16-electron complex is well known for allowing the insertion of many unsaturated substrates into the Zr-H bond.[21.26] Thus alkynes insert, the less bulky regio-isomer being usually preferred. The σ-vinylzirconium complexes obtained in this way from terminal alkynes are transmetallated onto palladium, which allows the same catalysis of C-C bond formation as that described above with organozinc compounds.

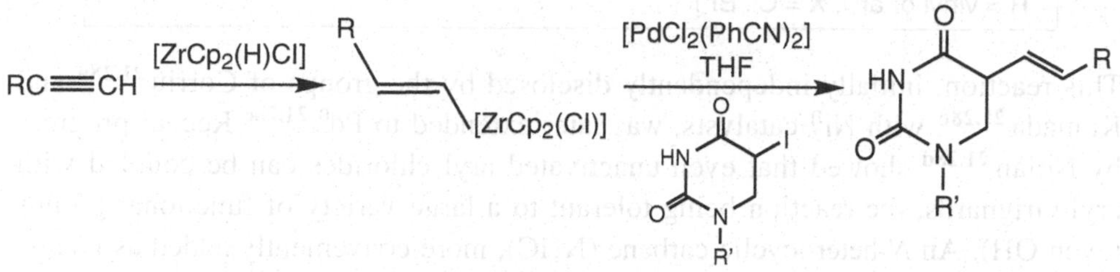

On the other hand, the σ-vinylzirconium complexes obtained from internal alkynes do not react with Pd complexes. The reaction becomes possible upon addition of $ZnCl_2$ because of the transmetallation: Zr ⟶ Zn ⟶ Pd.

3.5. STILLE REACTION: COUPLING WITH ORGANOTIN REAGENTS

Any substrate that undergoes oxidative addition onto Pd^0 can in principle be coupled with any transferable group using an organotin compound. The Stille coupling is the most used example in synthesis. It is a very efficient reaction that allows to couple a vinyl- or an alcynyltin derivative with a iodoalkene or iodoarene, the reaction being catalyzed by Pd^0: [21.27]

$$RI + Bu_3SnR' \xrightarrow[\text{THF}]{PdCl_2L_2,\ L = PPh_3 \text{ or } MeCN} R\text{-}R' + Bu_3SnI$$

R = vinyl or aryl R' = vinyl or alcynyl

54%
Zearalenone

The reaction leads to conjugated systems, the enyne fragment being particularly useful, because it is frequently found in antitumor antibiotics. In spite of the infatuation for this reaction, it is no longer advised because of the toxicity of organotin reagents, and it tends to be replaced by other coupling methods with zinc (above) and boron reagents (see below).

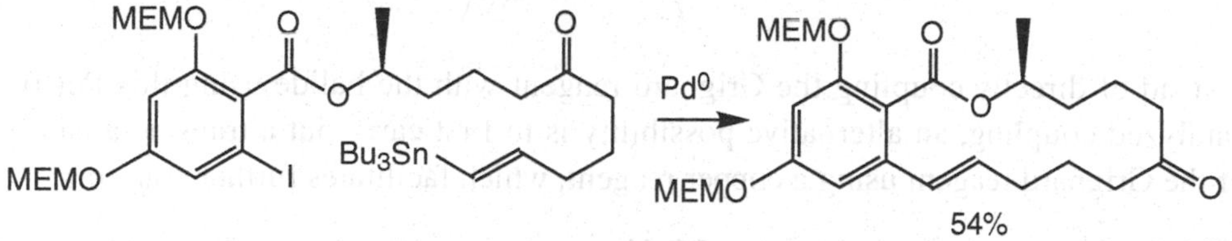

3.6. THE CORRIU-KUMADA REACTION: COUPLING WITH GRIGNARD'S REAGENTS

The reaction of Grignard with alkenyl or aryl halides can be catalyzed either by Ni or Pd:

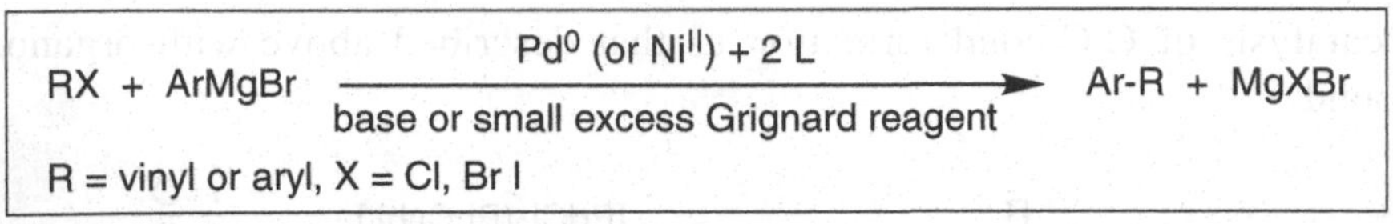
RX + ArMgBr
Pd0 (or NiII) + 2 L
base or small excess Grignard reagent
Ar-R + MgXBr
R = vinyl or aryl, X = Cl, Br I

This reaction, initially independently disclosed by the groups of Corriu [21.28a] and Kumada [21.28b] with Ni^{II} catalysts, was later extended to Pd^0.[21.28c] Recent progress by Nolan [21.28d] showed that even unactivated aryl chlorides can be coupled with aryl Grignards, the reaction being tolerant to a large variety of functional groups (even OH). An *N*-heterocyclic carbene (NHC), more conveniently added as its air-stable hydrochloride salt serves as Pd^0 ligand instead of a phosphine. The Grignard reagent is added in slight excess instead of the base in order to deprotonate this imidazolium salt providing the NHC Pd^0 ligand *in situ*.

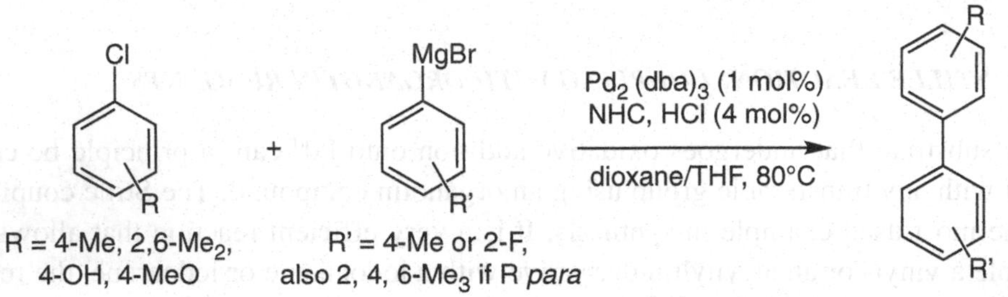
Cl
MgBr
+
R
R'
Pd2 (dba)3 (1 mol%)
NHC, HCl (4 mol%)
dioxane/THF, 80°C
R
R'
R = 4-Me, 2,6-Me2,
4-OH, 4-MeO
R' = 4-Me or 2-F,
also 2, 4, 6-Me3 if R para

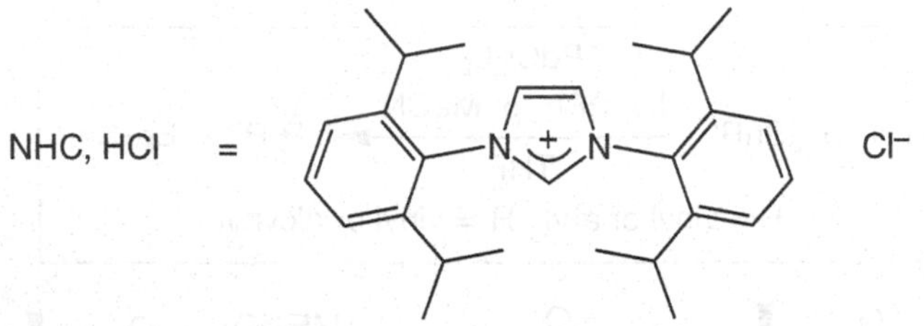
NHC, HCl =
N
+
N
Cl−

Instead of directly coupling the Grignard reagent with the halide using this Pd(0)-catalyzed coupling, an alternative possibility is to first carry out a transmetallation of the Grignard reagent using a copper reagent, which facilitates further coupling:

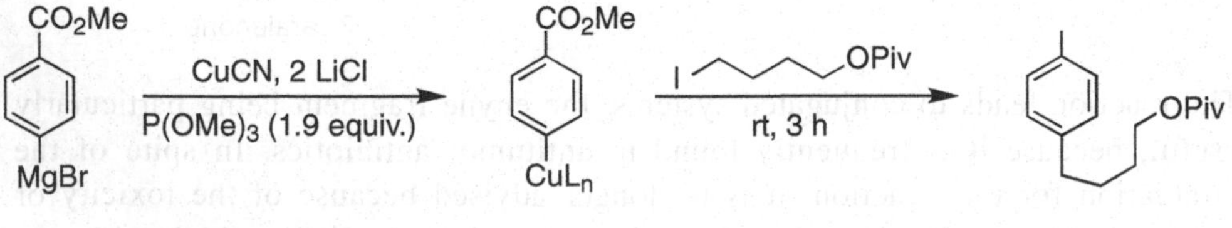
CO2Me
MgBr
CuCN, 2 LiCl
P(OMe)3 (1.9 equiv.)
CO2Me
CuLn
I
OPiv
rt, 3 h
OPiv

3.7. THE HIYAMA REACTION: COUPLING WITH TRIMETHOXYSILANES

The use of silicon-derived compounds for transmetallation [21.28e] would obviously be attractive, because of their low toxicity, low cost and availability. The reaction is known with aryl- and vinyltrimethoxysilane.

$$\boxed{\begin{array}{l} ArX\ +\ RSi(OMe)_3\ \xrightarrow{\ \ Pd^0\ \ }\ Ar\text{-}R\ +\ Si(X)(OMe)_3 \\ X = Br\ or\ Cl,\ R = vinyl\ or\ aryl \end{array}}$$

It is not common, however, probably because it requires high catalyst loading and yields both hetero- and homocoupled products. The amount of homocoupled product can be decreased by using excess (such as 3/1) $PhSi(OMe)_3$ and reduced temperature (60°C). For instance, the bulky NHC (above) was used to couple aryl- and heteroarylbromides in high yields without formation of homocoupled arene product. Even activated chloroarenes such as *para*-chloroacetophenone and *para*-chlorobenzonitrile could be coupled. Vinyltrimethoxysilane also quantitatively yields the substituted styrene upon reaction with *para*-chloroacetophenone for 18 h at 80°C in 1,4-dioxane/THF in the presence of TBAF.[21.28f]

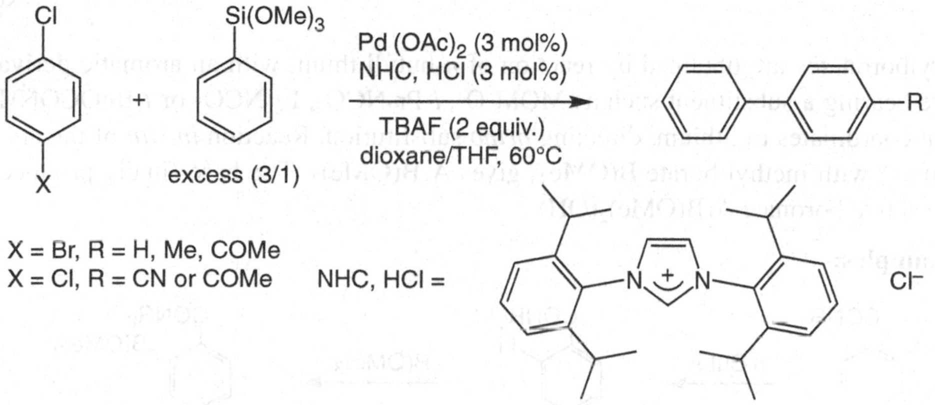

X = Br, R = H, Me, COMe
X = Cl, R = CN or COMe   NHC, HCl =

3.8. *THE SUZUKI REACTION: COUPLING WITH BORON REAGENTS*

The Suzuki reaction allows the coupling of vinyl and arylboranes, *via* boronates, with vinyl or aryl halides, using a Pd^0 catalyst.[21.29]

$$\boxed{\begin{array}{l} RI\ +\ R'BR''_2\ \xrightarrow[\text{base}]{\ \ Pd^0\ \ }\ R\text{-}R'\ +\ BR''_2I \\ R,\ R' = vinyl\ or\ aryl;\ R'' = OH,\ OR'''\ or\ alkyl\ (9\text{-}BBN) \end{array}}$$

This reaction is far-reaching, because a wide variety of air- and thermally stable organoboron reagents are commercially available or easily synthesized,[21.29b] and it has indeed found extensive use in natural product synthesis.[21.29c] The Suzuki reaction allows the introduction of alkenes, alkynes and arenes in C-C coupling processes, the hydroboration of alkenes and alkynes being very well known. It is the addition of a base in the medium of the Suzuki reaction that leads to the transformation of a borane BR_3 into a boronate $BR_3(OH)^-$ that is the active species for transmetallation towards Pd:

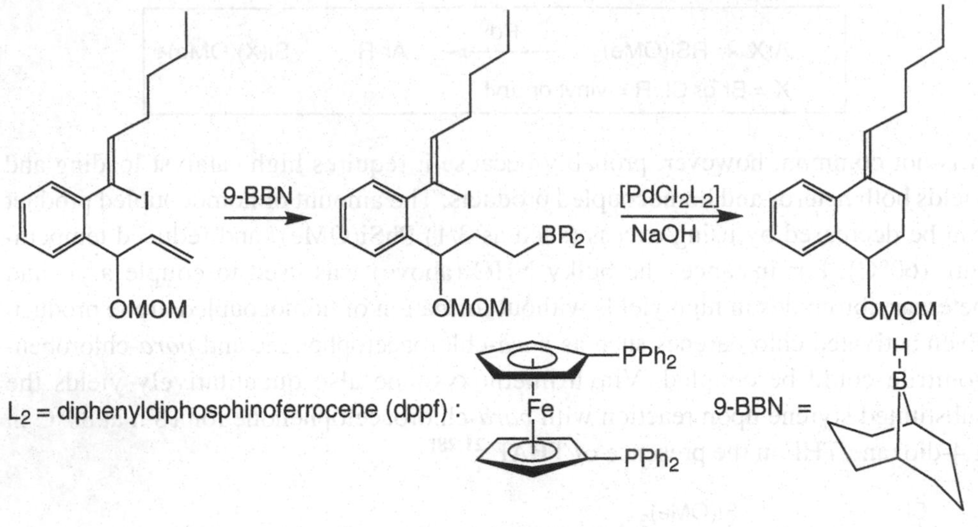

L₂ = diphenyldiphosphinoferrocene (dppf) :

9-BBN =

Arylboronates are obtained by reaction of *n*-butyllithium with an aromatic derivative bearing a substituent such as MOM-O-, *i*-Pr₂NCO-, Et₂NCO- or *t*-BuOCONH- that coordinates to lithium, directing *ortho* substitution. Reaction *in situ* of the aryllithium with methyl borate $B(OMe)_3$ gives $ArB(OMe)_2$. The base finally produces the active boronate $ArB(OMe)_2(OH)^-$.

Examples:

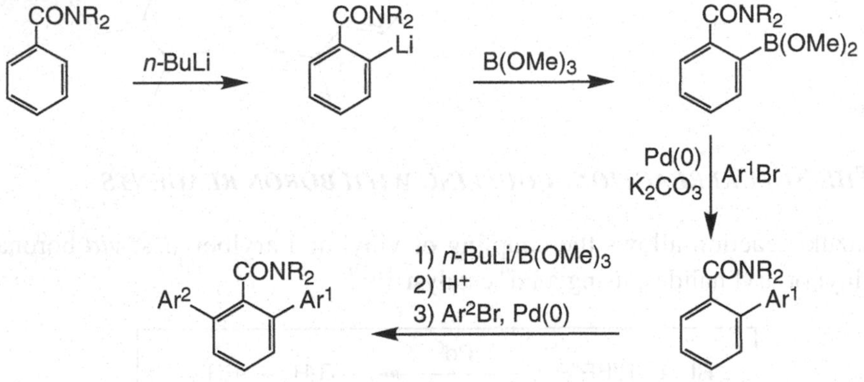

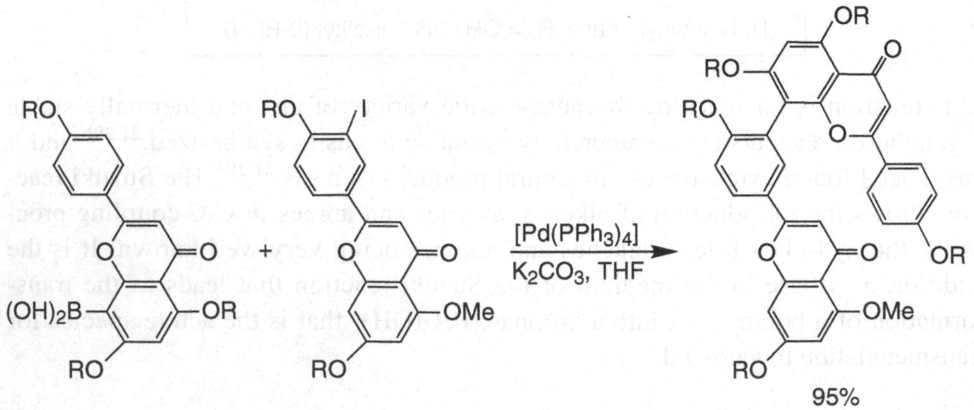

95%

The above examples are restricted to aryl iodides for which oxidative addition on Pd^0 is much more facile than with aryl chlorides (the most difficult to activate among the aryl halides) or even aryl bromides. Hartwig, Buchwald and Fu [21.29d] (*inter alia*) have shown in the late 1990s, however, that it is also possible to couple, for instance, bulky chloroarenes with $PhB(OH)_2$ and even bulky $ArB(OH)_2$ by using specially designed Pd^0 ligands:

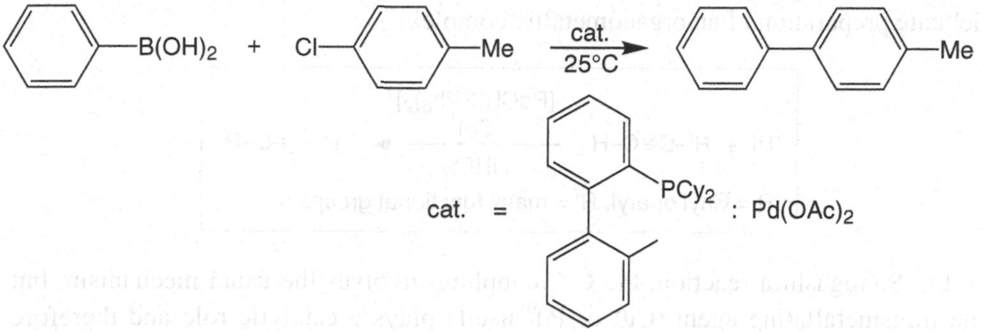

The use of dendritic diphosphines allows to recover and re-use the catalyst several times.[21.30a] The use of phosphines, however, can be avoided using Pd nanoparticles under mild conditions, or again NHC ligands.[21.30b] For instance, loading only with $Pd(OAc)_2$, bis-mesityl NHC, HCl and Cs_2CO_3, adding 1,4-dioxane and heating to 80°C for 30 min followed by cooling, addition of substrates, then cooling for a further 2 h afforded complete conversion for 4-chlorotoluene and phenylboronic acid. The best ligand/Pd ratio being 1/1, the catalytically active species is proposed to be 3-coordinate or 4-coordinate if solvent stabilization or a halide bridged dimer is involved. The activation period seems to be required for reaction of the base with the Pd^{II} salt and bis-mesityl NHC, HCl, generating the active NHC-Pd^0 species. A large variety of *para*-substituted functional aryl chlorides and triflates (easily accessible from phenols) were coupled with phenyl or *p*-anisyl boronic acid using this system. This procedure is excellent for pharmaceutical targets such as Fenbufen, an anti-inflammatory drug possessing analgesic properties, which acts by inhibition of prostaglandin synthesis:

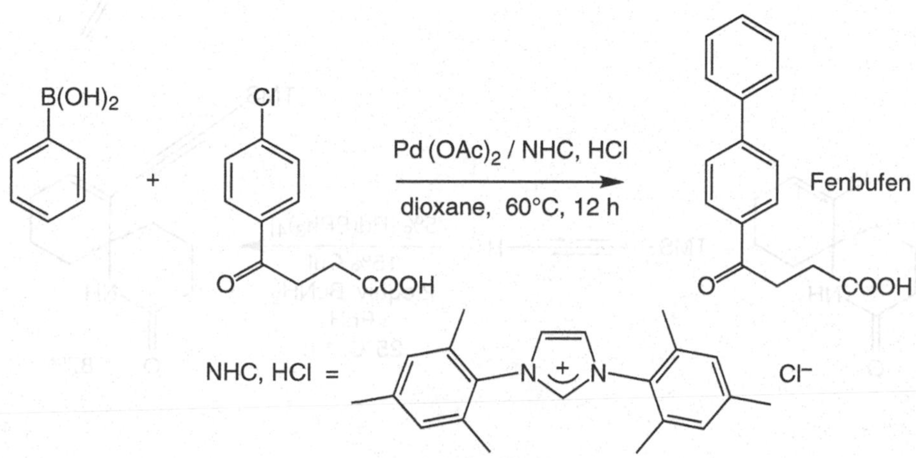

3.9. THE SONOGASHIRA REACTION: COUPLING TERMINAL ALKYNES WITH HALIDES CATALYZED BY Cu^I AND Pd^0 OR Pd^0 ALONE: METALLATION OF ALKYNES IN SITU BEFORE COUPLING

The Sonogashira reaction allows to couple a vinyl or aryl halide with a terminal alkyne.[21.31a-d] It is probably, together with the Heck reaction, the most spectacular among the coupling methods, because neither of these procedures requires the delicate preparation of an organometallic complex.

$$RI + R'\text{–}C{\equiv}C\text{–}H \xrightarrow[\text{NHEt}_2]{\substack{[PdCl_2(PPh_3)_2] \\ CuI}} R\text{–}C{\equiv}C\text{–}R'$$

R = vinyl or aryl; R' = many functional groups

In the Sonogashira reaction, the C-C coupling involves the usual mechanism, but the transmetallating agent (CuI or Pd^0 itself) plays a catalytic role and therefore needs not be isolated. Some electron-rich, bulky diphosphine Pd catalysts including dendritic ones can selectively couple even activated chloroarenes with phenylacetylene in the absence of copper cocatalyst and be recovered and re-used. Applications of these latter catalysts to star-shaped molecules are known.[21.31b] NHCs allow the activation of bromoarenes even in the absence of Cu cocatalyst with trimethylsilylacetylenes instead of terminal alkynes in order to avoid homocoupling. Application of rods made in this way and converted to molecular materials is of interest [21.31c] (see example p. 509 [21.31d]).

Mechanism and examples:

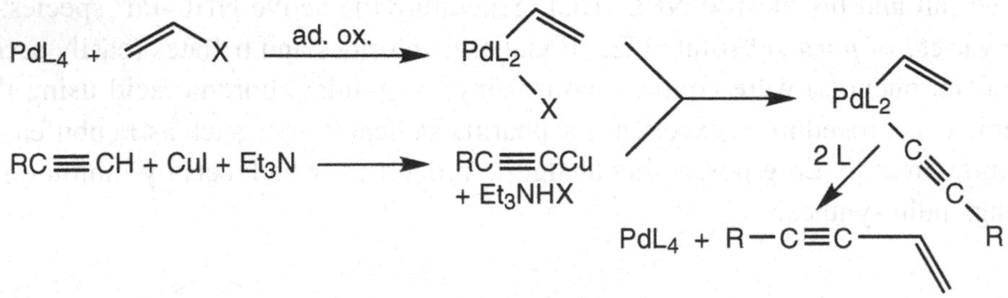

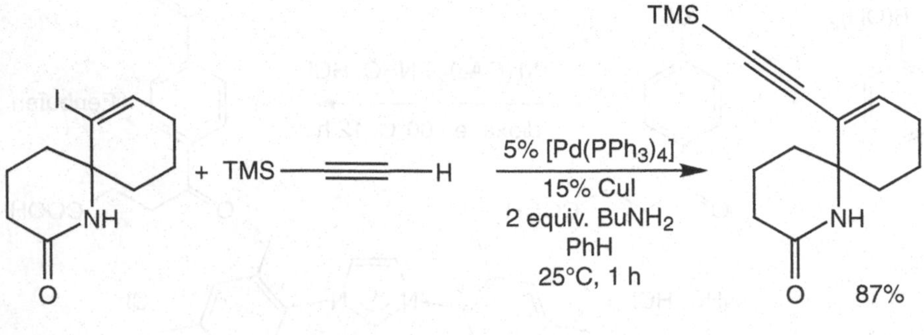

Oligophenylacetylenes can be synthesized using the Sonogashira reaction from iodoaromatics: [21.31d]

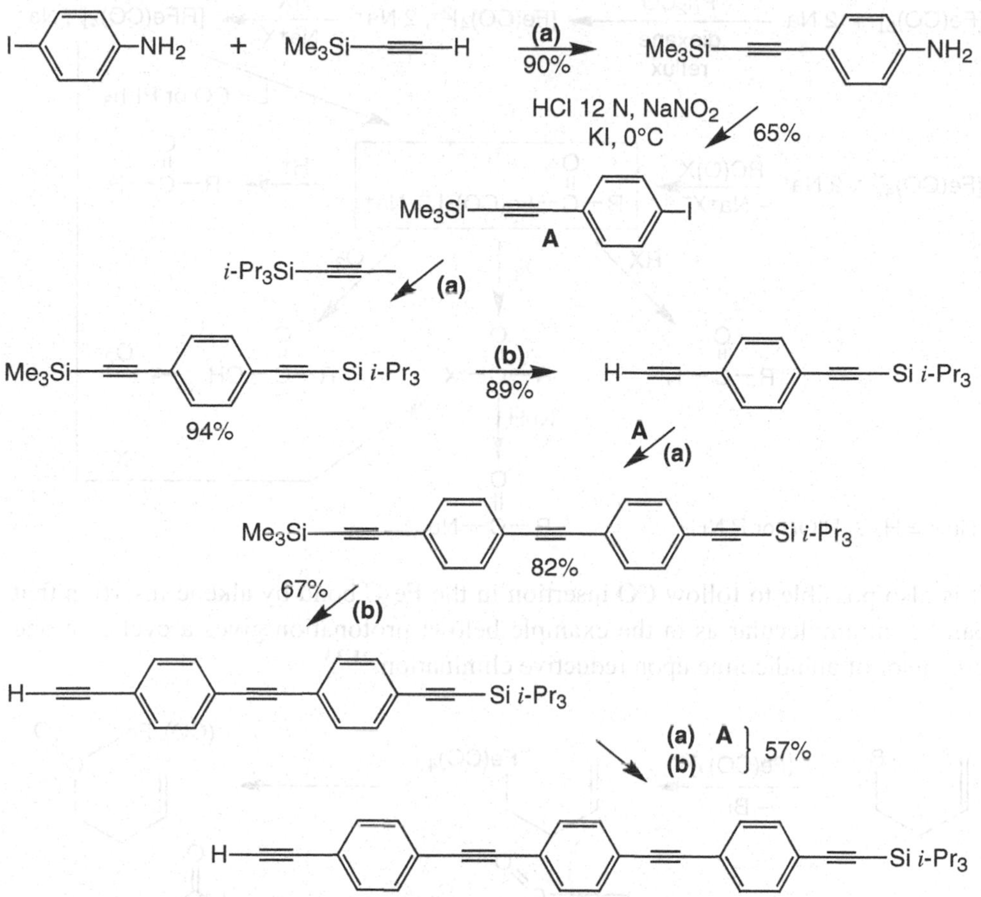

(a) - coupling: [PdCl₂(PPh₃)] (1 mol%), CuI (1 mol%), NEt₃ (solvent), RT, 18 h
(b) - deprotection: aqueous NaOH M in a THF/EtOH: 1/1 mixture, RT, 30 min

3.10. COLLMAN'S REAGENT Na₂Fe(CO)₄: COUPLING WITH CO INSERTION

CO insertion into a metal-alkyl bond, introduced in Chap. 6, is a well-known and useful method used to introduce a carbonyl into an organic substrate, for example, stoichiometrically with Collman's reagent[21.32] or catalytically with palladium acetate and PPh₃.[21.16,21.17]

Collman's reagent being a dianion, it can undergo two oxidative additions with organic halides: one before CO insertion, and the second after. This second reaction with an electrophile leads to an instable neutral Fe^{II} complex that undergoes fast (irreversible) reductive elimination of the coupled organic product. The large variety of organic electrophiles allows to obtain a whole range of functional organic compounds in this way:

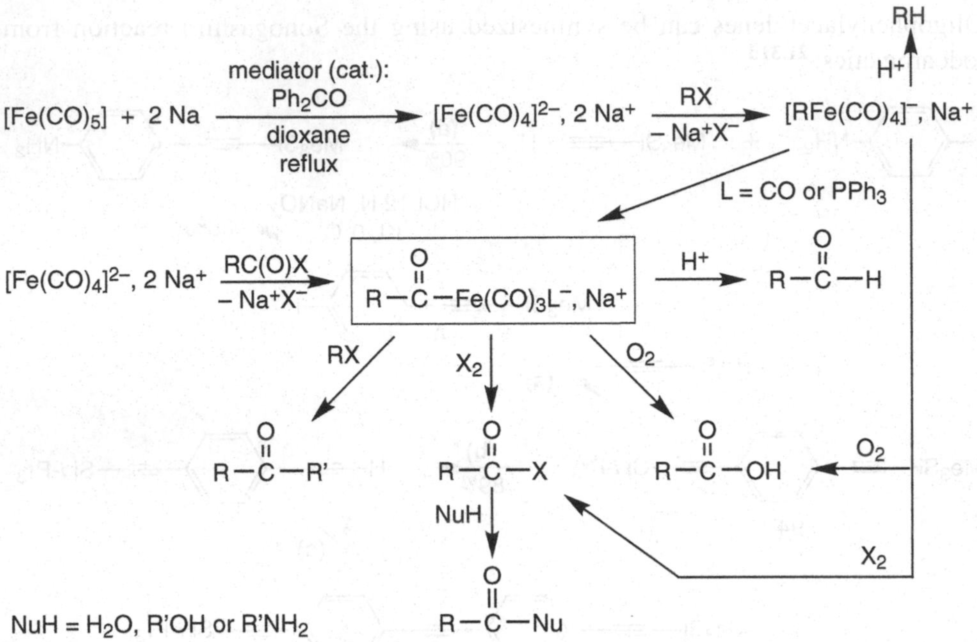

$[Fe(CO)_5]$ + 2 Na $\xrightarrow[\substack{\text{dioxane} \\ \text{reflux}}]{\substack{\text{mediator (cat.):} \\ Ph_2CO}}$ $[Fe(CO)_4]^{2-}$, 2 Na$^+$ $\xrightarrow[- Na^+X^-]{RX}$ $[RFe(CO)_4]^-$, Na$^+$

RH

H$^+$

L = CO or PPh$_3$

$[Fe(CO)_4]^{2-}$, 2 Na$^+$ $\xrightarrow[- Na^+X^-]{RC(O)X}$ $\boxed{\substack{O \\ \| \\ R-C-Fe(CO)_3L^-, \text{ Na}^+}}$ $\xrightarrow{H^+}$ $\substack{O \\ \| \\ R-C-H}$

RX $\swarrow$ X$_2$ $\downarrow$ O$_2$ $\searrow$

$\substack{O \\ \| \\ R-C-R'}$ $\substack{O \\ \| \\ R-C-X}$ $\substack{O \\ \| \\ R-C-OH}$ $\xleftarrow{O_2}$

NuH $\downarrow$ X$_2$

NuH = H$_2$O, R'OH or R'NH$_2$ $\substack{O \\ \| \\ R-C-Nu}$

It is also possible to follow CO insertion in the Fe-C bond by alkene insertion that can be intramolecular as in the example below; protonation gives a cyclic ketone precursor of aphidicoline upon reductive elimination.[21.33]

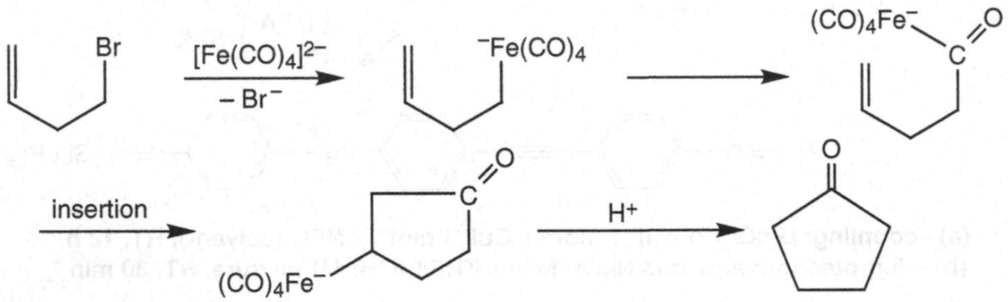

$\xrightarrow[- Br^-]{[Fe(CO)_4]^{2-}}$

$\xrightarrow{\text{insertion}}$

$\xrightarrow{H^+}$

Example:

CH$_2$OTs

30%

HO   CH$_2$OH

HO

HO

aphidicoline

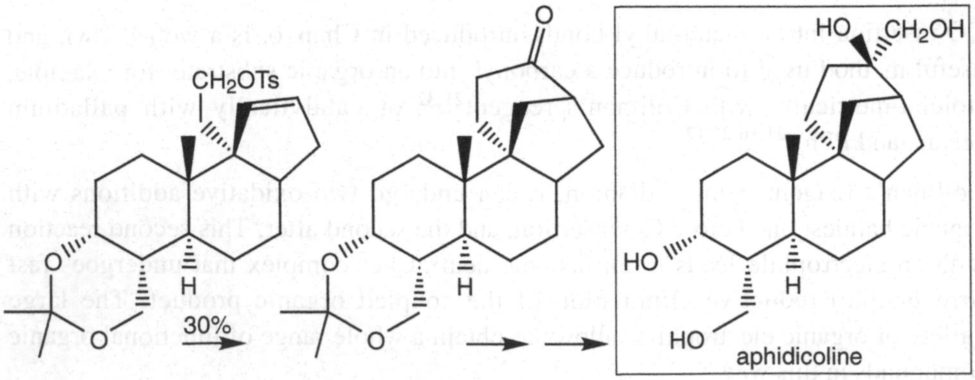

3.11. CARBONYLATION OF ORGANIC HALIDES CATALYZED BY Pd^0

Subsequent to oxidative addition of an organic halide on Pd^0 and CO insertion into the Pd^{II}-alkyl bond, the intermediate Pd^{II}-acyl species can be trapped by a nucleophile NuH which leads to the carbonylated organic product RC(O)Nu and regenerates Pd^0. This process represents a catalytic version of the carbonylation of organic halides using Collman's reagent:

$$R + CO + NuH + Et_3N \xrightarrow{\ Pd^0\ } RCONu + Et_3NH^+X^-$$

R = vinyl, aryl; NuH = alcohol, amine, etc.; ligands : PPh_3, etc.

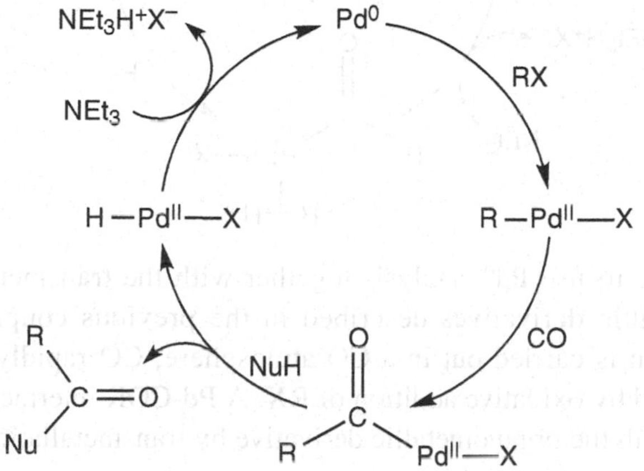

Various nucleophiles work: alcohols, amines and even H_2. Numerous applications have been exploited in organic synthesis, as for the Heck reaction.

Examples (catalyst = $Pd(OAc)_2/PPh_3$; solvent = DMF):

$$RX + CO + R'OH + NEt_3 \longrightarrow RCO_2R' + NHEt_3^+X^-$$

$$RX + CO + R'NH_2 + NEt_3 \longrightarrow RCONHR' + NHEt_3^+X^-$$

$$RX + CO + H_2 + NEt_3 \longrightarrow RC(O)H + NHEt_3^+X^-$$

The mechanisms are not always totally cleared up, in particular concerning the nucleophilic attack. One may envisage an oxidative addition of NuH onto Pd^{II} giving $Pd^{IV}(H)(Nu)$ or a direct nucleophilic attack onto the carbon atom bonded to Pd in the intermediate Pd^{II}-acyl species.

The intramolecular versions are especially efficient and very useful. In the following example, even the oxygen atom of a conveniently located enolate plays the role of the nucleophile for the synthesis of isocoumarines: [21.34]

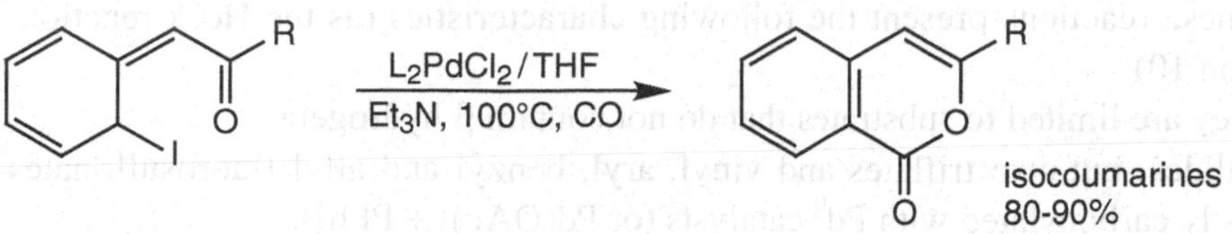

isocoumarines
80-90%

In the case of H_2, a Pd η^2-H_2 complex is a probable intermediate, and it could be deprotonated by NEt_3 to give $NHEt_3^+X^-$ and a palladium-hydride-acyl precursor of the aldehyde:

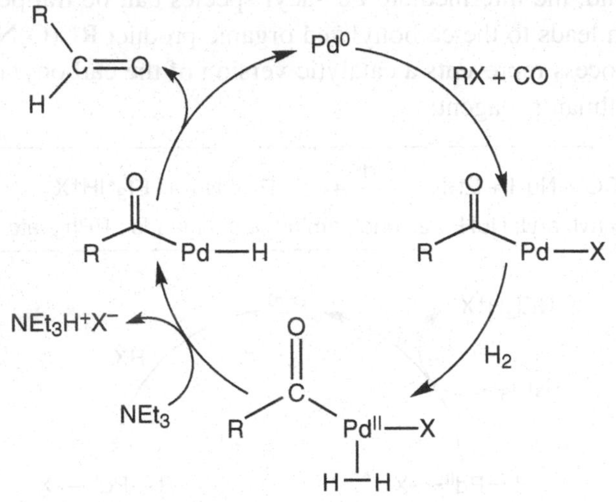

It is also possible to use Pd^0 catalysis together with the transmetallation systems from organometallic derivatives described in the previous coupling procedures. When the reaction is carried out in a CO atmosphere, CO rapidly inserts into the Pd-R bond formed by oxidative addition of RX. A Pd-COR intermediate is then formed and reacts with the organometallic derivative by transmetallation: [21.16,21.17,21.34]

$$RX + CO + MR' \xrightarrow{Pd^0} RCOR' + MX$$

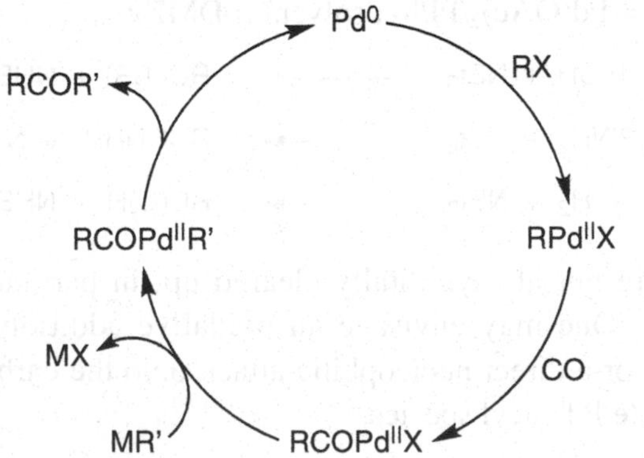

There are also other ways to trap this Pd-acyl intermediate; they are indicated at the end of the following section devoted to the very richly studied Heck reaction.

All these reactions present the following characteristics (as the Heck reaction, see section 10):

‣ They are limited to substrates that do not contain β hydrogens.
‣ Halides, but also triflates and vinyl, aryl, benzyl and allyl fluorosulfonates are easily carbonylated with Pd^0 catalysts (or $Pd(OAc)_2 + PPh_3$).

▸ In the reaction medium, Pd^{II} is reduced to Pd^0.

▸ NEt_3 used in order to neutralize HX also probably plays the role of a reducing agent of Pd^{II} to Pd^0. The reduction potential being low, the single-electron-transfer mechanism to Pd^{II} is a viable pathway. Another possible reduction route is β-elimination subsequent to coordination:

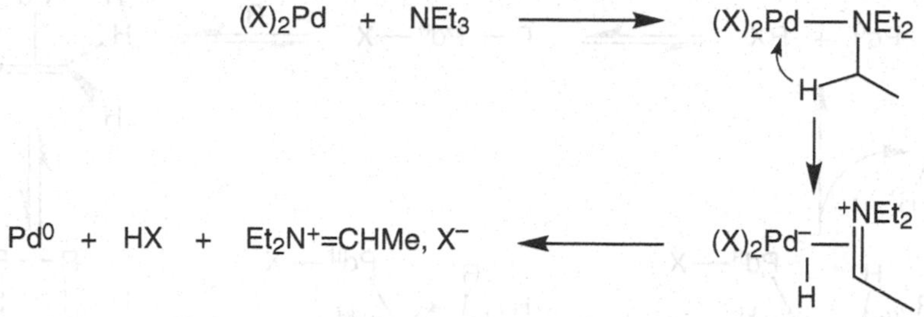

$$(X)_2Pd \; + \; NEt_3 \; \longrightarrow \; (X)_2Pd\text{---}NEt_2$$

$$Pd^0 \; + \; HX \; + \; Et_2N^+{=}CHMe, \; X^- \; \longleftarrow \; (X)_2Pd^-\text{---}\overset{+NEt_2}{\underset{H}{|}}$$

3.12. HECK REACTION: CATALYSIS BY Pd^0 OF THE COUPLING OF AN ORGANIC HALIDE WITH AN OLEFIN BY OLEFIN INSERTION

The Heck reaction allows to couple a vinyl, aryl or benzyl halide with an olefin using Pd^0 catalysis.[21.1,21.16] This is the best known among the C-C coupling reactions. It is of extensive use, especially because it tolerates a large variety of functional groups. As the Sonogashira reaction, it does not require the isolation of an organometallic derivative RM, which is an enormous advantage:

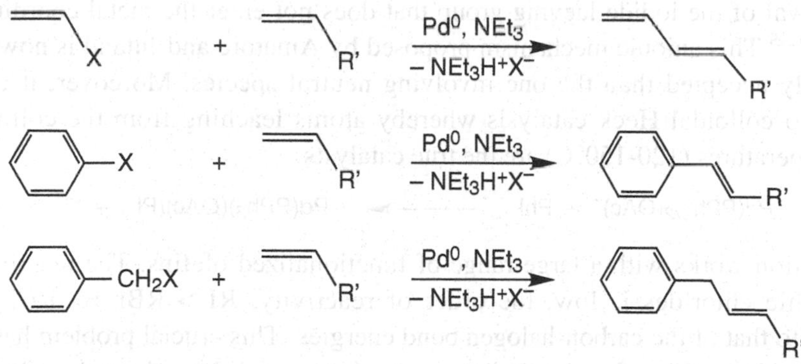

The Heck reaction is different from all the other Pd^0-catalyzed coupling reactions, because it involves insertion after the usual oxidative addtion step. In addition, it ends up by a β-elimination, whereas the last step of all other Pd^0-catalyzed coupling reactions is a reductive elimination.

The RX substrate must not contain a flexible β C-H bond so that the metal-alkyl species resulting from its oxidative addition be prevented from undergoing β-elimination before the insertion step (which would result in an overall dehydrohalogenation of RX). Thus, the R group must be a vinyl, aryl (the β position is not flexible in these groups) or benzyl radical. The reaction is carried out upon heating RX (organic halide or triflate) with the alkene, a catalytic amount of Pd^{II} and an excess of tertiary amine in acetonitrile.

The use of mild conditions is possible with more polar solvents such as DMF and in the presence of quaternary ammonium chloride in order to keep a high chloride concentration, inhibiting Pd^0 precipitation. The insertion and β-elimination occur with *cis* stereochemistry. The classic mechanism is the following:

$$Pd^0 + RX \rightleftharpoons R-Pd^0-X \rightleftharpoons \cdots$$

The nature of the ligands on the catalytically active Pd species varies with the conditions and nature of the substrates used at the beginning of the reaction. The presence of an anionic ligand on Pd^0 such as halide or acetate favors the oxidative addition by increasing the electron density at the metal center. For example, Amatore and Jutand have shown that if one starts from $Pd(OAc)_2$ in the presence of PPh_3, the active species is the anion $Pd(PPh_3)_2(OAc)^-$. The oxidative addition of iodobenzene then follows a SN_2 mechanism, i.e. the formation of the Pd-Ph bond goes along with the removal of the iodide laeving group that does not enter the metal coordination sphere. [21.35] This anionic mechanism proposed by Amatore and Jutand is now more commonly accepted than the one involving neutral species. Moreover, it can be applied to colloidal Heck catalysis whereby atoms leaching from the colloids at high temperatures (120-150°C) are the true catalysts.

$$Pd(PPh_3)_2(OAc)^- + PhI \longrightarrow Pd(PPh_3)(OAc)(Ph) + I^-$$

The reaction works with a large range of functionalized olefins. The reactivity of the organic chlorides is low, the order of reactivity: RI > RBr >> RCl being opposite to that of the carbon-halogen bond energies. This crucial problem has been recently addressed by chemists using some electron-rich ligands such as $P(t\text{-Bu})_3$ and some diphosphines.

The regioselectivity of the reaction depends in particular on the olefin substituent (steric effects favoring the formation of the C-C bond on the non-substituted carbon as shown on the preceding page), the nature of the leaving group X in RX and ligands on palladium. For example, in the presence of the cation Ag^+ or Tl^+ capable of abstracting the anion X^-, a cationic catalytically active intermediate $Pd(PPh_3)_2$ $(R)(\eta^2\text{-olefin})^+$ is formed, which enhances the electronic effects of the substituent. Thus, the electron-withdrawing substituents (CN, CO_2R, $CONH_2$) lead to the regioselectivity indicated above. On the contrary, with subtituents that have a donor effect such as OBu, the electronic effects dominate over the steric effects, reversing the regioselectivity that becomes in favor of the C-C bond formation on the substituted olefin carbon. It is even possible to make this regioselectivity specific with chelating diphosphine ligands such as $PPh_2(CH_2)_3PPh_2$.

Thousands of examples of the Heck reaction are known in organic synthesis with organic iodides and bromides, the most useful ones being those resulting from intramolecular reactions.[21.36]

Example: [21.37]

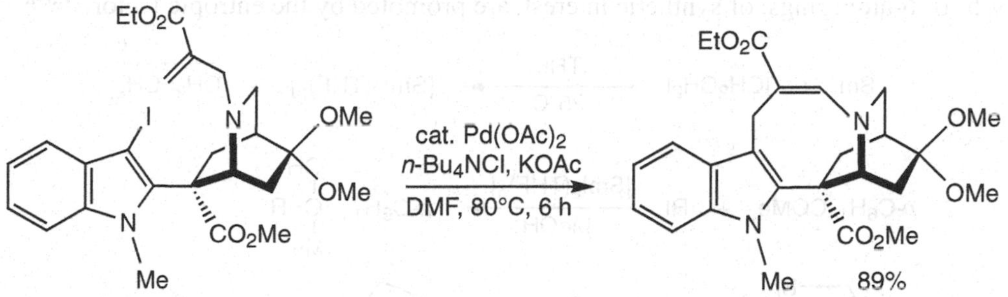

It is possible to achieve several successive olefin or CO insertions before the last step, the β-elimination, of the Heck reaction. The following example is particularly remarkable: [21.38a]

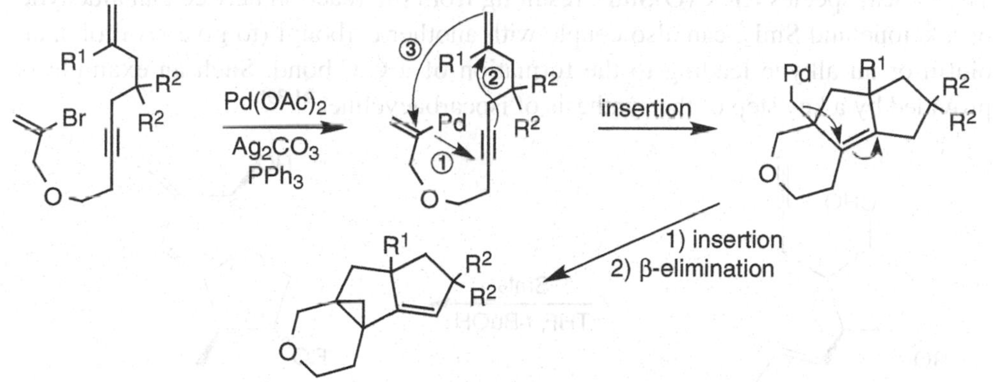

In the presence of double and triple C-C bonds in the same molecule, the triple bond usually inserts before the double bond. Both insertions have about the same energy, but alkyne insertion is entropically favored. If only a triple bond is intramolecularly inserted, the resulting vinyl-palladium species cannot β-eliminate. It can be trapped by transmetallation as in the preceding methods or by a nucleophile:[21.38b]

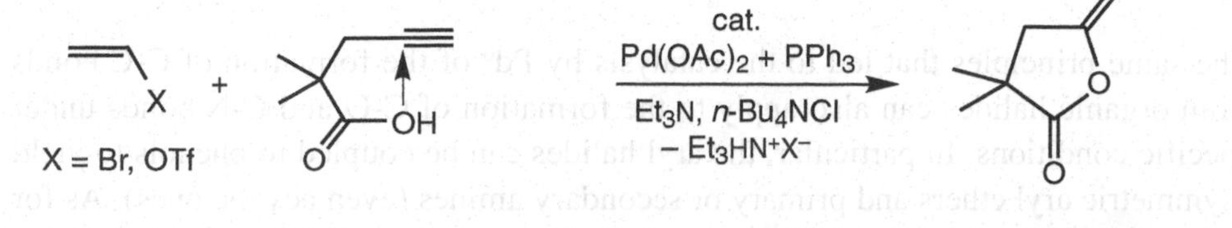

In conclusion, the Heck reaction and its derived versions are very efficient and of great potential in organic synthesis due to the multiple variations and the ease of implementation.

3.13. APPLICATIONS OF KAGAN'S REAGENT, SmI₂

The radicals R• generated by reduction of the halides RX (R = alkyl or aryl; X = Cl, Br, I, etc.) by SmI$_2$ (see Chap. 12.4.2) can be reduced (formation of RH by trapping an H atom from the solvent) or couple onto a carbonyl (aldehydes, ketones: Barbier reaction), an olefin or an alkyne. For coupling reactions, the competitive reduction to RH indicated above must be avoided. The intramolecular coupling with formation of 5- or 6-atom rings, of synthetic interest, are promoted by the entropic factor.[21.39a]

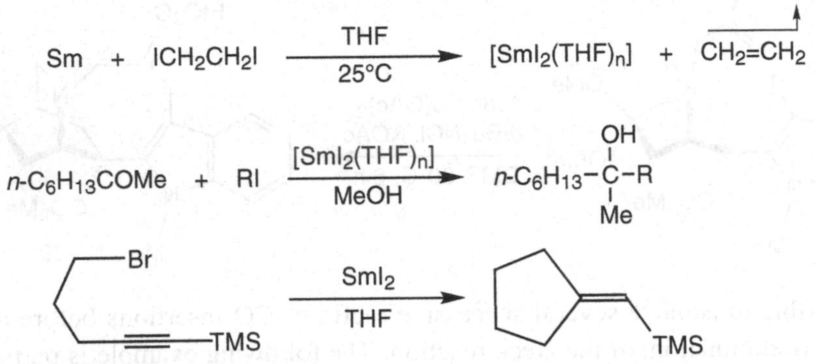

$$Sm \ + \ ICH_2CH_2I \ \xrightarrow[25°C]{THF} \ [SmI_2(THF)_n] \ + \ CH_2{=}CH_2 \ \longrightarrow$$

$$n\text{-}C_6H_{13}COMe \ + \ RI \ \xrightarrow[MeOH]{[SmI_2(THF)_n]} \ n\text{-}C_6H_{13}{-}\underset{Me}{\overset{OH}{\underset{|}{\overset{|}{C}}}}{-}R$$

The radical species RR'C(O)SmI$_2$, resulting from the reaction between an aldehyde or a ketone and SmI$_2$, can also couple with another carbonyl (to give *cis*-diols), an olefin or an alkyne leading to the formation of a C-C bond. Such an example is provided by a key step of the synthesis of isocarbacycline: [21.39b]

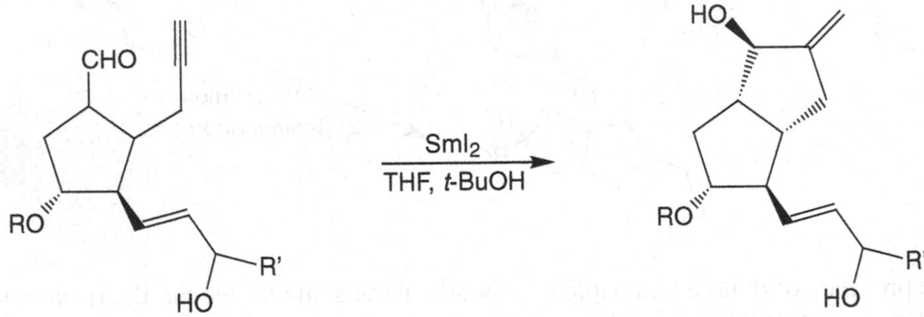

precursor of isocarbacycline

4. EXTENSION OF PALLADIUM CATALYSIS
TO THE FORMATION OF C-O AND C-N BONDS

The same principles that led to the catalysis by Pd^0 of the formation of C-C bonds from organic halides can also apply to the formation of C-O and C-N bonds under specific conditions. In particular, the aryl halides can be coupled to phenols to yield asymmetric aryl ethers and primary or secondary amines (even acyclic ones). As for the catalytic formation of C-C bonds, the nature of the halide in aryl halides is crucial, the aryl chlorides being the most difficult to activate. The use of particularly electron-rich and bulky phosphines favors Pd^0 catalysis and allows the use of various aryl chlorides as shown in the pioneering work of the Hartwig [21.40a] and Buchwald groups.[21.40c]

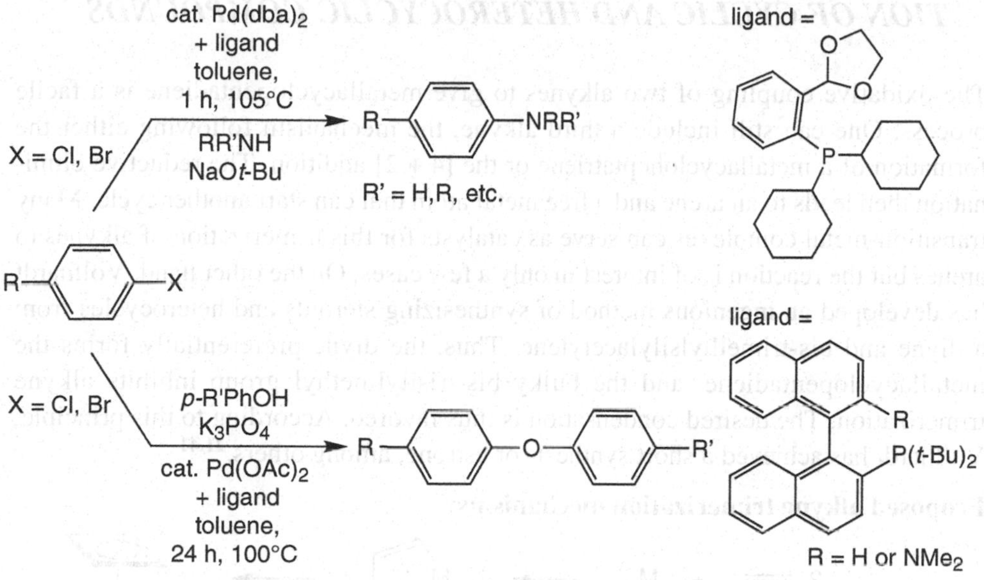

Mechanism:

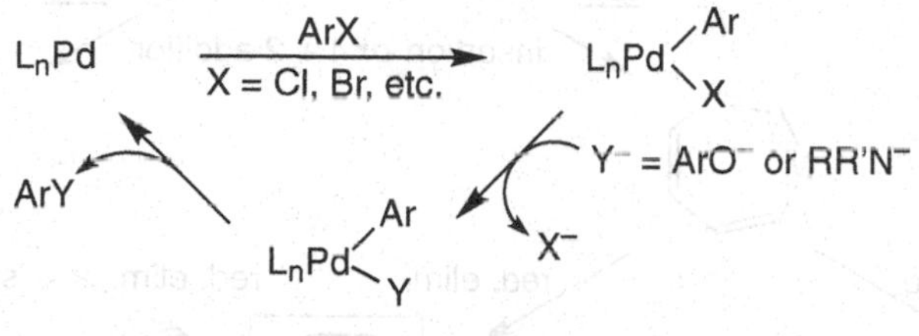

The use of NHCs in 1/1 ligand/Pd ratio here too permitted the catalytic coupling of aryl halides in dioxane (X = Br and I at RT, Cl at 100°C). KOt-Bu was used to deprotonate the imidazolium precursor of NHC and neutralize HX produced in the reaction. Even 4-chlorotoluene, 4-chloroanisole, *ortho*-substituted aryl halides and 2-chloropyridine could be efficiently aminated with primary and secondary alkyl-amines or anilines: [21.26b,21.40d]

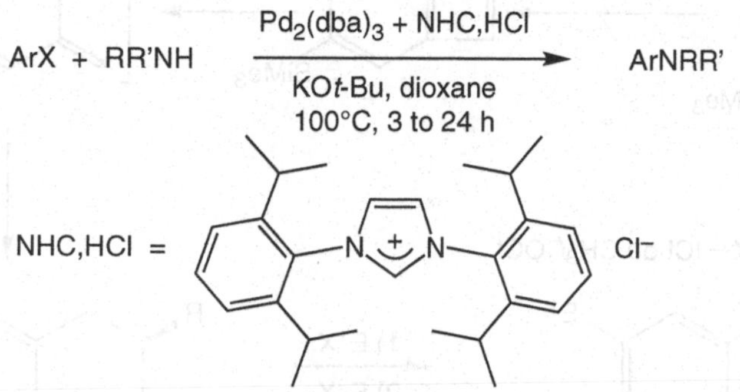

5. OXIDATIVE COUPLING REACTIONS OF ALKYNES WITH OTHER UNSATURATED FRAGMENTS FOR THE FORMATION OF CYCLIC AND HETEROCYCLIC COMPOUNDS

The oxidative coupling of two alkynes to give metallacyclopentadiene is a facile process. One can still include a third alkyne, the mechanism following either the formation of a metallacycloheptatriene or the [4 + 2] addition. The reductive elimination then leads to an arene and a free metal atom that can start another cycle. Many transition-metal complexes can serve as catalysts for this trimerization of alkynes to arenes but the reaction is of interest in only a few cases. On the other hand, Vollhardt has developed an ingenious method of synthesizing steroids and heterocycles from a diyne and bis-trimethylsilylacetylene. Thus, the diyne preferentially forms the metallacyclopentadiene, and the bulky bis-trisilylmethyl group inhibits alkyne trimerization. The desired condensation is thus favored. According to this principle, Vollhardt has achieved a short synthesis of estrone, among others.[21,41]

Proposed alkyne trimerization mechanisms:

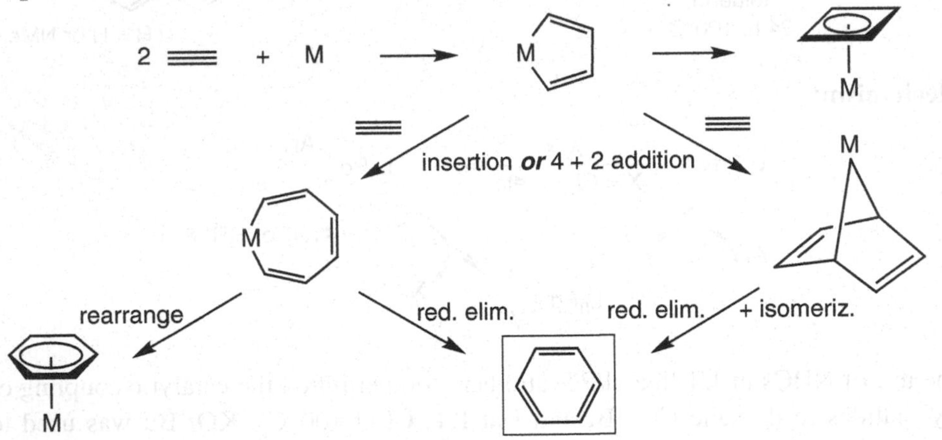

Example:

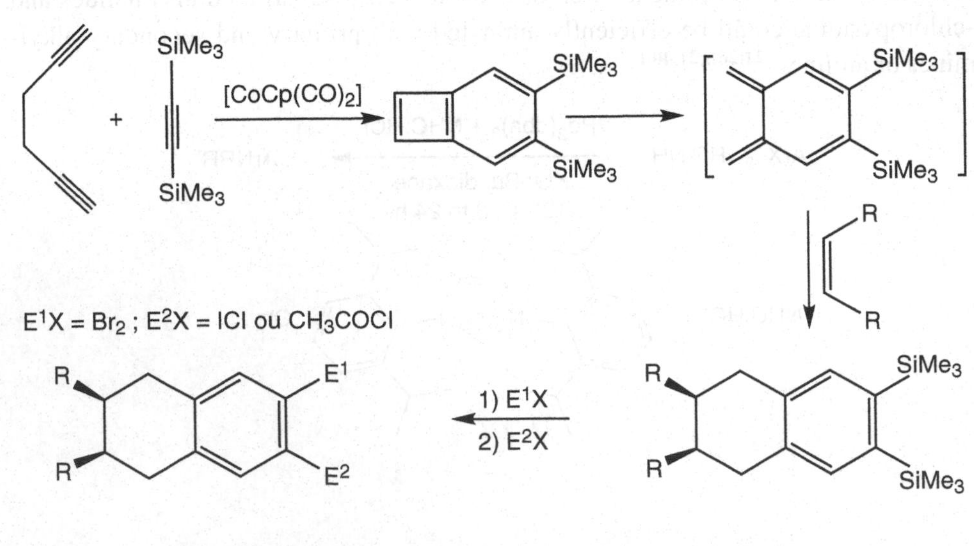

$E^1X = Br_2$; $E^2X = ICl$ ou CH_3COCl

Application to a short synthesis of estrone:

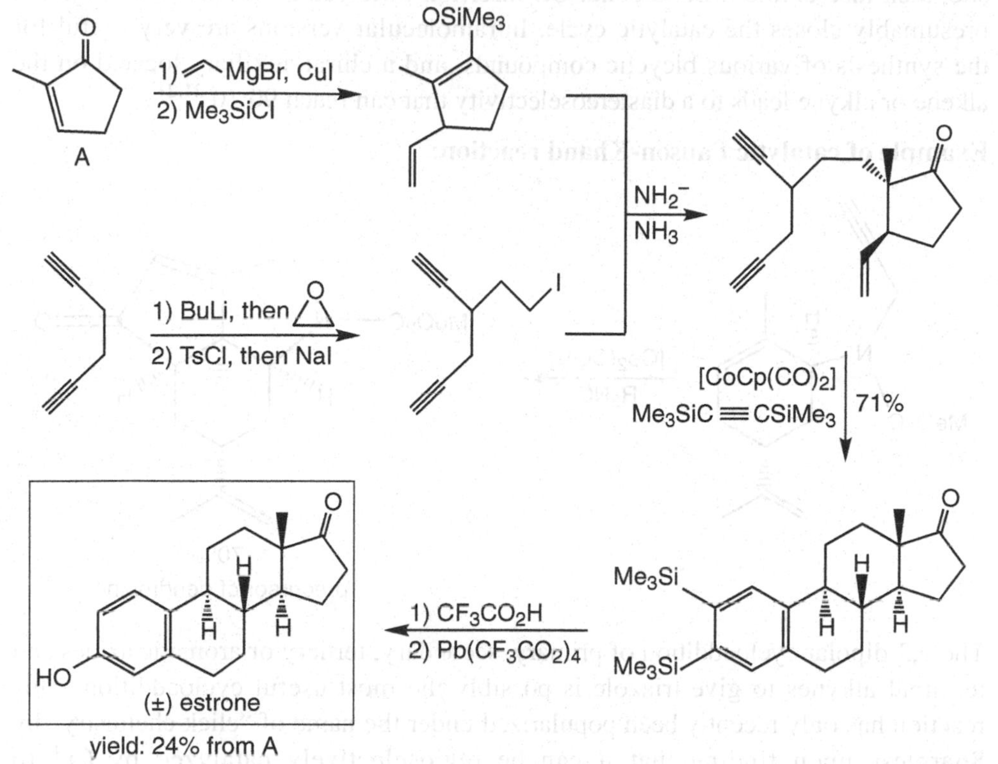

yield: 24% from A

The same condensation can also occur if the third unsaturated molecule is an alkene, a nitrile or an isocyanate, which leads respectively to a cyclohexadiene, a pyridine or a pyridone. There are thus a large variety of synthetic possibilities that have been exploited by Vollhardt and others, sometimes in combination with CO insertion.[21.41]

A related process is the Pauson-Khand reaction [21.42a] that couples an alkyne, an alkene and CO. The alkyne is coordinated to the $Co_2(CO)_6$ fragment in [$Co_2(CO)_6$ (μ^2,η^2-alkyne)] as in the protection procedure, forming the Co_2C_2 tetrahedron.

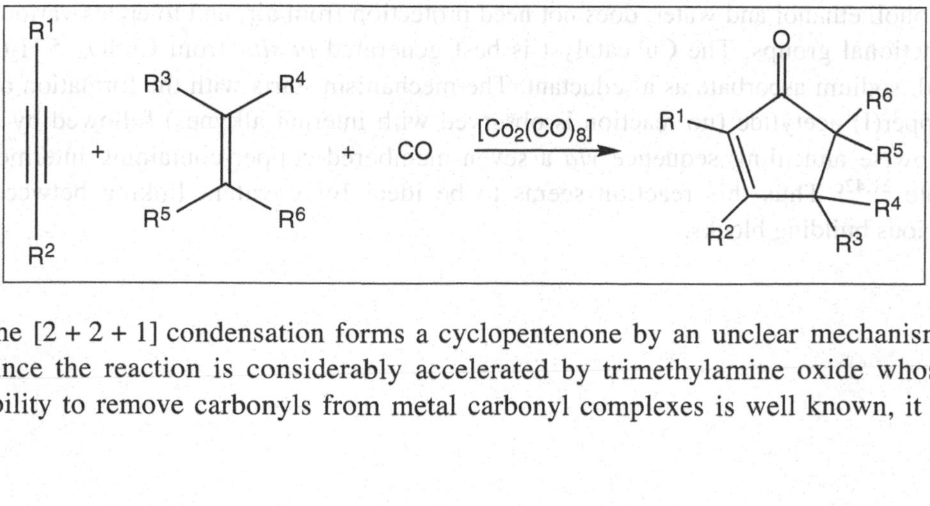

The [2 + 2 + 1] condensation forms a cyclopentenone by an unclear mechanism. Since the reaction is considerably accelerated by trimethylamine oxide whose ability to remove carbonyls from metal carbonyl complexes is well known, it is expected that the alkene coordinates to a metal center to fill the free coordination site, then inserts into a Co-C bond. CO insertion followed by reductive elimination presumably closes the catalytic cycle. Intramolecular versions are very useful for the synthesis of various bicyclic compounds, and a chiral auxiliary located on the alkene or alkyne leads to a diastereoselectivity that can reach 90:10.[21.21]

Example of catalytic Pauson-Khand reaction:

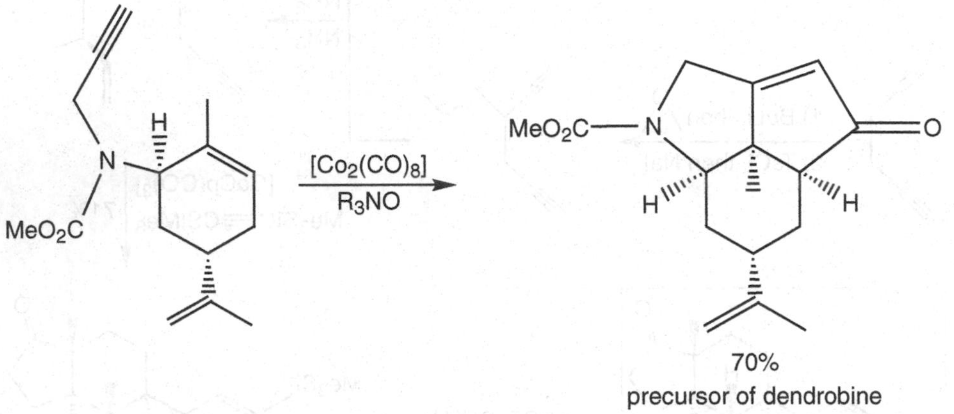

70%
precursor of dendrobine

The 1,3-dipolar cycloaddition of primary, secondary, tertiary or aromatic azides and terminal alkynes to give triazole is possibly the most useful cycloaddition. This reaction has only recently been popularized under the name of "click chemistry" by Sharpless upon finding that it can be regioselectively catalyzed by Cu^I to quantitatively provide only 1,4-disubstituted 1,2,3-triazoles.[21.42b]

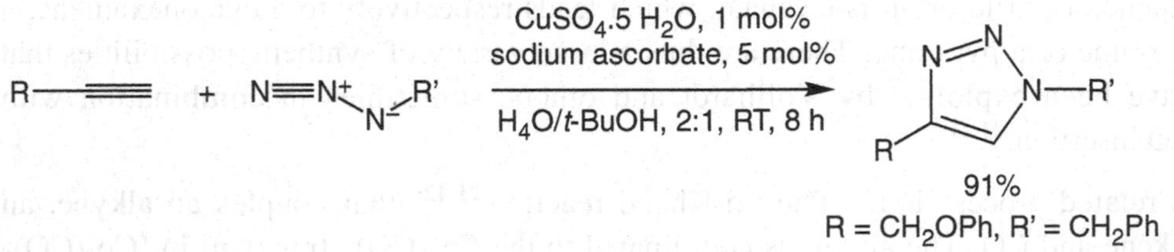

91%
R = CH₂OPh, R' = CH₂Ph

It can be carried out at room temperature in a variety of solvents including *tert*-butyl alcohol, ethanol and water, does not need protection from air, and tolerates various functional groups. The Cu^I catalyst is best generated *in situ* from $CuSO_4 \cdot 5\,H_2O$ with sodium ascorbate as a reductant. The mechanism starts with the formation of copper(I) acetylide (no reaction is observed with internal alkynes) followed by a stepwise annealing sequence *via* a seven-membered copper-containing intermediate.[21.42c] Thus this reaction seems to be ideal for covalent linking between various building blocks.

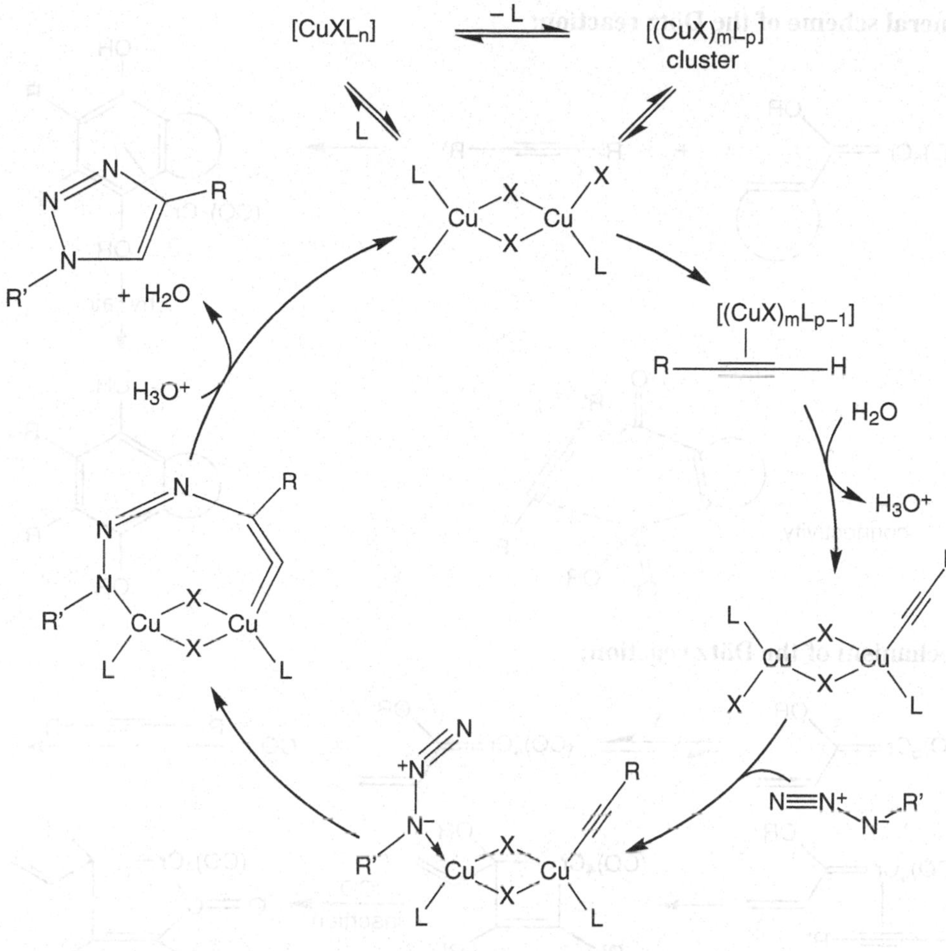

**Proposed mechanism for the Cu^I-catalyzed cycloaddition
of an azide and a terminal alkyne (click reaction)** [21.42c]

6. METAL-CARBENE COMPLEXES IN ORGANIC SYNTHESIS

6.1. APPLICATIONS OF FISCHER-TYPE METAL CARBENE COMPLEXES:
THE DÖTZ REACTION

Among the many reactions of Fischer-type metal-carbene complexes[21.43] (see Chap. 9), the most remarkable and most applied to organic synthesis is the Dötz reaction.[21.44] This complex reaction involves an unsaturated Fischer-type carbene complex with an alkyne yielding polycyclic hydroquinones. The principle of the scheme, the mechanism and an example [21.45] are represented p. 522.

General scheme of the Dötz reaction:

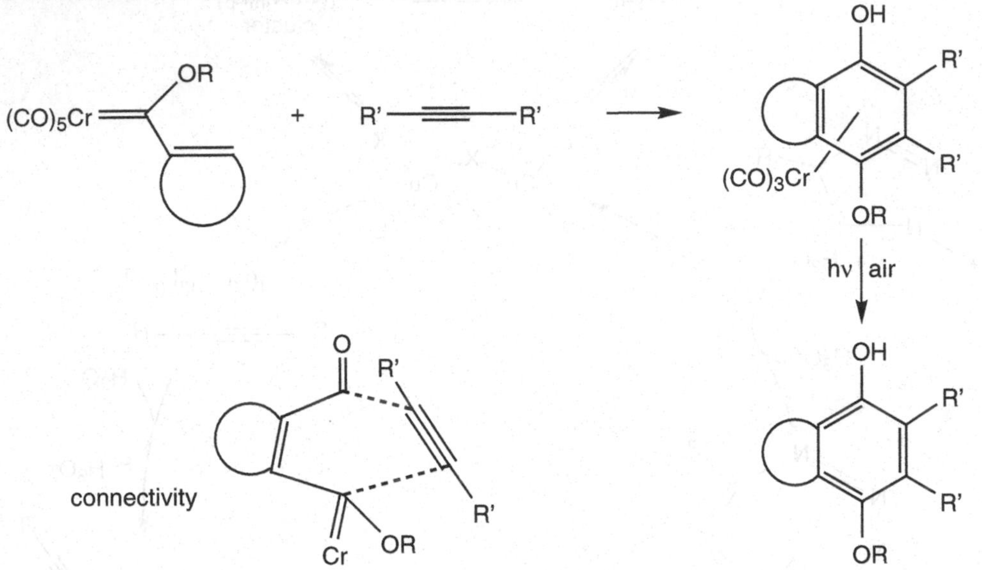

connectivity

Mechanism of the Dötz reaction:

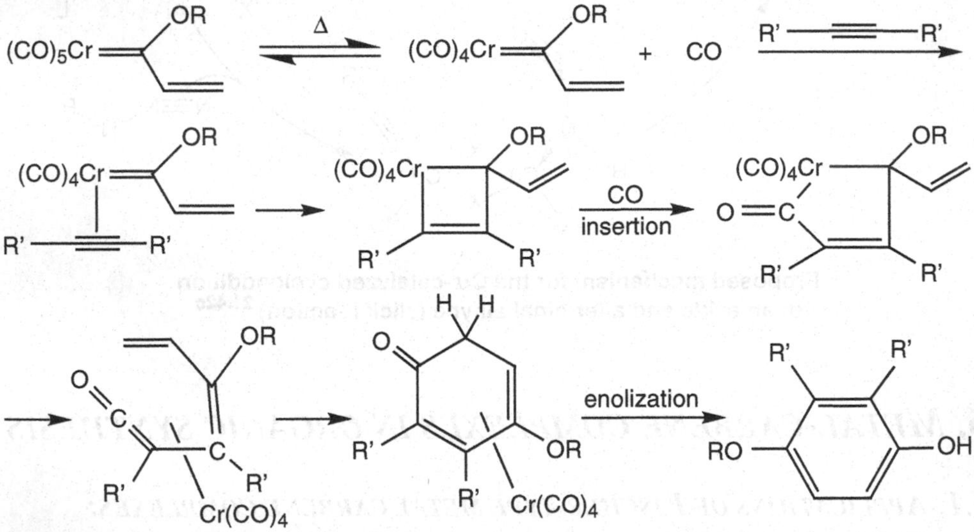

Example of the Dötz reaction:

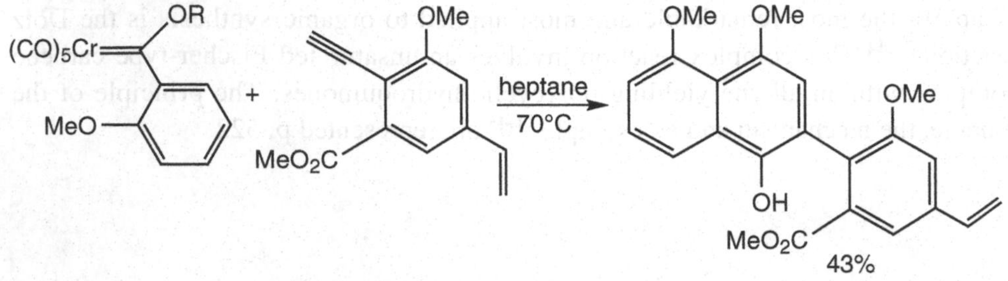

43%

6.2. *METHYLENATION OF CARBONYLS*
USING SCHROCK-TYPE METAL-METHYLENE COMPLEXES

Early transition-metal methylene (and alkylidene) are super-Wittig reagents, as shown by Schrock with the first example of these methylene complexes, $[TaCp_2Me(=CH_2)]$.[21.46] Tebbe's complex, $[TiCp_2(\mu_2\text{-}CH_2)_2AlMe_2]$, is a source of $[TiCp_2(=CH_2)]$; it has been much used because it is simply generated from $[TiCp_2Cl_2]$, $AlMe_3$ and pyridine. It reacts with carbonyl derivatives to give an oxatitanacyclobutane intermediate that yields $[TiCp_2(O)]$ and the methylenated organic derivative.[21.47]

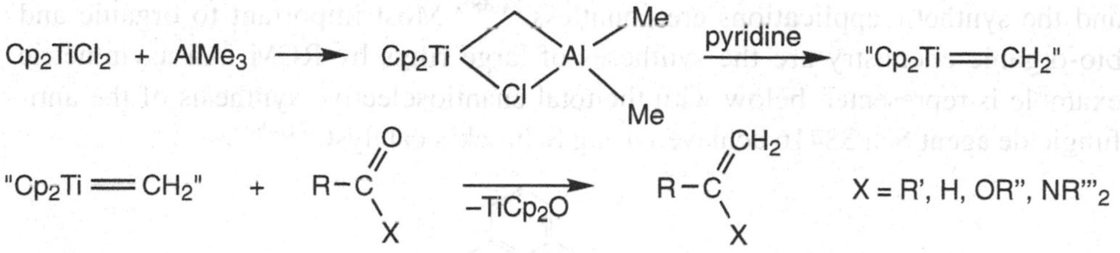

$$Cp_2TiCl_2 + AlMe_3 \longrightarrow Cp_2Ti \underset{Cl}{\overset{Me}{<}} Al \overset{Me}{\underset{Me}{<}} \xrightarrow{\text{pyridine}} \text{"}Cp_2Ti = CH_2\text{"}$$

$$\text{"}Cp_2Ti = CH_2\text{"} + R\overset{O}{\underset{X}{-}C} \xrightarrow{-TiCp_2O} R\overset{CH_2}{\underset{X}{-}C} \qquad X = R', H, OR'', NR'''_2$$

Example:

90%

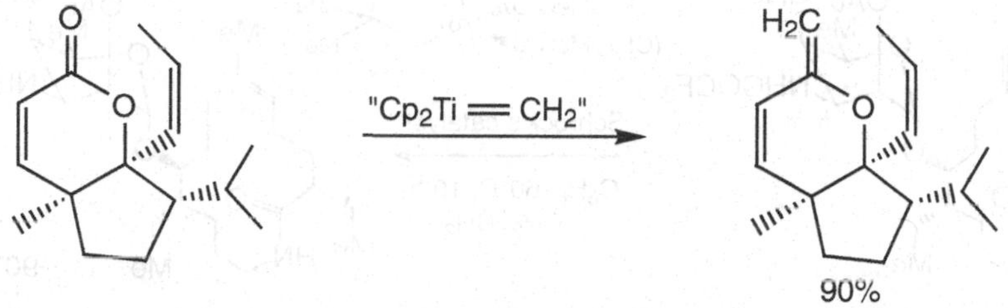

The complex $[TiCp_2Me_2]$, even more simply synthesized from $[TiCp_2Cl_2]$ and MeLi, generates $[TiCp_2(=CH_2)]$ by heating at 60°C. It is a clean methylenation reagent for ketones, aldehydes and esters avoiding the fragmentation of ethers, contrary to Tebbe' complex.[21.48a]

7. *EXAMPLES OF ASYMMETRIC CATALYSIS* [21.30-21.32]

The principle of asymmetric catalysis, developed from the pioneering work of Kagan with asymmetric hydrogenation (see Chap. 15), has now been extended to the overall field of catalysis. The asymmetric hydrogenation of prochiral olefins has itself given rise to various applications in the area of the synthesis of amino acids such as *L*-DOPA (see Chap. 15), antibiotics of the family of carbapenemes and isoquinoline alkaloids such as morphine and benzomorphanes. Asymmetric hydrogenation of prochiral C=O bonds (i.e., those in which the two carbon substituents are different) using Noyiori's catalysts of the type $[RuCl_2BINAP]$ leads to secondary alcohols with e.e.s close to 100%, and has been used in particular for the synthesis of prostaglandins.[21.50] Some representative examples of other reactions are shown here. Research in this field is rapidly expanding given the multiple applications, especially in the pharmaceutical and agricultural area.

7.1. OLEFIN METATHESIS CATALYZED BY MOLYBDENUM-ALKYLIDENE OR RUTHENIUM-BENZYLIDENE COMPLEXES

Some transition-metal-carbene complexes, in particular Schrock's molybdenum alkylidene complexes such as [Mo{CF₃}₂MeCO}₂(=CHCMe₂Ph)(=N-o-iPr₂Ph)], and Grubbs' ruthenium benzylidene complexes such as [RuCl₂(PCy₃)L(=CHPh)] (L = PCy₃ or cyclic bis(mesitylamino)carbene) catalyze various types of olefin metathesis: ring closing metathesis (RCM) and its reverse, ring opening metathesis (ROM), ring opening metathesis polymerization (ROMP), cross-metathesis (CM) and enzyme metathesis (EYM). The mechanistic aspects are discussed in Chap. 15 and the synthetic applications are countless.[21.48b] Most important to organic and bio-organic chemistry are the syntheses of large rings by RCM. An asymmetric example is represented below with the total enantioselective synthesis of the anti-fungicide agent Sch 38416 achieved using Schrock's catalyst:[21.48c]

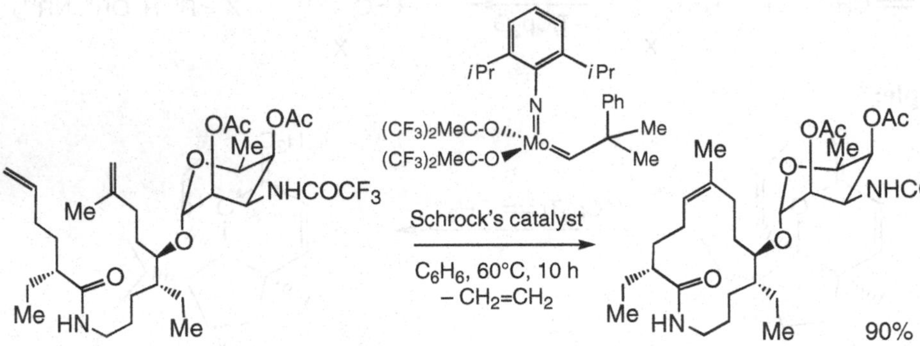

7.2. EPOXIDATION (see Chap. 16)

Optically active (+) or (–) endo- brevicomine is a pheromone representing a popular target whose synthesis has been achieved by various routes. Sharpless' asymmetric epoxidation (see Chap. 16) is one of them:[21.51]

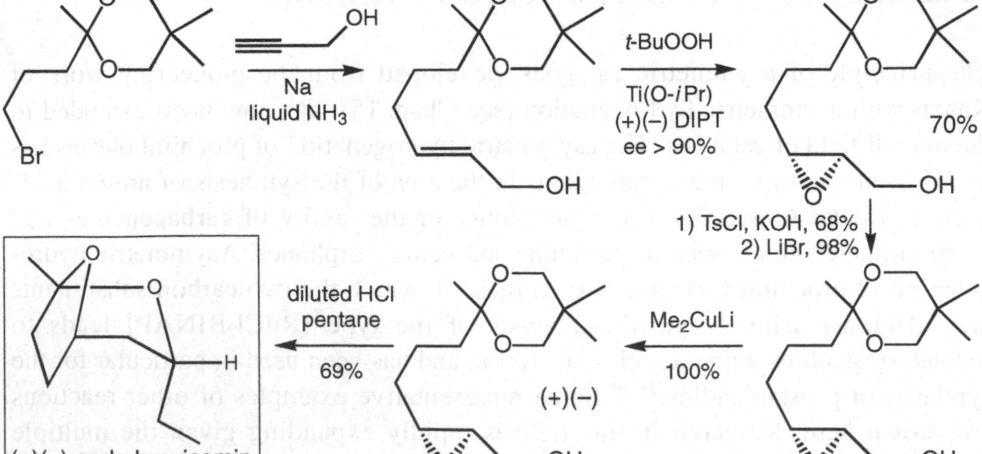

As shown above, the need to have an alcohol functionality in allylic position in order to obtain an efficient catalysis must be paid by additional steps for the introduction, then the removal of this functional group. This drawback has been bypassed using new epoxidation catalysts that do not require an OH group in the olefin. Thus, epoxidation of simple olefins such as indene using the chiral Jacobsen catalyst (see Chap. 16) is a key step in the synthesis of L-736524, a promising inhibiting agent of HIV protease (*anti*-AIDS drug).[21.52]

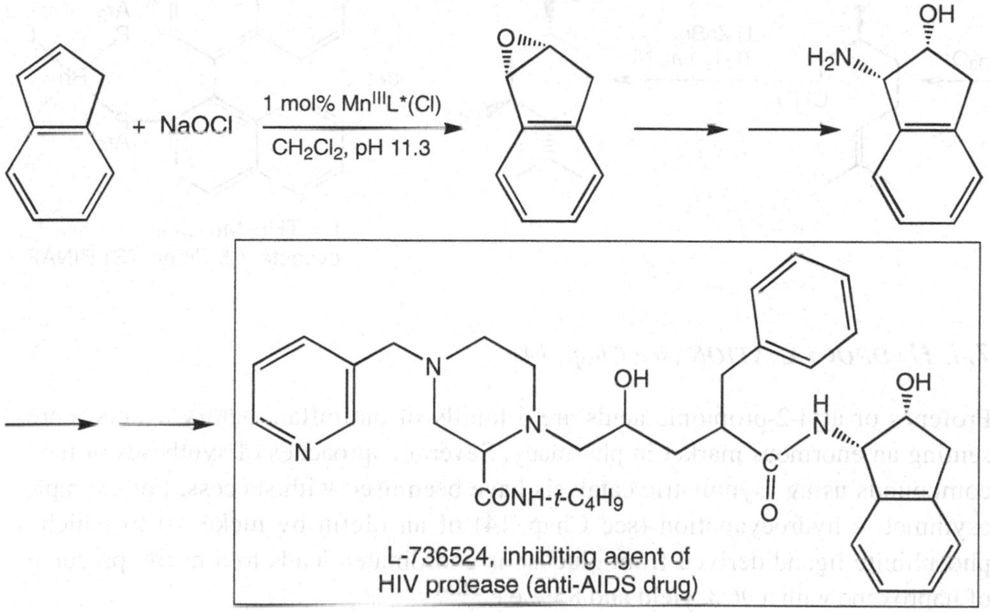

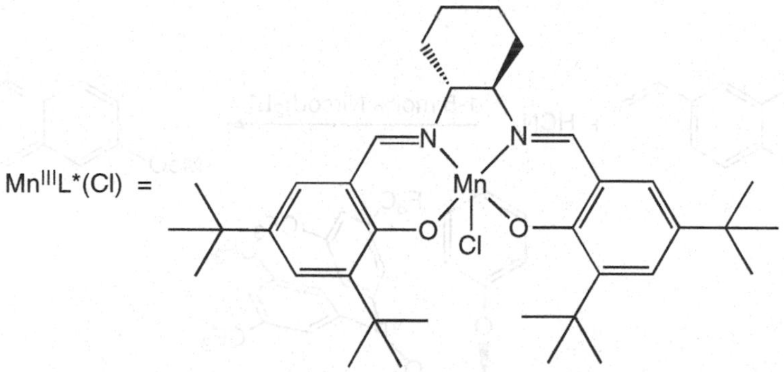

L-736524, inhibiting agent of
HIV protease (anti-AIDS drug)

$Mn^{III}L^*(Cl) =$

7.3. ISOMERIZATION OF ALLYLIC AMINES

Noyori has achieved the asymmetric catalytic isomerization of allylic amines to optically active enamines using his Rh^I-BINAP complex (see Chap. 14). This reaction, isomerization of geranyldiethylamine to (*E*)-enamine of (*R*)-citronellal, is used in Japan as a key step in the synthesis of 1500 tons per year of (–)menthol starting from myrcene. The catalyst is recycled, which still increases its efficiency, and the total chiral multiplication reaches 400,000 mol product *per* mol catalyst. The optical purity of citronellal obtained (96-99%) is far larger than that of natural citronellal (82%).[21.53]

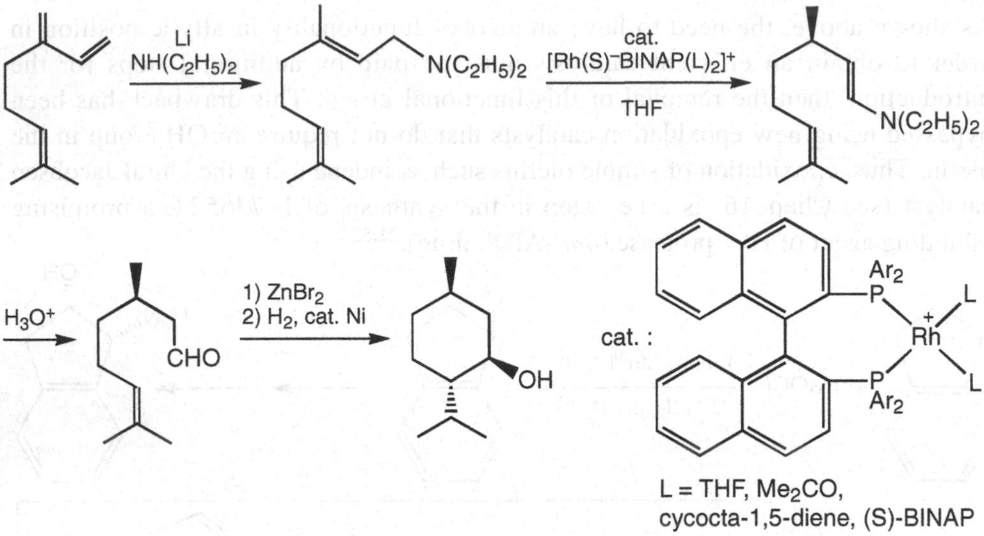

L = THF, Me$_2$CO,
cycocta-1,5-diene, (S)-BINAP

7.4. HYDROCYANATION (see Chap. 14)

Profenes or aryl-2-propionic acids are a family of antiinflammatory agents representing an enormous market in pharmacy. Several approaches of syntheses of these compounds using asymmetric catalysis have been used with success. For example, asymmetric hydrocyanation (see Chap. 14) of an olefin by nickel(0) to which a phosphinite ligand derived from glucose is coordinated leads to a nitrile precursor of naproxene with a 96% yield and 85% e.e.[21,54]

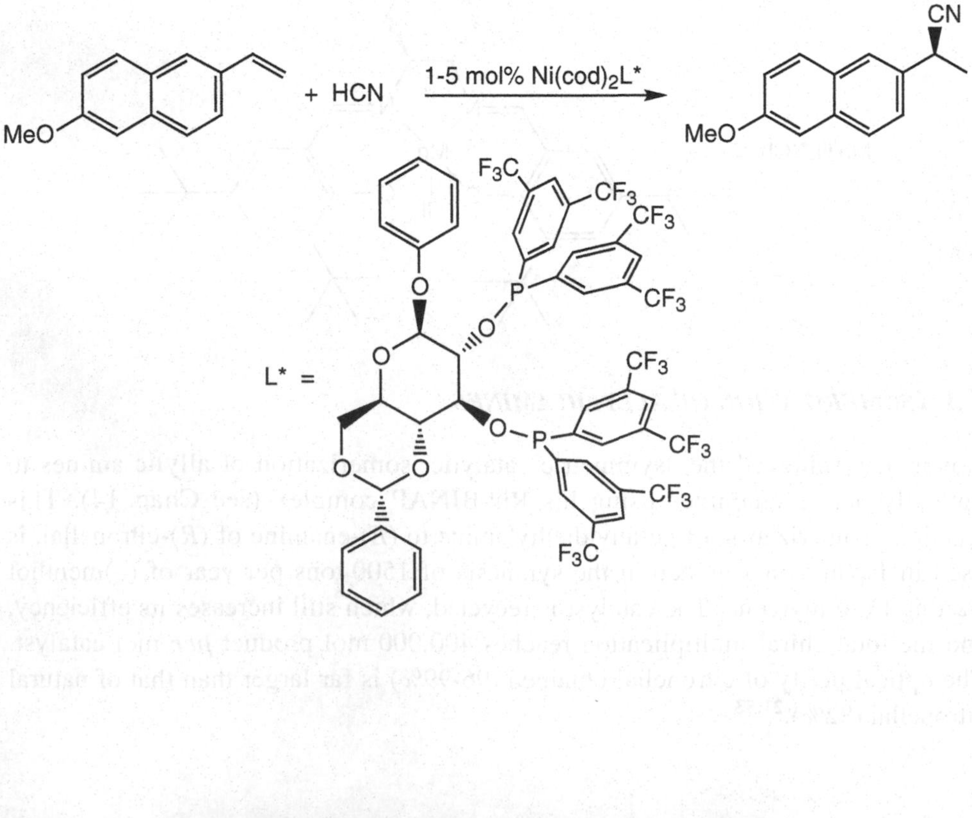

7.5. CARBENE TRANSFER (see Chap. 9)

Pyrethroids are a category of insecticides that are remarkable because of their high selectivity and low toxicity. Some members of this family, chrysanthemic esters, are obtained by asymmetric cyclopropanation by diazoacetates in the presence of a Cu^{II}-Schiff-base catalyst. Among other applications, one of these esters subsequently affords access to cilastatine, an excellent stabilizer of the imipenem antibiotic *in vivo*: [21.55]

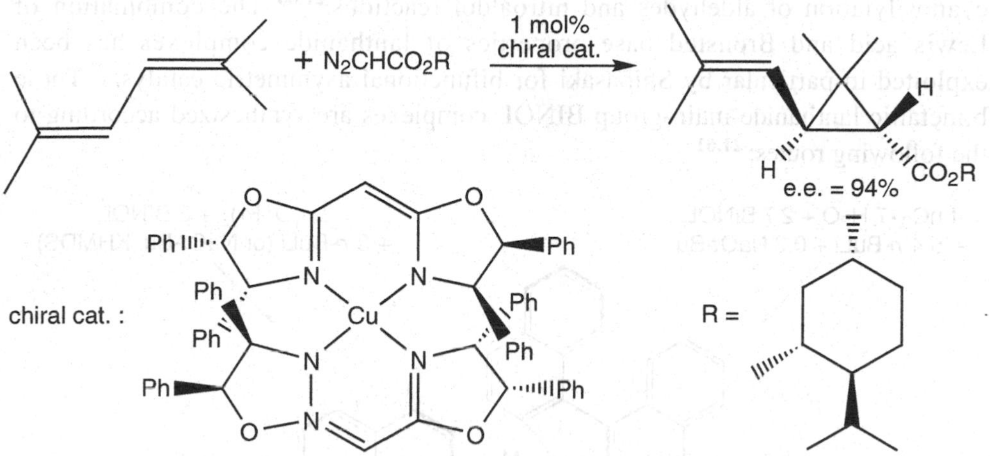

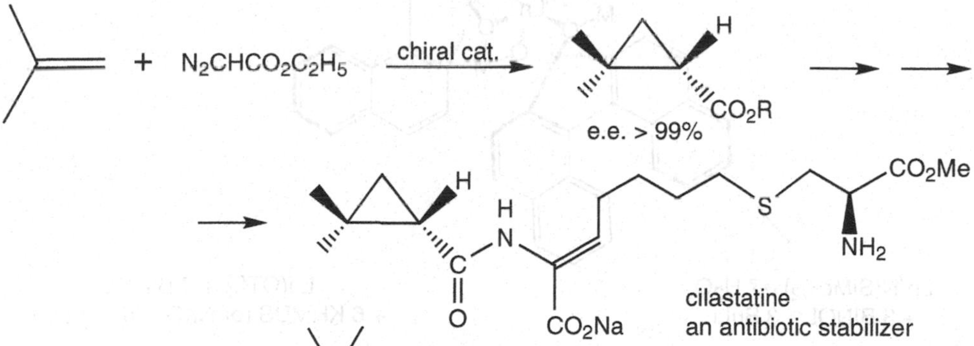

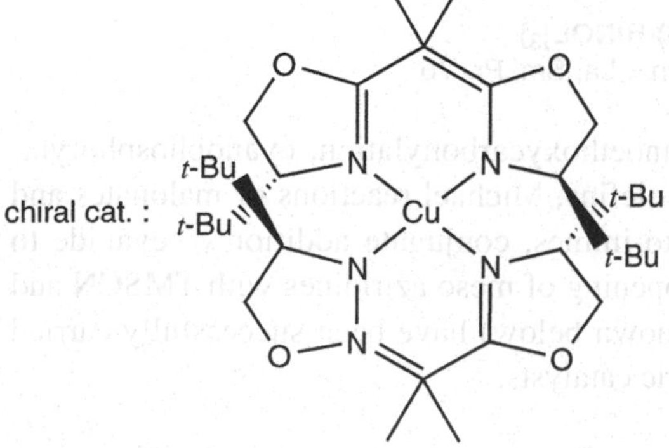

Many other reactions (metathesis, alkylations allylic, Diels-Alder, etc.) are rapidly being developed in asymmetric catalysis [21.56,21.57] and will continue to appear at a fast pace.

7.6. LANTHANIDE 1,1'-BI-2-NAPHTHOL (BINOL) COMPLEXES IN ASYMMETRIC SYNTHESIS EXAMPLE OF THE CYANOSILYLATION OF KETONES

The use of lanthanide complexes in asymmetric catalysis was pioneered by Danishefsky's group with the hetero-Diels-Alder reaction,[21.58] and their utility as chiral Lewis acid catalysts was shown by Kobayashi.[21.59] The Brønsted base character of lanthanide-alkoxides has been used by Shibasaki for aldol reactions, cyanosilylation of aldehydes and nitroaldol reactions.[21.60] The combination of Lewis acid and Brønsted base properties of lanthanide complexes has been exploited in particular by Shibasaki for bifunctional asymmetric catalysis. These bimetallic lanthanide-main-group BINOL complexes are synthesized according to the following routes: [21.61]

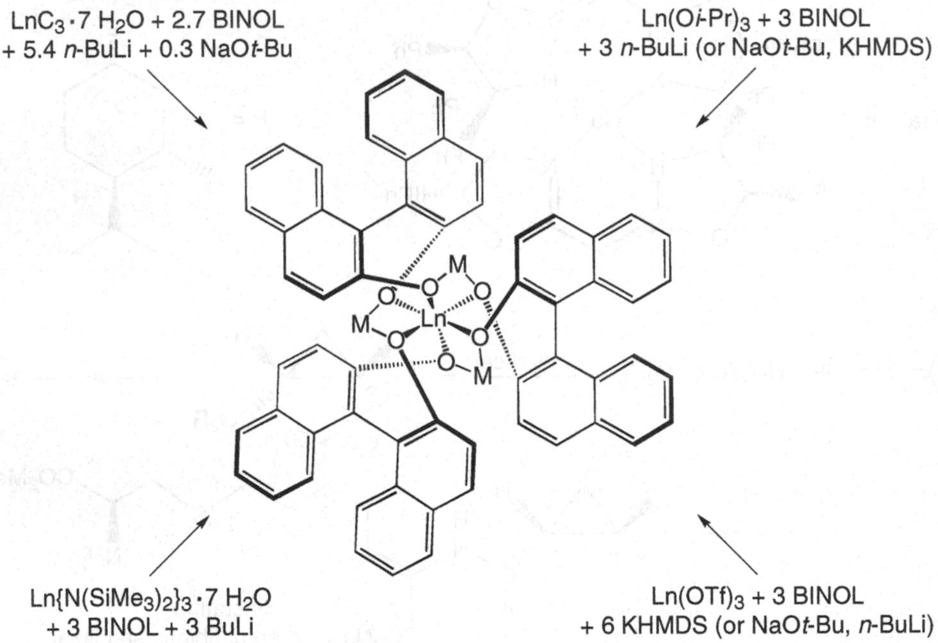

$LnC_3 \cdot 7\ H_2O$ + 2.7 BINOL
+ 5.4 n-BuLi + 0.3 NaOt-Bu

$Ln(Oi\text{-}Pr)_3$ + 3 BINOL
+ 3 n-BuLi (or NaOt-Bu, KHMDS)

$Ln\{N(SiMe_3)_2\}_3 \cdot 7\ H_2O$
+ 3 BINOL + 3 BuLi

$Ln(OTf)_3$ + 3 BINOL
+ 6 KHMDS (or NaOt-Bu, n-BuLi)

M{Ln[(S)-BINOL]₃}
M = Li, Na, K; Ln = La, Sm, Pr, Yb

Thus, the asymmetric catalysis of cyanoethoxycarbonylation, cyanophosphorylation, epoxidation of electron-deficient olefins, Michael reactions of malonates and β-keto-esters, Strecker reaction of keto-imines, conjugate addition of cyanide to α, β-unsaturated pyrrole amides, ring opening of meso aziridines with TMSCN and cyanosilylation of ketones (example shown below) have been successfully carried out using these complexes as asymmetric catalysts.

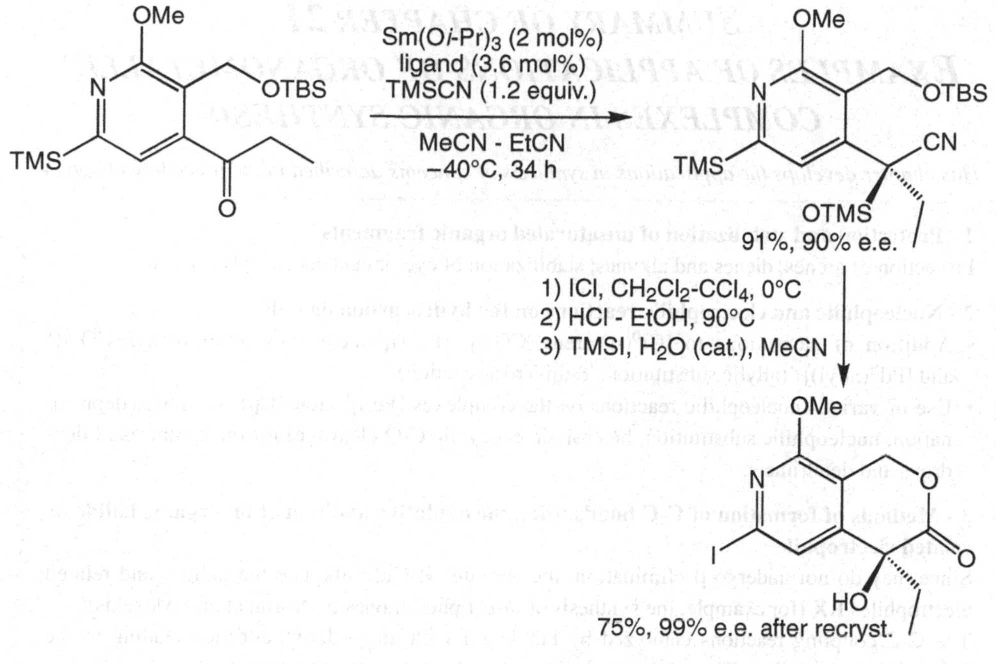

**S-selective, catalytic cyanosilylation of ketones (top),
and application to the synthesis of a key intermediate for camptothecin (bottom)**

SUMMARY OF CHAPTER 21
EXAMPLES OF APPLICATIONS OF ORGANOMETALLIC
COMPLEXES IN ORGANIC SYNTHESIS

This chapter develops the applications in synthesis of concepts described in the preceding chapters.

1 - Protection and stabilization of unsaturated organic fragments

Protection of arenes, dienes and alkynes; stabilization of cyclobutadiene, *o*-xylylene, etc.

2 - Nucleophilic and electrophilic reactions on the hydrocarbon ligands

- Addition of carbanions to $[Cr^0(\eta^6\text{-arene})(CO)_3]$, $[Fe^{II}(\eta^5\text{-methoxycyclohexadienyl})(CO)_3]^+$ and $[Pd^{II}(allyl)]^+$ (allylic substitution: Tsuji-Trost reaction).
- Use of various nucleophilic reactions on the complexes $[Fe(\eta^6\text{-arene})Cp]^+$ (benzylic deprotonation, nucleophilic substitution, heterolytic exocyclic C-O cleavage) for the synthesis of dendrons and dendrimers.

3 - Methods of formation of C-C bonds using the oxidative addition of an organic halide or related electrophile

Since they do not undergo β-elimination, the cuprates R_2CuLi alkylate the halides and related electrophiles RX (for example, the synthesis of insect pheromones by Normant and Alexakis).

The C-C coupling reactions catalyzed by Pd^0 begin with an oxidative addition leading to the $Pd^{II}(R)(X)$ intermediate. The relevant subsequent mechanims are of two types:

- transmetallation of an organometallic R'M (M = BR''$_2$, SnR''$_3$, CuR'', ZnR'') on the intermediate $Pd^{II}(R)(X)$ followed by reductive elimination: Suzuki, Stille, Sonogashira, Corriu-Kumada and Hiyama reactions, etc.;
- insertion of an olefin followed by β-elimination: the Heck reaction that, like the Sonogashira reaction, does not require the preparation of an organometallic compound.

Heck reaction:
$$R^1X + CH_2{=}CHR^2 + NEt_3 \xrightarrow{\text{cat. : } Pd^0} R^1CH{=}CHR^2 + NEt_3H^+X^- \quad \begin{cases} R^1 = \text{aryl,} \\ \text{vinyl, benzyl,} \\ \text{but not alkyl} \end{cases}$$

Sonogashira reaction:
$$ArX + HC{\equiv}CR' + NEt_3 \xrightarrow{\text{cat. : } Pd^0} ArC{\equiv}CR' + NEt_3H^+X^-$$

For these Pd-catalyzed reactions, activation of aryl chlorides requires the use of specific bulky phosphines (or a bulky NHC carbene).

4 - Extension of Pd catalysis to the formation of C-O and C-N bonds

$$ArX + NuH + B \xrightarrow{\text{cat. : } Pd^0} ArNu + BH^+X^- \qquad \text{with NuH = RR'NH or ArOH; B = base}$$
cat: specific bulky phosphines required (or bulky NHC carbene)

5 - Coupling reactions of alkynes with other unsaturated fragments

- Vollhardt reaction (cyclization 2+2+2: diyne + alkyne): example of the short synthesis of estrone.
- Pauson-Khand reaction (alkyne + alkene + CO $\longrightarrow$ cyclopentenone).
- Regioselective click reaction of terminal alkynes with azido derivatives catalyzed by Cu^I in water.

6 - Metal-carbene complexes in organic synthesis

Dötz reaction, methylenation of ketones and alkene metathesis (dimerization, cyclization).

7 - Some examples of applications in asymmetric catalysis

Schrock's asymmetric metathesis catalyst, Sharpless' epoxidation of allylic alcohols, Noyori's isomerization of allylic amines; hydrocyanation and carbene transfer: Lewis acid catalysis.

EXERCISES

21.1. What is the product of the oxidation of E-AcO-CH$_2$-CH=CH-(CH$_2$)$_3$-CH=CH$_2$ by O$_2$ in the presence of a PdCl$_2$/CuCl$_2$ catalyst?

21.2. What is the product of the isomerization of CH$_2$=CH-CH$_2$-C(O)O-CH$_2$-CH=CH$_2$ catalyzed by Ni[P(O-i-Pr)$_3$]$_4$?

21.3. What is the product of the reaction of 1,3-butadiene with a catalytic amount of [Ni(PPh$_3$)$_2$Br$_2$] and NaBH$_4$?

21.4. What is the product of the reaction of naphthalene with styrene in the presence of [Pd(acac)$_2$]?

21.5. What is the product of the reaction of Me$_2$NH with Z-CH$_2$Cl-CH=CH-CH$_2$Cl in the presence of a catalytic amount of [Pd(PPh$_3$)$_4$]?

21.6. What is the product of the reaction of 4-octene with [ZrCp$_2$(H)Cl], followed by the reaction with Br$_2$?

21.7. Propose a synthetic route for E-Ph-CH=CH-CH$_3$ starting from benzene and propene.

21.8. Is it possible to indiscriminately use Pd^{II} (for example [Pd(acac)$_2$]) or Pd^0 (for example [Pd(PPh$_3$)$_4$]) in Heck reactions? Why? Explain using a mechanistic scheme.

Answers to exercises

References

Abbreviations

Index

Atomic weights of the elements

Periodic table of the elements

ANSWERS TO EXERCISES

CHAPTER 1

1.1. [FeCp$_2$]$^+$, sandwich structure with parallel rings, [FeL$_4$X$_2$]$^+$, 17, 5, 3, 6; [RhCl(PPh$_3$)$_3$], square planar, [RhL$_3$X], 16, 8, 1, 4;

[Ta(CH$_2$CMc$_3$)$_3$(CHCMe$_3$)], tetrahedral, [TaX$_5$], 10, 0, 5, 4;

[ScCp*$_2$(CH$_3$)], bent sandwich with CH$_3$ in the equatorial plane, [ScL$_4$X$_3$], 14, 0, 3, 7;

[HfCp$_2$Cl$_2$], bent sandwich with both Cl ligands in the equatorial plane, [HfL$_4$X$_4$], 16, 0, 4, 8;

[W(H)(CO)$_5$]$^-$, octahedral, [WL$_5$X]$^-$, 18, 6, 0, 6;

[Os(NH$_3$)$_5$(η^2-C$_6$H$_6$)]$^{2+}$, octahedral, [OsL$_6$]$^{2+}$, 18, 6, 2, 6;

[Ir(CO)(PPh$_3$)$_2$(Cl)], square planar, [IrL$_3$X], 16, 8, 1, 4;

[ReCp(O)$_3$], pseudo-octahedral, [ReL$_2$X$_7$], 18, 0, 7, 6;

[PtCl$_4$]$^{2-}$, square planar, [PtX$_4$]$^{2-}$, 16, 8, 2, 4;

[PtCl$_3$(C$_2$H$_4$)]$^-$, square planar, [PtLX$_3$]$^-$, 16, 8, 2, 4;

[CoCp$_2$], sandwich structure with parallel rings, [CoL$_4$X$_2$], 19, 7, 2, 6;

[Fe(η^6-C$_6$Me$_6$)$_2$], sandwich structure with parallel rings, [FeL$_6$], 20, 8, 0, 6;

[AuCl(PPh$_3$)], linear, [AuLX], 14, 10, 1, 2;

[Fe(η^4-C$_8$H$_8$)(CO)$_3$], trigonal bipyramid (piano stool), [FeL$_5$], 18, 8, 0, 5;

[Ru(NH$_3$)$_5$(η^1-C$_5$H$_5$N)]$^{2+}$, octahedral, [RuL$_6$]$^{2+}$, 18, 6, 2, 6;

[Re(CO)$_4$(η^2-phen)]$^+$, octahedral, [(ReL$_6$]$^+$, 18, 6, 1, 6;

[FeCp*(CO)(PPh$_3$)(CH$_2$)]$^+$, pseudo-octahedral (piano stool), [(FeL$_4$X$_3$]$^+$, 18, 6, 4, 6;

[Ru(bpy)$_3$]$^{2+}$, pseudo-octahedral, [RuL$_6$]$^{2+}$, 18, 6, 2, 6.

1.2. In the form [FeCp*(η^2-dtc)$_2$], both dithiocarbamate ligands are chelated to iron, [FeL$_4$X$_3$], 19, 5, 3, 7. In the form [FeCp*(η^2-dtc)(η^1-dtc)], one of the two dithio-carbamate ligands is chelated to iron whereas the other is monodentate, [FeL$_3$X$_3$], 17, 5, 3, 6. The two forms having respectively 19 and 17 electrons are both to some extent destabilized, being both one electron away (in excess and shortage respectively) with respect to the robust 18-electron structure. The main reaction of 19-electron complexes in the absence of other substrate is to partially decoordinate giving a 17-electron complex, and this reaction is very fast. The main reaction of 17-electron complexes in the presence of a ligand L (in the absence of reducing agent) is also very fast, addition giving the 19-electron complex. Thus, the dissociation of 19-electron complexes and re-association of the 17-electron complex formed with the dissociated ligand make a rapid equilibrium between these two species that only ends up by the reaction of either form, if any.

1.3.   It is known that O_2 can coordinate to a transition metal in a side-on mode (as ethylene does), which is the case in this complex, or end-on (with an angle) as in hemoglobin and myoglobin. In the case of the side-on coordination, the ligand is an L ligand if the metal is electron poor, the length of the double O=O bond being relatively little changed by the coordination. If, on the other hand, the metal center has a high electron density, as inthis case, the O_2 ligand undergoes an oxidative addition leading to the lengthening of the O-O bond that can, at the limit, reach the length of a single O-O bond. The metal-O_2 species should then be formally described as an Ir^{III} metallocycle, the O_2 ligand now being a peroxo X_2 ligand in the Ir-O_2 triangle structure.

1.4. $[CoCp_2]^+$: organometallic 18-electron, thus stable complex.

$[CoCp_2]^-$: 20-electron organometallic sandwich complex (triplet state, having two unpaired electrons with the same spin) isoelectronic to nickelocene (a commercial compound). The stabilities of these two complexes are weakened by the double occupancy of the doubly degenerate antibonding orbital (two orbitals having the same energy are occupied by two electrons). The orbital energy level is not too high, however, in spite of its antibonding character, which allows the isolation and crystallization of these two reactive complexes.

$[V(CO)_7]^+$: the addition of two disadvantages, a coordination number larger than 6 and the cationic nature of a binary metal carbonyl leads to the non-inexistence of this complex in spite of its 18-electron structure.

$[Cr(H_2O)_6]$: non-existent as all the binary water complexes in which the metal has a low oxidation state such as (0).

$[Ni(H_2O)_6]^{2+}$: relatively stable 20-electrons complex as $[CoCp_2]^-$, *vide supra*, and in addition because of the ideal coordination number (6) and oxidation state (II).

$[ReH_9]^{2-}$: stable 18-electron complex in spite of its very high coordination number, because hydrides, the only ligands present in the coordination sphere, are very small.

$[Zr(CH_2Ph)_4]$: in principle an 8-electron complex that is thermodynamically stable but kinetically very reactive with many substrates. The stability, in spite of the very low NVE, is due partly to the absence of β H, and partly to the π interactions between the phenyl rings and the metal decreasing the Zr-C-Ph angle below 109° (the value of the NVE = 8, only is conventional and does not take this π interactions into account).

$[Zr(CH_2Ph)_4(CO)_2]$: does not exist as almost all the mono- and bimetallic metal-carbonyl structures in which the metals do not have non-bonding electrons (NNBE = 0) (the M-CO bond needs π backbonding from some filled metal d orbitals). There are only exceptions in lanthanide chemistry in which the metal-CO bond is very weak (R. Anderson's work).

$[Cr(CO)_5]^{2-}$: stable 18-electron complex as all the binary mono- or polyelectronic metal-carbonyl 18-electron complexes in which the metal is at the 0, −1 or −2 oxidation state.

$[Cu(Cl)(PPh_3)]$: stable; Cu, Ag and Au commonly gives complexes having the MLX 14-electron structure;

$[Ir(CO)_2(PPh_3)(Cl)]$, $[Pt(PPh_3)_3]$: 16-electron complexes of noble metals of groups 9 and 10 respectively, very stable as usually the group 8-10 noble metal 16-electron complexes are if the oxidation state is also adequate.

$[ScCp*_2(CH_3)]$: stable (but reactive) 14-electron complex (classic structure also with the lanthanides, these complexes being initiators of olefin polymerization)

$[NbCp_2(CH_3)_3]$: stable (but reactive) 18-electron organometallic complex.

$[FeCp*(dtc)_2]^+$: stable 18-electron complex in spite of its odd coordination number (7), because the high oxidation state (IV) is ideal in a mixed inorganic-organometallic complex with dominant inorganic components.

1.5. The optimal stability of this type of complex corresponds to a NVE = 18, the number of electrons provided by the ambivalent, versatile ligand Ph_2C_2 is 4 for two of them and 2 for the 3rd one.

1.6. The order of decreasing thermal stability of these metallocenes is dictated by the 18-electron rule (NEV): $[FeCp_2]$ (18e) > $[MnCp_2]$ (17e) and $[CoCp_2]$ (19e) > $[CrCp_2]$ (16e) and $[NiCp_2]$ (20e) > $[VCp_2]$ (15e) > $[TiCp_2]$ (14e).

1.7. The splitting of these 5 d orbitals into 3 degenerate bonding levels (of identical energy) t_{2g} and 2 degenerate antibonding e_g levels is represented page 40. Manganese has 5 d electrons, one in each orbital (the semi-filling of these 5 orbitals stabilizes this electronic structure). The total spin S is 5/2 (high-spin structure, multiplicity 2S + 1: sextuplet state). $\mu_{theor.} = [5\,(5+2)]^{1/2} = 5.916\ \mu_B$. With Cp*, the 5 electron-releasing methyl substituents strengthen the ligand field, forcing the electrons to pair in a low-spin state S = 1/2 (doublet state: $\mu_{theor.} = [1\,(1+2)]^{1/2} = 1.732\ \mu_B$).

1.8. The electronic density on manganese increases as the π acceptor character of the ligands decreases: CO < CS < C_2H_4 < THF < PPh_3. Detection by infrared spectroscopy or cyclic voltammetry: the absorption frequency of the carbonyls and the oxidation potential of the complex decrease as the electronic density increases. It is the complex $[MnCp(CO)_2L]$, L = CO, that is the most difficult to oxidize in this series.

1.9. Bent sandwich complex, the two carbonyls being in the equatorial plane. The two rings are connected by a dimethylene bridge that hinders the free rotation. One of the Cp's is pentahapto, whereas the other Cp has to be trihapto in order to obey the 18-electron rule.

CHAPTER 2

2.1.

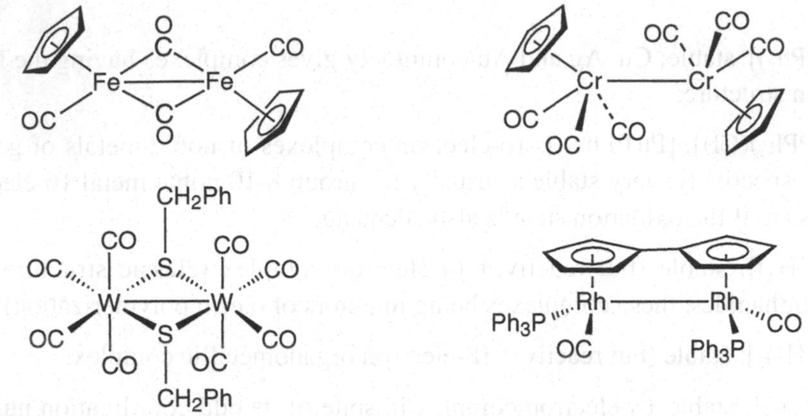

[FeCp(CO)(μ-CO)]$_2$: [FeL$_4$X$_2$], 18, 6, 6; [CrCp(CO)$_3$]$_2$: [CrL$_5$X$_2$], 18, 4, 7;

[W(CO)$_4$(μ$_2$-SCH$_2$Ph)]$_2$: [WL$_5$X$_2$], 18, 4, 7;

{[Rh(CO)(PPh$_3$)]$_2$(η^5,η^5,μ$_2$-fulvalenyl)}: [RhL$_4$X], 18, 8, 5.

2.2. 2, 3, 2.

2.3. [Re$_4$H$_4$(CO)$_{12}$]: NEC = 56 (localized count: counting, to simplify, H as bridging 1 electron; 17 on each Re, i.e. 68 for 4 Re, minus 12 for the 6 Re-Re bonds of the Re$_4$ tetrahedron);

{Fe$_6$[C(CO)$_{16}$]}$^{2-}$: NEC = 86. For the localized count, one can attribute 3 CO's for each of the 4 equatorial Fe's and 2 for the 2 apical Fe's that are formally bridged by the C ligand that is doubly bonded to each of the two Fe's. This gives 18 for each Fe, but the 2 electrons of the negative charges make too many for a localized count with 18 electrons *per* Fe. This count gives (6 × 18) + 2 = 110 less the 24 of the 12 formal Fe-Fe bonds of the octahedron.

[NiCp]$_6$: NEC = 90. For the localized count, one has 6 × NiL$_2$X$_5$, i.e. 6 × 19 = 114 less the 24 of the 12 formal Ni-Ni bonds of the Ni$_6$ octahedron.

In conclusion, the determination of the localized NVE on each metal does not work for any of these three diamagnetic clusters (otherwise, one would have found NEV = 18).

2.4. [Re$_4$H$_4$(CO)$_{12}$]: each ReCp(H) species brings an electronic doublet to the cluster, as Fe(CO)$_3$ or BH also does. The analogous borane would be B$_4$H$_4$, which does not fit Wade's rule.

{Fe$_6$[C(CO)$_{16}$]}$^{2-}$: if the central C is assimilated to a doubly bonded allenic carbon to each of the two apical Fe's, this ligand is assimilated to 2 CO ligands, and the cluster is equivalent to [Fe(CO)$_3$]$_6^{2-}$ or B$_6$H$_6^{2-}$ that follows Wade's rule (*closo* cluster with 6 summits, octahedral).

$[NiCp]_6$: each NiCp is equivalent to BH$^-$ bringing 3 electrons to the cluster that is thus equivalent to $B_6H_6^{6-}$. It is an *arachno* cluster with n = 1/2 (6 + 6 + 6) – 3 = 6 summits.

2.5. The number of atoms of each layer being 10 n^2 + 2, the numbers of atoms of the clusters containing 1, 2, 3, 4, 5, 6 and 7 layers are 13, 55, 147, 309, 561, 923 and 1415, respectively.

2.6. **a.** 6x – 14 = –2 → x = 2;
   b. (6 × 24) + 2 = 26;
   c. NEC = 26 + (14 × 2) = 54.

CHAPTER 3

3.1. **a.** The reaction is endergonic (thermodynamically unfavorable): $\Delta G° \sim 0.1$ V
   (2.3 kcal·mol^{-1} or ~ 10 kJ·mol^{-1});
   b. The reaction is exergonic (favorable): $\Delta G° \sim -1.8$ V
   (~ – 41 kcal·mol^{-1} or ~ – 173 kJ·mol^{-1}).

3.2. **a.** The reaction is impossible: $\Delta G° = 2$ V ~ 46 kcal·mol^{-1} ~ 190 kJ·mol^{-1};
   b. The photo-oxidation is favorable:
   $\Delta G° = -0.13$ V ~ – 3 kcal·mol^{-1} ~ – 12.5 kJ·mol^{-1}.

3.3. The protonation of the iron-hydride complex [Fe^{II}Cp(PMe$_3$)$_2$(H)] leads to the H$_2$ complex [Fe^{II}Cp(PMe$_3$)$_2$(H$_2$)]$^+$, because the iron atom in the starting complex is relatively electron poor, and the oxidation state Fe^{II} is much more favorable than Fe^{IV}. On the other hand, the protonation of the osmium complex leads to the protonation of Os^{II} that is an electron-rich metal center, i.e. the oxidative addition occurs to yield the dihydride [Os^{IV}Cp(PMe$_3$)$_2$(H)$_2$]$^+$, the oxidation state Os^{IV} being much more common than Fe^{IV}.

3.4. This is the σ-bond metathesis reaction in [LuCp*$_2$CH$_3$]; this reaction leads to the deuteration of the methyl group bonded to lutecium.

CHAPTER 4

4.1. 1) benzene; 2) cyclohexadienyl on an external coordinated C leading to cyclohexadiene.

4.2. Ph$_3$COH and the metal-carbene complex [FeCp(dppe)(CHCH$_3$)]$^+$.

4.3. No, because ethylferrocene is itself very reactive towards EtCl + AlCl$_3$. It is necessary to acetylate ferrocene, then to reduce the ketone by the Clemensen method, i.e. using Zn/HCl.

CHAPTER 5

5.1. A 21-electron intermediate or transition state would be of much too high energy, whereas the partial decoordination to the 17-electron species is almost barrierless.

5.2. The only selective method consists in carrying out an electrocatalytic ligand substitution reaction. It is necessary to use a reducing initiator, for example the electron-reservoir complex $[Fe^ICp(C_6Me_6)]$ in catalytic amount.

5.3. Electrocatalytic reaction with an oxidant such as ferrocenium as an initiator in catalytic amount.

CHAPTER 6

6.1. **a.** $[FeCp(CO(FcPPh_2)(COCH_3)]$;
     b. add a catalytic amount of a reducing agent.

6.2. $[ZrCp_2(Cl)]_2)(\mu_2\text{-}C_5H_8)]$.

6.3. The phenyl complexes decompose by β-elimination and the accessible complexes with the other aryl ligands are stable.

6.4. $[TaMe_5]$ and $[Ta(CH_2Ph)_5]$ are relatively stable (thermally) because of the absence of β H. $[Ta(CH_2CMe_3)_5]$ cannot be synthesized because of the steric bulk favoring α-elimination to Me_4C and $[Ta(CH_2CMe_3)_3(=CHCMe_3)]$ (see section 5).

6.5. Methane is produced by α-elimination (see section 5) and ethane is formed by reductive elimination of two CH_3 ligands (see section 3).

CHAPTER 7

7.1. The molecule contains both terminal and bridging carbonyl ligands (note that these ligands are not rigidly coordinated to the metal, however; they interconvert more slowly than the infrared time scale of 10^{-13} s, the molecule being fluxional): $[FeCp(CO)(\mu^2\text{-}CO)]_2$. This phenomenon is very common in polynuclear binary metal carbonyls.

7.2. $[FeCp(\mu_1\text{-}CO)]_2$.

7.3. Fluxional molecule; the carbonyl ligands rapidly "jump" from one metal to the other via a bridging situation ($\mu^2\text{-}CO$).

7.4. $[(FeCp)_2(\mu_2\text{-}CO)_3]$.

7.5. $[FeCp(\mu^3\text{-}CO)]_4$; NEV = 18; NEC = 60.

CHAPTER 8

8.1. [MoCp(CO)$_3$(CH$_2$CH$_3$)] and [MoCp$_2$(CH$_2$CH$_2$CH$_3$)$_2$] (NEV = 18); Cr(CH$_2$Ph)$_6$ (NEV = 12, no β H) and [Nb(CH$_3$)$_5$] (NEV = 10, no β H are stable (moderately for the latter); [Zr(CH$_2$CH$_2$Ph)$_4$] (NEV = 8); [Re(CH$_2$CH$_3$)$_5$] (NEV = 12); [TiCp$_2$(CH$_2$CH$_3$)$_2$] (NEV = 16) and [TaPh$_5$] (NEV = 10) are unstable and decompose by β-elimination, because they have both less than 18 valence electrons on the metal and one H atom on the β C atom.

8.2. $$[CH_3M(CO)_5] \longrightarrow CH_3^{\bullet} + {}^{\bullet}Mn(CO)_5$$

8.3. The Re complex is more stable than that of Mn for two reasons. The first one is that the electronic density in the M-alkyl bond is higher for Re than for Mn because of the additional electronic layer when going from Mn to Re. The second reason is that, Re being larger than Mn, the radical Re(CO)$_5$$^{\bullet}$ is less sterically stabilized than Mn(CO)$_5$$^{\bullet}$; thus, its formation requires more energy. The overall result is that the Re-CH$_3$ bond (52 kcal·mol^{-1}; 217 kJ·mol^{-1}) is stronger than the Mn-CH$_3$ bond (36 kcal·mol^{-1}; 150 kJ·mol^{-1}) in the complexes [CH$_3$M(CO)$_5$] (M = Mn vs. Re).

8.4. [TiPh$_4$] is very unstable because of the fast decomposition by β-elimination that is not possible with R = CH$_3$, CH$_2$Ph and OCH$_3$ giving more or less stable complexes. The order of stability is the following: CH$_3$ < CH$_2$Ph < OCH$_3$ < OPh. The complex [Ti(CH$_3$)$_4$] is not very stable, because its NVE is very low (8). This low NVE is compensated with R = Ph, CH$_2$Ph, OCH$_3$ and OPh by an additional coordination (attraction of the π electrons of the aromatic ring for Ph, donation of the non-bonding *p* doublet of the oxygen atom to the metal for the ligands OR). With OCH$_3$ there is the possibility of β-elimination upon heating, but not with OPh.

8.5. The geometry around the oxygen atom in the alkoxy ligand shows a more open angle than around the carbon atom of an alkyl ligand. This is all the more true as the donation of one or even two oxygen *p* doublets to the metal increases the angle of the M-O-C chain that becomes sometimes linear in the case of the donation of two oxygen *p* doublets (bulky alkoxy). The β H atoms are thus further from the metal in an alkoxy ligand than in an alkyl ligand, and thus less easily subjected to β-elimination than in the alkyl ligands. Thus for instance, the complex [TaCp*(C$_2$H$_5$)$_4$] very rapidly decomposes at low temperature by β-elimination, whereas [TaCp*(OMe)$_4$] is stable at ambient temperature.

8.6. The equilibrium is displaced from the preferred H$_2$ complex towards the dihydride complex upon going down the periodic table. Indeed, the metal becomes more electron rich upon adding electronic layers, thus more able to electronically support two M-H bonds (oxidative addition of H$_2$).

CHAPTER 9

9.1. [FeCp*(PPh$_3$)(CO)(CH$_2$)]$^+$ + PMe$_3$ $\longrightarrow$ [FeCp*(PPh$_3$)(CO)(CH$_2$PMe$_3$)]$^+$

[TaCp$_2$(Me)(CH$_2$)] + AlMe$_3$ $\longrightarrow$ [TaCp$_2$(Me)$^+$(CH$_2$AlMe$_3$)$^-$]

9.2. The first complex is almost a primary carbocation, thus a very powerful and very reactive electrophile, whereas the electron-releasing effects of the five methyl substituents on the Cp ligand and the increase of electronic density resulting from the replacement of CO by PPh$_3$ decrease the positive charge on the carbenic carbon in the second complex compared to that in the first one.

9.3. These reactions are of the same type, the Schrock carbene being eventually considered as an ylid:

$$Me_2C=O + PMe_3CH_2 \longrightarrow Me_2C=CH_2 + Me_3P=O$$
$$Me_2C=O + [TaCp_2 (Me)(CH_2)] \longrightarrow Me_2C=CH_2 + 1/x [(-TaCp_2 (CH_3)(O-)x]$$

9.4. $[(MnCp(CO)_2)_2(\mu_2\text{-}CH_2)]$.

9.5. The three oxides are isoelectronic, but CrO_4^{2-} and MnO_4^- are more stable than FeO_4, because the Fe^{VIII} is rendered very electron poor by the very electronegative oxygen atoms in the third complex due to the absence of negative charge. One notices that the number of charges in going from CrO_4^{2-} to FeO_4 goes along with a shortening of the bonds (the double bond character increases).

9.6. We know that double oxidative addition of an alkene to a metal to produce a metal-bis-carbene or triple oxidative addition of an alkyne to produce a metal-bis-carbyne complex is not possible. Olefin insertion into a metal-hydride giving a metal-alkyl, however, can be followed by α-elimination to yield a hydrido-metal-carbene or even further double α-elimination can give a hydrido-metal-carbyne. Such a case is known: the reaction of styrene or phenylacetylene with $[Os(H)Cl_2L_2]$ (L = Pi-Pr$_3$) yields the carbyne complex $[OsHCl_2(CCH_2Ph)L_2]$ via the metal-carbene intermediate. PhEt is also formed as a result of hydride consumption (K.G. Caulton, *J. Organomet. Chem.* **2001**, *617-618*, 56).

9.7. Thermolysis of $[Cp_2Zr(NHAr)_2]$ generates ArNH$_2$ and the catalyst $[Cp_2Zr=NAr]$ that reacts with the alkyne to provide the azazirconacyclobutene. Addition of ArNH$_2$ yields the bis-amino-Zr intermediate resulting from σ-bond methathesis, then reductive elimination gives the hydroamination product and the catalyst (R.G. Bergman *et al.*, *J. Am. Chem. Soc.* **1992**, *114*, 1708; A.L. Odon *et al.*, *J. Am. Chem. Soc.* **2004**, *126*, 1794).

CHAPTER 10

10.1. In the first two cases, the positive charge or the electron-withdrawing ligands decreases the electron density on the metal, favoring the true olefin complex. In the 3rd case, the electron-releasing character of the ligands and the electron-rich metal (3rd row, located on the right of the periodic table) provoke the oxidative addition to the platinacyclopropane complex. In the 4th example, the ancillary electron-releasing ligand and the very electron poor olefin (four fluorine substituents) favor the oxidative addition to the nickelacyclopropane.

10.2. a. $[WCp(CO)_3]_2 + [Cp_2Fe]^+PF_6^- + C_2H_4$

b. $[WCp(CO)_3Cl] + AgBF_4 + C_2H_4$

10.3. *Syn* (Me at the outside) and *anti* (Me inside).

10.4. The reaction starts by the substitution of two CO ligands by the diene to give $[Fe(CO)_3(\eta^4\text{-diene})]$, then continues by the substitution of a 3rd CO ligand and the intramolecular oxidative addition of the *endo* C-H bond leading to $[FeCp(CO)_2H]$. It ends by the dimerization with formation of the Fe-Fe bond and the loss of H_2. This last step is not mechanistically clear-cut, but it could be taken into account by partial decoordination of the Cp ring (from η^5- to η^3-), coordination of the Fe-H bond of another molecule on the vacant site and binuclear reductive elimination of H_2.

10.5. The fragment $Cr(CO)_3$ is not sufficiently activating to carry out this type of reaction. On the other hand, with the isoelectronic fragment $Mn(CO)_3^+$, the electron-withdrawing effect is considerably stronger because of the positive charge, which makes the reaction possible in spite of the low availability of the nitrogen doublet that is engaged in the conjugation with the aromatic ring.

CHAPTER 11

11.1. ▸ $[FeCp_2]^+$ is a 17-electron d^5 Fe^{III} complex: $(a_{1g})^1$; $S = 1/2$; $\mu_{theor} = 1.73$ μ_B

▸ $[CoCp_2]^-$ is a 20-electron d^8 Co^I complex: $(e^*_{1g})^2$; $S = 1$; $\mu_{theor} = 2.83$ μ_B

▸ $[Co(C_6Me_6)_2]^+$ is 19-electron Co^I d^7 complex $(e^*_{1g})^1$; $S = 1/2$; $\mu_{theor} = 1.73$ μ_B

11.2. The variations of the numbers of electrons concern the d metal orbitals. The metal is protected from the point of view of the reactivity by the cage of ligands, especially when these ligands are permethylated. The metal orbitals being of higher energy with the 2nd and 3rd row, the electron deficiencies and excesses then concern ligands whose radical reactivity is high even when they are permethylated.

11.3. The decaisopropylmetallocenes of the first row of transition metals (Fe, Co^+, etc.) cannot exist because of the steric bulk (mutual hindrance of the two rings that are too close to each other). On the other hand, the larger size of the elements of the 2nd and 3rd row of transition metals than with that of the 1st row metals locates the two Cp rings sufficiently remote from each other, and decaisopropylrhodocenium can be synthesized from decamethylrhodocenium (PF_6^- salts) by reaction of excess KOH and CH_3I.

11.4. Ferrocene first reacts with $CO_2 + AlCl_3$ to undergo carboxylation of one ring. Then $AlCl_3$ complexes the free ring (the more electron-rich one), provoking the weakening of the Fe-Cp bond, then its cleavage and substitution by the arene ring. The cation formed is insensitive towards electrophiles.

11.5. In the first case, the fulvene form is more stable due to the coordination of the exo-cyclic double bond, because the iron center reaches NVE = 18 in this way, whereas, in the second example, the p orbital of the exocyclic carbon and the d orbitals of iron push each other away (the metal orbitals are all filled, the metal having NVE = 18).

11.6. The 1,2, the 1,3 and the 1,1' isomers, but the two first isomers also have metallocenic planar chirality (the two substituents being different) and are thus racemic compounds that can be split into enantiomers. This is not the case for the 1,1' isomer due to free ring rotation about the Fe-ring axis.

11.7. Ferrocene is easily oxidized. Indeed, a positive charge makes the complex much more difficult to oxidize if the structure is otherwise the same. The difference of oxidation potentials between the two compounds is almost 2 volts, which is very large: ferrocene is oxidized at 0.4 V *vs.* the saturated calomel electrode (SCE) and $[FeCp(C_6H_6)]^+$ at 2.2 V *vs.* SCE.

CHAPTER 12

12.1. $LaMe_8^{3-} > KMe_8^{3-} > NaMe > LiMe = UMe_8^{3-} > MgMe_2 > TaMe_5 > ZnMe_2 > CdMe_2$.

12.2. The four first ones (ionic), but not the four last ones (partially covalent).

12.3. The acidity of a hydrogen atom bound to a carbon atom increases when the % of atomic *p* orbital decreases in the hybridization of this carbon: acetylene (*sp* C) is thus much more acidic ($pK_a = 24$) than benzene (*sp* C^2, $pK_a = 37$), itself more acidic than butane (sp^3 C, $pK_a = 44$). The acido-basic equilibrium between butyl lithium and acetylene (or benzene) is displaced towards the direction that is favorable for the reaction producing lithium acetylide (or phenyllithium) and butane (or benzene).

12.4. Dimer in which the hydride ligands bridge the two metallic atoms. In the monomer, the metal is a Lewis acid because of the low number of *d* electrons (the valence layer is incomplete); there is thus a strong tendency to accept the doublet of a M-H bond of another identical molecule.

CHAPTER 13

13.1. In organic synthesis, Cu^{II} is much used as a coupling agent of terminal alkynes and arenes (see text). In catalysis, $CuCl_2$ plays an essential role as a cocatalyst in the Wacker process (see Chap. 16). In biology, the redox Cu^I/Cu^{II} couple is important as active site of some metallo-enzymes catalyzing reactions that use molecular oxygen as shown in Chap. 19 (reduction of O_2 by Cu^I, as in the Wacker process).

13.2. The lithium and magnesium reagents being very polar, the carbanionic character of these alkylating agents makes them strong reductants. $TaCl_5$ is thus very easily reduced to $TaCl_4$ even at $-80°C$ by this type of alkylating agent. On the other hand, with dialkylzinc reagents that are almost apolar, this problem does not intervene, and alkylation of $TaCl_5$ can proceed smoothly at ambient temperature without side reaction.

13.3. In the absence of steric effect, Lewis acids have a tendency to dimerize, the E-R bond formally serving as ligand for the element E of another molecule to fit the octet rule. This is well known with BH_3 (see Chap. 2). With three alkyl substituents, the steric bulk of the substituents inhibits this dimerization in the case of boron, but not in the case of aluminum, a larger element than boron because of the additional electronic layer. The Al-Al bond is longer than the B-B distance and can thus form in spite of the bulk.

13.4. For about the same reason for which trialkylboranes (isoelectronic to carbocations) are stable as monomers, whereas trialkylalanes (isoelectronic to silicenium cations) are not. See the preceding exercise. Silicenium cations, however, do not dimerize whereas alanes do, because of the charge repulsion. They are extraordinarily strong electrophiles.

13.5. ^{119}Sn Mössbauer spectroscopy: the quadrupole splitting QS (distance between the two spectral lines) is very sensitive to the geometry and coordination environment of this metal. Tin (II) is known, for instance in monomeric bulky decasubstituted stannocenes $[Sn(C_5R_5)_2]$ such as decaphenylstannocene (R = Ph).

CHAPTER 14

14.1. Not necessarily. The proportion of enantiomers in the product formed also depends on the kinetic barrier. By the way, in the case of the asymmetric hydrogenation of the olefin precursor of *L*-DOPA by the Rh-DIOP catalyst (see text), the most stable diastereoisomer, formed in quantities of the order of 100 times larger than the other diastereoisomer, reacts of the order of 10,000 times more slowly than the other diastereoisomer, so that it leads to the corresponding enantiomer in only about 1% in the enantiomeric mixture.

14.2. According to the principle of micro-reversibility, such a process must be theoretically feasible, because the reverse process, hydrogenation of alkenes to alkanes, is well known. In order to be possible, the reaction must be made thermodynamically favorable overall, which is usually not the case. If one adds to the reaction medium an olefin such as isobutylene whose hydrogenation is thermodynamically very favorable, the coupling between alkane dehydrogenation with isobutylene hydrogenation becomes thermodynamically slightly favorable, thus possible if the catalyst is appropriate. The kinetic problem then resides in the fact that very few catalysts can give rise to oxidative addition of an alkane C-H bond. Such a system could be realized by Crabtree with iridium catalysts (iridium is one of the only metals which gives oxidative addition of alkane C-H bonds, see Chap. 4.3.2).

14.3. The cone angle of the phosphite directly determines the kinetics of dissociation from and association to the metal center that is itself directly connected to the kinetics of hydrocyanation catalysis. Indeed, the bulkier the ligand, the more easily it decoordinates. When the ligand becomes too large, however, a colloidal metal forms, then agglomerates and precipitates.

14.4. A metallic center should be chosen in such a way that it be electron-rich enough to give oxidative addition of H_2 to the dihydride. One of the two M-H bonds then formally undergoes the insertion of a double bond of benzene producing an intermediate cyclohexadienyl ligand, then the reductive elimination of the second hydride with the cyclohexadienyl ligand gives the cyclohexadiene ligand. A 2nd oxidative addition of H_2 can intervene to continue the reduction according to the same mechanism; the intermediate metal-cyclohexene species formed finally undergoes the 3rd oxidative addition of H_2 to complete the hydrogenation to cyclohexane. Mutterties (an exceptional chemist) has shown that the catalyst $[Co(\eta^3\text{-allyl})(P(OMe)_3)_3]$ is active. The mechanism, however, can be either of the classic type (the allyl ligand can liberate a coordination site while becoming monohapto), or of colloidal type (some colloids are known to readily hydrogenate benzene to cyclohexane).

14.5. Yes, because the decoordination of a 6-electron arene ligand leading to a very unsaturated 14-electron species is then much easier from such a 20-electron complex than from a robust 18-electron complex (arene-iron hapticities lower than 6 are unknown unlike with Ru and Os). Indeed, the presence of two electrons in antibonding orbitals of the 20-electron complex considerably weakens the Fe-arene bond compared to 18-electron complexes.

CHAPTER 15

15.1. It is probable that oxygen from air gives rise to double oxidative addition onto the transition metal of the catalyst leading to a dioxo metal species. The oxo ligands, isoelectronic to carbenes, can play the same role as the latter in the initiation of olefin metathesis. On the other hand, the double oxidative addition of an olefin to give a bis-carbene metal species is unknown (contrary to the simple oxidative addition giving a metallacyclopropane). This means that it is not possible to initiate metathesis with only the olefin if the metal does not bear a doubly bonded ligand such as carbene or oxo (initiation with $M=SiR_2$, $M=NR$ or $M=PR$ are unknown, however). One of the only known alternatives is an alkyl complex MCH_2R giving a metal-hydride-alkylidene species $M(H)(=CHR)$ by α-H elimination.

15.2. Terminal olefins that are very common and yield, as one of the metathesis products, ethylene whose volatility allows to shift the metathesis equilibrium towards the products. Note that the olefins $RCH=CHCH_3$ yield, upon metathesis, 2-butene, another gas whose formation also shifts the metathesis equilibrium towards products.

15.3. See text of Chapter 14.2.3 (ROMP mechanism).

15.4. The metal must be electron rich in order to be oxidized by two oxidation state units upon oxidative coupling, but it must also at the beginning, liberate two coordination sites in order to allow the coordination of two olefin molecules. The complex of a noble metal bearing PPh_3 ligands such as $[Pt(PPh_3)_3]$, correctly meets these criteria.

15.5. β-H elimination of the tantalacyclopentane followed by re-insertion in the opposite direction yields the substituted tantalacyclobutane (S.J. McLain, J. Sancho, R.R. Schrock, *J. Am. Chem. Soc.* **1979**, *101*, 5451 and **1980**, *102*, 5610; *Pure Appl. Chem.* **1980**, *52*, 5610).

CHAPTER 16

16.1. (J.A. Labinger, A.M. Herring and J.E. Bercaw in the general ref. indicated above in **16.2**, Chapter 15 of this ref.):

$$Pt^{2+} + R\text{-}H \longrightarrow [Pt(\eta^2\text{-}RH)]^{2+}$$
$$[Pt(\eta^2\text{-}R\text{-}H)]^{2+} \longrightarrow [Pt\text{-}R]^+ + H^+$$
$$[Pt\text{-}R]^+ + H_2O \longrightarrow ROH + Pt^0 + H^+$$
$$Pt^0 + Pt^{4+} \longrightarrow 2\,Pt^{2+}$$

16.2.
$$M^+\text{-}O^- + H_2 \longrightarrow H\text{-}M\text{-}OH$$

16.3. In the same way as protonation of a neutral metal-hydride complex yields a cationic H_2 complex, protonation of methane yielding transient CH_5^+ may be considered as protonation of a C-H bond leading to a transient H_2 complex of the carbocation CH_3^+, a strong Lewis acid.

16.4. (J.E. Bäckwall *et al.*, *J. A. C. S.* **1979**, *101*, 2411)

$$1/2\,O_2 + FePc \longrightarrow PcFe=O$$
$$PcFe=O + QH_2 \longrightarrow PcFe + Q + H_2O$$
$$Pd^0 + Q + 2H^+ \longrightarrow Pd^{II} + QH_2$$

16.5. The nucleophile is acetate instead of H_2O for the attack of Pd-coordinated ethylene in the Wacker process. It is thus vinyl acetate that is produced:

$$\text{CH}_2=\text{CH}_2 + HOAc + 1/2\,O_2 \xrightarrow{\;[Pd^{II}, Cu^{II}]\;} \text{CH}_2=\text{CHOAc}$$

16.6. In the catalytic cycle of the Wacker process with palladium, CO insertion occurs into the Pd-C bond, which leads to the acyl ligand, then to acrylic acid subsequent to tautomerization along the decomplexation (D. Fenton *et al.*, *Chemtech* **1972**, *2*, 220).

$$\text{CH}_2=\text{CH}_2 + CO + 1/2\,O_2 \xrightarrow{\;[Pd^{II}, Cu^{II}]\;} \text{CH}_2=\text{CHCO}_2\text{H}$$

CHAPTER 17

17.1. $Pd(CF_3CO_2)_3$ is much more electrophilic than $Pd(OAc)_3$, because the electron-withdrawing CF_3 groups remove electron density from Pd, thus electrophilic attack is much faster with $Pd(CF_3CO_2)_3$ than with $Pd(OAc)_3$. The stereoelectronic effect of the CF_3 groups also largely increases the *para* regioselectivity compared to CH_3 in OAc.

17.2. Electrophilic substitution: $M^+ + H_3C\text{-}H \longrightarrow M\text{-}CH_3 + H^+$

σ-bond metathesis: $M\text{-}R + H_3C\text{-}H \longrightarrow M\text{-}CH_3 + R\text{-}H$

See also Chap. 3 for the Shilov system.

17.3. The Lewis acid character of boron is responsible for the regioselective σ-bond metathesis of metal-boryl complexes with alkanes. The unoccupied p orbital at boron triggers the C-H activation process, so that intramolecular arrangement places the boryl group *cis* to the alkyl group.

17.4. Co(OAc)$_3$ is a stronger oxidant than Mn(OAc)$_3$ and can oxidize both alkanes and arenes, whereas only arene that have lower oxidation potential than alkanes, can be oxidized by Mn(OAc)$_3$.

17.5. The repulsion between the filled d metal orbital and filled p oxygen orbitals of the M=O bond is all the more important as the metals have more d electrons, thus as one moves to the right of the periodic table. When one moves to the left, the M=O bond is stronger and the metal has less tendency to deliver its oxo ligand onto a substrate. This is why the V and Cr complexes are much less active in oxo transfer than the Mn and Fe complexes.

CHAPTER 18

18.1. The major product is CH$_3$(CH$_2$)$_4$CHO (linear) and the minor one is CH$_3$(CH$_2$)$_3$CH(CHO)CH$_3$ (branched), the reaction being more or less regioselective in linear product depending on the catalyst.

18.2. Upon cathodic reduction (2 electrons are consumed and Cl$^-$ is produced, platinum cathode, iron anode, propylene carbonate, applied potential: -900 mV *vs*. Ag-AgCl). The catalysis uses a pressure of 10 atm (CO-H$_2$: 1-1) (A. Mortreux and F. Petit in *Homogeneous Transition Metal Catalyzed Reactions*, Eds W.R. Moser and D.W. Slocum, ACS Series, Washington, DC, 1992, Chap. 18).

18.3. The substitution of CO by PPh$_3$ leads to an increased electronic density on the metal centers, which should accelerate the catalytic reaction. The suppression of the catalytic activity means that other transformations are then occurring. In fact, a second PPh$_3$ ligand is coordinated and reacts with activation of a P-C bond leading to the liberation of benzene to give a PPh$_2$ ligand bridging two Ru atoms. This phosphido ligand undergoes oxidative addition of a C-H bond of a phenyl group (orthometallation) on the 3rd Ru atom with liberation of CO and formation of a Ru-aryl bond and a hydride bridging two Ru atoms. Thus, the transformed ligand caps the three Ru atoms, which makes it rigid and inert (G. Süss-Fink, Chap. 28 of the ref. indicated above in **18.2**).

18.4. The reaction is regio- and stereospecific and leads to the (Z)-1-silyl-2-formyl-1-hexenes, i.e. to the silylformylation products. In the most probable catalytic cycle, the insertion of alkyne occurs into a Rh-SiR$_3$ bond leading to an intermediate

2-silyl-1-alkylethynyl-Rh. Then, insertion of CO leads to the intermediate 3-silyl-2-alkylacryloyl-Rh. Finally, oxidative addition of the silane generates the metal-hydride species before reductive elimination of the acyl and hydride ligands liberating the reaction product (I. Ojima *et al.*, in the same ref. indicated above in **16.2**, Chap. 19 of this ref.).

18.5. Mechanism of the Monsanto process, but the oxidative addition of H_2 occurs more rapidly than the reductive elimination of the acetyl and iodo ligands that is inhibited by the chelating diphosphine ligand. Finally, the cycle is closed by the reductive elimination of the acyl and hydride ligand (K.N.G. Moloy and R.W. Wegman in the general ref. indicated above in **16.2**, Chap. 22 of this ref.).

18.6. The cocatalyst catalyzes the hydrogenation of acetaldehyde to ethanol (see the mechanism of olefin hydrogenation but replace the olefin C=C bond by the C=O aldehyde bond). Same ref. as **16.5**.

CHAPTER 19

19.1. The cytochromes have lower (less positive) Fe^{III}/Fe^{II} redox potentials than many organometallic complexes with the same oxidation state because, in the numerous polyene- or polyenyl complexes (ferrocene, iron tricarbonyl dienyl cations, etc.), the π backbonding of iron decreases its electronic density which increases the redox potential (i.e. the complex is more difficult to oxidize, the HOMO being essentially metal-based). On the contrary, in the cytochromes, the redox system is defined by the FeN_4 (N_4 = porphyrin) or Fe-S (S = cystein) core, the inorganic ligands having antibonding orbitals that are too high in energy to give rise to the π backbonding. The ligands being only σ donors, the metal is more electron rich, thus the redox potential is less positive (or more negative) than in the organometallic complexes.

19.2. The Co-C bond exists which fulfills the 18-electron configuration. Note that with porphyrins, however, the delocalization is important, and the 17-electron radical [Co(L)(porphyrin)] is very stabilized, so that it does not give a Co-alkyl bond. Consequently, the conjugation should not be too strong (corrin or dmg ligand) in order to allow the Co-alkyl bond to exist. A modest conjugation, added to the steric effect, restricts the energy of the Co-alkyl bond to 29 ± 5 kcal·mol^{-1} (121.2 kJ·mol^{-1}), which makes vitamin B_{12} a source or reservoir of alkyl radicals.

19.3. Because the Fe=O species is fragile due to the strong repulsion between the *d* metal orbitals and the *p* orbital of the oxene (oxo) ligand, these orbitals being filled and in front of one another.

19.4. Dinitrogen being a weak π acceptor, the negative charge located on the β nitrogen atom is weak. The possibility to protonate dinitrogen (which occurs on this β atom) depends on this charge, however. This step is the first one of the reduction of coordinated dinitrogen. Dinitrogen is thus a mediocre ligand that is readily displaced by all sorts of other ligands. The few complexes in which dinitrogen is reduced are those for which the π backbonding and consequently the negative charge on the β nitrogen atom is more important. A simple criterion is the infrared absorption frequency of the N-N bond (usually located around 1920 and 2150 cm^{-1}) that is all the weaker as the metal-ligand bonding interaction is stronger. The dinitrogen ligand is protonable if this frequency is low enough (for example [TiCp*$_2$(N$_2$)], in which the dinitrogen ligand is reducible, absorbs at 2023 and 2056 cm^{-1}). Finally note that, if protonation occurs on the metal, π backbonding of the metal to dinitrogen is considerably weakened, canceling any chance of dinitrogen reduction.

CHAPTER 20

20.1. Late transition-metal surfaces better chemisorb substrates than early transition-metal surfaces, because their larger number of valence d electrons facilitate dissociation by backbonding into the antibonding orbitals of the substrates.

20.2. The Al-O bond is more polarized Al$^{\delta+}$-O$^{\delta-}$ than Si-O because, having one less electron than Si, Al is more electropositive than Si. In zeolites, the AlO$_4^-$ tetrahedron is negatively charged, unlike the SiO$_4$ tetrahedron, thus the oxygen atoms become more basic as the Al content increases or more acidic as the Si content increases.

20.3. The reaction of C with CO$_2$ is the less costly as long as coal is available (CO$_2$ being abundant).

20.4. In the Ziegler-Natta process for olefin polymerization, the chain growth proceeds by olefin insertion into the metal-alkyl bond. A saturated alkane polymer is ultimately formed. In the Fischer-Tropsch process, the chain growth probably proceeds by surface methylene insertion into a surface-alkenyl bond. The common point is that an olefin is formed in the Fischer-Tropsch process, an olefin being also the substrate of the Ziegler-Natta process. Another common point is that paraffins are formed in both reactions, but the ones formed in the Ziegler-Natta polymerization are considerably longer than in the Fischer-Tropsch process. A last common point is that the chain growth proceeds by insertion into a metal-carbon bond in both cases, but a two-carbon fragment inserts in the Ziegler-Natta process, whereas a one-carbon fragment inserts in the Fischer-Tropsch reaction.

20.5. The common point that makes the Schrock and Basset catalysts active in alkene metathesis is the alkoxy or aryloxy ligand in Schrock's catalysts and the silyloxy ligand in Basset's catalysts. Also note that Basset's catalysts are in fact Schrock's complexes or catalysts that are grafted on silica in order to link the metal by a Si-O-M bond and make catalysts heterogeneous. Schrock's high oxidation state compounds also give σ-bond metathesis upon α-H elimination, but they do not catalyze alkane disproportionation.

CHAPTER 21

21.1. E-AcO-CH_2-CH=CH-$(CH_2)_3$-C(O)Me. The reaction is regiospecific (J. Tsuji *et al.*, *Tetrahedron Lett.* **1978**, 1817).

21.2. Isomerization occurs by oxidative addition of the O-allyl bond giving an intermediate Ni(η^3-allyl) that leads to the mixture of CH_2=CH-CH_2-CH=CH-CH_2-CO_2H and CH_2=CH-CH_2-CH_2-CH=CH-CO_2H (G.P. Chiusoli *et al.*, *Chem. Commun.* **1977**, 793).

21.3. 1,3,6-Octatriene E,E,E-CH_2=CH-CH=CH-CH_2-CH=CH-CH_3 (C.U. Pitman *et al.*, *J. Am. Chem. Soc.* **1975**, *97*, 341).

21.4. E-Styryl-1-naphthyl-2 ethylene (Y. Fugiwara, *Chem. Lett.* **1977**, 1061).

21.5. Z-Me_2N-CH_2-CH=CH-CH_2-NMe_2.

21.6. The insertion reaction of an olefin into the Zr-H bond is immediately followed by the complete isomerization of the branched alkyl chain to the linear chain on zirconium, then bromation gives [$ZrCp_2$(Cl)Br] and 1-bromooctane.

21.7. Chloromercuration yields PhHgCl, then reaction with $PdCl_2$ leads to transmetallation, i.e. formation of PhPdCl. Finally, reaction of the latter with propene provokes the coordination and insertion of propene into the Pd-Ph bond followed by β-elimination finally giving E-Ph-CH=CH-CH_3.

21.8. Yes, because one uses NEt_3 that rapidly reduces Pd(II) to Pd(0) *in situ* by single electron transfer or according to the scheme indicated in the text at the end of section 3.9.

GENERAL

Organometallic Chemistry Books

▸ J.P. Collman, L.S. Hegedus, J.R. Norton, R.G. Finke - *Principles and Applications of Organotransition Metal Chemistry*, University Science Books, Mill Valley, Ca, 2nd ed., **1987**. Complement by S.E. Kegley, R. Pinhas: *Problems and Solutions in Organometallic Chemistry*, University Science Books, Mill Valley, Ca, **1986**.

▸ Ch. Elschenbroich, A. Salzer - *Organometallics*, VCH, Weinheim, 4th ed., **2006**.

▸ R.II. Crabtree - *The Organometallic Chemistry of the Transition Metals*, Wiley, New York, 3rd ed., **2005**.

▸ J.K. Kochi - *Organometallic Mechanisms and Catalysis*, Academic Press, New York, **1978**.

▸ S. Komiya Ed. - *Synthesis of Organometallic Compounds*, Wiley, New York, **1997**.

▸ J.D. Atwood - *Inorganic and Organometallic Reaction Mechanisms*, Brooks/Cole, Belmont, CA, **1985**.

▸ A. Yamamoto - *Organotransition Metal Chemistry*, Wiley, New York, **1990**.

▸ C.M. Lukchart - *Fundamental Organometallic Chemistry*, Brooks, Cole, Monterey, **1985**.

▸ M. Bochmann - *Organometallics 1 and 2*, Oxford Science Publications, Oxford, **1994**.

▸ I. Haiduc, J.J. Zuckerman - *Basic Organometallic Chemistry*, Walter de Gruyter, Berlin, **1985**.

▸ A.W. Parkins, R.C. Poller, - *An Introduction to Organometallic Chemistry*, Palgrave Macmillan, Oxford, **1987**.

▸ J.S. Thayer - *Organometallic Chemistry, An Overview*, VCH, New York, **1988**.

▸ D.F. Shriver, M.A. Drezdzon - *The Manipulation of Air-Sensitive Compounds*, 2nd ed., Wiley, New York, **1986**.

▸ G.O. Spessard, G.L. Miessler - *Organometallic Chemistry*, Prentice Hall, Upper Saddle River, NJ, **1997**.

▸ P. Powell - *Principle of Organometallic Chemistry*, 2nd ed., Chapman and Hall, London, **1988**.

Encyclopedia: G. Wilkinson, F.G.A. Stone, E.W. Abel Eds - *Comprehensive Organometallic Chemistry* (9 vol.), Pergamon Press, Oxford, **1982**; 2nd ed. (*ibid.*), **1995**.

Applications of Organometallic Chemistry to Organic Synthesis

▸ F.J. McQuillin, D.G. Parker, G.R. Stephenson - *Transition Metal Organometallics for Organic Synthesis*, Cambridge University Press, Cambridge, **1991**.

▸ M. Schlosser Ed. - *Organometallics in Synthesis*, Wiley, New York, **1994**.

▸ A. de Mejere, H. tom Dieck - *Organometallics in Organic Synthesis 1*, Springer Verlag, Berlin, **1987**.

▸ H. Werner, G. Erker Eds - *Organometallics in Organic Synthesis 2*, Springer Verlag, Berlin, **1987**.

▶ L.S. Hegedus - *Transition Metals in the Synthesis of Complex Organic Molecules*, University Science Books, Mill Valley, Ca, **1994**.

▶ P.H. Harrington - *Transition Metals in Total Synthesis*, Wiley, New York, **1990**.

▶ S.G. Davies - *Organotransition Metal Chemistry: Application to Organic Synthesis*, Pergamon Press, Oxford, **1982**.

▶ A.J. Pearson - *Metallo-organic Chemistry*, Wiley, Chichester, **1985**.

▶ H. Alper - *Transition Metal Organometallics in Organic Synthesis*, Academic Press, New York, **1978**, Vol. I and II.

Inorganic Chemistry

▶ F. Mathey, A. Sevin - *Introduction to the Molecular Chemistry of the Transition Metal Elements*, Wiley, New York, **1998**.

▶ F.A. Cotton, G. Wilkinson, C.A. Murillo, M. Bochmann - *Advanced Inorganic Chemistry*, 6th ed., Wiley, New York, **1999**.

▶ J.E. Huheey - *Inorganic Chemistry, Principles of Structure and Reactivities*, 4th ed., Harper and Row, New York, **1993**.

▶ D.F. Shriver, P.W. Atkins, C.H. Langford - *Inorganic Chemistry*, 2nd ed., Freeman, New York, **1994**.

▶ K.F. Purcell, J.C. Kotz - *Inorganic Chemistry*, Saunders, Philadelphia, **1985**.

▶ F. Basolo, R.G. Pearson - *Mechanisms of Inorganic Reactions*, 2nd ed., Wiley, New York, **1967**.

Encyclopedia: G. Wilkinson, R.D. Guillard, J.E. Mc Cleverty Eds - *Comprehensive Coordination Chemistry* (7 vol.), Pergamon Press, Oxford, **1987**.

Catalysis

▶ A. Mortreux, F. Petit Eds - *Industrial Applications of Homogeneous Catalysis*, D. Reidel, Amsterdam, **1988**.

▶ B. Cornils, W.A. Herrmann Eds - *Applied Homogeneous Catalysis with Organometallic Compounds*, VCH, Weinheim, **1996**.

▶ B. Cornils, W.A. Herrmann, R. Schlögl, C.-H. Wong Eds - *Catalysis from A to Z. A Concise Encyclopedia*, Wiley-VCH, Weinheim, **2000**.

▶ P.A. Chaloner - *Handbook of Coordination Catalysis in Organic Chemistry*, Buttersworth, London, **1986**.

▶ R.A. Sheldon, J.K. Kochi - *Metal-Catalyzed Oxidations of Organic Compounds*, Academic Press, New York, **1981**.

▶ G.W. Parshall, S.D. Ittel - *Homogeneous Catalysis*, Wiley, New York, **1992**.

▶ W.R. Moser, D.W. Slocum Eds - Homogeneous Transition Metal Catalyzed Reactions, *Adv. Chem. Ser.* **230**, Washington DC, **1992**.

▶ A. Pfaltz, H. Yamamoto Eds - *Comprehensive Asymmetric Catalysis* (3 vol.), E.N. Jacobsen, Springer Verlag, Berlin, **1999**.

▶ H. Brunner, W. Zettlmeier - *Handbook of Enantioselective Catalysis with Transition-metal Compounds* (2 vol.), VCH, Weinheim, **1993**.

Electron Transfer, Inorganic and Organometallic Radicals

▶ D. Astruc - *Electron-Transfer and Radical Processes in Transition-Metal Chemistry. Part I: Theory, Techniques (Electrochemistry), Structures and Molecular Electronics. Part II: Mechanisms, Catalysis and Applications to Organic Synthesis.* VCH, New York, **1995**.

▶ W.C. Trogler Ed. - Organometallic Radical Processes, *J. Organomet. Chem. Library*, Elsevier, New York, **1990**, Vol. 22.

▸ M. Chanon, M. Juliard, J.-C. Poite Eds - *Paramagnetic Species in Activation, Selectivity, Catalysis*, NATO ASI Series C257, Kluwer, Dordrecht, **1989**.

▸ P. Zanello - *Inorganic Electrochemistry. Theory, Practice and Applications*, RSC, Cambridge, **2003**.

▸ A.J.L. Pombeiro, C. Amatore Eds - *Trends in Molecular Electrochemistry*, Marcel Dekker, New York, **2004**.

Biochemistry and Bioinorganic Chemistry

▸ L. Stryer - *Biochemistry*, 5th ed., Freeman, New York, **2002**.

▸ S.J. Lippard, J.M. Berg - *Principles of Bioinorganic Chemistry*, University Science Books, Mill Valley, Ca, **1994**.

▸ I. Bertini, H.B. Gray, S.J. Lippard, J.S. Valentine - *Bioinorganic Chemistry*, University Science Books, Mill Valley, Ca, **1994**.

Supramolecular Chemistry

▸ J.-M. Lehn - *Supramolecular Chemistry. Concepts and Perspectives*, VCH, Weinheim, **1995**.

CITED REFERENCES

Introduction - Historical aspects

0.1. C.A. Russell - *Edward Frankland*, Cambridge University Press, Cambridge, **1996**.

0.2. J.S. Thayer - Organometallic Chemistry. A Historical Perspective, *Adv. Organomet. Chem.* **13**, 1, **1975**.

0.3. G. Wilkinson - The Iron Sandwich. A Recollection of the First Four Months. *J. Organomet. Chem.* **100**, 273, **1975**.

Chapter 1 - Monometallic transition-metal complexes

1.1. P.R. Mitchell, R.V. Parish - The 18 Electron Rule, *J. Chem. Ed.* **46**, 8111, **1969**.

1.2. C.A. Tolman - The 16 and 18-Electron Rule in Organometallic Chemistry and Homogeneous Catalysis, *Chem. Soc. Rev.* **1**, 337, **1972**.

1.3. R.H. Crabtree - Transition Metal Complexes of σ Bonds, *Angew. Chem. Int. Ed. Engl.* **32**, 789, **1993**.

1.4. R. Hoffmann - Building Bridges Between Inorganic and Organic Chemistry (Nobel Lecture), *Angew. Chem. Int. Ed. Engl.* **21**, 711, **1982**.

1.5. S. Trofimenko - Tris(pyrazolyl)borates, *Chem. Rev.* **72**, 497, **1972** and *Polyhedron* **23**, 197, **2004**.

1.6. P. Mountford *et al.* - *Coordination, Organometallic and Related Chemistry of Tris(pyrazolyl)methane*, Dalton (Perspective), **2005**, 635.

1.7. D. Bourissou, O. Guerret, F.P. Gabbaï, G. Bertrand - Stable Carbenes, *Chem. Rev.* **100**, 39, **2000**; Y. Canac, M. Soleilhavoup, S. Conejero, G. Bertrand, Stable non-N-heterocyclic carbenes, *J. Organomet. Chem.* **689**, 3857, **2004**.

1.8. E. Peris, R.H. Crabtree - Recent Homogeneous Catalytic Applications of Chelate and Pincer N-heterocyclic Carbenes, *Coord. Chem. Rev.* **248**, 2239, **2004**.

1.9. P.P. Power - Some Highlight in the Development and Use of Bulky Monodentate Ligands, *J. Organomet. Chem.* **689**, 3904, **2004** (review).

Chapter 2 - Bimetallic transition-metal complexes and clusters

2.1. a) F.A. Cotton - Quadruple Bonds and Other Metal to Metal Bonds,
Chem. Soc. Rev. **4**, 27, **1975**; *Acc. Chem. Res.* **2**, 240, **1969**;
b) F.A. Cotton, R.A. Walton - *Multiple Bonds Between Metal Atoms*,
Wiley, New York, **1982**.

2.2. P. Chini - Large Metal Carbonyl Clusters, *J. Organomet. Chem.* **200**, 37, **1980**;
Adv. Organomet. Chem. **14**, 285, **1976**.

2.3. a) D.M.P. Mingos, D.J. Wales - *Introduction to Cluster Chemistry*,
Prentice Hall, Englewood Cliffs, **1990**;
d) P. Zanello - *Struct. Bond.* **79**, 101, **1992**.

2.4. D.F. Shriver, H.D. Kaez, R.D. Adams Eds - *The Chemistry of Metal Cluster
Complexes*, VCH, Weinheim, **1990**.

2.5. G. Schmid Ed. - *Transition Metal Clusters and Colloids*, VCH, Weinheim, **1994**;
G. Schmid - *Nanoparticles. From Theory to Application*, Wiley-VCH, Weinheim, **2004**.

2.6. a) Polyoxometallate issue, *Chem. Rev.* **98**, 1, **1998**;
b) A. Proust, P. Gouzerh - Main-Group Element, Organic and Organometallic
Derivatives of Polyoxometallates, *Chem. Rev.* **98**, 77, **1998**;
c) M.T. Pope, A. Müller - Polyoxometalate Chemistry: An Old Field with New
Dimensions in Several Disciplines, *Angew. Chem. Int. Ed. Engl.* **30**, 34, **1991**.

2.7. R. Chevrel, M. Hirrien, M. Sergent - Superconducting Chevrel Phases:
Prospects and Perspectives, *Polyhedron* **5**, 87, **1996**.

2.8. C. Perrin - Octahedral Clusters in Transition Element Chemistry, *J. of Alloys and
Compounds* **10**, 262-263, **1997**; A. Perrin, M. Sergent - Rhenium Clusters in
Inorganic Chemistry: Structures and Metal-Metal Bonding, *New J. Chem.* **12**, 337,
1988; A. Perrin, C. Perrin, M. Sergent - Octahedral Clusters in Mo^{II} and Re^{III}
Chalcohalide Chemistry, *J. Less Common Met.* **137**, 241, **1988**;
A. Simon - Condensed Metal Clusters, *Angew. Chem. Int. Ed. Engl.* **20**, 1, **1981**.

2.9. K. Wade - *Electron Defficient Compounds*, Thomas Nelson, London, **1971**
(great historical and pedagogical value).

2.10. M.F. Hawthorne - Dicarbollyl Derivatives of the Transition Metals. Metallocene
Analogs, *J. Am. Chem. Soc.* **90**, 862, 879, 896, **1968**; Ten Years of Metallocarboranes,
Adv. Organomet. Chem. **14**, 145, **1976**; *Acc. Chem. Res.* **1**, 281, **1968**.

2.11. a) P. Braunstein, L.A. Oro, P.R. Raithby Eds - *Metal Clusters in Chemistry* (3 vol.),
Wiley-VCH, Weinheim, **1999**;
b) G. Lavigne - Effects of Halide and Related Ligands on Reactions of Carbonyl
Ruthenium Complexes (Ru^0-Ru^{II}), *Eur. J. Inorg. Chem.*, 917, **1999** (review).

2.12. E. Alonso, J. Ruiz, D. Astruc - The Use of the Electron-reservoir Complexes
[Fe^{I}Cp(arene)] for the Electrocatalytic Synthesis of Redox-reversible Metal-Carbonyl
Clusters Containing Ferrocenyldiphenylphosphine, *J. Clust. Sci.* **9**, 271, **1998**.

2.13. Y.-C. Lin, A. Mayr, C.B. Knobler, H.D. Kaesz - Synthesis of Hydrodotriosmium
Complexes Containing Isomeric Carboxamido, Formamido, and Iminyl Groups,
J. Organomet. Chem. **272**, 207, **1984**.

2.14. D.F. Shriver *et al.* - Adduct Formation and Carbonyl Rearrangement of Polynuclear
Carbonyls in the Presence of Group III Halides, *Inorg. Chem.* **13**, 499, **1974**;
J. Am. Chem. Soc. **100**, 5239, **1978**.

2.15. A.J. Deeming, M. Underhill - Reactions of Dodecarbonyl-Triangulo-Triosnium with
Alkenes and Benzene, *J. Chem. Soc., Dalton Trans.: Inorg. Chem.* **13**, 1415, **1974**.

2.16. M.G. Thomas, B.F. Beier, E.L. Muetterties - Catalysis Using the M$_3$(CO)$_{12}$ Clusters
(M = Ru, Os, Ir), *J. Am. Chem. Soc.* **98**, 1296, **1976**; J.S. Bradley - *J. Am. Chem. Soc.*
101, 7419, **1979**; G. Süss-Fink - *Adv. Chem. Ser.* **230**, 1992, **1992**.

2.17. D. Méry, C. Ornelas, J. Ruiz, S. Cordier, C. Perrin, D. Astruc - Mo_6Br_8-Cluster-Cored Organometallic Stars and Dendrimers, in *Inorganic Clusters*, C. Perrin Ed., *C.R. Chimie* **8**, 1789, **2005**.

2.18. J. Ruiz, C. Belin, D. Astruc - Assembly of Dendrimers with Redox-Active Clusters $[CpFe(\mu_3\text{-}CO)]_4$ at the Periphery and Application to Oxo-anion and Adenosine-5'-Triphosphate (ATP) Sensing, *Angew. Chem. Int. Ed. Engl.* **45**, 132, **2006**.

Chapter 3 - Redox reactions, oxidative addition and σ-bond metathesis

3.1. H. Taube - *Electron-Transfer Reactions of Complex Ions in Solution*, Academic Press, New York, **1970**.

3.2. J.S. Miller *et al.* - Molecular Ferromagnets, *Chem. Rev.* **88**, 201, **1988**; *Acc. Chem. Res.* **21**, 114, **1988**; *Science* **240**, 40, **1988**.

3.3. D. Astruc - *Electron Transfer and Radical Processes in Transition Metal Chemistry*, VCH, New York, **1995**.

3.4. R.A. Marcus, N. Sutin - Electron Transfer in Chemistry and Biology, *Biochim. Biophys. Acta* **811**, 265, **1985**.

3.5. J.K. Kochi, T.M. Bockman - The Role of Electron Transfer and Charge Transfer in Organometallic Chemistry, *Pure Appl. Chem.* **52**, 571, **1980**; Organometallic Ions and Ion Pairs, *Adv. Organomet. Chem.* **33**, 51, **1991**.

3.6. H. Taube - Electron Transfer Between Metal Complexes. A Retrospective View, *Angew. Chem. Int. Ed. Engl.* **23**, 329, **1984** (Nobel Lecture).

3.7. D. Astruc - From Organo-Transition Metal Chemistry Towards Molecular Electronics: Electronic Communication Between Ligand-Bridged Metals, *Acc. Chem. Res.* **30**, 383, **1997**.

3.8. L. Vaska - Dioxygen-Metal Complexes: Towards a Unified View, *Acc. Chem. Res.* **9**, 175, **1976**.

3.9. a) R.G. Bergman - Activation of Alkanes with Organotransition Metal Complexes, *Science* **223**, 902, **1984** and **270**, 1970, **1995**; *J. Organomet. Chem.* **400**, 273, **1990**; *Acc. Chem. Res.* **28**, 154, **1995**; S.A. Blum, K.L. Tan, R.G. Bergman - Application of Physical Organic Methods to the Investigation of Organometallic Reaction Mechanisms. *J. Org. Chem. (Perspective)* **68**, 4127, **2003**; W.D. Jones, F.J. Feher - *J. Am. Chem. Soc.*, 1171, **1984**;
b) A.E. Shilov, G.B. Shulpin - Activation of C-H Bonds by Metal Complexes, *Chem. Rev.* **97**, 2879, **1997**.

3.10. J.-Y. Saillard, R. Hoffmann - C-H and H-H Activation by Transition Metal Complexes and Surfaces, *J. Am. Chem. Soc.* **106**, 2006, **1984** (very important article).

3.11. J. Halpern - Activation of C-H Bonds by Metal Complexes: Mechanistic, Kinetic and Thermodynamic Considerations, *Inorg. Chim. Acta* **100**, 41, **1985**.

3.12. J.A. Osborn - *Organotransition-Metal Chemistry*, Y. Ishii, M. Tsutsui Eds, Plenum, New York, 69, **1978**.

3.13. P.L. Watson, G.W. Parshall - Organolanthanides in Catalysis, *Acc. Chem. Res.* **18**, 51, **1985**.

3.14. a) A.E. Shilov, G.B. Shulpin - *Activation and Catalytic Reactions of Saturated Hydrocarbons in the Presence of Metal Complexes*, Kluwer, Dordrecht, **2000**;
b) J.A. Labinger, J.E. Bercaw - Mechanistic Aspects of C-H Activation by Pt Complexes, *Nature* **417**, 507, **2002**; R.H. Crabtree - Organometallic Chemistry of Alkanes, *Chem. Rev.* **85**, 245, **1985**; Aspects of Methane Chemistry, *Chem. Rev.* **95**, 987, **1995**; Organometallic Alkane Activation, *J. Organomet. Chem.* **689**, 4083, **2004**; P.E.M. Siegbahn, R.H. Crabtree - *J. Am. Chem. Soc.* **118**, 4442, **1996**;

    c) D.D. Wick, K. Goldberg - *J. Am. Chem. Soc.* **119**, 10235, **1997**
       and **121**, 11900, **1999**; see also K. Goldberg *et al.* - *J. Am. Chem. Soc.* **121**, 252,
       1999; **123**, 2576, **2001** and **125**, 8614 and 9442, **2003**;

    d) for discussion of the kinetic effects on this and related reactions,
       see W.D. Jones - *Acc. Chem. Res.* **36**, 140, **2003**;

    e) W.D. Jones - *J. Am. Chem. Soc.* **120**, 2843, **1998 and**; **123**, 9718, **2001**;
       Organometallics **120**, 2843 and 4784, **1998**.

3.15. M. Brookhart, M.L.H. Green - Carbon-Hydrogen-Transition Metal Bonds,
      J. Organomet. Chem. **250**, 395, **1983**.

3.16. **a)** Review on C-C oxidative addition:
       R. Rybtchinski, D. Milstein - *Angew. Chem. Int. Ed.* **38**, 870, **1999**;

    b) Review on pincer ligand complexes:
       M. Albrecht, G. Van Koten - *Angew. Chem. Int. Ed.* **40**, 3750, **2001**;
       M.E. Van der Boom, D. Milstein - *Chem. Rev.* **103**, 1759, **2003**;

    c) D. Milstein *et al.* - *Nature* **364**, 699, **1993**; *J. Am. Chem. Soc.* **117**, 9774, **1995**;
       118, 12406, **1996**; **120**, 13415, **1998**; **121**, 4528 and 6652, **1999**; **117**, 7723 and
       9774, **2000**; **123**, 9064, **2001**.

3.17. S.J. McLain, R.R. Schrock - Selective Olefin Dimerization *via* Tantallocyclopentane
      Complexes, *J. Am. Chem. Soc.* **100**, 1315, **1978**; J.D. Fellmann, G.A. Rupprecht,
      R.R. Schrock - Rapid Selective Dimerization of Ethylene to 1-Butene by a Tantalum
      Catalyst and a New Mechanism for Ethylene Oligomerization, *J. Am. Chem. Soc.* **101**,
      5099, **1979**; S.J. McLain, J. Sancho, R.R. Schrock - Metallacyclopentane to
      Metallacyclobutane Ring Contraction, *J. Am. Chem. Soc.* **101**, 5451, **1979**;
      R.R. Schrock, S.J. McLain, J. Sancho - *Pure Appl. Chem.* **52**, 729, **1980**.

3.18. H. Nishihara *et al.* - Synthesis of a New Bis(ferrocenyl)ruthenacyclopentatriene
      Compound with a Significant Inter-Metal Electronic Communication, *J. Organomet.*
      Chem. **637-639**, 80, **2001**; E. Singleton *et al.* - *Chem. Commun.*, 1682, **1986**.

3.19. S. Lippard *et al.* - Formation of Highly Functionalized Metal-Bound Acetylenes by
      Reductive Coupling of CO and Methyl Isocyanide, *J. Am. Chem. Soc.* **116**, 4166, **1992**.

Chapter 4 - Reactions of nucleophiles and electrophiles with complexes

4.1. S.G. Davies, M.L.H Green, D.M. Mingos - Nucleophilic Addition to Organo-
      Transition Metal Cations Containing Unsaturated Hydrocarbon Ligands,
      Tetrahedron **34**, 20, **1978**.

4.2. D. Astruc, P. Michaud, A.M. Madonik, J.-Y. Saillard, R. Hoffmann -
      The Regiospecific Reaction of Nucleophiles with [$Fe^{II}(\eta^5$-cyclohexadienyl)
      (η^6-benzene)]$^+$: a Frontier Orbital Controlled Route to [$Fe^0(\eta^4$-cyclohexadiene)
      (η^6-benzene)], *Nouv. J. Chim.* **9**, 41, **1985**.

4.3. E.O. Fischer - On the Way to Metal Carbene and Carbyne Complexes,
      Adv. Organomet. Chem. **1**, 14, **1976**.

4.4. **a)** C.P. Casey, W.M. Miles - The Preparation of Methoxyalkyliron Complexes:
       Precursors of Secondary Carbeneiron Complexes,
       J. Organomet. Chem. **254**, 333, **1983**; see also ref. **20.11**;

    b) B. Crociani, P. Uguagliati, U. Belluco - Steric Role of Aromatic Ring Ortho-
       Substituents in the Mechanism of Carbene Formation from Palladium(II)
       Arylisocyanide Complexes and Anilines, *J. Organomet. Chem.* **117**, 189, **1976**.

4.5. M.F. Semmelhack - Arene-Metal Complexes in Organic Synthesis, *Annals New York*
      Acad. Sci., 36, **1977**; B.M. Trost Ed. - *Comprehensive Organic Chemistry*,
      Pergamon Press, **1991**, Vol. 4, 517; *J. Organomet. Chem. Library* **1**, 361, **1976**.

4.6. M.F. Semmelhack *et al.* - Addition of Carbon Nucleophiles to Arene-Chromium
      Complexes, *Tetrahedron* **37**, 3957, **1981**; M.F. Semmelhack, A. Yamashita -
      J. Am. Chem. Soc. **102**, 5926, **1980**; G. Jaouen, A. Meyer, G. Simonneaux -
      J. Chem. Soc., Chem. Commun., 813, **1975**.

4.7. H. Des Abbayes, M.-A. Boudeville - Alkylation of Arylacetic Ester by Phase-Transfer Catalysis and Sodium Hydride: Activation and Stereochemical Effects of the Cr(CO)₃ Group, *J. Org. Chem.* **42**, 4104, **1977**.

4.8. A. Solladié-Cavallo, G. Soladié, E. Tsamo - Chiral Arene-Tricarbonyl-Chromium Complexes: Resolution of Aldehydes, *J. Org. Chem.* **44**, 4189, **1979**;
M.C. Sénéchal-Tocquer, D. Sénéchal, J.-Y. Le Bihan, D. Gentric, B. Caro - *Bull. Soc. Chim. Fr.* **129**, 121, **1992**.

4.9. E.P. Kündig *et al.* - Thermodynamic Control of Regioselectivity in the Addition of Carbanions to Arene-Tricarbonyl-Chromium Complexes, *J. Am. Chem. Soc.* **111**, 1804, **1989**.

4.10. D. Astruc - Organo-Iron Complexes of Aromatic Compounds. Application in Organic Synthesis, Tetrahedron Report N° 157, *Tetrahedron* **39**, 4027, **1983**.

4.11. A. Alexakis, F. Rose-Münch, E. Rose *et al.* - Resolution and Asymmetric Synthesis of Ortho-Substituted Benzaldehyde-Tricarbonyl-Chromium Complexes, *J. Am. Chem. Soc.* **114**, 8288, **1992**.

4.12. a) A.F. Cunningham Jr - Mechanism of Mercuration of Ferrocene: General Treatment of Electrophilic Substitution of Ferrocene Derivatives, *Organometallics* **16**, 1114, **1997**;
 b) J. Weber *et al.* - Density Functional Study of Protonated, Acetylated, and Mercurated Derivatives of Ferrocene: Mechanism of the Electrophilic Substitution Reaction, *Organometallics* **17**, 4983, **1998**.

Chapter 5 - Ligand substitution reactions

5.1. J.A.S. Howell, P.M. Buckinshow - Ligand Substitution Reactions at Low-Valent Four-, Five-, and Six-Coordinate Transition-Metal Complexes, *Chem. Rev.* **83**, 557, **1983**.

5.2. D.J. Darensbourg - Mechanistic Pathway for Ligand Substitution Processes in Metal Carbonyls, *Adv. Organomet. Chem.* **21**, 113, **1982**.

5.3. F. Basolo *et al.* - Rates and Mechanism of Substitution Reactions of Nitrosyltetracarbonyl-manganese (O), Trinitrosylcarbonyl-manganese (O) and Nitrosyltricarbonyl-cobalt (O), *J. Am. Chem. Soc.* **85**, 3929, **1966** and **89**, 4626, **1967**.

5.4. M.S. Wrighton - The Photochemistry of Metal Carbonyls, *Chem. Rev.* **74**, 401, **1974**.

5.5. G.L. Geoffroy, M.S. Wrighton - Organometallic Photochemistry, Academic Press, New York, **1975**; *J. Chem. Ed.* **60**, 861, **1983**.

5.6. D. Astruc - 19-Electron Complexes and their Role in Organometallic Mechanisms, *Chem. Rev.* **88**, 1189, **1988**.

5.7. R.L. Rich, H. Taube - Catalysis by Pt^III of Exchange Reactions of PtCl₄⁻ and PtCl₆⁻, *J. Am. Chem. Soc.* **76**, 2608, **1954**.

5.8. A.J. Pöe - Systematic Kinetics of Associative Mechanisms of Metal Carbonyls, *Pure Appl. Chem.* **60**, 1209, **1988**.

5.9. D. Astruc - Electron-Transfer-Chain Catalysis in Organotransition Metal Chemistry, *Angew. Chem. Int. Ed. Engl.* **27**, 643, **1988**; *Electron Transfer and Radical Processes in Transition Metal Chemistry*, VCH, New York, **1975**, Chap. 6 (Chain Reactions).

5.10. J.K. Kochi - Electron Transfer and Transient Radicals in Organometallic Chemistry, *J. Organomet. Chem.* **300**, 139, **1986**.

5.11. M. Chanon - Electron-Transfer Induced Chain Reactions and Catalysis Building Bridges Between Inorganic, Organic and Organometallic Substrates, *Acc. Chem. Res.* **20**, 214, **1987**.

5.12. C. Moinet *et al.* - Electrochemical Reduction of [FeCp(h⁶-arene)]⁺ in Basic Media, *J. Electroanal. Chem. Interfac. Electrochem.* **241**, 121, **1981**.

5.13. P. Boudeville, A. Darchen - Thermochemical and Kinetic Studies of the Electron-Transfer Catalysis of Arene Replacement by P(OMe)$_3$ Ligands in [Fe^{II}Cp(arene)] Cations, *Inorg. Chem.* **30**, 1663, **1991**.

5.14. **a)** J. Ruiz *et al.* - Arene-Exchange by P donors in the 19-Electron Complexes [Fe^{I}Cp(arene)]: Kinetics, Mechanisms and Salt Effects, *J. Am. Chem. Soc.* **112**, 5471, **1990**;

     b) E. Alonso, D. Astruc - Introduction of the Cluster Fragment Ru$_3$(CO)$_{11}$ at the Periphery of Phosphine Dendrimers Catalyzed by the Electron-Reservoir Complex [Fe^{I}Cp(C$_6$Me$_6$)], *J. Am. Chem. Soc.* **122**, 3222, **2000**.

5.15. C. Roger, C. Lapinte - Synthesis and Characterization of a Stable Iron-Methylene Complex, Methylene Activation by Electron-Transfer Catalysis, *J. Chem. Soc., Chem. Commun.*, 1598, **1989**.

5.16. D. Touchard, J.-L. Fillaut, H. Le Bozec, C. Moinet, P.H. Dixneuf - Monoelectronic Processes in Iron Complexes and Controlled Reactions by Ancillary PR$_3$ Ligands, in *Paramagnetic Species in Activation, Selectivity, Catalysis*, M. Chanon, M. Juliard, J.-C. Poite Eds, Kluwer, Dordrecht, **1989**, 311.

5.17. D. Astruc - Transition Metal Radicals: Chameleon Structure and Catalytic Functions, *Acc. Chem. Res.* **24**, 36, **1991**.

5.18. B.H. Byers, T.L. Brown - Transition Metal Substitution *via* a Radical Chain Pathway, *J. Am. Chem. Soc.* **97**, 947, **1975** and **99**, 2527, **1977**.

5.19. M. Chanon, M.L. Tobe - ETC: A Mechanistic Concept for Inorganic and Organic Chemistry, *Angew. Chem. Int. Ed. Engl.* **21**, 1, **1982**.

Chapter 6 - Insertion and extrusion reactions

6.1. F. Calderazzo - Synthetic and Mechanistic Aspects of Inorganic Insertion Reactions. Insertion of Carbon Monoxide, *Angew. Chem. Int. Ed. Engl.* **16**, 299, **1977**.

6.2. D.F. Shriver *et al.* - Activation of Coordinated CO Toward Alkyl Migratory Insertion by Molecular Lewis Acids, *J. Am. Chem. Soc.* **102**, 5093, **1980**; Surface Induced Alkyl Migration in Metal Carbonyls, *J. Am. Chem. Soc.* **102**, 5112, **1980**.

6.3. T.J. Marks *et al.* - CO Activation by Organoactinides. Migratory CO Insertion into M-H Bonds to Produce Nanonuclear Formyls, *J. Am. Chem. Soc.* **103**, 6959, **1981**.

6.4. R.H. Magnuson, W.P. Giering *et al.* - Detection and Characterization of Radical Cations Resulting from the Oxidation of Methyl and Acetyl Iron Complexes, *J. Am. Chem. Soc.* **102**, 6887, **1980**.

6.5. B.A. Narayanan, C. Amatore, C.P. Casey, J.K. Kochi - Novel Chain Reactions for the Formylmetal to Hydridometal Conversion. Free Radicals, Photochemical and Electrochemical Methods of Initiation, *J. Am. Chem. Soc.* **105**, 6351, **1983**.

6.6. M. Brookhart, M.L.H. Green, L.L. Wang - Carbon-Hydrogen-Transition-Metal Bonds, *Prog. Inorg. Chem.* **36**, 1, **1988**.

6.7. R.H. Crabtree - Iridium Compounds in Catalysis, *Acc. Chem. Res.* **12**, 331, **1979**.

6.8. J. Schwarz - Organozirconium Compounds in Organic Synthesis: Cleavage Reactions of Carbon-Zirconium Bonds, *Pure Appl. Chem.* **52**, 733, **1980**.

6.9. J.C.W. Chien - *Coordination Polymerization*, Academic Press, New York, **1975**.

6.10. P.L. Watson, D.C. Roe - β-Alkyl Transfer in a Lanthanide Model for Chain Termination, *J. Am. Chem. Soc.* **104**, 6471, **1982**.

6.11. R.H. Grubbs *et al.* - Olefin Insertion in a Metal Alkyl in a Ziegler Polymerization System, *J. Am. Chem. Soc.* **107**, 3377, **1985**.

6.12. H.H. Brintzinger *et al.* - Isotope Effect Associated with α-Olefin Insertion in Zirconocene-Based Polymerization Catalysts: Evidence for α-Agostic Transition State, *Angew. Chem. Int. Ed. Engl.* **29**, 1412, **1990**.

6.13. M. Brookhart *et al.* - Implication of Three-Center, Two-Electron M-H-C Bonding for Related Alkyl Migration Reactions: Design and Study of an Ethylene Polymerization Catalyst, *J. Am. Chem. Soc.* **107**, 1443, **1985**; Mechanism of Rh(III) Catalyzed Methyl Acrylate Dimerization, *ibid.* **114**, 4437, **1992**.

6.14. H.H. Brintzinger, D. Fischer, R. Mulhaupt, B. Rieger, R.M. Waymouth - Stereospecific Olefin Polymerization with Chiral Metallocene Catalysts, *Angew. Chem. Int. Ed. Engl.* **34**, 1143, **1995**.

6.15. **a)** W. Kaminsky - Polymerization, Oligomerization and Copolymerization of Olefins in *Applied Homogeneous Catalysis with Organometallic Compounds*, B. Cornils, W.A. Herrmann Eds, VCH, Weinheim, **1996**, Vol. 1, 220;
M. Bochmann - Kinetic and Mechanistic Aspects of Metallocene Polymerization Catalysts. *J. Organomet. Chem.* **689**, 3982, **2004** (review);
b) J.H. Teuben *et al.* - *J. Mol. Cat.* **62**, 277, **1990**;
L. Resconi *et al.* - *J. Am. Chem. Soc.* **114**, 1025, **1992**.

6.16. R.R. Schrock - Alkylidene Complexes of Niobium and Tantalum, *Acc. Chem. Res.* **12**, 98, **1979**.

6.17. **a)** M.L.H. Green - Studies of Synthesis, Mechanism and Reactivity of Some Organo-Molybdenum and Tungsten Compounds, *Pure Appl. Chem.* **50**, 27, **1978**;
b) J.E. Bercaw *et al.* - *J. Am. Chem. Soc.* **108**, 5347, **1986**.

6.18. G.M. Whitesides *et al.* - Ring Strain in Bis(triethylphosphine)-3,3 Dimethyl Platinacyclobutane is Small, *J. Am. Chem. Soc.* **103**, 948, **1981**.

6.19. P.R. Sharp, D. Astruc, R.R. Schrock - Niobium and Tantalum Mesityl Complexes and the Role of the Mesityl Ligand in Hydrogen Abstraction Reactions, *J. Organomet. Chem.* **182**, 477, **1979**.

6.20. T.J. Marks *et al.* - Intra- and Intermolecular Organoactinide Carbon-Hydrogen Activation Pathways. Formation, Properties, and Reactions of Thoracyclobutanes, *J. Am. Chem. Soc.* **104**, 7357, **1982**.

Chapter 7 - Metal carbonyls and complexes of other monohapto L ligands

7.1. F.A. Cotton - Metal Carbonyls: Some New Observations in an Old Field, *Prog. Inorg. Chem.* **21**, 1, **1976**.

7.2. J.E. Ellis - Highly Reduced Metal Carbonyl Anions: Synthesis, Characterization and Chemical Properties, *Adv. Organomet. Chem.* **31**, 1, **1990**.

7.3. G.R. Dobson - Trends in Reactivity for Ligand Exchange Reactions of Octahedral Metal Carbonyls, *Acc. Chem. Res.* **9**, 300, **1976**.

7.4. F. Basolo - Kinetics and Mechanisms of CO Substitution of Metal Carbonyls, *Polyhedron* **9**, 1503, **1990**.

7.5. H. Werner - Complexes of Carbon Monoxide: An Organometallic Family Celebrating its Birthday, *Angew. Chem. Int. Ed. Engl.* **29**, 1077, **1990**.

7.6. **a)** R. Poli - Open-Shell Organometallics as a Bridge Between Werner-Type and Low-Valent Organometallic Complexes. The Effect of the Spin State on the Stability, Reactivity and Structure, *Chem. Rev.* **95**, 2135, **1995**;
Molybden Open-Shell Organometallics: Spin State Changes and Pairing Energy Effects, *Acc. Chem. Res.* **30**, 494, **1997**; *J. Organomet. Chem.* **689**, 4291, **2004**;
b) A.E. Stiegman, D.R. Tyler - Reactivity of 17- and 19-Electron Complexes in Organometallic Chemistry, *Comments Inorg. Chem.* **5**, 215, **1986**.

7.7. L.G. Hubert-Pfalzgraf *et al.* - Preparation and Structure of a Mixed Nobium (I) Isocyanide Carbonyl Complex with a Bent C-N-C Linkage, *Inorg. Chem.* **30**, 3105, **1991**.

7.8.   I.S. Butler - Coordination Chemistry of the Thiocarbonyl Ligand,
        Pure Appl. Chem. **88**, 991, **1988**.

7.9.   P.V. Broadhurst - Transition Metal Thiocarbonyl Complexes,
        Polyhedron **4**, 1801, **1985**.

7.10. A.D. Allen, F. Bottomley - Inorganic Nitrogen Fixation: Nitrogen Compounds
        of the Transition Metals, *Acc. Chem. Res.* **1**, 360, **1968**.

7.11. R.R. Schrock - High Oxidation State Complexes [W(η^5-C$_5$Me$_5$)Me$_2$X] (X = OAr or
        SAr) and [MoCp*Me$_3$]$_2$ (μ-N$_2$), *J. Am. Chem. Soc.* **112**, 4331 and 4338, **1990**.

7.12. A.E. Shilov - *Metal Complexes in Biomimetic Chemical Reactions*,
        CRC Press, New York, **1996**, Chap. I (Nitrogen Fixation).

7.13. G.B. Richter-Addo, P. Legzdins - Organometallic Nitrosyl Chemistry,
        Chem. Rev. **88**, 991, **1988**.

7.14. C.A. Tolman - Steric Effects of Phosphorous Ligands in Organometallic Chemistry
        and Catalysis, *Chem. Rev.* **77**, 313, **1977**.

Chapter 8 - Metal-alkyl and -hydride complexes and other complexes of monohapto X ligands

8.1.   **a)** R.R. Schrock, G.W. Parshall - σ-Alkyl and -Aryl Complexes
        of the Group 4-7-Transition Metals, *Chem. Rev.* **76**, 243, **1976**;

        b) P.J. Davison, M.F. Lappert, R. Pearce - Metal σ-Hydrocarbyls, MR$_n$.
        Stoichiometry, Structures, Stabilities, and Thermal Decomposition Pathways,
        Chem. Rev. **76**, 219, **1976**.

8.2.   **a)** G. Wilkinson, *Pure Appl. Chem.* **71**, 627, **1959**.

        b) G.M. Whitesides *et al.* - Mechanism of Thermal Decomposition of
        [Pt^{II}(*n*-Bu)$_2$(PPh$_3$)$_2$], and of Pt^{II}(PR$_3$)$_2$ Metallocycles,
        J. Am. Chem. Soc. **94**, 5258, **1972** and **98**, 6521, **1978**;

8.3.   T.J. Marks - Bonding Energetics in Organometallic Compounds,
        ACS Symp. Series, Washington DC, **1990**.

8.4.   R. Usón, J.Forniés - Synthesis and Structures of Novel Types of Pt-M
        (M = Ag, Sn or Pb) Neutral or Anionic Organometallic Complexes,
        Inorg. Chim. Acta **198**, 165, **1992**.

8.5.   M. Brookhart, M.L.H. Green - Carbon-Hydrogen-Transition Metal Bonds,
        J. Organomet. Chem. **250**, 395, **1983**.

8.6.   **a)** V. Vidal, A. Theolier, J. Thivolle Cazat, J.-M. Basset - Metathesis of Alkanes
        Catalyzed by Silica Supported Transition Metal Hydrides, *Science* **276**, 99, **1997**;

        b) J. Corker, F. Lefebvre, C. Lecuyer, V. Dufaud, F. Quignard, A. Choplin, J. Evans,
        J.-M. Basset - Catalytic Cleavage of the C-H and C-C Bond of Alkanes by
        Surface Organometallic Chemistry: An EXAFS and IR Characterization of
        a Zr-H Catalyst, *Science* **271**, 966, **1996**.

8.7.   K.G. Caulton, L.G. Hubert-Pfalzgraf - Synthesis, Structural Principles and Reactivity
        of Heterometallic Alkoxides, *Chem. Rev.* **90**, 969, **1990**.

8.8.   P.R. Sharp, D. Astruc, R.R. Schrock - Niobium and Tantalum Mesityl Complexes
        and the Role of the Mesityl Ligand in Hydrogen Abstraction Reactions,
        J. Organomet. Chem. **182**, 477, **1979**.

8.9.   **a)** S.J. Mc Lain, R.R. Schrock, P.R. Sharp *et al.* - Synthesis of the Monomeric
        Niobium- and Tantalum-Benzyne Complexes and the Molecular Structure of
        [Ta(η^5-C$_5$Me$_5$)(C$_6$H$_4$)Me$_2$], *J. Am. Chem. Soc.* **101**, 263, **1979**;

        b) S.L. Buchwald, Q. Fang - *J. Org. Chem.* **194**, 45, **1989**.

8.10. **a)** T. Mallah, S. Thiébaut, M. Verdaguer, P. Veillet - High TC Molecular-Based Magnets: Ferrimagnetic Mixed-Valence Chromium(III)-Chromium(II) Cyanides with TC = 240 and 190K, *Science* **262**, 1554, **1993**;
b) M. Verdaguer - Molecular Electronics Emerges in Molecular Magnetism, *Science* **272**, 698, **1996**.

8.11. N. Le Narvor, L. Toupet, C. Lapinte - Elemental Carbon Bridging Two Iron Centers. Syntheses and Spectroscopic Properties of $Cp^*(dppe)FeC_4\text{-}FeCp^*(dppe)^{n+}nPF_6^-\cdot$X-Ray Crystal Structure of the Mixed Valence Complex (n = 1), *J. Am. Chem. Soc.* **117**, 7129, **1995**.

8.12. J.A. Gladysz *et al.* - Consanguineous of Coordinated Carbon: a ReC_4Re Assembly that is Isolable in Three Oxidation States Including Cristallographically Characterized $ReC\equiv CC\equiv CRe$ and $^+Re=C=C=C=C=Re^+$ Adducts and a Radical Cation in Which Charge is Delocalized Between Rhenium Termini, *J. Am. Chem. Soc.* **119**, 775, **1997**; S. Szafert, J.A. Gladysz - Carbon in One Dimension. Structural Analysis of the Higher Conjugated Polyynes, *Chem. Rev.* **103**, 4175, **2003**.

8.13. T.D. Tilley - in *The Chemistry of Organosilicon Compounds*, S. Patai Ed., Wiley, New York, **1989**, Vol. 2, 1415.

8.14. A. Dedieu Ed. - *Transition Metal Hydrides*, VCH, Weinheim, **1992**.

8.15. R. Schollhorn - Intercalation Compounds, in *Inclusion Compounds*, Academic Press, London, **1984**, 249-349.

8.16. G.J. Kubas - Molecular Hydrogen Complexes: Coordination of a σ Bond to Transition Metal, *Acc. Chem. Res.* **21**, 190, **1988**.

8.17. **a)** T.Y. Lee, L. Messerle - Utility of Hydridotributyltin as Both Reductant and Hydride Transfer Reagent in Organotransition Metal Chemistry I. A Convenient Synthesis of the Organoditantalum (IV) Hydride $Cp^*_2Ta_2(\mu^2\text{-}H)_2Cl_4$ from Cp^*TaCl_4, and the Probes of the Possible Reaction Pathways, *J. Organomet. Chem.* **553**, 397, **1998**;
b) V.D. Parker, M. Tilset *et al.* - Electrode Potentials and the Thermodynamics of Isodesmic Reactions, *J. Am. Chem. Soc.* **113**, 7493, **1991** and **115**, 7493, **1993**.

8.18. L.M. Venanzi - Transition Metal Complexes with Bridging Hydride Ligands, *Coord. Chem. Rev.* **43**, 251, **1982**.

8.19. R.H. Crabtree - Dihydrogen Complexes: Some Structural and Chemical Studies, *Acc. Chem. Res.* **23**, 95, **1990**; *Angew. Chem. Int. Ed. Engl.* **32**, 789, **1993**.

8.20. R.H. Crabtree, O. Eisenstein, A.L. Rheingold *et al.* - A New Intermolecular Interaction: Unconventional Hydrogen Bonds with Element-Hydride Bonds as Proton Acceptor, *Acc. Chem. Res.* **29**, 348, **1996**.

8.21. **a)** Y. Ouari, S. Sabo-Etienne, B. Chaudret - Exchange Coupling Between a Hydride and a Stretched Dihydrogen Ligand in Ruthenium Complexes, *J. Am. Chem. Soc.* **120**, 4228, **1998**;
b) M.A. Esteruelas, L. Oro - Dihydrogen Complexes as Homogeneous Reduction Catalysts, *Chem. Rev.* **98**, 577, **1998**.

8.22. P. Hamon, L. Toupet, J.-R. Hamon, C. Lapinte - Novel Diamagnetic and Paramagnetic Fe^{II}, Fe^{III} and Fe^{IV} Classical and Non-Classical Hydrides. X-Ray Crystal Structure of $[Fe(C_5Me_5)(dppe)D]PF_6$, *Organometallics* **11**, 1429, **1992**.

8.23. R.D. Simpson, R.G. Bergman - Synthesis, Structure, and Exchange-Reactions of Rhenium Alkoxide and Aryloxide Complexes. Evidence for Both Proton and H-Atom Transfer in the Exchange Transition State, *Organometallics* **12**, 781, **1993**.

8.24. R.G. Bergman *et al.* - Formation, Reactivity and Properties of Nondative Late Transition Metal-Oxygen and Nitrogen Bonds, *Acc. Chem. Res.* **35**, 44, **2002**.

Chapter 9 - Metal-carbene and -carbyne complexes and multiple bonds with transition metals

9.1.   K.H. Dötz *et al.* - *Transition Metal Carbene Complexes*,
Verlag Chemie, Weinheim, **1983**.

9.2.   R.R. Schrock - Alkylidene Complexes of Niobium and Tantalum,
Acc. Chem. Res. **12**, 98, **1979**; *Science* **13**, 219, **1983**.

9.3.   **a)** E.O. Fischer, U. Schubert, H. Fisher - Selectivity and Specificity in Chemical
Reactions of Carbene and Carbyne Metal Complexes,
Pure Appl. Chem. **50**, 857, **1978**;

b) P.J. Brothers, W.R. Roper - Transition-Metal Dihalocarbene Complexes,
Chem. Rev. **88**, 1293, **1988**.

9.4.   V. Guerchais, D. Astruc - New Entry to (C$_5$Me$_5$)Fe Chemistry and the Methylene
Complexes [Fe(η^5-C$_5$Me$_5$)(CO)(L)(=CH$_2$)]$^+$, L = CO or PPh$_3$,
J. Chem. Soc., Chem. Commun., 835, **1985**.

9.5.   M. Brookhart, W.B. Studabaker - Cyclopropanes from Reactions of Transition-
Metal-Carbene Complexes and Olefins, *Chem. Rev.* **87**, 411, **1987**.

9.6.   T.E. Taylor, M.B. Hall - Theoretical Comparison Between Nucleophilic and
Electrophilic Transition-Metal Carbenes Using Generalized MO and CI Methods,
J. Am. Chem. Soc. **106**, 1576, **1984**.

9.7.   **a)** W.A. Herrmann - The Methylene Bridge, *Adv. Organomet. Chem.* **20**, 159, **1982**;

b) F.N. Tebbe *et al.* - Olefin Homologation with Titanium Methylene Compounds,
J. Am. Chem. Soc. **100**, 3611, **1978**.

9.8.   **a)** M.I. Bruce - Transition-Metal Complexes Containing Allenylidene, Cumul-
enylidene and Related Ligands, *Chem. Rev.* **91**, 197, **1991** and **98**, 2797, **1998**;

b) V. Cardieno, M.P. Gamasa, J. Gimeno - *Eur. J. Inorg. Chem.*, 571, **2001**;

c) H. Werner - *Chem. Commun.*, 903, **1997**; Y. Wakatsuki - Mechanistic Aspects
Regarding the Formation of Metal-Vinylidenes from Alkynes and Related
Reactions, *J. Organomet. Chem.* **689**, 4092, **2004** (review);

d) R.H. Grubbs *et al.* - *J. Am. Chem. Soc* **114**, 3974, **1992** and **115**, 9858, **1993**;
For the Synthesis of Grubbs-type Ru-Alkylidene and Benzylidene Complexes that
Catalyze Olefin Metathesis, see ref. **15.1**, **15.3**, **15.11** and **15.12** by the Grubbs,
Nolan, Herrman and Fürstner groups;

e) S. Rigaut, D. Touchard, P.H. Dixneuf - Ruthenium-Allenylidene Complexes and
their Specific Behavior, *Coord. Chem. Rev.* **248**, 1585, **2004**;

f) S. Szafert, J.A. Gladysz - *Chem. Rev.* **103**, 4175, **2003**,; M.P. Cifuentes,
M.G. Humphrey - Alkynyl Compounds and Nonlinear Optics,
J. Organomet. Chem. **689**, 3968, **2004** (review);

g) J.P. Selegue - *Organometallics* **1**, 217, **1982**.

9.9.   **a)** R.R. Schrock *et al.* - Synthesis of Molybdenum Imido Alkylidene Complexes and
some Reactions Involving Acyclic Olefins, *J. Am. Chem. Soc.* **112**, 3875 **1990**;
J. Feldman, R.R. Schrock - *Prog. Inorg. Chem.* **39**, 1, **1991**;

b) R.R. Schrock - *Chem. Rev.* **101**, 145, **2001**.

9.10. D. Mansuy - Dichlorocarbene Complexes of Iron (II) Porphyrins-Crystal and
Molecular Structure of [Fe(TPP)(CCl$_2$)(H$_2$O)],
Angew. Chem. Int. Ed. **17**, 781, **1978**.

9.11. **a)** J.P. Collman - Disodium Tetracarbonylferrate, a Transition Metal Analog of a
Grignard Reagent, *Acc. Chem. Res.* **8**, 342, **1975**;

b) M.F. Semmehack, R. Tamura - *J. Am. Chem. Soc.* **105**, 4099, **1983**;

c) M.Y. Daresbourg - *Prog. Inorg. Chem.* **33**, 221, **1985**;

d) A. Loupy, B. Tchoubar, D. Astruc - Salt Effects Resulting from Exchange
Between Two Ion Pairs and their Crucial Role in Reaction Pathways,
Chem. Rev. **92**, 1141, **1992**.

9.12. a) M.F. Lappert *et al.* - Synthetic Routes to Electron-Rich Olefin-Derived Monocarbene-Rh^I Neutral and Cationic Complexes, *J. Chem. Soc., Dalton Trans.*, 883 and 893, **1984**;

   b) H. Le Bozec, A. Gorgues, P.H. Dixneuf - Novel Route to Iron-Carbene Complexes *via* η^2-CS_2 Derivatives. 1, 3-Dithiolium Species as Precursors for Dithiolene-Iron Complexes and Tetrathiafulvalenes, *J. Am. Chem. Soc.* **100**, 3946, **1978**; *Inorg. Chem.* **20**, 2486 and 3929, **1981**.

9.13. a) H. Rudler *et al.* - Dihydropyridines in Organometallic Synthesis. Formation of Pyridine and Dihydropyridine-Stabilized Alkylidene Complexes of W^0 and Cr^0 from Fischer Carbene Complexes: Structure and Reactivity, *J. Am. Chem. Soc.* **118**, 12045, **1996**;

   b) K.H. Dötz - Carbene Complexes in Organic Syntheses, *Angew. Chem. Int. Ed. Engl.* **23**, 587, **1984**;

   c) R.R. Schrock *et al.* - *J. Am. Chem. Soc.* **110**, 1423, **1988**;

   d) K.G. Caulton - Factors Governing the Equilibrium Between Metal-alkyl, Alkylidene and Alkylidyne: MCX_2R , X MCXR and X_2MCR, *J. Organomet. Chem.* **617-618**, 56, **2001**.

9.14. a) R.H. Grubbs - [$TiCp_2(CH_2)$] Complexes in Synthetic Applications, *Pure Appl. Chem.* **55**, 1733, **1983**;

   b) S. Nlate, V. Guerchais, C. Lapinte *et al.* - One-Electron Reduction of the Iron-Carbene Complexes [FeCp*$(CO)_2$ {C=(OMe)R}]$^+$ (R = H, Me) and H-Atom vs. Alkyl Transfer in the 19-Electron Intermediates, *J. Organomet. Chem.* **428**, 49, **1992**.

9.15. H. Fisher *et al.* - *Carbyne Complexes*, VCH, Weinheim, **1988**.

9.16. R.R. Schrock *et al.* - Metathesis of Acetylenes by W^{VI}-Alkylidyne Complexes, *J. Am. Chem. Soc.* **103**, 3932, **1981** and **102**, 4291, **1982**; ref. **9.15**, Chap. 5.

9.17. W.A. Nugent, J.M. Mayer - *Metal-Ligand Multiple Bonds. The Chemistry of the Transition-Metal Complexes Containing Oxo, Nitrido or Alkylidene Ligands*, Wiley, New York, **1988**.

9.18. D.E. Wigley - Organoimido Complexes of the Transition Metals, *Prog. Inorg. Chem.* **42**, 239, **1994**.

9.19. F. Mathey - The Formal Analogy Between the Species R_2C:, R_2Si:, RN: and RP:, *Angew. Chem. Int. Ed. Engl.* **26**, 275, **1987**.

9.20. A.H. Cowley *et al.* - Cleavage of a P = C Bond and Formation of a Linear Terminal Phosphinidene Complex, *J. Am. Chem. Soc.* **112**, 6734, **1990**.

9.21. G. Huttner, K. Evertz - Phosphinidene Complexes and their Higher Homologues, *Acc. Chem. Res.* **19**, 406, **1986**.

9.22. M.F. Lappert *et al.* - The First Stable Transition Metal (Mo or W) Complexes Having a Metal-P^{III} Double Bond: the Phosphorous Analogue of Metal Aryl- and Alkyl-imides: X-Ray Structure of [$MoCp_2$(=P-$C_6H_2Bu^t_3$-2,4,6)], *Chem. Commun.*, 1282, **1987**.

Chapter 10 - π Complexes of mono- and polyenes and enyls

10.1. M.J.S. Dewar - A Review of the π-Complex Theory, *Bull. Soc. Chim. Fr.* **18**, C79, **1951**.

10.2. R. Hoffmann *et al.* - Ethylene Complexes. Bonding, Rotational Barriers, and Conformational Preferences, *J. Am. Chem. Soc.* **101**, 3801, **1979**.

10.3. P.J. Fagan *et al.* - Metal Complexes of Buckminsterfullerene (C_{60}), *Science* **252**, 1160, **1991**; *Acc. Chem. Res.* **25**, 134, **1992** (the whole Acc. Chem. Res. issue is dedicated to fullerenes).

10.4. M. Rosenblum - Organoiron Complexes as Potential Reagents in Organic Synthesis, *Acc. Chem. Res.* **7**, 122, **1974**.

10.5. A.L. Balch *et al.* - Ir (η^2-C$_{70}$)(CO)(Cl)(PPh$_3$)$_2$: the Synthesis and Structure of an Iridium Organometallic Derivative of a Higher Fullerene, *J. Am. Chem. Soc.* **113**, 8953, **1991**.

10.6. M.B. Trost - New Rules of Selectivity: Allylic Alkylations Catalyzed by Palladium, *Acc. Chem. Res.* **13**, 385, **1980**.

10.7. M.L.H. Green *et al.* - Diene and Arene Compounds of Zirconium and Hafnium, *J. Chem. Soc., Dalton Trans.*, 2641, **1992**.

10.8. A.M. Madonik, D. Astruc - Electron-Transfer Chemistry of the 20-Electron Complex [Fe(C$_6$Me$_6$)$_2$] and its Strategic Role in C-H Bond Activation, *J. Am. Chem. Soc.* **106**, 3381, **1984**.

10.9. L. Ricard, R. Weiss *et al.* - Binding and Activation of Enzymatic Substrates by Metal Complexes. Structural Evidence for Acetylene as a Four-Electron Donor in [W(O)(C$_2$H$_2$)(S$_2$CNEt$_2$)$_2$], *J. Am. Chem. Soc.* **100**, 1318, **1978**.

10.10. C. Floriani *et al.* - Structure and Properties of [Ti(CO)Cp$_2$(PhC$_2$Ph)], *J. Chem. Soc., Dalton Trans.*, 1398, **1978**.

10.11. J.L. Templeton - Four-Electron Alkyne Ligands in Molybdenum (II) and Tungsten (II) Complexes, *Adv. Organomet. Chem.* **29**, 1, **1989**.

10.12. S.J. Mc Lain, R.R. Schrock, P.R. Sharp *et al.* - Synthesis of the Monomeric Niobium- and Tantalum-Benzyne Complexes and the Molecular Structure of [Ta(η^5-C$_5$Me$_5$)(C$_6$H$_4$)Me$_2$], *J. Am. Chem. Soc.* **101**, 263, **1979**.

10.13. M.A. Bennett *et al.* - Stabilization of 1,2,4,5-Tetrahydrobenzene by Complexation at Two Ni^0 Centers, *Angew. Chem. Int. Ed. Engl.* **97**, 941, **1988**.

10.14. D. Seyferth - Chemistry of Carbon-Functional Alkylidenetricobalt Nonacarbonyl Cluster Complexes, *Adv. Organomet. Chem.* **120**, 97, **1976**.

10.15. P. Maitlis *et al.* - The Pd^{II}-Induced Oligomerization of Acetylenes: an Organometallic Detective Story, *Acc. Chem. Res.* **9**, 93, **1976**.

10.16. R. Baker - Allyl Metal Derivatives in Organic Synthesis, *Chem. Rev.* **73**, 487, **1973**.

10.17. M.L.H. Green - Allyl Metal Complexes, *Adv. Organomet. Chem.* **2**, 325, **1964**.

10.18. M.L.H. Green *et al.* - Cycloheptatriene and -enyl Complexes of the Early Transition Metals, *Chem. Rev.* **95**, 439, **1995**.

10.19. F.A. Cotton, G. Wilkinson, P.L. Gauss - *Basic Inorganic Chemistry*, 3rd ed., Wiley, New York, **1995**, Chap. 18-20.

10.20. J. Okuda - Transition Metal Complexes of Sterically Demanding Cyclopentadienyl Ligands, *Top. Curr. Chem.* **160**, 97, **1992**.

10.21. B. Gloaguen, D. Astruc - Chiral Pentaisopropyl- and Pentaisopentyl Cyclopentadienyl Complexes: One-Pot Synthesis by Formation of Ten Carbon-Carbon Bonds from Pentamethylcobaltocenium, *J. Am. Chem. Soc.* **112**, 4607, **1990**.

10.22. F. Mathey - From Phosphorous Heterocycles to Phosphorous Analogues of Unsaturated Hydrocarbon-Transition Metal π Complexes, *J. Organomet. Chem.* **400**, 149, **1990**; Phospha-Organic Chemistry: Panorama and Perspectives, *Angew. Chem. Int. Ed.* **42**, 1578, **2003**.

10.23. H. Taube - Chemistry of Ru^{II} and Os^{II} Amines, Pure Appl. Chem. **51**, 901, **1979**; W.D. Harman, H. Taube, Synthesis of [Os(NH$_3$)$_5$(η^2-TMB)]$^{2+}$: a Valuable Precursor for Pentaammineosmium (II) Chemistry - *Inorg. Chem.* **26**, 2917, **1987**; *J. Am. Chem. Soc.* **109**, 1883, **1987** and **108**, 8223, **1986**.

10.24. F. Rose *et al.* - [Cr(η^6-arene)(CO)$_3$] and [Mn(η^5-cyclohexadienyl)(CO)$_3$] Complexes: Indirect Nucleophilic Substitutions, *Coord. Chem. Rev.* **178-179**, 249, **1998**.

Chapter 11 - Metallocenes and sandwich complexes

11.1. G. Wilkinson - The Iron Sandwich. A Recollection of the First Four Months,
  J. Organomet. Chem. **100**, 273, **1975**.

11.2. A. Togni, T. Hayashi Eds - *Ferrocenes*, VCH, Weinheim, **1995**.

11.3. D. Astruc - *Electron Transfer and Radical Processes in Transition Metal Chemistry*,
  VCH, New York, **1995**, 142.

11.4. O. Kahn - *Molecular Magnetism*, VCH, New York, **1994**, 294.

11.5. D. Astruc - The Use of Organoiron Compounds in Aromatic Chemistry,
  Top. Curr. Chem. **160**, 47, **1992**.

11.6. J.H. Ammeter *et al.* - On the Low-Spin-High-Spin Equilibrium of Manganocene and
  Dimethylmanganocene, *J. Am. Chem. Soc.* **96**, 7833, **1974**.

11.7. Y. Mugnier, C. Moïse, J. Tirouflet, E. Laviron - Réduction électrochimique du
  ferrocène, *J. Organomet. Chem.* **186**, C49, **1980**.

11.8. A.J. Bard, V.V. Strelets *et al.* - Electrochemistry of Metallocenes at Very Negative
  and Very Positive Potentials. Electrogeneration of Cp_2Co^{2+}, Cp_2Co^{2-} and Cp_2Ni^{2+}
  Species, *Inorg. Chem.* **32**, 3528, **1993**.

11.9. T.J. Keally, P.L. Pauson - A New Type of Organoiron Compound,
  Nature **168**, 1039, **1951**.

11.10.a) H. Sitzmann *et al.* - Pentaisopropylcyclopentadienyl: Singlet Anion, Doublet
  Radical and Triplet Cation of a Carbocyclic π System,
  J. Am. Chem. Soc. **115**, 12003, **1993**;
  b) G.E. Herberich *et al.* - [FeCp*CpR] Derivatives. A New and Highly Selective
  Synthesis, *Organometallics* **17**, 5931, **1998**.

11.11. S.A. Miller, J.A. Tebboth, J.F. Tremaine - Dicyclopentadienyliron,
  J. Chem. Soc., 632, **1952**.

11.12. Gmelin Handbook of Inorganic and Organometallic Chemistry, 8th ed., Fe Organic
  Compounds, Vols. A1-A8 (**1974**, **1977**, **1978**, **1980**, **1981**, **1985**, **1989**, **1991**).

11.13. B.W. Rockett, G. Marr - Ferrocene Chemistry,
  J. Organomet. Chem. **416**, 327, **1991** and cited references.

11.14. G.E. Herberich - A Novel and General Route to Metal (η^5-Borole)Complexes:
  Compounds of Manganese, Ruthenium and Rhodium,
  Angew. Chem. Int. Ed. Engl. **22**, 996, **1983**.

11.15.a) D. Buchholz *et al.* - Mono- and Bis-Pentaisopropylcyclopentadienyl Cobalt and
  Rhodium Sandwich Complexes and other Decabranched Cyclopentadienyl
  Complexes, *Chem. Eur. J.* **1**, 374, **1995**;
  b) E. Carmona *et al.* - *Angew. Chem. Int. Ed. Engl.* **39**, 1949, **2000**; *Chem. Commun.*,
  2916, **2000**; Synthesis, Solid-State Structure, and Bonding Analysis of the Octa,
  Nona, and Decamethylberyllocenes, *Chem. Eur. J.* **9**, 4452, **2003**.

11.16.a) H. Werner, A. Salzer - The Triple -Decker Sandwich [Ni_2Cp_3]$^+$,
  Angew. Chem. Int. Ed. Engl. **11**, 930, **1972**;
  b) W. Siebert - 2,3-Dihydro-1,3-Diborole-Metal Complexes with Activated
  C-H Bonds: Building Blocks for Multilayered Compounds,
  Angew. Chem. Int. Ed. Engl. **24**, 943, **1985**;
  c) J.E. Bercaw *et al.* - Synthesis of Decamethyltitanocene and its Reactions with
  Dinitrogen, *J. Am. Chem. Soc.* **100**, 3078, **1978**.

11.17. E.L. Muetterties, T.A. Albright *et al.* - Structural, Stereochemical and Electronic
  Features of Arene-Metal Complexes, *Chem. Rev.* **82**, 499, **1982**.

11.18. a) Ch. Elschenbroich - μ-(η^6,η^6-Biphenyl)-bis[Cr(η^6-C$_6$H$_6$)] and
Bis [μ-(η^6,η^6-biphenyl)]-dichromium. Novel Species to Explore Mixed-Valence
Sandwich Complex Chemistry, *J. Am. Chem. Soc.* **101**, 6773, **1979**;

b) M.A. Bennett - *J. Organomet. Chem.* **175**, 87, **1979**;

c) V. Boekelheide *et al.* - *J. Am. Chem. Soc.* **108**, 3324 and 7010, **1986**.

11.19. a) J. Ruiz *et al.* - First 17-18-19-Electron Triads of Stable Isostructural Organometal-
lic Complexes. The 17-Electron Complexes [Fe(C$_5$R$_5$)(arene)]$^{2+}$ (R = H or Me), a
Novel Family of Strong Oxidants: Isolation, Characterization, Electronic
Structure and Redox Properties, *J. Am. Chem. Soc.* **120**, 11693, **1998**;

b) H. Trujillo *et al.* - Thermodynamics of C-H Activation in Multiple Oxidation
States: Compared Benzylic C-H Acidities and C-H Bond Dissociation Energies in
the Isostructural 16 to 20-Electron Complexes [Fe^x(η^5-C$_5$R$_5$)(η^6-arene)]^n,
x = 0-IV, R = H or Me; n = –1 to +3, *J. Am. Chem. Soc.* **121**, 5674, **1999**;

c) D. Astruc - Transition-Metal Sandwiches as Reservoirs of Electrons, Protons,
Hydrogen Atoms and Hydrides, in *Mechanism and Processes in Molecular
Chemistry*, D. Astruc Ed., Gauthier-Villars, Paris, *New J. Chem.* **16**, 305, **199**;
Acc. Chem. Res. **19**, 377, **1986**; **24**, 36, **1991**; **30**, 383, **1997** and **33**, 287, **2000**.

11.20 a) P. Michaud *et al.* - Electron-Transfer Pathways in the Reduction of d^6 and d^7
Organo-Iron Cations by LiAlH$_4$ and NaBH$_4$, *J. Am. Chem. Soc.* **104**, 3755, **1982**;

b) D. Astruc *et al.* - Organometallic Electron Reservoirs. A Novel Mode of
C-H Activation Using Dioxygen *via* Superoxide Radical Anion in Solution and
in the Solid State with C$_5$R$_5$Fe^{I}C$_6$R'$_6$. Subsequent Bond Formation with C, Si, P,
Mn, Fe, Cr, Mo and Halogens, *J. Am. Chem. Soc.* **103**, 7502, **1981**;

c) J.-R. Hamon *et al.* - Syntheses, Characterization and Stereoelectronic
Stabilization of Organometallic Electron-Reservoirs: the 19-Electron d^7 Redox
Catalysts [Fe^I(η^5-C$_5$R$_5$)(η^6-C$_6$R'$_6$)], *J. Am. Chem. Soc.* **103**, 758, **1981**.

11.21 a) C. Valério *et al.* - The Dendritic Effect in Molecular Recognition: Ferrocene
Dendrimers and their Use as Supramolecular Redox Sensors for the Recognition
of Small Inorganic Anions, *J. Am. Chem. Soc.* **119**, 2588, **1997**;

b) C. Valério *et al.* - A Polycationic Metallodendrimer with
24 [FeCp*(η^6-N-Alkylaniline)]$^+$ Termini That Recognizes Chloride
and Bromide Anions, *Angew. Chem. Int. Ed. Engl.* **38**, 1747, **1999**.

11.22 a) D.S. Brown *et al.* - Electron-Reservoir Complexes Fe^{I}Cp(η^6-arene) as Selective
Initiators for a Novel Electron-Transfer-Chain Catalyzed Reaction: General
Synthesis of Fulvalene-Bridged Homo- and Heterobimetallic Zwitterions,
Angew. Chem. Int. Ed. Engl. **33**, 661, **1994**;

b) A. Buet *et al.* - *J. Chem. Soc., Chem. Commun.*, 447, **1979**;

c) S. Rigaut *et al.* - Triple C-H/N-H Activation by O$_2$ for Molecular Engineering:
Heterobifunctionalization of the 19-Electron Redox Catalysts Fe^{I}Cp(η^6-arene),
J. Am. Chem. Soc. **119**, 11132, **1997**;

d) J. Ruiz Aranzaes *et al.* - Metallocenes as References for the Determination
of Redox Potentials by Cyclic Voltammetry, *Can. J. Chem.* **84**, 228, **2006**.

11.23 a) D. Astruc *et al.* - Design, Stabilization and Efficiency of Organometallic
"Electron Reservoirs". 19-Electron Sandwiches [Fe^I(η^5-C$_5$R$_5$)(η^6-C$_6$R'$_6$)],
a Key Class Active in Redox Catalysis, *J. Am. Chem. Soc.* **101**, 5445, **1979**;

b) J-R. Hamon *et al.* - Multiple Formation and Cleavage of C-C Bonds
in CpFe$^+$(η^6-arene) Sandwiches and the Unusual C$_6$Et$_6$ Geometry in the X-Ray
Crystal Structure of [Fe^{II}(η^6-C$_6$Et$_6$)]$^+$PF$_6^-$, *J. Am. Chem. Soc.* **104**, 7549, **1982**;

c) J.-L. Fillaut *et al.* - Single-Step Six-Electron Transfer in a Heptanuclear Complex:
Isolation of Both Redox Forms, *Angew. Chem. Int. Ed. Engl.* **33**, 2460, **1994**.

Chapter 12 - Ionic and polar metal-carbon bonds: alkali and rare earth complexes

12.1. M.E. O'Neil, K. Wade - *Structural and Bonding Relationships among Main Group Organometallic Compounds,* in *Comprehensive Organometallic Chemistry,* G. Wilkinson *et al.* Eds, Pergamon, Oxford, **1982,** Vol. 1.1.

12.2. a) A. Loupy, B. Tchoubar, D. Astruc - Salt Effects in Organic and Organometallic Chemistry, *Chem. Rev.* **92,** 1141, **1992;**

b) W. Jenneskens *et al.* - Water-Soluble Poly(4,7,10,13-Tetraoxatetradecylmethyl silane): Enhanced Yield and Improved Purity *via* Polymerization Using C_8K as Reducing Agent, *J. Chem. Soc., Chem. Commun.,* 329, **1997;**

c) A. Fürstner - Chemistry of and with Highly Reactive Metals, *Angew. Chem. Int. Ed.* **32,** 164, **1993.**

12.3. G. Fraenkel, H. Hsu, B.M. Su - The Structure and Dynamic Behavior of Organolithium Compounds in Solution, ^{13}C, ^6Li and ^7Li NMR, in *Lithium Current Applications in Science, Medecine and Technology,* R. Balch Ed., Wiley, New York, **1985.**

12.4. a) O.A. Reutov, I.P. Beletskaya, K.P. Butin - *C-H Acids,* Pergamon, Oxford, **1978;**

b) J.R. Jones - *The Ionization of Carbon Acids,* Academic Press, London, **1973.**

12.5. H. Normant - Chemistry of Organomagnesiums after V. Grignard, *Pure Apl. Chem.* **30,** 463, **1972;** Alkenylmagnesium Halides, *Adv. Org. Chem. Methods Results* **2,** 1, **1960.**

12.6. J. Villeras - α-Haloalkyl Grignard Reagents, *Organomet. Chem. Rev.* **A(7),** 81, **1971.**

12.7. G. Courtois, L. Miginiac - Allyl Derivatives of Mg, Zn and Cd, *J. Organomet. Chem.* **69,** 1, **1974.**

12.8. P. Knochel *et al.* - Preparation and Reactions of Polyfunctional Organozinc Reagents in Organic Synthesis, *Chem. Rev.* **93,** 2117, **1993;** *J. Am. Chem. Soc.* **114,** 3983, **1992;** *J. Organomet. Chem.* **680,** 126, **2003.**

12.9. H.M. Walborsky - Mechanism of Grignard Reagent Formation. The Surface Nature of the Reaction, *Acc. Chem. Res.* **23,** 286, **1990;** M. Chanon, J.-C. Negrel, N. Bodineau, J.-M. Mattalia, E. Peralez, A. Goursot - Electron Transfer at the Solid Liquid Interface; New Insights on the Mechanism of Formation of the Grignard Reagent, *Macromol. Symp.* **134,** 13, **1998.**

12.10. P. Knochel *et al.* - Highly Functionalized Organomagnesium Reagents Prepared through Halogen-Metal Exchange, *Angew. Chem. Int. Ed.* **42,** 4302, **2003.**

12.11. E.C. Ashby *et al.* - Stereochemistry of Organometallic Compound Addition to Ketones, *Chem. Rev.* **75,** 521, **1975;** The Mechanism of Grignard Reagent Addition to Ketones, *Acc. Chem. Res.* **7,** 272, **1974.**

12.12. T.J. Marks - Chemistry and Spectroscopy of f-Elements Organometallics, *Prog. Inorg. Chem.* **25,** 223, **1979;** *Science* **217,** 989, **1982;** K.N. Raymond, C.W. Eigenbrot, Jr. - Structural Criteria for the Mode of Bonding of Organo-Actinides and -Lanthanide Compounds, *Acc. Chem. Res.* **13,** 276, **1980.**

12.13. P. Mountford, B.D. Ward - Recent Development in the Non-cyclopentadienyl Organometallic and Related Scandium Chemistry, *Chem.Commun.* (Feature Article), 1797, **2003.**

12.14. H. Schumann - Homoleptic Rare-Earth Compounds, *Comments Inorg. Chem.* **2,** 247, **1983;** Organolanthanides, *Angew. Chem. Int. Ed. Engl.* **23,** 474, **1984.**

12.15. a) W.J. Evans - Organolanthanides, J. Organomet. Chem. **250,** 217, **1982;**

b) Perspective in Reductive Lanthanide Chemistry, *Coord. Chem. Rev.* **206-207,** 2000;

c) W.J. Evans *et al.* - Reaction Chemistry of Sterically Crowded [SmCp*$_3$], *J. Am. Chem. Soc.* **120,** 9273, **1998.**

12.16. M. Ephritikhine - Synthesis, Structure and Reactions of Hydride, Borohydride and Aluminohydride Compounds of the f-Elements, *Chem. Rev.* **97**, 3469, **1997**.

12.17. S. Arndt, J. Okuda - Mono-cyclopentadienyl Complexes of the Rare-Earth Metals, *Chem. Rev.* **102**, 1953, **2002**; J. Okuda - *Rare-Earth Metal Complexes that Contain Linked Amido-cyclopentadienyl Ligands: Ansa-metallocene Mimics and "Constrained Geometry" Catalysts*, Dalton (Perspective) **2003**, 2367.

12.18. S. Kobayashi - *Chem. Lett.*, 2183, **1991**;
A Novel Chiral Lead(II) Catalyst for Enantioselective Aldol Reactions in Aqueous Media, *J. Am. Chem. Soc.* **122**, 11531, **2000** and **125**, 2989, **2003**.

12.19. P. Girard, J.L. Namy, H. Kagan - Divalent Lanthanide Derivatives in Organic Synthesis. I. Mild Preparation of SmI_2 and YbI_2 and their Use as Reducing Agents or Coupling Agents, *J. Am. Chem. Soc.* **102**, 2693, **1980**; *Tetrahedron* **42**, 6573, **1980**; in *Mechanisms and Processes in Molecular Chemistry*, D. Astruc Ed., Gauthier-Villars, Paris; *New J. Chem.* **16**, 89, **1992**.

12.20. J.L. Luche - Lanthanides in Organic Chemistry. I. Selective 1,2 Reductions of Conjugated Ketones, *J. Am. Chem. Soc.* **100**, 2226, **1978**; see refs **19.57**, **7** and **45**.

Chapter 13 - Covalent chemistry of the organoelements of frontier (11, 12) and main (13-16) groups

13.1. J.F. Normant - Stoichiometric *vs.* Catalytic Use of Cu^I Salts in Main-Group Organometallic Synthesis, *Pure Appl. Chem.* **50**, 709, **1978** and **56**, 91, **1984**.

13.2. J.F. Normant - Organocopper Reagents in Organic Synthesis, *J. Organomet. Chem. Library* **1**, 219, **1976**; *Synthesis*, 63, **1972**.

13.3. G.H. Posner - *An Introduction to Synthesis Using Organocopper Reagents*, Wiley, New York, **1980**.

13.4. T. Kaufmann - Oxidative Coupling *via* Organocopper Compounds, *Angew. Chem. Int. Ed. Engl.* **13**, 291, **1974**.

13.5. J.G. Noltes *et al.* - Organosilver Chemistry, *Organomet. Chem. Rev.* **A(5)**, 215, **1970**.

13.6. R.J. Pudephatt - *The Chemistry of Gold*, Elsevier, Amsterdam, **1978**, Chap. 7 (Organogold Chemistry).

13.7. a) H. Schmidbauer - Organogold Chemistry, *Angew. Chem. Int. Ed. Engl.* **15**, 728, **1976**;
b) H. Schmidbauer Ed. - *Progress in Gold Chemistry, Biochemistry and Technology*, Wiley, New York, **1999**;
c) A. Laguna - Gold Compounds of the Heavy Group V Elements, in ref. **13.7b**, 349-427.

13.8. a) H.E. Simmons *et al.* - Halomethylzinc-Organic Compounds in the Synthesis of Cyclopropanes, *Org. React.* **20**, 1, **1973**;
b) P. Knochel *et al.* - Organozinc Mediated Reactions, *Tetrahedron* **54**, 8275, **1998** (Tetrahedron Report N° 459); R. Giovannini, P. Knochel - Ni^{II}-Catalyzed Cross-Coupling Between Polyfunctional Arylzinc Derivatives and Primary Alkyl Iodides, *J. Am. Chem. Soc.* **120**, 11186, **1998**.

13.9. a) T.P. Hanusa - Ligand Influences on Structure and Reactivity in Organoalkaline Earth Chemistry, *Chem. Rev.* **93**, 1023, **1993**; P. Jutzi, N. Burford - *Chem. Rev.* **99**, 969, **1999**; P. Jutzi, N. Burford - in *Metallocenes*, A. Togni, R.L. Haterman Eds, Wiley-VCH, Weinheim, **1998**, Chap. 2;
b) I. Resa, E. Carmona, E. Gutiérrez-Puebla, A. Monge - *Science* **305**, 1136, **2004**; D. del Rio, A. Galindo, I. Resa, E. Carmona - Theoretical and Synthetic Studies on $[Zn(\eta^5\text{-}C_5Me_5)_2]$: Analysis of the Zn-Zn Bonding Interaction. *Angew. Chem. Int. Ed.* **44**, 1244, **2005**.

13.10. P.R. Jones, P.J. Desio - Less-Familiar Reactions of Organocadmiums, *Chem. Rev.* **78**, 491, **1978**.

13.11. R. Larock - *Organomercury Compounds in Organic Synthesis*,
Springer Verlag, New York, **1985**; *Tetrahedron* **38**, 1713, **1982**.

13.12. D.L. Rabenstein - The Chemistry of Methylmercury Toxicology,
Acc. Chem. Res. **11**, 101, **1978**; *J. Chem. Ed.* **55**, 292, **1978**.

13.13. H.C. Brown - Organoboranes, *Pure Appl. Chem.* **47**, 49, **1976** and **55**, 1387, **1983**.

13.14. a) D.S. Matteson - Preparation and Use of Organoboranes in Organic Synthesis,
in ref **13.14b**, Chap. 3;
  b) F.R. Hartley, S. Patai Eds - *The Chemistry of the Metal-Carbon Bond*,
  Wiley, New York, **1987**, Vol. 4.

13.15. E.L. Muetterties Ed. - *The Chemistry of Boron and its Compounds*,
Wiley, New York, **1967** (contains several chapters on organoboranes).

13.16. H. Schmidbauer - Organophosphane-Borane Chemistry,
J. Organomet. Chem. **200**, 287, **1980**; R.T. Paine - Recent Advances in
Phosphinoboranes, *Chem. Rev.* **95**, 343, **1995**.

13.17. W. Siebert - Boron Heterocycles as Transition-Metal Ligands, *Adv. Organomet.
Chem.* **18**, 301, **1980**.

13.18. T. Mole, E.A. Jeffrey - *Organoaluminum Compounds*, Elsevier, Amsterdam, **1972**.

13.19. a) P.A. Chaloner - Preparation and Use of Organoaluminum Compounds in Organic
Synthesis, in ref. **13.14b**, Chap. 4;
  b) G. Courtois, L. Miginiac - Allyl Derivatives of Aluminum,
  J. Organomet. Chem. **69**, 1, **1974**.

13.20. H.W. Roesky, S.S. Kumar - Chemistry of Aluminum(I), *Chem. Commun*, 4027, **2005**;
H. Roesky, S. Singh, V. Jancik - A Paradigm Change in Assembling OH
Functionalities on Metal Centers, *Acc. Chem. Res.* **37**, 969, **2004**;
H.W. Roesky, S. Singh, K.K.M. Yusuff - Organometallic Hydroxides of Transition
Elements, *Chem. Rev.* **106**, 3813, **2006**.

13.21. H. Sinn, W. Kaminsky - Ziegler-Natta Catalysis,
Adv. Organomet. Chem. **18**, 99, **1980**.

13.22. a) S. Uemura - Preparation and Use of Organothallium Compounds in Organic
Synthesis, in ref. **13.14b**, Chap. 5;
  b) L. Gade - *Molecular Structures, Solid-State Aggregation and Redox
  Transformations of Low-valent Amido-Thallium Compounds*,
  Dalton (Perspectives) **2003**, 267.

13.23. E.W. Colvin - Preparation and Use of Organosilicon Compounds in Organic
Synthesis, in ref. **14b**, Chap. 6.

13.24. a) R.J.P. Corriu, C. Guérin, J. Moreau - Stereochemistry at Silicon,
Top. Stereochem. **15**, 43, **1984**;
  b) R.J.P. Corriu - Hypervalent Species at Silicon: Structure and Reactivity,
  J. Organomet. Chem. **400**, 81, **1990**; *Chem. Rev.* **93**, 1371, **1993**;
  c) H. Sakurai - *Synlett.* **1**, 1, **1989**.

13.25. I. Fleming - The Chemistry of an Allylsilane: the Synthesis of a Prostaglandin
Intermediate and of Loganin, *Chem. Soc. Rev.* **10**, 83, **1981**.

13.26. R.J.P. Corriu - Ceramics and Nanostructures from Molecular Precursors.
Angew. Chem. Int. Ed. **39**, 1376, **2000**; R.D.Miller, J. Michl - Polysilane
High Polymers, *Chem. Rev.* **89**, 1359, **1989**; M. Birot, J.-P. Pillot, J. Dunoguès -
Comprehensive Chemistry of Polycarbosilanes, Polysilazanes and
Polycarbosilazanes as Precursors of Ceramics. *Chem Rev.* **95**, 1443, **1995**.

13.27. H. Mimoun - Selective Reduction of Carbonyl Compounds by
Polymethylhydrosiloxane in the Presence of Metal Hydride Catalysts.
J. Org. Chem. **64**, 2582, **1999**.

13.28. S. Masamune - Strain-Ring and Double-Bond Systems Consisting of the Group 14 Elements Si, Ge and Sn, *Angew. Chem. Int. Ed. Engl.* **30**, 920, **1991**.

13.29. R. West - Chemistry of the Silicon-Silicon Double Bond, *Angew. Chem. Int. Ed. Engl.* **26**, 1201, **1987**.

13.30. P. Jutzi - Cyclopentadienyl Complexes with Main-Group Elements as Central Atoms, *J. Organomet. Chem.* **400**, 1, **1990**; The Versatility of the Pentamethylcyclopentadienyl Ligand in Main-Group Chemistry, *Comments Inorg. Chem.* **6**, 123, **1987**; *Pure Appl. Chem.* **75**, 483, **2003**.

13.31. J.B. Lambert - Modern Approaches to Silylium Cation in the Condensed Phase. *Chem. Rev.* **95**, 1191, **1995**; J.B. Lambert *et al.* - Crystal Structure of a Silyl Cation with No Coordination to Anion and Distant Coordination to Solvent, *Science*, **260**, 1917, **1993**; C.A. Reed *et al.* - Closely Approaching the Silylium Ion R_3Si^+, *Science* **162**, 402, **1993**.

13.32. R.J.P. Corriu, M. Henner - The Silicon Ion Question, *J. Organomet. Chem.* **74**, 1, **1974**; *Angew. Chem. Int. Ed. Engl.* **33**, 1097, **1994**; G. Cerveau, R.J.P. Corriu, E. Framery - Nanostructured Organic-Inorganic Hybrid Materials: Kinetic Control of the Texture. *Chem. Mater.* **13**, 3373, **2001**.

13.33. J. Satgé - Reactive Intermediates in Organogermane Chemistry, *Pure Appl. Chem.* **56**, 137, **1984**; *Adv. Organomet. Chem.* **21**, 241, **1982**; *J. Organomet. Chem.* **400**, 121, **1990**.

13.34. a) Organic Aspects: M. Pereyre, J.P. Quintard, A. Rahm - *Organotin Chemistry for Synthetic Applications*, Butterworth, London, **1987**;
b) Inorganic aspects: Ch. Elschenbroich, A. Salzer - *Organometallics*, VCH, Weinheim, 2nd ed., **1992**, Chap. 8.

13.35. H. Schmidbauer - Pentaalkyls and Alkylidene Trialkyls of the Group V Elements, *Adv. Organomet. Chem.* **14**, 205, **1976**.

13.36. A.H. Cowley - Stable Compounds with Double Bonds Between the Heavier Main-Group Elements, *Acc. Chem. Res.* **17**, 386, **1984**; From Multiple Bonds to Materials Chemistry, *J. Organomet.* **400**, 71, **1990**; Some Past Vignettes and Future Prospects for Main Group Chemistry, *J. Organomet. Chem.* **600**, 168, **2000**; From Group 13-Donor-acceptor Bonds to Triple-decker Cations, *Chem. Commun.* (Feature Article), 2369, **2004**.

13.37. H.J. Reich - Functionnal Group Manipulation Using Organoselenium Reagents, *Acc. Chem. Res.* **12**, 22, **1979**.

IV - Introduction to catalysis

IV.1. K.P.C. Vollhardt - *Organic Chemistry*, Freeman, New York, **1987**, 752.

IV.2. B. Cornils, W.A. Herrmann, R. Schlögl, C.-H. Wong Eds - *Catalysis from A to Z. A Concise Encyclopedia*, Wiley-VCH, Weinheim, **2000**, 391.

IV.3. D. Astruc - Electron-transfer-Chain Catalysis in Organotransition-Metal Chemistry, *Angew. Chem. Int. Ed. Engl.* **27**, 643, **1988**.

IV.4. V. Balzani, F. Scandola - *Supramolecular Photochemistry*, Ellis Horwood, New York, **1991**, Chap. 12.

IV.5. J.-M. Savéant - Catalysis of Chemical Reactions by Electrodes, *Acc. Chem. Res.* **13**, 323, **1980** and **26**, 455, **1993**.

IV.6. D. Astruc - *Electron-Transfer and Radical Processes in Transition Metal Chemistry*, VCH, New York, **1995**, Chap. 7 (Redox Catalysis and its Applications to Synthesis), 413.

IV.7. a) B. Cornils, W.A. Herrmann Eds - *Applied Homogeneous Catalysis with Organometallic Compounds*, 2nd ed., Wiley-VCH, Weinheim, **2002**;
b) J. Gladysz - *Pure Appl. Chem.* **73**, 1319, **2001**; *Chem. Rev.* **102**, 3215, **2002**.

IV.8. G. Ertl, H. Knözinger, J. Weitkamp Eds - *Handbook of Heterogeneous Catalysis* (4 vol.), Wiley-VCH, Weinheim, **1997**.

IV.9. **a)** J.M. Thomas - Design, Synthesis, and In Situ Characterization of New Solid Catalysts, *Angew. Chem. Int. Ed.* **38**, 3588, **1999**; **b)** E. Lindner *et al.* - *Angew. Chem. Int. Ed.* **38**, 2154, **1999**.

IV.10. N.E. Leadbeater, M. Marco - Preparation of Polymer-Supported Ligands and Metal Complexes for Use in Catalysis, *Chem. Rev.* **102**, 3217, **2002**.

IV.11. D. Astruc, F. Chardac - Dendritic Catalysts and Dendrimers in Catalysis, *Chem. Rev.* **101**, 2991, **2001**.

IV.12. **a)** J.W. Knappen, A.W. Van der Made, J.C. de Wilde, P.W.N.M. Van Leeuwen, P. Wijkens, D.M. Grove, G. Van Koten *et al.* - Homogeneous Catalysts Based on Silane Dendrimers Functionalized with Arylnickel (II) complexes, *Nature* **372**, 659, **1994**; **b)** R. Van Heerbeek, P.C.J. Kamer, P.W.N.M. Van Leeuwen, J.N.H. Reek - Dendrimers as Recoverable Supports and Reagents, *Chem. Rev.* **102**, 3717, **2002**; **c)** H. Alper *et al.* - Supported dendritic catalysts, *J. Am. Chem. Soc.* **121**, 3035, **1999**; **122**, 956, **2000**, **123**, 2889, **2001**.

IV.13. **a)** E. Kunz - *Chemtech.* **17**, 570, **1987**; **b)** B. Cornils, W.A. Herrmann Eds - *Aqueous-Phase Organometallic Catalysis*, Wiley-VCH, Weinheim, **2004**.

IV.14. I.T. Horvath - Fluorous Biphase Chemistry, *Acc. Chem. Res.* **31**, 641, **1998**.

IV.15. L.P. Barthel-Rosa, J.A. Gladysz - Chemistry in Fluorous Media: A User's Guide to Practical Considerations in the Application of Fluorous Catalysts and Reagents, *Coord. Chem. Rev.* **190-192**, 587, **1999**.

IV.16. K. Drauz, H. Waldmann, S.M. Roberts Eds - *Enzyme Catalysis in Organic Synthesis*, 2nd ed., Wiley-VCH, Weinheim, **2000**.

IV.17. J. Dupont, R.F. de Souza, P.A. Suarez - Ionic Liquid (Molten Salt) Phase Organometallic Catalysis, *Chem. Rev.* **102**, 3667, **2002**.

IV.18. P.G. Jessop, T. Ikariya, R. Noyori - Homogeneous Catalysis in Supercritical Fluids, *Science* **269**, 1065, **1995**.

IV.19. **a)** P.T. Anastas, T.C. Williamson Eds - Green Chemistry, *Adv. Chem. Ser.* **626**, Washington DC, **1996**; **b)** P.T. Anastas, J.C. Warner - *Green Chemistry: Theory and Practice*, Oxford University Press, New York, **1998**. **c)** C. Copéret, B. Chaudret Eds - Surface and Interfacial Organometallic Chemistry and Catalysis, *Top. Organomet. Chem.*, Springer, Berlin, **2005**, Vol. 16.

IV.20. For Chauvin-type catalytic mechanisms involving metallosquares, see: **a)** D. Astruc - La métathèse : de Chauvin à la chimie verte, *L'Act. Chim.* **273**, 3, **2004**; **b)** D. Astruc - Metathesis Reactions: From a Historical Perspective to Recent Developments, *New J. Chem.* **29**(1), 42, **2005**.

IV.21. L. Mauron, L. Perrin, O. Eisenstein - DTF Study of CH_4 Activation by d^0Cl_2LnZ (Z = H, CH_3) Complexes, *J. Chem. Soc., Dalton Trans.*, 534, **2002**.

Chapter 14 - Hydrogenation and hydroelementation of alkenes

14.1. H. Brunner - *Hydrogenation*, in ref. **14.2**, Vol. 1, Chap. 2.2.2, 201-219.

14.2. B. Cornils, W.A. Herrmann Eds - *Applied Homogeneous Catalysis with Organometallic Compounds*, VCH, Weinheim, **1996**.

14.3. B.R. James - Homogeneous Catalytic Hydrogenation, in *Comprehensive Organometallic Chemistry*, G. Wilkinson *et al.* Eds, Pergamon, Oxford, **1982**; *Adv. Organomet. Chem.* **17**, 319, **1979**.

14.4. C. Larpent, R. Dabard, H. Patin - Rh^I Production During the Oxidation by Water of a Hydrosoluble Phosphine, *Inorg. Chem.* **26**, 2922, **1987**; *J. Mol. Cat.* **44**, 191, **1988**.

14.5. J. Halpern - Free Radical Mechanisms in Coordination Chemistry, *Pure Appl. Chem.* **51**, 2171, **1979**.

14.6. D. Evans, J.A. Osborn, F.H. Jardine, G. Wilkinson - Homogeneous Hydrogenation and Hydroformylation Using Ru Complexes, *Nature* **208**, 1203, **1965**.

14.7. T.J. Marks *et al.* - Highly Reactive Actinides. Synthesis, Chemistry and Structures of Group 4 Hydrocarbyls and Hydrides with Chelating Bis(polymethylcyclopentadienyl) Ligands, *J. Am. Chem. Soc.* **107**, 8103 and 8111, **1985**.

14.8. H. Kagan, T.P. Dang - Asymmetric Catalytic Reduction with Transition Metal Complexes. A Catalytic System of Rh^I with (-)-2,3-*o*-Isopropylidene-2,3-dihydroxy-1,4-bis (diphenylphosphino) butane, a New Chiral Diphosphine, *J. Am. Chem. Soc.* **94**, 6429, **1972**; *Chem. Commun.*, 481, **1971**.

14.9. W.S. Knowles - Asymmetric Hydrogenations, *Acc. Chem. Res.* **16**, 106, **1983**.

14.10. J.S. Speir - Hydrosilylation, *Adv. Organomet. Chem.* **407**, 407, **1979**; B. Marciniec - in ref. **14.2**, Chap. 2.6, 487-506.

14.11. N.L. Lewis - Hydrosilylation, a "Homogeneous" Reaction Really Catalalyzed by Colloids, in Homogeneous Transition-Metal Catalyzed Reactions, W.R. Moser, D.W. Slocum Eds, *Adv. Chem. Ser.* **230**, Washington DC, **1992**, Chap. 37, 541-549; *J. Am. Chem. Soc.* **112**, 5998, **1990**.

14.12. K. Burgess - Transition-Metal-Promoted Hydroboration of Alkenes: Emerging Methodology for Organic Transformations, *Chem. Rev.* **91**, 1179, **1991**.

14.13. K. Huthmacher, S. Krill - Hydrocyanation, in ref. **14.2**, Chap. 2.5, 465-486.

14.14. C.A. Tolman - Homogeneous Catalyzed Olefin Hydrocyanation, *Adv. Catal. Series*, Academic Press, New York, **1985**, Vol. 33; *J. Chem. Ed.* **3**, 199, **1986**.

Chapter 15 - Transformations of alkenes and alkynes

15.1. R.H. Grubbs Ed. - *Handbook of Metathesis*, Wiley-VCH, Weinheim, **2003**, Vol. 1-3.

15.2. R.R. Schrock - Recent Advances In Olefin Metathesis by Molybdenum and Tungsten Imido Alkylidene Complexes, *J. Mol. Cat. A: Chemical* **213**, 21, **2004**.

15.3. R.H. Grubbs - Olefin Metathesis, *Tetrahedron* **60**, 7117, **2004**; The Development of L$_2$X$_2$Ru=CHR Olefin Metathesis Catalysts: an Organometallic Success Story, *Acc. Chem. Res.* **34**, 18, **2001**.

15.4. a) K.J. Ivin, J.C. Mol - *Olefin Metathesis and Metathesis Polymerization*, Academic Press, New York, **1997**; **b)** T.J. Katz - The Olefin Metathesis Reaction, *Adv. Organomet. Chem.* **16**, 283, **1977**;

15.5. D. Astruc - The Metathesis Reactions: from a Historical Perspective to Recent Developments, *New J. Chem.* **29**, 42, **2005**.

15.6. R.L. Banks, G.C. Bailey - *Ind. Eng. Chem., Prod. Res. Dev.*, 170, **1964**; H.S. Eleuterio - Historical Account of the Early Days, *J. Mol. Cat.* **65**, 55, **1991**.

15.7. J.L. Hérisson, Y. Chauvin - Catalyse de transformation des oléfines par les complexes du tungstène: télomérisation des oléfines cycliques en présence d'oléfines acycliques, *Macromol. Chem.* **141**, 161, **1971**,.

15.8. R.R. Schrock - Catalysis by Transition Metals: Metal-Carbon Double Bonds and Triple Bonds, *Science* **219**, 13, **1983**.

15.9. R.R. Schrock *et al.* - *J. Am. Chem. Soc.* **102**, 4515 and 6236, **1980**; *J. Mol. Cat.* **8**, 73, **1980**.

15.10. J. Kress, M. Welosek, J.A. Osborn - W(IV) Carbene for the Metathesis of Olefins. Direct Observation and Identification of the Chain Carrying Carbene Complexes in a Highly Active Catalyst System, *J. Chem. Soc., Chem. Commun.*, 514, **1982**; *J. Am. Chem. Soc.* **105**, 6346, **1983**.

15.11. a) P. Schwab, M.B. France, J.W. Ziller, R.H. Grubbs - A Series of Well-Defined Metathesis Catalysts- Synthesis of [RuCl$_2$(=CHR')(PR$_3$)] and its Reactions, *Angew. Chem. Int. Ed. Engl.* **34**, 2039, **1995**; P. Schwab, R.H. Grubbs, J.W. Ziller *J. Am. Chem. Soc.* **118**, 100, **1996**; Ring-Closing Metathesis and Related Processes in Organic Synthesis, *Acc. Chem. Res.* **28**, 446, **1995**;
b) M. Scholl, T.M. Trnka, J.P. Morgan, R.H. Grubbs - *Tetrahedron Lett.* **40**, 2247, **1999**; J. Huang, E.D. Stevens, S.P. Nolan, J.L. Petersen - *J. Am. Chem. Soc.* **121**, 2647, **1999**; L. Ackermann, A. Fürstner, T. Weskamp, F.J. Kohl, W.A. Herrmann *Tetrahedron Lett.* **40**, 4787, **1999**; M. Scholl, S. Ding, C.W. Lee, R.H. Grubbs - *Org. Lett.* **1**, 953, **1999**.

15.12. A. Fürstner - Olefin Metathesis and Beyond, *Angew. Chem. Int. Ed.* **39**, 3012, **2000**.

15.13. C. Dietrich-Bucherer, G. Rapenne, J.-P. Sauvage - Efficient Synthesis of a Molecular Knot by Copper(I)-Induced Formation of the Precursor Followed by Ruthenium(II)-Catalyzed Ring Closing Metathesis, *Chem. Commun.*, 2053, **1997**.

15.14. M. Schuster, S. Blechert - *Angew. Chem. Int. Ed.* **36**, 2036, **1997**; S.J. Connor, S. Blechert - Recent Developments in Olefin Cross-Metathesis, *Angew. Chem. Int. Ed.* **42**, 1900, **2003**.

15.15. J.A. Gladysz *et al.* - Metal-catalyzed Olefin Metathesis in Metal Coordination Spheres, *Chem. Eur. J.* **7**, 3931, **2001**; in ref. **15.1**, Vol. 2, 403; *Angew. Chem. Int. Ed.* **43**, 5537, **2004**.

15.16. A.V. Chuchryukin, H.P. Dijkstra, B.M.J.M. Suijkerbuijk, R.J.M. Klein Gebbink, G.P.M. Van Klink, A.M. Mills, A.L. Spek, G. Van Koten - Macrocycle Synthesis by Olefin Metathesis on a Nanosized, Shape-Persistent Tricationic Platinum Template, *Angew. Chem. Int. Ed.* **42**, 228, **2003**.

15.17. P. Wang, C.N. Moorefield, G.R. Newkome - Nanofabrication: Reversible Self-Assembly of an Imbedded Hexameric Metallomacrocycle within a Macromolecular Superstructure, *Angew. Chem. Int. Ed.* **44**, 1679, **2005**.

15.18. N.G. Lemcoff, T.A. Spurlin, A.A. Gewirth, S.C. Zimmerman, J.B. Beil, S.L. Elmer, H.G. Vandeveer - Organic Nanoparticles Whose Size and Rigidity Are Finely Tuned by Cross-Linking the End Groups of Dendrimers, *J. Am. Chem. Soc.* **126**, 11420, **2004**; J. Beil, N.G. Lemcoff, S.C. Zimmerman - *J. Am. Chem. Soc.* **126**, 13576, **2004**.

15.19. a) V. Martinez, J.-C. Blais, D. Astruc - A Fast Organometallic Route from *p*-Xylene, Mesitylene, and *p*-Diisopropylbenzene to Organoiron and Polycyclic Aromatic Cyclophanes, Capsules and Polymers, *Angew. Chem. Int. Ed.* **42**, 4366, **2003**;
b) C. Ornelas, D. Méry, J.-C. Blais, E. Cloutet, J. Ruiz Aranzaes - Efficient Mono- and Bifunctionalization of Polyolefin Dendrimers by Metathesis, *Angew. Chem. Int. Ed.* **44**, 7852, **2005**.

15.20. R.R. Schrock, A.H. Hoveyda - Molybdenum and Tungsten Imido Alkylidene Complexes as Efficient Olefin-Metathesis Catalysts, *Angew. Chem. Int. Ed.* **42**, 4592, **2003**.

15.21. C. Coperet, F. Lefebvre, J.-M. Basset - Well-defined Metallocarbenes and Metallocarbynes Supported on Oxide Supports Prepared *via* Surface Organometallic Chemistry: a Source of Highly Active Alkane, Alkene and Alkyne Metathesis Catalysts, in ref. **15.1**, Vol. 1, 33 and 190; *J. Am. Chem. Soc.* **123**, 2062, **2001**.

15.22. A. Mortreux, M. Blanchard - Alkyne Metathesis Catayzed by Mo(CO)$_6$ + Phenol. *J. Chem. Soc., Chem. Commun.*, 786, **1974**; A. Mortreux, F. Petit, M. Blanchard - ^{13}C Tracer Studies of Alkynes Metathesis, *Tetrahedron Lett.* **49**, 4967, **1978**.

15.23. M. Blanchard, A. Mortreux - *J. Mol. Cat.* **1**, 101, **1975-6**.

15.24. R.R. Schrock - *Acc. Chem. Res.* **19**, 342, **1986**; High Oxidation State Multiple Metal-Carbon Bonds, *Chem. Rev.* **102**, 145, **2002**; in ref. **15.1**, Vol. 1, 173.

15.25. K. Grela, J. Ignatowska - An Improved Catalyst for Ring-Closing Alkyne Metathesis Based on Molybdenum Hexacarbonyl/2-Fluorophenol, *Org. Lett.* **4**, 3747, **2002**.

15.26. V. Huc, R. Weihofen, I. Martin-Jimenez, P. Oulié, C. Lepetit, G. Lavigne, R. Chauvin - Ether-Induced Rate Enhancement of Mo-Catalyzed Alkyne Metathesis Under Mild Conditions, *New J. Chem.* **27**, 1412, **2003**.

15.27. U.H.F. Bunz - *Acc. Chem. Res.* **34**, 998, **2001**; in *Modern Arene Chemistry*, D. Astruc Ed., Wiley-VCH, Weinheim, **2002**, 217.

15.28. a) A. Fürstner, C. Mathes, C.W. Lehmann, - Mo[N(t-Bu)(Ar)]₃ Complexes as Catalyst Precursors: *In Situ* Activation and Application to Metathesis Reactions of Alkynes and Diynes, *J. Am. Chem. Soc.* **121**, 9453, **1999**; in ref. **15.1**, Vol. 2, 432;
b) A. Mortreux, O. Coutelier - *J. Mol. Cat. A: Chemical* **254**, 96, **2006**.

15.29. C. Cummins - Reductive Cleavage and Related Reactions Leading to Molybdenum-Element Multiple Bonds: New Pathways Offered by Three-Coordinate Mo^III, *Chem. Commun.* **47**, 1777, **1998**; *Prog. Inorg. Chem.* **47**, 685, **199**.

15.30. W. Zhang, S. Kraft, J.S. Moore - A Reductive Recycle Strategy for the Facile Synthesis of Mo^VI Alkylidyne Catalysts for Alkyne Metathesis, *Chem. Commun.*, 832, **2003**; *J. Am. Chem. Soc.* **126**, 329, **2004**.

15.31. R.R. Schrock - Living Ring-Opening Metathesis Polymerization Catalyzed by Well-Characterized Transition Metal Alkylidene Complexes, *Acc. Chem. Res.* **23**, 158, **1990**.

15.32 R.H. Grubbs, W. Tumas - Polymer Synthesis and Organotransition Metal Chemistry, *Science* **243**, 907, **1989**.

15.33. M.-H. Desbois, D. Astruc - Coupling Electron-Transfer Catalysis with Organometallic Catalysis: Carbenic Polymerization of Alkynes, *New. J. Chem.* **13**, 595, **1989**.

15.34. A. Bre, Y. Chauvin, D. Commereuc - Mode of Decomposition of Titanacyclopentane Complexes. Model of Intermediate Species in the Dimerization of Olefins Catalyzed by Titanium Complexes, *Nouv. J. Chim.* **10**, 535, **1986**.

15.35. Y. Chauvin, H. Olivier - Dimerization and Codimerization, in ref. **15.36a**, Vol 1, Chap. 2.3.1.4, 258-268.

15.36. a) B. Cornils, W.A. Herrmann Eds - *Applied Homogeneous Catalysis with Organometallic Compounds,* VCH, Weinheim, **1996**;
b) in ref. **15.36a**, Vol 1, Chap. 2.3.1-2.3.6.

15.37. R.R. Schrock, S. McLain, J. Sancho - Tantalacyclopentane Complexes and Their Role in the Catalytic Dimerization of Olefins, *Pure Appl. Chem.* **52**, 729, **1980**; *J. Am. Chem. Soc.* **101**, 4558, **1979**.

15.38. R.H. Grubbs *et al.* - Metallacyclopentanes as Catalysts for the Linear and Cyclodimerization of Olefins, *J. Am. Chem. Soc.* **100**, 7416, **1978**.

15.39. P. Braunstein, F. Naud - Hemilability oh Hybrid Ligands and the Coordination Chemistry of Oxazoline-Based Systems, *Angew. Chem. Int. Ed.* **40**, 680, **2001**.

Chapter 16 - Oxidation of olefins

16.1. R.A. Sheldon, J.K. Kochi - *Metal-Catalyzed Oxidations of Organic Compounds*, Academic Press, New York, **1981**.

16.2. B. Meunier Ed. - *Biomimetic Oxidations Catalyzed by Transition Metal Complexes*, Imperial College Press, London, **2000**.

16.3. a) Ref. **16.3b**, Vol 1, Part. 2.4 (Oxidations);
b) B. Cornils, W.A. Herrmann Eds - *Applied Homogeneous Catalysis with Organometallic Compounds*, VCH, Weinheim, **1996**.

16.4. J.E. Bäckvall *et al.* - Stereochemistry and Mechanism for the Pd^{II}-Catalyzed Oxidation of Ethene in Water (the Wacker Process), *J. Am. Chem. Soc.* **101**, 2411, **1979**.

16.5. J.K. Stille, R. Divakaruni - Mechanism of the Wacker Process, Stereochemistry of the Hydroxypalladation, *J. Organomet. Chem.* **169**, 239, **1979**.

16.6. a) P.M. Henry - *Palladium Catalyzed Oxidations of Hydrocarbons*, Reidel, Dordrecht, **1980**;
b) E. Szromi, H. Shan, P.R. Sharp - Alkene Oxidation by a Platinum Oxo Complex and Isolation of a Platinaoxetane, *J. Amer. Chem. Soc.* **12**, 10522, **2003**.

16.7. K.B. Sharpless *et al.* - The First Practical Method for Asymmetric Epoxidation of Allylic Alcohols, *J. Am. Chem. Soc.* **102**, 5974, **1980**, in *Asymmetric Synthesis*, J.D. Morrison Ed., Academic Press, New York, **1985**, Vol. 5, Chap. 8; *Pure Appl. Chem.* **55**, 1823, **1983**.

16.8. H. Mimoun - in *Comprehensive Coordination Chemistry*, G. Wilkinson *et al.* Eds, Pergamon, Oxford, **1987**, Vol. 6, 317; Activation of Molecular Oxygen and Selective Oxidation of Olefins Catalyzed by Group VIII Transition Metal Complexes, *Pure Appl. Chem.* **53**, 2389, **1981**.

16.9. K.B. Sharpless *et al.* - Asymmetric Epoxidation of Allylic Alcohols, *J. Am. Chem. Soc.* **113**, 106, **1991**; *J. Org. Chem.* **49**, 3707, **1984**; **50**, 422, **1985**, and **51**, 1922, **1986**.

16.10. J.P. Collman, E. Rose *et al.* - An Efficient Asymmetric Epoxidation of Terminal Olefins, *J. Am. Chem. Soc.* **121**, 460, **1999**.

16.11. E.N. Jacobsen - Asymmetric Catalysis of Epoxide Ring-Opening Reactions, *Acc. Chem. Res.* **33**, 421, **2000**.

16.12. R.A. Sheldon, J.K. Kochi - Allylic Oxidation, in ref. **17.1**, 289-294; N. Rabjohn. - *Org. React.* **24**, 261, **1976**.

16.13. D. Mansuy, P. Battioni - Dioxygen Activation at Heme Centers in Enzymes and Synthetic Analogs, in *Bioinorganic Catalysis*, J. Reedijk Ed., Dekker, New York, **1993**, Chap. 12, 395-424; *Pure Appl. Chem.* **59**, 759, **1987** and **62**, 741, **1990**.

16.14. B. Meunier - Metalloporphyrins as Versatile Catalysts for Oxidation Reactions and Oxidative DNA Cleavage, *Chem. Rev.* **92**, 1411, **1992**; *Bull. Soc. Chim. Fr.* **4**, 578, **1986**.

16.15. K.B. Sharpless *et al.* - Diols *via* Catalytic Dihydroxylation, in ref. **17.3b**, Chap. 3.3.2, 1009-1024; *Chem. Rev.* **94**, 2483, **1994**.

16.16. a) H. Alper *et al.*, - Phase-Transfer Catalysis and Related Systems, in ref. **17.3**, Chap. 3.2.4, 844-865; *Aldrichim. Acta* **24**, 3, **1991**;
b) H. Alper, H. des Abbayes - Phase-Transfer Catalyzed Organometallic Chemistry. The Cobalt Carbonyl-Catalyzed Carbonylation of Halides, *J. Organomet. Chem.* **134**, C11, **1977**.

16.17. E.V. Dehmlov, S.S. Dehmlov - *Phase Transfer Catalysis*, 3rd ed., VCH, Weinheim, **1993**.

16.18. B. Meunier, E. Guilmet, M.-E. De Carvalho, R. Poilblanc - Sodium Hypochlorite: A Convenient Oxygen Source for Olefin Epoxidation Catalyzed by (Porphyrinato) Manganese Complexes, *J. Am. Chem. Soc.* **106**, 6668, **1984**.

Chapter 17 - C-H activation and functionalization of alkanes and arenes

17.1 a) For a leading reference, see K.I. Goldberg, A.S. Goldman Eds - *Activation and Functionalization of C-H Bonds*, ACS Symp. Series 885, Washington DC, **2004**;
b) D. Sames - in ref. **17.1a**, Chap. 9, 155-168;
c) V. Snieckus - in *Modern Arene Chemistry*, D. Astruc Ed., Wiley-VCH, Weinheim, **2002**, Chap. 10, 330-367.

17.2. S.H. Brown, R.H. Crabtree - *Chem. Commun.*, 970, **1997**; Making mercury-photosensitized dehydrodimerization into an organic synthetic method. Vapor-pressure selectivity and the behavior of functionalized substrates. *J. Am. Chem. Soc.* **111**, 2935, **1989** and 2946; *Chemtech.* **21**, 634, **1991**.

17.3. **a)** C.A. Tolman *et al.* - in *Activation and Functionalization of Alkanes*, C.L. Hill Ed., Wiley, New York, **1989**, 303-360;
b) D.T. Sawyer *et al.* - *J. Am. Chem. Soc.* **115**, 5817, **1993**.

17.4. J.L. McLain, J. Lee, J.T. Groves - in *Biomimetic Oxidations Catalyzed by Transition Metal Complexes*, B. Meunier Ed., Imperial College Press, London, **2000**, 91.

17.5. B. Meunier - Metalloporphyrins as Versatile Catalysts for Oxidation Reactions and Oxidative DNA Cleavage, *Chem. Rev.* **92**, 1411, **1992**.

17.6. D. Mansuy, P. Battioni - Dioxygen Activation at Heme Centers in Enzymes and Synthetic Analogs, in *Bioinorganic Catalysis*, J. Reedijk Ed., Dekker, New York, **1993**, Chap. 12, 395-424; *Pure Appl. Chem.* **59**, 759, **1987** and **62**, 741, **1990**.

17.7. L. Que *et al.* - Dioxygen Activation at Nonheme Iron Active Sites: Enzymes, Models, and Intermediates, *Chem. Rev.* **104**, 939, **2004**; *J. Biol. Inorg. Chem.* **10**, 87, **2005**.

17.8. J.M. Mayer - in *Biomimetic Oxidations Catalyzed by Transition Metal Complexes*, B. Meunier Ed., Imperial College Press, London, **2000**, 1;
Acc. Chem. Res. **31**, 441, **1998**; J.M. Mayer *et al.* - *Science* **294**, 2524, **2001**;
J.R. Bryant, J.M. Mayer - *J. Am. Chem. Soc.* **125**, 10351, **2003**.

17.9. A.E. Shilov *et al.* - *Nouv. J. Chim.* **7**, 729, **1983**; A.E. Shilov, A.A. Shteinman - Activation of Saturated Hydrocarbons by Metal Complexes in Solution, *Coord. Chem. Rev.* **24**, 97, **1977**; see also ref. **3.14a**.

17.10. S. Stahl, J.A. Labinger, J.E. Bercaw - Homogeneous Oxidation of Alkanes by Electrophilic Late Transition Metals, *Angew. Chem. Int. Ed.* **37**, 2181, **1998**; see also ref. **3.14b**.

17.11. D.D. Wick, K.I. Goldberg - Studies of Reductive Elimination Reactions to Form Carbon-Oxygen Bonds from Pt^{IV} Complexes, *J. Am. Chem. Soc.* **118**, 4442, **1996**;
B.S. Williams, K.I. Goldberg - *J. Am. Chem. Soc.* **123**, 2576, **2001**;
U. Fekl, K.I. Goldberg - *Adv. Inorg. Chem.* **54**, 259, **2003**.

17.12. M. Tilset - Mechanistic Aspects of C-H Activation by Pt Complexes, *Chem. Rev.* **105**, 2471, **2005**.

17.13. R.A. Periana *et al.* - Platinum Catalysts for the High-Yield Oxidation of Methane to a Methanol Derivative, *Science* **280**, 560, **1998**.

17.14. A. Sen - Catalytic Functionalization of Carbon-Hydrogen and Carbon-Carbon Bonds in Protic Media, *Acc. Chem. Res.* **31**, 550, **1998**; *Top. Organomet. Chem.* **3**, 81, **1999**.

17.15. Y. Fujiwara *et al.* - *Chem. Lett.*, 1687, **1989**; Catalytic Functionalization of Arenes and Alkanes *via* C-H Bond Activation, *Acc. Chem. Res.* **34**, 633, **2001**.

17.16. R.A. Periana *et al.* - Catalytic, oxidative condensation of CH_4 to CH_3COOH in One Step *via* CH Activation, *Science* **301**, 814, **2003**.

17.17. **a)** R.H. Crabtree *et al.* - Iridium Complexes in Alkane Dehydrogenation, *J. Am. Chem. Soc.* **101**, 7738, **1979**;
b) M.J. Burk, R.H. Crabtree, *J. Am. Chem. Soc.* **109**, 8125, **1987**.

17.18. D. Baudry, M. Ephritikhine, H. Felkin - The Activation of Carbon-Hydrogen Bonds in Cyclopentane by Bis(phosphine)rheniumheptahydrides, *J. Chem. Soc., Chem. Commun.*, 1243, **1980**.

17.19. **a)** A.S. Goldman *et al.* - Mechanism of Alkane Transfer-Dehydrogenation Catalyzed by a Pincer-Ligated Iridium Complex, *J. Am. Chem. Soc.* **125**, 7770, **2003**;
b) A.S. Goldman *et al.* - *Chem. Commun.*, 2060, **2003**.

17.20. T. Sakakura, M. Tanaka - Efficient Catalytic C-H Activation of Alkanes: Regioselective Carbonylation of the Terminal Methyl Group of *n*-Pentane by [RhCl(CO)(PMe₃)₂], *J. Chem. Soc., Chem. Commun.*, 758, **1987**.

17.21. A.J. Kunin, R. Eisenberg - Photochemical Carbonylation of Benzene by Ir^I and Rh^I Square-Planar Complexes, *J. Am. Chem. Soc.* **108**, 535, **1986**.

17.22. A.S. Goldman - Study of the Mechanism of Photochemical Carbonylation of Benzene Catalyzed by Rh(PMe₃)₂(CO)Cl, *J. Am. Chem. Soc.* **116**, 9498, **1994**; L.D. Field *et al.* - *J. Am. Chem. Soc.* **116**, 9492, **1994**.

17.23. W.D. Jones *et al.* - Carbon-Hydrogen Bond Activation by Ruthenium for the Catalytic Synthesis of Indoles, *J. Am. Chem. Soc.* **108**, 5640, **1986** and **109**, 5047, **1987**; *Organometallics* **9**, 718, **1990** and **13**, 385, **1994**.

17.24. a) K.M. Waltz, J.F. Hartwig - *Science* **277**, 211, **1997**;
    b) H. Chen, J.F. Hartwig - Catalytic, Regiospecific End-Functionalization of Alkanes: Rhenium-Catalyzed Borylation Under Photochemical Conditions, *Angew. Chem. Int. Ed.* **38**, 3391, **1999**;
    c) H. Chen, S. Schlecht, T.C. Semple, J.F. Hartwig - *Science* **287**, 1995, **2000**;
    d) P. Rablen, J.F. Hartwig - Accurate Borane Sequential Bond Dissociation Energies by High-Level ab Initio Computational Methods, *J. Am. Chem. Soc.* **118**, 4648, **1996**;
    e) J.F. Hartwig - in ref. **17.1**, Chap. 8.

17.25. a) J.M. Mayer - Hydrogen Atom Abstraction by Metal-Oxo Complexes: Understanding the Analogy with Organic Radical Reactions, *Acc. Chem. Res.* **31**, 441, **1998**;
    b) J.M. Mayer *et al.* - *Science* **294**, 2524, **2001**;
    c) J.M. Mayer *et al.* - *J. Am. Chem. Soc.* **124**, 10112, **2002**.

17.26. F.G. Bordwell *et al.* - Bond Dissociation Energies in DMSO Related to the Gas Phase Values, *J. Am. Chem. Soc.* **102**, 5751, **1980**; *J. Am. Chem. Soc.* **113**, 9790, **1991** and **118**, 8777, **1996**.

17.27. a) R. Taylor - *Electrophilic Aromatic Substitution*, Wiley, New York, **1990**;
    b) H.C. Brown, R.A. Wirkkala - *J. Am. Chem. Soc.* **88**, 1447, **1966**;
    c) G.B. Deacon, G.J. Farquharson - *J. Organomet. Chem.* **67**, C1, **1974**;
    d) Y. Fujihara *et al.* - *Chem. Lett.*, 345, **1995**; *J. Organomet. Chem.* **580**, 290, **1999**;
    e) V.V. Grushin *et al.* - *Adv. Synth. Catal.* **343**, 161, **2001**; in ref. **17.1**, Chap. 24.

17.28. a) V. Snieckus - Directed Ortho-Metalation. Tertiary Amide and *o*-Carbamate Directors in Synthetic Strategies for Polysubstituted Aromatics, *Chem. Rev.* **90**, 879, **1990**;
    b) S. Murai *et al.* - *Nature* **366**, 529, **1993**; *Chem. Lett.*, 681, **1995**; Density Functional Study on Activation of ortho-CH Bond in Aromatic Ketone by Ru Complex. Role of Unusual Five-Coordinated d⁶ Metallacycle Intermediate with Agostic Interaction, *J. Am. Chem. Soc.* **120**, 12692, **1998**;
    c) *J. Org. Chem.* **63**, 5129, **1998**; *J. Am. Chem. Soc.* **123**, 10935, **2001**;
    d) F. Kakiuchi, S. Murai - *Acc. Chem. Res.* **35**, 826, **2002**; V. Ritleng, C. Sirlin, M. Pfeffer - *Chem. Rev.* **35**, 826, **2002**.

17.29. H.D. Kaesz, R.B. Saillant - Hydride Complexes of the Transition Metals, *Chem. Rev.* **72**, 231, **1972**; T.F. Blackburn, J. Schwarz - *J. Chem. Soc., Chem. Commun.*, 157, **1977**; J.-E. Bäckwell *et al.* - *Tetrahedron Lett.* **34**, 5459, **1993**; *J. Chem. Soc., Chem. Commun.*, 1037, **1994**.

17.30. a) P. Shapley *et al.* - Heterobimetallic Catalysts for the Oxidation of Alcohols. [Os(N)R₂(CrO₄)]⁻ (R = Me, CH₂SiMe₃), *J. Am. Chem. Soc.* **110**, 6591, **1988**;
    b) P. Shapley *et al.* - *Inorg. Chem.* **27**, 976, **1988** and **32**, 5646, **1993**;
    c) P. Shapley *et al.* - *J. Am. Chem. Soc.* **122**, 1079, **2000**;
    d) M.E. Tess *et al.* - *Inorg. Chem.* **39**, 3942, **2000**; P. Shapley *et al.* - *Organometallics* **20**, 4700, **2001**.

Chapter 18 - Carbonylation and carboxylation reactions

18.1. I. Tkatchenko - Synthesis with Carbon Monoxide and a Petroleum Product, in *Comprehensive Organometallic Chemistry*, G. Wilkinson, F.G.A. Stone, E.W. Abel Eds, Pergamon Press, Oxford, **1982**, Vol. 8, Chap. 50.3, 101-224.

18.2. **a)** M. Gauss *et al.* - Synthesis of Acetic Acid and Acetic Anhydride from Methanol, in ref. **16.2b**, Vol. 1, Chap. 2.1.2, 104-138; D. Forster. - *Adv. Organomet. Chem.* **17**, 255, **1979**;

    b) B. Cornils, W.A. Herrmann Eds - *Applied Homogeneous Catalysis with Organometallic Compounds,* VCH, Weinheim, **1996**;

    c) B. Cornils, W.A. Herrmann - *Angew. Chem. Int. Ed.* **36**, 1048, **1997**.

18.3. C.D. Frohning, C.W. Kohlpainter - Hydroformylation (Oxo Synthesis, Roelen Reaction), in ref. **16.2b**, Vol. 1, Chap. 2.1.1, 29-104.

18.4. **a)** Ph. Kalck *et al.* - Hydroformylation Catalyzed by Ruthenium Complexes, *Adv. Organomet. Chem.* **32**, 121, **1991**; *Pure Appl. Chem.* **61**(5), 967, **1989**;

    b) E. Kunz - *Info Chimie* **421**, 51, **2000**.

18.5. **a)** A. Höhn - Synthesis of Propionic and Other Acids, in ref. **16.2b**, Vol. 1, Chap. 2.1.2.2;

    b) W. Clegg - *Chem. Commun.* 1877, **1999**;

    c) E. Drent, P.H.M. Budzelaar - *Chem. Rev.* **96**, 663, **1996**;

    d) E. Drent *et al.* - New Developments in the Carbonylation of Alkynes: a Clean Route to Methacrylates, in ref. **16.2b**, Chap. 3.3.9, 1119.

18.6. S. Inoue, N. Yamazaki Eds - *Organic and Bio-organic Chemistry of Carbon Dioxide*, Wiley, New York, **1982**.

18.7 R.P.A. Sneeden - Reactions of Carbon Dioxide, in *Comprehensive Organometallic Chemistry*, G. Wilkinson, F.G.A. Stone, E.W. Abel Eds, Pergamon Press, Oxford, **1982**, Vol. 8, Chap. 50.4, 225-284.

Chapter 19 - Bio-organometallic chemistry: enzymatic catalysis

19.1. L. Stryer - *Biochemistry,* 5th ed., Freeman, New York, **2002**.

19.2. J.J.R.F. Da Silva, R.J.P. Williams - *The Biological Chemistry of the Elements,* Oxford University Press, New York, **1991**.

19.3. R.H. Holm *et al.* - Structural and Functional Aspects of Metal Sites in Biology, *Chem. Rev.* **96**, 2239, **1996**.

19.4. J. Reedijk Ed. - *Bioinorganic Catalysis,* Dekker, New York, **1993**.

19.5. R.H. Crabtree - *The Organometallic Chemistry of the Transition Metals,* 2nd ed., Wiley, New York, **1994**, Chap. 16 (Bioorganometallic Chemistry), 428-456.

19.6. D. Dolphin - B_{12}, Wiley, New York, **1982**.

19.7. J. Halpern - Mechanistic Aspects of Coenzyme B_{12}- Dependent Rearrangements. Organometallics as Free Radical Precursors, *Pure Appl. Chem.* **55**, 1059, **1983**; *Acc. Chem. Res.* **15**, 231, **1982**.

19.8. G.N. Schrauzer *et al.* - The Mechanism of Coenzyme B_{12} Action in Dioldehydrase, *J. Am. Chem. Soc.* **93**, 1503 and 1505, **1971**.

19.9. Chap. 14 in *Comprehensive Organometallic Chemistry*, G. Wilkinson, F.G.A. Stone, E.W. Abel Eds, Pergamon Press, Oxford, **1982**; R.H. Homes, J.A. Ibers - in *Iron Sulfur Proteins,* W. Lovenberg Ed., Academic Press, New York, **1976**, Vol. 3, Chap. 7.

19.10.a) R.A. Sheldon, J.K. Kochi - *Metal-Catalyzed Oxidation of Organic Compounds,* Academic Press, New York, **1981**, Chap. 8 (Biological Oxidations), 216-270;

    b) D. Mansuy, P. Battioni - Dioxygen Activation at Heme Centers in Enzymes and Synthetic Analogues, in ref. **18.4**, Chap. 12, 395-424.

19.11. L. Que - Oxygen Activation at Nonheme Iron Centers, in ref. **19.4,** Chap. 11, 347-394; *Prog. Inorg. Chem.* **38**, 97, **1990**; M. Costas, M.P. Mehn, M.P. Jensen, L. Que Jr. - Dioxygen Activation at Mononuclear Nonheme Iron Active Sites: Enzyme, Models, and Intermediates, *Chem. Rev.* **104**, 939, **2004**; K.D. Koehntop, J.P. Emerson, L. Que Jr. - The 2-His-1-carboxylate Facial Triad: a Versatile Platform for Dioxygen Activation by Mononuclear Nonheme Iron (II) Enzymes, *J. Biol. Inorg. Chem.* **10**, 87, **2005**.

19.12. a) A.E. Shilov - *Metal Complexes in Biomimetic Chemical Reactions,* CRC Press, New York, **1996**, Chap. 1, 1-100 (N_2 Fixation); *Pure Appl. Chem.* **1409**, 1992, **1992**;

   b) M.D. Fryzuk, S.A. Johnson - *Coord. Chem. Rev.* **200-202**, 379, **2000**;

   c) C.M. Kozak, P. Mountford - Revelations in Dinitrogen Metal Complexes and Functionalization by Metal Complexes, *Angew. Chem. Int. Ed.* (Highlight) **43**, 1186, **2004**.

19.13. R.H. Holmes, E.I. Salomon Eds - Nitrogen Metabolism, in the special issue: Bioinorganic Enzymology, *Chem. Rev.* **96**(7), 2951-3030, **1996** (4 reviews); D.J. Evans *et al.* - Catalysis by Nitrogenase and Synthetic Analogs, in ref. **19.4**, Chap. 5, 89-130.

19.14. a) R.R. Schrock - *Acc. Chem. Res.* **30**, 9, **1997**; *Chem. Commun.* 2389, **2003**;

   b) D.V. Yandulov, R.R. Schrock - Reduction of Dinitrogen to Ammonia at a Well-Protected Reaction Site in a Molybdenum Triamidoamine Complex, *J. Am. Chem. Soc.* **124**, 6252, **2002**; *Science* **301**, 76, **2003**;

   c) T.A. Bazhenova, A.E. Shilov - *Coord. Chem. Rev.* **144**, 69, **1995**.

19.15. S.W. Ragsdale, M. Kumar - Nickel Containing Carbon Monoxide Dehydrogenase/Acetyl-Coenzyme A Synthase, *Chem. Rev.* **96**, 2515, **1996**.

19.16. J.R. Lancaster - *The Bioinorganic Chemistry of Nickel,* VCH, Weinheim, **1988**.

19.17. R. Cammack - Catalysis by Nickel in Bioinorganic Systems, in ref. **18.4**, Chap. 7, 189-225; *Adv. Inorg. Chem.* **32**, 297, **1988**.

19.18. R.H. Crabtree *et al.* - Functional Modeling of CO Dehydrogenase: Catalytic Reduction of Methylviologen by CO/H_2O with and N,O,S-Ligated Nickel Catalyst, *Angew. Chem. Int. Ed. Engl.* **32**, 92, **1993**.

19.19. R.H. Holm *et al.* - Reaction Sequence Related to that of Carbon Monoxide Dehydrogenase (Acetyl Coenzyme A Synthase): Thioester Formation Mediated at Structurally Defined Nickel Centers, *J. Am. Chem. Soc.* **112**, 5385, **1990**.

19.20. G. Jaouen, A. Vessieres, I.S. Butler - Bioorganometallic Chemistry. A Future Direction for Transition Metal Organometallic Chemistry?, *Acc. Chem. Res.* **26**, 361, **1993**; G. Jaouen Ed. - *Organometallics*, Wiley-VCH, Weinheim, **2006**.

19.21. R.H. Fish *et al.* - A New, Aqueous ^{1}H NMR Shift Reagent Based on Host-Guest Molecular Recognition Principles Using a Supramolecular Host: [Cp*Rh(2'-deoxyadenosine)]$_3$(OTf)$_3$, *J. Org. Chem.* **63**, 7151, **1998**.

Chapter 20 - Heterogeneous catalysis

20.1. a) For a leading reference, see G.A. Somorjai - *Introduction to Surface Chemistry and Catalysis*, Wiley, New York, **1994**;

   b) G. Ertl, H. Knözinger, J. Veitkamp Eds - *Handbook of Heterogeneous Catalysis*, Wiley-VCH, Weinheim, **1997**.

20.2. a) J.-F. Lamber, E. Bordes-Richard Eds - *Act. Chim.* **2002**, May-June;

   b) G. Martino, J.-P. Boîtiaux in ref. **20.2a**, 7;

   c) J.-C. Volta, J.-L. Portefaix - *Appl. Catal.* **18**, 1, **1985**; E. Bordes-Richard - Challenges in Heterogeneous Catalytic Oxidation, in ref. **20.2a**, 38.

20.3. **a)** A. Nielsen - Review of Ammonia Catalysis, *Catal. Rev.* **4**, 1, **1970**;
    b) J.H. Sinfelt - Catalytic Hydrogenolysis on Metals, *Catal. Lett.* **9**, 159, **1991**;
    c) C.J.H. Jacobson - *Chem. Commun.*, 1057, **2000**.

20.4. M.P. Kiskinova - Electronegative Additives and Poisoning in Catalysis,
Surf. Sci. Rep. **8**, 359, **1988**.

20.5. G.A. Somorjai - On the Mechanism of Sulfur Poisoning of Platinum Catalysts,
J. Catal. **27**, 453, **1972**.

20.6. S. Bhatia *et al.* - Deactivation of Zeolite Catalysts,
Catal. Rev. Sci. Eng. **31**, 431, **1989/1990**.

20.7. O.V. Krylov - *Catalysis by Non-Metals*, Academic Press, New York, **1970**.

20.8. G.A. Somorjai - Active Sites in Heterogeneous Catalysis, *Adv. Catal.* **26**, 1, **1977**.

20.9. R.G. Herman - Classical and Non-classical Routes for Alcohol Synthesis,
in *New Trends in CO Activation. Studies in Surface Science and Catalysis*,
L. Guczi Ed., Elsevier, Amsterdam, **1991**, Vol. 64.

20.10. D.A. Hickman, L.D. Schmidt - Production of Syngas by Direct Catalytic Oxidation
of Methane, *Science* **259**, 343, **1993**.

20.11. C. Lapinte, D. Catheline, D. Astruc - $NaBH_4$ Reduction of CO in the Cationic
Carbonyl Complexes $[C_5Me_5Fe(CO)_2L][PF_6]$, *Organometallics* **7**, 1683, **1988**.

20.12. **a)** G.A. Somorjai *et al.* - Potassium Coadsorption Induced Dissociation of CO on the
    Rh(111) Crystal Face: an Isotope Mixing Study, *J. Phys. Chem.* **89**, 1598, **1985**;
    b) P. J. Flory - *Principles of Polymer Chemistry*,
    Cornell University Press, Itahaca, New York, **1953**.

20.13. J.T. Kummer, P.T. Emmet - Fischer-Tropsch Synthesis Mechanistic Studies,
J. Am. Chem. Soc. **75**, 5177, **1953**.

20.14. D.J. Dwyer, G.A. Somorjai - The Role of Readsorption in Determining the Product
Distribution During CO Hydrogenation over Fe Single Crystals,
J. Catal. **56**, 249, **1979**.

20.15. R.B. Anderson - *The Fischer-Tropsch Synthesis*, Academic Press, London, **1984**.

20.16. P.M. Maitlis *et al.* - Investigations by ^{13}C NMR Spectroscopy of Ethene-Initiated
Catalytic CO Hydrogenation, *J. Am. Chem. Soc.* **124**, 10456, **2002**;
J. Mol. Cat. A: Chemical **204-205**, 55, **2003**; *J. Organomet. Chem.* **689**, 4366, **2004**.

20.17. Z.-P. Liu, P. Hu - A New Insight into Fischer-Tropsch Synthesis,
J. Am. Chem. Soc. **124**, 11568, **2002**.

20.18. I.M. Ciobica *et al.* - Mechanisms for Chain Growth in Fischer-Tropsch Synthesis
over Ru(0001), *J. Catal.* **212**, 136, **2002**.

20.19. R.D. Adams, I.T. Horvath - The Unusual Structures, Bonding, and Reactivity of
Some Sulfido-Bridged Tungsten-Osmium Carbonyl Cluster Compounds,
Prog. Inorg. Chem. **33**, 127, **1985**; J.S. Bradley - *Adv. Organomet. Chem.* **22**, 1, **1983**.

20.20. F.G.A. Stone *et al.* - Chemistry of Polynuclear Metal Complexes with Bridging
Carbene or Carbyne Ligands, *J. Chem. Soc., Dalton Trans.*, 1201, **1987**.

20.21. P.H. Emmet *et al.* - Mechanism Studies of the Fischer-Tropsch Synthesis. The
Addition of Radioactive Alcohol, *J. Am. Chem. Soc.* **73**, 564, **1951**; **82**, 1027, **1960**.

20.22. H. Schulz, M. Claeys - Kinetic Modelling of Fischer-Tropsch Product Distributions,
Appl. Catal. A: General **186**, 91, **1999**.

20.23. M. Ichikawa *et al.* - Heterogenized Bimetallic Clusters. Their Structures and
Bifunctional Catalysis, *J. Mol. Cat.* **62**, 15, **1990**;
J. Chem. Soc., Chem. Commun., 458, **1988**.

20.24. P.M. Maitlis - New Models for the Formation of Oxygenates in the Fischer-Tropsch
Reaction, *J. Organomet. Chem.* **334**, C14 and C17, **1987**;
J. Chem. Soc., Dalton Trans., 2193, **1992**.

20.25. G.A. Somorjai - Surface Science and Catalysis,
Philos. Trans. R. Soc., London, **A318**, 81, **1986**.

20.26. G.A. Somorjai *et al.* - A Comparison of Gas-Phase and Electrochemical
Hydrogenation of Ethylene at Platinum Surfaces, *J. Am. Chem. Soc.* **107**, 5910, **1985**.

20.27. G.A. Somorjai, B.E. Bent - The Structure of Adsorbed Monolayers. The Surface
Chemical Bond, *Prog. Colloid Polym. Sci.* **70**, 38, **1985**.

20.28. S.J. Thompson, G. Webb - Catalytic Hydrogenation of Olefins on Metals, a New
Interpretation, *J. Chem. Soc., Chem. Commun.* **13**, 526, **1976**.

20.29. J.M. Thomas - *Stud. Surf. Sci. Catal.* **140**, 1, **2001**; Structural Elucidation of
Microporous and Mesoporous Catalysts and Molecular Sieves by High-Resolution
Electron Microscopy, *Acc. Chem. Res.* **34**, 583, **2001**;
Angew. Chem. Int. Ed. Engl. **38**, 3588, **1999**; *Top. Catal.* **15**, 85, **2001**.

20.30. M. Bowker - *The Basis and Applications of Heterogeneous Catalysis*,
Oxford University Press, New York, **1998**.

20.31. H. Akawara, M. Aresta, J.N. Armor, M.A. Barteau, E.J. Beckman, A.T. Bell,
J.E. Bercaw, C. Creuz, E. Dinjus, D.A. Dixon, K. Domen, D.L. Dubois, J. Eckert,
E. Jujita, D.H. Gibson, W.A. Goddart, D.W. Goodman, J. Keller, G.J. Kubas,
H.H. Kung, J.E. Lyons, L.E. Manzer, T.J. Marks, K. Kormuka, K.M. Nicholas,
R. Periana, L. Que, J. Rostrup-Nielson, W.H.M. Sachtler, L.D. Schmidt, A. Sen,
G.A. Somorjai, P.C. Stair, B.R. Stults, W. Tomas - Catalysis Research of Relevance
to Carbon Management: Progress, Challenges, and Opportunities,
Chem. Rev. **101**, 953, **2001**.

20.32. D. Bazin, C. Mottet, G. Treglia - New Opportunities to Understand Heterogeneous
Catalysis Processes on Nanoscale Bimetallic Particles Through Synchrotron
Radiation and Theoretical Studies, *Appl. Catal. A* **200**, 47, **2000**.

20.33. A. Corma - From Microporous to Mesoporous Molecular Sieve Materials and Their
Use in Catalysis, *Chem. Rev.* **97**, 2373, **1997**.

20.34. J.-C. Volta - Site Isolation for Light Hydrocarbons Oxidation,
Top. Catal. **15**, 93, **2001**.

20.35. J.-M. Millet, J.C. Védrine - Importance of Site Isolation in the Oxidation of
Isobutyric Acid to Methacrylic Acid on Iron Phosphate Catalysts,
Top. Catal. **15**, 139, **2001**.

20.36. K.L. Fudala, T.D. Tilley - Design and Synthesis of Heterogeneous Catalysts:
the Thermodynamic Molecular Approach, *J. Catal.* **216**, 265, **2003**.

20.37. T.D. Tilley *et al.* - Tris(tert-butoxy)siloxy Complexes as Single-Source Precursors to
Homogeneous Zirconia- and Hafnia-Silica Materials. An Alternative to the Sol-Gel
Method, *J. Am. Chem. Soc.* **119**, 9745, **1997**; *Chem. Mater.* **3**, 1001, **1991**.

20.38. R. Rulkens, T.D. Tilley - A Molecular Precursor Route to Active and Selective
Vanadia-Silica-Zirconia Heterogeneous Catalysts for the Oxidative Dehydrogenation
of Propane, *J. Am. Chem. Soc.* **120**, 9959, **1998**.

20.39. H.H. Kung - *Adv. Catal.* **40**, 1, **1994**.

20.40. a) B.K. Warren, S.T. Oyama - Heterogeneous Hydrocarbon Oxidation,
in *ACS Symp. Ser.*, Washington DC, **1996**;
b) S. Albonetti, F. Cavani, F. Trifirò - *Key Aspects of Catalyst Design for the
Selective Oxidation of Paraffins*, Dekker, New York, **1996**;
c) B.K. Hondett - *Heterogeneous Catalytic Oxidation*, Wiley, Chichester, **2000**.

20.41. T.D. Tilley *et al.* - Silica-Supported, Single-Site Titanium Catalysts for Olefin
Epoxidation. A Molecular Precursor Strategy for Control of Catalyst Structure,
J. Am. Chem. Soc. **124**, 8380, **2002**; *J. Am. Chem. Soc.* **119**, 9745, **1997**.

20.42. M. Chabanas, A.E. Quadrielli, C. Copéret, J. Thivolle-Cazat, J.M. Basset, A. Lesage, L. Emslet - Molecular Insight into Surface Organometallic Chemistry Through the Combined Use of 2D HETCOR Solid-State NMR Spectroscopy and Silsesquioxane Analogues, *Angew. Chem. Int. Ed.* **40**, 4493, **2001**.

20.43. C. Copéret, M. Chabanas, R.P. Saint-Arroman, J.-M. Basset - Homogeneous and Heterogeneous Catalysis: Bridging the Gap Through Surface Organometallic Chemistry, *Angew. Chem. Int. Ed.* **42**, 156, **2003**.

20.44. D. Soulivong, C. Copéret, J. Thivolle-Cazat, J.-M. Basset, B. Maunders, R.B.A. Pardy, G.J. Sunley - Cross-Metathesis of Propane and Methane: a Catalytic Reaction of C-C Bond Cleavage of a Higher Alkane by Methane, *Angew. Chem. Int. Ed.* **43**, 5366, **2004**.

20.45. R.L. Burnett, T.R. Hughes - Mechanism and Poisoning of the Molecular Redistribution Reaction of Alkanes with a Dual-Functional Catalyst System, *J. Mol. Cat.* **31**, 55, **1973**.

20.46. a) J.-M. Basset *et al.* - Characterization of Surface Organometallic Complexes Using High Resolution 2D Solid-State NMR Spectroscopy. Application to the Full Characterization of a Silica Supported Metal Carbyne: SiO-Mo(C-Bu-t)(CH$_2$-Bu-t)$_2$, *J. Am. Chem. Soc.* **123**, 3820, **2001**;

    b) F. Blanc, J. Thivolle-Cazat, J.-M. Basset, C. Copéret, A.S. Hock, Z.J. Tonzetich, R.R. Schrock - *J. Am. Chem. Soc.* **129**, ASAP, **2007**.

20.47. C. Copéret, F. Lefebvre, J.-M. Basset - Well-Defined Metallocarbenes and Metallocarbynes Supported on Oxide Supports Prepared *via* Surface Organometallic Chemistry: a Source of Highly Active Alkane, Alkene, and Alkyne Metathesis Catalysts, in *Handbook of Metathesis*, R.H. Grubbs Ed., Wiley-VCH, Weinheim, **2003**, Vol. 1, Chap. 1.12, 190-2004.

20.48. J. Li, X. Li, H.-J. Zhai, L.-S. Wang - Au$_{20}$: a Tetrahedral Cluster, *Science* **299**, 864, **2003**.

20.49. R.G. Finke *et al.* - Polyoxoanion- and Tetrabutylammonium-Stabilized Rh^0_n Nanoclusters: Unprecedented Nanocluster Catalytic Lifetime in Solution, *J. Am. Chem. Soc.* **121**, 8803, **1999** and **124**, 5796, **2002**.

20.50. D. Astruc, F. Lu, J. Ruiz - Nanoparticles as Recyclable Catalysts: the Frontier Between Homogeneous and Heterogeneous Catalysis, *Angew. Chem. Int. Ed.* **44**, 7399, **2005**, *Inorg. Chem.* **46**, 1884, **2007**.

20.51. M. Haruta - Nanoparticulate Gold Catalysts for Low Temperature CO Oxidation, *J. New. Mat. Electrochem. Systems* **7**, 163, **2004**; C. Louis - in *Nanoparticle Catalysis*, D. Astruc Ed., Wiley, Weinheim, **2007**.

20.52. M.-C. Daniel, D. Astruc - Gold Nanoparticles: Assembly, Supramolecular Chemistry, Quantum-Size-Related Properties, and Applications Towards Biology, Catalysis and Nanotechnology, *Chem. Rev.* **104**, 293, **2004**.

20.53. a) M.T. Reetz, J.G. de Vries - Ligand-Free Heck Reactions Using Low Pd-Loading, *Chem. Commun.*, 1559, **2004**; J.G. de Vries - *Dalton Trans.*, 421, **2006**;

    b) *Nanoparticle Catalysis*, D. Astruc Ed., Wiley, Weinheim, **2007**; N.T. Phan, M. Van der Sluys, C.W. Jones - On the Nature of the Active Species in Palladium Catalyzed Mizoroki-Heck and Suzuki-Miyaura Couplings- Homogeneous or Heterogeneous Catalysis, A Critical Review, *Adv. Syn. Catal.* **348**, 609, **2006**.

20.54. B.H. Lipshutz, S. Tasler, W. Chrisman, B. Spliethoff, B. Tesche - On the Nature of the 'Heterogeneous' Catalyst: Nickel-on-Charcoal, *J. Org. Chem.* **68**, 1177, **2003**; B.H. Lipshutz, H. Ueda - *Angew. Chem. Int. Ed.* **39**, 4492, **2000**.

20.55. S. Pröckl, W. Kleist, M.A. Gruber, K. Köhler - In Situ Generation of Highly Active Dissolved Palladium Species from Solid Catalysts - A Concept for the Activation of Aryl Chlorides in the Heck Reaction, *Angew. Chem. Int. Ed. Engl.* **43**, 1881, **2004**; L. Djakovitch, M. Wagner, C.G. Hartung, M. Beller, K. Köhler - *J. Mol. Cat. A: Chemical* **219**, 121, **2004**.

20.56. J. Grunes, J. Zhu, G.A. Somorjai - Catalysis and Nanoscience, *Chem. Commun.* (Focus Article) 2257, **2003**; P. Serp, P. Kalk, R. Feurer - Chemical Vapor Deposition for the Controlled Preparation of Supported Catalytic Material, *Chem. Rev.* **102**, 3085, **2002**; G. Renaud, J. Jupille *et al.* - Real-time Monitoring of Growing Nanoparticles, *Science* **300**, 1416, **2003**.

20.57. a) J.-H. He, I. Ichinose, T. Kunitake, A. Nakao, Y. Shiraishi, N. Toshima - Facile Fabrication of Ag-Pd Bimetallic Nanoparticles in Ultrathin TiO$_2$-Gel Films: Nanoparticle Morphology and Catalytic Activity, *J. Am. Chem. Soc.* **125**, 11034, **2003**;
b) N. Toshima, T. Yonezawa - *New J. Chem.* **22**, 1179, **1998**.

Chapter 21 - Organometallic complexes in organic synthesis
Examples of applications

21.1. L.S. Hegedus - *Transition Metals in the Synthesis of Complex Organic Molecules*, University Science Books, MillsValley, Ca, **1994**.

21.2. F.J. McQuillin, D.G. Parker, G.R. Stephenson - *Transition Metal Organometallics for Organic Synthesis*, Cambridge University Press, Cambridge, **1991**.

21.3. J. Harrington - *Transition Metals in Total Synthesis*, Wiley, New York, **1990**.

21.4. S.G. Davies - *Organotransition Metal Chemistry: Applications to Organic Synthesis*, Pergamon Press, Oxford, **1982**.

21.5. K.M. Nicolas, R. Pettit - An Alkyne Protecting Group, *Tetrahedron Lett.*, 3475, **1971**.

21.6. D.R.H. Barton, H. Patin *et al.* - Synthetic Uses of Steroidal Ring Protection: 22,23-Dihydroergosterol, *J. Chem. Soc., Perkin Trans. I.*, 821, **1976**.

21.7. M. Franck-Neumann, D. Martina - Reaction of Diazoalkanes with Tropone-Iron Carbonyl: a New Synthesis of 2,3-Tropones, *Tetrahedron Lett.*, 1759, **1975**.

21.8. D. Seyferth, A.T. Wehman - Aromatic Electrophilic Substitution Reactions of Phenylmethylidynecobalt Nonacarbonyl and Diphenylacetylene Hexacarbonyl. Preparatively Useful Reactions of Coordinated Organic Ligands, *J. Am. Chem. Soc.* **92**, 5520, **1970**.

21.9. K.M. Nicholas *et al.* - Total Synthesis of Cyclocolorenone, *Tetrahedron Lett.* **27**, 915, **1986**; *J. Org. Chem.* **51**, 1960, **1986**.

21.10. D. Astruc - Use of Organoiron Compounds in Organic Synthesis, in *The Chemistry of the Metal-Carbon Bond*, F.R. Hartley, S. Patai Eds, Wiley, New York, **1987**, Vol. 4, 625-731.

21.11. H. Alper - (Dihydropyridine) Iron Tricarbonyl Complexes, *J. Organomet. Chem.* **96**, 95, **1975**.

21.12. M.F. Semmelhack, A. Yamashita - Arene-Metal Complexes in Organic Synthesis, Synthesis of Acorenone and Acorenone B, *J. Am. Chem. Soc.* **102**, 5924, **1980**.

21.13. V. Sartor, L. Djakovitch, J.-L. Fillaut *et al.* - Organoiron Route to a New Dendron for Fast Dendritic Syntheses Using Divergent and Convergent Methods, *J. Am. Chem. Soc.* **121**, 2929, **1999**; *New J. Chem.* **24**, 351-370, **2000**.

21.14.a) N. Ardoin, D. Astruc - Molecular Trees: from Syntheses Towards Applications, *Bull. Soc. Chim.* **132**, 875, **1995**;

  b) D. Astruc - Research Avenues on Dendrimers Towards Molecular Biology: from Biomimetism to Medicinal Engineering, *C.R. Acad. Sci.* **322**, Série IIb, 757, **1996**.

21.15. A.J. Pearson, D.C. Rees - Total Synthesis of (±) and (+) Limaspermine and Formal Synthesis of (±)-Aspidospermine Using Organoiron Complexes, *J. Chem. Soc., Perkin Trans. I.*, 2467, **1982**; Synthetic Approaches to C-18 Oxygenated Aspidosperma Alkaloids *via* Organoiron Complexes, *J. Chem. Soc., Perkin Trans. I.*, 619, **1983**.

21.16.a) R.F. Heck - *Palladium Reagents in Organic Synthesis*, Academic Press, New York, **1985**;

  b) A. de Mejere, F.E. Meyer - Fine Feathers Make Fine Birds: the Heck Reaction in Modern Garb, *Angew. Chem. Int. Ed. Engl.* **33**, 2379, **1994**;

  c) W.A. Herrmann - Catalytic C-C Coupling by Pd Complexes: Heck Reaction, in ref. **15.2**, Vol. 2, Chap. 3.1.6, 712-732;

  d) W. Cabri, I. Candiani - Recent Developments and New Perspectives in the Heck Reaction, *Acc. Chem. Res.* **28**, 2, **1995**; N.J. Whitcombe, K.K. Hii, S.E. Gibson - Advances in the Heck Chemistry of Aryl Bromides and Chlorides, *Tetrahedron* **57**, 7449, **2001** (Tetrahedron Report N° 582);

  e) A.F. Little, G.C. Fu - *Angew. Chem. Int. Ed.* **41**, 4176, **2002**;

  f) R.B. Bedford, C.S.J. Casin, D. Holder - *Coord. Chem. Rev.* **248**, 2283, **2004**;

  g) I.P. Beletskaya, A.V. Cheprakov - *Chem. Rev.* **100**, 3009, **2000**.

21.17. J. Tsuji - Organic Synthesis with Palladium Compounds, Springer Verlag, Berlin, **1980**; *Synthesis*, 369, **1984**.

21.186. J. Tsuji *et al.* - Total Synthesis of Monarch Butterfly Pheromone, *Tetrahedron Lett.* **22**, 2575, **1981**.

21.19.a) A. Alexakis, J.-F. Normant *et al.* - Reactivity of RCu, BF$_3$ and R$_2$CuLi, BF$_3$ towards the Ether Linkage. Epoxides Acetals and Orthoformiates, *Tetrahedron Lett.* **25**, 3075, **1984**;

  b) A. Alexakis, J.-F. Normant *et al.* - Acetal as Chiral Auxiliaries: Asymmetric Synthesis of γ, δ-Ethylenic Aldehydes. An Approach to the California Red Scale Pheromone, *Tetrahedron Lett.* **28**, 2363, **1987**.

21.20. A. Alexakis, J.-F. Normant - Chirals Acetals in Enantio- and Diastereoselective Substitution or Elimination Reactions, *Pure Appl. Chem.* **60**, 49, **1988**.

21.21. S.H. Bertz *et al.* - New Copper Chemistry. V. New Heterocuprates with Greatly Improved Thermal Stability, *J. Am. Chem. Soc.* **104**, 5824, **1982**; *J. Org. Chem.* **49**, 1119, **1984**.

21.22. S.F. Martin, R.A. Jones, A.H. Cowley *et al.* - Structure and Reactivity of Novel Lithium Di-*tert*-butylphosphido(alkyl)cuprates, *J. Am. Chem. Soc.* **110**, 7226, **1988**.

21.23. I. Marek, A. Alexakis, J.-F. Normant - γ-Tributylstannyl-β-Metallated Acrolein: a Versatile Synthon, *Tetrahedron Lett.* **32**, 6337, **1991**.

21.24. J.P. Quintard *et al.* - Regio- and Stereocontrolled Stannylmetallation of 3,3-diethoxy-prop-1-yne and 4-4-diethoxy-but-1-yne: an Efficient Access to the Corresponding Vinyltins with Fixed Configurations, *Tetrahedron Lett.* **32**, 6333, **1991**.

21.25. M. Tius *et al.* - C-Glycosylanthraquinone Synthesis: Total Synthesis of Vineomycinone B$_2$ Methyl Ester, *J. Am. Chem. Soc.* **113**, 5775, **1991**.

21.26.a) W.A. Herrmann *et al.* - N-Heterocyclic Carbenes (NHCs) in Catalysis, *Chem. Eur. J.* **2**, 772, **1996**; reviews: W.A. Herrmann - *Angew. Chem. Int. Ed.* **41**, 1291, **2002**;

  b) S.P. Nolan *et al.* - *J. Organomet. Chem.* **653**, 69, **2002**;

  c) J. Schwartz - Carbene Complexes in Organic Synthesis, *Pure Appl. Chem.* **53**, 733, **1980**.

21.27. a) J.K. Stille *et al.* - Stereospecific Pd-Catalyzed Coupling Reaction of Vinyl Iodide with Acetylenic Tin Reagents, *J. Am. Chem. Soc.* **109**, 2138 et 5478, **1987**;

b) A. Kalivretenos, J.K. Stille, L.S. Hegedus - Synthesis of β-Resorcyclic Macrolides *via* Organo-Pd Chemistry. Application to the Total Synthesis of (S)-Zearalenone, *J. Org. Chem.* **56**, 2883, **1991**.

21.28. Coupling Grignard reagents with aryl halides:

a) R.J.P. Corriu, J.P. Masse - *Chem. Commun.*, 144, **1972**;

b) K. Tamao, K. Sumitani, M. Kumada - *J. Am. Chem. Soc.* **94**, 4374, **1972**; review: M. Kumada - *Pure Appl. Chem.* **52**, 669, **1980**;

c) M. Yamamura, I. Moritani, S.J. Murahashi - *J. Organomet. Chem.* **91**, C39, **1975**;

d) J. Huang, S.P. Nolan - *J. Am. Chem. Soc.* **121**, 9889, **1999**;

e) K. Gouda, E. Hagiwara, Y. Hatanaka, T. Hiyama - *J. Org. Chem.* **61**, 7232, **1996**; Hiyama *et al.* - *Tetrahedron Lett.* **38**, 439, **1997**; M.E. Mowery, P. DeShong - *J. Org. Chem.* **63**, 3156, **1998** and **64**, 1684 and 3266, **1999**;

f) H.M. Lee, S.P. Nolan - *Org. Lett.* **2**, 2053, **2000**.

21.29. a) A. Suzuki, N. Miyaura - Palladium Catalyzed Cross-Coupling Reactions of Organoboron Compounds, *Chem. Rev.* **95**, 2457, **1995**; A. Suzuki - in *Modern Arene Chemistry*, D. Astruc Ed., Wiley-VCH, Weinheim, **2002**, 53-106;

b) A. Suzuki *et al.* - Availability of Various Functional Boron Reagents for the Suzuki Reaction, *J. Org. Chem.* **58**, 2201, **1993**;

c) S.P. Stanforth - *Tetrahedron* **54**, 263, **1998**;

d) A.F. Littke, G.C. Fu - *J. Org. Chem.* **64**, 10, **1999**.

21.30. a) F. Lu, J. Ruiz, D. Astruc - Palladium-Dodecanethiolate Nanoparticles as Stable and Recyclable Catalysts for the Suzuki-Miyaura Reaction of Aryl Halides Under Ambient Conditions, *Tetrahedron Lett.*, 9443, **2004**;

b) J. Lemo, K. Heuze, D. Astruc - Efficient Dendritic Diphosphino Pd^{II} Catalysts for the Suzuki Reaction of Choroarenes, *Org. Lett.* **7**, 2253, **2005**.

21.31. a) K. Sonogashira *et al.* - *Tetrahedron Lett.*, 4467, **1975**; K. Sonogashira *et al.* - A Convenient Synthesis of Ethynylarenes and Diethynylarenes, *Synthesis*, 627, **1980**;

b) D. Méry, K. Heuze, D. Astruc - *Chem. Commun.*, 1934, **2003**; K. Heuze, D. Méry, D. Gaus, D. Astruc - *Chem. Commun.*, 2274, **2003**; Copper-free Monomeric and Dendritic Pd Catalysts for the Sonogashira Reaction: Substituent Effects, Synthetic Applications, and the Recovery and Re-use of the Catalysts, *Chem. Eur. J.* **10**, 3936, **2004**;

c) A.L. Rusanov, I.A. Khotina, M.M. Begretov - *Russ. Chem. Rev.* **66**, 1053, **1997**;

d) O. Lavastre, L. Ollivier, P.H. Dixneuf - Sequential Catalytic Synthesis of Rod-Like Conjugated Poly-ynes, *Tetrahedron* **52**, 5495, **1996**.

21.32. J.P. Collman *et al.* - Disodium Tetracarbonylferrate - A Transition Metal Analog of a Grignard Reagent, *Acc. Chem. Res.* **8**, 342, **1975**; Lewis Acid Catalyzed [RFe(CO)₄]⁻ Alkyl Migration Reactions. A Mechanistic Investigation, *J. Am. Chem. Soc.* **100**, 4766, **1978**.

21.33. a) J.E. Mc Murry *et al.* - Stereospecific Total Synthesis of Aphidicolin, *J. Am. Chem. Soc.* **101**, 1330, **1979**;

b) E.-I. Negishi *et al.* - Synthesis of Enol Esters and Enol Lactones *via* Palladium-Catalyzed Carbonylation of Aryl and Alkenyl Halides, *Tetrahedron Lett.* **31**, 2841, **1990**.

21.34. L. Hegedus - in *Organometallics in Synthesis*, M. Schlosser Ed., Wiley, New York, **1994**, Chap. 5.

21.35. C. Amatore, A. Jutand *et al.* - Rates and Mechanism of the Formation of Pd^0 Complexes from Mixtures of Pd(OAc)$_2$ and PR$_3$ and their Reactivity in Oxidative Additions, *Organometallics* **14**, 1818 and 5605, **1995**; *Acc. Chem. Res.* **33**, 314, **2000**; A. Jutand - Mechanisms of Palladium-Catalyzed Reactions: Role of Chloride Ions, *Appl. Organomet. Chem.* **18**, 574, **2004**.

21.36. R.F. Heck - Pd-Catalyzed Reactions of Organic Halides with Olefins, *Acc. Chem. Res.* **12**, 146, **1979**; G. Davis, A. Hallberg - *Chem. Rev.* **89**, 1433, **1989**.

21.37. R.J. Sunberg, R.J. Cherney - Synthesis of Analogs of Iboga Alkaloids. Investigation of Electrophilic, Pd-Catalyzed, and Radical Cyclisations for the Preparation of 5,6-Homoiboga Derivatives, *J. Org. Chem.* **55**, 6028, **1990**.

21.38. a) A. de Meyere *et al.* - Pd-Catalyzed Polycyclization of Dienynes: Surprisingly Facile Formation of Tetracyclic Systems Containing a Three-Membered Ring, *J. Org. Chem.* **56**, 6487, **1991**;
  b) A. Arcadi, S. Cacchi *et al.* - Pd-Catalyzed Reaction of Vinyl Triflates and Vinyl/Aryl Halides with 4-Alkynoic Acids: Regio- and Stereoselective Synthesis of (E)-d-Vinyl/Aryl-γ-methylene-γ-butyrolactones, *J. Org. Chem.* **57**, 976, **1992**.

21.39. a) M. Sasaki, J. Colin, H.B. Kagan - Divalent Lanthanide Derivatives in Organic Synthesis: New Developments in SmI$_2$ Mediated Barbier Type Reactions, *New J. Chem.* **16**, 89, **1992** (see also ref. **12.18**); S.M. Bennett *et al.* - *Synlett.*, 805, **1991**;
  b) K. Banai *et al.* - *Tetrahedron* **46**, 6689, **1990**.

21.40. a) J.F. Hartwig - Carbon-Heteroatom Bond-Forming Reductive Eliminations of Amines, Ethers and Sulfides, *Acc. Chem. Res.* **31**, 852, **1998**; J.F. Hartwig - Palladium-Catalyzed Amination of Aryl Halides and Sulfonates, in *Modern Arene Chemistry*, D. Astruc Ed., Wiley-VCH, Weinheim, **2002**, 107-168;
  b) A.S. Gram *et al.* - P, O and P, N Ligand-Catalyzed Aminations of Aryl Halides, *Organometallics* **18**, 1840, **1999**;
  c) S.L. Buchwald *et al.* - Rational Development of Practical Catalysts for Aromatic Carbon-Nitrogen Bond Formation, *Acc. Chem. Res.* **31**, 805, **1998**; S.L. Buchwald *et al.* - Electron-Rich Bulky Phosphine Ligands Facilitate the Pd-Catalyzed Preparation of Diaryl Ethers, *J. Am. Chem. Soc.* **121**, 4369, **1999**;
  d) S.P. Nolan *et al.* - Aryl C-N Coupling Catalyzed by NHC-Pd, *Org. Lett.* **1**, 1307, **1999**; *J. Org. Chem.* **66**, 7729, **2001**.

21.41. a) K.P.C. Vollhardt - Transition-Metal Catalyzed Acetylene Cyclizations in Organic Synthesis, *Acc. Chem. Res.* **10**, 1, **1977**;
  b) K.P.C. Vollhardt - Cobalt-Mediated [2+2+2]-Cycloadditions: a Maturing Synthetic Strategy, *Angew. Chem. Int. Ed.* **40**, 2004, **2001**.

21.42. a) S. Takano *et al.* - A Facile Chiral Approach to the Dendrobine Skeleton by Intramolecular Pauson-Khand Reaction, *Chem. Lett.*, 443, **1992**; N.E. Shore - *Org. React.* **40**, 1, **1991**;
  b) K.B. Sharpless *et al.* - *Angew. Chem. Int. Ed. Engl.* **23**, 539, **1984**;
  c) V.D. Bock, H. Hiemstra, J.H. Van Maarseveen - *Eur. J. Org. Chem.*, 51, **2006**.

21.43. K.H. Dötz - *Transition Metal Carbene Complexes*, Verlag Chemie, Weinheim, **1983**.

21.44. K. H. Dötz - Carbene Complexes in Organic Synthesis, *Angew. Chem. Int. Ed. Engl.* **23**, 587, **1984**.

21.45. K.A. Parker, C.A. Coburn - A Strategy for the Convenient Synthesis of Gilvo-Carcins *via* Chromium Carbene Benzannulation, *J. Org. Chem.* **56**, 1666, **1991**.

21.46. R.R. Schrock - Alkylidene Complexes of Niobium and Tantalum, *Acc. Chem. Res.* **12**, 98, **1979**.

21.47. a) L.F. Cannizo, R.H. Grubbs - Reactions of [Cp$_2$Ti=CH$_2$] with Acid Anhydrides and Imides, *J. Org. Chem.* **50**, 2316, **1985**; In Situ Preparation of [TiCp$_2$(μ_2-Cl)(μ_2-CH$_2$)AlMe$_2$] (Tebbe's Reagent), *ibid.*, 2386;

  b) H.J. Kang, L.A. Paquette - Claisen-Based Strategy for the de Novo Construction of Basmane Diterpenes. Enantiospecific Synthesis of (+)-7, 8-Epoxy-2-basmen-6-one, *J. Am. Chem. Soc.* **112**, 3252, **1990**.

21.48. a) N.A. Petasis *et al.* - Titanium-Mediated Carbonylations. Benzylidenation of Carbonyl Compounds with Dibenzyltitanocene, *J. Org. Chem.* **57**, 1327, **1992**; *Synlett.*, 5, **1992**;

  b) R. Roy, S.K. Das - Recent Applications of Olefin Metathesis and Related Reactions in Carbohydrate Chemistry, *Chem. Commun.*, 519, **2000**;

  c) A.F. Houri - *J. Am. Chem. Soc.* **117**, 2943, **1995**.

21.49. H. Kagan - in *Comprehensive Organometallic Chemistry*, G. Wilkinson *et al.* Eds, Pergamon Press, Oxford, **1982**, Vol. 8, Chap. 53.

21.50. a) R. Noyori - *Asymmetric Catalysis in Organic Synthesis*, Wiley, New York, **1994**; in ref. **32b**, Chap. 2.9, 552-571;

  b) B. Cornils, W.A. Herrmann Eds - *Applied Homogeneous Catalysis with Organometallic Compounds*, VCH, Weinheim, **1996**.

21.51. K.B. Sharpless *et al.* - Asymmetric Epoxidation of Allylic Alcohols, *Pure Appl. Chem.* **55**, 1823, **1983**; A.C. Oehlschlager, B.D. Johnston - *J. Org. Chem.* **52**, 940, **1987**.

21.52. C.H. Senayake, E. Roberts *et al.* - The Behavior of Indene Oxide in the Ritter Reaction: a Simple Route to cis-Aminoindanol, *Tetrahedron Lett.* **36**, 3993, **1995**.

21.53. R. Noyori, S. Otsuka *et al.* - Highly Enantioselective Isomerization of Prochiral Allylamines Catalyzed by Chiral Diphosphine Rh^I Complexes. Preparation of Optically Active Enamines, *J. Am. Chem. Soc.* **106**, 5208, **1984**.

21.54. A.L. Casalnuovo *et al.* - Ligand Electronic Effects in Asymmetric Catalysis: Enhanced Enantioselectivity in the Asymmetric Hydrocyanation of Vinylarenes, *J. Am. Chem. Soc.* **116**, 9869, **1994**.

21.55. T. Aratani - Catalytic Asymmetric Synthesis of Cyclopropane-Carboxylic Acids: an Application of Chiral Copper Carbenoid Reaction, *Pure Appl. Chem.* **57**, 1839, **1985**.

21.56. B.M. Trost - Asymmetric Transition-Metal Catalyzed Allylic Alkylation, *Chem. Rev.* **95**, 395, **1995**; B.M. Trost, M.L. Crawley - Asymmetric Transition-Metal Catalyzed Allylic Alkylations: Applications in Total Synthesis, *Chem. Rev.* **103**, 2921, **2003**; B.M. Trost *et al.* - New Insights into the Mechanism of Mo-Catalyzed Asymmetric Alkylation, *Pure Appl. Chem.* **76**, 625, **2003**.

21.57. a) L.F. Tieztze, F. Haunert - Domino Reactions in Organic Synthesis. An Approach to Efficiency, Elegance, Ecological Benefit, Economic Advantage and Preservation of our Resources in Chemical Transformations, in *Stimulating Concepts in Chemistry*, F. Vögtle, J.F. Stoddart, M. Shibasaki Eds, Wiley-VCH, Weinheim, **2000**, 39-64.

  b) A.C. Humphries, A. Pfaltz - Enantioselective Catalysis Using Electronically Unsymmetrical Ligands, *ibid.*, 89-103;

  c) M. Shibasaki - Asymmetric Two-Center Catalysis, *ibid.*, 105-121;

  d) A.H. Hoveyda - Metal-Catalyzed Enantioselective Reactions, *ibid.*, 145-162.

ABBREVIATIONS

Δ	heat (thermal reaction)
ε	dielectric constant
δ	NMR chemical shift (most often given *vs*. TMS)
λ	redox reorganization energy parameter when $\Delta G° = 0$
ν_{CO}	carbonyl infrared frequency
$\Delta G^{\dagger}$	reaction activation enthalpy
$\Delta G°$	standard free enthalpy of the reaction
μ_n	indicates the number n of metals to which the ligand is bonded
$\Delta S*$	reaction activation entropy
18e	18 valence electrons
9-BBN	9-bora [3,3,1] bicyclononane
Å	Angstrom
Ac	acetyl, $CH_3C(O)-$
ADN	deoxyribonucleic acid
AIBN	2,2'-azobisisobutyronitrile, $Me_2C(CN)N=NC(CN)Me_2$
AIDS	acquired immuno-deficiency syndrome
ARN	ribonucleic acid
Ar	aryl
ATP	5-adenosyl triphosphate
acac	acetylacetonato (LX ligand), $MeCOCH^{\bullet}COMe$
aq.	aqueous
atm	atmosphere (pressure)
Bu or *n*-Bu	*n*-butyl
Bu^t or *t*-Bu	*ter*-butyl
bipy	bipyridine
C	coordinance of the metal
Chap.	chapter
CoA	coenzyme A
Cp	cyclopentadienyl, C_5H_5
Cp*	pentamethylcyclopentadienyl, C_5Me_5
Cy	cyclohexyl, $C_6H_{11}-$
cat	catalyst
co	cofactor

cod   cyclo-octadiene, C_8H_{12}

cot   cyclo-octatetraene C_8H_8

d.e.   diastereoisomeric excess

DIOP   chiral diphosphine used by Kagan for asymmetric hydrogenation

DiPT   diisopropyltartrate

DME   1,2-dimethoxyethane, $MeOCH_2CH_2OMe$

DMF N,N-dimethylformamide, Me_2NCHO

DMSO   dimethylsulfoxide, Me_2SO

DOPA   3,4-dihydroxyphenylalanine

dba   dibenzylidene acetone (L_2 ligand)

dmg   dimethylglyoximato

dmgH   dimethylglyoxime

dmpe   dimethyldiphosphinoethane ($Me_2PCH_2CH_2PMe_2$, L_2 ligand)

dppe   diphenyldiphosphinoethane ($Ph_2PCH_2CH_2PPh_2$, L_2 ligand)

dppf   1,1'-diphenyldiphosphinoferrocene ($[Fe(\eta^5-C_5H_4PPh_2)_2]$, L_2 ligand)

dppm   diphenyldiphosphinomethane ($Ph_2PCH_2PPh_2$, L_2 ligand)

dtc   dialkyldithiocarbamato S_2CNR_2 (R = Me or Et)

E, E^+ or El^+ electrophile

eu   entropy unit

$E°$ standard redox potential

E_{pa} anodic peak potential

E_{pc} cathodic peak potential

EPR   paramagnetic electron resonance

equiv.   molar equivalent

ET   electron transfer

EXAFS   Extended X-ray Absorption Fine Structure

e   electron

e.e.   enantiomeric excess

eV   electron volt

FAD   flavine-adenine dinucleotide

$FADH_2$ reduced form of FAD

Fc   ferrocenyl radical, $[FeCp(\eta^5-C_5H_4-)]$

Fp   cyclopentadienyl iron dicarbonyl radical, $FeCp(CO)_2-$

g   gram

H_2ase   hydrogenase enzyme

H_4folate   tetrahydrofolate

HMPA   hexamethylphosphoramide, $(Me_2N)_3PO$

HOMO	Highest Occupied Molecular Orbital
Hz	hertz, s^{-1}
HIV	Human Immunodeficiency Virus
hv	radiative energy
h^n or η^n	hapticity (indicates the number n of atoms of the ligand bonded to the metal)
In•	radical initiator
IR	infrared
i-Bu	isobutyl, $-CH_2CHMe_2$
i-Pr	isopropyl, $-CHMe_2$
irrev.	irreversible
K	equilibrium constant of the reaction
k	rate constant of the reaction
kcal	kilocalorie
kJ	kilojoule
L	ligand molecule (in its neutral form) giving an electron pair to the metal
LUMO	Lowest Unoccupied Molecular Orbital
M	metal or molar
MAO	methylaluminoxane
MEM	β-methoxyethoxymethyl, $-CH_2OCH_2CH_2OCH_3$
Me	methyl, CH_3
Mes	mesityl (2,4,6-trimethylphenyl)
MMO	methane monoxygenase
min	minute
N	solution containing one mol-equivalent *per* liter
N_2ase	nitrogenase enzyme
NADP+	nicotinamide adeninedinucleotide phosphate, oxidized form
NADPH	reduced form of NADP+
NBS	*N*-bromosuccinimide
EAN	number of electrons of the cluster
NNBE	number of non-bonding electrons (also indicated as d^n)
NVE	number of valence electrons
Nu or Nu−	nucleophile
OAc	acetate, $-O-COCH_3$
OS	metal oxidation state
ox	oxidized form of a redox system
ox. ad.	oxidative addition
Pc	phthalocyanine

Ph	phenyl, C_6H_5-
Pr	*n*-propyl
PVA	polyvinyl alcohol
PVC	polyvinylchloride
Py	pyridine, C_6H_5N
phen	*o*-phenanthroline (1,10-phenanthroline)
ppm	part *per* million (10^{-6})
R	constant of ideal gases
red	reduced form of a redox system
red. el.	reductive elimination
RM	organometallic reagent
RNR	ribonucleotide reductase
RT	room temperature
SCE	saturated calomel electrode (reference electrode)
SN_1	1st order nucleophilic substitution (monomolecular)
SN_2	2nd order nucleophilic substitution (bimolecular)
SN_{Ar}	nucleophilic aromatic substitution
s	second
T	absolute temperature in Kelvin
TBHP	*ter*-butyl hydroperoxide
TCNE	tetracyanoethylene, $(NC)_2C=C(CN)_2$
TEM	Transmission Electron Microscopy
Tf	trifluoromethanesulfonate (triflate) CF_3SO_3-
THF	tetrahydrofuranne, C_4H_8O
TMEDA	tetramethylethylenediamine, $Me_2NCH_2CH_2NMe_2$
TMP	*meso*-tetramesitylporphyrinato
TMS	tetramethylsilane, $SiMe_4$
TPP	*meso*-tetraphenylporphyrinato
Tp	tris-pyrazolato
Ts	toluenesulfonate (tosylate), p-Me-C_6H_4-SO_3-
TyO$^\bullet$	tyrosyl radical
tmt	tetramethylthiophene
UV	ultraviolet
V	volt
X	radical ligand (in its neutral form) giving one electron to the metal
[e]	electron in catalytic amount
°C	Celsius degree

Printed in the United States
by Bookmasters

Printed in the United States
By Bookmasters